SMOOTHING OF MULTIVARIATE DATA

WILEY SERIES IN PROBABILITY AND STATISTICS

Established by WALTER A. SHEWHART and SAMUEL S. WILKS

A complete list of the titles in this series appears at the end of this volume.

SMOOTHING OF MULTIVARIATE DATA

Density Estimation and Visualization

Jussi Klemelä

WILEY

A JOHN WILEY & SONS, INC., PUBLICATION

Published by John Wiley & Sons, Inc., Hoboken, New Jersey.
Published simultaneously in Canada.

Library of Congress Cataloging-in-Publication Data:

Klemelä, Jussi,
Smoothing of multivariate data: density estimation and visualization /
Jussi Klemelä.
p. cm.
Includes bibliographical references and index.
ISBN 978-0-470-29088-0 (cloth)
1. Smoothing (Statistics) 2. Estimation theory
QA278.K584 2009
519.5 - - dc22
Printed in the United States of America.

10 9 8 7 6 5 4 3 2 1

CONTENTS

PREFACE

This book is about analyzing multivariate data through smoothing. We take smoothing to mean multivariate density estimation. We think that it is important to combine density estimation with visualization because we believe that multivariate density estimation cannot be fully useful without visualization tools. On the other hand, smoothing can be considered a visualization tool. The book is divided into three parts: visualization, mathematical and algorithmic tools, and a toolbox of density estimators.

The first part presents visualization tools that are based on shape isomorphic transforms and related ideas. These visualization tools can be used to visualize density estimates, but they can also be used to visualize quite general functions and sets. In addition the first part gives a new look at some classical visualization tools and at cluster analysis.

The second part of the book provides a mathematical and algorithmic background for the third part. Modern statistics is based on probability theory, approximation theory, theory of algorithms and computation, information theory, and so on. Thus we have to try to provide a glimpse at the spectrum of these disciplines.

The third part discusses several multivariate density estimators: anisotropic kernel estimators, minimization estimators, multivariate adaptive histograms, wavelet estimators, best basis selection, and stagewise minimization. Asymptotic theory is applied to characterize the properties of the estimators. Several versions of the estimators are presented, and some are more suited for implementation. We do not present a single estimator that can adapt to all kinds of data. Instead, we present a toolbox of estimators, and try to provide guidance on the choice of the right kind of estimator.

The book was written at the University of Heidelberg, Institute of Applied Mathematics, University of Mannheim, Department of Economics, and University of Oulu, Department of Mathematical Sciences. The writing was financed by Deutsche Forschungsgemeinschaft under projects MA1026/6-2, MA1026/8-1, MA1026/8-2, MA1026/9-2.

JUSSI KLEMELÄ

Oulu, Finland
June, 2009

INTRODUCTION

We will analyze data that are given as an $n \times d$ matrix of real numbers. The number in the ith row and in the jth column is the measurement of the jth property of the ith object. For example, the objects might be companies and the properties might be the stock price, debt, number of employees, earnings, or the objects might be persons and the measurements might be height, weight, age.

I.1 SMOOTHING

A fundamental idea is to *smooth* the data. Smoothing means that we interpret the data as n realizations of d-dimensional identically distributed random vectors and estimate the density function of the observations. A density function is a function $\mathbf{R}^d \to \mathbf{R}$ that describes the distribution of the probability mass in the d-dimensional Euclidean space.

The invention of the Cartesian coordinate system made it possible to visualize two-dimensional data with scatter plots. One may interpret the $n \times d$ data matrix as n points in the d-dimensional Euclidean space, and when $d = 2$ to plot the points in the Cartesian coordinate system. Scatter plots may be used, for example, to find the regions where most of the observations are concentrated. Finding regions where the observations are concentrated translates into the problem of finding regions where the density function takes large values, since a density function is a function that describes

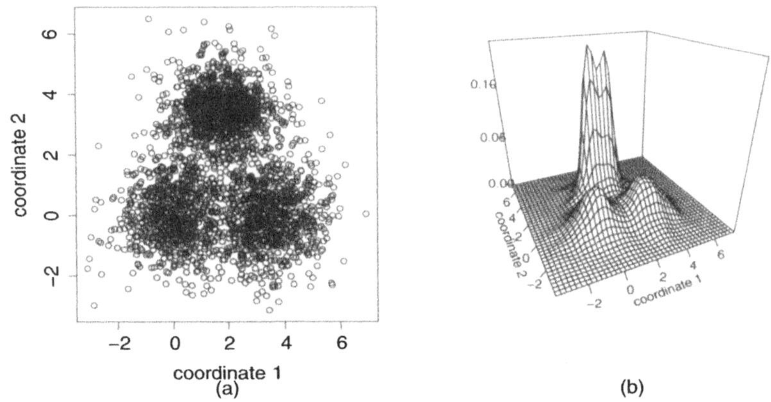

Figure I.1 Shown are (a) a scatter plot and (b) a kernel estimate of data of size $n = 3000$.

the denseness of the observations in the regions of the sample space. Figure I.1(a) shows a scatter plot of data of size $n = 3000$, and panel (b) shows a perspective plot of a kernel estimate. The figure illustrates the fact that the scatter plot makes it possible to identify individual points but the perspective plot of the density estimate visualizes the overall denseness of the observations.

I.2 VISUALIZATION

Functions $\mathbf{R}^d \to \mathbf{R}$ are much more complex objects than $n \times d$ data matrices. Thus it would seem that smoothing multivariate data is not useful in visualization. Is it possible to extend the success story of smoothing from the cases $d = 1$ and $d = 2$ to the cases $d \geq 3$? In our opinion only the very first steps have been made in finding visualization tools for multivariate functions, sets, and data.

The usual graphs seem simple to us, but the idea did not occur to the Greeks or Romans, or to Newton and Leibniz. Lambert (1779) used bivariate function graphs to analyze physical data, and Playfair (1786, 1801) invented the histogram, the pie chart, and the line graph. Still the progress in using these graphs in scientific reporting was slow, and even the scientifically trained readers had to learn how to cope with the new methods. (Spence and Lewandowsky 1990, pp. 13-14). Visualization makes the data visible, and seeing is one of the basic ways for humans to perceive reality. This does not mean that visualization is trivial and that new tools cannot be developed.

Humans can see only one-, two-, or three-dimensional objects. Thus visualization of multidimensional objects is possible only by transforming multidimensional objects to one, two, or three-dimensional objects. Furthermore science is communicated through paper and the computer screen, and this puts emphasis on the two-dimensional case. How to transform multidimensional objects to one, two, or

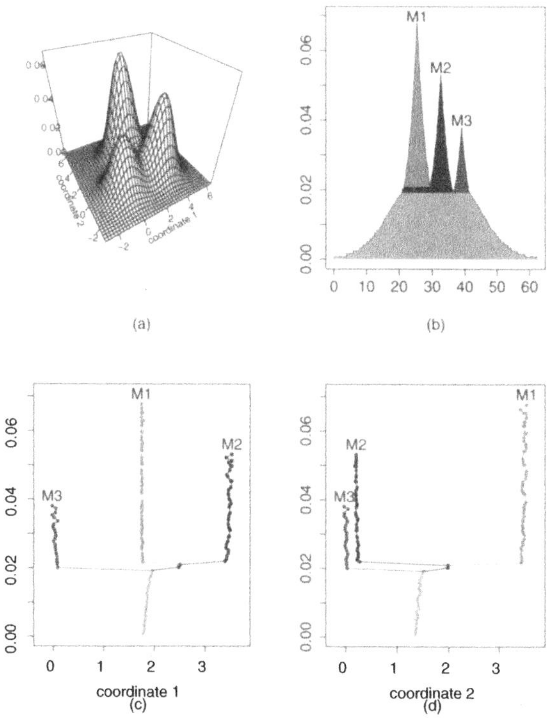

Figure I.2 Shown are (a) a perspective plot, (b) volume plot, and (c–d) barycenter plots of a three-modal density.

three-dimensional objects? A useful method is to apply projections and slices, but there are other possibilities.

Figure I.2, Figure I.3, and Figure I.4, show visualizations of objects of three different types: a function, a set, and data. These three objects have something in common: they are all three-modal objects. The visualizations in the figures reveal the modality of the objects by way of shape isomorphic transforms. These visualizations are one of the main subjects of the book.

In topology one says that two sets are topologically equivalent if they are diffeomorphic or homeomorphic. For example, a donut and a coffee cup may be said to be topologically equivalent. The definition of topological equivalence in terms of diffeomorphisms or homeomorphisms applies to objects of same dimension, but we are interested in the similarity of objects of different dimensions; a multivariate function may be visualized by a one- or two-dimensional function if these functions are similar in some sense.

"Visual geometry is like an experienced doctor's savvy in reading a patient's complexion, charts, and X-rays. Precise analysis is like the medical test results–the

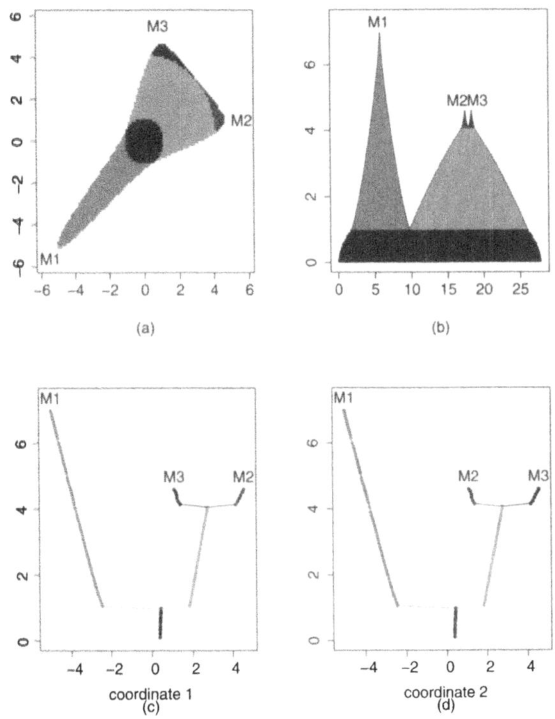

Figure I.3 Shown are (a) a standard plot, (b) radius plot, and (c–d) location plots of a level set of a density with Clayton copula, and Student marginals.

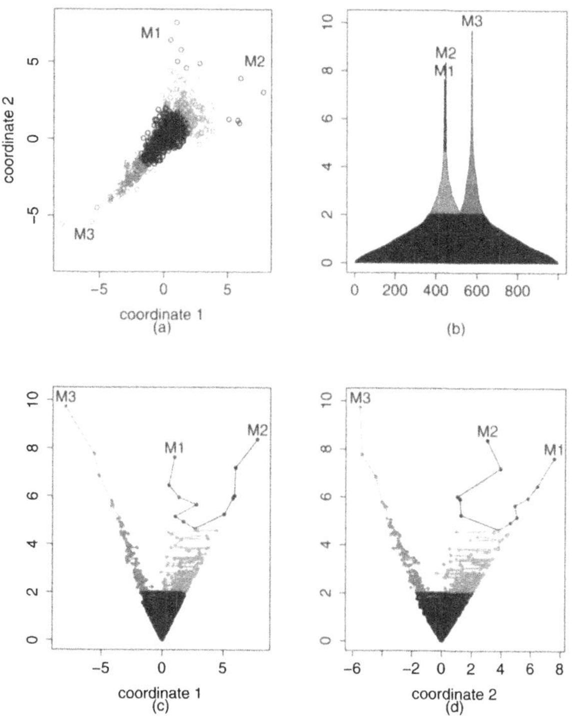

Figure I.4 Shown are (a) a scatter plot, (b) tail frequency plot, and (c–d) tail tree plots of a sample of size 1000 from a density with Clayton copula and Student marginals.

raw numbers of blood pressure and chemistry." (Mandelbrot and Hudson 2004, the prelude). Visualization cannot replace probabilistic inference, but probabilistic inference cannot alone be sufficient for scientific inference, without the support of visualization. Sometimes graphical methods are the only tool we need. For example, when the sample size is very large, we do not need to worry about the random fluctuation part of the experiment.

At least three categories of research on visualization have been classified: statistical visualization or data visualization, scientific visualization, and information visualization. Data visualization studies the direct visualization of data matrices, including the visualization of categorical data. Scientific visualization has concentrated on the visualization of 3D objects, functions, and processes, addressing the issues of industrial design and the medical, chemical, and meteorological visualization. Information visualization has addressed the visualization of various kinds of abstract data structures, like networks and text corpuses. Our main emphasis is on the visualization of multivariate functions. The research on the visualization of functions can be seen as a part of the discipline of information visualization. Our basic setting is to analyze statistical data and thus our research could also be seen as belonging to statistical visualization.

I.3 DENSITY ESTIMATION

Multivariate density estimation is difficult. Parametric Gaussian models fail because they have $2d + d(d-1)/2$ parameters (d parameters for the mean, d parameters for the diagonal of the covariance matrix, and $d(d-1)/2$ parameters for the off-diagonals of the symmetric covariance matrix). Nonparametric estimators that use local averaging fail because local neighborhoods are almost empty of observations in high-dimensional Euclidean spaces. Although some classical methods fail, this does not mean that some other methods could not work. There exists a rich and growing population of density estimators that add to the toolbox of fully parametric and fully nonparametric methods.

The additions to the toolbox could include structured nonparametric methods that utilize structural restrictions in the underlying function. Consider, for example, estimating a multivariate density with a product density, or estimating a regression function with an additive function. Consider imposing shape restrictions like unimodality or imposing structural restrictions on the level sets of the density. Recent additions to the toolbox of density estimators include the estimators based on semiparametric models and mixture models. Infinite mixture models are convex hulls of a base class of densities. This leads to the use of ensemble methods, like bootstrap aggregation, boosting, and stagewise minimization estimators.

Density estimation is a high-precision tool for statistical inference. It can give detailed knowledge about the distribution. Functions defined in moderate-dimensional Euclidean spaces, say four- or five-dimensional spaces, can be extremely complex, and it can be almost impossible to detect all features of a joint distribution of four random variables. Sometimes the data contain hundreds of variables, and there is no hope to reach detailed knowledge about the full joint distribution with a finite amount of measurements. In this case it may be useful to apply dimension reduction techniques before continuing the analysis. Just like a sculptor starts with a hammer and a chisel to create the first contours of the sculpture, and then proceeds with high precision instruments to create the final details, a scientist could start with dimension reduction techniques and then proceed with density estimation. Statistics needs different kinds of tools to be used for different purposes. The right tools are chosen taking into account the available material and taking into account the aims of the work.

I.4 PLAN OF THE BOOK

Part I of the book covers visualization of multivariate functions, sets, data, and scales of multivariate density estimates. Part II gives basic mathematical tools to analyze asymptotically the behavior of multivariate density estimators and describes algorithms that are needed in visualization and in estimation of multivariate densities. Part III presents a toolbox of multivariate density estimators.

I.5 WEB PAGE AND THE CODE

Our hope is that the book satisfies the requirements of reproducible research. We provide software packages to reproduce the main figures and experiments of the book. The R-packages "denpro"and "delt" may be downloaded from the Web page http://www.denstruct.net or from the Web page http://r-project.org. The Web page of the book contains instructions for applying the packages. The Web page of the book contains also the colored figures of the book and the code for reproducing the figures.

I.6 BIBLIOGRAPHIC NOTES

The classic introductions to multivariate density estimation are those by Tapia and Thompson (1978), who discuss penalized likelihood density estimation, Silverman (1986), who considers applications of kernel density estimation, and Scott (1992), who addresses issues of visualization. Kernel estimation is studied in Wand and Jones (1995). A mathematical exposition with the L_1 view is given by Devroye and Györfi (1985) and Devroye (1987). An applied view is given by Simonoff (1996). Efromovich (1999) covers curve estimation with an emphasis on series methods. Tsybakov (2004) covers asymptotic minimax theory of density estimation.

A semiological study of graphics is given by Bertin (1967, 1981). Tukey (1977) gives a foundation for exploratory data analysis. Visualization of information is considered in Tufte (1983, 1990, 1997). Cleveland (1993*b*, 1994) considers principles of graph construction and strategies for data analysis, treating curve fitting as a visualization tool. The topological concepts of scientific visualization are presented in Fomenko and Kunii (1997). Information visualization, as visualization of graphs, trees, knowledge domains, and virtual environments, is discussed in Chen (2004). Spence (2001) treats general information visualization and includes also classical statistical visualization from an information visualization viewpoint.

PART I

VISUALIZATION

CHAPTER 1

VISUALIZATION OF DATA

Our data are given as an $n \times d$ matrix of real numbers, which is called a data matrix. The rows represent the cases and the columns represent the variables: the number in the ith row and in the jth column gives the measurement of the ith object with respect to the jth variable. A data matrix is often called a "data table" in information visualization. Bertin (1981) calls the cases "objects" and the variables "characteristics". We may interpret the rows of the data matrix as points in the d-dimensional Euclidean space. Thus the elementary data visualization task is to visualize a set $\{x_1, \ldots, x_n\} \subset \mathbf{R}^d$.

Variables come in three basic types: nominal, ordinal, and quantitative. Nominal variables allow only meaningful use of the identity relation, ordinal variables allow the use of order relation, and quantitative variables allow the use of arithmetic on them. We will ignore the important types of nominal and ordinal variables and restrict ourselves to the quantitative variables. In addition we will ignore the crucial phase of transforming a raw data to a data matrix.

One can make a distinction between the raw information that is contained in a visualization and the information that can be extracted with a human perception system. For example, a scatter plot of 2D data and the corresponding data matrix of numbers contain the same raw information (up to rounding errors and overplotting);

we can make a computer program that maps a scatter plot to a data matrix, and conversely maps a data matrix to a scatter plot. On the other hand, a human can see patterns from a 2D scatter plot, whereas looking at the numbers of a data matrix printed on a sheet of paper does not usually reveal much to a human.

Some visualizations reduce the raw information but increase the information that may be perceived by a human. For example, kernel estimates do not possess the original information; we may not recover the original data from a kernel estimate. The empirical distribution function contains exactly the same information as the original data; we may recover the original data from the distribution function. However, kernel estimates visualize local concentrations of data well, whereas distribution functions are inefficient in visualizing these local concentrations.

Pixel-based visualizations require at least one pixel to represent one observation. When the size of the data grows, at some stage the pixel based visualizations fail because a human perception system cannot separate pixels when the resolution grows too fine. However, when the size of the data is large, one may use smoothing as a visualization tool.

Thus smoothing arises in visualization in two ways: (1) as a tool that leads to enhanced visualizations taking into account the properties of human perception and (2) as a replacement of pixel based visualizations for large data sets.

Smoothing leads to the task of visualization of functions (visualization of density estimates). Visualization of functions turns out to be related with the visualization of trees. Visualization of functions is discussed in Chapter 2 and visualization of trees is discussed in Chapter 3. This chapter gives an overview of some important data visualization methods.

- Section 1.1 discusses scatter plots, projections, dynamic scatter plots, slices, prosections, and subsetting.
- Section 1.2 discusses visualization of univariate data. Covered are line plots, 1D scatter plots, index plots, time series plots, empirical distribution functions, tail plots, PP-plots, QQ-plots, Box-plots, and kernel estimates.
- Section 1.3 discusses parallel level plots. Parallel level plots are defined as a series of 1D scatter plots.
- Section 1.4 discusses graphical matrices. Graphical matrices are $n \times d$ matrices of graphical elements.
- Section 1.5 discusses visualization with graphical elements. Parallel coordinate plots, Andrew's curves, and Chernoff faces are mentioned.
- Section 1.6 discusses linking across dimensions.
- Section 1.7 discusses descriptive statistics that characterize distributions. Mean, median, mode, and dispersion matrices and vectors are defined.
- Section 1.8 discusses dimension reduction. Principal components, projection pursuit, self-organizing maps, and multidimensional scaling are mentioned.

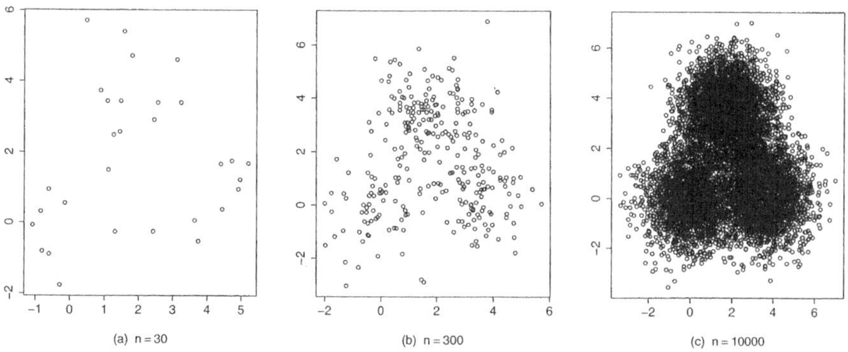

Figure 1.1 *(Scatter plots.)* Showing data of different sample sizes.

1.1 SCATTER PLOTS, PROJECTIONS, AND SLICES

1.1.1 Scatter Plots

The plot of a two-dimensional point cloud $\{x_1, \ldots, x_n\} \subset \mathbf{R}^2$ in the Cartesian coordinate system is called a scatter plot.

Figure 1.1 shows three scatter plots of data of three different sizes, when the data is simulated from the same density. The sample sizes are $n = 30$, $n = 300$, and $n = 10,000$. The data are generated from the three-modal density later shown in Figure 2.7. When the sample size is small, then the point cloud is too sparse to reveal the three modes of the density. When the sample size is large, then the scatter plot is mostly black, so the visuality of density of the points in different regions is obscured. When the sample size is moderate, a more accurate impression of the distribution of the probability mass is possible. Smoothing provides a better way to visualize the density of the observation than a scatter plot; see Figure I.1.

1.1.2 Projections

When $d \geq 3$, we may project data to the 2D Euclidean space and then visualize with a scatter plot this two-dimensional point cloud. We may choose to look only at the projections on the coordinate planes. A *scatter plot matrix*, or a *draftman's plot*, shows scatter plots of all two-dimensional projections to the coordinate planes. A scatter plot matrix consists of

$$\binom{d}{2} = \frac{d(d-1)}{2}$$

scatter plots. A difficulty with the use of scatter plot matrices comes from the fact that single observations cannot be identified across the different windows. A *brush* is a rectangle that can be moved with the mouse. The observations inside the rectangle are highlighted or colored in every window of the scatter plot matrix.

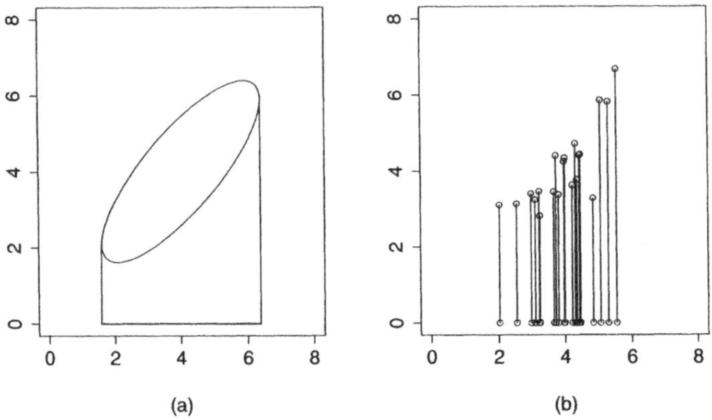

Figure 1.2 *(Projection.)* Shown are one-dimensional projections of two-dimensional sets. (a) Projection to the x_1 axis of an ellipse; (b) projection to the x_1 axis of an ellipsoidal point cloud.

Projections are defined in Appendix B.7. We may also construct a projection by first making a rotation and then projecting to the coordinate planes. See Appendix B.7 for the definition of rotations.

Figure 1.2 illustrates 1D projections of 2D sets. Panel (a) shows how an ellipse is projected on the x_1-axis. Panel (b) shows how an ellipsoidal point cloud is projected on the x_1-axis.

1.1.3 Dynamic Scatter Plots

Dynamic scatter plots show a series of linked 2D projections. Making a series of 2D projections that are close to each other conveys an impression of 3D projections. Smooth transitions between 2D projections make it possible to identify observations in different projections, and brushing may be used to additionally ease the identification of the observations across different projections.

A *grand tour* is a sequence of 2D projections that goes through all possible angles. A dense set of 2D planes is needed so that one can move smoothly through the set; a continuous space-filling path thus moves through the 2D planes in the d-dimensional space. Asimov (1985) introduces the grand tour and Wegman and Solka (2002) discusses algorithms for the grand tour.

1.1.4 Slices

A two-dimensional slice, or section, of a multivariate set $A \subset \mathbf{R}^d$, along the coordinate axes, with the conditions $x_3 = x_{30}, \ldots, x_d = x_{d0}$, is the set

$$\{(x_1, x_2) : (x_1, x_2, x_3, \ldots, x_d) \in A,\ x_3 = x_{30}, \ldots, x_d = x_{d0}\} \subset \mathbf{R}^2.$$

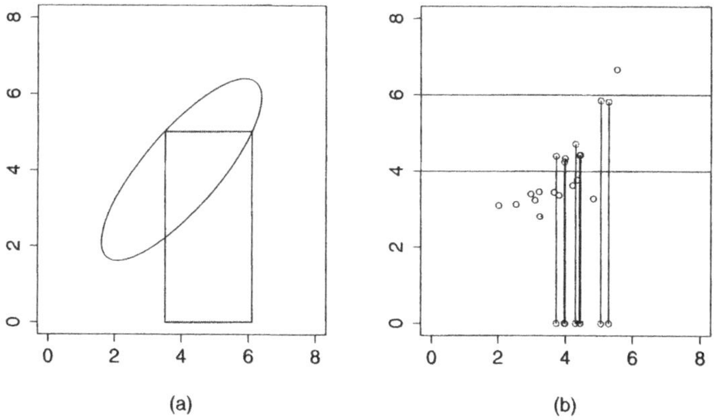

Figure 1.3 *(Slice.)* One-dimensional slices of two-dimensional sets. (a) A slice of an ellipse; (b) a thickened section of an ellipsoidal point cloud.

When set $A \subset \mathbf{R}^d$ is a point cloud, almost all slices are empty. Thus we expand the condition on the coordinates $3, \ldots, d$ to be an inclusion on an interval. This leads to a thickened section

$$\{(x_1, x_2) : (x_1, x_2, x_3, \ldots, x_d) \in A,\ (x_3, \ldots, x_d) \in [a_3, b_3] \times \cdots \times [a_d, b_d]\}.$$

A scatter plot of a thickened section shows two of the coordinates when the other coordinate values are restricted to an interval. A scatter plot of a thickened section is called also a *conditioning plot*, or *coplot*. A coplot visualizes conditional dependence of the two variables, conditioning on the chosen interval of the values of the other variables.

Thickened sections may be useful in visualizing dependence, but there exists a huge number of possible slices to consider: for the d-dimensional data one has $d(d-1)/2$ coordinate directions, but in choosing slices, one has to consider also other than the coordinate directions in making 2D projections. Finally, for each 2D projection one needs to make a series of slices.

Figure 1.3 illustrates 1D slices of 2D sets. Panel (a) shows a slice along the line $x_2 = 5$ of an ellipse. Panel (b) shows a thickened section of an ellipsoidal point cloud when the condition is $x_2 \in [4, 6]$.

1.1.5 Prosections

A limitation of projections is that a projection of an object to a low-dimensional space can fill this space. For example, a projection of a plane in 3D space to a 2D space fills the 2D space. More generally, a projection of an m-dimensional plane in the d-dimensional space to the k-dimensional space fills the k-dimensional space when $m \geq k$.

A limitation of slices is that they may be empty with high probability. For example, a slice with a plane in the 3D space will miss a point in the 3D space almost surely. More generally, a slice of an m-dimensional plane in the d-dimensional space to the k-dimensional space misses the plane when $d - m > k$. One calls the number $d - m$ the codimension of the m-dimensional plane.

Thus 2D projections and slices can reveal planes of dimension 0, 1 and $d-2, d-1$, d. When $d \leq 4$, all cases are covered. When $d > 4$, a combination of a projection and a slice may be used, and these combinations are called prosections. For example, if we have an m-plane in the d-dimensional space, make an $(d - m + 1)$-dimensional section, and then project the section to 2D space, then we will get a line. This is discussed more in detail by Furnas and Buja (1994).

1.1.6 Subsetting

Scatter plots suffer from overplotting when the number of observations is large; see Figure 1.1 and Figure I.1. The same holds for all the other pixel-based visualization tools (parallel coordinate plots, graphical matrices, tail tree plots, etc.). The human perception system has a limited capability to distinguish objects at fine resolution levels. Wegman and Luo (2002) estimate that with single pixel coding less than 10^6 data items can be visualized, and this is a generous upper bound.

One may solve the problem of overplotting by switching to visualization methods based on smoothing; see Figure I.1. However, one may also visualize a large number of observations with pixel-based tools by using *subsetting*. Subsetting means that one chooses subsets of the data and visualizes only one subset at a time. Subsetting is a concept that has been used particularly in earth sciences to describe the process of retrieving just the parts of large files that may arise, for example, when collecting data with earth-observing satellites. The classical subsetting techniques include the following:

1. The subset chosen randomly.

2. The subset consists only of the central data points, with outliers removed by taking all inside a box or inside a sphere.

3. When data contain some categorical variables, these variables can be used to choose a subset.

4. When data does not contain categorical variables it is possible to define such variables using cluster analysis, for example.

5. It is possible to use one variable to define the groups (different age groups or income groups), for example, by taking all on a halfspace (thresholding).

6. With time series data subsets can be chosen temporally by date/time ranges.

7. With spatial data subsets can be chosen by latitude/longitude.

We illustrate a method that may be called *likelihood subsetting*. Likelihood subsetting is a method that lies conceptually between visualizing the complete data and smoothing the data. In likelihood subsetting one chooses the observations that are inside a level set of a density estimate. Thus we consider subsets

$$A_\lambda = \left\{ x_1, \ldots, x_n : \hat{f}(x_i) \geq \lambda \right\},$$

where $\hat{f} : \mathbf{R}^d \to \mathbf{R}$ is a density estimate and $0 \leq \lambda \leq \max\{\hat{f}(x_i) : i = 1, \ldots, n\}$. There are $n-1$ distinct sets A_λ as λ varies, but it is reasonable to expect that for most data sets it is enough to try only few values of λ. Likelihood subsetting is typically computationally expensive, but it is much less expensive than the full program of estimating and visualizing densities, where one needs typically to evaluate density estimates on a multivariate grid.

We may implement the likelihood subsetting with a nearest neighbor density estimate, defined in Section 14.4. First we need to calculate pairwise distances between observations, which takes $O(d \cdot n^2)$ steps. Evaluation of the estimate at the data points takes $O(n^2)$ steps (for each observation, find the distance to the kth nearest neighbor).

Figure 1.4 shows scatter plots of exchange rates of Brazilian Real and Mexican new Peso between 1995-01-05 and 2007-09-26. The rates are with respect to one U.S. Dollar and transformed to returns (we use the transformation $r_i \mapsto (r_i - r_{i-1})/r_{i-i}$). Panel (a) shows 3197 observations and panel (b) shows a subset of size 1769. We applied the nearest-neighbor likelihood subsetting with the number of neighbors $k = 30$. Subsetting reveals the positive correlation between Brazilian Real and Mexican Peso. The data is provided by Federal Reserve Economic Data (http://research.stlouisfed.org).

1.2 UNIVARIATE DATA

We mention five types of visualization methods for univariate data: (1) a line plot, 1D scatter plot, index plot, time series plot; (2) empirical distribution function, tail plot; (3) PP-plot and QQ-plot; (4) box plot; and (5) kernel estimate.

1.2.1 Line Plot, 1D Scatter Plot, Index Plot, Time Series Plot

Visualizing 2D data with scatter plots is a natural visualization technique. At first sight the 1D case seems more difficult: a naive 1D scatter plot would place the observations on the coordinate axis and the overplotting would make the plot useless. Figure 1.5 shows a *line plot* of a sample of size 100 from the two-modal density, whose density is shown in Figure 1.9 and in Figure 1.13(a) as a solid line. Each observation is plotted as a vertical line, placed at the location of the observation in the coordinate axis. A line plot suffers heavily from overplotting.

Lifting is the the main idea for making useful 1D scatter plots. Lifting means that one adds a new variable to the data and makes a 2D scatter plot of the original

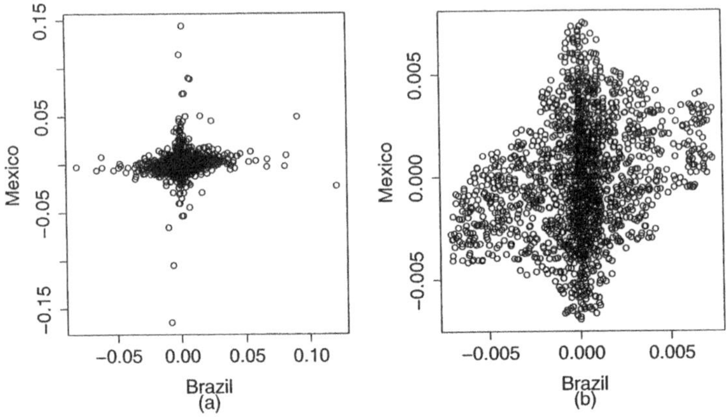

Figure 1.4 *(Likelihood subsetting.)* (a) A scatter plot of exchange rates for Brazilian Real and Mexican Peso ($n = 3197$); (b) a subset of data (size 1769) obtained by a likelihood subsetting.

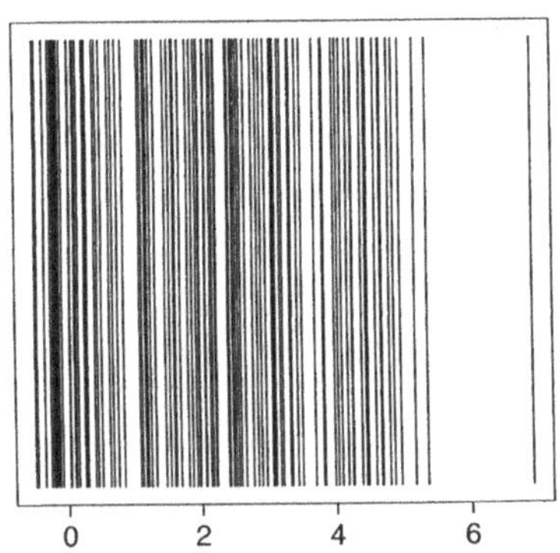

Figure 1.5 *(Line plot.)* A line plot of data of size 100.

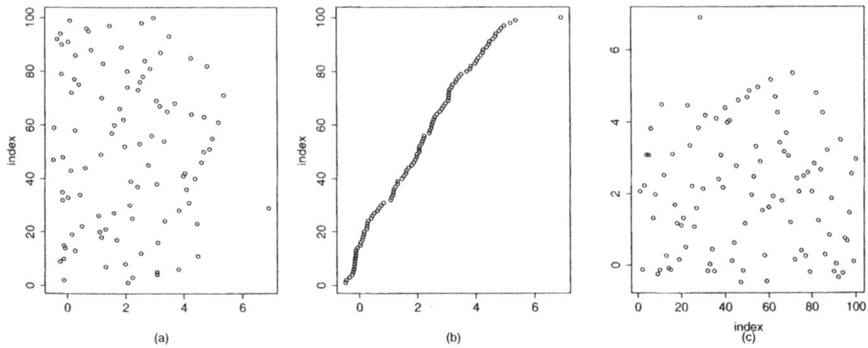

Figure 1.6 *(1D scatter plots.)* 1D scatter plots of data in Figure 1.5 when three different methods of lifting are applied.

observations together with the synthetic measurements. The new variable may be called level because its role is to spread the observations in the direction of the new coordinate. In the simplest case the level may be chosen to be the index of the data points: level$(x_i) = i$, $i = 1, \ldots, n$. In the case of time series data it is natural to choose the level to be the time index of the observations. In many cases one is free to choose the level so that the plot reveals as much information as possible. We will generalize 1D scatter plots to the multivariate case and call them *parallel level plots*. In the multivariate case one makes a 1D scatter plot from each coordinate and the level identifies the observations among the d scatter plots. Parallel level plots are defined in Definition 1.2.

Figure 1.6 shows three 1D scatter plots. In panel (a) the level is chosen to be the index. In panel (b) the level is chosen to be the number of observations which are smaller or equal to this observation,

$$\text{level}(x_i) = \#\{x_j : x_j \leq x_i,\ j = 1, \ldots, n\}.$$

In panel (c) the level is chosen to be the index, but the level is plotted as an x-coordinate and the observed values as a y-coordinate, unlike in panel (a).

One may enhance 1D scatter plots by adding connections to the plot. Figure 1.7 shows the scatter plots of Figure 1.6 enhanced. In the scatter plot of panel (a) each observation is connected with a line to the arithmetic mean of the data. We may call this plot an *index plot*. The scatter plot of panel (b) is the same as the empirical distribution function of Figure 1.9, up to the normalizing factor n^{-1}. Figure 1.7(c) shows a *time series plot*; now the observations are joined to each other with lines, and the level is shown at the vertical axis

Time series plots may be used to visualize the spread of the data. Figure 1.8 shows time series plots of three simulated data of size 500. Panel (a) shows data whose distribution is the Student distribution with degrees of freedom 3, panel (b) shows

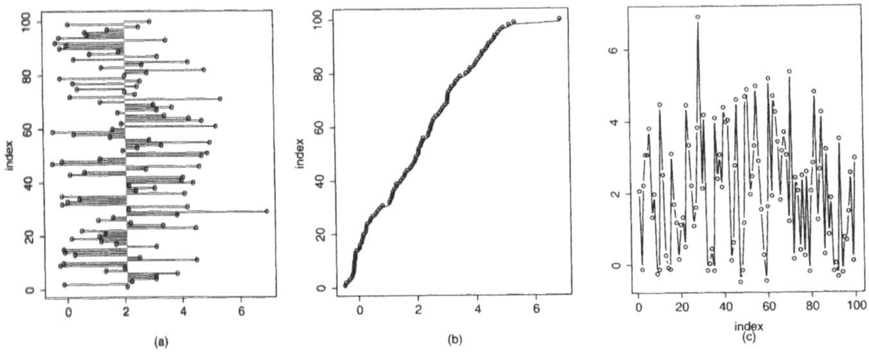

Figure 1.7 *(Enhanced 1D scatter plots.)* Shown are an index plot, an empirical distribution function, and a time series plot corresponding to the 1D scatter plots in Figure 1.6.

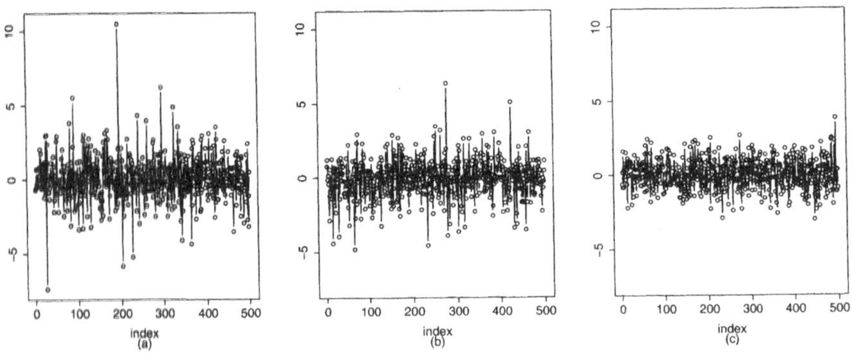

Figure 1.8 *(Time series plots.)* Time series plots used in visualizing the spread of data. (a) A sample from the Student distribution with degrees of freedom 3; (b) a sample from the standard double exponential distribution; (c) a sample from the standard Gaussian distribution. Sample sizes are 500.

data from the standard double exponential distribution, and panel c) shows data from the standard Gaussian distribution.

1.2.2 Empirical Distribution Function and Tail Plot

Empirical distribution functions may be used in visualizing the heaviness of the tails. In contrast, kernel estimates are better in mode detection.

The empirical distribution function is defined as

$$F_n(t) = n^{-1}\#\{i : i = 1, \ldots, n,\ x_i \le t\}, \qquad t \in \mathbf{R}.$$

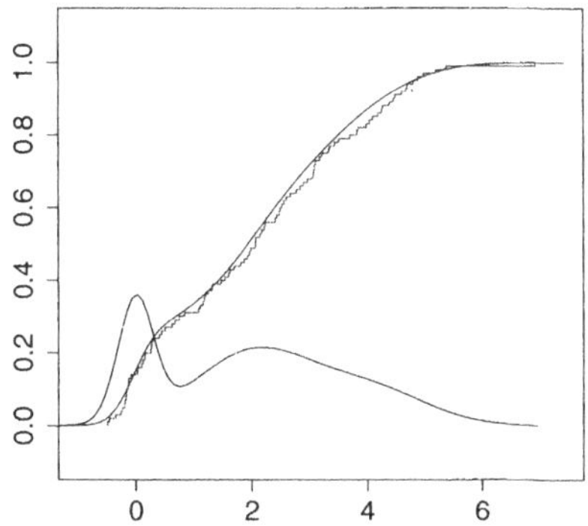

Figure 1.9 *(Distribution function versus density function.)* Empirical distribution function compared with the true distribution function and the density function. The empirical distribution function is the step function, the true distribution function is the smooth monotonically increasing function, and the density function is the two-modal function.

The empirical distribution function may be thought of as a modification of the 1D scatter plot. The level of an observation is equal to the number of observations that are smaller or equal to this observation, and we normalize the level by dividing by the sample size:

$$\text{level}(x_i) = n^{-1}\#\{x_j : x_j \le x_i,\ j = 1, \ldots, n\}. \tag{1.1}$$

In addition the observations are joined with lines to make a step function from the scatter plot.

Figure 1.9 shows the empirical distribution function of the data in Figures 1.5 through 1.7. The true distribution function and the density function are included in the same plot. The true distribution function is the smooth monotonically increasing function and the density function is the two-modal function. The figure illustrates the difficulty of detecting modes with a distribution function.

When we want to visualize the heaviness of the tails of the underlying distribution, it useful to use *tail plots* of the data. We divide the data to the left tail and to the right tail, and visualize separately the two tails. The right tail is $\mathcal{R} = \{x_i : x_i > \mu,\ i = 1, \ldots, n\}$, where μ may be chosen to be the median of the observations. Now we choose the level

$$\text{level}(x_i) = \#\{x_j : x_j \ge x_i, x_j \in \mathcal{R}\}, \qquad x_i \in \mathcal{R}.$$

The scatter plot of $\mathcal{R}$ with this choice of level visualizes the heaviness of the right tail. It is useful to use a logarithmic axis for the y-axis. When the data are centered around 0, we may choose $\mu = 0$, and we may use a logarithmic scale also for the x-axis.

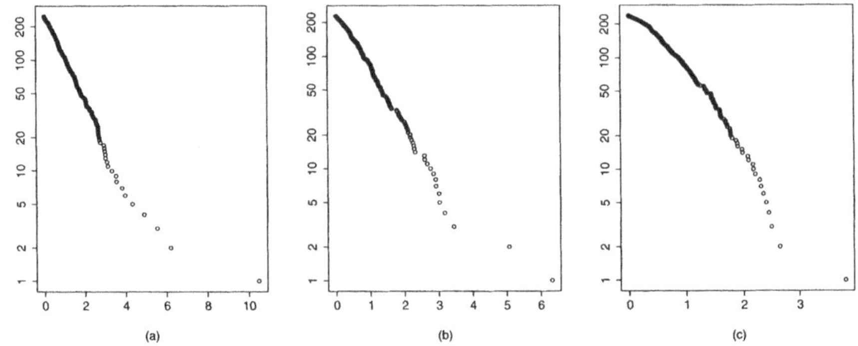

Figure 1.10 *(Right-tail plot.)* Right-tail plots of the data in Figure 1.8: (a) Sample from the Student distribution with degrees of freedom 3; (b) sample from the standard double exponential distribution; (c) sample from the standard Gaussian distribution. Sample sizes are 500.

Figure 1.10 shows right tail plots of the data in Figure 1.8. We us a logarithmic scale only for the y-axis. Tail plots have been used by Mandelbrot (1963), Bouchaud and Potters (2003), Sornette (2003).

1.2.3 PP-Plot and QQ-Plot

PP-plots and QQ-plots visualize the empirical distribution function or its inverse (quantile function) by comparing them to a reference distribution. The reference distribution might represent a null-hypothesis, so the PP-plot and QQ-plot are used to assess the goodness of fit. The reference distribution may have a parametric form and the parameters are estimated from the data. One may also scan through a collection of reference distributions in the spirit of exploratory data analysis.

We denote with $F : \mathbf{R} \to [0,1]$ a reference distribution function and with $F^{-1} : [0,1] \to \mathbf{R}$ the generalized inverse of F:

$$F^{-1}(p) = \inf\{t \in \mathbf{R} : F(t) \geq p\}, \qquad p \in [0,1].$$

Definition 1.1 (Probability plot, Quantile plot.) *A* PP-plot, *or a* probability plot, *of the data $x_1, \ldots, x_n \in \mathbf{R}$, associated to a reference distribution function $F : \mathbf{R} \to [0,1]$, is a plot of points*

$$(F_n(x_i), F(x_i)), \qquad i = 1, \ldots, n,$$

where F_n is the empirical distribution function. A QQ-plot, *or a* quantile plot, *of the data $x_1, \ldots, x_n \in \mathbf{R}$, associated to a reference distribution function $F : \mathbf{R} \to [0,1]$, is a plot of points*

$$\left(F^{-1}(p_i), F_n^{-1}(p_i)\right), \qquad i = 1, \ldots, n.$$

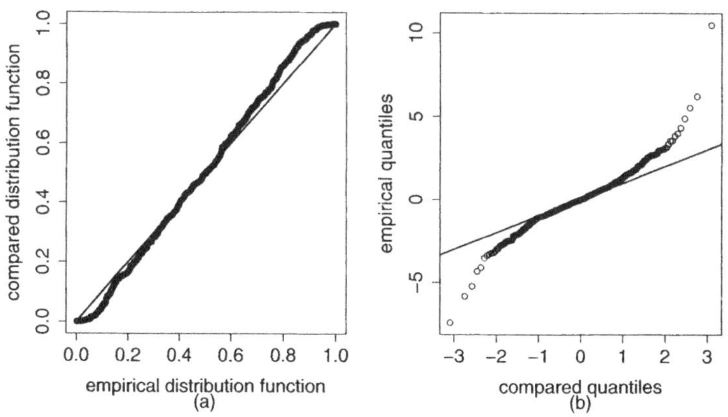

Figure 1.11 *(PP-plot, QQ-plot.)* (a) A PP-plot of a sample from a Student distribution with degrees of freedom 3; (b) a QQ-plot of the sample. The reference distribution is the standard Gaussian distribution.

where $p_i = (i - 1/2)/n$. *The graph of the line* $y = x$ *(*$x_1 = x_2$*) is also included in the plots.*

Note that $F_n(x_{(i)}) = i/n$, where $x_{(1)} < \cdots < x_{(n)}$ is the ordered sample, and thus the PP-plot is the plot of the points

$$\left(i/n, F(x_{(i)})\right), \qquad i = 1, \ldots, n.$$

Note also that

$$F_n^{-1}(p) = \inf\{t \in \mathbf{R} : F_n(t) \geq p\} = \begin{cases} x_{(1)}, & 0 < p \leq 1/n, \\ x_{(2)}, & 1/n < p \leq 2/n, \\ \vdots & \end{cases}$$

Thus the QQ-plot is the plot of the points

$$\left(F^{-1}(p_i), x_{(i)}\right), \qquad i = 1, \ldots, n.$$

Figure 1.11 shows a PP-plot and a QQ-plot for a sample of size 500 from the Student density with degrees of freedom 3, when the reference distribution is the standard Gaussian distribution. Panel (a) shows a PP-plot and panel (b) shows a QQ-plot.

When the sample is originating from the reference distribution F, the points in the PP-plot lie near the line $x_1 = x_2$. Location differences will lead to plots where the points lie above or below this line. Scale differences will lead to plots where the points lie first on the one side of the line $x_1 = x_2$, next intersect this line, and then lie on the other side of the line.

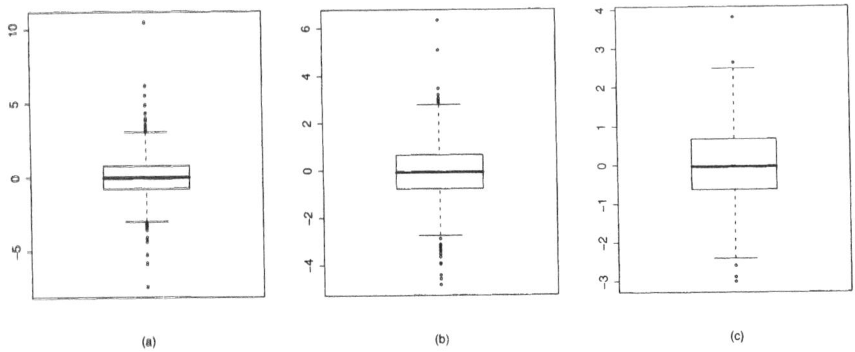

Figure 1.12 *(Box plot.)* Box plots of the data in Figure 1.8. (a) A sample from the Student distribution with degrees of freedom 3; (b) a sample from the standard double exponential distribution; (c) a sample from the standard Gaussian distribution. Sample sizes are 500.

A PP-plot is related to the Kolmogorov-Smirnov test statistic

$$KS_n(G) = \sup_{x \in \mathbf{R}} |G(x) - F_n(x)|,$$

which is used to test the null-hypothesis $F = G$, where F is the true distribution function. The asymptotic distribution of the normalized test statistic $\sqrt{n} KS_n(F)$ is the distribution of $\sup_{p \in [0,1]} |B(p)|$, where B is the standard Brownian bridge. Thus the asymptotic distribution, when the null-hypothesis holds, is independent of the true distribution F. The formal testing procedure is useful when we accompany it with the PP-plot: for the case where the null-hypothesis is rejected, we need a graphical tool that indicates the direction of the deviation from the null-hypothesis.

1.2.4 Box Plot

A *box plot* shows a box with a center line at the median, the upper bound of the box at the 75% quantile, and the lower bound of the box at the 25% quantile. Thus the box shows the interval where 50% of the data lies. In addition lines are drawn at 1.5 times the inter-quartile range, and the extreme observations beyond these lines are drawn as circles. Figure 1.12 shows box plots of the data in Figure 1.8. The previous plots did not reduce the information content of the sample (except that a right-tail plot and a left-tail plot visualize only half of the data); one may reconstruct the data up to the effects of overplotting. A box plot reduces the information in the sample to the 5 numbers: median, 2 quartiles, and 2 extremes. A box plot may be more suited for the presentation of the data than for the exploration of the data. Note, however, that the extreme observations are plotted individually in a box plot, and often only the extreme observations need to be identified, whereas the central observations may be aggregated in visualizations.

1.2.5 Kernel Estimates

The previous visualization methods are not effective for the visualization of multimodality. We may use smoothing to visualize the modes. Thus we consider density estimates, like kernel estimates, as a tool for visualizing data. Kernel estimators are defined in Section 14.3.

Unlike 1D scatter plots, empirical distribution functions, and many other visualization methods, a kernel estimate looses some information: it is not possible to recover the original data when we have available only a kernel estimate. However, less information is lost when the smoothing parameter becomes smaller. In the limit, as the smoothing parameter of kernel estimates approaches zero, a kernel estimate shows delta peaks at each observation. This leads to the idea of visualizing the data by showing a scale of kernel estimates corresponding to a sequence of smoothing parameter values. A scale of kernel estimates shows simultaneously a summary of the data (when the smoothing parameter is large or moderate) and idiosyncratic features (when the smoothing parameter is small). We do not have to choose a single best kernel estimate but we may visualize the data with a scale of density estimates. The scale space view of curve estimation is developed in Chaudhuri and Marron (2000); Chaudhuri and Marron (1999) introduced SiZer as an exploratory tool in the spirit of the scale space view.

Figure 1.13(a) shows a scale of kernel estimates. The true density is shown in panel (b) as the solid line, and also in Figure 1.9. The sample size is 100 and the smoothing parameter is in the interval $h \in [0.2, 1]$. The Gaussian kernel is used. Figure 1.13(b) shows the kernel estimates with $h = 0.1$ and $h = 1$ as dashed lines, in addition to showing the true density with the solid line. We discuss in Chapter 19 a stagewise minimization estimator which estimates this target density better; see Figure 19.1(a).

We can combine a kernel estimate and a 1D scatter plot when the level of an observation is chosen to be the value of the kernel estimate at the observation:

$$\text{level}(x_i) = \hat{f}(x_i), \qquad i = 1, \ldots, n,$$

where $\hat{f}$ is a kernel estimate. Figure 1.14 shows two 1D scatter plots with the level determined by kernel estimates with $h = 0.2$ and $h = 1$. The standard Gaussian kernel function is used in the kernel estimates. The data are shown also in Figures 1.5 through 1.7.

1.3 PARALLEL LEVEL PLOTS

In *parallel level plots* we introduce a new dimension to the point cloud, called a level. We attach each observation with a level and make d parallel scatter plots. Each coordinate gives rise to a scatter plot and the level is the y-coordinate of the scatter plots. Thus we identify the observations across scatter plots through the value of the level of the observations. Linking across dimensions is discussed also in Section 1.6 and parallel level plots will be enhanced in Chapter 6 by introducing tail tree plots.

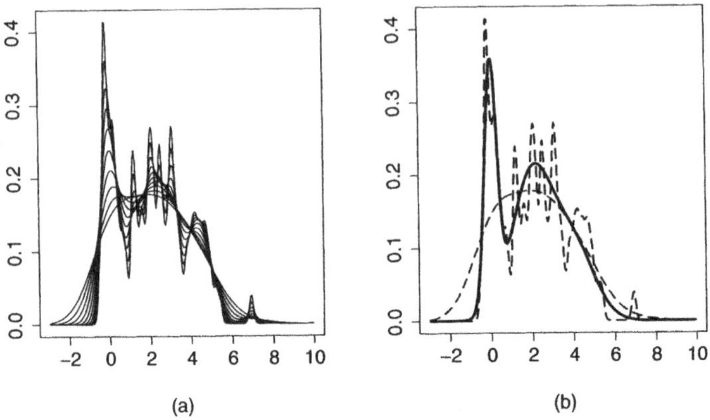

Figure 1.13 *(Scale of kernel estimates.)* (a) A scale of 10 kernel estimates with $h \in [0.2, 1]$. (b) The true density with a solid line, and the estimates with $h = 0.2$ and $h = 1$ with dashed lines. The data of size $n = 100$ is shown in Figures 1.5 through 1.7.

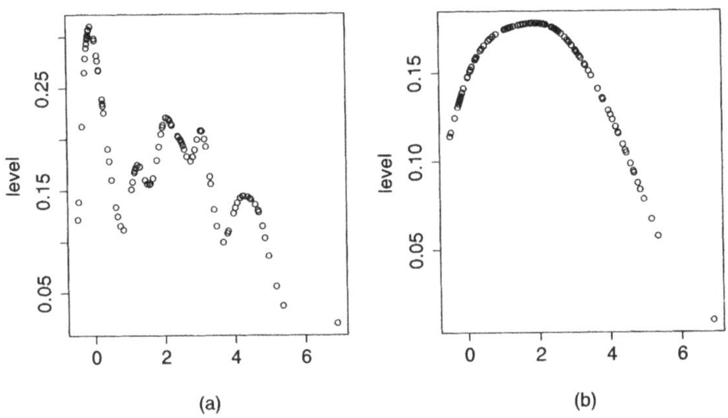

Figure 1.14 *(1D scatter plots via kernel estimates.)* Shown are 1D scatter plots of the data in Figures 1.5 through 1.7. The level of scatter plots is the value of a kernel estimate at the observations. (a) $h = 0.2$; (b) $h = 1$.

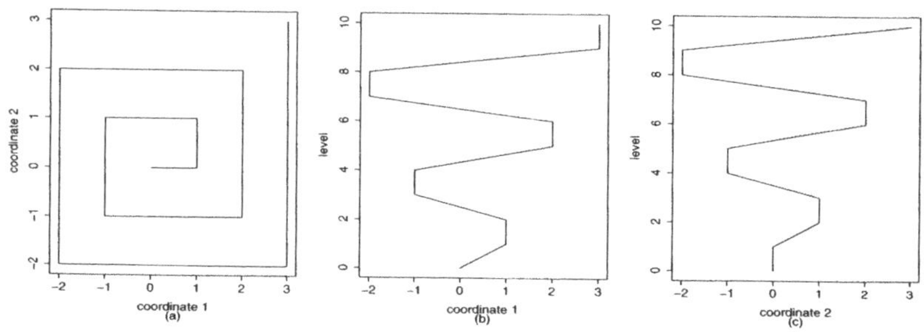

Figure 1.15 *(Parallel level plot, 2D.)* (a) A rectangular spiral in the 2D Euclidean space; (b-c) a parallel level plot of the spiral.

Definition 1.2 (Parallel level plot.) A parallel level plot *of data* $x_1, \ldots, x_n \in \mathbf{R}^d$, *associated with a level mapping* $\text{level}(x_i) \in \mathbf{R}$, *consist of d windows, where the kth window,* $k = 1, \ldots, d$, *shows a scatter plot of points* $(x_i^k, \text{level}(x_i))$, $i = 1, \ldots, n$. *We use the notation* $x_i = (x_i^1, \ldots, x_i^d) \in \mathbf{R}^d$.

When the level mapping satisfies $\text{level}(x_i) \neq \text{level}(x_j)$, for $i \neq j, i, j = 1, \ldots, n$, the unique identification of the observations across the windows is possible.

Parallel level plots generalize the 1D scatter plot, shown in Figures 1.6 and 1.7, to the multivariate case. All these plots introduce a new variable (level) to the data. In the multivariate case the level is the same for each coordinate of a single observation, and the level is used to identify the observations across the windows.

1.3.1 Multivariate Time Series

Parallel level plots have been common in visualizing multivariate time series: one takes the time index of the data points to be their level. If we have d time series, and each time series contains measurements at the same n time points, then the plot of these times series in d parallel windows is a parallel level plot. Becker, Clark and Lambert (1994) uses a visualization, called a cave plot, which resembles the index plot of Figure 1.7(a) to visualize a multivariate time series.

1.3.2 One-dimensional Curves

One-dimensional curves in the d-dimensional Euclidean space may be visualized naturally with parallel level plots by taking the parameter of the curve to be the level: if we have a curve $z(t) = (z_1(t), \ldots, z_d(t)) \in \mathbf{R}^d$, $t \in [0, \infty)$, we may visualize it with a parallel level plot by choosing a fine grid $0 \leq t_1 < \cdots < t_n < \infty$, defining $x_i = z(t_i)$, and $\text{level}(x_i) = t_i$, $i = 1, \ldots, n$. We will illustrate parallel level plots in visualizing spirals.

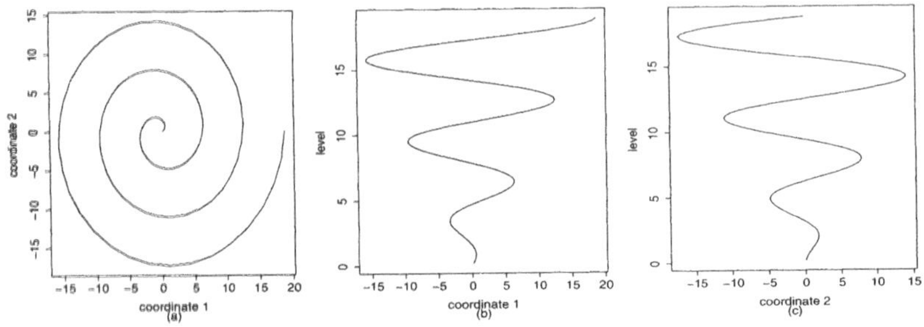

Figure 1.16 *(Parallel level plot, 2D.)* (a) An Archimedean spiral in the 2D Euclidean space; (b-c) a parallel level plot of the spiral.

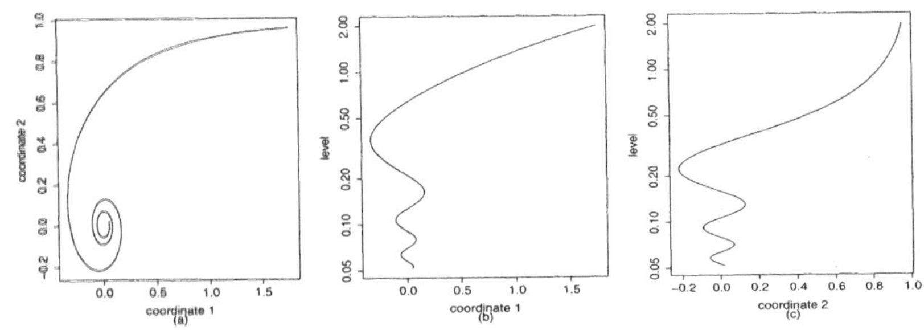

Figure 1.17 *(Parallel level plot, 2D.)* (a) A hyperbolic spiral in the 2D Euclidean space; (b-c) a parallel level plot of the spiral.

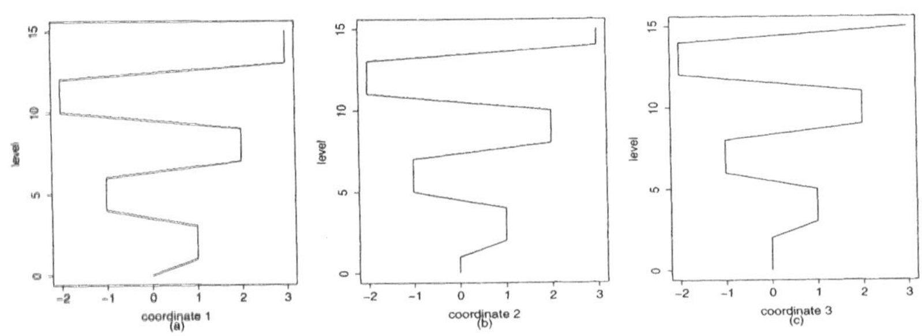

Figure 1.18 *(Parallel level plot, 3D.)* A parallel level plot of a 3D rectangular spiral.

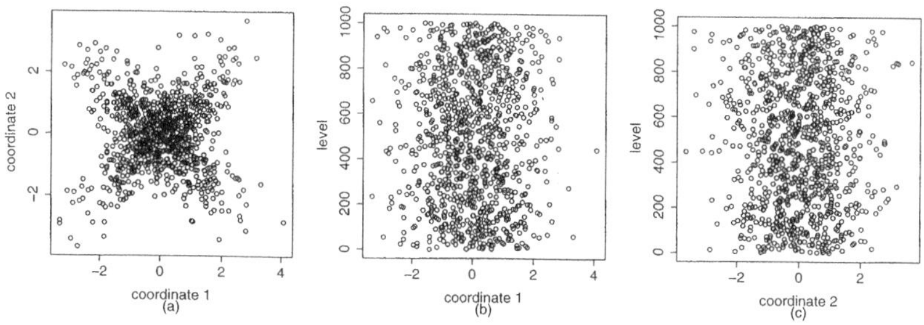

Figure 1.19 *(Parallel level plot, point cloud.)* (a) A scatter plot of size 1000 from a mixture of two Gaussian distributions. (b-c) A parallel level plot of the data in (a) when the level is the index of observations.

1. Figure 1.15(a) shows a 2D rectangular spiral and Figure 1.15(b)-(c) shows a parallel level plot of the spiral. We have taken the level equal to the length of the spiral, starting at the origin.

2. Figure 1.16(a) shows a 2D Archimedean spiral and Figure 1.16(b)-(c) shows a parallel level plot of the spiral. The Archimedean spiral is defined by $r \mapsto r(\cos r, \sin r)$, $r \geq 0$. We have taken the level equal to the radius r.

3. Figure 1.17(a) shows a 2D hyperbolic spiral and Figure 1.17(b)-(c) shows a parallel level plot of the spiral. The hyperbolic spiral is defined by $r \mapsto (r\cos(1/r), r\sin(1/r))$, $r > 0$. We have taken the level equal to the radius r, and $r \geq 0.05$.

4. Figure 1.18 shows a parallel level plot of a 3D rectangular spiral.

1.3.3 Point Clouds

When visualizing point clouds in general, the usefulness of a parallel level plot depends on the choice of level. A natural choice for the level is the distance from a center point, say, from the arithmetic mean. Define

$$\text{level}(z) = \|z - \mu\|, \qquad \mu = \frac{1}{n}\sum_{i=1}^{n} x_i. \tag{1.2}$$

An other choice for the level is the value of the first coordinate, say,

$$\text{level}(z) = z^1, \qquad z = (z^1, \ldots, z^d) \in \mathbf{R}^d. \tag{1.3}$$

We illustrate the effect of the choice of the level with four figures.

1. Figure 1.19 shows a 2D point cloud which is a sample of size 1000 from a mixture of two Gaussians. Panels (b)-(c) show the same data with a parallel

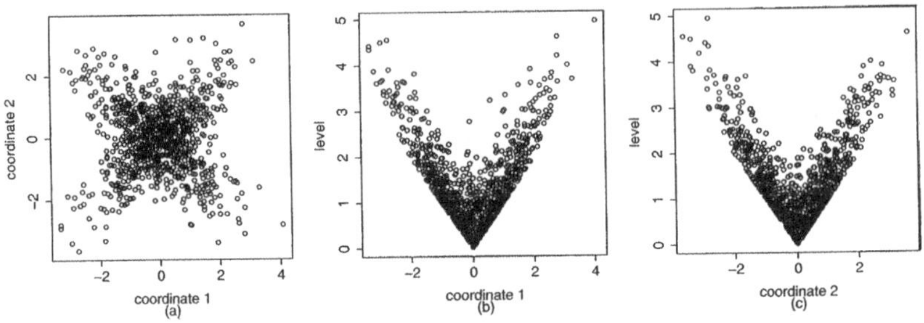

Figure 1.20 *(Parallel level plot, point cloud.)* (a) The same data as in Figure 1.19(a). (b-c) A parallel level plot when the level is the distance from the arithmetic mean.

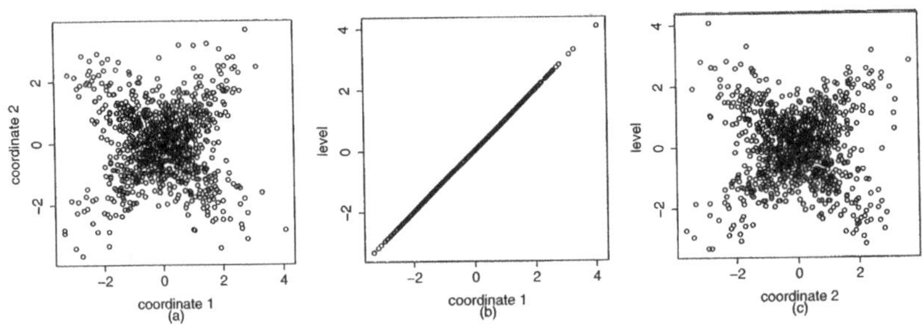

Figure 1.21 *(Parallel level plot, point cloud.)* (a) The same data as in Figure 1.19(a). (b-c) A parallel level plot when the level is the first coordinate.

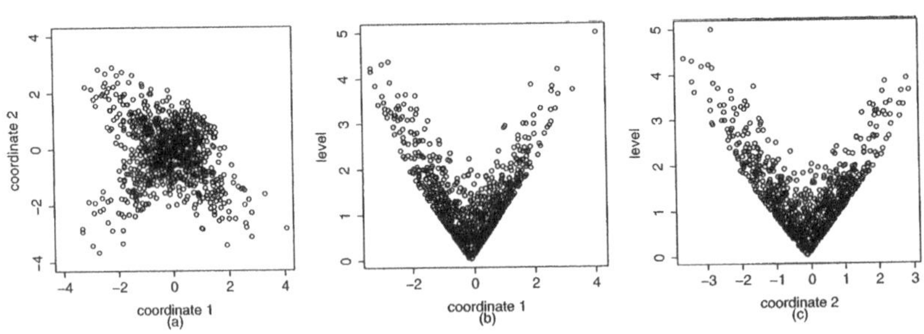

Figure 1.22 *(Parallel level plot, point cloud.)* (a) A scatter plot of the data in Figure 1.19 when one tail is removed. (b-c) A parallel level plot when the level is the distance from the arithmetic mean.

level plot where we have attached a unique random level in $\{1, \ldots, 1000\}$ to each observation (the level is in fact the index of the observations). The scatter plot and the parallel level plot contain the same information (for a computer program), but the scatter plot lets a human eye to detect the pattern of four tails in the data.

2. Figure 1.20(b)–(c) shows a parallel level plot of the data in panel (a) when we choose the level as the distance to the arithmetic mean, as in equation (1.2). The parallel level plot indicates that there are at least two tails in the data. We define in Section 6.2 a *tail tree plot*. A tail tree plot adds a sufficient structure to a parallel level plot making it possible for humans to detect the four tails in the data, as is shown in Figure 6.9.

3. Figure 1.21(b)–(c) shows a parallel level plot when the level is the first coordinate, as in (1.3). In this case the second window of the parallel level plot is a scatter plot where the coordinates have been switched.

4. Figure 1.22 shows the data in Figures 1.19 through 1.21 when one tail is cut away. Panel (a) shows a scatter plot and panels (b–c) show a parallel level plot when the level is the Euclidean distance from the arithmetic mean, as in equation (1.2). The parallel level plot of the data does not appear to differ from the parallel level plot in Figure 1.20(b)–(c), where the data has four tails. This demonstrates further the need for tail tree plots.

1.4 GRAPHICAL MATRICES

A graphical matrix is a $n \times d$ matrix of graphical elements; a $n \times d$ data matrix of real numbers is transformed by representing each real number by a graphical element. Graphical matrices were studied by Bertin (1967, 1981). There are many ways to code real numbers with graphical elements. When the data consists only of binary values, one may represent these values with white and black rectangles. Figure 1.23 shows a graphical matrix containing data about townships. The variables take binary values, and the values are coded with black and white. Figure 1.24 shows the same data when the columns and rows have been reordered, so that the patterns are revealed. Real values can be represented with bars or lines, whose length is proportional to the value. Bertin (1981) mentions seven visual variables that can be used to code numerical values: form, position, size, value, texture, color, and orientation. Further developments of a graphical matrix include the Data Image as in Minnotte and West (1999).

1.4.1 Bar Matrix

Figure 1.25(a) shows a scatter plot of five observations. Panel (b) shows a graphical matrix of these data, where the measurements are pictured as bars whose heights are proportional to the values of measurement (relative to the range of the variable). We

Figure 1.23 *(Unordered table.)* The unordered table of Bertin is reproduced from Bertin (1981).

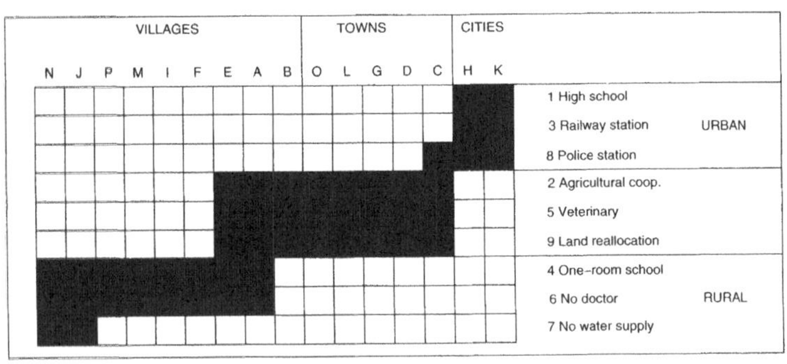

Figure 1.24 *(Ordered table.)* The ordered table of Bertin is reproduced from Bertin (1981).

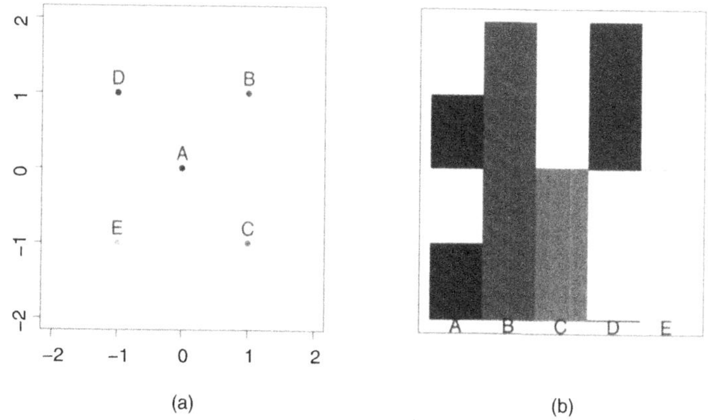

Figure 1.25 *(Bar matrix, definition.)* (a) A scatter plot of five points, and (b) a graphical matrix of the data with bars as graphical elements (bar matrix).

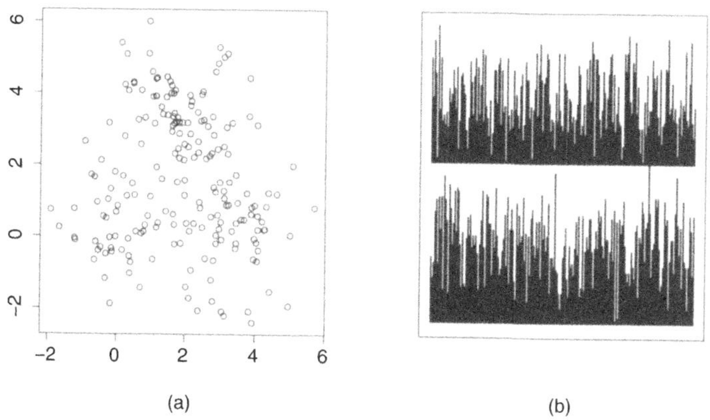

Figure 1.26 *(Bar matrix, illustration.)* (a) A scatter plot of 200 points and (b) a graphical matrix of the data with bars as graphical elements (bar matrix).

have labeled the bars to show the correspondence with the points in the scatter plot. This kind of a graphical matrix may be called a *bar matrix*.

Figure 1.26 shows an other example of a scatter plot and a bar matrix. Panel (a) shows a data of size 200 (the data are generated from the density shown in Figure 2.7), and panel (b) shows a graphical bar matrix with a random permutation.

The usefulness of a graphical matrix depends on the choice of an informative permutation of the rows and the columns of the matrix. Similarly to the parallel level plot in Figure 1.19(b)–(c), and similarly to the Bertin's table in Figure 1.23, a

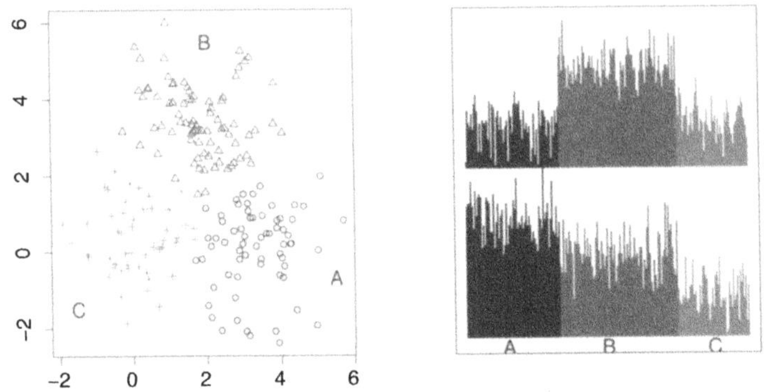

Figure 1.27 *(Bar matrix, ordered.)* (a) A scatter plot of 200 points when the k-means clustering algorithm has been applied to label the data points. (b) A graphical matrix of the data with bars as graphical elements (bar matrix). The observations belonging to a same cluster are grouped together.

random permutation leads typically to useless graphical matrices. A suggestion of Bertin (1981) is to order indexes so that the values of one variable are in a monotonic order. The permutation could also be done with the help of a clustering algorithm. The observations belonging to the same cluster are grouped together in the graphical matrix.

Figures 1.26 and 1.27 illustrate the usefulness of a permutation. Figure 1.26(b) shows an unordered graphical matrix and Figure 1.27(b) shows an ordered graphical matrix. The ordering has been done by a permutation where the observations belonging to a same cluster are grouped together. The clusters were obtained by the k-means algorithm with $k = 3$; see Section 8.2 for the definition of k-means clustering.

1.4.2 Index Plot Matrix

Figure 1.28(a) shows a scatter plot of five observations. Panel (b) shows a graphical matrix of these data, where each measurement is drawn as a circle that is connected with a line to the arithmetic mean of the coordinate. This graphical matrix is identical with a parallel level plot, where the level is the index of the observation, and we delete the lines and keep only the circles. (Another possibility would be to delete the circles and keep only the lines.) Note also that this graphical matrix generalizes the 1D index plot, as shown in Figure 1.7(a).

Figure 1.29 shows an other example of an index plot matrix. Panel (a) shows a scatter plot of data of size 200 that contains four tail clusters and panel (b) shows an index plot matrix with random ordering of the observations.

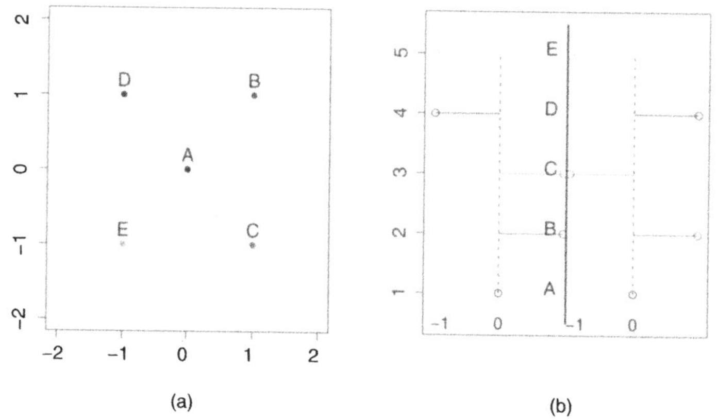

Figure 1.28 *(Index plot matrix, definition.)* (a) A scatter plot of five points, and (b) a graphical matrix of the data with lines as graphical elements (index plot matrix).

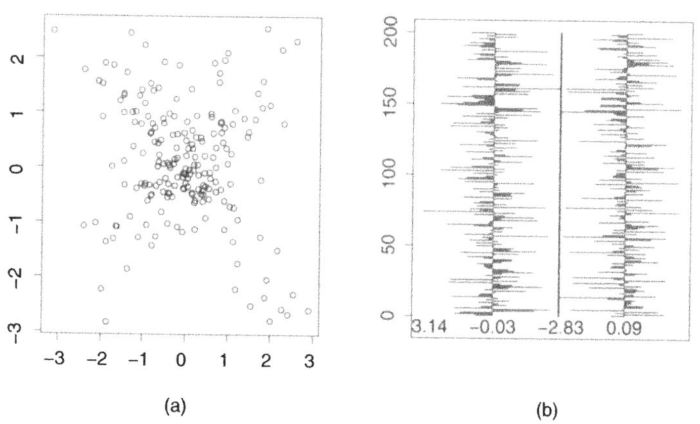

Figure 1.29 *(Index plot matrix, illustration.)* (a) A scatter plot of 200 points and (b) a graphical matrix of the data with lines as graphical elements (index plot matrix).

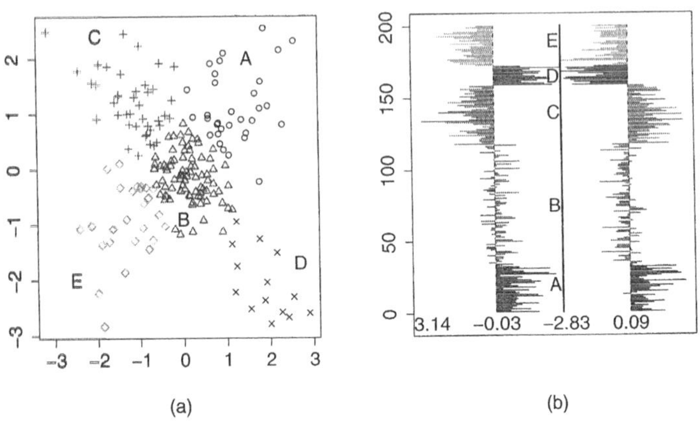

Figure 1.30 *(Index plot matrix, ordered.)* (a) A scatter plot of 200 points when the k-means clustering algorithm has been applied to label the data points. (b) A graphical matrix of the data with lines as graphical elements (index plot matrix). The observations belonging to a same cluster are grouped together.

Figures 1.29 and 1.30 illustrate the usefulness of a permutation based on a clustering. Figure 1.29(b) shows an unordered matrix, and Figure 1.30(b) shows an ordered matrix for groups that were obtained by the k-means algorithm with $k = 5$, The observations belonging to a same cluster are grouped together. Cluster analysis is discussed more in detail in Chapter 8. Cluster analysis may be used to visualize data with graphical matrices, or conversely graphical matrices may be used to visualize and validate the results of a cluster analysis.

1.5 OBSERVATIONS AS OBJECTS

Parallel coordinate plots and Andrew's curves represent observations as 1D curves. Chernoff faces represent observations with schematic human faces. We mention also other possibilities for visualizing data by representing the data points as objects. These methods work typically for small data sets. Also, the methods are usually sensitive for the ordering of the variables.

1.5.1 Parallel Coordinate Plots

Parallel coordinate plots may be used to visualize a multivariate point cloud by looking at the one-dimensional projections of the data to the coordinate axes, which are drawn parallel to each other. One needs to identify the observations across the d projections. Parallel coordinate plots identify the observations across the coordinate axes by joining them with straight lines.

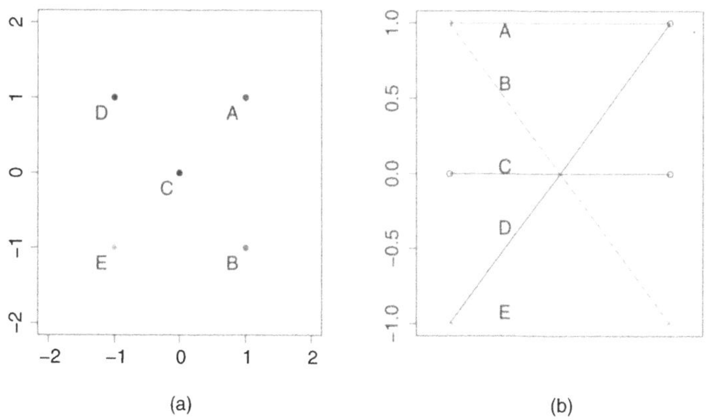

Figure 1.31 *(Parallel coordinate plot, definition.)* (a) A scatter plot of five points and (b) a parallel coordinate plot of these points.

In a parallel coordinate plot the number of the coordinate is mapped onto the horizontal axis and the value is mapped onto the vertical axis, and then these points are linearly interpolated. Observation x_i, whose index is $i \in \{1, \ldots, n\}$, is represented as the curve that linearly interpolates the points

$$(1, x_i^1), (2, x_i^2), \ldots, (d, x_i^d).$$

Parallel coordinate plots were introduced by Inselberg (1985). See also Inselberg and Dimsdale (1990), Wegman (1990), and Inselberg (1997).

Parallel coordinate plots resemble parallel level plots: in both cases one visualizes 1D projections of the data at the coordinate axis, and uses a method to identify the observations among the d projections. Parallel coordinate plots provide a more compact visualization of the data than parallel level plots. Consequently parallel coordinate plots suffer more from overplotting. Indeed the lines connecting the observations introduce more ink to the plot (bad "signal-to-ink-ratio"). In contrast, parallel level plots introduce a new variable to the plot (level), and spread the observations with the help of the new variable, thus diminishing the problem of overplotting. Miller and Wegman (1991) introduces a parallel coordinate density plot which that may be used to solve the problem of black ink.

1. Figure 1.31 illustrates the definition of a parallel coordinate plot. Panel (a) shows a scatter plot of five points, and panel (b) shows a parallel coordinate plot of these five points.

2. Figure 1.32 shows a scatter plot and the corresponding parallel coordinate plot. We have generated 500 observations from an equal mixture of two Gaussian distributions. The distribution is the same as in Figure 1.19, but the sample size is only half of that in Figure 1.19.

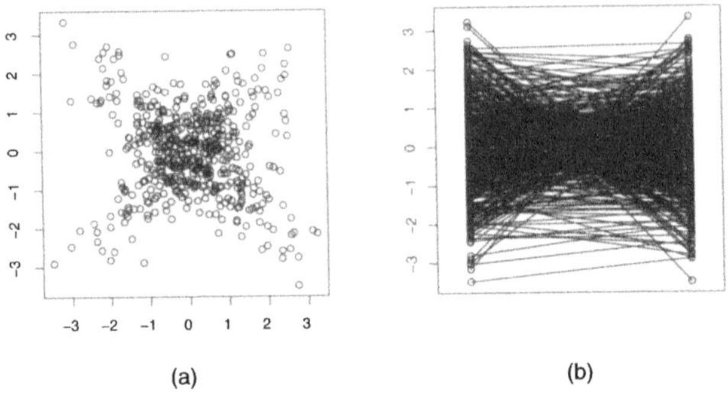

Figure 1.32 *(Parallel coordinate plot, illustration.)* (a) A scatter plot and (b) a corresponding parallel coordinate plot. The sample of size 500 is generated from a mixture of two Gaussian distributions. The distribution is the same as in Figure 1.19, where the sample size was 1000.

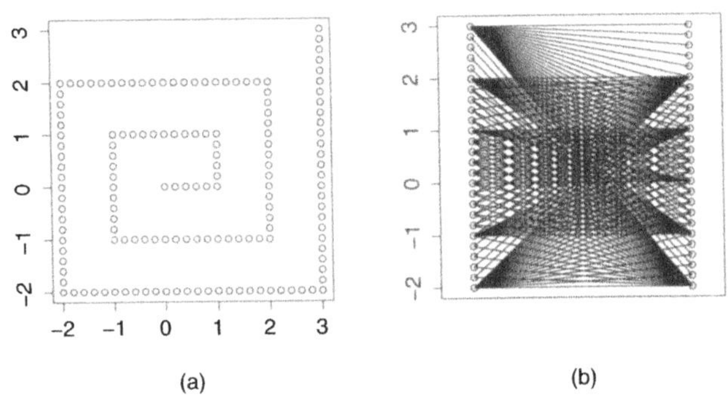

Figure 1.33 *(Parallel coordinate plot, rectangular spiral.)* (a) A scatter plot of points lying on a rectangular spiral and (b) a parallel coordinate plot of the points in panel (a).

3. Figure 1.33 visualizes a 2D rectangular spiral with a parallel coordinate plot.

4. Figure 1.34 visualizes a 2D Archimedean spiral with a parallel coordinate plot.

5. Figure 1.35 visualizes a 2D hyperbolic spiral with a parallel coordinate plot.

6. Figure 1.36 visualizes a 3D rectangular spiral with a parallel coordinate plot.

1.5.2 Multivariate Time Series

If we have n time series, and each time series is a vector of measurements at the same d time points, then the plot of these time series in a single window, joining the

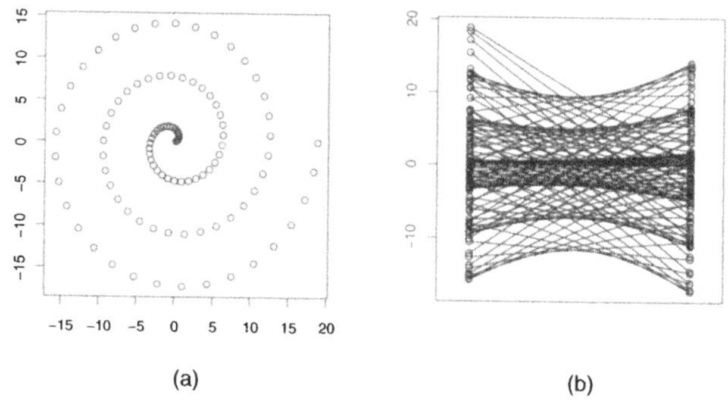

Figure 1.34 *(Parallel coordinate plot, Archimedean spiral.)* (a) A scatter plot and (b) a parallel coordinate plot.

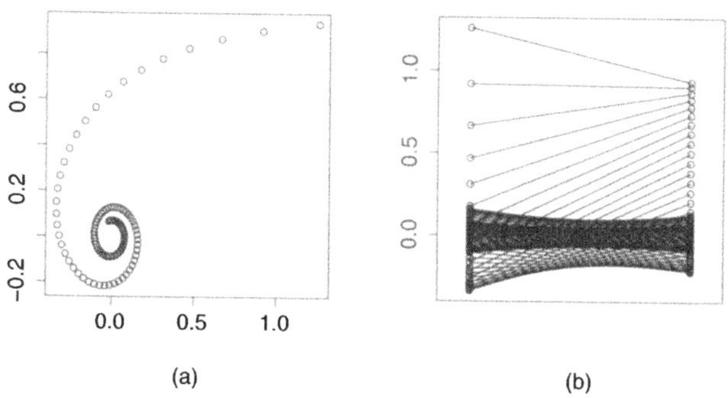

Figure 1.35 *(Parallel coordinate plot, hyperbolic spiral.)* (a) A scatter plot and (b) a parallel coordinate plot.

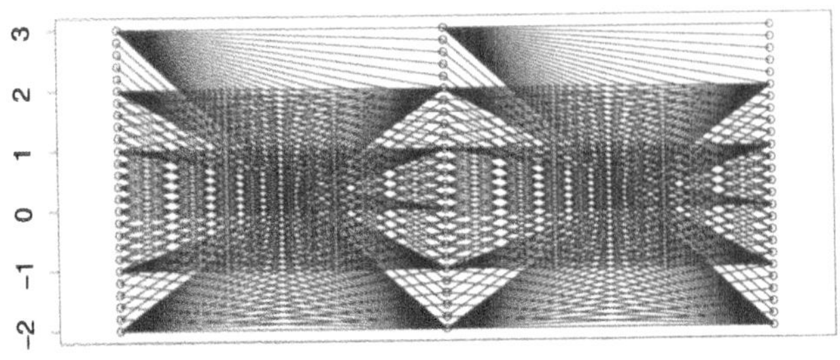

Figure 1.36 *(Parallel coordinate plot, 3D rectangular spiral.)* A parallel coordinate plot of points lying on a 3D rectangular spiral.

observations of a single time series with straight lines, is a parallel coordinate plot. Figure 1.7(c) shows a time series plot of a single time series.

1.5.3 Andrew's Curves

Andrew's curves, introduced in Andrews (1972), represent observations as one-dimensional curves. The curves are Fourier series whose coefficients are equal to the the observation values. The definition of the ith Andrew's curve is

$$f_i(t) = \begin{cases} 2^{-1/2} x_i^1 + \sum_{k=1}^{(d-1)/2} x_i^{2k} \sin(kt) + x_i^{2k+1} \cos(kt), \\ \qquad \text{when } d \text{ is odd}, \\ 2^{-1/2} x_i^1 + \sum_{k=1}^{d/2-1} \left[x_i^{2k} \sin(kt) + x_i^{2k+1} \cos(kt) \right] + x_i^d \sin(dt/2), \\ \qquad \text{when } d \text{ is even}, \end{cases}$$

for $i = 1, \ldots, n$, $d \geq 3$, where $t \in [-\pi, \pi]$. The ordering of the variables affects the visualization: the last variables will have only a small contribution to the visualization. Andrews (1972) suggests to use the ordering of the variables given by the principal component analysis; see Section 1.8.1 for the definition of the principal component analysis.

1.5.4 Faces

Faces are graphical representations of a data matrix where the size of face elements are assigned to variables (the columns of the data matrix). Chernoff (1973) introduced faces, and one often uses the term "Chernoff faces". Flury and Riedwyl (1981) and Flury and Riedwyl (1988) have further developed the face technique, and defined the following characteristics: right eye size, right pupil size, position of right pupil, right eye slant, horizontal position of right eye, curvature of right eyebrow, density of right eyebrow, horizontal position of right eyebrow, vertical position of right eyebrow, right upper hair line, right lower hair line, right face line, darkness of right hair, right hair slant, right nose line, right size of mouth, right curvature of mouth, and the same for the left side of the face.

Each observation is represented by one face. We have defined together 36 characteristics of a face. Thus we may visualize data up to 36 variables, but we are not able to visualize many observations with faces. With faces we may find clusters from data by looking groups of similar looking faces.

1.5.5 Other Possibilities

- *Profiles.* Profiles represent each observation by d vertical bars. Each bar has a height proportional to the value of the corresponding variable. Sometimes profiles are drawn as polygonal lines. Bertin (1967) uses profiles to make graphical matrices, as discussed in Section 1.4. Bar matrices shown in Figures 1.25 through 1.27 are related to profiles: a transposed bar matrix would show n rows of profiles.

- *Stars.* Stars are similar to profiles. The values of the variables are represented by equally spaced rays emanating from a fixed point. The endpoints of the rays are usually connected in a polygon. This definition of stars can be found in Kleiner and Hartigan (1981).

- *Glyphs.* Glyphs are similar to profiles and stars. The values of the variables are represented by not equally spaced rays emanating from a boundary of a circle of a fixed positive radius. Anderson (1960) defines glyphs.

- *Boxes.* Boxes represent each variable as a length in one of the three dimensions of a box. If there are more than three variables, then there will be several segments within each dimension. Hartigan (1975*b*) defines boxes.

- *Trees.* To draw trees, one makes first a tree with d nodes; each node corresponds to a variable (a hierarchical clustering algorithm is applied to construct the tree). All the observations have the same tree structure, and the values of the variables are represented with the widths of the branches of the tree. Using a tree structure for representing points makes the method less sensitive to the problem of the ordering of the variables. Kleiner and Hartigan (1981) introduce trees.

Härdle and Simar (2003) discuss parallel coordinate plots, Andrews's curves, and Chernoff faces. Kleiner and Hartigan (1981) introduce trees and castles and discuss profiles, stars, glyphs, faces, and boxes.

1.6 LINKING ACROSS DIMENSIONS

We have shown many ways to visualize multivariate data by making d univariate visualizations and linking these visualizations together, by identifying observations

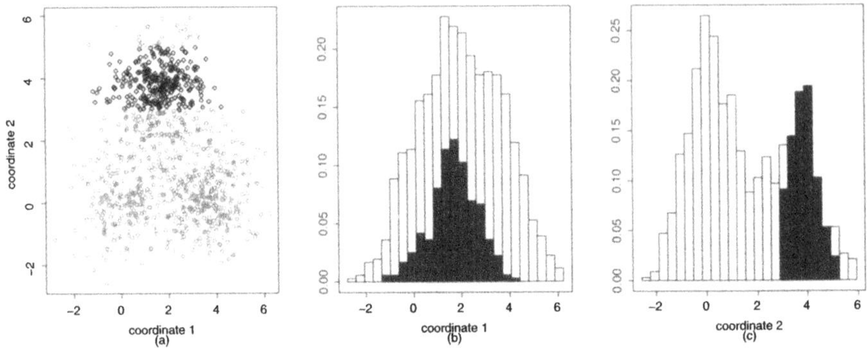

Figure 1.37 *(Linking across dimensions.)* Linking across dimensions illustrated with histograms.

across the dimensions. Parallel level plots identify observations across the dimensions by using an additional height-variable. Parallel coordinate plots identify observations across the dimensions by connecting them with straight lines. Graphical matrices make a layout of the data so that the attributes of a single observations are in the same row (or in a single column, depending on the arrangement). Tail tree plots, to be introduced in Chapter 6, enhance parallel level plots by coloring the observations with the help of a tail tree, and this helps identify observations across the 1D scatter plots.

When one makes several 2D projections and shows the scatter plots of the projected data, it is useful to apply brushing to identify observations across the scatter plots.

Brushing techniques can be used with all 1D visualization tools introduced in Section 1.2. For d one-dimensional visualizations brushing is used to connect observations across the dimensions. For example, one may select observations lying in an interval in one dimension and look how these observations are distributed in the other dimensions. It is useful to have a dynamic brushing tool to be able to try various subsets and refine the selections during the progress of exploration.

A particular example of linking across the dimensions is provided by the Attribute Explorer, introduced in Tweedie, Spence, Williams and Bhogal (1994) and Spence and Tweedie (1998), and whose use is illustrated in Spence (2001, Section 5.4). The Attribute Explorer makes d one-dimensional histograms, and each observation is identified with a rectangular region of a histogram bar. Figure 1.37 shows an example of using histograms in linking across dimensions. Panel (a) shows a data of size 1000. The data are generated from the same density as the data in Figure 1.1. Panel (b) shows a histogram of the first coordinate and panel (c) shows a histogram of the second coordinate, when the number of bins is 25. The observations whose second coordinate lies in the interval $[3, 5]$ are brushed.

1.7 DESCRIPTIVE STATISTICS

We will discuss descriptive statistics that characterize the distribution. The location of the data is characterized by a d dimensional vector that gives a center point for the data. We can make a very condensed visualization of the data by calculating a center point of the data. We can also characterize the dispersion of the data with various descriptive statistics. For example, box plots, shown in Figure 1.12, utilize descriptive statistics to make a visualization of data.

1.7.1 Location

Mean The expected value of random variable $X \in \mathbf{R}^d$ is written as EX. We use the notation EF for the mean of the distribution with distribution function

$$F(x) = P((-\infty, x]), \qquad x \in \mathbf{R}^d,$$

where $(-\infty, x] = (-\infty, x_1] \times \ldots \times (-\infty, x_d]$. When the distribution has density $f : \mathbf{R}^d \to \mathbf{R}$, we have

$$EF = \int_{\mathbf{R}^d} x f(x)\, dx,$$

assuming the integral is finite. A data version of the expected value may be taken to be the arithmetic mean:

$$\bar{X}_n = \frac{1}{n} \sum_{i=1}^{n} X_i.$$

Univariate Median A median of a univariate distribution F may be defined as such value $\text{med}(F) \in \mathbf{R}$ that

$$F(\text{med}(F)) = 0.5.$$

In terms of the density function f we have

$$\int_{-\infty}^{\text{med}(F)} f(y)\, dy = 0.5.$$

This equation may not uniquely define the median, since the distribution function F might not be genuinely increasing inside the support. The not genuinely increasing distribution function happens when the support is not an interval, or when the distribution is discrete. In general, we define

$$\text{med}(F) = \inf\{x \in \mathbf{R} : F(x) \geq 0.5\}.$$

A data version of median is a midpoint of the data, for example,

$$\text{med}_n = X_{([n/2])},$$

where $X_{(1)} < \cdots < X_{(n)}$ is the ordered sample. Then $\text{med}_n \approx \text{med}(F_n)$, where F_n is the empirical distribution function. Unlike the mean, the median always exists.

Multivariate Median In the multivariate case it is not straightforward to define a median. We give several definitions for *data depthness*. A depth function $D : \mathbf{R}^d \to [0, \infty)$ corresponding to a distribution function $F : \mathbf{R}^d \to \mathbf{R}$, is such that when $D(x)$ is large, x is close to the center of the distribution (it is deep inside the distribution), and when $D(x)$ is small (close to 0), x is far away from the center. A median would be a value maximizing D. We may define a data version of the depth function by defining the depth function corresponding to the empirical distribution function.

1. *Mahalabonis depth.* Define the Mahalabonis distance, corresponding to the distribution function F, by

$$d_M^2(x) = (x - EF)^T \Sigma_F^{-1} (x - EF),$$

where EF is the mean vector of the distribution and Σ_F is the covariance matrix of the distribution, defined in (1.4). Define the Mahalabonis depth function as

$$D_M(x) = (1 + d_M(x))^{-1}.$$

The data version is obtained by applying the sample mean and covariance.

2. *Half-space depth.* The half-space depth function is defined as

$$D_h(x) = \inf\{P_F(H) : H \text{ is a closed half-space in } \mathbf{R}^d \text{ such that } x \in H\}.$$

The data version is obtained by applying the empirical distribution function. We may also define the sample version in the following way: The depth of a point $x \in \mathbf{R}$, with respect to a one-dimensional data set, is the minimum of the number of data points in one side of x. The depth of a point $x \in \mathbf{R}^d$, with respect to a high-dimensional data set, is the smallest depth in any one-dimensional projection of the data set. That is, the depth of a point $x \in \mathbf{R}^d$, with respect to a data set $\{X_1, \ldots, X_n\} \subset \mathbf{R}^d$, is

$$D_{h,n}(x) = n^{-1} \min\{\#\{i : u^T X_i \le u^T x\} : \|u\| = 1\}.$$

The half-space depth was defined by Hodges (1955) and Tukey (1975).

3. *Data depths based on simplexes.* Let

$$d_O(x) = E_F \text{volume}[S(x, X_1, \ldots, X_d)],$$

where $S(x, X_1, \ldots, X_d) \subset \mathbf{R}^d$ is the closed simplex with vertices x and $X_1, \ldots, X_d$, and $X_1, \ldots, X_d$ are i.i.d. with distribution F. We define the Oja depth by

$$D_O(x) = (1 + d_O(x))^{-1},$$

as in Oja (1983). We define the simplicial depth by

$$D_s(x) = E_F I_{S(X_1, \ldots, X_{d+1})}(x),$$

where $S(X_1, \ldots, X_{d+1}) \subset \mathbf{R}^d$ is the closed simplex with $X_1, \ldots, X_{d+1}$ as vertices, and $X_1, \ldots, X_{d+1}$ are i.i.d. with distribution F. Simplicial depth was defined by Liu (1990).

Data based versions may be calculated with U-statistics. Define $d_n(x)$ as the average of $h(x, x_{i_1}, \ldots, x_{i_m})$ taken over all subsets $\{i_1, \ldots, i_m\}$ of size m from $\{1, \ldots, n\}$. That is,

$$d_n(x) = \binom{n}{m}^{-1} \sum_{*} h(x, x_{i_1}, \ldots, x_{i_m}),$$

where the summation is taken over all m-lets $(i_1, \ldots, i_m)$ such that $1 \le i_1 \le \cdots \le i_m \le n$. For the Oja depth we choose $m = d$ and h as the volume of the simplex formed by x and d data points: $h(x) = \text{volume}[S(x, X_{i_1}, \ldots, X_{i_d})]$, and $D_{O,n}(x) = (1 + d_n(x))^{-1}$. For the simplicial depth we choose $m = d+1$ and h as the indicator of a simplex formed by $d + 1$ data points: $h(x) = I_{S(X_{i_1}, \ldots, X_{i_{d+1}})}(x)$, and $D_{s,n}(x) = d_n(x)$. In the case of the simplicial depth one computes the fraction of the sample random simplexes containing the point x.

4. *Data depths based on the L_1 norm.* Define

$$D_{L_1}(x) = (1 + d_{L_1}(x))^{-1},$$

where we may consider the following possibilities:

$$d_{L_1}(x) = \begin{cases} E\|x - X\| \\ E\|x - X\|_{\Sigma^{-1}}, \end{cases}$$

where Σ is the covariance matrix and $\|x\|^2_{\Sigma^{-1}} = x^T \Sigma^{-1} x$. These depth functions were studied, for example, by Zuo and Serfling (2000). Note that after defining med(F) with an L_1 depth function, the Mahalabonis distance may be modified to $d_{M,2}(x) = \|x - \text{med}(F)\|_{\tilde{\Sigma}^{-1}}$, where $\tilde{\Sigma}$ is a suitable covariance matrix. This distance was proposed in Serfling (2002*a*).

5. *Convex hull peeling depth.* The convex hull peeling depth has only the data version, and this depth is defined only at the data points $X_1, \ldots, X_n$, but not at arbitrary points $x \in \mathbf{R}^d$. Let the data points at the perimeter of the convex hull of the data have the smallest depth 1. Remove these data points and form the convex hull of the remaining data points. The points at the perimeter have the one step larger depth 2. Continue until a nested sequence of convex hulls is formed and all data points have been reached. Convex hull peeling depth was defined by Barnett (1976).

Liu, Parelius and Singh (1999) give a review of applications of data depthness. Many of the notions of a data depth are related to level sets or to estimates of the level sets; consider the depth regions

$$\{x \in \mathbf{R}^d : D(x) \ge \lambda\}.$$

Depth regions are often close to level sets of the density f of a unimodal distribution. Indeed one way to define a depth function is to take the depth equal to the density: $D(x) = f(x)$. We may call this the density depth, or the likelihood depth.

When the density is elliptical, as defined in Section 10.1.3, then the depth regions of Mahalabonis depth are equal to the level sets of the density. The simplicial depths measure the local concentration of the probability mass and are thus related to the density depth. The depth regions of convex peeling depth estimate the level sets of densities whose level sets are convex.

Mode A mode may be defined to be a point at which the density has a local maximum. In the univariate case the unimodality is relatively straightforward to define.

Definition 1.3 (Unimodal 1D distribution.) *A distribution is* unimodal *about a mode* $\mu \in \mathbf{R}$, *if the distribution function* $F(t) = P((-\infty, t])$ *is convex on* $(-\infty, \mu)$ *and concave on* (μ, ∞).

In the multivariate case there are many alternative definitions of unimodality. A very general definition is the following:

Definition 1.4 (Star unimodality.) *A distribution with density* f *is* star unimodal *about* $\mu \in \mathbf{R}^d$, *if for every* $\lambda > 0$ *the level set* $\{x : f(x) \geq \lambda\}$ *is star-shaped about* μ.

Set $A \subset \mathbf{R}^d$ is called star shaped when there exists a reference point $\mu \in A$ so that for every $x \in A$ the line segment joining x and μ belongs to A: $\{\mu + p(x - \mu) : p \in [0, 1]\} \subset A$; see also Definition 5.2. The set of reference points about which A is star shaped is called the kernel of A. The star unimodality is equivalent to the following two conditions:

1. A distribution with density f is star unimodal about $\mu \in \mathbf{R}^d$ if $f(\mu + ux) \leq f(\mu + tx)$ for every $0 < t < u < \infty$ and $x \neq 0$. That is, the density is decreasing on rays emanating from μ.

2. A distribution is star unimodal about $\mu \in \mathbf{R}^d$ if it belongs to the closed convex hull of the set of all uniform distributions on sets that are star shaped about μ. Here the topology on the set of probability measures is the topology of weak convergence. This criterion does not require that the density exists.

The equivalence of these two conditions with the star unimodality is proved in Dharmadhikari and Joag-dev (1988, p. 38). Figure 1.38(a) shows a density that is not star unimodal. We may define the unimodality so that also this density is unimodal. (Figure 1.38(b) is discussed in the connection of Lemma 2.1.)

Definition 1.5 (Level set unimodality.) *A distribution with density* f *is* level set unimodal, *if for every* $\lambda > 0$ *the level set* $\{x : f(x) \geq \lambda\}$ *is connected.*

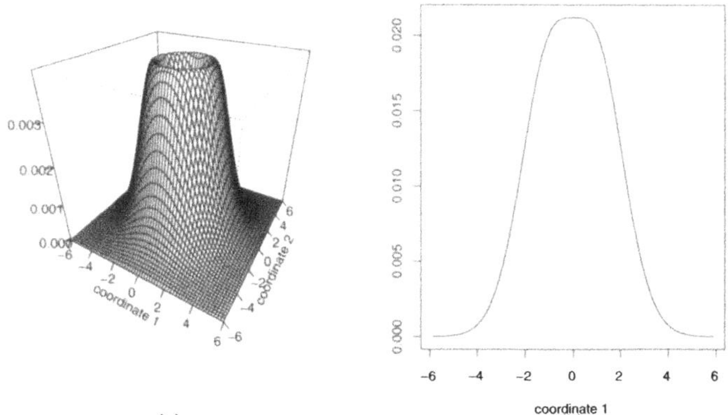

(a) (b)

Figure 1.38 *(Star unimodality and hat function.)* (a) A density which is not star unimodal. (b) A marginal density of the density in (a). The density is defined in (2.5).

A density that is unimodal according to Definition 1.4 or Definition 1.5 may have many maxima, and we may define the "mode" to be the set of maxima. For a uniform distribution with a star-shaped or connected support the whole interior of the support could be defined as the mode. It is possible to generalize Definition 1.4 to the definition of *α-unimodality.*

Definition 1.6 (α-unimodality.) *A distribution P is* α-unimodal *about $\mu \in \mathbf{R}^d$, for the index $\alpha \geq 0$ if, for $X \sim P$, $t^\alpha E[g(t(X - \mu))]$ is nondecreasing in $t \in (0, \infty)$, for every bounded, nonnegative, Borel measurable function $g : \mathbf{R}^d \to \mathbf{R}$.*

If a distribution is α-unimodal, then the distribution is β-unimodal for $\beta > \alpha$. A distribution with density f is α-unimodal about $\mu \in \mathbf{R}^d$, if $t^{d-\alpha} f(\mu + tx)$ is non-increasing on $t \in (0, \infty)$, for every $x \neq 0$. This is proved in Dharmadhikari and Joag-dev (1988, Thm. 3.6). A star unimodal distribution is α-unimodal with $\alpha = d$. An interest in α-unimodality comes from the fact that the marginals of α-unimodal distributions are α-unimodal. This does not hold for the other two concepts of multivariate unimodality which we have defined; see Lemma 2.1.

We have followed Dharmadhikari and Joag-dev (1988) in the exposition of unimodality, except for Definition 1.5.

Estimation of the modes (the location of the modes) may be done with the help of density estimators: one estimates the density and finds the modes of the estimate. This was discussed in Parzen (1962). The estimation of the location of the mode is as difficult as the estimation of the gradient (first derivative) of the density: for s-smooth density the rate of convergence is $n^{(s-1)/(2s+d)}$.

1.7.2 Dispersion

Dispersion Matrix A dispersion matrix or a scatter matrix is defined as a functional that maps d-dimensional random vectors to $(d \times d)$-dimensional matrices. A dispersion matrix $\Sigma(X)$ of a random vector $X = (X_1, \ldots, X_d)$ is a symmetric and positive semi-definite matrix that is affine invariant. Affine invariance means that

$$\Sigma(a + BX) = B\Sigma(X)B^T,$$

for every d-vector a and invertible $d \times d$-matrix B.

In order to express the shape of an elliptical distributions with a dispersion matrix it is useful to write the matrix in a canonical form. A positive semi-definite symmetric matrix Σ can be written as

$$\Sigma = \lambda UCU^T,$$

where $\lambda > 0$, U is an orthogonal matrix of eigenvectors, and C is a diagonal matrix of standardized eigenvalues, so that $|\det C| = 1$. We may call λ the *scale*, λ^2 the *generalized variance*, C the *shape*, and U the *orientation*.

Below are some examples of dispersion matrices.

- The *covariance matrix* of distribution F, or the random vector $X \sim F$, is defined by

  $$\Sigma = \mathrm{Cov}(X) = E\left[(X - EF)(X - EF)^T\right]. \quad (1.4)$$

  The covariance matrix can be estimated by

  $$\Sigma_n = \frac{1}{n}\sum_{i=1}^{n}(X_i - \bar{X}_n)(X_i - \bar{X}_n)^T, \quad (1.5)$$

  where $\bar{X}_n$ is the arithmetic mean.

- The positive semi-definite symmetric matrix Σ in the definition of an elliptical distribution in (10.4) is a dispersion matrix.

- *Modal dispersion* is defined for a random vector X with density f_X by

  $$\Sigma(X) = -\left(D^2 f_X(\mathrm{mode}(f_X))\right)^{-1},$$

  where $D^2(f)$ is the Hessian matrix of the second derivatives of f and $\mathrm{mode}(f)$ is the mode of f; see Meucci (2005), page 368. The modal dispersion may be motivated by the observation: the larger the absolute value of the second derivative, the thinner is the density function around its mode and thus the less disperse is the distribution. Clearly, modal dispersion is a local concept (unlike the covariance matrix). Also, in order for modal dispersion to be defined, we need to assume that the density is unimodal.

Note that we define various measures for dependency in Section 9.3.2. If we have a measure $\rho_{i,j}$ for the dependency of $X_i, X_j \in \mathbf{R}$, then we may construct the $d \times d$ matrix R with the elements $\rho_{i,j}$, and a similar matrix may be constructed for data versions. These matrices R are not always dispersion matrices.

Dispersion Vector We may use the generalized correlation, defined in (9.5), to define a dispersion vector. The generalized correlation is the correlation between a real valued random variable and a vector. Denote with Z_i the vector $X = (X_1, \ldots, X_d)$ without the ith component: $Z_i = (X_1, \ldots X_{i-1}, X_{i+1}, \ldots X_d)$. Define similarly to (9.5),

$$\rho_{G,i} = \max_{\phi} \rho(X_i, \phi^T Z_i),$$

where ρ is the linear correlation defined in (9.3), and the maximum is over k-vectors. Then we may use the vector $(\rho_{G,1}, \ldots, \rho_{G,d})$ to describe the spread of $(X_1, \ldots, X_d)$. However, a typical characterization of the spread is the covariance matrix, or more generally a dispersion matrix.

1.7.3 Higher Order Moments

We may use higher moments to characterize the distribution. Let $i_1, \ldots, i_k \in \{1, \ldots, d\}$. The kth central moment is defined as

$$E\left[(X_{i_1} - EX_{i_1}) \cdots (X_{i_k} - EX_{i_k})\right].$$

For a univariate distribution we may define the kth standardized moment as

$$\frac{\mu_k}{\sigma^k},$$

where $\mu_k = \int_{-\infty}^{\infty} (x - EF)^k f(x)\, dx$ is the kth moment about the mean and σ is the standard deviation, $\sigma^2 = \int_{\mathbf{R}^d} (x - EF)^2 f(x)\, dx$, and where f is the density and EF is the expectation. Value μ_3/σ^3 is called the skewness and μ_4/σ^4 is called the kurtosis. The kurtosis measures the heaviness of the tails of the density. Often the kurtosis is defined as $\mu_4/\sigma^4 - 3$, to make the kurtosis of the normal distribution equal to zero. For a definition of skewness and kurtosis in terms of cumulants, see (1.9).

1.8 DIMENSION REDUCTION OF DATA

1.8.1 Principal Components

Principal component analysis tries to find linear combinations of the variables that will contain as much information as possible. We reduce the dimension by choosing only the most important linear combinations.

Definition of the Principal Component Transformation In the theoretical version of the principal component transformation we use the spectral representation of the covariance matrix Σ of random vector $X \in \mathbf{R}^d$, as defined in (1.4). In the empirical version the covariance matrix Σ is replaced by the empirical covariance matrix Σ_n, defined in (1.5).

1. The *spectral representation*, or the *eigen decomposition*, of Σ is given by

$$\Sigma = A \Lambda A^T, \tag{1.6}$$

where Λ is the $d \times d$ diagonal matrix of the *eigenvalues* of Σ and A is the orthogonal $d \times d$ matrix of the *eigenvectors*, or the *principal components*, of Σ (the columns of A are the eigenvectors); see Appendix B.7. If Σ has rank d, then all λ_i are positive. Principal component transformation is defined by

$$Y = A^T(X - EX) \in \mathbf{R}^d. \tag{1.7}$$

The covariance matrix of Y is the matrix Λ of eigenvalues: $E(YY^T) = \Lambda$.

2. The empirical spectral representation can be calculated by finding the *singular value decomposition* of the centering of the $n \times d$ data matrix $\mathbb{X}_n$:

$$\mathbb{X}_n - 1_{n\times 1}\bar{X}_n^T = UDV^T,$$

where U is a $n \times d$ orthogonal matrix of left singular vectors, V is a $d \times d$ orthogonal matrix of right singular vectors, and D is a $d \times d$ diagonal matrix of singular values; see Appendix B.7. Here we denote with $1_{n\times 1}$ the $n \times 1$ column vector of ones and $\bar{X}_n$ is the $d \times 1$ column vector of arithmetic means. Then

$$\Sigma_n = \frac{1}{n}\left(UDV^T\right)^T \cdot UDV^T = \frac{1}{n} VD^2V.$$

Thus $A = V$ and $\Lambda = n^{-1}D^2$.

Data sphering is the principal component transformation followed by the normalization of the marginal variances to unity. Examples of data sphering are given in Section 9.2.1.

Optimality of the Principal Component Transformation Let Y_1 and Y_2 be the first two components of Y in (1.7). There is no other linear combination with larger variance than Y_1, and Y_2 is the linear combination, uncorrelated with Y_1, with the second largest variance. More precisely, let

$$a_1 = \text{argmax}_{a:\|a\|=1}\text{Var}(a^T X). \tag{1.8}$$

Then a_1 is the column of A corresponding to the largest eigenvalue. Let

$$a_2 = \text{argmax}_{a:\|a\|=1, a\perp a_1}\text{Var}(a^T X).$$

Then a_2 is the column of A corresponding to the second largest eigenvalue.

Dimension Reduction with the Principal Component Transformation We can define the dimension reduction for the theoretical version or for the empirical version.

1. The theoretical dimension reduction proceeds as follows: choose random vector $(Y_1, \ldots, Y_{d'})^T$, where $1 \le d' \le d$ and Y_i are the elements of vector Y defined in (1.7). Now d' could be chosen so that the ratio

$$\frac{\sum_{k=1}^{d'} \lambda_k}{\sum_{k=1}^{d} \lambda_k}$$

is sufficiently close to 1, for example, larger than 0.95, where λ_i, $i = 1, \ldots, d$, are the eigenvalues (the diagonal elements of Λ); we assume that the eigenvalues are in order $\lambda_1 \ge \lambda_2 \ge \cdots \ge \lambda_d > 0$.

2. The empirical dimension reduction finds new data matrix $\mathbb{Y}_n$ of dimension $n \times d'$,

$$\mathbb{Y}_n = \left(\mathbb{X}_n - 1_{n\times 1}\bar{X}_n^T\right) \times (a_1 \cdots a_{d'}),$$

where $\mathbb{X}_n$ is the original $n \times d$ data matrix, and $(a_1 \cdots a_{d'})$ is the $d \times d'$ matrix of the d' first eigenvectors of Σ_n. When principal component analysis is used to reduce dimension, we obtain new random variables that are linear combinations of the original random variables. Thus it may be difficult to interpret the new random variables.

1.8.2 Projection Pursuit

Projection pursuit tries to find interesting low-dimensional projections of high-dimensional data by maximizing a projection index. In the theoretical version of projection pursuit we find a projection matrix A that maximizes $Q(AX)$, where $X \in \mathbf{R}^d$ is a random vector and Q is the projection index. Huber (1985) mentions the following projection indexes, which are defined for the case of 1D projections. In all cases one can show that $Q(X) \ge 0$ and $Q(X) = 0$ if X is normal.

1. *(Cumulant.)* Let

$$Q(X) = |c_m(X)|/c_2(X)^{m/2}, \qquad m > 2,$$

where c_m is the mth cumulant:

$$c_m = \frac{1}{i^m}\frac{\partial^m}{\partial t^m} \log E\exp\{itX\}\bigg|_{t=0}, \qquad m \ge 2, \tag{1.9}$$

where i is the imaginary unit. For $m = 3$ we obtain the absolute skewness and for $m = 4$ the absolute kurtosis.

2. *(Fisher information.)* Let

$$Q(X) = \mathrm{Var}(X)I(X) - 1,$$

where $I(X)$ is the Fisher information

$$I(X) = \frac{\int_{-\infty}^{\infty}(f_X')^2}{f_X},$$

where f_X is the density of X.

3. *(Shannon entropy.)* Let

$$Q(X) = -S(x) + \log\left(\mathrm{std}(X)\sqrt{2\pi e}\right),$$

where $S(X)$ is the Shannon entropy

$$S(x) = -\int_{-\infty}^{\infty} f_X \log f_X,$$

where f_X is the density of X.

Note that principal component analysis can be seen as projection pursuit, where the projection index is the variance, see (1.8).

Huber (1985) gives a review of projection pursuit. The term "projection pursuit" was coined by Friedman and Tukey (1974). Further studies of projection pursuit include Cook, Buja and Cabrera (1993), Cook, Buja, Cabrera and Hurley (1995).

1.8.3 Self-organizing Maps

Self organizing map (SOM) is a discretized scatter plot of a two-dimensional data that is obtained from the original data by a nonlinear mapping. SOM finds k cluster centers, or prototypes, and each prototype is associated with a two-dimensional index in $I = \{i_1, \ldots, i_N\} \times \{j_1, \ldots, j_M\}$. Each observation in $\{X_1, \ldots, X_n\}$ is mapped to the closest cluster center, and thus each observation is associated to an index in I. The plot of those indexes is a self-organizing map of the data. Kohonen (2001) gives an exposition of SOM. We present the batch version of SOM following Hastie, Tibshirani and Friedman (2001).

1. Choose k cluster centers $\mathcal{M} = \{m^1, \ldots, m^k\} \subset \mathbf{R}^d$. Each cluster center is associated with an index $l(m) \in I$.

2. Iterate the following steps until convergence.

   (a) Go through the cluster centers $m \in \mathcal{M}$.

   (b) Consider cluster center m. Find the set of cluster centers in the neighborhood of m: $\mathcal{M}_m = \{m' : \delta(l(m'), l(m)) \le \rho\}$, where δ is, for example, the Euclidean distance.

   (c) Find the observations $\mathbb{X}_m$ that belong to the Voronoi neighborhoods of the cluster centers in $\mathcal{M}_m$:

   $$\mathbb{X}_m = \bigcup \{\mathcal{X}_{m'} : m' \in \mathcal{M}_m\},$$

   where $\mathcal{X}_{m'}$ is the Voronoi neighborhood of m' (the set of those observations that are closer to m' than to any other cluster center):

   $$\mathcal{X}_{m'} = \{X_i : \|X_i - m'\| = \min\{\|(X_i - m''\| : m'' \in \mathcal{M}\}\}.$$

(d) Update m to be the weighted average of the observations in $\mathbb{X}_m$:

$$m = \frac{\sum_{X \in \mathbb{X}_m} w_X X}{\sum_{X \in \mathbb{X}_m} w_X},$$

where $w_X > 0$ are weights.

Define $\mu : \{X_1, \ldots, X_n\} \to \mathcal{M}$ to be the mapping of observations to the closest cluster center:

$$\|\mu(X_i) - X_i\| = \min\{\|m' - X_i\| : m' \in \mathcal{M}\}.$$

(Then we may write $\mathcal{X}_m = \{X_i : m = \mu(X_i)\}$, where $\mathcal{X}_m$ is the Voronoi neighborhood of m.)

Definition 1.7 (Self-organizing map.) *) A* Self-organizing map *of data* $\{X_1, \ldots, X_n\}$ *is the scatter plot of* $\{l(\mu(X_i)) : i = 1, \ldots, n\}$.

The application of the algorithm requires a choice for the radius $\rho > 0$ and the weights w_X. When ρ is so small that $\mathcal{M}_m = \{m\}$, then the cluster centers are equal to the cluster centers of the k-means algorithm, defined in Chapter 8.2.

1.8.4 Multidimensional Scaling

In multidimensional scaling one tries to find such mapping of the d-dimensional data to a two-dimensional Euclidean space that the distances of the observations remain as close to the original distances as possible; that is, one tries to find such a mapping $Q : \{X_1, \ldots, X_n\} \to \mathbf{R}^2$, that the stress functional

$$\sum_{1 \le i < j \le n} \left(\|X_i - X_j\| - \|Q(X_i) - Q(X_j)\|\right)^2$$

is minimized. Sammon's mapping uses the stress functional

$$\sum_{1 \le i < j \le n} \frac{\left(\|X_i - X_j\| - \|Q(X_i) - Q(X_j)\|\right)^2}{\|X_i - X_j\|}.$$

This stress functional emphasizes small distances. For example, the Newton–Raphson algorithm can be used for the minimization of the stress functional.

CHAPTER 2

VISUALIZATION OF FUNCTIONS

A one-dimensional function may be visualized by plotting the graph of the function (scatter plot of the graph). A two-dimensional function may be visualized by drawing a perspective plot of the graph of the function or a contour plot showing the level curves of the function. A three-dimensional function may be visualized by drawing a perspective plot of level surfaces. In order to use perspective plots and contour plots for a high-dimensional function one has to reduce the dimension of the function. The use of slices and projections is classical. For example, four-dimensional functions may be visualized by visualizing 3D level curves of a series of slices as the fourth variable is changed over its range. These issues are discussed Section 2.1.

There are many ways to transform a function to a lower dimensional function. Section 2.2 discusses useful transforms in order to visualize the spread of a multivariate distribution. The concept of a level set of a density function appears as a central concept, anticipating level set trees and shape trees discussed in later chapters. The spread of a multivariate distribution is visualized with the help of univariate functions. One may distinguish between density type visualizations and distribution function type visualizations. Density type visualizations depict the functional relationship between the level and the volume of the level sets of a density, whereas distribution function type visualizations apply general sequences of nested sets and visualize the probability content of the sets as function of the volume of the sets.

A multivariate function is given (stored in a computer) typically by either giving its values on a multivariate grid or by defining it as a linear combination of some simple functions, for example, as an expansion in an orthonormal system or as a mixture of Gaussians. Kernel estimates are mixtures of scaled kernel functions. Orthogonal series estimators are expansions with basis functions. Boosting and bootstrap aggregation give density estimates that have a mixture form, with a large number of mixture members. Even when there are only few terms in the linear combination, it is typically difficult to grasp the shape of the function if we are only given the coefficients of the expansion. That is why we need visualization tools.

2.1 VISUALIZATION OF LOW-DIMENSIONAL FUNCTIONS

2.1.1 One-dimensional Functions

Graphs and ff-Plots A one-dimensional function may be visualized by plotting the graph of the function. An ff-plot may be used to compare two 1D functions.

Graph The graph of the function $f : \mathbf{R} \to \mathbf{R}$ is the set of points

$$\{(t, f(t)) : t \in \mathbf{R}\}. \tag{2.1}$$

Figure 2.1 shows examples of graphs of 1D densities: the double exponential, Gaussian, Student, and Bartlett densities. The densities are defined in Section 10.1.1. In Figure 2.1 the standard Gaussian density is used as the reference density. With the help of the reference density, the Student density and the Bartlett density may be compared although they are in different windows. Using a reference function is important since the aspect ratio affects the visual appearance of the functions.

A graph is defined in (2.1) as a set of infinite cardinality. In practice, we have available the values of the function on a finite grid. The function is drawn (on a finer grid) by interpolating the function values from the courser grid to the finer grid. An alternative way to plot a nonregular function is to use a rather course grid, not use interpolation, but show the behavior of the function between the grid points by a vertical line. Figure 2.2 shows a time series of the prices of the NASDAQ stock index in the period of three months between 3 April 2006 and 28 June 2006 (61 observations). The opening, closing, highest, and lowest daily prices are shown. The height of the vertical line shows the span between the lowest and the highest price, the horizontal line that points to the left shows the opening price, and the horizontal line that points to the right shows the closing price.

ff-Plot We may visualize function $f : \mathbf{R} \to \mathbf{R}$ by plotting the points

$$\{(\phi(t), f(t)) : t \in \mathbf{R}\},$$

where $\phi : \mathbf{R} \to \mathbf{R}$ is a reference function. This plot may be called an ff-plot, in analogy to the PP-plot and QQ-plot of Definition 1.1. An ff-plot is used to highlight

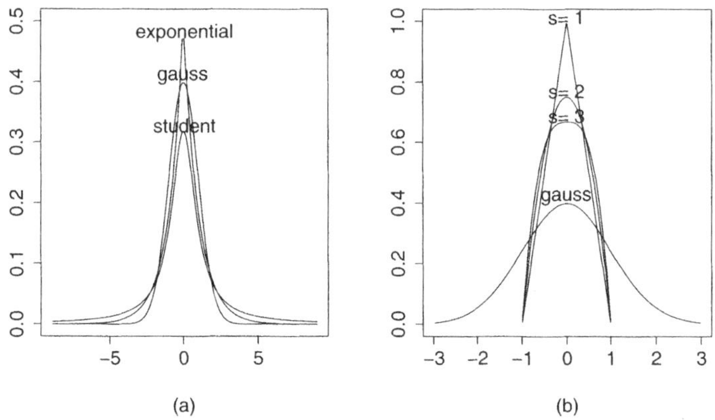

Figure 2.1 *(Graphs of density functions.)* (a) The double exponential, standard Gaussian, and Student density with degrees of freedom $\nu = 1$. (b) The polynomial densities with $s = 1, 2, 3$, and the standard Gaussian density. The densities are defined in Section 10.1.1.

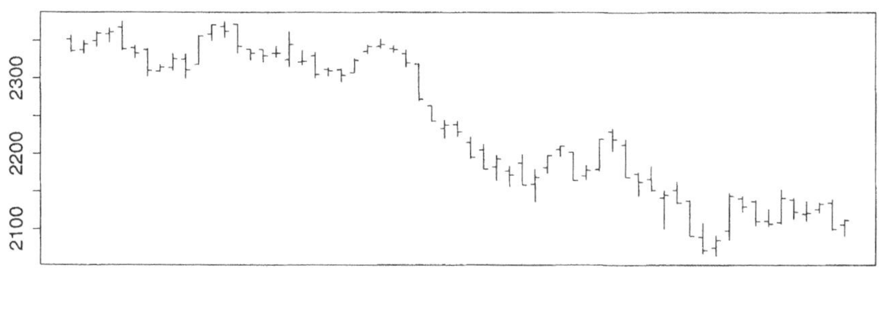

Figure 2.2 *(Bar chart of stock prices.)* Time series of prices of the NASDAQ stock index during the period 3 April 2006 to 28 June 2006.

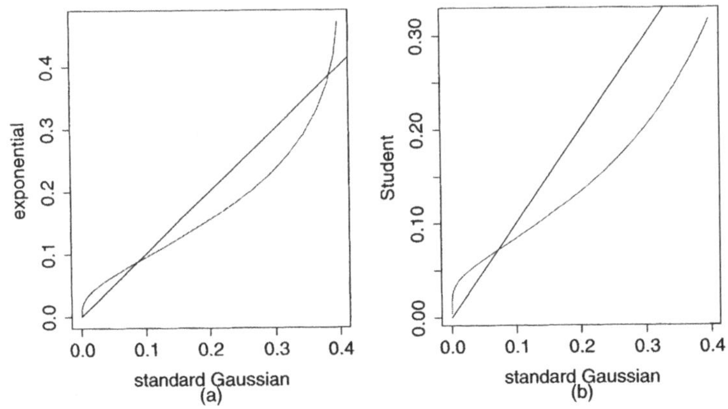

Figure 2.3 *(ff-plots of densities.)* (a) An ff-plot of the double exponential versus the standard Gaussian. (b) An ff-plot of the Student density with degrees of freedom $\nu = 1$ versus the standard Gaussian.

the differences between the function and a reference function. We plot also the line $x = y$ at the window: if f and ϕ are equal, then an ff-plot shows only the line $x = y$. If ϕ is not monotonic, then the points $(\phi(t), f(t))$ and $(\phi(u), f(u))$ may coincide although $t \neq u$. Thus we might not be able to recover the function when we have available only an ff-plot. Note that if ϕ is monotonic, then an ff-plot amounts to plotting the graph of the function with a rescaled x-coordinate, and thus we can recover the function when we have available an ff-plot with a monotonic reference function ϕ.

Figure 2.3 shows ff-plots of the double exponential density and a Student density. Figure 2.3(a) shows an ff-plot of the double exponential with relation to the standard Gaussian density. We can see that the double exponential takes larger values at the tails and at the center, but otherwise the standard Gaussian takes larger values. Indeed the ff-plot stays above the curve $x = y$ for the low values of the functions, then crosses this line, but goes again above the line for large values of the functions. Figure 2.3(b) shows an ff-plot of the Student density with degrees of freedom $\nu = 1$ with relation to the standard Gaussian. The Student density dominates at the tails, but the standard Gaussian dominates at the center.

Aspect Ratio The aspect ratio of a graph is the height of the data rectangle divided by the width (Cleveland 1993*b*, p. 88). The judgment of the curve is based on the aspect ratio. Figure 2.4 shows the daily closing prices of the SP500 stock index during the four year period 1 June 2005 to 20 May 2009 (there are 1000 observations). The prices are interpolated with straight lines. Three aspect ratios are shown. The changes in the price seem more moderate (the volatility appears to be small) when the width of the data rectangle is the largest (the aspect ratio is small).

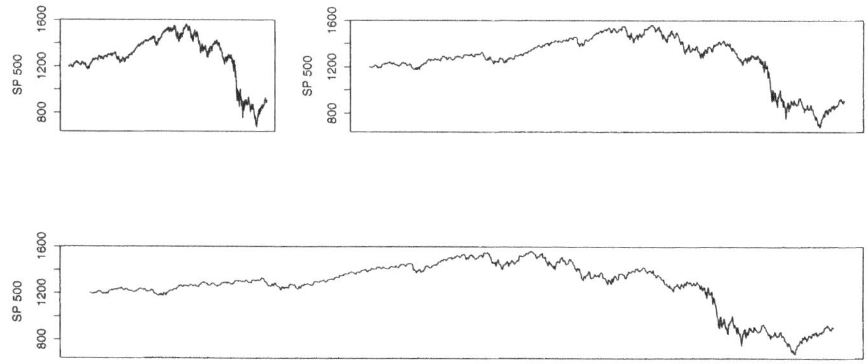

Figure 2.4 *(Three different aspect ratios.)* The value of the SP500 stock index is shown during the period 1 June 2005 to 20 May 2009.

Let us assume that a curve is made up of line segments. According to Cleveland, McGill and McGill (1988) and Cleveland (1993*a*), the judgments of a curve are optimized when the absolute values of the orientations are centered on 45°. One calls *banking* choosing the aspect ratio of a graph to enhance the perception. Banking to 45° is achieved when

$$\frac{\sum_{i=1}^{k} \theta_i l_i}{\sum_{i=1}^{l} l_i} = 45^o,$$

where k is the number of line segments making up the curve, θ_i is the orientation of the ith line segment, and l_i is the length of the ith line segment (Cleveland 1993*b*, p. 90).

Distributions A one-dimensional distribution may be visualized by visualizing the density function or by visualizing the distribution function.

Density Functions Figure 2.1 shows graphs of densities and Figure 2.3 shows ff-plots of densities.

Distribution Functions A distribution function is defined by $F(t) = \int_{-\infty}^{t} f(u)\, du$, $t \in \mathbf{R}$, where f is the density function. Figure 2.5(a) shows graphs of the distribution functions of the double exponential, standard Gaussian, and Student distribution with degrees of freedom $\nu = 1$. Note that the Student distribution function does not reach the value 1, since we have plotted the functions only in the range $[-8, 8]$. Panel (b) shows the distribution functions of the polynomial densities with $s = 1, 2, 3$, and the distribution function of the standard Gaussian density. The distributions are defined in terms of densities in Section 10.1.1. The plotting range varies between the panels, which affects the visual appearance. We have plotted the standard Gaussian

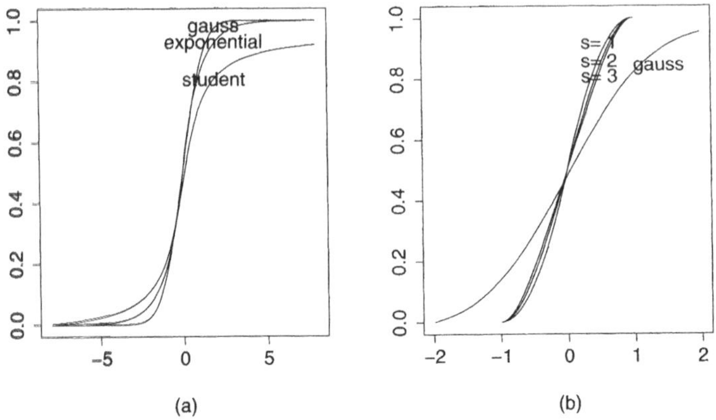

Figure 2.5 *(Graphs of distribution functions.)* (a) The double exponential, standard Gaussian, and Student distribution with degrees of freedom $\nu = 1$. (b) The polynomial distributions with $s = 1, 2, 3$, and the standard Gaussian distribution.

distribution function in both panels, in order to make the comparison between the distributions easier.

An ff-plot is called an PP-plot when the functions are distribution functions (and the other distribution function is an empirical distribution function). The PP-plot is defined in Definition 1.1. Figure 2.6(a) shows an PP-plot of the double exponential with relation to the standard Gaussian distribution function. Note that the line of the PP-plot crosses the line $x = y$ three times. This behavior is difficult to detect from the graphs of the distribution functions. Figure 2.6(b) shows an PP-plot of the Student distribution with degrees of freedom $\nu = 1$ set against the standard Gaussian distribution. Now the line of the PP-plot crosses the line $x = y$ only once.

Multimodality Distribution functions are useful in visualizing the tail behavior of distributions. However, they are not helpful in visualizing multimodality of distributions, since modes are local maxima of densities. Modes of a distribution may be detected from a graph of a distribution function by looking at convex and concave parts of the distribution function: a univariate distribution is unimodal with mode μ if the distribution function is convex on $(-\infty, \mu)$ and concave on (μ, ∞). Figure 1.9 shows both the density function and the distribution function of a distribution with multimodal density (mixture of Gaussians).

2.1.2 Two- and Three-dimensional Functions

Perspective plots of functions may be applied to visualize two-dimensional functions. Perspective plots of contours may be applied to visualize three-dimensional functions. Another tool for visualizing two-dimensional functions is a contour plot.

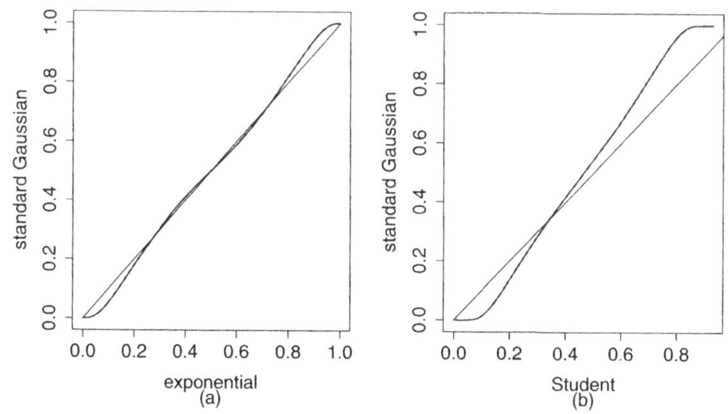

Figure 2.6 *(ff-plots of distribution functions.)* (a) An ff-plot of the double exponential versus the standard Gaussian. (b) An ff-plot of the Student distribution function with degrees of freedom $\nu = 1$ versus the standard Gaussian.

Perspective Plots The graph of function $f : \mathbf{R}^d \to \mathbf{R}$ is the set of points $(x, f(x))$, $x \in \mathbf{R}^d$. A graph is a set in the $d+1$-dimensional Euclidean space. When $d = 2$, we may make a perspective plot of this 3D set.

Figure 2.7(a) shows a perspective plot of a mixture of three standard Gaussian densities. The mixture weights are $p_1 = 0.45$, $p_2 = 0.35$, and $p_3 = 0.25$, and the mixture means are $m_1 = (1.75, 3.5)$, $m_2 = (3.5, 0.2)$, and $m_3 = (0, 0)$.

Figure 2.7(a) is a wireframe plot, with hidden line removal. The 3D effect is provided by the perspective and by the hiding of the features in the background. The box around the plot enhances the perception of depth. The function is evaluated on a grid, which makes a 3D point cloud, and the wireframe plot shows the lines that connect the points. Terms "surface plot", "mesh plot", or "wire mesh surface" are in use. (In a wire surface graph secondary lines are placed between the grids.)

Cleveland (1993*b*, p. 249) lists the following choices to be made in wireframe rendering: grid density (the grid density cannot be too large since then the lines connecting the points merge and the perception of the depth is lost), the orientation of the surface, viewing distance, ratios of the lengths of the three dimensions of the plot, and overall size of the display. It is also possible to use an orthogonal view, instead of rendering the surface in perspective. In an orthogonal view parallel lines moving from foreground to background stay a constant distance apart.

Another way to make a perspective plot is to use a shaded surface. Cleveland (1993*b*, p. 268) lists the following choices to be made in shading: reflective properties of the surface (matte or shiny finish), color of the surface, number, color, intensity, and location of point light sources, color and intensity of ambient light. Foley, van Dam, Feiner and Hughes (1997) and Watt (2000) give more details on computer graphics. Rock (1984) and Marr (1992) discuss principles of vision and perception.

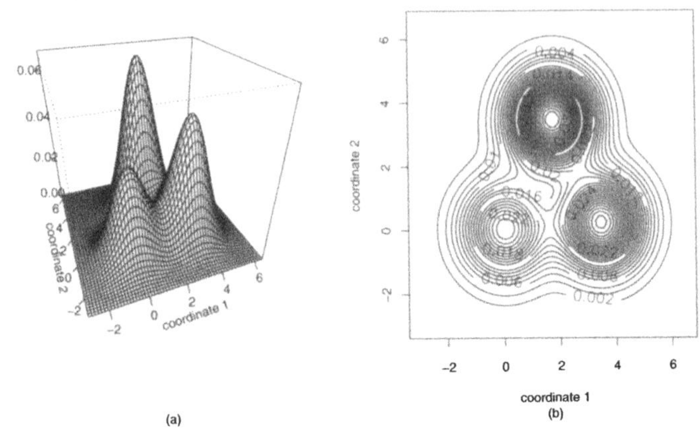

Figure 2.7 *(A mixture of three standard Gaussian densities.)* (a) A perspective plot and (b) a contour plot. The mixture weights are $p_1 = 0.45$, $p_2 = 0.35$, and $p_3 = 0.25$, and the mixture means are $m_1 = (1.75, 3.5)$, $m_2 = (3.5, 0.2)$, and $m_3 = (0, 0)$.

Contour Plots The λ-level curve of function $f : \mathbf{R}^d \to \mathbf{R}$ is defined by

$$\Gamma(f, \lambda) = \{x \in \mathbf{R}^d : f(x) = \lambda\}, \tag{2.2}$$

where $\lambda \in \mathbf{R}$. A contour plot shows a collection of contours for equispaced levels of a 2D function. To visualize 3D functions, one can apply perspective plots of level curves (level surfaces).

Figure 2.7(b) shows a contour plot corresponding to the perspective plot of Figure 2.7(a). The numbers show the levels of the level curves. 2D contour plots do not suffer from hiding effects like perspective plots of the function.

Contour plots have some drawbacks:

1. Contour plots have difficulties to visualize functions that have flat regions and sharp jumps. A contour plot is suitable for smooth functions, but for piecewise constant functions a perspective plot is usually preferred.

2. Figure 2.8 illustrates a difficulty when one visualizes heavy tailed densities with contour plots. We visualize the density with the Gaussian copula with the correlation coefficient $\rho = 0.8$. The density has Student marginals with degrees of freedom 1. Panel (a) shows a contour plot with 10 contours, panel (b) has 50 contours, and panel (c) has 500 contours. The shape of the tails of the density is revealed only with a large number of equispaced contours.

3. Figure 2.9 illustrates an other difficulty with contour plots. A contour plot does not make a visual difference between local minima and local maxima. The information whether a local extreme is a minimum or a maximum is obtained by looking at the level numbers attached to the contour lines. Also, to find out

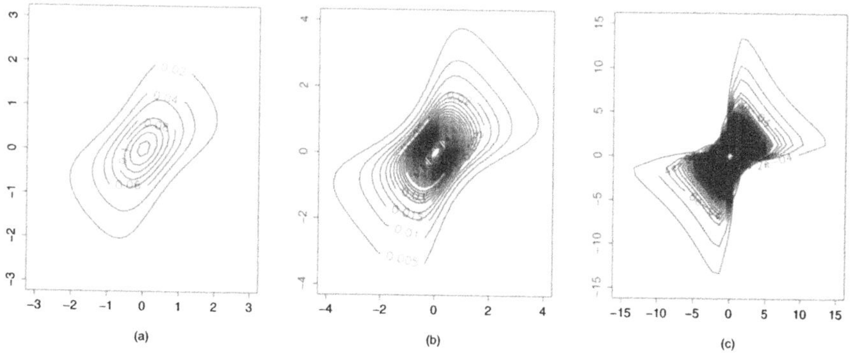

Figure 2.8 *(Three contour plots.)* The density has the Gaussian copula with the correlation coefficient $\rho = 0.8$ and Student marginals with degrees of freedom 1. (a) A contour plot with 10 contours, (b) with 50 contours, and (c) with 500 equispaced contours.

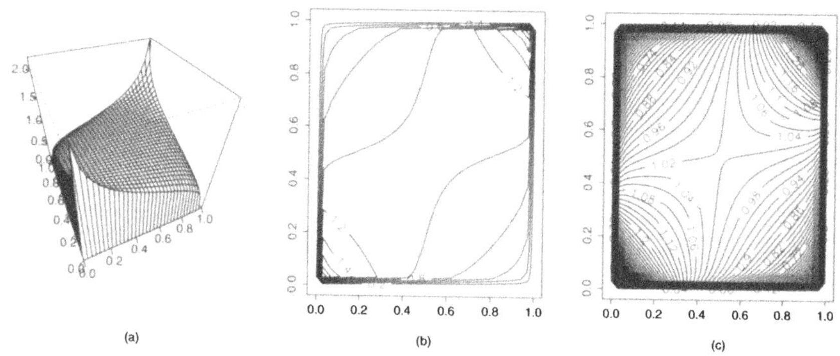

Figure 2.9 *(A perspective plot and two contour plots.)* The density has the Gaussian copula with the correlation coefficient $\rho = 0.2$ and the uniform marginals. (a) A perspective plot, (b) a contour plot with 10 contours, and (c) a contour plot with 100 equispaced contours.

whether the function is increasing or decreasing in a region requires looking at the level numbers. It is helpful to draw additionally a perspective plot to get the right visual impression. Figure 2.9 visualizes the Gaussian copula density with the correlation parameter $\rho = 0.2$. The density has two local maxima and two local minima. Panels (b) and (c) illustrate the effect of the number of levels to the contour plot.

2.1.3 Dimension Reduction of Functions

Multivariate functions may be visualized by looking at 1D to 3D dimensional slices and Radon transforms. When the function is a density, a Radon transform is called a marginal density. In the case of densities one may look also at the low-dimensional conditional densities, but this does not seem to bring advances as compared to looking at slices.

The most straightforward application of dimension reduction is to reduce the dimension of a d-dimensional function to a two-dimensional function. However, to visualize 4D functions, a sliding technique may be applied, where one goes through 3D slices of the function and for each 3D slice visualize a series of 3D density contours, as suggested by Härdle and Scott (1992). Scott (1992, sec. 1.4.3) discusses visualization of functions up to five dimension with slices.

Slice A slice of a function $f : \mathbf{R}^d \to \mathbf{R}$ is a lower dimensional function which we get by fixing the values of some variables. An example of a two-dimensional slice of function f is

$$g(x_1, x_2) = f(x_1, x_2, a_3, \ldots, a_d), \tag{2.3}$$

where we have fixed variables $3 - d$ to take the values $a_3, \ldots, a_d \in \mathbf{R}$. We may take slices along other directions than along the coordinate axes, and even more generally, we may define slice $g : \mathbf{R}^{d'} \to \mathbf{R}$ by

$$g(x_1, \ldots, x_{d'}) = f(\phi(x_1, \ldots, x_{d'})),$$

where $\phi : \mathbf{R}^{d'} \to \mathbf{R}^d$, with $1 \le d' < d$. For example, in the slice of (2.3) we have chosen

$$\phi(x_1, x_2) = (x_1, x_2, a_3, \ldots, a_d).$$

A potentially useful example is to take ϕ equal to a parametrization of a 2D part of the sphere with radius r:

$$\phi(\theta_1, \theta_2) = r \cdot \Psi(\theta_1, \theta_2, a_3, \ldots, a_{d-1}),$$

where $r > 0$, $\Psi : \Theta \to \mathbf{S}_{d-1}$, $\Theta = [0, 2\pi] \times [0, \pi]^{d-2}$ is a parametrization of the unit sphere $\mathbf{S}_{d-1} = \{x \in \mathbf{R}^d : \|x\| = 1\}$, defined in (B.8). For example, when $d = 3$, we may take

$$\phi(\theta_1, \theta_2) = r \cdot (\sin\theta_1 \sin\theta_2, \cos\theta_1 \sin\theta_2, \cos\theta_2).$$

When a density is spherically symmetric, it is constant on these slices.

Difficulties with Slices To find all features of a function, it may be necessary to go through a huge number of slices. In the d-dimensional Euclidean space there are $d(d-1)/2$ ways to choose two coordinate directions, but we need also to consider other than coordinate directions, and for each direction we need a grid of slices. A series of slices is also called a *trellis* of plots.

After looking at a number of slices, we may not be sure whether all important features have been found. It may also be very difficult to keep track of all slices and to get a mental picture of the shape of the function based on a series of slices.

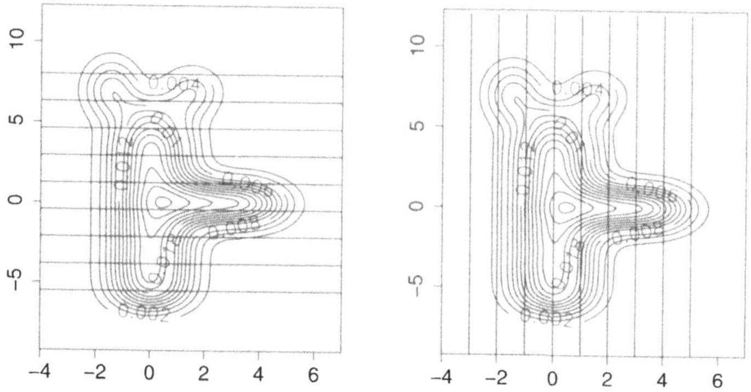

Figure 2.10 *(The positions of the slices.)* (a) Slices parallel to the x-axis; (b) slices parallel to the y-axis.

Examples of Slices We draw slices of the 2D density shown in Figure 2.10. Note that our example is the simplest illustrative case, since we use 1D slices to visualize the shape of a two-dimensional density. It is essentially more difficult to make inference from, say, a five-dimensional density using two-dimensional slices. The density has *multimodal level sets*, in the sense defined in Definition 5.5. With skillfully chosen slices it is possible to visualize the multimodality of level sets. However, the main problem is to find the right slices and to keep records of the information provided by the slices.

Slices along the Coordinate Axes Figure 2.11 shows the slices parallel to the x-axis and Figure 2.12 shows the slices parallel to the y-axis. The last two panels in Figure 2.11 visualize the two ears of the level sets (two modes of the level sets at the north of the contour plot). Figure 2.10 shows a contour plot of the 2D density and the positions of the slices.

Rotated Slices Rotating the slices is equivalent to rotating the density. We rotate the density 45 degrees and look at the slices parallel to the coordinate axes. Figure 2.13 shows the positions of the slices. Figure 2.14 shows the slices parallel to the x-axis, and Figure 2.15 shows the slices parallel to the y-axis. The two ears of the level sets are not visible in any of the rotated slices.

Radon Transform and Marginal Density

Radon Transform Marginal densities are a special case of Radon transforms. A Radon transform is defined as the integrals of a function over lower dimensional hyperplanes: a Radon transform of a d-dimensional function is the function obtained by integrating over all d'-dimensional hyperplanes, $1 \leq d' < d$. Define the d'-

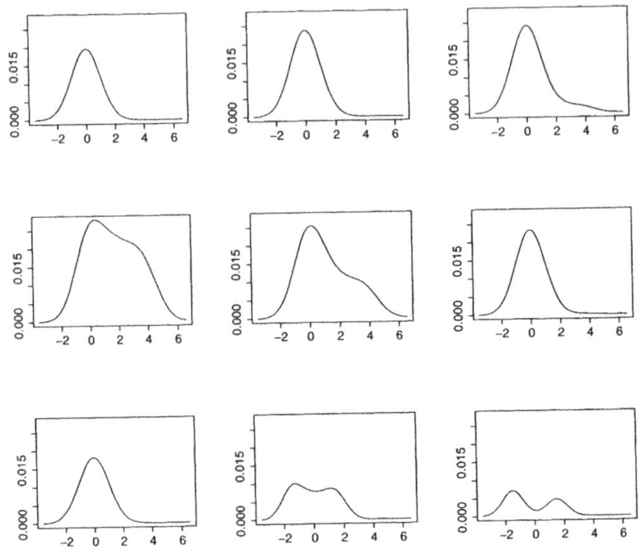

Figure 2.11 *(Slices parallel to the x-axis.)* We go through the slices in Figure 2.10(a) starting from the bottom.

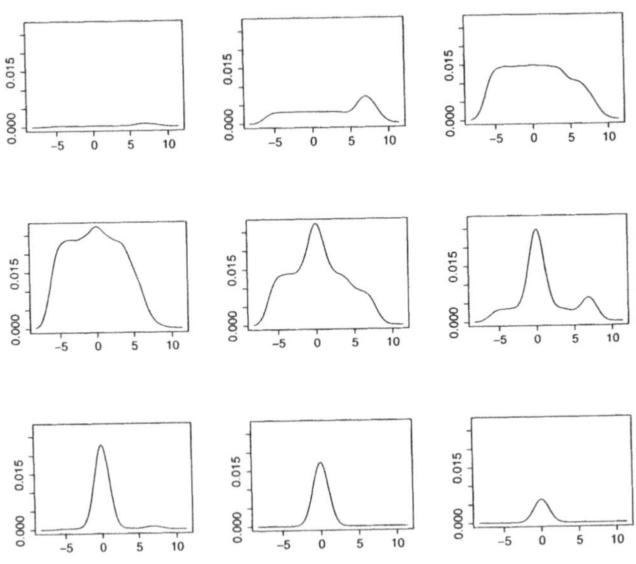

Figure 2.12 *(Slices parallel to the y-axis.)* We go through the slices in Figure 2.10(b) starting from the left.

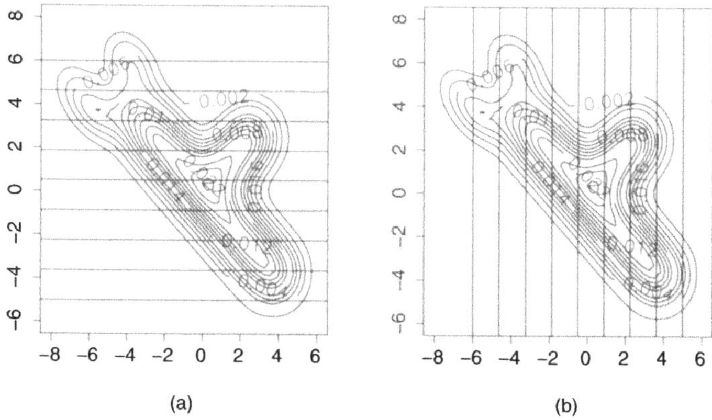

Figure 2.13 *(The positions of the slices when the density is rotated.)* (a) Slices parallel to the x-axis; (b) slices parallel to the y-axis.

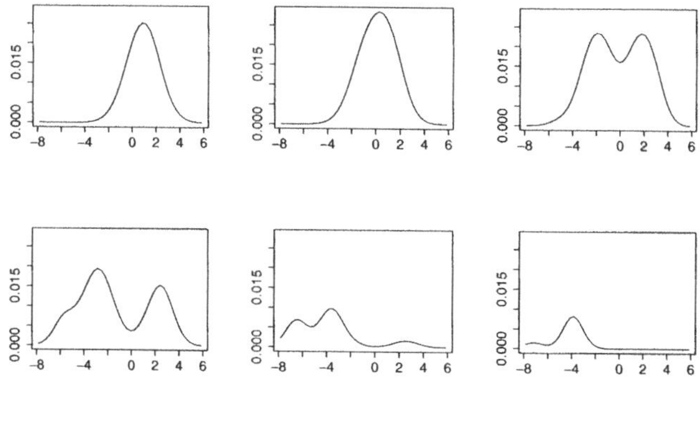

Figure 2.14 *(Slices parallel to the x-axis when the density is rotated.)* We go through the slices in Figure 2.13(a) starting from the bottom.

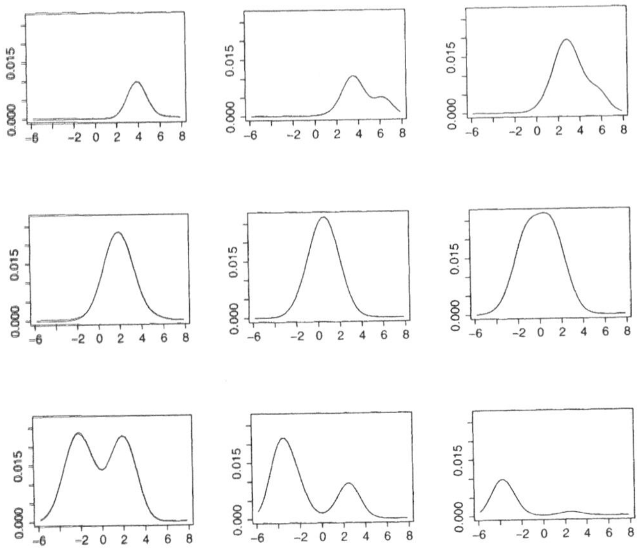

Figure 2.15 *(Slices parallel to the y-axis when the density is rotated.)* We go through the slices in Figure 2.13(b) starting from the left.

dimensional Radon transform as the function

$$(Rf)(P) = \int_P f, \qquad P \in \mathcal{P},$$

where $\mathcal{P}$ is the set of d'-dimensional hyperplanes, and the integration is with respect to the d'-dimensional Lebesgue measure on P. This is equivalent to the definition in (9.26).

Marginal Density A marginal density is obtained by integrating out some variables of a density. A two-dimensional marginal density $g : \mathbf{R}^2 \to \mathbf{R}$ of density $f : \mathbf{R}^d \to \mathbf{R}$ is defined by

$$g(x_1, x_2) = \int_{-\infty}^{\infty} \cdots \int_{-\infty}^{\infty} f(x_1, x_2, x_3, \ldots, x_d)\, dx_3 \cdots dx_d. \tag{2.4}$$

A two-dimensional marginal density g is a restriction of the two-dimensional Radon transform. Let for $(x_1, x_2) \in \mathbf{R}^2$,

$$P_{x_1,x_2} = \left\{(x_1, x_2, x_3, \ldots, x_d) : (x_3, \ldots, x_d) \in \mathbf{R}^{d-2}\right\}$$

be the $(d-2)$-dimensional hyperplane, orthogonal to the $(x_1$-$x_2)$-plane, intersecting the point (x_1, x_2). When $d = 3$, then P_{x_1,x_2} is a one-dimensional ray. Let $\mathcal{P}_{1,2} =$

$\{P_{x_1,x_2} : (x_1, x_2) \in \mathbf{R}^2\}$ be the set of such $(d-2)$-dimensional hyperplanes so that $\mathcal{P}_{1,2} \subset \mathcal{P}$. Then we can write the marginal density defined in (2.4) as

$$g(x_1, x_2) = (Rf)(P_{x_1,x_2}), \qquad P_{x_1,x_2} \in \mathcal{P}_{1,2}.$$

More generally, we may first make a rotation, and then integrate out some of the variables. Matrix A is a rotation matrix if it is orthogonal and $\det(A) = 1$. When A is a $d \times d$ rotation matrix,

$$g(x_1, x_2) = \int_{-\infty}^{\infty} \cdots \int_{-\infty}^{\infty} f(Ax)\, dx_3 \cdots dx_d$$

is a marginal density. Density g is the density of the first two variables of random vector $A^{-1}(X_1, \ldots, X_d)$, where $(X_1, \ldots, X_d) \sim f$. See Appendix B.7 for the definition of a rotation.

Marginal densities may be defined equivalently with the help of projections. See Appendix B.7 for the definition of a projection. When P is a projection to the subspace S of $\mathbf{R}^d$, we can write $\mathbf{R}^d = S \oplus S^{\perp}$, and the marginal density on subspace S is defined by

$$f_S(s) = \int_{S^{\perp}} f(s+y)\, dy, \qquad s \in S.$$

When $e_1, \ldots, e_{d'}$ is a basis of S, then we may define the marginal density by

$$g(x_1, \ldots, x_{d'}) = f_S(x_1 e_1 + \cdots + x_{d'} e_{d'}), \qquad x_1, \ldots, x_{d'} \in \mathbf{R}.$$

Kernel Estimator When we are interested only in marginal densities of a kernel estimate, we may first project the data into two dimensions and use a two-dimensional kernel estimate to estimate the marginal distribution. Indeed, let $f_n(x) = n^{-1}\sum_{i=1}^{n} K_h(x - X_i)$ be a kernel density estimate, where $h > 0$ is the smoothing parameter, $K : \mathbf{R}^d \to \mathbf{R}$ is the kernel function, and $K_h(x) = K(x/h)/h^d$. Let K_{X_1,X_2} be the marginal density of the kernel K in the first two dimensions. Then

$$\begin{aligned}\int_{-\infty}^{\infty} \cdots \int_{-\infty}^{\infty} & f_n(x_1, x_2, x_3, \ldots, x_d)\, dx_3 \cdots dx_d \\ &= \frac{1}{nh^2} \sum_{i=1}^{n} K_{X_1,X_2}\left(\frac{x_1 - X_i^1}{h}, \frac{x_2 - X_i^2}{h}\right).\end{aligned}$$

This fact was noted in Wegman and Luo (2002).

Difficulties with Marginal Densities The difficulties with marginal densities are similar to the difficulties with slices: (1) there are a huge number of possible projections, and (2) it is difficult to get a mental picture from the shape of a density based only on low dimensional views. There are fewer marginal densities than slices, since for each projection one has a unique marginal density but many slices.

The collection of all slices contains the complete information on the function. The collection of all marginal densities may lose some information. Multimodality

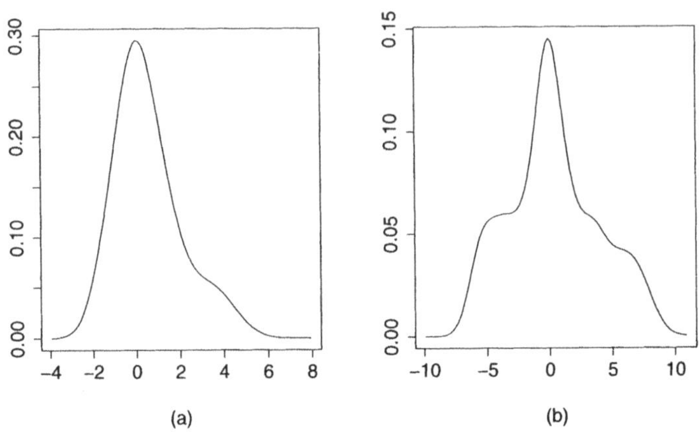

Figure 2.16 *(Marginal densities.)* (a) The marginal density on the x-axis of the density in Figure 2.10. b) The marginal density on the y-axis.

of densities may sometimes be detected by visualizing marginal densities, problems arising when the modes are close to each other or when there are a large number of modes. On the other hand, detecting "multimodality of level sets" with marginal densities is fundamentally more difficult. We illustrate the difficulty of visualizing multimodality of level sets with a two-dimensional density. Difficulties with visualizing multimodality with marginal densities when the modes are close to each other is illustrated with a three- and a four-dimensional example.

Two-dimensional Example *(Multimodality of level sets.)* Figure 2.16 shows the marginal densities along the coordinate axes of the density shown in Figure 2.10. The precise shape would be difficult to reconstruct even with a large number of marginal densities.

Three-dimensional Example *(Modes that are close to each other.)* We consider an equal mixture of standard Gaussian densities, located at the vertices of a tetrahedron. Let

$$f(x) = \frac{1}{4}\sum_{i=1}^{4} \phi(x - m_i), \qquad x \in \mathbf{R}^3,$$

where ϕ is the standard Gaussian density, and

$$\begin{aligned} m_1 &= D \times (1/2, 0, 0), \\ m_2 &= D \times (-1/2, 0, 0), \\ m_3 &= D \times (0, \sqrt{3}/2, 0) \\ m_4 &= D \times (0, 1/(2\sqrt{3}), \sqrt{2/3}), \end{aligned}$$

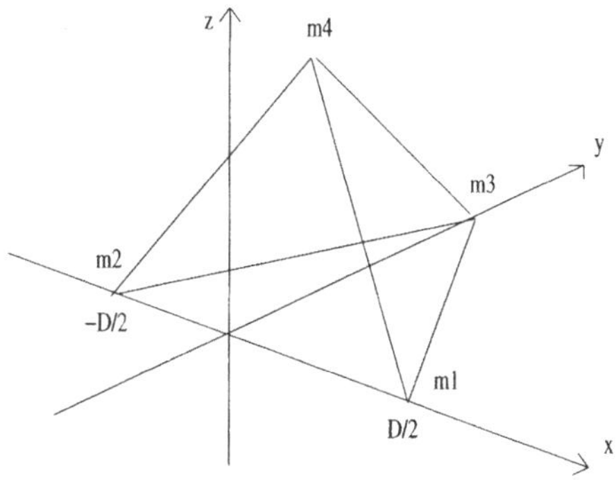

Figure 2.17 *(Tetrahedron.)* The vertices of the tetrahedron are $m1$, $m2$, $m3$, and $m4$. The distance between the vertices is D.

where $D > 0$. Points m_1 through m_4 lie on the vertices of a tetrahedron and the distance between vertices is D. See Figure 2.17.

When D is large, f has four modes, but when D is small, f is unimodal. Even when we have an analytical expression for f, it is difficult to know the exact value of D that makes the function unimodal. When D is small, two-dimensional projections do not reveal the modes of the density even if the function were multimodal. In Figure 2.18(a)–(c) we show the marginal densities on the three coordinate planes of $\mathbf{R}^3$ when $D = 3$. Notice that there are no better projections than the coordinate planes. When the distance between the components of the mixture is large, the projection to the x-y plane reveals all four modes: in Figure 2.18(d) we show the marginal density on the x-y plane when $D = 4$.

Four-dimensional Example *(Modes close to each other.)* We consider an equal mixture of standard Gaussian densities. The means of the components of the mixture lie on the vertices of a pentahedron. We choose as means

$$
\begin{aligned}
m_1 &= D \times (1/2, 0, 0, 0), \\
m_2 &= D \times (-1/2, 0, 0, 0), \\
m_3 &= D \times (0, \sqrt{3}/2, 0, 0), \\
m_4 &= D \times (0, 1/(2\sqrt{3}), \sqrt{2/3}, 0), \\
m_5 &= D \times (0, 1/(2\sqrt{3}), 1/(2\sqrt{6}), \sqrt{15/24}).
\end{aligned}
$$

Points m_1 through m_5 lie on the vertices of a pentahedron, and the distance between vertices is D. When we choose m_i sufficiently close to each other, the two-dimensional projections do not reveal the modes of the density. In Figures 2.19(a)–(f) we show marginal densities on the six coordinate planes of $\mathbf{R}^4$ of the equal mixture of standard Gaussian densities, with $D = 4$. We denote coordinate directions by (x, y, z, u). Even when the distance between components of the mixture is large, there exists no projection to the coordinate planes that would reveal all five modes at once.

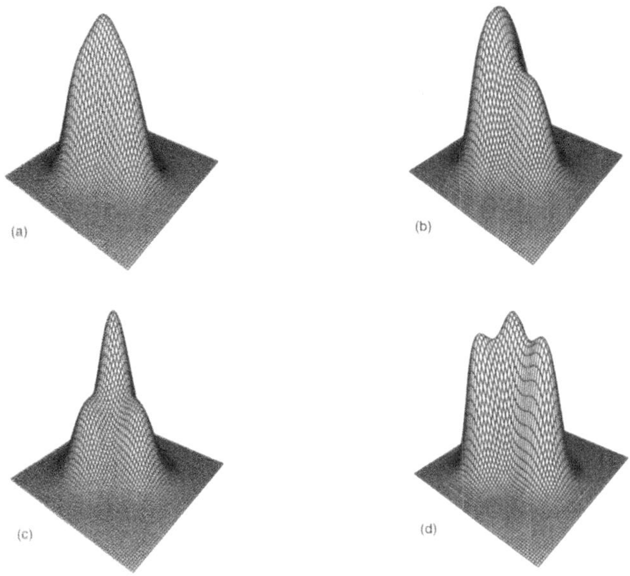

Figure 2.18 *(Marginal densities of a 3D function.)* The marginal densities are shown on the coordinate planes of a mixture of four standard Gaussian densities whose domain is $\mathbf{R}^3$. (a) x-y plane, $D = 3$; (b) x-z plane, $D = 3$; (c) y-z plane, $D = 3$; (d) x-y plane, $D = 4$, where D is the distance between the mixture modes.

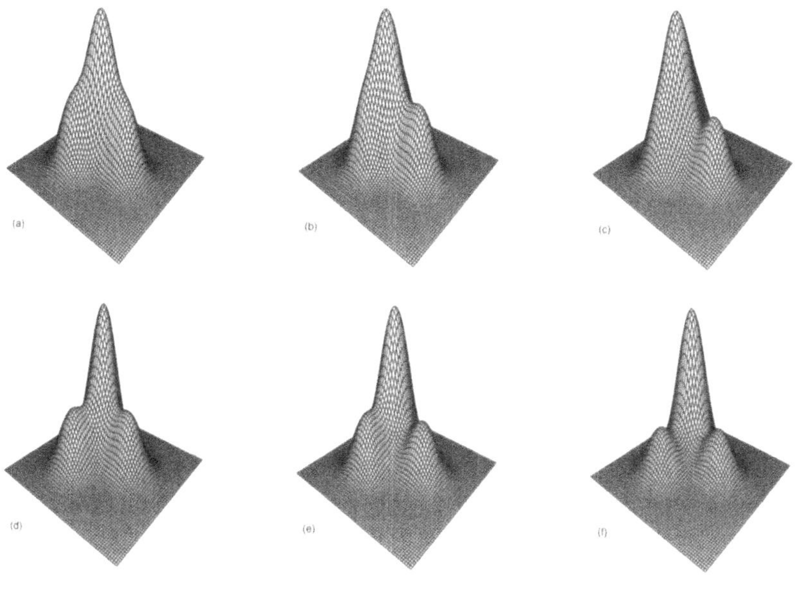

Figure 2.19 *(Marginal densities of a 4D function.)* The marginal densities are shown on the coordinate planes of a mixture of five standard Gaussian densities whose domain is $\mathbf{R}^4$, when the distance between the mixture modes is $D = 4$. (a) x-y plane; (b) x-z plane; (c) x-u plane; (d) y-z plane; (e) y-u plane; (f) z-u plane.

Marginals and Unimodality It is of interest to ask in which cases a unimodal multivariate distribution has unimodal marginals. We defined three notions of unimodality in Definition 1.4, Definition 1.5, and Definition 1.6. We obtain a general result only for α-unimodal distributions.

Lemma 2.1 *(1) There exists a star unimodal distribution that has a multimodal marginal. (2) There exists a distribution that is not star unimodal but whose marginals are unimodal. (3) There exists a level set unimodal distribution that has multimodal marginals. (4) There exists a distribution that is not level set unimodal but whose marginals are unimodal. (5) The marginals of an α-unimodal distribution are α-unimodal.*

Proof. (1) Let P be the uniform distribution on $[0,1]^2 \cup [-1,0]^2$. Distribution P is star unimodal about 0, but if $X \sim P$, then $X_1 + X_2$ has a bi-modal distribution.

(2) Let

$$f(x) = C\left[\exp\left\{-\frac{a^2\|x\|^2}{2}\right\} - b\exp\left\{-\frac{\|x\|^2}{2}\right\}\right], \qquad x \in \mathbf{R}^d. \tag{2.5}$$

where $C^{-1} = (2\pi)^d(a^{-d} - b)$, and parameters a and b are chosen so that f is nonnegative. The first marginal is

$$f_1(x_1) = C(2\pi)^{(d-1)/2}\left[a^{1-d}\exp\left\{-\frac{a^2x_1^2}{2}\right\} - b\exp\left\{-\frac{x_1^2}{2}\right\}\right], \qquad x_1 \in \mathbf{R}.$$

When $d = 2$, parameters a and b may be chosen so that marginal f_1 is unimodal, and f is not star unimodal. Figure 1.38 shows the case $a = 0.5$ and $b = 1$.

(3) We give two examples. (i) The density of item (1) is also level set unimodal, and serves as an example. (ii) The density in (2.5) is always level set unimodal, but parameters a and b may be chosen so that its marginals are two-modal.

(4) The three-dimensional density whose marginals are shown in Figure 2.18 and the four-dimensional density whose marginals are shown in Figure 2.19 serve as examples.

(5) When $g : \mathbf{R} \to \mathbf{R}$ is bounded and nonnegative, $g^* : \mathbf{R}^d \to \mathbf{R}$, $g^*(x) = g(x_1)$ is bounded and nonnegative. Thus $t^\alpha E[g(t(X_1 - \mu_1))]$ is nondecreasing if $t^\alpha E[g^*(t(X - \mu))]$ is nondecreasing. □

Conditional Density We obtain conditional density

$$f_{X_1,X_2|X_3=a_3,\ldots,X_d=a_d} : \mathbf{R}^2 \to \mathbf{R}$$

of density $f : \mathbf{R}^d \to \mathbf{R}$ by dividing the slice with the marginal density:

$$f_{X_1,X_2|X_3=a_3,\ldots,X_d=a_d}(x_1,x_2) = \frac{f_{X_1,X_2,X_3=a_3,\ldots,X_d=a_d}(x_1,x_2)}{f_{X_1,X_2}(x_1,x_2)},$$

where

$$f_{X_1,X_2,X_3=a_3,\ldots,X_d=a_d}(x_1,x_2) = f(x_1,x_2,a_3,\ldots,a_d)$$

is the slice and

$$f_{X_1,X_2}(x_1,x_2) = \int_{-\infty}^{\infty} \cdots \int_{-\infty}^{\infty} f(x_1,x_2,x_3,\ldots,x_d)\, dx_3 \cdots dx_d.$$

is the marginal density. In this definition we assume that $f_{X_1,X_2}(x_1,x_2) > 0$. There are as many conditional densities as there are slices. The normalization of a slice with the marginal density does not seem to bring advantages.

2.2 VISUALIZATION OF THE SPREAD

We have discussed the visualization of the spread of univariate data in Section 1.2. In particular, PP-plot and QQ-plot were mentioned. In this section we discuss the visualization of the spread of multivariate distributions.

In the visualization of the spread we concentrate on the noncentral regions of the distribution, and we assume that the distribution is unimodal or nearly unimodal. The methods are, however, usually robust with respect to small deviations from the unimodality; the density may have several small local extremes.

We define transformations of a multivariate distribution to a univariate function in order to visualize a multivariate distribution. In the case of two-dimensional distributions we may apply perspective plots or contour plots to draw the graph of the density or the graph of the distribution function, but also in the two-dimensional case it is useful to apply 1D curves to visualize the spread of the distribution, since perspective plots and contour plots do not give a transparent visualization of the tail behavior of the distribution. We apply two types of visualizations: (1) density type visualizations and (2) distribution and quantile function type visualizations.

In density type visualizations, we look at how the volumes of the level sets of a density are changing as function of the level. This leads to one-dimensional curves that visualize the spread of a multivariate distribution. In multivariate cases the volumes of the level sets are often exploding when we move to the lower levels, and they are shrinking fast to zero when we move to the higher levels. In addition the volumes of some multivariate sets vary irregularly as a function of the dimension. For example, the volumes of unit balls vary irregularly as a function of the dimension. To solve these problems, we apply a dimension normalization.

In the distribution and quantile function type visualizations, we can apply other sequences of sets than the sequence of the level sets of the density. For example, we could look at the depth regions defined by a depth function. We visualize the functional relation between the probability content and the volume of the sets in the sequence. We consider a nested sequence of sets and choose a fixed reference point inside these sets. This is different from the univariate case where the distribution function visualizes the cumulation of the probability mass as one travels from left to right. Again, it is useful to apply a dimension normalization to get more easily conceivable one-dimensional spread functions.

We may be interested in the spread of a distribution, but we have available only a sample of observations from the distribution. To make density type visualizations, we need to estimate the level sets. Applying distribution and quantile function type visualizations is sometimes easier, since we may need only to calculate the empirical probabilities over some sequence of sets. However, it may be useful to apply also in distribution and quantile function type visualizations a sequence of estimates of level sets.

2.2.1 Density Type Visualizations

Density type visualizations apply the sequence $(\Lambda_\lambda)_{\lambda\in[0,\infty)}$ of the level sets of a density. The level set of a density $f : \mathbf{R}^d \to [0, \infty)$ with level λ is defined by

$$\Lambda(f,\lambda) = \{x \in \mathbf{R}^d : f(x) \geq \lambda\}, \qquad \lambda \in [0,\infty). \tag{2.6}$$

Density type visualizations show the functional relation between the level and the volume of the level sets in the sequence. When the density is unknown, we have to estimate the level sets with a sample of observations from the distribution of the density. We may estimate the level sets by estimating the density function, or we may estimate the level sets directly.

Unimodal Volume Function We call a *level-to-volume* function a function that maps levels to the volumes of the level sets in the sequence. A *volume-to-level* function is the generalized inverse of this function. These univariate functions characterize the spread of a multivariate density.

Definition 2.1 (Level-to-volume function, Volume-to-level function.) *The* level-to-volume *function* $V : [0,\infty) \to [0,\infty)$, *associated to a multivariate density* $f : \mathbf{R}^d \to \mathbf{R}$, *is defined by*

$$V(\lambda) = \text{volume}(\Lambda(f,\lambda)), \qquad \lambda \in [0,\infty). \tag{2.7}$$

The volume-to-level *function* $V^{-1} : [0,\infty) \to [0,\infty)$, *is defined by*

$$V^{-1}(v) = \sup\{\lambda \in [0,\infty) : V(\lambda) \geq v\}, \qquad v \in [0,\infty),$$

where we use the convention $\sup \emptyset = 0$.

A volume-to-level function seems to be more natural, since we can modify this function so that the univariate symmetric densities remain unchanged through the transform. We reflect a volume-to-level function with respect to the origin and scale it with the factor 2 to get a symmetric density. We call this function a *unimodal volume function*.

Definition 2.2 (Unimodal volume function.) *A* unimodal volume function, *associated to a multivariate density* $f : \mathbf{R}^d \to \mathbf{R}$, *is any translation of function* $W : \mathbf{R} \to [0,\infty)$,

$$W(t) = \begin{cases} V^{-1}(2t), & t \geq 0 \\ V^{-1}(-2t), & t < 0, \end{cases}$$

where V^{-1} is the volume-to-level function.

We have $\int_{-\infty}^{\infty} W = \int_0^{\infty} V^{-1} = 1$ and $W \geq 0$. We may use the term *unimodal volume transform*, since we have defined a transformation of a multivariate density to a univariate (unimodal symmetric) density. The unimodal volume function of Definition 2.2 is related to the volume function defined in Definition 4.5. When the density $f : \mathbf{R}^d \to \mathbf{R}$ is unimodal, the definitions coincide. For multimodal densities Definition 4.5 leads to more informative visualizations where the volumes of the separated regions of level sets are visualized, and thus relative largeness of the modes are visualized.

Illustrations We illustrate Definition 2.1 and Definition 2.2.

1. For $f(x) = (1/2)I_{[-1,1]}(x)$ as the uniform density on $[-1, 1]$, we have $V(\lambda) = 2$ when $\lambda \in [0, 1/2]$ and $V(\lambda) = 0$ when $\lambda > 1/2$. Function V does not have an inverse, but the generalized inverse is $V^{-1}(v) = 1/2$ when $v \in [0, 2]$ and $V^{-1}(v) = 0$ when $v > 2$.

2. Figure 2.20 illustrates the concepts of a level-to-volume function, volume-to-level function, and unimodal volume function. Paanel (a) shows a level-to-volume function of the univariate Bartlett density $t \mapsto (3/4)(1 - t^2)_+$, where $(t)_+ = \max\{0, t\}$. Panel (b) shows the inverse of the level-to-volume function of the Bartlett density (volume-to-level function). Panel (c) shows the unimodal volume function. We have positioned the unimodal volume function so that the left boundary of the support is at the origin.

3. *(Spherically symmetric densities.)* Let $f(x) = g(\|x\|^2)$ be a spherically symmetric density, where $g : [0, \infty) \to \mathbf{R}$. We call g the density generator. We have

$$\{x : f(x) \geq \lambda\} = \{x : \|x\| \leq r_\lambda\}, \qquad r_\lambda = \sqrt{g^{-1}(\lambda)}, \tag{2.8}$$

when g is monotonically decreasing. The level-to-volume function is thus $V(\lambda) = \text{volume}(B_{r_\lambda})$, where $B_r = \{x \in \mathbf{R}^d : \|x\| \leq r\}$.

Examples We give examples of unimodal volume functions of spherically symmetric densities. We show that unimodal volume functions of functions of the same functional form may have different shapes in different dimensions. This is because the volumes of multivariate sets can vary irregularly as the dimension changes. To be specific, let us consider the volume of a ball. The volume of a ball with radius $r > 0$ is $\text{volume}(B_r) = C_d r^d$, where $C_d = \text{volume}(B_1)$ is defined in (B.10). Figure 2.21 shows the volumes of balls with radii $r = 1/2$ and $r = 1$ as a function of the dimension. When $r = 1/2$, the ball is inside a unit square and its volume is always less than one. As the dimension grows, the dimension of the ball becomes negligible. When $r = 1$, the volume is first increasing as the dimension increases but then starts

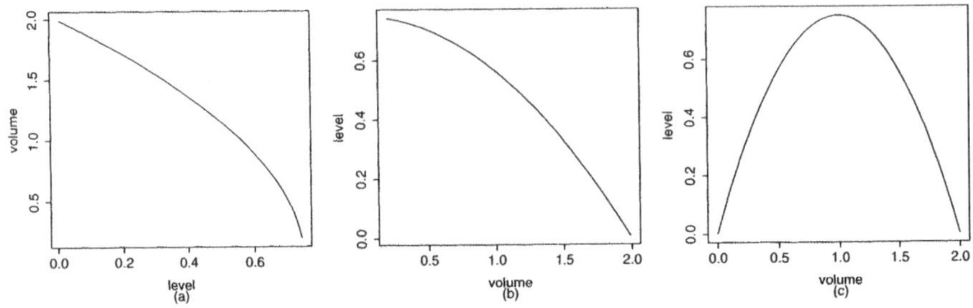

Figure 2.20 *(Definitions 2.1 and 2.2 illustrated.)* (a) The level-to-volume function, (b) the inverse of the level-to-volume function, and (c) a unimodal volume function of the Bartlett density.

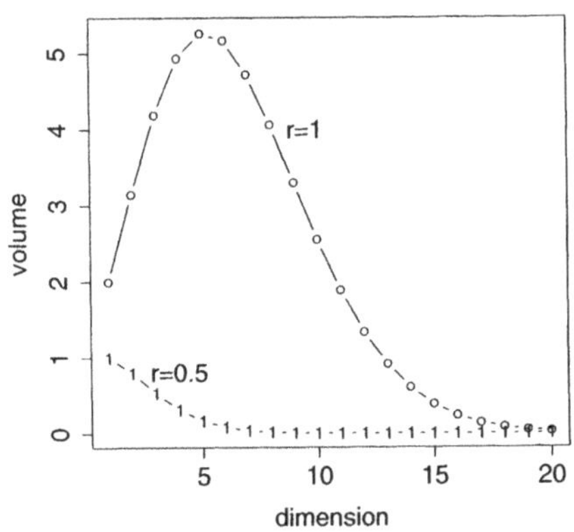

Figure 2.21 Volumes of balls with radii $r = 1/2$ and $r = 1$ shown as a function of the dimension.

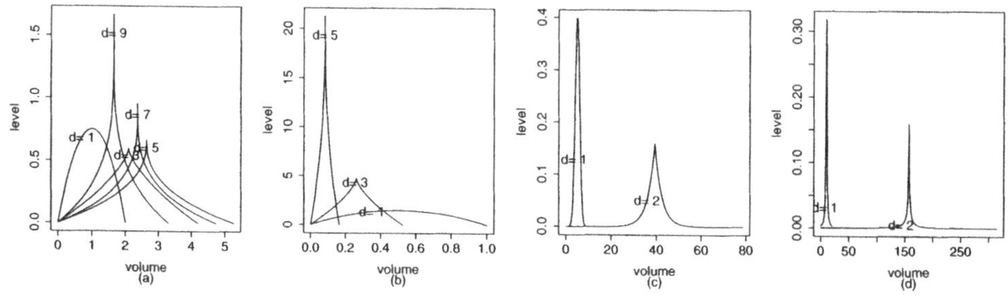

Figure 2.22 *(Unimodal volume functions.)* (a) Unimodal volume functions of Bartlett densities with $\sigma = 1$ for dimensions $1, 3, 5, 7, 9$. (b) The case of Bartlett densities with $\sigma = 1$ and $d = 1, 3, 5$. (c) The case of standard Gaussian densities for dimensions $1, 2$. (d) The case of Student densities with degrees of freedom $\nu = 1$ for dimensions $1, 2$.

decreasing. When $r > (\Gamma(3/2)\sqrt{\pi})^{-1} \approx 0.6366198$, the volume is first increasing before it starts decreasing.

1. *(Bartlett densities.)* The Bartlett density generator g is defined in (10.7). The multivariate scaled Bartlett density is defined by $x \mapsto \sigma^{-d} g(\|x/\sigma\|^2)$. Figure 2.22(a) shows unimodal volume functions for dimensions $1, 3, 5, 7, 9$ of the Bartlett density when $\sigma = 1$. Since $\sigma = 1$, the volume of the support is increasing until $d = 5$, but then starts decreasing, as suggested by Figure 2.21.

   Figure 2.22(b) shows unimodal volume functions for dimensions $1, 3, 5$ of the Bartlett density with $\sigma = 1/2$. Since $\sigma = 1/2$, the volume of the support is decreasing as function of the dimension.

2. *(Gaussian densities.)* The Gaussian density generator is defined in (10.5). Figure 2.22(c) shows unimodal volume functions of the standard Gaussian density for dimensions $1 - 2$. The volume of the lower level sets is increasing when the dimension is increasing.

3. *(Student densities.)* The Student density generator is defined in (10.6). Figure 2.22(d) shows unimodal volume functions of the Student density with degrees of freedom $\nu = 1$ for dimensions $1 - 2$. The volume of the lower level sets is increasing when the dimension is increasing.

Problems The previous examples have brought up some problems with the unimodal volume function.

1. *(Concentration effect.)* The volume of the ball with radius r contains the term r^d. When r is small (smaller than 1), the volume of the ball with radius r is very small, and when r is large (larger than 1), the volume of the ball with radius r is very large. Thus a unimodal volume function has often a sharp peak

at the center and its tails are spreading out. Note that in the case of visualizing multimodal densities with a volume transform, zooming may be used to see the details, as in Section 4.5.

2. *(Dimension non-invariance.)* The volume of a ball contains multiplier C_d that is not monotonic with respect to dimension d, and thus the shapes of unimodal volume functions may vary irregularly when the dimension varies. For example, the unimodal volume functions of Bartlett densities have different shapes for various dimensions.

To address these problems, we define a dimension normalized version of a unimodal volume function.

Dimension Normalized Unimodal Volume Function We define a dimension normalized unimodal volume function that visualizes the shape of a multivariate density in a dimension insensitive way.

Definition 2.3 (Dimension normalized level-to-volume function, Dimension normalized volume-to-level function.) *The* dimension normalized level-to-volume function $V^* : [0,\infty) \to [0,\infty)$, *associated to a multivariate density* $f : \mathbf{R}^d \to \mathbf{R}$, *is defined by*

$$V^*(\lambda) = \left(\frac{1}{C_d} \text{volume}(\Lambda(f,\lambda)) \right)^{1/d}, \qquad \lambda \in [0,\infty),$$

where C_d is defined in (B.10). The dimension normalized volume-to-level function $(V^*)^{-1} : [0,\infty) \to [0,\infty)$, *is defined by*

$$(V^*)^{-1}(v) = \sup\{\lambda \in [0,\infty) : V^*(\lambda) \geq v\}, \qquad v \in [0,\infty),$$

where we use convention $\sup \emptyset = 0$.

As in the case of non-dimension normalized functions we prefer the volume-to-level function and we symmetrize this function. In addition we normalize the function to integrate to one which makes the function more dimension insensitive.

Definition 2.4 (Dimension normalized unimodal volume function.) *A* dimension normalized unimodal volume function, *associated to a multivariate density* $f : \mathbf{R}^d \to \mathbf{R}$, *is any translation of function* $W^* : \mathbf{R} \to [0,\infty)$,

$$W^* = c\, W_0^*, \qquad W_0^*(t) = \begin{cases} (V^*)^{-1}(t), & t \geq 0 \\ (V^*)^{-1}(-t), & t < 0, \end{cases}$$

where $(V^*)^{-1}$ is the dimension normalized volume-to-level function, and c is the normalization constant: $c^{-1} = \int_{-\infty}^{\infty} W_0^*$.

As in the non-dimension normalized case symmetric univariate densities remain unchanged through the transform, up to a translation.

Proposition 2.2 *When $d = 1$ and density f is symmetric unimodal, $W^* = f(\cdot - \mu)$, for some $\mu \in \mathbf{R}$.*

Proof. We have $C_1 = 2$, and thus when $d = 1$, we have $W_0^* = W^*$ is a density whose level sets have lengths equal to the lengths of the level sets of f. □

Elliptical Densities Let $f(x) = |\det\Sigma|^{-1/2} g(x^T\Sigma^{-1}x)$ be an elliptical density, where $g : [0,\infty) \to \mathbf{R}$ is a density generator and Σ is a symmetric positive semi-definite dispersion matrix. We have

$$\{x : f(x) \geq \lambda\} = \{x : x^T\Sigma^{-1}x \leq r_\lambda^2\}, \qquad r_\lambda = \sqrt{g^{-1}\left(|\det\Sigma|^{1/2}\lambda\right)}, \quad (2.9)$$

when g is monotonically decreasing. The volume of the ellipsoid in (2.9) is equal to

$$\text{volume}\left(\{x : x^T\Sigma^{-1}x \leq r_\lambda^2\}\right) = |\det\Sigma|\, C_d\, r_\lambda^d,$$

where C_d is defined in (B.10). The dimension normalized level-to-volume function is $\lambda \mapsto |\det\Sigma|^{1/d} r_\lambda$, where r_λ is defined in (2.9), and the dimension normalized volume-to-level function is $r \mapsto |\det\Sigma|^{-1/2} g(|\det\Sigma|^{-2/d} r^2)$, $r \geq 0$. A dimension normalized unimodal volume function is $t \mapsto c\, g(|\det\Sigma|^{-2/d} t^2)$, $t \in \mathbf{R}$, where $c^{-1} = |\det\Sigma|^{1/d} 2 \int_0^\infty g(u^2)\,du$. Thus a dimension normalized unimodal volume function depends on the dimension only through g and $|\det\Sigma|^{1/d}$. We have proved the following proposition:

Proposition 2.3 *Let $f_i : \mathbf{R}^{d_i} \to \mathbf{R}$, $i = 1, 2$, be elliptical densities, $f_i(x) = |\det\Sigma_i|^{-1/2} g_i(x^T\Sigma_i^{-1}x)$. If $g_i = c_i g$ for some $g : [0,\infty) \to [0,\infty)$, $i = 1, 2$, and $|\det\Sigma_1|^{1/d_1} = |\det\Sigma_2|^{1/d_2}$, then the dimension normalized unimodal volume functions of the corresponding elliptical densities are equal, up to a translation.*

In particular, for spherically symmetric Gaussian and Bartlett densities the dimension normalized unimodal functions are equal for all dimensions, up to translation. Note that the shape of the Student density generator depends on the dimension, in order the density to be integrable in all dimensions when $\nu > 0$. Figure 2.23 shows dimension normalized unimodal volume functions for the Bartlett, standard Gaussian, and Student densities with $\nu = 1$ and $d = 1 - 3$.

2.2.2 Distribution Function Type Visualizations

In Section 2.2.1 we considered only sequences of level sets of densities. When we define quantile and distribution function type visualizations, it makes sense to consider also other sequences of sets. We visualize the relationship between the probability content and the volume of the sets in the sequence.

Sequences of Sets Let $A_\lambda \subset \mathbf{R}^d$, $\lambda \in [0,\infty)$, be a collection of sets indexed with parameter λ. We assume that the collection of sets is nested and decreasing:

$$A_{\lambda_1} \supset A_{\lambda_2} \quad \text{when} \quad 0 \leq \lambda_1 \leq \lambda_2 < \infty. \quad (2.10)$$

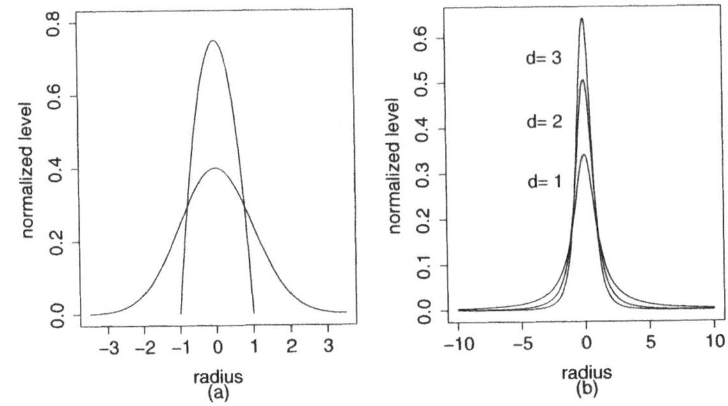

Figure 2.23 Dimension normalized unimodal volume functions are shown for the Bartlett, standard Gaussian, and Student densities with degrees of freedom $\nu = 1$ and $d = 1 - 3$.

We index the sequence of sets $(A_\lambda)_{\lambda \in [0,\infty)}$ with the probabilities:

$$C_p = A_{\lambda_p}, \qquad p \in [0, 1], \tag{2.11}$$

where

$$\lambda_p = \sup\{\lambda \in [0, \infty) : P(A_\lambda) \geq p\}. \tag{2.12}$$

The sequence A_λ is decreasing as λ is increasing and thus the level λ_p corresponding to probability p is well defined. We may consider the following cases:

1. The sequence of sets depends on the underlying distribution.
   (a) The sequence is constructed assuming knowledge of the underlying distribution.
      i. Level sets as defined in (2.6)
      ii. Depth regions
      iii. Minimum volume sets
      iv. Central regions of a quantile function
   (b) The sequence is estimated based on a sequence of identically distributed random vectors.
2. The sequence of sets does not depend on the underlying distribution.

Depth Regions We may choose the collection of sets to be the depth regions associated with various depth functions $D : \mathbf{R}^d \to [0, \infty)$: $A_\lambda = \{x \in \mathbf{R}^d : D(x) \geq \lambda\}$. A depth function D corresponding to a distribution function $F : \mathbf{R}^d \to \mathbf{R}$ is such that when $D(x)$ is large, x is close to the center of the distribution. One way to define a depth function is to take the depth equal to the density: $D(x) = f(x)$ and many of the notions of data depth regions may be related to level sets. Examples of depth functions are given in Section 1.7.1.

Minimum Volume Sets We may choose the sets C_p to be the minimum volume sets with a given probability content when we perform the minimization over a given collection of sets. Let $\mathcal{S}$ be a class of measurable sets and define

$$C_p = \text{argmin}_{C \in \mathcal{S}} \{\text{volume}(C) : P(C) \geq p\}. \tag{2.13}$$

These sets where considered, for example, by Einmahl and Mason (1992) and Polonik (1999). Again, we may find a connection to level sets. Indeed a level set is the minimum volume set with a given probability content. When P has density $f : \mathbf{R}^d \to \mathbf{R}$, and class $\mathcal{S}$ is the class of Borel sets of $\mathbf{R}^d$, the minimum in the definition of C_p is achieved by a level set of f if f has no flat parts: volume$(\{x : f(x) = \lambda\}) = 0$ for $\lambda > 0$. If f has no flat parts, then we may find λ_p such that $P_f((\Lambda(f, \lambda_p)) = p$ and it holds

$$C_p = \text{volume}(\Lambda(f, \lambda_p)).$$

In general, the following lemma holds:

Lemma 2.4 *For all measurable $C \subset \mathbf{R}^d$ with $P_f(C) \geq p$, $0 < p < 1$, we have that*

$$\text{volume}(C) \geq \text{volume}(\Lambda(f, \lambda_p)).$$

where $\lambda_p = \inf\{\lambda : P_f(\Lambda(f, \lambda)) \leq p\}$.

Proof. Let $C \subset \mathbf{R}^d$ be such that $P_f(C) \geq p$. Let λ be large enough so that $P_f(\Lambda(f, \lambda)) \leq p$. Denote $\Lambda = \Lambda(f, \lambda)$. We have

$$p = \int_{C \setminus \Lambda} f + \int_{C \cap \Lambda} f \leq \lambda \, \text{volume}(C \setminus \Lambda) + \int_{C \cap \Lambda} f.$$

Also

$$p \geq \int_{\Lambda \setminus C} f + \int_{C \cap \Lambda} f \geq \lambda \, \text{volume}(\Lambda \setminus C) + \int_{C \cap \Lambda} f.$$

Thus we have

$$\text{volume}(C \setminus \Lambda) \geq \text{volume}(\Lambda \setminus C)$$

and correspondingly

$$\text{volume}(C) \geq \text{volume}(\Lambda).$$

We have proved the lemma. □

Central Regions of a Quantile Function A multivariate quantile function may be defined to be a function $Q : \mathbf{S}_{d-1} \times [0,1] \to \mathbf{R}^d$, where $\mathbf{S}_{d-1} = \{x \in \mathbf{R}^d : \|x\| = 1\}$ is the unit sphere. These types of quantile functions are "center-oriented" quantile functions. In the univariate case $d = 1$ we may define $\mathbf{S}_0 = \{-1, 1\}$ and $Q(-1,p) = F^{-1}((1-p)/2)$ as the left quantile and $Q(1,p) = F^{-1}(1-(1-p)/2)$ as the right quantile, where F is the distribution function. This is a median-oriented quantile function. Define the pth central region as

$$C_p = \{Q(u,q) : u \in \mathbf{S}_{d-1},\ 0 \le q \le p\}, \qquad p \in [0,1]. \tag{2.14}$$

These sets were proposed by Serfling (2002*a*).

It is left to define a multivariate quantile function to be used in (2.14). In the univariate case quantiles are defined through the inverse of the distribution function. In the univariate case a quantile is the solution to the minimization problem

$$\begin{aligned} Q(p) &= \text{argmin}_{Q \in \mathbf{R}} E\left[p\,(X-Q)^+ + (1-p)\,(X-Q)^-\right] \\ &= \text{argmin}_{Q \in \mathbf{R}} E\Phi(2p-1, X-Q), \end{aligned}$$

where $(a)^+ = \max\{0, a\}$ is the positive part, $(a)^- = \max\{0, -a\}$ is the negative part, and $\Phi(u,t) = |t| + ut$; see Ferguson (1967). The sample pth quantile $\hat{Q}(p)$, $0 < p < 1$, is the solution to the minimization problem

$$\hat{Q}(p) = \text{argmin}_{Q \in \mathbf{R}} \sum_{i=1}^{n} \Phi(2p-1, X_i - Q).$$

Chaudhuri (1996) and Koltchinskii (1997) propose to consider in the multivariate case the minimizers

$$Q(x) = \text{argmin}_{Q \in \mathbf{R}^d} E\Phi(x, X-Q),$$

where $x \in \mathbf{R}^d$, $\|x\| < 1$ and

$$\Phi(x,y) = \|y\| + (x,y), \qquad x, y \in \mathbf{R}^d,$$

where (x,y) is the Euclidean inner product. Numbers $Q(x)$ are called multivariate quantiles. Note that since $\|x\| < 1$, we can write $x = pu$ for a unique $0 < p < 1$ and $u \in \mathbf{S}_{d-1}$. The sample version is

$$\hat{Q}(x) = \text{argmin}_{Q \in \mathbf{R}^d} \sum_{i=1}^{n} \Phi(x, X_i - Q).$$

We have now defined a sequence of sets as the sequence of central regions of a quantile function but it is possible to go in the other direction and define a quantile function based on a sequence of sets. Assume that we have a sequence of sets $(A_\lambda)_{\lambda \in [0,\infty)}$ that are nested and decreasing as in (2.10). Let $\mu \in \mathbf{R}^d$ such that $\mu \in A_\lambda$ for all nonempty A_λ. For example, when A_λ are the level sets of a unimodal density,

μ is the mode of the density, and when A_λ are the depth regions associated with a depth function, μ is the multivariate median associated with the depth function. We index the sequence with the probabilities and denote $(A_{\lambda_p})_{p\in[0,1]}$. When in addition sets A_λ are star shaped, we may associate every point in the boundary of A_λ with a unique direction $u \in \mathbf{S}_{d-1}$ from the center μ. Denote with $Q(u,p)$ the point in the boundary of A_{λ_p} which is at direction $u \in \mathbf{S}_{d-1}$ from the center point μ. Function Q is called a multivariate quantile function. Zuo and Serfling (2000) consider almost sure convergence of the sample versions, and Nolan (1992) considers asymptotic distribution theory for the depth region based quantile function estimation.

A more complete review of several notions of multivariate quantile functions is given in Serfling (2002*b*).

Sequences Independent of the Distribution We may take the sequence to be independent of the underlying distribution. A simple and useful example is provided by the complements of balls, or the balls itself: $A_\lambda = \mathbf{R}^d \setminus \{x : \|x-\mu\| \le \lambda\}$ or $A_\lambda = \{x : \|x-\mu\| \le 1/\lambda\}$, where μ is a center point. Note that the center point μ has typically to be chosen based on the data.

Quantile and Distribution Function A quantile function maps probabilities to the volumes and a distribution function is the generalized inverse of a quantile function.

Definition 2.5 (Probability-to-volume function, Volume-to-probability function.) *A* probability-to-volume function, *or a* quantile function, *corresponding to distribution P, is defined by*

$$Q(p) = \text{volume}\left(A_{\lambda_p}\right), \qquad p \in [0,1],$$

where λ_p is defined in (2.12). A volume-to-probability function, *or a* distribution function, *is defined by*

$$F(v) = \inf\{p \in [0,1] : Q(p) \ge v\}, \qquad v \in [0,\infty).$$

Alternatively, we may index the sequence of sets with volumes:

$$\lambda_v = \sup\{\lambda \in [0,\infty) : \text{volume}(A_\lambda) \ge v\}, \qquad v \in [0,\infty), \tag{2.15}$$

and define the distribution function as $v \mapsto P(A_{\lambda_v})$. The quantile function is then defined as the generalized inverse of the distribution function.

Liu et al. (1999) call a probability-to-volume-function a "scale curve" or a "scalar form of scale/dispersion", in the case sets A_λ are depth regions and Serfling (2002*b*) considers the case where the sets A_λ are the central regions determined by a multivariate quantile function.

Examples We illustrate probability-to-volume functions for the case of Bartlett, standard Gaussian, and Student distributions, defined in (10.7), (10.5), and (10.6).

1. *(Bartlett distribution.)* Figure 2.24(a) shows probability-to-volume functions for the Bartlett distribution for dimensions $1, 3, 5, 7, 9$ and for $\sigma = 1$. Here σ

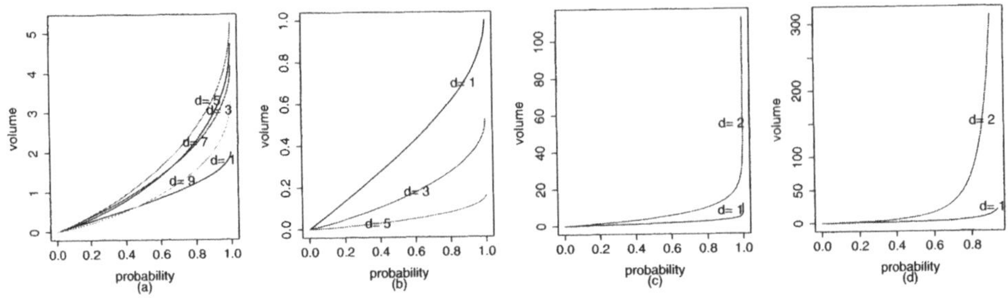

Figure 2.24 *(Probability-to-volume functions.)* (a) Probability-to-volume functions of Bartlett distributions with $\sigma = 1$ for dimensions $1, 3, 5, 7, 9$. (b) The case of Bartlett distributions with $\sigma = 0.5$ and $d = 1, 3, 5$. (c) The case of standard Gaussian distributions for dimensions $1, 2$. (d) The case of Student distributions with degrees of freedom $\nu = 1$ for dimensions $1, 2$. The corresponding unimodal volume functions are shown in Figure 2.22.

means the scaling factor defined by $x \mapsto \sigma^{-d} g(\|x/\sigma\|^2)$. In the range $d = 1-5$ a quantile function corresponding to a higher dimension dominates a quantile function corresponding to a lower dimension, but in the range $d = 6-10$ this relation reverses.

Figure 2.24(b) shows dimensions $1, 3, 5$ for the Bartlett distribution when $\sigma = 1$. Because $\sigma = 0.5$, a quantile function corresponding to a lower dimension dominates a quantile function corresponding to a higher dimension. The dominating relations are explained by Figure 2.21.

2. *(Gaussian distribution.)* Figure 2.24(c) shows probability-to-volume functions for the standard Gaussian distribution for dimensions $1, 2$. A quantile function corresponding to a higher dimension dominates a quantile function corresponding to a lower dimension.

3. *(Student distribution.)* Figure 2.24(d) shows probability-to-volume functions for the Student density with degrees of freedom $\nu = 1$ for dimensions $1, 2$. As in the Gaussian case a quantile function corresponding to a higher dimension dominates a quantile function corresponding to a lower dimension.

Dimension Normalized Versions Quantile and distribution functions have similar problems as unimodal volume functions: the concentration effect and the dimension non-invariance. The concentration effect for quantile functions means that the function is near zero at $[0, 1-\epsilon]$ but then explodes to take large values at $[1-\epsilon, 1]$. Thus we define a dimension normalized quantile and distribution function. A dimension normalized quantile function maps probabilities to the normalized volumes.

Definition 2.6 (Dimension normalized probability-to-volume function, Dimension-normalized volume-to-probability function.) *A* dimension normalized probability-to-volume function, *or a* dimension normalized quantile function, *corresponding to distribution P, is defined by*

$$Q^*(p) = \left[\frac{1}{C_d}\,\text{volume}\left(A_{\lambda_p}\right)\right]^{1/d}, \qquad p \in [0,1],$$

where C_d is defined in (B.10) and λ_p is defined in (2.12). A dimension-normalized volume-to-probability function, *or a* dimension normalized distribution function, *is defined by*

$$F^*(v) = \inf\{p \in [0,1] : Q^*(p) \geq v\}, \qquad v \in [0,\infty).$$

As an example, consider the case where the sequence of sets is a sequence of balls. The dimension normalized volume of a ball is equal to its radius. Thus the dimension normalized quantile function is equal to $p \mapsto r_p$, where r_p is the radius of a ball with probability content p.

Examples We illustrate dimension normalized quantile functions and distribution functions for the case of Bartlett, standard Gaussian, and Student distributions, defined in (10.7), (10.5), and (10.6).

1. *(Bartlett distribution.)* Figure 2.25(a) shows the dimension normalized probability-to-volume functions for the Bartlett density, for dimensions 1 to 20. Figure 2.26(a) shows the dimension normalized volume-to-probability functions.

2. *(Gaussian distribution.)* Figure 2.25(b) shows the dimension normalized probability-to-volume functions for the standard Gaussian density, for dimensions 1 to 20. Figure 2.26(b) shows the dimension normalized volume-to-probability functions.

3. *(Student distribution.)* Figure 2.25(c) shows the dimension normalized probability-to-volume functions for the Student density with degrees of freedom $\nu = 1$, for dimensions 1 to 20. Figure 2.26(c) shows the dimension normalized volume-to-probability functions for the same parameters. We plot the functions only up to radius 20, and thus the probability content of the sets does not reach 1.

2.3 BIBLIOGRAPHIC NOTES

2.3.1 Visualization of High-dimensional Functions

We will discuss later the approach of level set trees to visualize high dimensional functions. We will mention only two examples from the book Card, Mackinlay and Shneiderman (1999) that address the visualization of high-dimensional functions. (1) Feiner and Beshers (1990) presents a slicing scheme to visualize multivariate

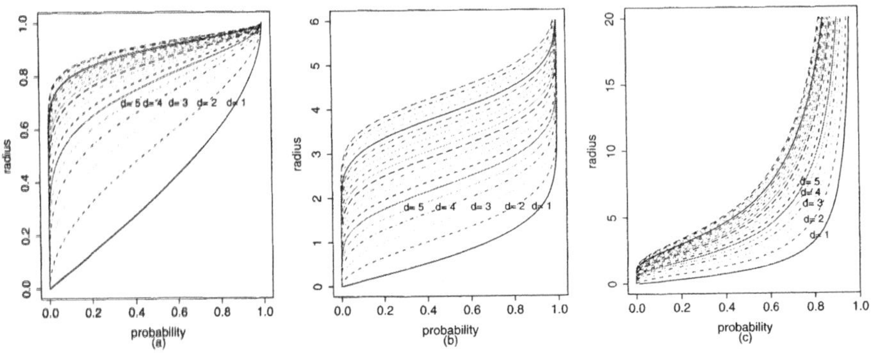

Figure 2.25 *(Dimension normalized quantile functions for dimensions* 1-20.*)* (a) The Bartlett distribution; (b) the standard Gaussian distribution; (c) the Student distribution with degrees of freedom 1.

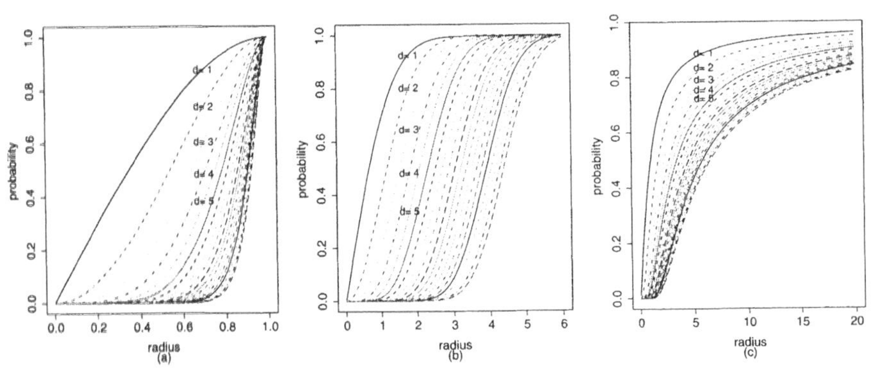

Figure 2.26 *(Dimension normalized distribution functions for dimensions* 1-20.*)* (a) The Bartlett distribution; (b) the standard Gaussian distribution; (c) the Student distribution with degrees of freedom 1.

functions: one shows perspective plots of 2D slices and manipulates the choices of the slices interactively. (2) Mihalisin, Timlin and Schwegler (1991) considers schemes to go through a dense collection of slices and considers also the possibility of nonlinear slicing.

2.3.2 Visualization of the Spread of Multivariate Densities

A further possibility to visualize the spread of a density f is to apply the real valued random variable $f(X)$, where $X \sim f$. Note that the density of random variable $f(X)$ is $g : [0, \infty) \to [0, \infty)$, $g(\lambda) = -\lambda V'(\lambda)$, where $V(\lambda)$ is the volume of the level set with level λ, as defined in (2.7). This was proved in Troutt (1991).

Liu et al. (1999) propose four univariate curves to visualize kurtosis, relative-spread, and heavy tailedness.

1. They propose to plot the Lorenz curve of $f(X)$. The Lorenz curve of a random variable $Y \in \mathbf{R}$, with distribution function F, is defined to be

$$L(p) = \frac{1}{EY} \int_0^p F^{-1}(t)\, dt = \frac{p}{EY} E(Y \,|\, Y \le F^{-1}(p)), \qquad p \in [0, 1].$$

The graph $(p, L(p))$, $p \in [0, 1]$ is equal to the graph of $(F(y), \Omega(y))$, $y \in \mathbf{R}$, where

$$\Omega(y) = \frac{1}{EY} \int_{-\infty}^{y} t\, dF(t), \qquad y \in \mathbf{R}.$$

Thus the Lorenz curve of $f(X)$ is equal to

$$L(p) = \frac{\int_{C_p^c} f\, dF}{\int f\, dF} = \frac{p}{Ef(X)} E(f(X) \,|\, X \in C_p^c), \qquad p \in [0, 1],$$

where C_p is defined in (2.11) with A_λ the level set of the density, and C_p^c is the complement of C_p.

2. They propose an other Lorenz curve type plot

$$L_n(p) = \frac{\sum_{i=1}^{[np]} Z_i}{\sum_{i=1}^{n} Z_i}, \qquad p \in [0, 1],$$

where Z_i is the Mahalabonis distance from the deepest point μ_n, determined by a data depth: $Z_i = (X_i - \mu_n)^T \Sigma_n^{-1} (X_i - \mu_n)$, where Σ_n is a sample dispersion matrix.

3. A shrinkage plot is defined to be the plot of

$$a_n(p) = n^{-1} \#\{X_i \in C_{n,p}^s\}, \qquad p \in [0, 1],$$

where $C_{n,p}^s$ is the shrinkage of $C_{n,p}$ towards the center, $C_{n,p}^s = s(C_{n,p} - \mu_n)$, $0 < s < 1$, and $C_{n,p}$ is the pth central hull.

4. A fan plot is defined by

$$b_n(t) = \frac{\text{volume}(C_{n,p}(t))}{\text{volume}(C_{n,p})}, \qquad t \in [0, 1],$$

where $C_{n,p}(t)$ is the convex hull of the $100t\%$ central-most data points in $C_{n,p}$.

CHAPTER 3

VISUALIZATION OF TREES

We discuss visualization of trees when the nodes of the tree are carriers of information. We are interested not only in the visualization of the underlying tree structure but want to visualize the additional information associated with the nodes.

Trees are often visualized with plots as in Figure 3.1(a). Nodes are drawn as circles or bullets and parent–child relations are expressed with straight lines joining a child to the parent. The horizontal position (height) of a node is equal to the depth of the node. There are many other ways to visualize a tree. Figure 3.1(b) shows the tree of panel (a) with a visualization where the nodes are represented with nested sets. The root node is the largest set, and parent–child relations are expressed with set inclusion. This visualization looks like a contour plot of a function. In fact we will later go into the other direction; we start with a multivariate function and represent this function with a tree structure.

Figure 3.1 visualizes only the tree structure. Often trees are carriers of information: every node is annotated with an information content. In these cases we are seldom interested only in the visualization of the tree structure but want to visualize the information associated with the nodes. We consider two decorations of nodes with information:

Smoothing of Multivariate Data: Density Estimation and Visualization. By Jussi Klemelä

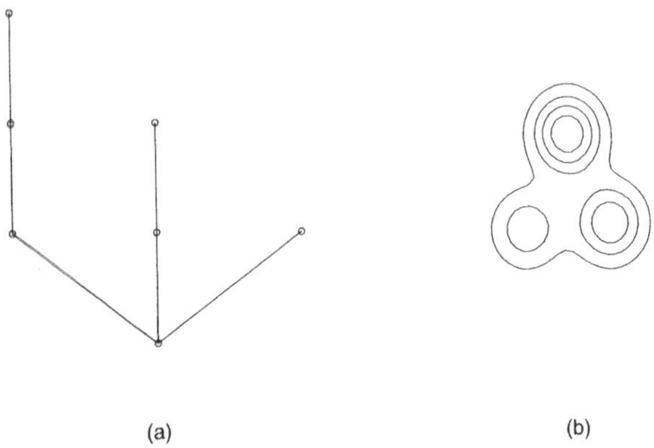

Figure 3.1 *(Tree.)* (a) The nodes of a tree as circles and (b) the nodes of a tree as sets.

1. Section 3.1 considers the case where every node is associated with a vector of d real numbers,

2. Section 3.2 considers the case where every node is associated with a positive real number.

A *tree* is a set of nodes together with parent–child relations between the nodes. A tree is defined more abstractly in Appendix D.1. Graph drawing algorithms make it possible to draw trees automatically. Graph drawing algorithms are given, for example, in Tollis, Battista, Eades and Tamassia (1998).

3.1 VISUALIZATION OF SPATIAL TREES

3.1.1 Spatial Tree

We consider visualization of trees whose nodes are annotated with vectors of real numbers. We interpret the vectors as points in the Euclidean space; d-dimensional vectors give the locations of the nodes in the d-dimensional Euclidean space. When the nodes that are close in the tree structure are also close in the Euclidean space, it is natural to call this kind of tree a *tree that generates a branching curve*. We use the term *spatial tree* for shortness. We assume also that the nodes are associated with a real-valued height. When this is not the case, we can always define the height of a node to be its depthness in the tree structure. (The depth of the root is defined to be 0, and the depth of a child is defined to be one more than the depth of the parent.)

Definition 3.1 (Spatial tree.) *A* spatial tree *is a tree satisfying the following properties.*

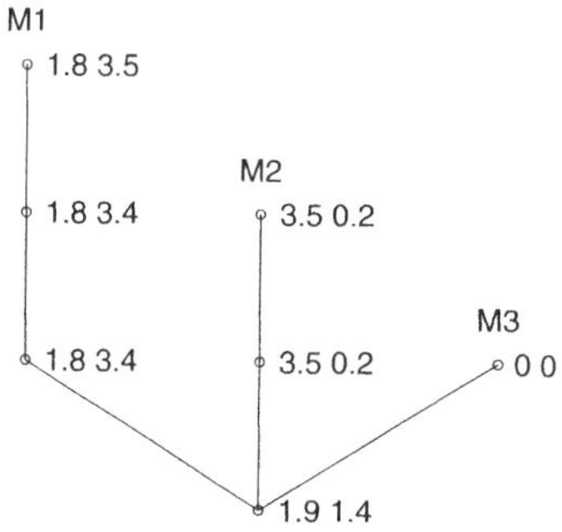

Figure 3.2 *(Spatial tree.)* Each node of the spatial tree is associated with a vector in $\mathbf{R}^2$.

1. *Each node is associated with a vector of d real numbers.*
2. *Each node is associated with a real valued height and the height of a child is larger than the height of the parent.*

3.1.2 Spatial Tree Plot

A spatial tree may be interpreted as as a discrete curve in the d-dimensional Euclidean space, having branches, and thus one may visualize this tree by plotting each coordinate separately.

Definition 3.2 (Spatial tree plot.) *A* spatial tree plot *consists of d windows. The nodes of the tree are drawn as bullets.*

1. *The horizontal position of a node in the ith window, $i = 1, \ldots, d$, is equal to the ith coordinate of the vector associated with the node.*
2. *The vertical position of a node is equal to the height of the node.*
3. *The parent–child relations are expressed by the line joining a child with the parent.*

Examples We give examples of spatial tree plots:

1. Figure 3.2 shows a spatial tree, where each node is annotated with a vector in $\mathbf{R}^2$. Figure 3.3 shows a corresponding spatial tree plot. The leafs of the tree are labeled with M1 to M3.

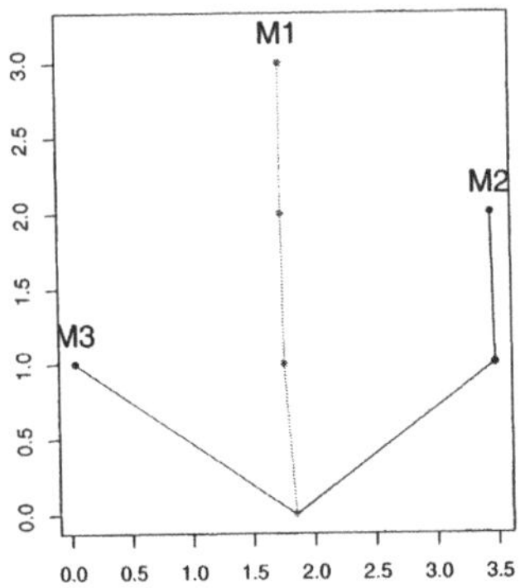

(a) coordinate 1

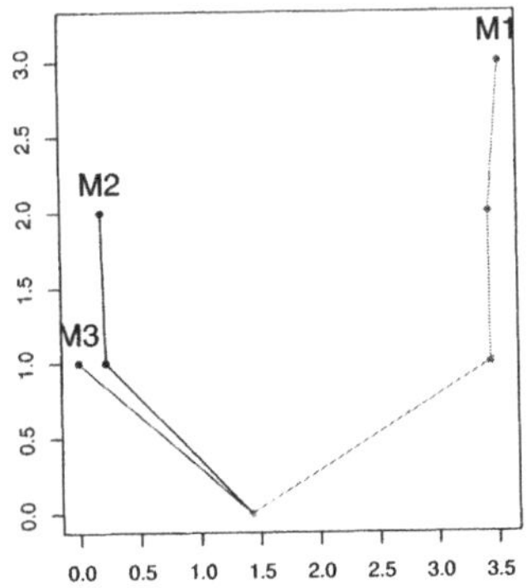

(b) coordinate 2

Figure 3.3 *(Spatial tree plot.)* A spatial tree plot of the spatial tree in Figure 3.2.

2. Figure 3.4(a) shows a hyperbolic spiral with three branches. We may interpret the curves as a spatial tree, whose root node is at the origin. Panels (b) and (c) show the corresponding spatial tree plot. The hyperbolic spiral is defined by $r \mapsto (r\cos(1/r), r\sin(1/r))$.

   Section 1.3 introduces parallel level plots, which may be used to visualize 1D curves in the d-dimensional Euclidean space, for example. Spatial tree plots are closely related to parallel level plots: the use of colors in a spatial tree plot makes it possible to visualize 1D curves that have branches. We may say that a spatial tree plot is a colored parallel level plot. Note that Figure 1.17 shows the hyperbolic spiral without the branches of Figure 3.4.

3. Examples of spatial tree plots include a barycenter plot of a level set tree as defined in Section 4.3, a location plot of a shape tree as defined in Section 5.4, a tail tree plot of data as defined in Section 6.2, a multiframe mode graph of a scale of estimates as defined in Section 7.1, and a parallel dendrogram plot defined in Section 8.1.

3.1.3 Colors and Labels

When the leaf nodes are labeled uniquely, it is possible in principle to identify the nodes across the d windows. Adding colors to the plot eases the identification of the nodes across different windows. Besides the nodes, it is useful to color also the lines joining a child and a parent. The color of a line will be the same as the color of the child node. Often it is enough to be able to identify different branches across the windows, and thus it is enough to choose different colors for the branches. The segmentation of a tree to branches is defined in Section D.3.1.

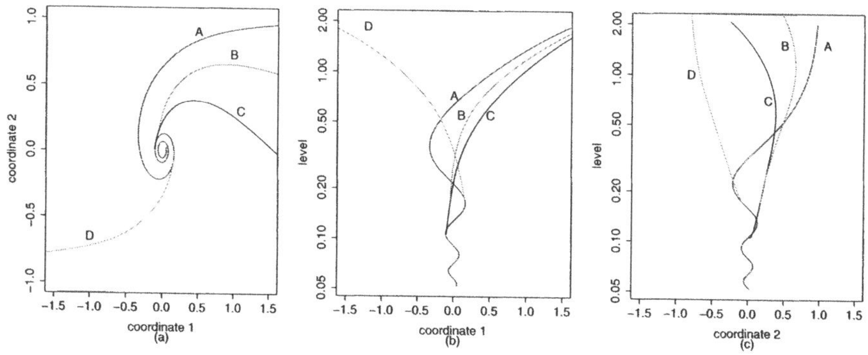

Figure 3.4 *(Hyperbolic spiral with three branches.)* (a) A hyperbolic spiral with three branches and (b)–(c) a corresponding spatial tree plot. We have interpreted the hyperbolic spiral and its branches as a tree whose root is at the origin. The height of the nodes is taken to be the distance from the origin.

We may apply the *leafs-first coloring* or the *root-first coloring*. In the leafs-first coloring we choose first distinct colors for the leaf nodes and then travel toward the root node, changing the color always when two branches are merging. In the root-first coloring we start by choosing a color for the root. We travel toward the leafs, and when the tree is branching, we choose new colors for the children. One of the children may have the same color as the parent.

The leafs-first coloring is used when we want to highlight the leaf nodes. Then we may choose the most distinguished colors to separate the leaf nodes (red, blue, green, orange, etc.). After a while it is difficult to find colors that are distinguishable, and thus one may not separate the nodes easily near the root. The leafs-first coloring will be used for level set trees and shape trees.

When we want to highlight the nodes near the root nodes, we use the root-first coloring. After reaching a certain level we may also fix the colors and choose always the color of the parent for the children. This type of coloring is used for spatial tree plots of dendrograms, since then we see which observations belong to the chosen clusters. Also with multiframe mode graphs we use the root-first coloring.

3.2 VISUALIZATION OF FUNCTION TREES

3.2.1 Function Tree

A *function tree* is a tree whose nodes are annotated with lengths (positive real numbers). We may visualize a function tree so that every node is drawn as a line whose length is equal to the length of the node. As in the case of spatial trees we assume furthermore that every node is associated with a real valued height, which determines the vertical positions of the nodes.

When the tree has a large number of nodes, it is useful to visualize the tree as a function. We construct a 1D piecewise constant function by considering every node as a level set of a function; the length and height of the level set is determined by the the length and height associated with the node. For this reason the term "function tree" is used.

We define a function tree to be a tree with several root nodes, so it is in fact a collection of trees (a forest).

Definition 3.3 (Function tree.) A function tree *is a collection of trees satisfying the following properties:*

1. *Each node is associated with a positive length and the sum of the lengths of children is smaller than or equal to the length of the parent.*
2. *Each node is associated with a real valued height and the height of a child is larger than the height of the parent.*
3. *The root nodes and the children of each node have been ordered.*

3.2.2 Function Tree Plot

Next we define the function that is generated by a function tree. A function tree plot is a plot of this function. As a first step in the construction we associate the nodes of a function tree with intervals. The lengths of the intervals are determined by the lengths associated with the nodes. We start by associating the root nodes with intervals, and then we proceed by creating a nested family of intervals so that the intervals associated with the children are inside the interval associated with the parent. The child intervals are disjoint and equally spaced.

Definition 3.4 (Intervals of a function tree.) Intervals associated with the nodes of a function tree, *using the equal spacing method, are defined with the following rules:*

1. *Assume that the tree has N root nodes that are associated with lengths L_i, $i = 1, \ldots, N$. Choose a distance $\delta_r > 0$. The ith root node is associated with interval $[a_i, b_i]$, where $a_1 = 0$, $b_1 = L_1$, and for $i = 2, \ldots, N$,*

$$a_i = b_{i-1} + \delta_r, \qquad b_i = a_i + L_i. \tag{3.1}$$

2. *Let a node be associated with interval $[a, b]$, and assume that this node has $M \geq 1$ children associated with lengths L_i, $i = 1, \ldots, M$. Define the distance between the child intervals as*

$$\delta = \frac{(b-a) - \sum_{i=1}^{M} L_i}{M+1}.$$

The ith child is associated with interval $[a_i, b_i]$, where $a_1 = a + \delta$, $b_1 = a + L_1$, and for $i = 2, \ldots, M$,

$$a_i = b_{i-1} + \delta, \qquad b_i = a_i + L_i. \tag{3.2}$$

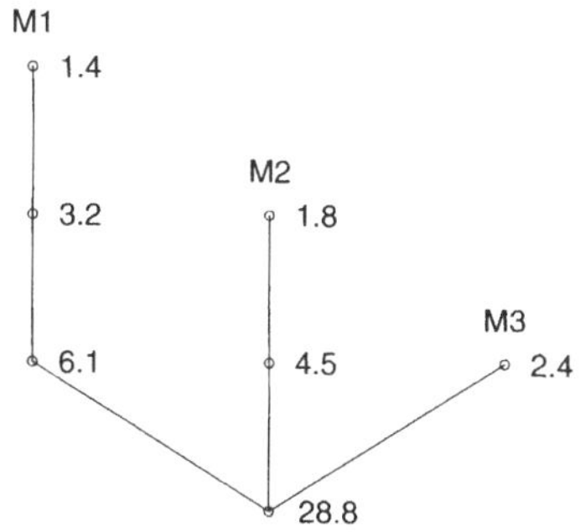

Figure 3.5 *(Function tree.)* Each node of the function tree is associated with a positive real number.

The distance $\delta_r > 0$ between the intervals associated with the root nodes may be chosen arbitrarily. For example, in package "denpro" it is chosen as $\delta_r = \text{volume}(\text{supp}(f))/[4(N+1)]$, where N is the number of root nodes.

Now we are ready to give a definition of the 1D function associated with a function tree. The basic idea is that a level set of the 1D function is formed by the union of those intervals that are associated with the nodes whose height is larger than, or equal to the level of the level set.

Definition 3.5 (Function generated by a function tree, Function tree plot.) *The function $g : \mathbf{R} \to \mathbf{R}$, generated by function tree T, is such that for level $\lambda \in \mathbf{R}$, the level set*

$$\{x \in \mathbf{R} : g(x) \geq \lambda\} = \bigcup \{\mathcal{I}_m : m \textit{ is such node of } T \textit{ that } H_m \geq \lambda\}, \qquad (3.3)$$

where H_m is the height and $\mathcal{I}_m$ is the interval associated to node m, as defined in Definition 3.4. A function tree plot *is a plot of the function generated by the function tree.*

Examples We give examples of function trees and functions generated by a function tree.

1. Figure 3.5 shows a function tree, as defined in Definition 3.3. The intervals of the tree are associated with lengths (positive real numbers).

   Figure 3.6(a) shows the intervals that are associated with the nodes of the tree shown in Figure 3.5, when the intervals associated with the nodes of a function tree are as defined in Definition 3.4. We have drawn these intervals at the heights associated with the nodes.

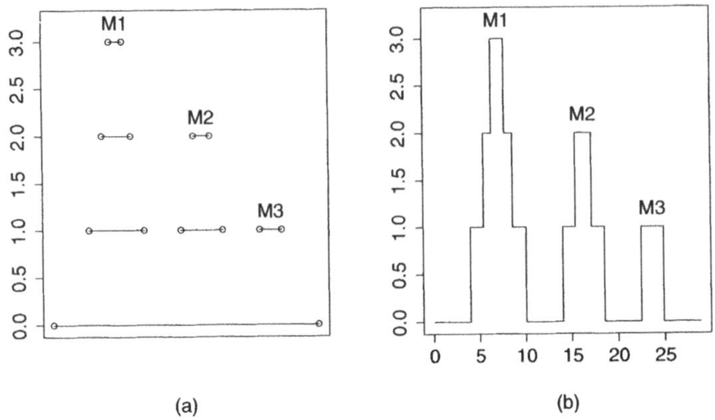

Figure 3.6 *(Function tree plot.)* (a) The intervals associated with the nodes of the function tree shown in Figure 3.5. (b) The function generated by the function tree.

Figure 3.6(b) shows the corresponding function, generated by the function tree, as defined in Definition 3.5 The graph of the function may be thought to be constructed so that we draw vertical lines at the end points of the intervals, joining a child to the parent, but delete the intervals themselves.

2. Figure 3.7(a) shows a tree. We have labeled the 22 leaf nodes with M1 to M22. The nodes are associated with positive lengths, to make the tree to a function tree. (The tree is a level set tree of a kernel estimate, and the lengths of the nodes are equal to the volumes of the sets associated with the nodes, see Section 4.2.) Panel (b) shows a corresponding function tree plot. The length of a node can be interpreted as measuring the "importance" of the node, and the function tree plot visualizes the importance of the nodes, whereas the usual tree plot treats every node equally.

3. Examples of function tree plots include a volume plot of a level set tree as defined in Section 4.2, a radius plot, probability content plot, and tail probability plot of a shape tree as defined in Section 5.3, and a tail frequency plot of a tail tree as defined in Section 6.3.

3.3 BIBLIOGRAPHIC NOTES

We mention three additional methods of visualizing trees:

1. *(Tree map.)* Drawing a tree map starts with a large rectangle and within that rectangle are smaller rectangles, one for each child. This construction is repeated until all nodes are included. Tree maps are based on the same idea

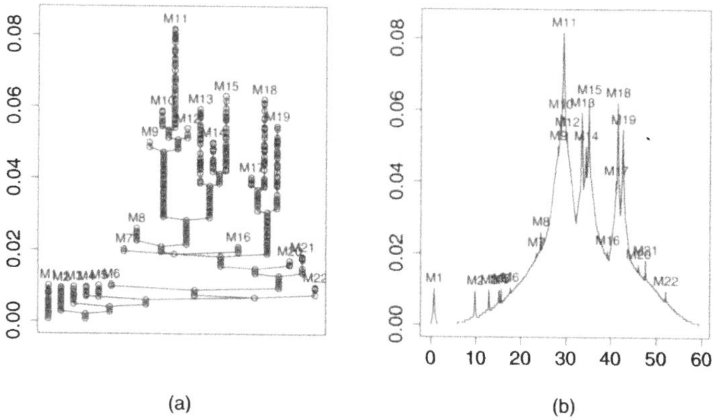

Figure 3.7 *(Function tree, Function tree plot.)* (a) A tree with 22 leaf nodes and (b) the corresponding tree function with 22 local maxima.

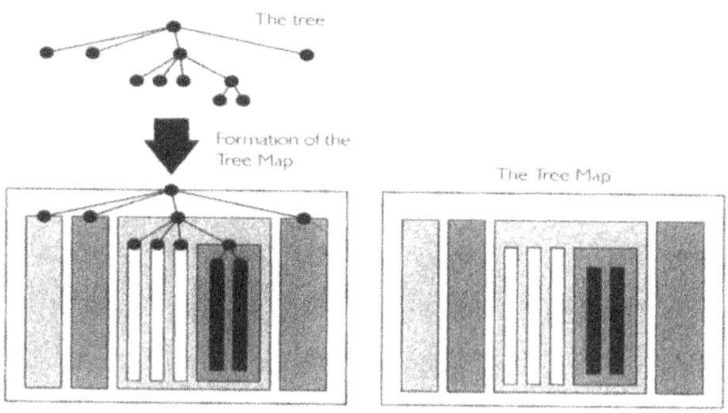

Figure 3.8 *(Tree map.)* Shown is Figure 8.26 from Spence (2001).

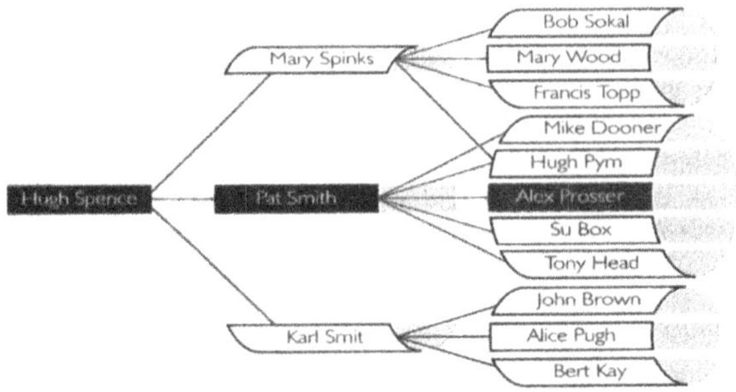

Figure 3.9 *(Cone tree.)* Shown is Figure 6.7 from Spence (2001).

as Figure 3.1. Figure 3.8 shows a construction of a tree map. Figure 3.8 is Figure 8.26 from Spence (2001). Tree maps were introduced by Johnson and Shneiderman (1991) and Shneiderman (1992).

2. *(Cone tree.)* A cone tree remodels a tree in three-dimensional space so that all children of a given node are arranged in a circle that forms a cone. Cone trees should support rotation of the cones due to the occlusion of the nodes. Figure 3.9 shows a cone-tree. Figure 3.9 is Figure 6.7 from Spence (2001). Cone trees were introduced by Robertson, Mackinlay and Card (1991).

3. *(Hyperbolic tree.)* A hyperbolic tree draws a tree so that large trees can be kept within the boundaries of a circular area. A hyperbolic geometric transformation is applied to arrange the nodes and the lines connecting them. The root is at the center, and the children are arranged around it. Hyperbolic browsers were introduced by Lamping and Rao (1994), Lamping, Rao and Pirolli (1995), and Lamping and Rao (1996).

CHAPTER 4

LEVEL SET TREES

A *level set tree* of a function is a tree structure of the separated components of the level sets of the function. A *level set* of a function is the set of those points where the function exceeds a given value. The level set of function $f : \mathbf{R}^d \to \mathbf{R}$ at level $\lambda \in \mathbf{R}$ is defined as

$$\Lambda(f, \lambda) = \left\{ x \in \mathbf{R}^d : f(x) \geq \lambda \right\}.$$

A level set tree is a tree whose root nodes are the separated components of the lowest level set of the function. The child nodes of a given node correspond to the separated components of a part of the level set whose level is one step higher than the level of this parent node.

Level set tree based visualization methods may be applied in exploratory data analysis, to make graphical inference on the existence of the modes, and to apply graphical methods to the smoothing parameter selection. It is possible to visualize important shape characteristics of a multivariate function by studying its level sets for a series of levels. Important shape characteristics of a multivariate density that we visualize with level set trees are the number, location, and relative largeness of the modes of the density. With a mode we mean a local extreme of the density (or the location of a local extreme of the density). The relative largeness of a mode will be defined to mean the probability mass associated with the mode (excess mass associated with the mode). Other shape characteristics that we may visualize with level set trees include skewness and kurtosis.

With the help of level set trees we may define a transformation of a multivariate function to a univariate function. We call this transform a *volume transform*. A volume transform preserves the mode structure of the function, and we say that the volume transformed function is *mode isomorphic* to the original multivariate function. We call a *volume function* the function that we obtain by a volume transform. A volume function visualizes certain pure shape characteristics, and we need also to visualize how the shape is located in the multivariate Euclidean space. We use a *barycenter plot* to visualize the locations of the barycenters of the separated components of the level sets of a function. In particular, a barycenter plot visualizes the locations of the modes.

Contours of level sets of 3D functions may be visualized with perspective plots, as explained in Chapter 2. Level set trees may be used as a graphical user interface to this type of visualization. However, level set trees provide through volume functions and barycenter plots a new edge on the visualization; the use of level set tree based visualizations is less dimension-sensitive than the use of slices and projections. The definition of a level set tree does not depend on the dimension of the function, and the barycenter plot is a plot which consists only of d windows.

To further motivate level set trees, note that we may look at functions in two different ways.

1. Define a function as a mapping which assigns real values to the points of the d-dimensional Euclidean space:

$$x \mapsto f(x) \in \mathbf{R}, \qquad x \in \mathbf{R}^d.$$

   This is the classical definition of a function.

2. Define a function as a mapping that assigns sets to the real values:

$$\lambda \mapsto \Lambda(f, \lambda) \subset \mathbf{R}^d, \qquad \lambda \in \mathbf{R}.$$

   That is, we assign level sets to the levels. Function $\tilde{f} : \text{Range}(f) \to \{A : A \subset \mathbf{R}^d\}$,

$$\tilde{f}(\lambda) = \Lambda(f, \lambda) \tag{4.1}$$

   is an equivalent way of defining function f, where $\text{Range}(f) = \{f(x) : x \in \mathbf{R}^d\}$ is the range of f. See Exercise 4.1.

4.1 DEFINITION OF A LEVEL SET TREE

A level set tree of a function is a tree whose nodes correspond to connected components of level sets of the function. We will define level set trees for functions defined on a Riemannian manifold $\mathbf{M}$ with distance function $D : \mathbf{M} \times \mathbf{M} \to [0, \infty)$. The most important example is the Euclidean space $\mathbf{M} = \mathbf{R}^d$ with the Euclidean distance

$D(x, y) = \|x - y\|$. An other example is the unit sphere $\mathbf{M} = \mathbf{S}_{d-1} = \{x \in \mathbf{R}^d : \|x\| = 1\}$ with the distance $D(x, y) = \arccos(x^T y)$. We will denote with ν the Lebesgue measure on $\mathbf{M}$. Appendix B.4 contains the definition of a Riemannian manifold.

A level set with level $\lambda \in \mathbf{R}$ of a function $f : \mathbf{M} \to \mathbf{R}$ is the set of those points where the function exceeds λ:

$$\Lambda(f, \lambda) = \{x \in \mathbf{M} : f(x) \geq \lambda\}. \quad (4.2)$$

We may call $\Lambda(f, \lambda)$ defined in (4.2) an *upper level set*. Level set trees may be used to visualize the local maxima of a function. Visualizing the local minima of a function may be done with the help of a tree of *lower level sets*

$$\Lambda^-(f, \lambda) = \{x \in \mathbf{M} : f(x) \leq \lambda\}. \quad (4.3)$$

One may also consider *level curves* (*isosurfaces, exact level sets*)

$$\Gamma(f, \lambda) = \{x \in \mathbf{M} : f(x) = \lambda\}. \quad (4.4)$$

In addition to a level set tree we may also consider a *contour tree*, which is a graph of connected components of level curves. A contour tree combines the information in a tree of upper level sets and in a tree of lower level sets.

The definition of a level set tree depends on the concepts of separated sets and a connected set.

Definition 4.1 (Separated sets, Connected set.) *Let* $\mathbf{M}$ *be a Riemannian manifold with distance* D.

1. *Sets* $A, B \subset \mathbf{M}$ *are* separated, *if* $\inf\{D(x, y) : x \in A,\ y \in B\} > 0$.

2. *Set* $A \subset \mathbf{M}$ *is* connected, *if for each nonempty* $B, C \subset \mathbf{M}$ *such that* $A = B \cup C$, *sets* B *and* C *are not separated.*

Thus two sets are separated if there is some space between them, and a set is connected if it cannot be written as a union of two separated sets. To write a set as a union of such separated sets where each set in the union is connected implies that we have decomposed the set in a maximal way: since each component is connected, it cannot be decomposed to finer separated components.

A level set tree may have any finite number of root nodes, and every node may have any finite number of child nodes. The root nodes of a level set tree correspond to the connected components of the lowest level set of the function. The child nodes of a given parent node correspond to those connected components of the level set with one step higher level than the level of the parent node, that are contained in the set corresponding to the parent node. We associate to every node a level and a set. The set associated with the node is the corresponding connected component of the level set and the level associated with the node is the level of the level set. Since a level set tree may have several root nodes, it is in fact a collection of trees (forest), but we will use the term "tree".

Definition 4.2 (Level set tree, Lower level set tree.) A level set tree *of function $f : \mathbf{M} \to \mathbf{R}$, associated with set of levels $\mathcal{L} = \{\lambda_1 < \cdots < \lambda_L\}$, where $\lambda_L \leq \sup_{x \in \mathbf{M}} f(x)$, is a tree whose nodes are associated with subsets of* $\mathbf{M}$ *and levels in* $\mathcal{L}$ *in the following way:*

1. *Write*

$$\Lambda(f, \lambda_1) = A_1 \cup \cdots \cup A_K,$$

where sets A_i are pairwise separated, and each is connected. The level set tree has K root nodes that are associated with sets A_i, $i = 1, \ldots, K$, and each root node is associated with the same level λ_1.

2. *Let node m be associated with set $B \subset \mathbf{M}$ and level $\lambda_l \in \mathcal{L}$, $1 \leq l < L$.*

   (a) *If $B \cap \Lambda(f, \lambda_{l+1}) = \emptyset$, then node m is a leaf node.*

   (b) *Otherwise, write*

$$B \cap \Lambda(f, \lambda_{l+1}) = C_1 \cup \cdots \cup C_M,$$

where sets C_i are pairwise separated, and each is connected. Then node m has M children, which are associated with sets C_i, $i = 1, \ldots, M$, and each child is associated with the same level λ_{l+1}.

The definition of a lower level set tree *is analogous to the definition of a level set tree. The definition uses the concept of a lower level set, defined in (4.3). The growing of a lower level set tree starts at the highest level set $\Lambda(f, \lambda_L)$ and proceeds to the lowest level λ_1.*

Illustrations We will make several illustrations of Definition 4.2. Figures 4.1 through 4.5 show standard tree plots of level set trees. In addition to the tree structure these plots show the levels of the nodes: the height of a node is equal to the level of the corresponding level set. In particular, the standard tree plot of a level set tree shows the number of local maxima and their heights. Later we will construct further visualizations of level set trees by annotating level set trees with volumes and barycenters of the sets associated to the nodes.

1. Figure 4.1 shows a contour plot and a level set tree of the 3 modal function shown in Figure 2.7. We have applied a grid of 5 levels.

2. Figure 4.2 shows again a contour plot and a level set tree of the function shown in Figure 2.7, but now we have applied a grid of 105 equidistant levels.

   A level set tree is associated with a certain finite set of levels. When we increase the number of levels, the level set tree gives a better representation of the function, at least for a function that has an infinite number of distinct level sets.

3. Figure 4.3 illustrates the growing of a level set tree. We consider again the function shown in Figure 2.7. The lowest level sets have only one connected

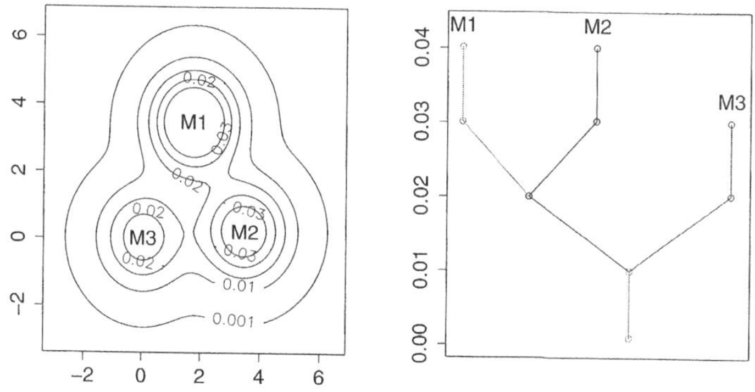

(a) (b)

Figure 4.1 *(Level set tree, sparse.)* A level set tree of the three-modal function shown in Figure 2.7. (a) A contour plot with five levels and (b) the corresponding level set tree with the grid of five levels.

component. Thus the level set tree has only one root node, and the level set tree is not branching at the low levels. When we proceed to the higher levels, there are two branching of level sets, and thus the final level set tree has tree leafs.

4. Figure 4.4 shows a perspective plot, a contour plot, and a level set tree of a function with 22 local maxima. The function is a kernel estimate with the standard Gaussian kernel and smoothing parameter $h = 0.28$. The data of size 200 is generated from the distribution whose density is shown in Figure 2.7. The level set tree has two roots and 22 leafs.

5. Figure 4.5 shows a perspective plot, a level set tree, and a lower level set tree of a function that has three local maxima and two local minima. The level set tree shows the local maxima and the lower level set tree shows the local minima.

Remarks We make several comments concerning Definition 4.2.

1. *(Piecewise constant functions.)* We say that a function $f : \mathbf{M} \to \mathbf{R}$ is piecewise constant if its range is a finite set:

$$\text{range}(f) \stackrel{def}{=} \{f(x) : x \in \mathbf{M}\} = \{\lambda_1, \ldots, \lambda_N\}, \tag{4.5}$$

where $\lambda_1 < \cdots < \lambda_N$. If the function is piecewise constant, as defined in (4.5), then we may choose the grid of levels of a level set tree to be $\lambda_1, \ldots, \lambda_N$.

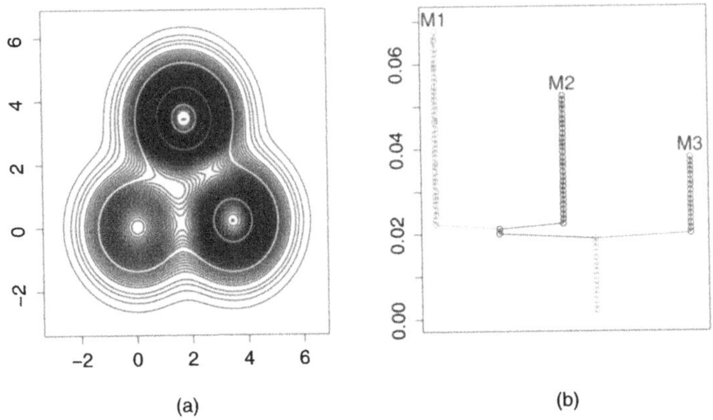

Figure 4.2 *(Level set tree, dense.)* A level set tree of the three-modal function shown in Figure 2.7. (a) A contour plot with 105 levels and (b) the corresponding level set tree with the grid of 105 levels.

Definition 4.3 (Saturated level set tree.) *If a function is piecewise constant as defined in (4.5), then the level set tree with the grid of levels $\lambda_1, \ldots, \lambda_N$ is called the* saturated level set tree *of the piecewise constant function.*

The saturated level set tree of a piecewise constant function is such that adding more levels to the grid does not lead to a more accurate representation of the function with its level set tree.

2. *(Existence of level set trees.)* Definition 4.2 of a level set tree assumes that each level set corresponding to levels in $\mathcal{L} = \{\lambda_1 < \cdots < \lambda_L\}$ has a finite maximal decomposition in the following sense.

Definition 4.4 The finite maximal decomposition to connected components *of set $A \subset \mathbf{M}$ is the collection $\mathcal{C}$ of sets so that*

(a) $A = \cup_{C \in \mathcal{C}} C$,

(b) for each $C, C' \in \mathcal{C}$ such that $C \neq C'$, C and C' are separated,

(c) each $C \in \mathcal{C}$ is connected.

We write $\mathcal{C} = \mathcal{C}(A)$ when we want to specify that we have made a decomposition of set A.

This kind of partition $\mathcal{C}$ of A is maximal. That is to say, if we try to make a larger partition by splitting some member of $\mathcal{C}$ into two parts, then by the condition 3 of connectedness, condition 2 will become violated.

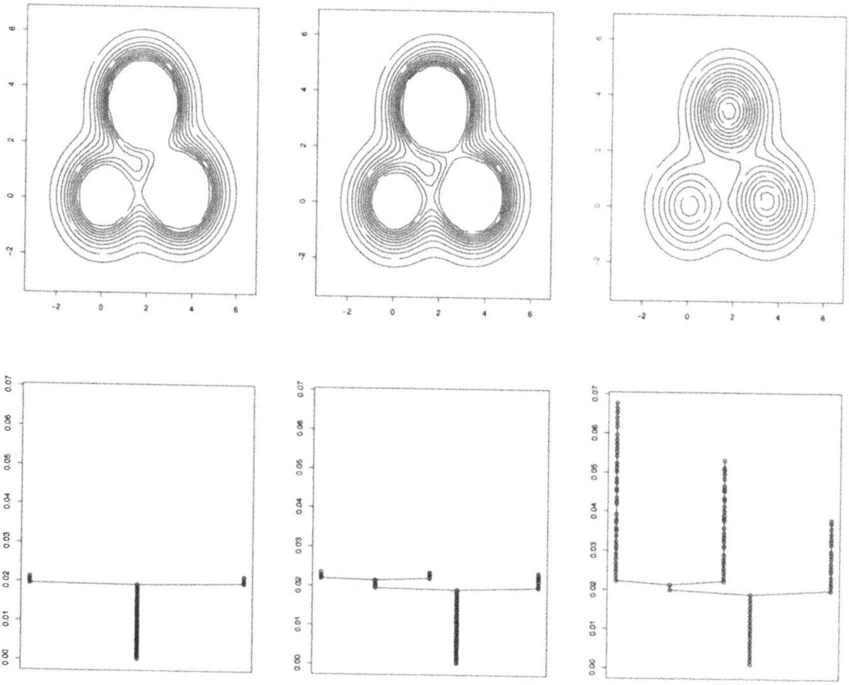

Figure 4.3 *(Growing of a level set tree.)* Illustrated is the growing of a level set tree of the three-modal function shown in Figure 2.7.

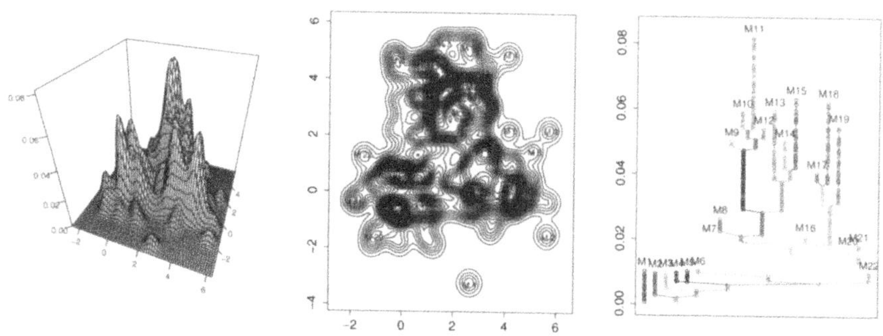

Figure 4.4 *(Level set tree, many modes.)* Shown are (a) a perspective plot, (b) a contour plot, and (c) a level set tree of a function with 22 local maxima.

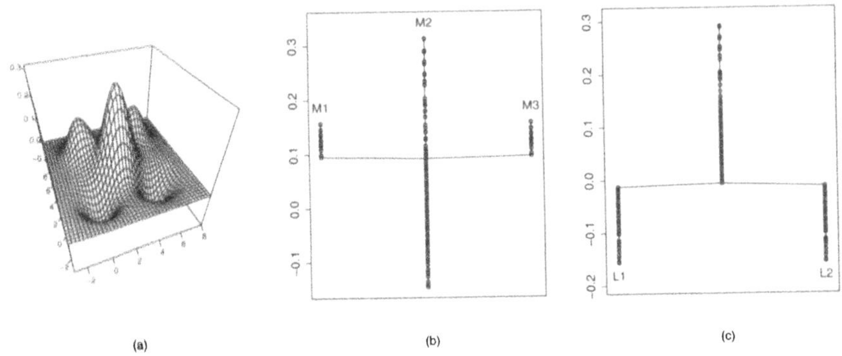

Figure 4.5 *(Level set tree, Lower level set tree.)* (a) A perspective plot of a function having three local maxima and two local minima. (b) A level set tree of the function. (c) A lower level set tree of the function.

There exist functions for which a level set tree, as defined in Definition 4.2, does not exist. It is difficult to formulate general regularity conditions that guarantee the existence of a level set tree, without being overly restrictive. Sufficient regularity conditions for the existence of level set trees are further discussed in Chapter 4.6.1, where Morse functions are defined. Also functions of the form $f(x) = \sum_{i=1}^{M} a_i I_{A_i}(x)$, where $a_i \in \mathbf{R}$ and A_i are rectangles, have always a level set tree.

3. *(Ordered level set trees.)* We have not defined a level set tree as an ordered tree. (In an ordered tree the children of a given node are ordered.) We do not include a definition for the ordering in the proper definition of a level set tree, given in Definition 4.2, for two reasons: (1) there are several almost equally natural ways to define an ordering for the siblings, and (2) with such an ordering we may express less useful information than otherwise with level set trees. However, to define a volume transform in Section 4.2, we need to define a level set tree as an ordered tree. We may use the methods of Section D.3.2 to define an ordering for the children of a given parent (siblings). To do this, we derive first a spatial tree from a level set tree by associating each node with the barycenter of the set associated with a node. The barycenter of a set is defined in (4.7). We have used the descendant based ordering rule of Definition D.3 in this book.

4. *(Alternative definitions of a level set tree.)* Definition 4.13 in Exercise 4.2 defines the same concept of a level set tree as Definition 4.2, but the definition is made by using set inclusion of the connected components of level sets.

   Definition 4.14 in Exercise 4.3 address a subtle issue concerning the definition of a level set tree. Definition 4.14 (for piecewise constant functions) builds a level set tree only from distinct components of level sets; when two components of level sets, corresponding to different levels, are equal, we delete the node corresponding to the lower level. In contrast, Definition 4.2 builds a level set tree from all separated subsets of all level sets corresponding to the given grid of levels.

5. Level set trees contain certain ambiguity because the grid of levels of a level set tree is arbitrary and not determined by the underlying function. We will apply level set trees to describe quantitative features of a function and define a *volume function*, associated to a level set tree, in Chapter 4.2. We can remove to a certain extent the ambiguity in a volume function by defining a limit volume function; see Definition 4.6 below.

4.2 VOLUME TRANSFORM

The *volume transform* is the basic *shape isomorphic transform* that we will discuss. The basic motivation for searching for shape isomorphic transforms comes from the fact that it is possible to visualize a multidimensional object with a low-dimensional object if these objects have the same shape. A volume transform is a transform that is used to visualize local extremes of a function. A volume transformed function has the same shape as the original function in the sense that the number and the sizes of the local extremes (either minima or the maxima) are equal. We will use the term *mode isomorphism* to describe this type of similarity in shape. The volume of an Euclidean set $A \subset \mathbf{R}^d$ is

$$\text{volume}(A) = \int_A dx.$$

We will continue to work in the framework of Riemannian manifolds $\mathbf{M}$, and in this setting the volume of a Borel set $A \subset \mathbf{M}$ is defined as

$$\text{volume}(A) = \int_A d\nu(x),$$

for a Borel measure ν on $\mathbf{M}$.

4.2.1 Volume Transform and Volume Function

We will first give a quick definition of a volume transform. After that, the precise definition is given in terms of function generating trees. Let $f : \mathbf{M} \to \mathbf{R}$ be a function, let ν be a Borel measure on $\mathbf{M}$, and let T be a level set tree of f. The volume transform is a 1D function $v(f;T) : \mathbf{R} \to \mathbf{R}$. We will define $v(f;T)$ in terms of its level sets:

- Annotate each node m of the level set tree T with an interval $\mathcal{I}_m \subset \mathbf{R}$. Let the length of an interval be equal to the ν-volume of the set annotated with the node. Let the intervals be nested according to the tree structure of the level set tree. We comment below in page 102 on the exact definition of the intervals.

- Volume transform $v(f;T) : \mathbf{R} \to \mathbf{R}$ is such that for each level $\lambda \in \mathbf{R}$,

$$\begin{aligned} &\{x \in \mathbf{R} : v(f;T)(x) \geq \lambda\} \\ &= \bigcup \{\mathcal{I}_m : m \text{ is such node of } T \text{ that } \lambda_m \geq \lambda\}, \end{aligned}$$

where λ_m is the level and $\mathcal{I}_m$ is the interval associated to node m.

We can define the volume transform using the concept of a function tree discussed in Section 3.2; a volume transform is defined as the function generated by a function tree (Definition 3.5). The function tree is a level set tree whose nodes are associated with lengths and heights in a natural way. To be precise, a level set tree is not a function tree (it is not an ordered tree), but we make a remark below concerning the ordering of a level set tree.

Definition 4.5 (Volume function, Volume transform.) *A volume function* $v(f;T)$: $\mathbf{R} \to \mathbf{R}$ *of a function* $f : \mathbf{M} \to \mathbf{R}$, *associated with a level set tree* T *of* f, *is a 1D function defined through a function tree as in Definition 3.5. We get the function tree, in the sense of Definition 3.3, from the level set tree* T *by defining the height and the length of the nodes in the following way:*

1. *The height associated to a node of the level set tree is equal to the level associated with the node.*

2. *The length associated to node* m *of the level set tree is equal to* volume(A_m), *where* A_m *is the set associated to node* m.

A volume transform *is the mapping of* f *to* $v(f;T)$.

Remarks We make several observations concerning Definition 4.5:

1. *(Locations of the intervals.)* We recall the equal spacing method, as defined in Definition 3.4, which is used to specify the locations of the intervals $\mathcal{I}_m$ associated to the nodes of a level set tree. Choose first an interval $[0, L]$, where L is greater than the sum of the volumes of the sets associated to the root nodes. Then the intervals associated to the root nodes are positioned inside $[0, L]$ in a symmetric way. After that, one positions the intervals recursively, making a nested collection of intervals according to the tree structure, and positioning the intervals symmetrically. Note that we have not excluded the case where some of the level sets of the function have infinite volume. Note also that we have not defined a level set tree as an ordered tree so that the positioning of the sibling intervals may be done in an arbitrary order.

2. *(Ordering of the siblings.)* A level set tree was not defined as an ordered tree, but we may define an ordering for siblings for example by the rules given in Section D.3.2, see the comment in page 100.

3. *(2D functions.)* A volume transform is interesting because it can be used to visualize local maxima or minima of a function of any dimension. However, the

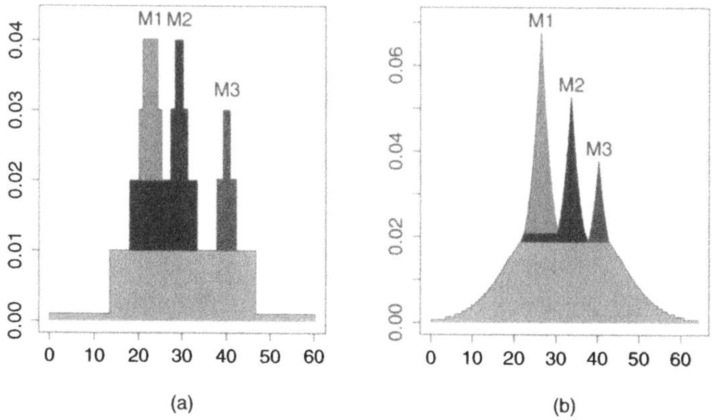

Figure 4.6 *(Volume functions.)* These volume functions correspond to the level set trees in Figure 4.1 and Figure 4.2, and to the 2D three-modal function in Figure 2.7. The volume function in panel (a) is based on a level set tree with few levels and the volume function in panel (b) is based on a level set tree with many levels.

transform is also useful in visualizing 2D functions. To be efficient, visualizing 2D functions with perspective plots requires motion and interaction, since 3D display suffers from distortions arising from a perspective view and occlusion. On the other hand, visualizing 2D functions with contour plots does not always convey information about the largeness of the local extrema, and separating local maxima from local minima is not visually intuitive with contour plots. Furthermore, visualizing functions defined on the unit sphere of the three-dimensional Euclidean space can be inconvenient with perspective plots or contour plots, and here a volume transform provides a very useful additional tool.

4. *(Densities.)* When $\lambda_1 > 0$ and f is a density, the level set $\Lambda(f, \lambda_1)$ has finite volume. Indeed, since densities are integrable, they may be bounded away from zero only on a set of finite volume. When f is a density, then we may without loss of generality choose $\lambda_1 > 0$ in Definition 4.2, since for densities we may, without loss of generality, assume that $\Lambda(f, 0) = \mathbf{R}^d$. Indeed, when density f is defined only on $A \subset \mathbf{R}^d$, we can set $f(x) = 0$ for $x \in A^c$ to get a density with the same distribution.

Illustrations We will make several illustrations of Definition 4.5.

1. Figure 4.6 shows volume functions of the 2D three modal function in Figure 2.7. The volume function in panel (a) is based on the level set tree in Figure 4.1. The volume function in panel (b) is based on the level set tree in Figure 4.2.

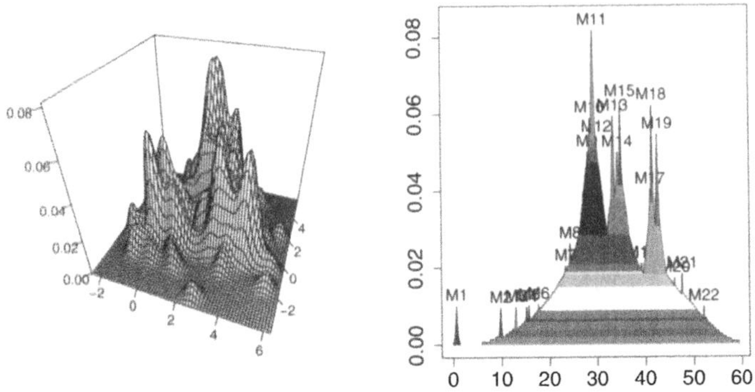

(a) (b)

Figure 4.7 *(Volume function, many modes.)* (a) A perspective plot of a multimodal function, and (b) a volume function of the multimodal function. The multimodal function is the same as in Figures 4.4(a) and 4.4(b), and a level set tree of the function is shown in Figure 4.4(c).

The area under the graphs has been colored. The coloring has been chosen identical with the coloring of the corresponding barycenter plots in Figure 4.12 and Figure 4.13. The coloring helps to find the corresponding nodes of the level set trees in the volume functions and in the barycenter plots.

2. Figure 4.7 shows a perspective plot and a volume function of a function with 22 modes. The function is the same as in Figure 4.4(a) and 4.4(b) and the corresponding level set tree is shown in Figure 4.4(c).

3. Figure 4.8(a) shows a perspective plot of a function with 3 local maxima and 2 local minima. Panel (b) shows a volume function corresponding to an (upper) level set tree, and panel (c) shows a volume function corresponding to a lower level set tree. The level set trees are shown in Figure 4.5.

4. Figure 4.9 shows a hat function and its volume function. The volume function looks like a volume function of a typical unimodal function, and the shape of the hat function is not revealed.

4.2.2 A Limit Volume Function

Definition 4.5 of a volume function depends on a level set tree, as defined in Definition 4.2, that depends in turn on a finite grid of levels used to construct this level set tree. It is of interest to define a limit volume function that is independent of any grid of levels. A natural way to define a limit function is to choose a sequence of piecewise constant functions f_k that converge to f as $k \to \infty$ (e.g., in $L_1(\mathbf{M})$), calculate a volume function $v(f_k; T_k)$ for each k, and define a limit volume function

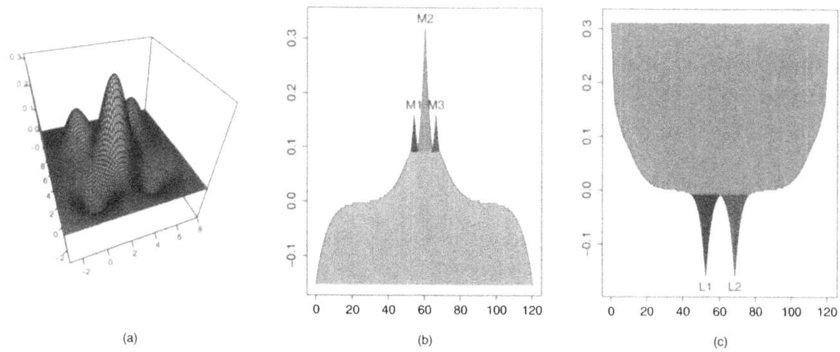

Figure 4.8 *(Volume functions, local maxima, and minima.)* Shown are volume functions of a function with three local maxima and two local minima. (a) A perspective plot of the function. (b) A volume function corresponding to a level set tree. (c) A volume function corresponding to a lower level set tree. The underlying function is the same as in Figure 4.5(a) and the level set tree and the lower level set tree are shown in Figures 4.5(b) and 4.5(c).

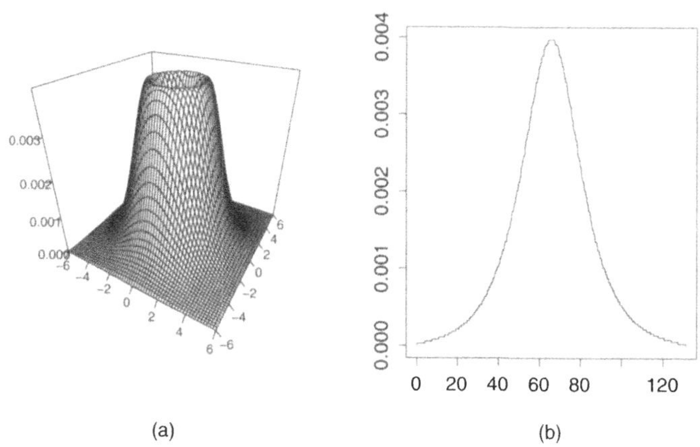

Figure 4.9 *(Hat function.)* (a) A perspective plot of a hat function and (b) its volume function.

$v(f)$ as the limit of $v(f_k; T_k)$ as $k \to \infty$ (e.g., in $L_1(\mathbf{R})$). The naive version of this approach does not work because the ordering of the nodes in the level set trees affects the volume function, and we want to make the definition independent of this ordering. Thus we are led to the following definition:

Definition 4.6 (Limit volume function.) *Let $f : \mathbf{M} \to \mathbf{R}$ be a bounded function. Function $v(f) : \mathbf{R} \to \mathbf{R}$ is a* limit volume function *of f if for each sequence of grids of levels $\mathcal{L}_k = \{\lambda_{k,1} < \cdots < \lambda_{k,N_k}\}$, $k = 1, 2, \ldots$, such that*

1. *$\lambda_{k,N_k} \le \sup_{x \in \mathbf{M}} f(x)$, and*

2. *$f_k \to f$ in $L_1(\mathbf{M})$, where*

$$f_k(x) = \sum_{i=1}^{N_k} \lambda_{k,i} I_{A_{k,i}}(x), \qquad x \in \mathbf{M}, \tag{4.6}$$

with

$$A_{k,i} = \Lambda(f, \lambda_{k,i}) \setminus \Lambda(f, \lambda_{k,i+1}), \;\; i = 1, \ldots, N_k - 1, \;\; A_{k,N_k} = \Lambda(f, \lambda_{k,N_k}),$$

we can find a sequence of level set trees T_k and volume functions $v(f_k; T_k)$ such that

a. *$\mathcal{L}_k$ is the grid of levels of level set tree T_k, and*

b. *$v(f_k; T_k)$ converges to $v(f)$ in $L_1(\mathbf{R})$, as $k \to \infty$.*

Note that a limit volume function is not unique because we can choose in many different ways the volume functions $v(f_k; T_k)$ from the equivalence class of all volume functions, whose level set tree is T_k; there exist many choices for the converging sequence. Figure 4.10 shows two volume functions that belong to the different converging sequences of volume functions. The sequences are converging to a limit volume function of the 2D three modal function in Figure 2.7. Volume functions in Figure 4.10 belong to the same equivalence class of volume functions.

4.3 BARYCENTER PLOT

A barycenter plot visualizes the "skeleton" of the function. A barycenter plot shows the 1D curves that go through the barycenters of all separated components of the level sets. The *barycenter* of set $A \subset \mathbf{R}^d$ is defined as

$$\text{barycenter}(A) = \frac{1}{\text{volume}(A)} \int_A x \, dx \in \mathbf{R}^d. \tag{4.7}$$

A barycenter is a d-dimensional point that is located at the center of mass of the set. It is the expectation of the random vector that is uniformly distributed on the set. We

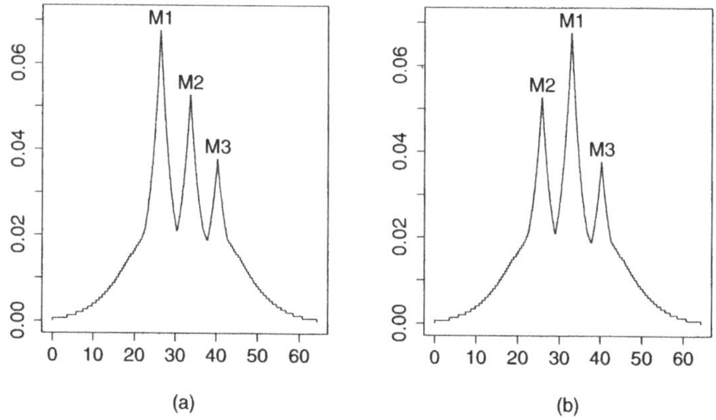

Figure 4.10 *(Equivalent volume functions.)* Shown are two different volume functions of the same 2D function in Figure 2.7.

can define the barycenter also for $A \subset \mathbf{M}$, where $\mathbf{M} \subset \mathbf{R}^d$ is a Riemannian manifold with Lebesgue measure ν:

$$\text{barycenter}(A) = \frac{1}{\text{volume}(A)} \int_A x \, d\nu(x) \in \mathbf{R}^d.$$

The barycenter can lie outside set A and it can lie outside manifold $\mathbf{M}$, unless $\mathbf{M} = \mathbf{R}^d$. We will give examples of barycenter plots only for the Euclidean case.

A Spatial Tree of Barycenters A spatial tree of barycenters (branching curve of barycenters) is shown in Figure 4.11. A spatial tree of barycenters is a spatial tree in the sense of Definition 3.1, when every node of a level set tree is associated with the barycenter of the set associated with the node. When $f : \mathbf{M} \to \mathbf{R}$ is unimodal, the spatial tree is the one-dimensional curve $b_c : [B_1, B_2] \to \mathbf{M}$, where $b_c(\lambda) = \text{barycenter}(\Lambda(f, \lambda))$, $B_1 = \inf_{x \in \mathbf{M}} f(x)$, $B_2 = \sup_{x \in \mathbf{M}} f(x)$, and $\Lambda(f, \lambda)$ is the level set of f at level λ. In the general multimodal case, level set $\Lambda(f, \lambda)$ may have many separated components, and the spatial tree corresponds to the mapping $b_c : [B_1, B_2] \to \mathbf{M}^\infty$, $b_c(\lambda) = (b_{\lambda,1}, \ldots, b_{\lambda,l})$ where $b_{\lambda,i} \in \mathbf{M}$ is the barycenter of the ith separated component of $\Lambda(f, \lambda)$.

Barycenter Plot A barycenter plot is a plot of the spatial tree of barycenters. A barycenter plot consists of d windows when the function is defined in the d-dimensional Euclidean space. We have a window for each coordinate. Each window shows the positions of one coordinate of the barycenters for different levels. The nodes of the tree are drawn as bullets.

1. The horizontal position of a node in the ith window, $i = 1, \ldots, d$, is equal to the ith coordinate of the barycenter of the set associated with the node.

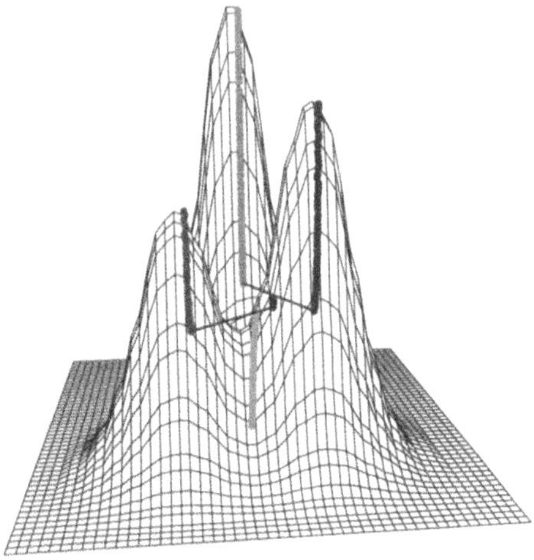

Figure 4.11 *(Skeleton of a function.)* Illustrated is a skeleton of a function. The corresponding barycenter plot is shown in Figure 4.13.

2. The vertical position of a node is equal to the level of the level set that contains the set associated with the node.

3. The parent–child relations are expressed by the line joining a child with the parent.

The definition of a barycenter plot can be given also in terms of spatial trees; we define a barycenter plot as a plot of a spatial tree.

Definition 4.7 (Barycenter plot.) *A barycenter plot is a plot of the spatial tree, as defined in Definition 3.2, when the spatial tree is a level set tree whose nodes are annotated with vectors and heights: (1) the vector is the barycenter of the set associated with the node, (2) the height is the level of the level set that contains the set associated with the node.*

Colors To identify the nodes between different windows of a barycenter plot, and between volume functions and a barycenter plot, we label the modes. The labeling of the modes will uniquely determine the correspondence of all nodes in different windows. To ease the identification of nodes and branches across different windows, we will also color the nodes. We will use the leafs-first coloring, as defined in Section 3.1, page 86. That is, we first choose distinct colors for the leaf nodes and then travel toward the root nodes by changing the color always when two branches are merging. We also color the lines joining a child and a parent. The color of a line will be the same as the color of the child node; that is, the color of the line will be

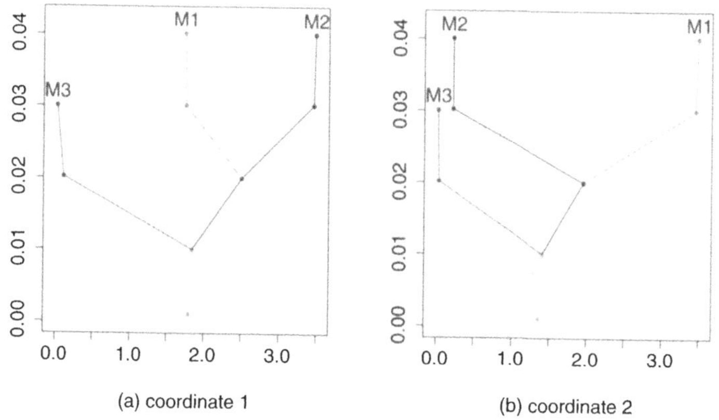

Figure 4.12 *(Barycenter plot, sparse.)* A barycenter plot corresponding to the volume function in Figure 4.6(a). The underlying function is the 2D three-modal density shown in Figure 2.7 and in Figure 4.11.

the same as the color of the node that is at the lower end of that line. The leafs-first coloring is appropriate because we want to highlight the modes of the function and thus we want to choose the colors for the modes in such a way that the modes are easy to distinguish from one another.

Illustrations We now give examples of barycenter plots.

1. Figure 4.12 shows the barycenter plot corresponding to the volume function in Figure 4.6(a).

2. Figure 4.13 shows the barycenter plot corresponding to the volume function in Figure 4.6(b).

3. Figure 4.14 shows the barycenter plot corresponding to the volume function in Figure 4.7.

4.4 INTERPRETATIONS

4.4.1 Mode Isomorphism

We justify the usefulness of a volume transform with the concept of a *mode isomorphism*: a volume transform is useful because it is a 1D function that is mode isomorphic to the original multivariate function. To define a mode isomorphism, first, we need to define the excess mass associated with a node of a level set tree and, second, we need to define the excess mass isomorphism of level set trees.

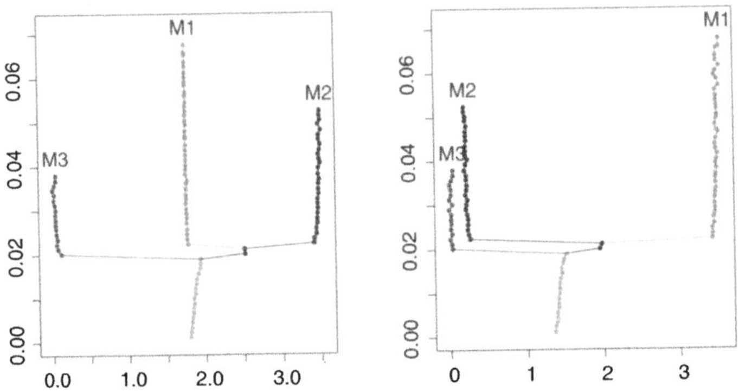

(a) coordinate 1 (b) coordinate 2

Figure 4.13 *(Barycenter plot, dense.)* A barycenter plot corresponding to the volume function in Figure 4.6(b). The underlying function is the 2D three-modal density shown in Figure 2.7 and in Figure 4.11.

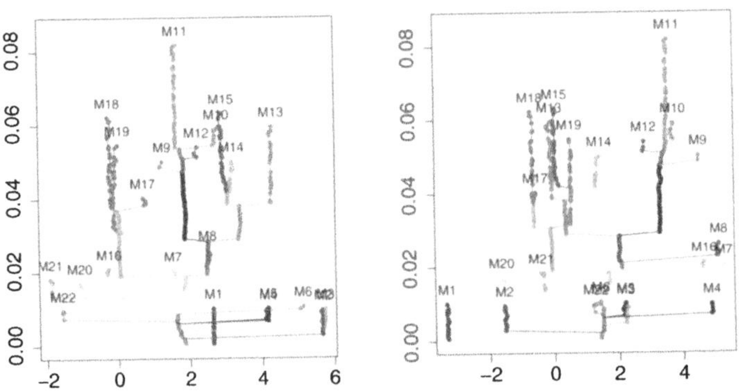

(a) coordinate 1 (b) coordinate 2

Figure 4.14 *(Barycenter plot, many modes.)* The barycenter plot corresponding to the volume function in Figure 4.7.

Excess Mass For piecewise constant functions, an excess mass associated with a node of a level set tree is equal to

$$\int_A (f - \lambda)\, d\nu, \tag{4.8}$$

where $A \subset \mathbf{M}$ is the set associated with the node, λ is the level associated with the node, and ν is the measure on $\mathbf{M}$; see (4.10) below for a precise expression. For functions that are not piecewise constant, the excess mass associated with a node of a level set tree will typically converge to the integral in (4.8) when the grid of levels of the level set trees becomes finer.

To define the excess mass associated with a node of a level set tree, we use the following notations: Assume that level λ and set A are associated with node m of a level set tree. Then we write

$$\text{set}(m) = A, \qquad \text{level}(m) = \lambda.$$

Furthermore, with parent(m), we mean the unique parent of node m. We say that a node is a descendant of node m if it is either a child of m or a child of an other descendant of m.

Definition 4.8 (Excess mass.) *The* excess mass, *associated with node m of a level set tree, is defined by*

$$\begin{aligned} &\text{excmass}(m) \\ &= \sum \{\nu(\text{set}(\text{parent}(m_0)))\ [\text{level}(m_0) - \text{level}(\text{parent}(m_0))] : \\ &\qquad m_0 \text{ is a descendant of } m\}. \end{aligned} \tag{4.9}$$

Remarks We can make some observations concerning the definition of the excess mass:

1. For piecewise constant functions, the sum in the right-hand side of (4.9) can be written as an integral. We assume that the level set tree T is saturated in the sense of Definition 4.3; that is, the grid of the levels of the level set tree is $\mathcal{L} = \text{range}(f)$. Then, for every node m of T,

$$\text{excmass}(m) = \int_{\text{set}(m)} (f(x) - \text{level}(m))\, d\nu(x). \tag{4.10}$$

   Equation (4.10) says that the excess mass is the volume of the area that the function delineates over a given level, in a given branch of the level set tree.

2. For a level set tree that has only one root node and the grid of levels is $\mathcal{L} = \{\lambda_1 < \cdots < \lambda_N\}$, the excess mass of the root node is equal to $\int_{\Lambda(f,\lambda_1)}(f - \lambda_1)_+$. When f is a density and $\lambda_1 = 0$, the excess mass of the root node is

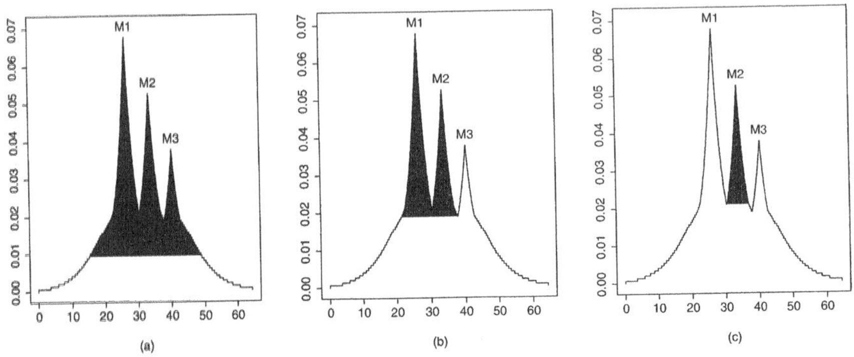

Figure 4.15 *(Excess mass.)* An illustration of the excess masses associated to nodes of a level set tree of the 2D three-modal function of Figure 2.7.

equal to one, and the excess masses of the other nodes are smaller fractions of the total probability mass.

3. An excess mass of a density function $f : \mathbf{R}^d \to \mathbf{R}$ is defined sometimes as the integral of f over a given level: $\int_{\mathbf{R}^d}(f - \lambda)_+$, where $\lambda \geq 0$. We have defined excess masses in a different way, restricting the integration to a set annotated to the node of a level set tree. This makes it possible to annotate excess masses to modes and to mode bundles.

Illustrations Figure 4.15 illustrates the excess masses corresponding to the nodes of a level set tree of the 2D three-modal function of Figure 2.7. Panel (a) shows as a shaded area the excess mass of a node that has as descendants all the leaf nodes (modes). Panel (b) shows the excess mass of a node that is just above the branching level between the mode bundles M1 and M2 and the mode M3. Panel (c) shows the excess mass of a node just above the branching level between mode M1 and mode M2. Figure 7.3(a) shows a level set tree where some of the nodes are augmented with the numerical values of the excess masses.

Excess Mass Isomorphism Level set trees are said to be excess mass isomorphic if the level set trees have isomorphic tree structures and if the excess masses of the corresponding nodes are equal. We say that trees T_1 and T_2 are isomorphic when there is a bijection I from the set of nodes of T_1 to the set of nodes of T_2 such that $I(m_0)$ is the parent of $I(m_1)$ if m_0 and m_1 are nodes of T_1 and m_0 is the parent of m_1.

Definition 4.9 (Excess mass isomorphism of level set trees.) *Level set trees T_1 and T_2 are* excess mass isomorphic *when*

1. *trees T_1 and T_2 are isomorphic,*

2. *for every node m of T_1,*

$$\text{excmass}(m) = \text{excmass}(I(m))$$

where I is the isomorphism between T_1 and T_2.

Mode Isomorphism Now we are ready to define the concept of a mode isomorphism. Mode isomorphism is defined between functions that are defined on possibly different manifolds $\mathbf{M}_1$ and $\mathbf{M}_2$. For example, the manifolds could be Euclidean spaces with different dimensions: $\mathbf{M}_1 = \mathbf{R}^d$ and $\mathbf{M}_2 = \mathbf{R}^{d'}$.

Definition 4.10 (Mode isomorphism.) *Let $f : \mathbf{M}_1 \to \mathbf{R}$ and $g : \mathbf{M}_2 \to \mathbf{R}$ be bounded functions. Denote*

$$\lambda_0^{(f)} = \sup_{x \in \mathbf{M}_1} f(x), \qquad \lambda_0^{(g)} = \sup_{x \in \mathbf{M}_2} g(x),$$

and

$$\Delta_{f,g} = \lambda_0^{(g)} - \lambda_0^{(f)}. \tag{4.11}$$

Functions f and g are mode isomorphic *when for all*

$$\lambda_1 < \cdots < \lambda_N \le \sup_{x \in \mathbf{M}_1} f(x),$$

level set trees T_f of f and T_g of g are excess mass isomorphic, where T_f has grid of levels $\{\lambda_1, \ldots, \lambda_N\}$, and T_g has grid of levels $\{\lambda_1 + \Delta_{f,g}, \ldots, \lambda_N + \Delta_{f,g}\}$.

Remarks Some observations concerning Definition 4.10 are as follows.

1. A first suggestion to define a mode isomorphism of functions might have been to say that f and g are mode isomorphic when their volume functions are identical. However, the definition of a volume function depends on the ordering of the siblings. We want to define a concept of isomorphism that is independent of the rule applied for the ordering of the siblings. Thus we have defined the concept of mode isomorphism using the concept of excess mass isomorphism.

2. We list some properties of mode isomorphism. Let f and g be mode isomorphic. Let $\Delta_{f,g}$ be defined in (4.11). Then

   (a) f and g have the same number of local maxima,

   (b) the level set of f with level λ has the same number of separated components as the level set of g with level $\lambda + \Delta_{f,g}$,

   (c) for all $\lambda \in \mathbf{R}$,

$$\int_{(f \ge \lambda)} f \, d\nu_2 = \int_{(g \ge \lambda + \Delta_{f,g})} g \, d\nu_2,$$

where we use the notation $(f \geq \lambda) = \{x \in \mathbf{M}_1 : f(x) \geq \lambda\}$.

3. We can also tie the concept of a mode isomorphism and the concept of a limit volume function together.

Proposition 4.1 *Let $f : \mathbf{M} \to \mathbf{R}$. Function f and a limit volume function $v(f)$ of f, defined in Definition 4.6, are mode isomorphic if $v(f)$ is such that $\sup_{x\in\mathbf{M}} f(x) = \sup_{t\in\mathbf{R}} v(f)(t)$.*

Proof. If f and $v(f)$ are not mode isomorphic, then there exists a grid $\mathcal{L}$ of levels such that the corresponding level set trees are not excess mass isomorphic. This implies that for some $\lambda \in \mathcal{L}$,

$$0 < \left| \int_{\mathbf{M}} (f-\lambda)_+ - \int_{-\infty}^{\infty} (v(f)-\lambda)_+ \right|. \tag{4.12}$$

Let $\mathcal{L}_k$ be such sequence of grids of levels that $\lambda \in \mathcal{L}_k$, for $k \geq k_0$, for some k_0, and satisfying the conditions 1 and 2 of Definition 4.6. Let T_k be the level set tree of f corresponding to grid $\mathcal{L}_k$, and let f_k be the quantization (4.6) of f corresponding to grid $\mathcal{L}_k$. We have

$$\begin{aligned}
&\left| \int_{\mathbf{M}} (f-\lambda)_+ - \int_{-\infty}^{\infty} (v(f)-\lambda)_+ \right| \\
&\leq \left| \int_{\mathbf{M}} (f-\lambda)_+ - \int_{\mathbf{M}} (f_k-\lambda)_+ \right| \\
&\quad + \left| \int_{\mathbf{M}} (f_k-\lambda)_+ - \int_{-\infty}^{\infty} (v(f;T_k)-\lambda)_+ \right| \\
&\quad + \left| \int_{-\infty}^{\infty} (v(f;T_k)-\lambda)_+ - \int_{-\infty}^{\infty} (v(f)-\lambda)_+ \right| \\
&\stackrel{def}{=} A_{1,k} + A_{2,k} + A_{3,k}.
\end{aligned}$$

We have that $\lim_{k\to\infty} A_{1,k} = 0$, since $f_k \to f$ in L_1, by the choice of $\mathcal{L}_k$. Also $A_{2,k} = 0$ for all $k \geq k_0$, since by the construction, f_k and $v(f;T_k)$ are mode isomorphic. (Indeed the level set trees of f_k and $v(f;T_k)$ are isomorphic by the construction. These level set trees are also excess mass isomorphic because the level sets of f_k and $v(f;T_k)$ have the same volumes and the same levels, and the definition of excess mass in (4.9) depends only on the volumes and levels of the level sets.) Finally, $\lim_{k\to\infty} A_{3,k} = 0$, since $v(f;T_k) \to v(f)$ in L_1, as $k \to \infty$, by the definition of a limit volume function in Definition 4.6. We have reached a contradiction with (4.12). □

4. Piecewise constant functions $f : \mathbf{M}_1 \to \mathbf{R}$ and $g : \mathbf{M}_2 \to \mathbf{R}$ are mode isomorphic when the saturated level set trees of f and g are excess mass isomorphic. Thus, if we want to prove that two piecewise constant functions are mode isomorphic, then we need to check the excess mass isomorphism only

for two level set trees. Saturated level set trees are defined in Definition 4.3, and we use the term "piecewise constant function" to mean a function satisfying (4.5).

4.4.2 Skewness and Kurtosis

Level set trees are a natural tool to visualize multimodality, but level set trees give also information on the spread (kurtosis) and the skewness. The spread and skewness are used to characterize unimodal distributions. However, it is useful to understand how the spread and skewness are visualized with volume functions and barycenter plots. The basic idea is that one may decompose a multimodal function to *unimodal components*.

We may construct a unimodal function by restricting a multimodal function to a region where it is unimodal. However, a more natural way to define unimodal components of a multimodal function is based on the level set tree structure of the function. This is a "horizontal" way of defining unimodal components. We have defined a segmentation of a tree in Section D.3.1. The segmentation of the nodes of a tree is such that each branch of a tree is made to a separate segment. This segmentation is used to define the unimodal components of a function.

We define unimodal components of a function by constructing a level set tree of a function, decomposing the level set tree to the branches, and finding the functions corresponding to each branch of the level set tree.

Definition 4.11 (Unimodal component of a function.) *Let U be a branch of a level set tree of a function $f : \mathbf{M} \to \mathbf{R}$, where the branches of a tree are defined in Section D.3.1. Write $U = \{m_1, \ldots, m_L\}$, where m_i are nodes of the tree. Let λ_i and A_i be the level and the set associated with node m_i. Assume that $\lambda_i < \lambda_{i+1}$, so that $A_i \supset A_{i+1}$. A unimodal component of the function, corresponding to U, is defined by*

$$f(x, U) = \lambda_L I_{A_L}(x) + \sum_{i=1}^{L-1} \lambda_i I_{A_i \setminus A_{i+1}}(x), \qquad x \in \mathbf{M}.$$

Figure 4.16 illustrates a composition of a multimodal function to unimodal components.

A volume function visualizes the spread of unimodal components and a barycenter plot visualizes the skewness of unimodal components.

- If the volumes of level sets of a density increase fast when one moves toward the lower levels, this indicates heavy tails of the distribution. Section 2.2 discusses visualization of the spread of unimodal functions more in detail.

- If the barycenters of the level sets are not the same for different levels, this indicates skewness of the function.

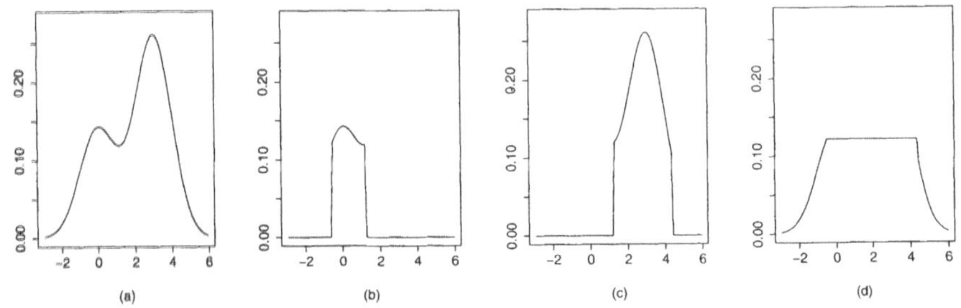

Figure 4.16 *(Unimodal components.)* Illustrated is a decomposition of a multimodal function to unimodal components. (a) A two-modal function; (b)–(d) the three unimodal components.

4.5 EXAMPLES OF LEVEL SET TREES

4.5.1 Three-dimensional Example

We constructed in Section 2.1.3, (page 62), an example of a three-dimensional density whose modes are so close to each other that marginal densities do not reveal the modes. With volume functions and barycenter plots we may reveal the modes. Figure 4.17(a) shows a volume function of the density. The volume function does not reveal the details of the density in the highest levels, and we show in Figure 4.17(b) a *zoomed volume function* that shows only the highest levels. Figure 4.18 shows the corresponding barycenter plot. The modes of the density are located at $M1 = (1.5, 0, 0)$ (orange), $M2 = (-1.5, 0, 0)$ (green), $M3 = (0, 2.6, 0)$ (blue), and $M4 = (0, 0.9, 2.4)$ (red), as can be seen from the barycenter plot, although some nodes overlay each other. We have discretized the function by evaluating it on a grid of 32^3 points.

4.5.2 Four-dimensional Example

We constructed in Section 2.1.3, (page 63), an example of a four-dimensional density whose modes are so close to each other that the marginal densities at two-dimensional coordinate planes do not reveal the modes. Figure 4.19(a) shows a volume function of the density. We have truncated the density to rectangle $[-4, 4]^4$. Figure 4.19(b) shows a zoomed volume function. Figure 4.20 shows the corresponding barycenter plot. The modes of the density are located at $M1 = (2, 0, 0, 0)$, $M2 = (-2, 0, 0, 0)$, $M3 = (0, 3.5, 0, 0)$, $M4 = (0, 1.2, 3.3, 0)$, and $M5 = (0, 1.2, 0.8, 3.2)$. We have discretized the function by evaluating the function on a grid of 16^4 points.

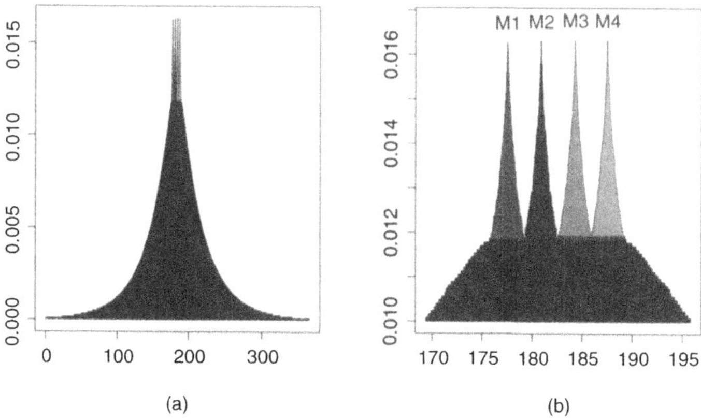

Figure 4.17 *(3D four modal example, volume function.)* Shown are a volume function and a zoomed volume function of a mixture of four 3D Gaussian densities. The density is defined in Section 2.1.3, page 62.

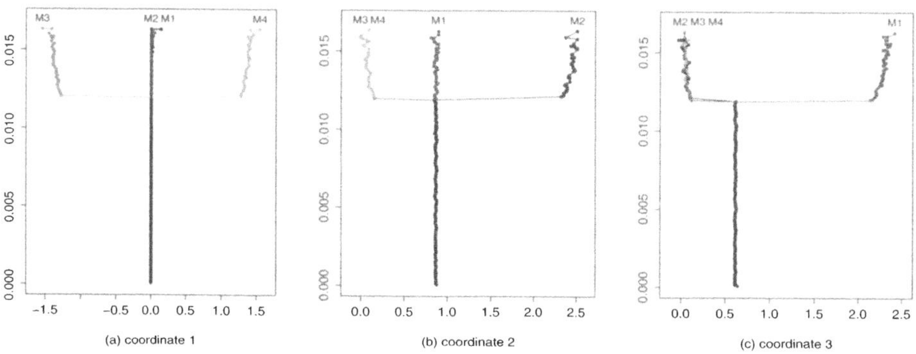

Figure 4.18 *(3D four modal example, barycenter plot.)* Shown is a barycenter plot of a mixture of four 3D Gaussian densities, corresponding to the volume function in Figure 4.17. The density is defined in Section 2.1.3, page 62.

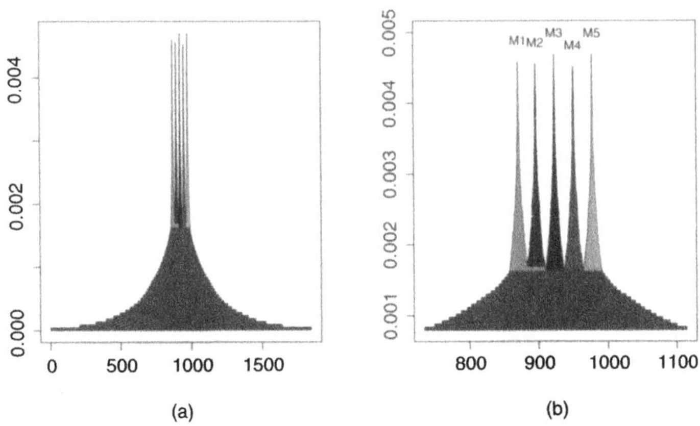

Figure 4.19 *(4D five modal example, volume function.)* Shown are a volume function and a zoomed volume function of a mixture of five 4D Gaussian densities. The density is defined in Section 2.1.3, page 63.

4.6 BIBLIOGRAPHIC NOTES

Volume functions and barycenter plots were introduced in Klemelä (2004). Related research has been done in topological data analysis, where Edelsbrunner, Letscher and Zomorodian (2000) introduce a persistence diagram or barcode to represent topological information about functions, and Singh, Memoli and Carlsson (2007) visualize data by simplicial complexes, where volume information is included as the largeness of the nodes of the complexes. Carlsson (2009) gives a review of concepts of topological data analysis. In cluster analysis Stuetzle (2003) defines a cluster tree, which is related to the concept of a level set tree. In the context of scientific visualization Weber, Bremer and Pascucci (2007) use volume information to make a 2D configuration of contour trees. We will give below a short review of Morse theory and Reeb graphs, since they are relevant to the concept of a level set tree.

4.6.1 Morse Theory

Morse theory finds information on the shape of a space by defining functions on the space. We follow Fomenko and Kunii (1997), which makes "a link between the theoretical aspects of modern geometry and topology, on the one hand, and experimental computer geometry, on the other" and Milnor (1963), Guillemin and Pollack (1974), and Matsumoto (2000), who give a more detailed mathematical discussion of differential topology. Appendix B.4 contains the definitions related to differential topology.

We start by defining a Morse function. A Morse function is sufficiently regular that we can construct a level set tree for it.

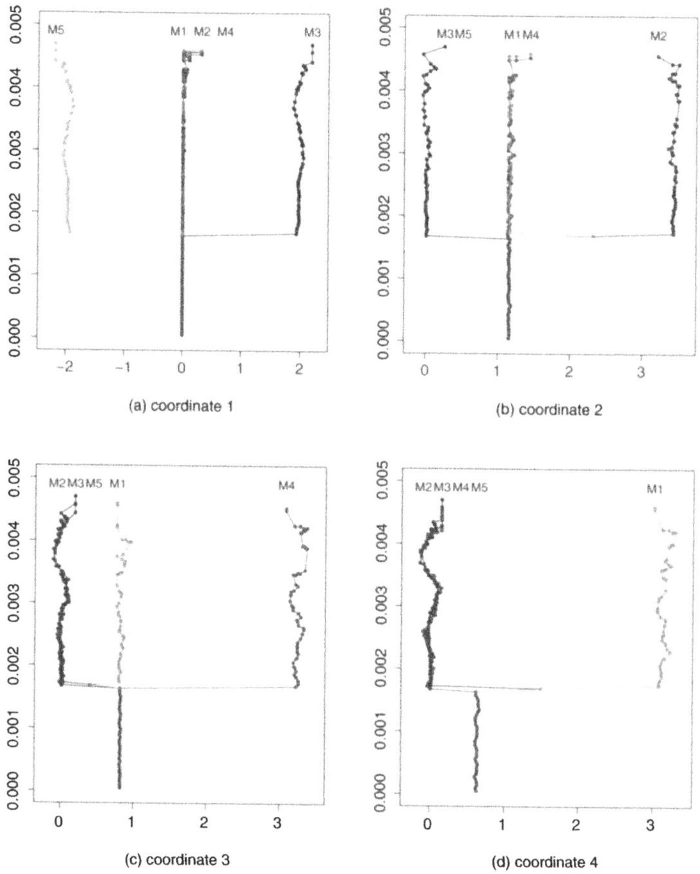

Figure 4.20 *(4D five modal example, barycenter plot.)* Shown is a barycenter plot of a mixture of five 4D Gaussian densities, corresponding to the volume function in Figure 4.19. The density is defined in Section 2.1.3, page 63.

Definition 4.12 (Morse function.) *Let* $f : \mathbf{M} \to \mathbf{R}$ *be a smooth function (infinitely times continuously differentiable).*

1. *Point* $x \in \mathbf{M}$ *is a* critical point *of* f *if all the partial derivatives of* f *vanish at* x*. If* x *is not a critical point, then it is said to be a* regular point.

2. *A critical point* x *is* nondegenerate *if* $\det(D^2 f(x)) \neq 0$*, where* $D^2 f(x)$ *is the Hessian matrix of the second partial derivatives of* f *at* x*.*

3. *Function* f *is a* Morse function *if all its critical points are nondegenerate.*

We say that λ*, with* $\min_{x \in \mathbf{M}} f(x) \leq \lambda \leq \max_{x \in \mathbf{M}} f(x)$*, is a* critical value *if* $f^{-1}(\lambda)$ *contains a critical point.*[1]

We make several remarks concerning Definition 4.12:

1. The preimage theorem states that if y is a regular value of a smooth map $f : \mathbf{M} \to \mathbf{R}$, then the preimage (isosurface) is a submanifold of $\mathbf{M}$, with $\dim f^{-1}(y) = \dim \mathbf{M} - 1$; see Guillemin and Pollack (1974, sec. 1.4). When y is a critical value, the isosurface can be rather complicated, but the condition of nondegeneracy brings regularity. Also, when y is a regular value, the preimage $\{x \in \mathbf{M} : f(x) \leq y\}$ is a smooth manifold with boundary.

2. Sard's theorem (Guillemin and Pollack 1974, sec. 7) states that if $f : \mathbf{M}_1 \to \mathbf{M}_2$ is a smooth map, then almost every point in $\mathbf{M}_2$ is a regular value (the set of critical values has measure zero). This implies that the regular values are dense in $\mathbf{M}_2$.

3. Morse functions on a closed manifold $\mathbf{M}$ are dense in the space of all smooth functions on $\mathbf{M}$; see Matsumoto (2000, th. 2.20) or Fomenko and Kunii (1997, th. 8.1.1).

4. Morse functions contain lots of information about the topology of their domain manifolds. Thus we can visualize a set by defining a Morse function on a set and visualizing this function. We develop this line of investigation in Chapter 5.

When the gradient vanishes at x, the Hessian can be used to test whether the point is a minimum, maximum, or a saddle point. Morse's lemma deals with this issue and describes the behavior of the function in a neighborhood of a nondegenerate critical point.

Lemma 4.2 (Morse's lemma.) *Let* $f : \mathbf{M} \to \mathbf{R}$*, where* $\mathbf{M}$ *is a d-dimensional manifold. Let* $x_0 \in \mathbf{M}$ *be a nondegenerate critical point of* f*. Then there exists an*

[1]More generally, for a smooth map of manifolds $f : \mathbf{M}_1 \to \mathbf{M}_2$, a point $y \in \mathbf{M}_2$ is called a regular value if $Df(x) : T_x(\mathbf{M}_1) \to T_y(\mathbf{M}_2)$ is surjective at every point x such that $f(x) = y$. The gradient (differential) $Df(x)$ of f at x is a linear map from the tangent space $T_x(\mathbf{M}_1)$ to the tangent space $T_y(\mathbf{M}_2)$. When $\mathbf{M}_2 = \mathbf{R}$ is one-dimensional, this is equivalent to saying that $Df(x)$ is the zero map.

open neighborhood U *of* x_0, *local coordinates* $y = (y_1, \ldots, y_d)$, $y : U \to \mathbf{R}^d$, *and index* $0 \le I \le d$ *such that*

$$f(x) = f(x_0) - \sum_{i=1}^{I} y_i^2(x) + \sum_{i=I+1}^{d} y_i^2(x), \qquad x \in U.$$

Index I is called the index of the critical point x_0.

A proof of Morse's lemma may be found in Milnor (1963, lmm. 2.2), or Matsumoto (2000, thm. 2.16). There are several points to regard concerning Lemma 4.2:

1. Morse's lemma says that the function can be expressed as a diagonalized quadratic function (in local coordinates). The expression is exact.
2. Let $x_0 \in \mathbf{M}$ be a nondegenerate critical point of f.
   (a) If the index of x_0 is 0, then x_0 is a minimum and
   $$f(x) = f(x_0) + \sum_{i=1}^{d} y_i^2(x), \qquad x \in U.$$
   If the index of x_0 is d, then x_0 is a maximum and
   $$f(x) = f(x_0) - \sum_{i=1}^{d} y_i^2(x), \qquad x \in U.$$
   (b) If $d = 2$, then we have only one further case: that index of x_0 is 1 and x_0 is a saddle point:
   $$f(x) = f(x_0) - y_1^2(x) + y_2^2(x), \qquad x \in U.$$
   (c) If $d = 3$, then x_0 may be a saddle point whose index is 1 or 2:
   $$f(x) = f(x_0) - y_1^2(x) + y_2^2(x) + y_3^2(x), \qquad x \in U$$
   or
   $$f(x) = f(x_0) - y_1^2(x) - y_2^2(x) + y_3^2(x), \qquad x \in U.$$
3. For a Morse function f on $\mathbf{M}$, the kth *Morse number* $\mu_k(f)$ of f is the number of critical points of f having index k. For example, $\mu_0(f)$ is the number of local minima of f and $\mu_d(f)$ is the number of local maxima of f. When $d = 2$, $\mu_1(f)$ is the number of saddle points of f.

Morse's lemma implies that a Morse function on a compact manifold has only finitely many critical points and these points are isolated (there cannot be a sequence of nondegenerate critical points converging to a nondegenerate critical point). Morse's lemma implies that for a smooth function on a compact smooth manifold M:

1. if λ is not a critical value, then there exists $\epsilon > 0$ such that $[\lambda - \epsilon, \lambda + \epsilon]$ does not contain a critical value, and

2. if λ is a critical value, then there exists $\epsilon > 0$ such that $[\lambda - \epsilon, \lambda + \epsilon]$ does not contain other critical values.

Now we can state the facts that imply the level set trees to be well defined for Morse functions. Fomenko and Kunii (1997, sec. 6.2) suggest the following facts for the two- and three-dimensional cases:

1. Let $f : \mathbf{M} \to \mathbf{R}$ be a Morse function. Assume that λ is not a critical value. Let $\epsilon > 0$ be such that $[\lambda - \epsilon, \lambda + \epsilon]$ does not contain a critical value. Then $f^{-1}(\lambda - \epsilon)$ and $f^{-1}(\lambda + \epsilon)$ are diffeomorphic and

$$\{x \in \mathbf{M} : f(x) \geq \lambda - \epsilon\}, \qquad \{x \in \mathbf{M} : f(x) \geq \lambda + \epsilon\},$$

   are diffeomorphic. In particular, $f^{-1}(\lambda - \epsilon)$ and $f^{-1}(\lambda + \epsilon)$ consists of the same number of connected components which are diffeomorphic to a ball.

2. Let $f : \mathbf{M} \to \mathbf{R}$ be a Morse function. Assume that λ is a critical value of index $1 \leq I < d$. Let $\epsilon > 0$ be such that $[\lambda - \epsilon, \lambda + \epsilon]$ does not contain other critical values. Then $\{x \in \mathbf{M} : f(x) \geq \lambda - \epsilon\}$ contains one connected component less than $\{x \in \mathbf{M} : f(x) \geq \lambda + \epsilon\}$.

   Fomenko and Kunii (1997, thm. 6.3.1) give a more precise statement: $\{x \in M : f(x) \geq \lambda - \epsilon\}$ is obtained from $\{x \in M : f(x) \geq \lambda + \epsilon\}$ by attaching an d-dimensional handle H_I of index I, defined as $H_I = D^I \times D^{d-I}$, where $D^k = \{x \in \mathbf{R}^k : \|x\| \leq 1\}$ is the closed k-disk. We define $H_0 = H_d = D^d$.

4.6.2 Reeb Graphs

We can combine information in a level set tree and in a lower level set tree and make a level curve tree. A level curve tree is a tree whose nodes correspond to connected components of level curves of the function. A level curve of $f : \mathbf{M} \to \mathbf{R}$ is $\Gamma(f, \lambda) = \{x \in \mathbf{M} : f(x) = \lambda\}$. Figure 4.21(b) illustrates a level curve tree. We draw a level curve tree of the function shown in panel (a). Note that a level set tree and a lower level set tree of the function are shown in Figure 4.5.

Fomenko and Kunii (1997, def. 8.1.2) define a *Reeb graph*. A Reeb graph of a function $f : \mathbf{M} \to \mathbf{R}$, where $\mathbf{M}$ is a compact manifold, is the space of connected components of the isosurfaces of the function: a Reeb graph is defined as a quotient space of $\mathbf{M}$ defined by the equivalence relation $x_1 \sim x_2$, where $x_1 \sim x_2$ means that (1) x_1 and x_2 belong to the same isosurface ($f(x_1) = f(x_2)$), and (2) x_1, x_2 are in the same connected component of the isosurface ($x_1, x_2 \in f^{-1}(f(x_1))$). Fomenko and Kunii (1997) refer to Reeb (1946) as the origin of the concept, and they refer to René Thom for inventing the term Reeb graph. Applications of Reeb graphs in computer geometry were described in Kunii and Shinagawa (1992).

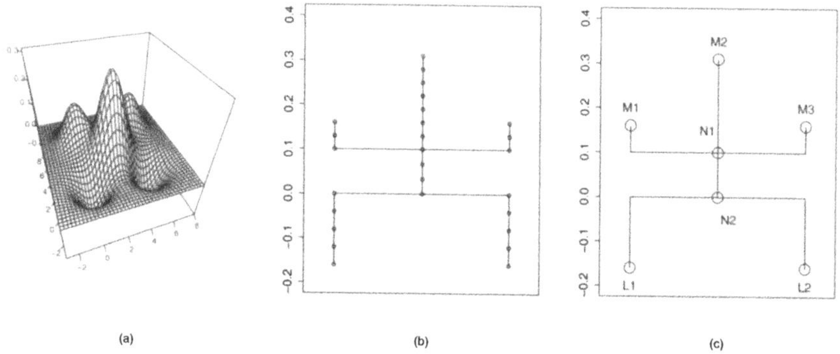

Figure 4.21 *(A level curve tree and the Reeb graph of a function.)* (a) A perspective plot of a function, (b) a level curve tree of the function, and (c) the Reeb graph of the function.

A Reeb graph is a graph whose nodes correspond to contours but only at critical levels, when contours are defined as connected components of level curves (isosurfaces). The nodes of a contour tree are only the contours that correspond to a change in the topology of the isosurface; these contours correspond to a minimum, a maximum, or a saddle point. The nodes or vertices of a Reeb graph correspond to the critical points of the function, and the nodes are connected with arcs or edges, which correspond to the connected components of the level curves that are situated strictly between two critical levels. Critical points and levels are defined in Definition 4.12. From a Reeb graph a directed graph can be constructed by ordering the nodes with respect to the values of the function.

Figure 4.21(c) illustrates the Reeb graph of a function shown in panel (a). The function has three local maxima and the nodes M1, M2, and M3 of the Reeb graph correspond to the local maxima. The function has two local minima and the nodes L1 and L2 of the contour tree correspond to the local minima. Nodes N1 and N2 of the contour tree correspond to the changes in the topology of the level sets.

Those Reeb graphs that are trees are called contour trees in computer graphics. Term "merge tree" or "join tree" has been used to denote a tree built from upper level sets, and the term "split tree" has been used to denote a tree built from lower level sets. Stuetzle (2003) uses the term "cluster tree" to denote the tree where only the nodes corresponding to the change of the topology of upper level sets are included. Carr, Snoeyink and Axen (2003) use the term "augmented contour tree" for a tree where the nodes corresponding to regular levels are also included.

EXERCISES

4.1 Prove the following facts:

(a) Level sets satisfy monotonicity: if $\lambda_1 \geq \lambda_0$, then $\Lambda(f, \lambda_0) \subset \Lambda(f, \lambda_1)$.

(b) For a function $f : \mathbf{R}^d \to \mathbf{R}$ we have

$$f(x) = \sup\{\lambda : x \in \Lambda(f, \lambda)\}.$$

(c) Level sets satisfy continuity: $\Lambda(f, \lambda) = \cap_{\lambda_0 < \lambda} \Lambda(f, \lambda_0)$.

(d) Let $u : \mathbf{R} \to \mathbf{R}$ be strictly increasing. Then $\Lambda(f, \lambda) = \Lambda(u \circ f, u(\lambda))$.

4.2 Show that Definition 4.13 below is equivalent to the Definition 4.2. We define a level set tree in Definition 4.13 as a tree whose nodes are the pairs where the first element is a level in the grid $\mathcal{L} = \{\lambda_1, \ldots, \lambda_L\}$ and the second element is a connected component of some level set whose level is in set $\mathcal{L}$. A parent–child relation may be defined by set inclusion. Let us denote with

$$\mathcal{C}_f^{(s)}(\lambda) = \mathcal{C}(\Lambda(f, \lambda))$$

the maximal decomposition of level set $\Lambda(f, \lambda)$, as defined in Definition 4.4. The set of nodes of a level set tree is the set of pairs

$$\mathcal{N}_f^{(s)}(\mathcal{L}) = \left\{(\lambda, S) : \lambda \in \mathcal{L},\ S \in \mathcal{C}_f^{(s)}(\lambda)\right\}.$$

Definition 4.13 (Level set tree.) A level set tree *of function* $f : \mathbf{M} \to \mathbf{R}$, *associated with set of levels* $\mathcal{L} = \{\lambda_1 < \cdots < \lambda_L\}$, *where* $\lambda_L \le \sup_{x \in \mathbf{M}} f(x)$, *is a tree whose nodes are the elements of the set* $\mathcal{N}_f^{(s)}(\mathcal{L})$.

1. *The root nodes of the tree are* $\{(\lambda, S) \in \mathcal{N}_f^{(s)}(\mathcal{L}) : \lambda = \lambda_1\}$.

2. *We say that node* $(\lambda, S) \in \mathcal{N}_f^{(s)}(\mathcal{L})$ *is the parent of node* $(\lambda', S') \in \mathcal{N}_f^{(s)}(\mathcal{L})$, *when*

   (a) $\lambda = \lambda_i$, $\lambda' = \lambda_{i+1}$, *for some* $i = 1, \ldots, N-1$, *and*

   (b) $S' \subset S$.

4.3 We define in Definition 4.14 below a reduced level set tree for piecewise constant functions. Piecewise constant function $f : \mathbf{M} \to \mathbf{R}$ is defined in Definition 4.5: $\{f(x) : x \in \mathbf{M}\} = \{\lambda_1, \ldots, \lambda_L\}$, where $\lambda_1 < \cdots < \lambda_L$.

(a) Give an example of a piecewise constant function $f : \mathbf{R} \to \mathbf{R}$ for which the level set tree (Definition 4.2) associated with levels $\lambda_1, \ldots \lambda_L$ and reduced level set tree (Definition 4.14 below) are the same.

(b) Give an example of a function $f : \mathbf{R} \to \mathbf{R}$ for which these trees are different.

Definition 4.14 (Reduced level set tree of a piecewise constant function.) A reduced level set tree *of piecewise constant function* $f : \mathbf{M} \to \mathbf{R}$ *is a tree whose nodes are associated with subsets of* $\mathbf{M}$ *and levels* $\lambda \in \mathbf{R}$ *in the following way:*

1. *Write*

$$\Lambda(f, \lambda_1) = A_1 \cup \cdots \cup A_K$$

where sets A_i are pairwise separated and each is connected The level set tree has K root nodes which are associated with sets A_i, $i = 1, \ldots, K$, and levels $\lambda_i = \min\{f(x) : x \in A_i\}$, $i = 1, \ldots, K$.

2. *Let node m be associated with set $B \subset \mathbf{M}$ and level $\lambda \in \mathbf{R}$.*

    (a) *If $\{x \in A : f(x) > \lambda\} = \emptyset$, then this node is a leaf node.*

    (b) *Otherwise, if $\{x \in A : f(x) > \lambda\} \neq \emptyset$, then write*

    $$\{x \in A : f(x) > \lambda\} = C_1 \cup \cdots \cup C_M$$

    where sets C_i are pairwise separated and each is connected. Then node m has M children, which are associated with sets C_i, $i = 1, \ldots, M$, and levels $c_i = \min\{f(x) : x \in C_i\}$, $i = 1, \ldots, M$.

CHAPTER 5

SHAPE TREES

Shape trees are an analogous tool for visualizing sets as level set trees are for visualizing functions. Shape trees can be defined as level set trees of distance functions of connected sets. A distance function gives the distance of each point of a set from a center point of the set.

With the help of level set trees one may visualize the mode structure of a function. Chapter 4 introduces a volume function for the visualization of the relative largeness of the modes and a barycenter plot for the visualization of the barycenters of the level sets, in particular, the locations of the modes. However, beyond the mode structure, these plots give only a limited insight into the shape of the function. We need additional visualization tools, since unimodal functions also have a large variety of shapes, and shape trees provide those additional tools.

Analogously to a volume function of a level set tree we can define a radius function of a shape tree. In fact, we can define a family of shape functions, where either the Lebesgue measure or the underlying probability measure is used to define the volumes. For example, a radius function reveals basically the following features of the set:

1. *Qualitative information.* The modes of a radius function correspond to the "extensions" or "tails" of the set, where by extensions of a set we mean those

parts of the set that do not fit inside a given ball. We call a connected set "multimodal" when a radius function is multimodal.

2. *Quantitative information.*

   (a) The lengths of the level sets of a radius function are equal to the volumes of the corresponding regions of the original set.

   (b) The levels of the level sets of a radius function are equal to the distances of the corresponding regions of the original set from the reference point.

Analogously to a barycenter plot of a level set tree we can define a location plot of a shape tree. A location plot visualizes the locations of the extensions of the set. In the location plot we plot the barycenters of the sets associated with the nodes of the shape tree.

Shape functions, like volume and radius functions, are 1D functions that visualize the shape of a multivariate object. It is natural to search for 2D functions that can visualize the shape of a multivariate object. We may visualize a multivariate function by constructing 1D shape functions from a series of its level sets and cluing these shape functions together to make a 2D function. These methods are able to visualize the anisotropic spread of a distribution: visualization tools show how the spread varies in different directions. Each mode of the 2D functions corresponds to a specific tail of the distribution.

Shape trees were introduced in Klemelä (2006).

5.1 FUNCTIONS AND SETS

We can analyze and visualize shapes of sets by first defining functions on the sets, and second by using the level set tree based methods of Chapter 4 for analyzing and visualizing these functions. This is similar to the fundamental idea behind the Morse theory, where one studies spaces by defining functions on the spaces. The two basic ways to define functions on a set use either a *distance function* or a *height function*.

Definition 5.1 (Distance function, Height function.) *Let* $A \subset \mathbf{R}^d$.

1. *A* distance function $f_A : A \to \mathbf{R}$ *with reference point* $\mu \in \mathbf{R}^d$ *is defined by* $f_A(x) = \|\mu - x\| I_A(x)$.

2. *A* height function $f_A : A \to \mathbf{R}$ *is the orthogonal projection of* A *onto a fixed line on* $\mathbf{R}^d$, *when the line is identified with* $\mathbf{R}$.

A third kind of function that can be used to visualize a set is the *boundary function* of a *star-shaped set*. A distance function and a height function are defined on any set, but a boundary function will be defined only for star-shaped sets.

Definition 5.2 (Star-shaped set, Boundary function.) *We define star-shaped sets first for* $A \subset \mathbf{R}^d$ *and then for* $A \subset \mathbf{S}_{d-1}$, *where* $\mathbf{S}_{d-1}$ *is the unit sphere* $\mathbf{S}_{d-1} = \{x \in \mathbf{R}^d : \|x\| = 1\}$.

1. *We say that set* $A \subset \mathbf{R}^d$ *is* star-shaped *if there is a reference point* $\mu \in A$ *and a* boundary function $f_A : \mathbf{S}_{d-1} \to [0, \infty)$ *so that we can write the set as*

$$A = \{\mu + r\xi : \xi \in \mathbf{S}_{d-1}, \ \ 0 \le r \le f_A(\xi)\}.$$

2. *We say that set* $A \subset \mathbf{S}_{d-1}$, $d \ge 2$, *is* star-shaped *if there is a reference point* $\mu \in A$ *and a* boundary function $f_A : T_\eta \to [0, \pi]$, $T_\eta = \{\xi \in \mathbf{S}_{d-1} : \xi \perp \eta\}$, *so that we can write the set as*

$$A = \{\mu \cdot \cos\theta + \xi \cdot \sin\theta : \xi \in T_\eta, \ \ 0 \le \theta \le f_A(\xi)\}.$$

Note that set T_η *is isomorphic with* S_{d-2}.

Remarks We make remarks concerning Definition 5.1 and Definition 5.2

1. A distance function is constructed by foliating the Euclidean space into a union of concentric spheres. A height function is constructed by foliating the Euclidean space into a union of parallel planes.

2. A distance function and a boundary function are related to each other: both can be used to describe how the set is extending in various directions around a center point. A natural choice for the center point is the barycenter, as defined in (4.7). The barycenter is the center of the mass of the set.

3. A difference between a distance function and a boundary function is that a distance function is just a function defined on a set, but a boundary function completely defines the set. Note also, that a boundary function is defined on the unit sphere, which is a lower dimensional manifold than than the Euclidean space where the set is defined. Thus we achieve a dimension reduction when we describe a star-shaped set with its boundary function.

4. A height function is based on a one dimensional projection and thus one would need several height functions to visualize all tails of a star-shaped set.

Illustration Definition 5.1 is illustrated in Figure 5.1, which shows a set in $\mathbf{R}^2$, its distance function, and its height function. The center point of the distance function is the origin. The height function is drawn by the orthogonal projection to the y-axis. Note that a distance function and a height function are defined only on set A, but the perspective plots show also the area outside set A.

5.2 DEFINITION OF A SHAPE TREE

We have two closely related approaches to define a shape tree of a set.

1. We could define a shape tree as a level set tree of a distance function of the set.

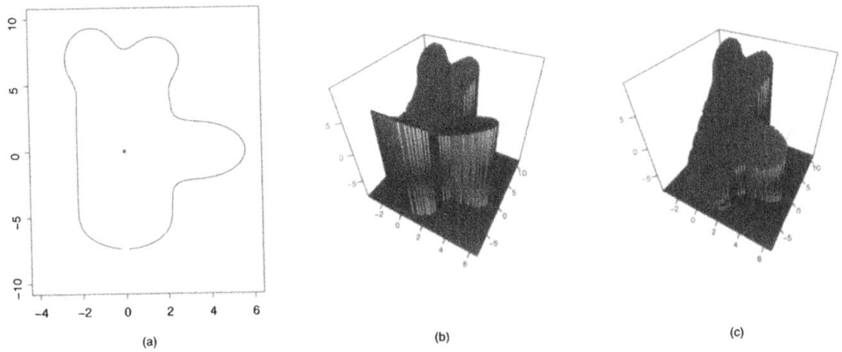

Figure 5.1 *(Distance function, Height function.)* Shown is (a) a set in $\mathbf{R}^2$, (b) its distance function, and (c) its height function.

2. We could define a shape tree as a level set tree of a boundary function of a star-shaped set.

We will follow the first approach and define a shape tree as a level set tree of a distance function. There are two reasons for this choice. First, boundary functions are defined only for star-shaped sets, and thus we gain some generality when we choose distance functions. Second, boundary functions are defined on the unit sphere, but it seems more convenient to work with functions defined on $\mathbf{R}^d$.

The following concepts are needed in the definition of a shape tree: a connected set is defined in Definition 4.1, a level set tree is defined in Definition 4.2, and a distance function is defined in Definition 5.1.

Definition 5.3 (Shape tree.) A shape tree *of a set $A \subset \mathbf{R}^d$, associated with reference point $\mu \in A$, and set of radii $\mathcal{R} = \{0 = r_0 < r_1 < \cdots < r_L\}$ is the level set tree of distance function f_A of A with the reference point μ. The grid of levels of the level set tree is $\mathcal{R}$.*

For transparency we will define shape trees also directly, without using the concept of a level set tree.

Definition 5.4 (Shape tree, direct definition.) A shape tree *of a set $A \subset \mathbf{R}^d$, associated with reference point $\mu \in A$, and set of radii $\mathcal{R} = \{0 = r_0 < r_1 < \cdots < r_L\}$, is a tree whose nodes are associated with subsets of A and radii in $\mathcal{R}$ in the following way.*

1. *Write*

$$A = A_1 \cup \cdots \cup A_K,$$

where sets A_i are pairwise separated, and each is connected. The shape tree has K root nodes that are associated with sets A_i, $i = 1, \ldots, K$, and each root node is associated with the same radius $r_0 = 0$.

2. *Let node m be associated with set $B \subset A$ and radius $r_l \in \mathcal{R}$, $0 \le l < L$.*

   (a) *If $B \setminus B_{r_{l+1}}(\mu) = \emptyset$, then node m is a leaf node.*

   (b) *Otherwise, write*

$$B \setminus B_{r_{l+1}}(\mu) = C_1 \cup \cdots \cup C_M \tag{5.1}$$

where sets C_i are pairwise separated, and each is connected. Then node m has M children, which are associated with sets C_i, $i = 1, \ldots, M$, respectively, and each child is associated with the same radius r_{l+1}.

A shape tree is a tree structure associated with a set, with a reference point inside this set, and with a grid of radii. The nodes of the tree are associated with subsets of the set and with the radii in the grid. The tree has a single root node and we associate the root with the set itself. To find the children of a given node, we increase the radius by one step, make the intersection of the set associated with the node with the complement of the ball with this new radius, and associate the children with the separated components of this intersection.

Illustrations We illustrate Definition 5.3.

1. Figure 5.2 shows the growing of a shape tree. The reference point of the shape tree is the origin. The root node (labeled with A) is associated with the the whole set. The child of the root node (labeled with B) is associated with a smaller set (a disk is removed from the original set). Node B has three children (C, D, and E). The node E has two children (F and G).

2. Figure 5.3 shows two shape trees with different grids of radii. Panel (a) shows the set. Panel (b) shows a shape tree with a sparse grid of radii. The grid is the same as that in Figure 5.2. Panel (c) shows a shape tree with a dense grid of radii. The reference point is the origin. The height of the nodes is equal to the radius associated with the nodes.

Remarks We make comments concerning the definition of a shape tree.

1. *(Modes of a Connected Set.)* The basic qualitative features of a set that are represented with a shape tree are the *modes of the set*.

   Definition 5.5 (Modes of a set.) *The* modes of a connected set, *associated with a reference point and a grid of radii, are those sets that are associated with the leafs of the corresponding shape tree of the set.*

   Figure 5.3 illustrates the definition of the modes of a set. As the grid of radii becomes finer, the modes of a connected set typically shrink to a single point of the Euclidean space. Using the term "modes of a set" is motivated by the fact

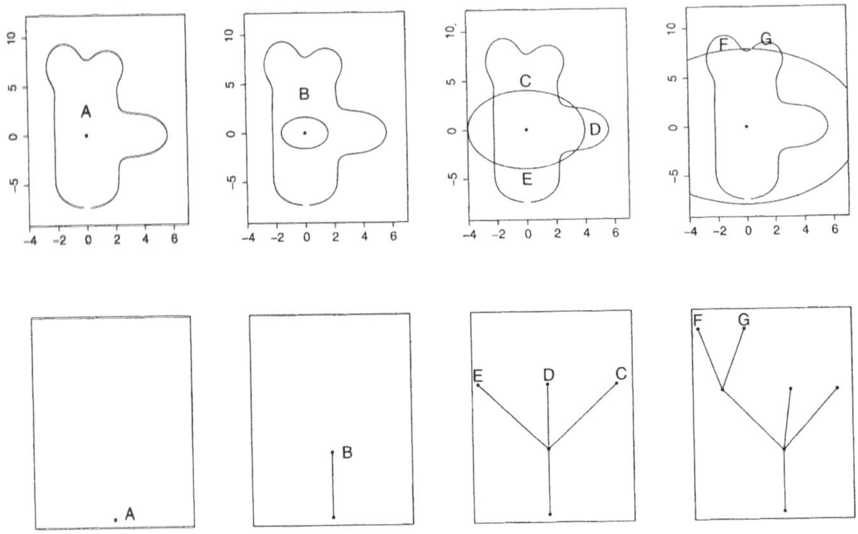

Figure 5.2 *(Growing of a shape tree.)* The growing of a shape tree is illustrated.

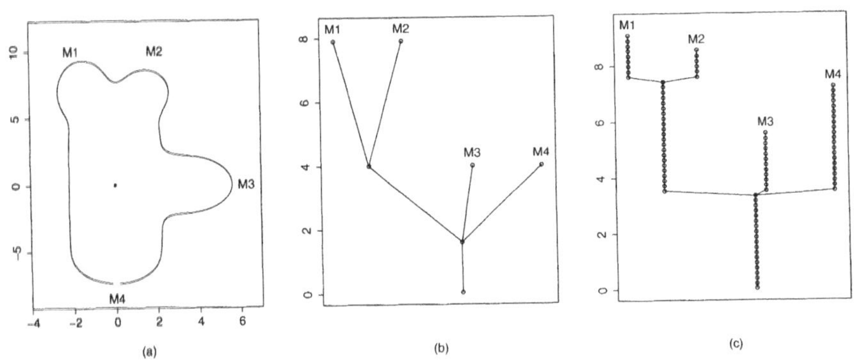

Figure 5.3 *(Shape trees with different grids of radii.)* (a) The set, (b) a shape tree with a sparse grid of radii, and (c) a shape tree with a dense grid of radii. The modes of the set are labeled M1 through M4.

that there is a unique correspondence between the modes of a shape function and with the modes of the original set. Shape functions and shape transforms are defined in Section 5.3.

2. *(Boundary Function.)* We will return to the idea of using a boundary function to define a shape tree. A boundary function of set A is a function $f_A : \mathbf{S}_{d-1} \to \mathbf{R}$. We state a lemma that says that for star-shaped sets this will lead essentially to the same concept of a shape tree.

Lemma 5.1 *Let $A \subset \mathbf{R}^d$ be a star-shaped set. The level set trees of a distance function of A and the level set trees of a boundary function of A are isomorphic as tree structures when the reference point of the distance function and the boundary function are the same.*

Proof. For $\lambda \geq 0$ the sets

$$\Lambda_\lambda^{(1)} = \{x \in A : \|x - \mu\| \geq \lambda\}, \qquad \Lambda_\lambda^{(2)} = \{\xi \in \mathbf{S}_{d-1} : f_A(\xi) \geq \lambda\}$$

have the same number of separated components, where f_A is the boundary function of A for the reference point $\mu \in A$. Let us make the induction hypothesis that the level set trees are isomorphic up to node m. Let the connected sets $B_1 \subset \Lambda_\lambda^{(1)}$ and $B_2 \subset \Lambda_\lambda^{(2)}$ correspond to node m. Then the sets

$$\{x \in B_1 : \|x - \mu\| \geq \lambda'\}, \qquad \{\xi \in B_2 : f_A(\xi) \geq \lambda'\},$$

for $\lambda' \geq \lambda$, have the same number of separated components. □

A parametrization of a sphere is defined in (B.8) as a function $\Psi : \Theta \to \mathbf{S}_{d-1}$, $\Theta = [0, 2\pi] \times [0, \pi]^{d-2}$. Note that function $\theta \mapsto f_A(\Psi(\theta))$ could have a level set tree that is not isomorphic to the level set tree of f_A.

5.3 SHAPE TRANSFORMS

A *shape transform* is a transform of a multidimensional set to a univariate function. Shape transforms are defined similarly as a volume transform of a multivariate function in Section 4.2. We can in fact define a shape transform as a volume transform of a distance function of a set. This particular shape transform will be called a *radius transform*. However, we can define several different kinds of shape transforms. Other shape transforms include the *tail probability transform* and the *probability content transform*.

All shape transforms are based on a construction where a shape tree is augmented to a function tree in the sense of Definition 3.3. A function tree defines in turn a function, according to Definition 3.5. A shape tree is augmented to a function tree by assigning a length and a height to the nodes. To define the length of a node, one finds the set associated with the node, and then one may use the volume of the set or the probability content of the set to define the length. To define the height of a node we use the radius of the node, but we can use also the probability content to define the radius.

Examples of shape transforms are given also in Figures 10.6, 10.11, 10.16, 10.21, and 10.26.

5.3.1 Radius Transform

Definition of Radius Transform A radius transform is defined for connected sets. To put it briefly, a radius transform is a volume transform of a distance function. A direct definition of a radius transform is made by describing how a shape tree is transformed to a function tree. The height of a node of a shape tree is defined to be equal to the radius associated with the node, and the length of a node is defined to be equal to the volume of the set associated with the node.

Definition 5.6 (Radius function, Radius transform.) *A radius function $r(A;T)$ of a set $A \subset \mathbf{R}^d$, associated with a shape tree T of A, is a 1D function defined through a function tree as in Definition 3.5. We get the function tree, in the sense of Definition 3.3, from the shape tree T by defining the height and the length of the nodes in the following way:*

1. *The height associated to a node of the shape tree is equal to the radius associated with the node.*
2. *The length associated to node m of the shape tree is equal to* volume(A_m), *where A_m is the set associated to node m.*

A radius transform *is a transformation of set A to a radius function.*

Illustrations We illustrate Definition 5.6.

1. Figure 5.4(a) shows a 2D set, and panels (b) and (c) show radius functions of this set. In the radius function of panel (b) the grid of radii is the same as in Figure 5.3(b). In the radius function of panel (c) the grid of radii is the same as in Figure 5.3(c).

   We have labeled the modes of the set M1 to M4. A mode of a set was defined in Definition 5.5. There is a unique correspondence between the modes of the set and the modes of the radius functions.

2. Section 10.1.4 contains more examples of radius functions.

Example Let $A = \{x \in \mathbf{R}^d : \|x\| \leq R\}$ be the ball of radius $R > 0$. We will find a limiting radius transform of A. A limiting volume transform of a function was defined in Definition 4.6. This definition can be modified to define a limiting radius transform. The volume of a ball with radius $r > 0$ is given in (B.9) as $C_d r^d$, where

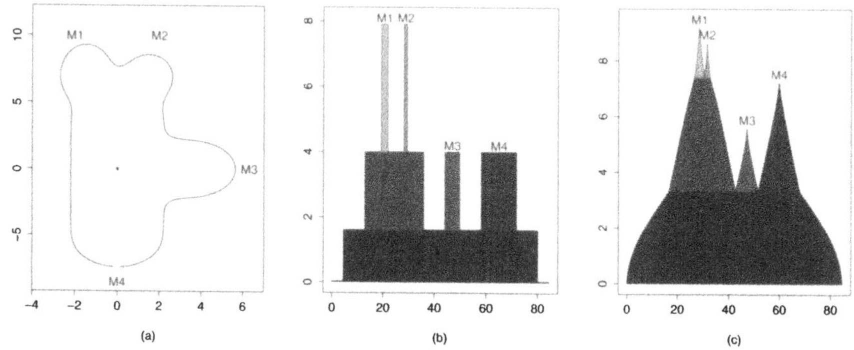

Figure 5.4 *(Radius functions with different grids of radii.)* (a) The set, (b) a radius function with a sparse grid of radii, and (c) a radius function with a dense grid of radii. The modes of the set are labeled with M1 througho M4.

$C_d = \pi^{d/2}/\Gamma(d/2+1)$. Thus the level set of a radius function of A with level r has length $C_d(R^d - r^d)$. We will invert the function $r \mapsto C_d(R^d - r^d)/2$ to get a limiting radius function g as

$$g(t) = \begin{cases} \left(R^d - 2t/C_d\right)^{1/d}, & 0 \le t \le R^d C_d/2 \\ \left(R^d + 2t/C_d\right)^{1/d}, & -R^d C_d/2 \le t \le 0. \end{cases} \qquad (5.2)$$

Figure 5.14(b) shows a radius transform of a ball for a finite grid of radii. Note that we usually apply the convention where the support of a radius function lies on the positive real axis, unlike in (5.2) where the support is centered at the origin.

Radius Transform with Boundary Functions The boundary function of a 2D ball $A = \{x \in \mathbf{R}^2 : \|x\| \le R\}$ is $f_A(\theta) = RI_{[0,2\pi]}(\theta)$. It would be of interest to define a radius transform in such a way that in the 2D case a limiting radius function of a ball would equal to its boundary function. This could be done by defining a shape tree as a level set tree of a boundary function, and not as a level set tree of a distance function as we have done. However, Lemma 5.1 states that this approach will lead to an essentially equivalent way of defining a shape tree. Thus we can leave the definition of a shape tree unchanged and instead modify the definition of a radius transform. In a radius transform we have defined the length associated with a node to be equal to volume(A_m), where A_m is the set associated with the node. We could also associate the node with length

$$\text{volume}\left(\{\xi \in \mathbf{S}_1 : \mu + r f_A(\xi) \in A_m \text{ for some } r \in [0, f_A(\xi)]\}\right),$$

where μ is the reference point of the boundary function f_A. Then the radius transform of a ball would be $RI_{[0,2\pi R]}(t)$, since the length of the perimeter of the ball of radius R is equal to $2\pi R$.

5.3.2 Tail Probability Transform

A tail probability transform is defined for a connected set when there is a probability measure defined on the set. For example, when the set is a level set of a density function, the density function defines a probability measure on the set. A tail probability transform is otherwise similar to a radius transform, but now the length of the nodes is equal to the probability content of the sets and not to the Lebesgue measure of the sets as in the case of a radius transform.

Definition 5.7 (Tail probability function, Tail probability transform.) A tail probability function *of a set* $A \subset \mathbf{R}^d$*, associated with a shape tree* T *of* A *and with a probability measure* P*, is a 1D function defined through a function tree as in Definition 3.5. We get the function tree, in the sense of Definition 3.3, from the shape tree* T *by defining the height and the length of the nodes in the following way:*

1. *The height associated to a node of the shape tree is equal to the radius associated with the node.*

2. *The length associated to node* m *of the shape tree is equal to the probability of the set associated with the node:* $P(A_m)$*, where* A_m *is the set associated to node* m*.*

A tail probability transform *is the transformation of set* A *to a tail probability function.*

Illustrations We illustrate Definition 5.7.

1. Figure 5.5 shows tail probability functions of the set shown in Figure 5.3(a). The set is the lowest level set in the contour plot shown in Figure 5.5(a), and the level of this level set is 0.002. The contour plot shows the density associated to the tail probability functions. Panel (b) shows the tail probability function where the grid of radii is the same as that in Figure 5.3(b). Panel (c) shows the tail probability function where the grid of radii is the same as that in Figure 5.3(c). The modes of the tail probability functions are thinner than the modes of the radius functions in Figure 5.4.

2. Section 10.1.4 contains more examples of the differences between a radius function and a tail probability function.

5.3.3 Probability Content Transform

Definition of Probability Content Transform A probability content transform is defined for a connected set when there is a probability measure defined on the set. A probability content transform is otherwise similar to a radius transform, but now the heights of the nodes are determined so that a probability content function visualizes the probability content inside the set. The length of a node is taken to be the volume of the associated set, like in a radius function.

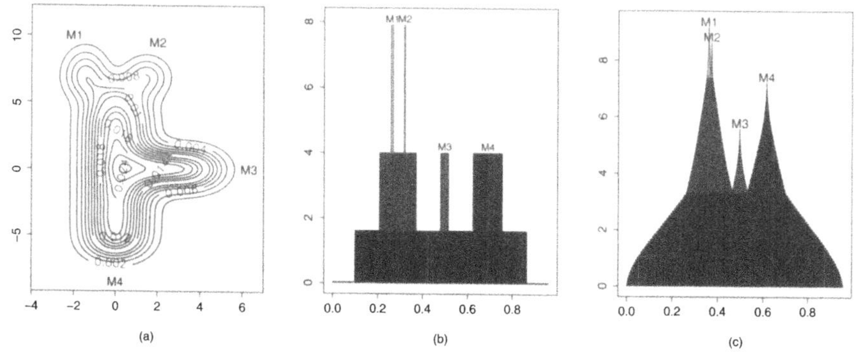

Figure 5.5 *(Tail probability functions.)* (a) A contour plot of a density. (b–c) Tail probability functions of the set shown in (a) as the lowest level set in the contour plot of the density. The same set is also shown in Figure 5.3(a). (b) A radius function with a sparse grid of radii, and (c) a radius function with a dense grid of radii.

Definition 5.8 (Probability content function, Probability content transform.) *A* probability content function *of set $A \subset \mathbf{R}^d$, associated with a shape tree T of A, and with a probability measure P, is a 1D function defined through a function tree as in Definition 3.5. We get the function tree, in the sense of Definition 3.3, from the shape tree T by defining the height and the length of the nodes in the following way.*

1. *(a) Let the height of the root node be* 0.

   (b) Let node m of the shape tree be a non-root node. Let the parent of node m be associated with height H, let node m be associated with set A, and let the children of node m be associated with sets $B_1, \ldots, B_M$. The height of node m is

$$H + \frac{P(A) - \sum_{i=1}^{M} P(B_i)}{\text{volume}(A)}. \tag{5.3}$$

2. *The length associated to node m of the shape tree is equal to* volume(A), *where A is the set associated to node m.*

A probability content transform *is the transformation of set A to a probability content function.*

Figure 5.6 shows probability content functions of the set shown in Figure 5.3(a). The set is the lowest level set in the contour plot shown in Panel (a), and the level of this level set is 0.002. The contour plot shows the density which is associated to the tail probability functions. Panel (b) shows the probability content function where the grid of radii is the same as in Figure 5.3(b). Panel (c) shows the probability content function where the grid of radii is the same as in Figure 5.3(c).

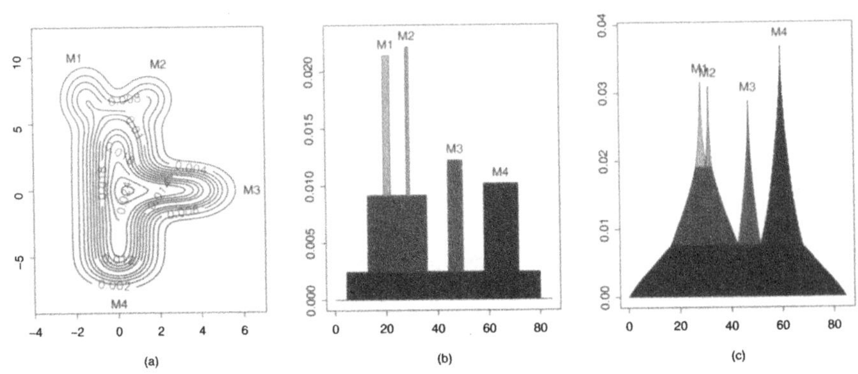

Figure 5.6 *(Probability content functions.)* (a) A contour plot of a density. (b–c) Probability content functions of the set shown in (a) as the lowest level set in the contour plot of the density. The set is also shown in Figure 5.3(a). (b) A probability content function with a sparse grid of radii, and (c) a probability content function with a dense grid of radii.

The probability content functions separate modes more clearly than the radius functions and the tail probability functions, in the sense that the valleys separating the modes are deeper. Differences between a radius function and a probability content function are further discussed in Section 5.5.2.

Probability Content In Lemma 5.2 we state that the tail probabilities on set A are equal to the excess masses of the probability content function of A. To formulate the lemma we use the following notations: When m is a node of a shape tree, we denote with A_m the set associated with node m. The shape tree is augmented to a function generating tree, and we denote with H_m the height and with $\mathcal{I}_m$ the interval (as defined in Definition 3.4) associated with node m. We denote by parent(m) the parent of node m.

Lemma 5.2 *Let m_0 be a node of a shape tree of set $A \subset \mathbf{R}^d$, associated to probability measure P. Let $g : \mathbf{R} \to \mathbf{R}$ be a probability content function of A, as defined in Definition 5.8. Then the probability $P(A_{m_0})$ of a tail region $A_{m_0} \subset A$ is equal to the excess mass of g, at interval $\mathcal{I}_{m_0}$ and at level $H_{\text{parent}(m_0)}$:*

$$P(A_{m_0}) = \int_{\mathcal{I}_{m_0}} \left(g - H_{\text{parent}(m_0)} \right),$$

where in the case that m_0 is the root node we denote $H_{\text{parent}(m_0)} = 0$.

Proof. We say that a node is a descendant of a node m if it is a child of m or a child of other descendant of m. We have that

$$P(A_{m_0}) = \sum \{ P(A_m) - \sum \{ P(A_{m'}) : m' \text{ is a child of } m \} : m \text{ is a descendant of } m_0 \text{ or } m = m_0 \} \qquad (5.4)$$

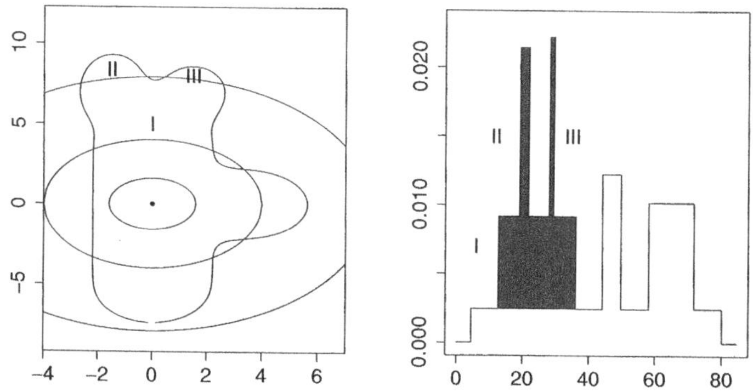

(a) (b)

Figure 5.7 *(Steps of the proof of Lemma 5.2.)* (a) Regions I, II, and III of the set. (b) The corresponding regions I, II, and III in the probability content function.

$$
\begin{aligned}
&= \sum \{\text{volume}(A_m) \left(H_m - H_{\text{parent}(m)} \right) : \\
&\qquad\qquad m \text{ is a descendant of } m_0 \text{ or } m = m_0\} && (5.5)\\
&= \int_{\mathcal{I}_{m_0}} \left(g - H_{\text{parent}(m_0)} \right). && (5.6)
\end{aligned}
$$

We illustrate the steps of the proof with Figure 5.7. Now $A_{m_0} = C$, where C is the set shown in Figure 5.2(c).

1. Step (5.4) in the proof of Lemma 5.2 follows from Figure 5.7(a): $P(A_{m_0})$ is equal to the sum of the probabilities of regions I, II, and III.

2. Step (5.5) follows from the definition of the height of the nodes given in (5.3).

3. Step (5.6) follows from Figure 5.7(b): the excess mass of function g in interval $\mathcal{I}_{m_0}$ is equal to the the volume of the colored region, and this volume is equal to the sum of the volumes of the three rectangles whose union is forming the colored region.

We have proved the lemma. □

As a corollary of Lemma 5.2 we have that the probability of set A is equal to the integral of the probability content transform: when g is the probability content transform of A, we have

$$
P(A) = \int_{-\infty}^{\infty} g.
$$

5.4 LOCATION PLOT

A location plot visualizes the locations of the sets associated with the nodes of a shape tree. A location plot is analogous to a barycenter plot of a level set tree, as defined in Definition 4.7. Both are plots of a spatial tree when the vector associated with a node is equal to the barycenter of the set associated with the node. Spatial trees are defined in Definition 3.1, and spatial tree plots are defined in Definition 3.2.

A location plot is always associated with a shape transform (radius transform, tail probability transform, or probability content transform). The height of a node in a location plot is equal to the height of the node in the associated shape transform. A location plot consists of d windows when the function is defined in the d-dimensional Euclidean space. There is a window for each coordinate. Each window shows the positions of one coordinate of the barycenters of the sets associated with the nodes of the shape tree. The leaf nodes of a shape tree correspond to the certain most extreme boundary points of the set, and the leaf nodes in a location plot show the locations of these extreme points. Thus a location plot visualizes a delineator of the set.

Definition 5.9 (Location plot.) *A location plot, associated with a shape transform, is a spatial tree plot, as defined in Definition 3.2. The spatial tree is defined by associating the shape tree with vectors and heights in the following way.*

1. *The vector associated with a node is equal to the barycenter of the set associated with the node.*

2. *The height associated with the node is equal to the height of the nodes in the associated shape plot.*

Illustrations We illustrate Definition 5.9.

1. Figure 5.8 shows a location plot of the set of Figure 5.4(a).The location plot corresponds to the radius function of Figure 5.4(b).

2. Figure 5.9 shows a location plot of the set of Figure 5.4(a). The location plot corresponds to the radius function of Figure 5.4(c).

3. Figure 5.10 illustrates further the definition of a location plot. We interpret the location plot of Figure 5.8. We have drawn as circles the seven barycenters of the sets associated with the seven nodes of the shape tree. These barycenters are joined with dotted lines to the corresponding nodes of the trees of the first and the second window of the location plot.

Colors It is enough to label the leaf nodes to identify the nodes between different windows of a location plot. The labels of the leaf nodes identify also the nodes of a location plot with the level sets of the associated shape function. To ease the identification of the nodes across different windows, we will use the leafs-first coloring, as defined in Chapter 3.1, page 86. The leafs-first coloring is appropriate because we want to highlight the extensions of the set, and thus we choose the best distinguishable colors for the extensions (for the modes of the shape function).

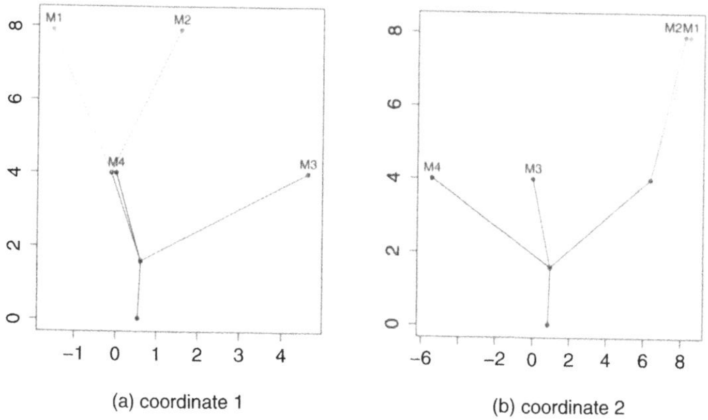

Figure 5.8 *(Location plot, sparse grid.)* Shown is a location plot of the set in Figure 5.4(a). The location plot corresponds to the radius function in Figure 5.4(b).

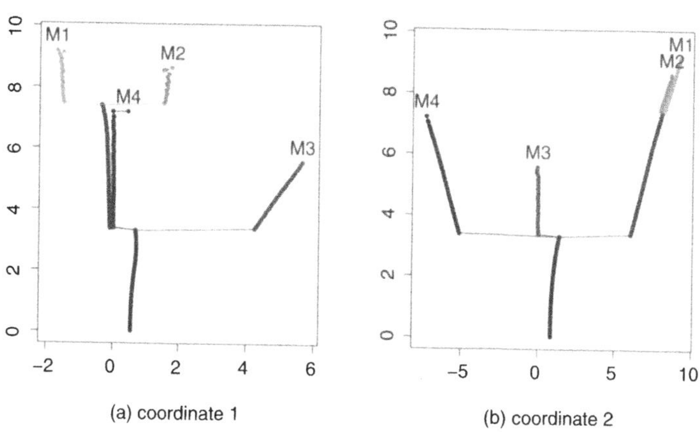

Figure 5.9 *(Location plot, dense grid.)* Shown is a location plot of the set in Figure 5.4(a). The location plot corresponds to the radius function in Figure 5.4(c).

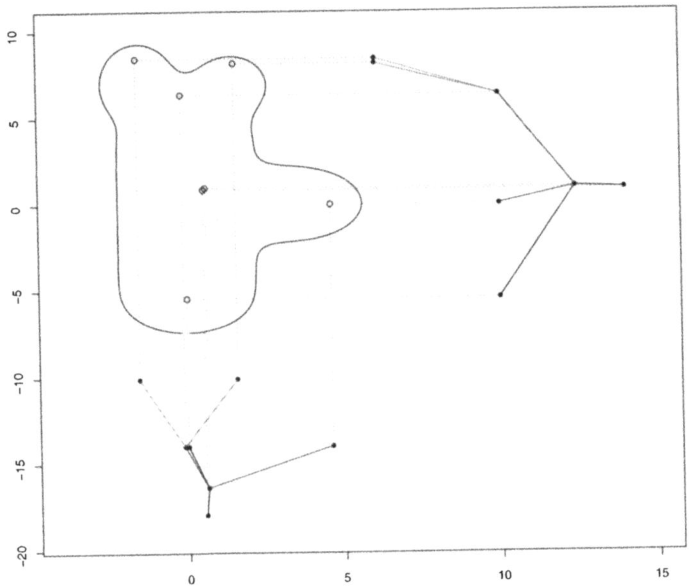

Figure 5.10 *(Location plot.)* Illustrated is the location plot of Figure 5.8.

5.5 CHOICE OF THE PARAMETERS

We discuss the choice of the reference point in the construction of a shape tree, differences between a radius function and a probability content function, and the choice of the metric in the definition of a shape tree.

5.5.1 Reference Point

The choice of the reference point of a shape tree has a substantial influence on the shape functions and on the location plots.

1. A natural choice for the reference point is the barycenter of the set. When the barycenter is the reference point, the shape plots seem to be easily interpretable.

2. When the set is a level set of a unimodal function, the mode of the function is a possible choice. The mode is the only point that belongs to every level set of a unimodal function, and it may be helpful to apply the same reference point for every level set when we are studying a series of level sets. The barycenter of a level set is not equal to the mode when the function is skewed. Before applying shape trees, it is useful to investigate the skewness of the function with a barycenter plot of a level set tree of the function, as defined in Section 4.3.

3. In the case of visualizing level sets of densities, any concept of location (expectation, median), as defined in Section 1.7.1, could be used if any of these reference points belong to the level set. The barycenter is the expectation of a uniformly distributed random variable on the set.

Illustrations We illustrate the effect of the reference point.

1. Figure 5.11 shows a set, its radius function, and the corresponding location plot. The reference point is marked with the square in panel (a). The radius function is unimodal, although the set is not a ball, because the reference point is far from the barycenter of the set. The skewness in the location plot reveals the fact that the chosen reference point is not equal to the barycenter.

2. Figure 5.12 shows the same set as in Figure 5.11, its radius function and the corresponding location plot. Now the reference point is the barycenter, marked with a dot in panel (a). The radius function is bimodal. The northeast extension of the set is closer to the barycenter, with more excess volume, than the the southwest extension. Thus the corresponding mode in the radius function is shorter and wider than the mode corresponding to the southwest extension.

5.5.2 Radius Function versus Probability Content Function

The definition of a radius function is easier to understand than the definition of a probability content function. However, a probability content function visualizes

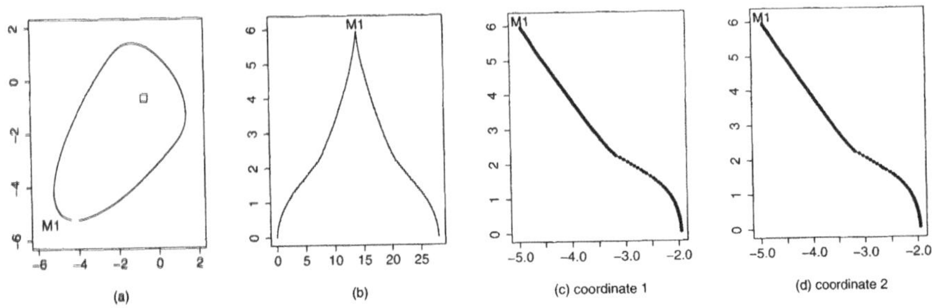

Figure 5.11 *(Radius function, the reference point is not equal to the barycenter.)* (a) The set, (b) its radius function, and (c–d) the corresponding location plot.

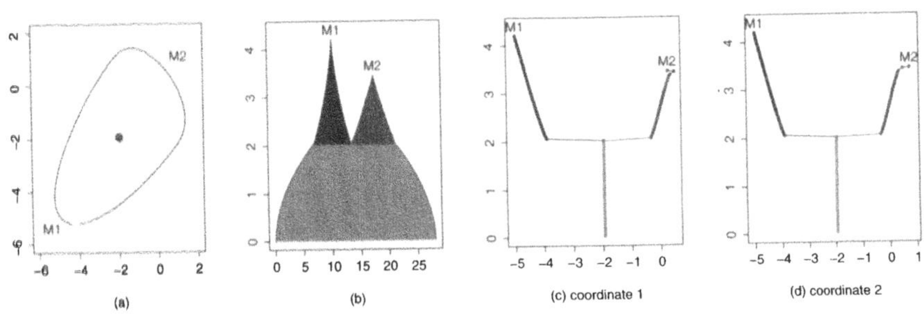

Figure 5.12 *(Radius function, the reference point is equal to the barycenter.)* (a) The set, (b) its radius function, and (c–d) the corresponding location plot.

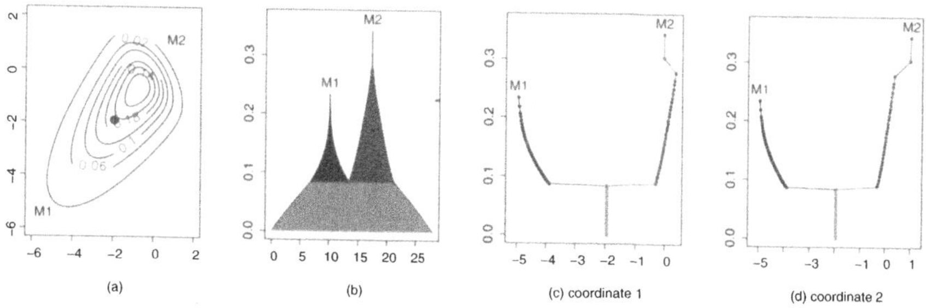

Figure 5.13 *(Probability content function, the reference point is equal to the barycenter.)* (a) A skewed Gaussian density, (b) a probability content function of the 0.02 level set, and (c–d) the corresponding location plot.

alternative information, showing the probability mass inside the set. A probability content plot may be useful in visualizing skewed densities.

Figure 5.13 shows a probability content function and a location plot for the 0.02 level set of the skewed Gaussian density, whose contour plot is shown in panel (a). Skewed Gaussian densities are defined in Section 10.1.5. The reference point is the barycenter of the level set, shown as a dot in panel (a). Note that the 0.02 level set is also shown in Figure 5.11 and in Figure 5.12. A skewed Gaussian density is defined in Section 10.1.5 (the parameters of the density are $\mu = (0, 0)$, $(\sigma_1, \sigma_2) = (3, 1)$, $\alpha = (6, 0)$, and then this density was rotated 135 degrees in the clockwork direction). We may compare the probability content function with the radius function in Figure 5.12(b). In the probability content function of Figure 5.13(b) the mode corresponding to the northeast extension of the level set is clearly bigger than the mode corresponding to the southwest extension. In the radius function of Figure 5.12(a) these extensions correspond to modes of roughly the same size. Thus the probability content function has visualized the fact that the northeast direction contains more probability mass.

5.5.3 Choice of the Metric

We have defined shape trees as level set trees of distance functions. A distance function is defined in Definition 5.1, using the Euclidean distance. We may consider other distances than the distance defined by the Euclidean norm. For example, we may take $\|x\|_{alt} = \max_{i=1,\ldots,d} |x_i|$, or $\|x\|_{alt}^2 = x^T Ax$, for some matrix A. In terms of Definition 5.4 the change of the metric means that we change the sequence of balls $B_r(\mu) = \{x \in \mathbf{R}^d : \|x - \mu\| \le r\}$, $r \in \mathcal{R}$, to the sequence of sets $S_r(\mu) = \{x \in \mathbf{R}^d : \|x - \mu\|_{alt} \le r\}$. With the norm $\|x\|_{alt} = \max_{i=1,\ldots,d} |x_i|$ the sequence consists of rectangles and with the norm $\|x\|_{alt}^2 = x^T Ax$ the sequence consists of ellipsoids. The use of alternative norms may be motivated when we want to visualize deviations from a certain shape with shape functions. The shape transforms of $A \subset \mathbf{R}^d$ are unimodal when A is a ball in the same metric as the metric

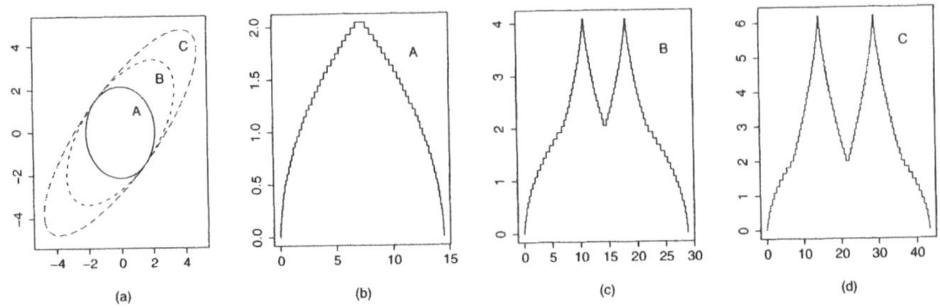

Figure 5.14 *(Radius functions.)* (a) A ball and two ellipses, (b) a radius function of the ball, (c) a radius function of the shorter ellipse, and (d) a radius function of the longer ellipse.

of the distance function. Thus we may visualize how A deviates from a rectangle if we choose the norm of the distance function to be $\|x\|_{alt} = \max_{i=1,\ldots,d} |x_i|$.

5.6 EXAMPLES OF SHAPE TREES

5.6.1 Uni- and Bimodality

Ellipsoidal Level Sets A shape function of a ball is a unimodal function. Shape functions of ellipses are bimodal functions when the reference point is not close to the boundary of the ellipse. Figure 5.14 shows three sets and their radius functions, when the reference point is at the origin, which is the barycenter of the sets. Panel (a) shows a ball and two ellipses, panel (b) shows a radius function of the ball, panel (c) shows a radius function of the shorter ellipse, and panel (d) shows a radius function of the longer ellipse. The modes of the radius transforms become more distinguished as the ellipses become longer.

Nonsymmetric Tail Dependence Figure 5.15 visualizes three level sets when the density is the Clayton density with standard Gaussian marginals. The Clayton copula with parameter $\theta > 0$ is defined in (10.21). Figure 5.15(a) shows the 10% level sets of Clayton densities with $\theta = 1$, $\theta = 2$, and $\theta = 4$, and panels (b), (c), and (d) show the corresponding radius functions. The densities in the Clayton family have nonsymmetric tail dependence; the dependence is larger in the negative orthant. The radius functions visualize the asymmetry of dependence: the mode corresponding to the area in the negative orthant is higher, but the mode corresponding to the area in the positive orthant is wider.

5.6.2 Multimodality of Level Sets

Figure 5.16 shows three level sets which have four-modal shape functions. In panel (a) the solid line delineates a level set of a mixture of two Gaussian densities, the dashed line delineates a level set of a density with Student copula and with standard Gaussian

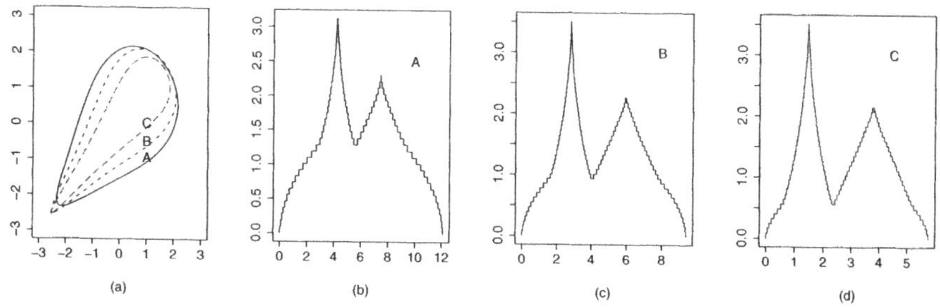

Figure 5.15 *(Radius functions.)* (a) Three Claytonian level sets and (b–d) the corresponding radius functions. Panel (b) corresponds to the set with the solid boundary, (c) to the set with the dashed boundary, and (d) to the set whose boundary is drawn with long dashes.

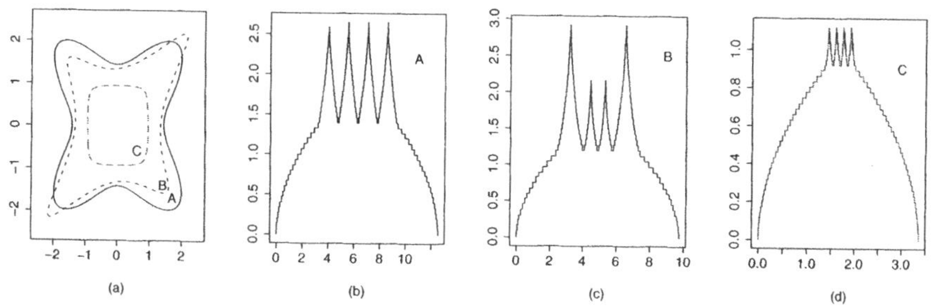

Figure 5.16 *(Radius functions.)* (a) Three level sets and (b–d) the corresponding radius functions. Panel (b) corrsponds to the set with the solid boundary, (c) to the set with the dashed boundary, and (d) to the set whose boundary is drawn with long dashes.

margins, and the line drawn with long dashes delineates the 10% level set of the Bartlett–Epanechnikov product density. The Student copula with parameters ν and ρ is defined in (10.20). In Figure 5.16 the parameter for degrees of freedom is $\nu = 3$ and the parameter for the correlation is $\rho = 0.3$. The Bartlett–Epanechnikov density is defined by $(x_1, \ldots, x_d) \mapsto (3/4)^d \Pi_{i=1}^d \max\{0, 1 - x_i^2\}$.

Panels (b) through (d) show the corresponding radius functions. The four tails of the mixture of Gaussians show up in the radius transform as four modes of equal size. The density with Student copula has also four tails but of unequal size, and they show up in the radius function as modes of unequal size. The 10% level set of Bartlett–Epanechnikov density has a slightly rectangular form and thus the radius function has four modes but these modes are of small size.

5.7 SHAPES OF DENSITIES

Shape functions and location plots visualize the shape, orientation, and location of sets. In particular, they visualize multimodality of sets. Multimodality of level sets of a density may arise in various ways. For example, mixtures of Gaussians with the same expectation but different covariance matrices are densities that may have multimodal level sets. (Analogously, mixtures of Gaussians with different expectations may be multimodal densities.) Product densities may have multimodal level sets, as shown in Figure 10.1. More complex dependency structures lead also to densities with multimodal level sets, as shown in Section 10.1.4 for various copula models.

The idea of visualizing densities through visualization of level sets appears at least in Scott (1992). This idea has been used to gain one more dimension: 3D functions may be visualized with perspective plots of 3D level sets. It is often enough to visualize only few level sets to get an accurate impression from the shape of the density. Also the shape tree based visualizations utilize the fact that the shape of the density is reflected by the shape of the level sets. However, besides the shape of the level sets one has to take into account the spacing of the level sets. The spacing of the level sets affects the "kurtosis" and "skewness" of the density.

1. Level sets may be unequally spaced in the vertical direction ("kurtosis"). For example, all spherically symmetric densities have ball-shaped level sets. Spherically symmetric densities may be written as $f(x) = g(\|x\|)$, where $g : [0, \infty) \to [0, \infty)$ and $\| \cdot \|$ is the Euclidean distance. The volumes of the level sets are, however, changing differently as the function of the level, depending on function g. Section 2.2 and Section 5.8 address the issues of spread visualization. In particular, a volume function of a level set tree may be used to diagnose kurtosis of the density.

2. Level sets may be unequally spaced in the horizontal direction. This happens for skewed densities. A barycenter plot of a level set tree may be used to detect the skewness of the density.

We may apply the following steps in exploratory data analysis:

1. Construct nonparametric density estimates from the data. One should calculate several estimates corresponding to a scale of smoothing parameters.

2. Apply the level set tree based techniques (the volume function and barycenter plot), as introduced in Chapter 4, to visualize the density estimates.

   (a) Make inference on the multimodality of the density. At this step we may also choose the values of the smoothing parameters of the estimator.

   (b) Make inference on the skewness and the kurtosis of the density. If the density is concluded to be multimodal, we study the skewness and the kurtosis of the density only on a region where the level sets are connected.

3. Apply the shape tree based techniques to visualize level sets of a density estimate.

   (a) If we have concluded that the density is unimodal, then we may apply shape trees to visualize the complete collection of level sets of the estimate.

   (b) If the density is concluded to be multimodal, we may restrict ourselves to a lower level set, or to a single mode (single connected component of a level set), and apply shape trees to make inference on the shape of the density restricted to this set.

5.8 2D SHAPE TRANSFORMS

Shape functions defined in Section 5.3 visualize a single level set of the function. We will define 2D shape functions that visualize simultaneously a series of level sets of the function. These 2D shape functions clue together several 1D shape functions, corresponding to several level sets of the original multivariate function. Thus a 2D shape function visualizes globally the underlying function. In addition to visualizing the shape of the level sets we are able to visualize the spread and the tail behavior with 2D shape functions.

Univariate density or distribution function type visualizations defined in Section 2.2 do not visualize an anisotropic tail behavior. We say that a distribution has anisotropic tails when the spread or dependence is different in various directions. 2D shape functions visualize anisotropic tails of a multivariate distribution because the modes of a 2D shape function correspond to the extensions of the level sets of the multivariate density. We apply a radius transform of level sets to get a density type visualization, and we apply a tail probability transform to get a distribution or quantile type visualization.

5.8.1 A 2D Volume Function

A 2D volume function is defined for a function, $f : \mathbf{R}^d \to \mathbf{R}$, $d \geq 2$, when the level sets of f are connected. A 2D volume function can be regarded as an extension of a unimodal volume function, as defined in Section 2.2.1.

A 2D volume function is based on the concept of a radius function, defined in Section 5.3.1. We calculate a series of radius functions from a series of level sets of the function, and clue these radius functions together in such a way that each radius function is a slice of the 2D volume function.

Radius functions are 1D functions whose support is an interval whose length is equal to the volume of the underlying set. We will normalize the radius functions so that their support is the interval $[0, 1]$. In addition we normalize the radius functions so that they integrate to the volume of the underlying set. The normalization makes it possible to construct a 2D function whose slices are identical with the normalized radius functions.

Let $f : \mathbf{R}^d \to \mathbf{R}$, $d \geq 2$, be a function whose level sets are connected. Let $r_\lambda : [0, v] \to [0, \infty)$ be a radius function of level set $\Lambda(f, \lambda)$, where $v = \text{volume}(\Lambda(f, \lambda))$. The normalized radius function is $h_\lambda : [0, 1] \to [0, \infty)$,

$$h_\lambda(t) = c r_\lambda(v t), \qquad t \in [0, 1],$$

where the normalization constant c is equal to $c = v / \int_0^1 r_\lambda(vt)\, dt$. Now a 2D volume function is defined to be the 2D function whose slices are the normalized radius transforms.

Definition 5.10 (2D volume function.) *A 2D volume function* $\mathcal{V} : [\lambda_0, \lambda_1] \times [0, 1] \to [0, \infty)$, *corresponding to function* $f : \mathbf{R}^d \to [\lambda_0, \lambda_1]$, *is the function whose slices are given by the normalized radius functions:*

$$\mathcal{V}(\lambda, t) = h_\lambda(t), \qquad t \in [0, 1],$$

for each level $\lambda \in [\lambda_0, \lambda_1]$.

Remarks We make several remarks concerning Definition 5.10.

1. Unimodal volume functions are defined in Section 2.2.1. Unimodal volume functions are 1D curves that visualize the spread of a multivariate function. We can interpret 2D volume functions as a generalization of unimodal volume functions. Note that the 1D volume functions defined in Section 4.2, based on level set trees, can also be seen as a generalization of unimodal volume functions. 2D volume functions are defined for unimodal functions and they visualize anisotropic spread and dependence, whereas 1D volume functions, based on level set trees, visualize multimodality of functions.

2. There are two additional ways to clue normalized radius transforms $(h_\lambda)_{\lambda \in [\lambda_0, \lambda_1]}$ together to make a 2D function: (1) we can let the graphs of the normalized radius transforms to be the level sets of a 2D function, or (2) we can let the normalized radius transforms to be the boundary functions of 2D sets and define these sets to be the level sets of a 2D function. We have chosen a third way, and have taken the normalized radius functions to be the slices of a 2D volume function, since it seems to lead to smooth 2D functions and this way of defining a 2D function does not require the nestedness of the graphs of the normalized radius transforms. The nestedness holds in typical cases, but we cannot guarantee the nestedness in all cases.

Illustrations We illustrate Definition 5.10.

1. Figure 5.17 visualizes a density with Gumbel copula with parameter $\theta = 2$ and the standard Gaussian marginals. Panel (a) shows a contour plot of the density and panel (b) shows a 2D volume function of the density. The 2D

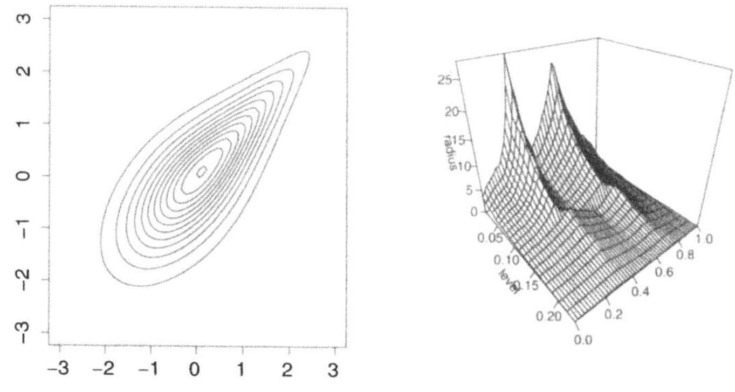

(a) (b)

Figure 5.17 *(Gumbel copula and standard Gaussian marginals, 2D volume function.)* Shown are (a) a contour plot of the density and (b) a corresponding 2D volume function.

volume function visualizes the eggshapes of the level sets: the radius functions have two modes of unequal size and shape. The shapes of the level sets do not change much when we move to the higher levels.

2. Figure 5.18 visualizes the density of the equal mixture of two 2D Gaussian densities, centered at the origin and whose marginal standard deviations are $(0.5, 1.5)$ and $(3.5, 0.5)$. Panel (a) shows a contour plot and panel (b) a 2D volume function. The density has four extensions.

3. Figure 5.19(a) shows a 2D volume function of a 3D density with Gumbel copula with parameter $\theta = 2$ and the standard Gaussian marginals. The shape of the function is similar to the 2D Gumbel density shown in Figure 5.17(b), but the 2D volume function of the 3D density has higher modes.

   Figure 5.19(b) shows a 2D volume function of the product of three 1D Student densities, with degrees of freedom $\nu = 1$. This 3D density has six extensions, which show up as six modes in the 2D volume function.

5.8.2 A 2D Probability Content Function

A 2D probability content function is associated to a sequence of connected sets $A_\lambda \subset \mathbf{R}^d$, $d \geq 2$, $\lambda \geq 0$, and to a probability measure P. A 2D probability content function can be regarded as an extension of a volume-to-probability function (distribution function), as defined in Definition 2.5.

Quantile and distribution functions as defined in Section 2.2.2 do not visualize anisotropic tail behavior, but a 2D probability content function addresses the visualization of anisotropic tails. Similar to quantile and distribution functions, a 2D

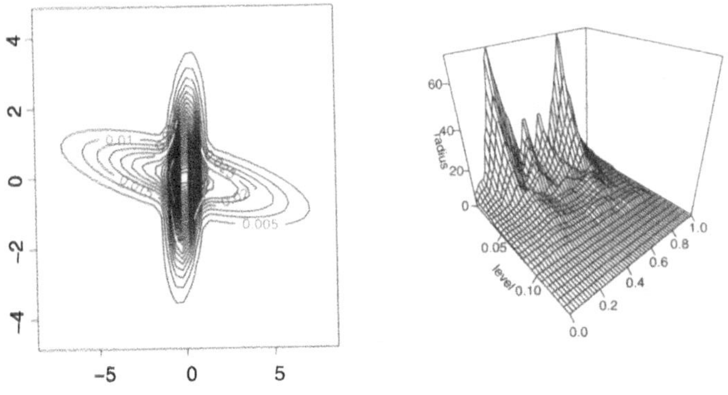

(a) (b)

Figure 5.18 *(Mixture of two 2D Gaussians, 2D volume function.)* Shown are (a) a contour plot of the density and (b) a corresponding 2D volume function.

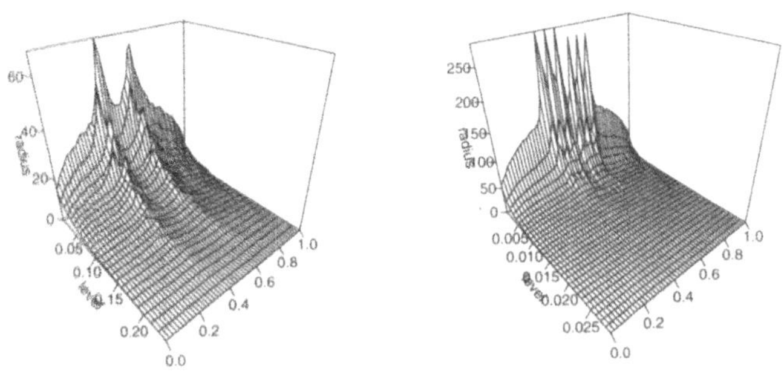

(a) (b)

Figure 5.19 *(2D volume functions of 3D densities.)* (a) A 2D volume function of a 3D Gumbel density and (b) a 2D volume function of the product of three univariate Student densities.

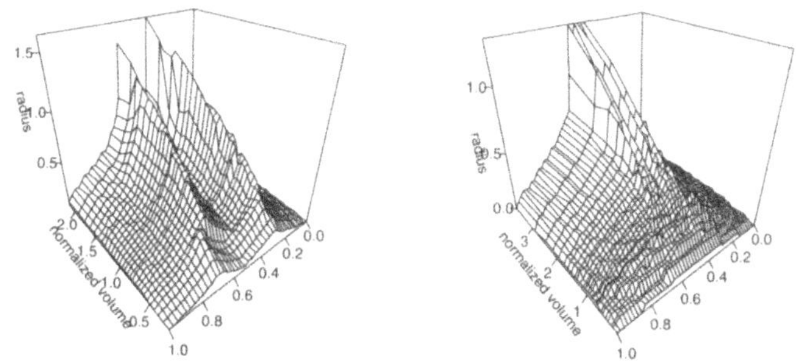

(a) (b)

Figure 5.20 *(2D probability content functions of 2D densities.)* (a) A 2D volume function of a distribution with a Gumbel copula and the standard Gaussian marginals. (b) A 2D volume function of a product of two Student densities.

probability content function is associated to a sequence of sets A_λ, $\lambda \geq 0$, where A_λ can be a level set, a depth region, a minimum volume set, or a central region of a quantile function. A useful example would be the case where A_λ is the level set of f with level λ, and f is the density of P.

A 2D probability content function is defined similarly to a 2D volume function, but now we clue together tail probability functions, as defined in Section 5.3.2, and not radius functions. Similar to radius functions, tail probability functions are shape isomorphic transforms of a multivariate connected set to a univariate function. However, the length of a disconnected part of a level set of a tail probability function is equal to the probability content of the corresponding tail, and not equal to its volume, as in the case of a radius function.

We define a normalized tail probability function so that a tail probability function is scaled to have support $[0, 1]$ and normalized to integrate to the probability content of the set. Let $g_\lambda : [0, p] \to [0, \infty)$ be a tail probability function of a connected set A_λ, where $p = P(A_\lambda)$. The normalized tail probability function is $h_\lambda : [0, 1] \to [0, \infty)$,

$$h_\lambda(t) = c\, g_\lambda(pt), \qquad t \in [0, 1],$$

where $c = p / \int_0^1 g_\lambda(pt)\, dt = p^2 / \int_0^p g_\lambda$. We clue normalized tail probability functions together indexing the sets A_λ with their volumes, or with the dimension normalized volumes.

Definition 5.11 (2D probability content function.) *A 2D probability content function $\mathcal{F} : (0, \infty) \times [0, 1] \to [0, \infty)$, associated to a sequence of connected sets $A_\lambda \subset \mathbf{R}^d$, $d \geq 2$, $\lambda \geq 0$, and to a probability measure P, is a 2D function whose slices are* *equal to the normalized tail probability functions:*

$$\mathcal{F}(v,t) = h_{\lambda_v}(t), \qquad t \in [0,1],$$

for $v \in (0,\infty)$, *when* λ_v *is either defined by*

$$\lambda_v = \sup\{\lambda \in [0,\infty) : \text{volume}(A_\lambda) \geq v\}, \qquad v \in [0,\infty),$$

as in (2.15), or we index the sets with the dimension normalized volumes

$$\lambda_v = \sup\{\lambda \in [0,\infty) : \left(C_d^{-1}\text{volume}(A_\lambda)\right)^{1/d} \geq v\}, \qquad v \in [0,\infty),$$

where C_d *is defined in (B.10).*

Illustrations We illustrate Definition 5.11. Sets A_λ are the level sets, and we index the sets with the dimension normalized volumes.

1. Figure 5.20(a) shows a 2D probability content function of the distribution with Gumbel copula with parameter $\theta = 2$ and the standard Gaussian marginals. Note that Figure 5.17(b) shows the 2D volume function of the same distribution.

2. Figure 5.20(b) shows the product of two univariate Student densities whose degrees of freedom are $\nu = 1$. Now the 2D probability content function has four modes.

CHAPTER 6

TAIL TREES

The concept of a *tail tree* replaces the concept of a shape tree for the sets of finite cardinality.

With tail trees one may visualize the shape, the location, and the orientation of a multivariate point cloud $x_1, \ldots, x_n \in \mathbf{R}^d$. The point cloud is interpreted as realizations of n random vectors having a common density function. The visualizations with tail trees are tailored to the case where the point cloud does not have clusters (it is connected in the sense of a single linkage hierarchical clustering). The assumption of the connectedness of the point cloud may be interpreted as arising from the connectedness of the support, or a low level set, of the density that has generated the observations.

The method is designed for visualizing dependency among several variables, whereas many of the other visualization tools are more directed to finding and visualizing clusters in the data. In some cases one can assume that the data do not contain clusters. This is typically the case for example for data consisting of financial returns. Then one is interested in the dependence structure and not interested in finding clusters. If one cannot exclude the possibility of clusters a priori, then one should explore the existence of clusters with other methods (projections, parallel

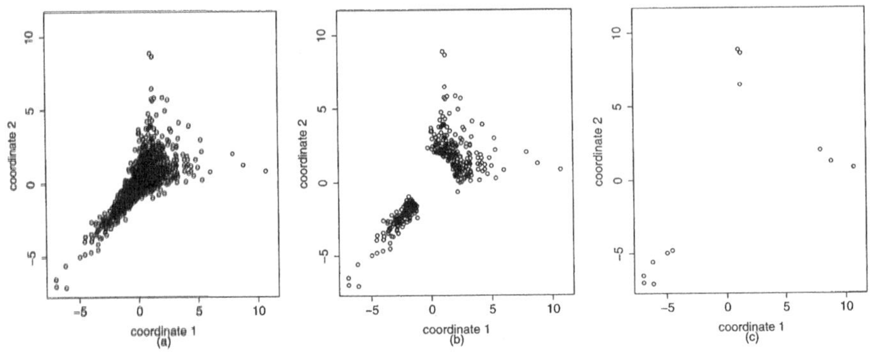

Figure 6.1 *(Tail clustering.)* (a) A data set with three tail clusters. (b–c) The data after removing central observations.

coordinate plots, graphical matrices), and possibly divide data into clusters before applying tail trees.

The basic idea is to define a tail tree among the observations. A tail tree is visualized with a tail tree plot and with a tail frequency plot, and a tail tree defines a segmentation of the data that may be used to enhance the grand tour, graphical matrices, and parallel coordinate plots.

1. A *tail tree* is a tree whose nodes are associated with the observations so that the root node corresponds to a central observation and the nodes in the upper levels of the tree correspond to the observations lying in the tails of the data. If the data is not spherically symmetric but has separated tails, then a tail tree has many branches, each branch corresponding to a separate tail of the data.

   The concept of a tail tree is closely related to the concept of a shape tree as defined in Chapter 5. A shape tree was defined for a connected set, when the connectedness was defined in terms of the topology of the Euclidean metric. We want to modify that concept in order to visualize sets of finite cardinality. A shape tree may be applied to visualize level sets of a density estimate, but with tail trees we may visualize the raw data, avoiding the density estimation step. Since density estimation is difficult in high-dimensional cases, we may be able to analyze higher dimensional data with tail trees than with shape trees. In addition sometimes a part of the data arises from a discrete distribution so that there does not exist an underlying density for the complete data.

2. A tail tree is defined with the help of a *tail clustering* of the data. We find tail clusters of the data when we remove central data points and then cluster the remaining observations. However, tails may appear at different distances, and there is no distinguished rule for choosing the number of central points to be removed. We may instead consider a sequence of balls of increasing radius,

and make a tree structure of clusters by clustering at each step the observations in the complement of a ball of the sequence.

Figure 6.1(a) shows a scatter plot of data of size 2000, generated from the distribution whose copula is the Clayton copula with parameter $\theta = 4$, and whose marginals have the Student distribution with degrees of freedom 4. This data may be said to contain three tails. Figure 6.1(b) shows the remaining observations when we have removed the observations inside a ball of radius 2. At this stage there are two tail clusters. Figure 6.1(c) shows the three most distant tail clusters. Now we have removed the observations inside a ball of radius 6.5.

A tail tree is defined for sets of finite cardinality, and we do not grow a sequence of balls with continuously increasing radius as in Figure 6.1: since the data has a finite cardinality, we may remove observations from the data set one at a time. We start at a center point of the data, and remove observations in the order determined by the closeness to the center point. At each step we cluster the remaining observations.

3. A *tail tree plot* consists of d scatter plots, where the x-coordinate of the ith scatter plot corresponds to the ith coordinate of the observations, and the y-coordinate of the scatter plots gives the Euclidean distance of the observations from the center point. We visualize in each scatter plot the tail tree by connecting the observations with straight lines, and use colors and labels to identify the branches of the tail tree among the d scatter plots. A tail tree adds a structure to the data that makes it possible for humans to detect important features in the data. A basic idea behind a tail tree plot is that a tail tree may be considered as a spatial tree (1D curve in the d-dimensional Euclidean space that has branches) and 1D curves in the d-dimensional space may be visualized with d projections to the coordinate axes.

   A tail tree plot may be seen as an enhancement of a parallel level plot defined in Section 1.3. The structure of the tail tree is used as an additional help to identify observations across the dimensions. A tail tree plot may also be seen as belonging to the family of linking techniques discussed in Section 1.6. A tail tree plot has some advantages over other methods: it is a dimension-insensitive tool, it may be used to visualize dependence and anisotropic tails, and it is not severely affected by overplotting. Projections may be excellent in visualizing clusters in the data. When we are interested in the dependence among the variables, tail tree plots may provide good alternatives to projections. A tail tree plot consists only of d windows, and it is a less dimension sensitive tool than the use of projections and slices.

4. A *tail frequency plot* visualizes the heaviness of the tails separately in each tail of the point cloud. This visualization is useful when the underlying distribution has anisotropic tails. A tail frequency plot visualizes some characteristics of a multivariate point cloud with the help of an 1D curve.

5. A tail tree defines a *segmentation of the data*. The segmentation may be used to enhance projections of the data (grand tour), graphical matrices, and parallel coordinate plots. We choose a distinct color for each segment and choose the color of the observation to be the color of the segment where it belongs. The coloring helps the orientation during the grand tour and makes the patterns in a parallel coordinate more visible. In addition the segmentation of the data may be used to define a permutation of the data that reveals patterns when one uses a graphical matrix to visualize the data.

Graphs may be used in two different ways: to communicate information to an audience and to analyze data. Graphs used for communications (presentation graphs) tend to be simple and are mainly intended for the display of discovered patterns. Graphs used for analysis may be quite detailed, complicated, or esoteric, containing most or even all of the original data, and are tools for the detection of important or unusual features in the data. (Spence and Lewandowsky (1990, p. 19–20)) Tail tree plots may be used to analyze data, and tail tree plots have intuitive appeal which makes them useful also for presentation purposes.

6.1 TAIL TREES

A tail tree is like a shape tree, defined in Section 5.2: we will define a tail tree as a level set tree of a distance function of a set. However, unlike in the case of a shape tree, we restrict ourselves to the case where the set has finite cardinality. Definition 4.2 of a level set tree was based on the concepts of separated sets and a connected set. Since the set that we consider has finite cardinality, we cannot use the same concepts of separated sets and a connected set as in Definition 4.1. We will first generalize the concepts of separated sets and a connected set in Section 6.1.1 and then define a tail tree in Section 6.1.2.

6.1.1 Connected Sets and Single Linkage Clustering

We define ρ-separated sets and a ρ-connected set, by modifying Definition 4.1. The definition may be used for sets of finite cardinality, but it applies for general sets in the multivariate Euclidean space, and not only for sets of finite cardinality.

Definition 6.1 (ρ-separated sets, ρ-connected set.)

1. *Sets $A, B \subset \mathbf{R}^d$ are* separated *for the resolution threshold $\rho \geq 0$ (ρ-separated), if for each $x \in A$ and $y \in B$, $\|x - y\| > 2\rho$, where $\|\cdot\|$ is the Euclidean norm.*

2. *Set $A \subset \mathbf{R}^d$ is* connected *for the resolution threshold $\rho \geq 0$ (ρ-connected), if for every nonempty B, C such that $A = B \cup C$, B and C are not separated for the resolution threshold ρ.*

In other words, a set is connected if it cannot be written as a union of two separated sets. When $\rho > 0$, set A is connected for the resolution threshold ρ if for each $x, y \in A$, there is a path $z_1, \ldots, z_m \in A$ so that $z_1 = x$, $z_m = y$ and $\|z_i - z_{i+1}\| \le 2\rho$ for $i = 1, \ldots, m-1$. A set of finite cardinality is ρ-connected if the union of balls, centered at the observations, with radius ρ, is 0-connected. (That is why we use distance 2ρ in the definition of separated sets.)

For resolution threshold $\rho = 0$ a set of finite cardinality is not connected. If a set is bounded, then for large enough resolution threshold the set is connected. Sets of finite cardinality are always bounded, and thus for large enough resolution threshold they are connected. For a set of cardinality n we have a finite sequence of intervals $[0, \rho_1), \ldots, [\rho_{n-1}, \infty)$ with the property that, when $\rho \in [\rho_{k-1}, \rho_k)$, $k = 1, \ldots, n$, $\rho_0 = 0$, $\rho_n = \infty$, the set has $n - k + 1$ connected components that are pairwise separated. We call values $\rho_1, \ldots, \rho_{n-1}$ the critical resolution thresholds.

The single linkage clustering partitions a finite set to separated sets in the sense of Definition 6.1. The single linkage clustering is an agglomerative hierarchical clustering algorithm, where the distance between clusters is defined to be the distance between the closest pair of data points in the clusters. In fact the hierarchical clustering tree (dendrogram) with the single linkage agglomeration gives the sequence $\rho_1, \ldots, \rho_{n-1}$ of the critical resolution thresholds. We build the hierarchical clustering tree by finding the 2 closest observations, and joining them together into a cluster. The distance between the two closest data points is equal to $2\rho_1$. At step k, $k = 1, \ldots, n-1$, we find the two closest clusters among the set of $n - k + 1$ previously found clusters, and join these clusters together. The distance between these two closest clusters is equal to $2\rho_k$. Finally, at step $n-1$ we merge together the remaining two clusters, and the distance between these clusters is equal to $2\rho_{n-1}$. Hierarchical clustering is discussed in Section 8.1.

6.1.2 Definition of a Tail Tree

We will define a tail tree for a finite set $A = \{x_1, \ldots, x_n\}$. A tail tree will be defined as a level set tree of a distance function of A. A distance function $f_A : A \to \mathbf{R}$ with reference point $\mu \in \mathbf{R}^d$, of a set $A \subset \mathbf{R}^d$, is defined in Definition 5.1 as $f_A(x) = \|\mu - x\| I_A(x)$. A level set tree depends on a grid of levels. In the case of a tail tree we will define the grid of levels to be the distances of the points $x_1, \ldots x_n$ from the reference point $\mu \in \mathbf{R}^d$.

Definition 6.2 (Tail tree.) *A* tail tree *of set $A = \{x_1, \ldots, x_n\} \subset \mathbf{R}^d$, associated with a resolution threshold $\rho \ge 0$, and with center point $\mu \in \mathbf{R}^d$, is a level set tree of a distance function f_A of A with the reference point μ. The level set tree is defined as in Definition 4.2, but with the concepts of ρ-separated sets and a ρ-connected set as in Definition 6.1. The grid of levels of the level set tree is $\mathcal{R} = \{r_1 < \cdots < r_n\}$, where*

$$r_i = \min\{\|x - \mu\| : x \in \mathcal{X}_i\},$$

with $\mathcal{X}_1 = A$ and for $i = 1, \ldots, n-1$,

$$\mathcal{X}_{i+1} = \{x \in A : \|x - \mu\| > r_i\}.$$

In addition we associate each node with a single observation: if a node is associated with level r_i, then we can associate it with observation x_i that satisfies $r_i = \|x_i - \mu\|$.

For transparency, we will give also a direct definition of a tail tree. Since we associate a node with the observation x_i that satisfies $r_i = \|x_i - \mu\|$, where r_i is the level associated with the node, we can drop the association of the nodes with the levels r_i. This association is redundant, and the reduction will be done in the following definition.

Definition 6.3 (Tail tree, direct definition) *A tail tree of set $A = \{x_1, \ldots, x_n\} \subset \mathbf{R}^d$, associated with a resolution threshold $\rho \geq 0$, and with center point $\mu \in \mathbf{R}^d$, is a tree where each node is associated with a subset of A and with one observation.*

1. *Write*
$$A = A_1 \cup \cdots \cup A_K,$$
where sets A_i are pairwise separated, and each is connected, with the resolution threshold ρ. The tail tree has K root nodes, which are associated with sets A_i and observations $a_i = \operatorname{argmin}_{y \in A_i} \|y - \mu\|$, $i = 1, \ldots, K$.

2. *Let node m be associated with set $B \subset A$.*

   (a) *If B contains only one observation $x \in A$, then node m is a leaf node, associated with set $B = \{x\}$ and observation x.*

   (b) *Otherwise, let $x = \operatorname{argmin}_{y \in B} \|y - \mu\|$ be the closest point in B to center point μ. Write*
$$B \setminus \{x\} = C_1 \cup \cdots \cup C_M, \tag{6.1}$$
where sets C_i are pairwise separated, and each is connected, with the resolution threshold ρ. Then node m has M children, which are associated with sets C_i and observations $c_i = \operatorname{argmin}_{y \in C_i} \|y - \mu\|$, $i = 1, \ldots, M$.

Illustrations We will illustrate the definition of a level set tree.

1. Figure 6.2(a) shows a data, and panel (b) shows a tail tree of the data. The data is the same data as in Figure 6.1(a). The resolution threshold is $\rho = 1.1$, and the center point is $(0.026, 0.031)$. The nodes in the tail tree are plotted at the height that is equal to the Euclidean distance of the corresponding observation from the center point. We have colored the data and the nodes of the tail tree so that the color of a node is the same as the color of the observation associated with the node.

2. Figure 6.3 gives a step by step illustration of Definition 6.3. We have drawn the data points as colored bullets. In the first window we have drawn a circle of radius $\rho = 0.75$ around each observation, and one sees that the set of observations is connected for resolution threshold ρ. The center point is chosen to be the origin. In the second window we have removed the circles

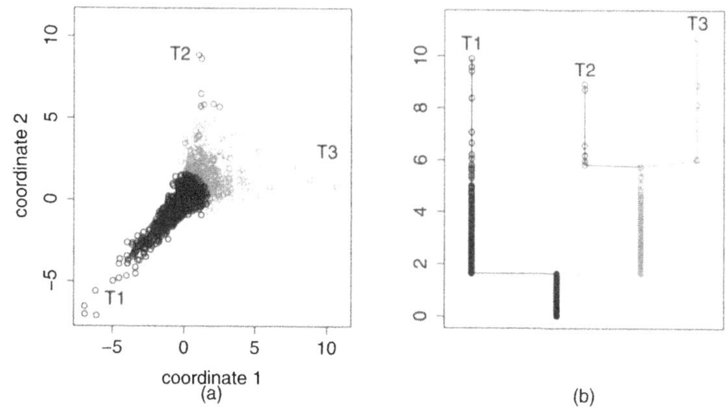

Figure 6.2 *(Tail tree.)* (a) A scatter plot of a data and (b) a tail tree of the data. The coloring of the nodes helps to find the corresponding observations in the scatter plot.

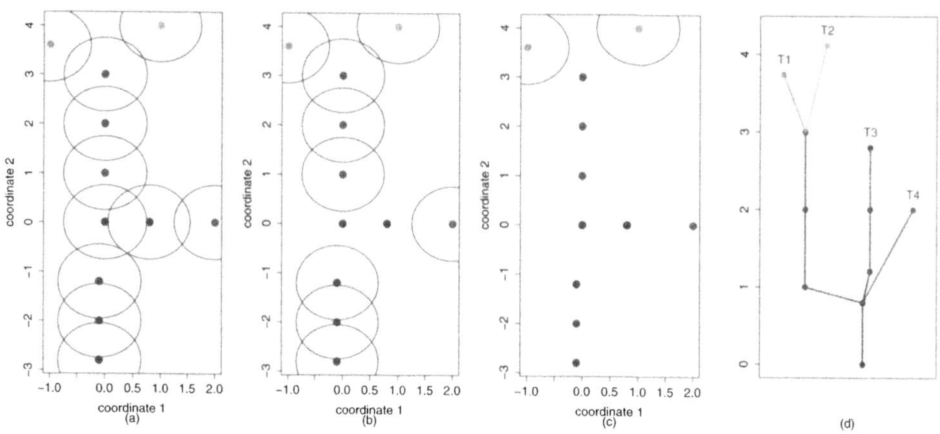

Figure 6.3 *(Tail tree.)* Illustrated is Definition 6.3 of a tail tree. (a–c) The data and (d) a tail tree of the data set.

around the two closest observations to the center, and one sees that the set of the remaining observations consists of three separated components. The rightmost component consists of a single red observation, and this observation corresponds to the leaf node T4. In the third window we have removed the circle around the other but the two furthest observations. These two furthest observations correspond to the leaf nodes T1 and T2. In the fourth window we show the tail tree of the data set.

Remarks We give remarks concerning the definition of a tail tree.

1. *(The center point.)* We choose in the most cases the center point to be the arithmetic mean: $\mu = n^{-1} \sum_{i=1}^{n} x_i$. An other possibility is to use some concept of a multivariate median, defined in terms of a data depth, see Section 1.7.1.

2. *(Algorithm.)* An algorithm for calculating a tail tree is presented in Section 13.2.1.

3. *(Decoration.)* A node of a tail tree is associated both with an observation and with a subset of the observations. We may always define one decoration in the terms of the other. Let us denote with set(m) the set associated with node m in a tail tree of the data, and let us denote with obs(m) the observation associated with the node. We may write a set associated with node m as

$$\text{set}(m) = \bigcup \{\text{obs}(c) : c = m \text{ or } c \text{ is a descendant of } m\}.$$

We may write the observation associated with a node m as

$$\text{obs}(m) = \begin{cases} \text{set}(m) \setminus \bigcup \{\text{set}(c) : c \text{ is a child of } m\}, & m \text{ a non-leaf node,} \\ \text{set}(m), & m \text{ a leaf node,} \end{cases}$$

where we use the convention $\{x\} = x$, where $x \in \mathbf{R}^d$. The decoration with observations is used to define a tail tree plot and the decoration with subsets of observations is used to define a tail frequency plot.

4. *(Other clustering distances.)* The definition of separated sets in Definition 6.1 may be rephrased in the following way: sets $A, B \subset \mathbf{R}^d$ are separated for the resolution threshold $\rho \geq 0$, if $D(A, B) > 2\rho$, where $D(A, B) = \inf\{\|x - y\| : x \in A,\ y \in B\}$ is the single linkage distance between two sets. The concept of separated sets is the fundamental underlying concept in the definition of a tail tree, and it is natural to ask whether the single linkage distance may be replaced by some other distance between sets in order to get other versions of a tail tree. Indeed each hierarchical agglomerative clustering algorithm is based on some concept of a distance $D(A, B)$ between sets $A, B \subset \mathbf{R}^d$, where the distance need to be defined only for sets of finite cardinality. For example, the complete linkage clustering uses the distance $D(A, B) = \sup\{\delta(x, y) : x \in A,\ y \in B\}$, where δ is typically the Euclidean distance, and the average linkage clustering uses the distance $D(A, B) = (\#A \cdot \#B)^{-1} \sum_{x \in A, y \in B} \delta(x, y)$, where A and B have finite cardinality. Section 8.1 discusses hierarchical clustering. The essential condition for the definition of a tail tree to be meaningful is the following.

$$\begin{aligned} &\text{When } A, B \subset \mathbf{R}^d \text{ are } \rho\text{-separated and } A_0 \subset A,\ B_0 \subset B, \\ &\text{sets } A_0 \text{ and } B_0 \text{ are } \rho\text{-separated.} \end{aligned} \tag{6.2}$$

Indeed, when we build a tail tree, we are at each step removing observations from the original set. If the condition (6.2) is violated, it may happen that after two branches have been separated, they may later reconnect after more observations have been removed. In this case we are building a graph that is not a tree. These kinds of graphs may indeed be useful in visualization and segmentation, but we don't study these graphs in this book. The distances of the complete linkage clustering and the average linkage clustering do not satisfy condition (6.2) when δ is the Euclidean distance.

5. *(Tail clustering.)* Tail clustering is discussed in Section 8.4.

6. *(Different scales.)* When the measurement scales are different for various coordinates of the observations, we may either (1) define a tail tree using a more flexible definition of separated sets given in Definition 6.4 or (2) normalize the data so that the measurement scales are compatible, and after that define a tail tree using the definition of separated sets given in Definition 6.1.

Definition 6.4 (Separated sets, Connected set.) *Sets $A, B \subset \mathbf{R}^d$ of finite cardinality are separated for a vector of resolution thresholds $R = (\rho_1, \ldots, \rho_d)$, where $\rho_i > 0$, if sets $\cup_{a \in A} B_R(a)$ and $\cup_{b \in B} B_R(b)$ are separated for resolution threshold $\rho = 0$, in the sense of Definition 6.1, where we denote $B_R(x) = \{y \in \mathbf{R}^d : \sum_{i=1}^{d} (y_i - x_i)^2 / \rho_i^2 \leq 1\}$.*

6.2 TAIL TREE PLOT

A tail tree plot visualizes the locations of the observations associated with the nodes of a tail tree. A tail tree plot is analogous to a barycenter plot of a level set tree, as defined in Definition 4.7, and to a location plot of a shape tree, as defined in Definition 5.9. Both are plots of a spatial tree when the vector associated with a node is taken to be the observation associated with the node. Spatial trees are defined in Definition 3.1, and spatial tree plots are defined in Definition 3.2.

6.2.1 Definition of a Tail Tree Plot

We have defined a tail tree as a tree whose nodes are associated with observations. We may consider a tail tree as a spatial tree (discrete curve, having branches, in the d-dimensional Euclidean space). Thus we may visualize this tree by plotting each coordinate separately, as in a spatial tree plot of Definition 3.2. An idea behind the definition of a spatial tree plot is that 1D curves in the d-dimensional Euclidean space may be visualized with coordinate functions (projections to the coordinate axes), and 1D branching curves may be visualized with projections when we add colors to the plot to identify different branches.

Definition 6.5 (Tail tree plot.) A tail tree plot *of data $A = \{x_1, \ldots, x_n\} \subset \mathbf{R}^d$, associated with a tail tree, is a spatial tree plot, as defined in Definition 3.2. The spatial tree is defined by associating the tail tree with vectors and heights in the following way:*

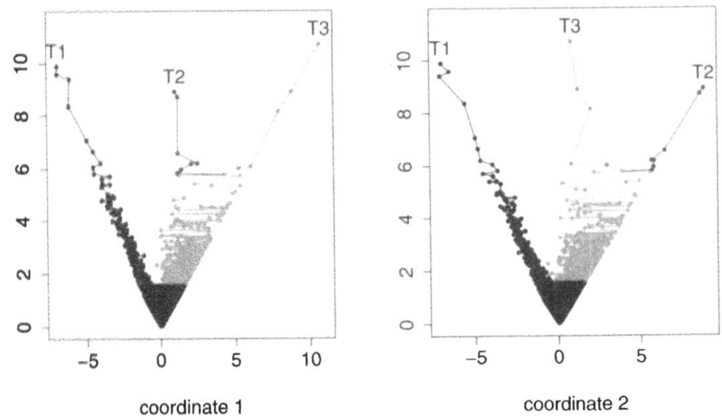

Figure 6.4 *(Tail tree plot.)* A tail tree plot of the data in Figure 6.2(a) and Figure 6.1(a).

1. *The vector associated with a node is equal to the observation associated with the node.*

2. *The height associated with the node is equal to the distance of the observation associated with the node from the center point.*

In summary, in a tail tree plot the horizontal position of a node in the ith window, $i = 1, \ldots, d$, is equal to the ith coordinate of the observation associated with the node. The vertical position of a node is equal to the distance of the observation associated with the node to the center point, The parent–child relations are expressed by lines joining the parent with its children.

Illustrations We illustrate tail tree plots.

1. Figure 6.4 shows a tail tree plot of the data in Figure 6.2(b) and in Figure 6.1(a). The tail tree plot shows the three tails of the data and one sees that the southwest tail (T1, blue branch) is much thinner than the northeast tail (T2-T3, orange branch). Branches T2 and T3 have different positions in the two windows, and the different positions visualize the different locations of these tails.

2. Figure 6.5 shows a tail tree plot of the data in Figure 6.3.

Remarks We make several remarks concerning the definition of a tail tree plot.

1. We have defined a tail tree plot differently than a location plot in Definition 5.9. Indeed a location plot visualizes the locations of the barycenters of the sets associated with the nodes of a shape tree. In a tail tree plot we have not used barycenters but visualized the locations of single observations (the locations of

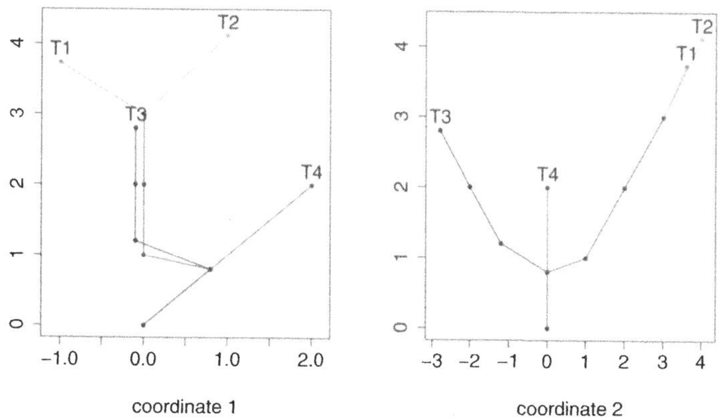

Figure 6.5 *(Tail tree plot.)* A tail tree plot of the data in Figure 6.3.

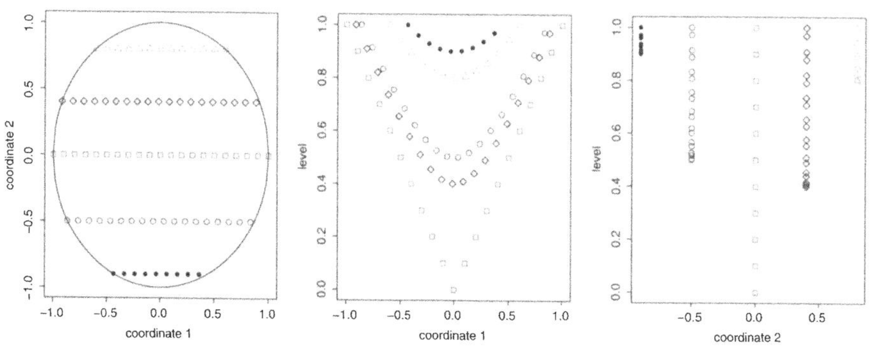

Figure 6.6 *(Correspondence with scatter plots.)* (a) Horizontal lines mapped onto (b) (smoothed) wedges in the x-window of a tail tree plot and onto (c) vertical lines in the y-window of a tail tree plot.

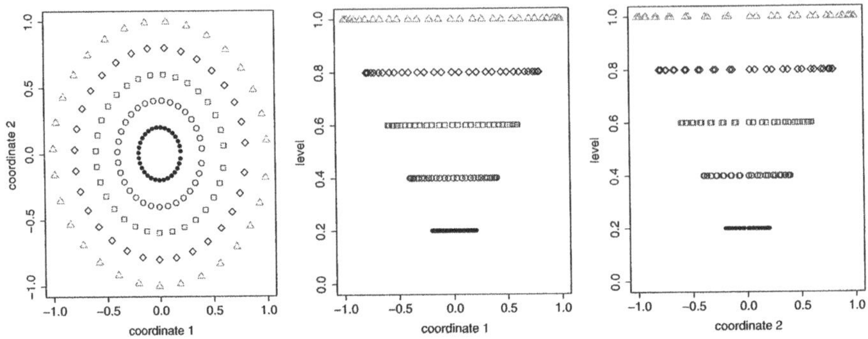

Figure 6.7 *(Correspondence with scatter plots.)* Spheres mapped onto horizontal lines.

the observations that are closest to the center point, in the sets associated with the nodes of a tail tree). This seems to be a more direct way of visualizing the data.

2. *(Colors.)* We add colors to the plot to identify the branches across different windows. We choose first distinct colors for the leaf nodes and then travel toward the root node, changing the color always when two branches are merging. This is called the leafs-first-coloring, as defined in Section 3.1, page 86, and it finds the best distinguishable colors for the leafs. We color also the lines joining a child and a parent. The color of a line will be the same as the color of the child node. As an additional help in identifying the nodes across the windows one may also label the leaf nodes with numbers or letters.

3. *(Correspondence with scatter plots.)* Figure 6.6 shows how points in a scatter plot correspond to nodes in a tail tree plot. Figure 6.6(a) shows points inside a unit ball, Figure 6.6(b) shows images of these points under the mapping $(x_1, x_2) \mapsto (x_1, (x_1^2 + x_2^2)^{1/2})$. The points on the x-axis (the squares in the first window) are mapped on a wedge in the x-window of the tail tree plot, and the points parallel to the x-axis are mapped to lifted smoothed wedges. Figure 6.6(c) shows images under the mapping $(x_1, x_2) \mapsto (x_2, (x_1^2 + x_2^2)^{1/2})$.

   Figure 6.7 shows that the points on spheres are mapped onto horizontal lines in a tail tree plot.

   The bullet corresponding to observation $x \in \mathbf{R}^d$ is drawn in the ith window of a tail tree point at the location $(x_i, \|x - \mu\|)$, where μ is the center point. We have $|x_i - \mu_i| \leq \|x - \mu\|$, where the equality is achieved when $x = (0, \ldots, 0, x_i, 0, \ldots, 0)$. Thus the bullets in the ith window always belong to the set $\{(x_i, \lambda) : |x_i - \mu_i| \leq \lambda, \lambda \geq 0\} \subset \mathbf{R} \times [0, \infty)$. Sometimes it is useful to plot the wedge $\{(x_i, \lambda) : |x_i - \mu_i| = \lambda, \lambda \geq 0\}$ as a dashed line in a tail tree plot; see Figure 6.10 below.

4. *(Overplotting.)* Any pixel based visualization (a scatter plot, parallel level plot, parallel coordinate plot, graphical matrix, tail tree plot) suffers from overplotting when the size of the data is large. Tail tree plots suffer from overplotting in the center of the data, but in the tails of the data single observations are distinguishable. Usually it is these "outliers" that are more interesting and in need to be identified, whereas central observations are not individually interesting.

   One may reduce overplotting in a tail tree plot by plotting only the bullets corresponding to the observations and leaving out the lines connecting the observations. Often one can visualize the tree structure sufficiently well only with the coloring of the observations. Figure 6.8 illustrates this fact. Panel (a) shows the first coordinate of a tail tree plot and panel (b) shows the same coordinate without the lines. The lines are hiding the fact that the triangle is not dense at the center. (The data come from Frank's copula and are the same as in Figure 6.17.)

   An additional way to reduce the overplotting is by subsetting, as was discussed in Section 1.1.6.

5. *(Comparison with parallel level plots.)* Parallel level plots are defined in Definition 1.2. Figure 6.9(a)–(b) shows a parallel level plot of the data in Figure 1.19 when we choose the level as in (1.2). (Figure 6.9(a)–(b) is the same as Figure 1.20(b)–(c)). Figure 6.9(a)–(b) indicates that there are at least two tails in the data, but the fact that there are four tails is only revealed when we draw the tail tree plot, as is shown in panels (c) and (d). The four tails of the data are shown as four branches of the tail tree, colored with black, red, blue, and green. The tail tree plot adds a sufficient structure to the visualization, making it possible for humans to detect the four tails in the data.

6.2.2 Examples of Tail Tree Plots

Spherically Symmetric Data We show tail tree plots of spherically symmetric data. For speherically symmetric data the shape of the point cloud is always the same, but the spread and the location of the data may vary. Figure 6.10(a) through (c) shows three scatter plots of spherically symmetric data. We generated a sample of size 1000 from the uniform distribution on the 2D ball, from the 2D standard Gaussian distribution, and from the 2D Student distribution with degrees of freedom 3. Figure 6.10(d) through (f) shows the first coordinates of tail trees of the data. We have chosen the resolution thresholds so that the data sets are connected, and this lead to thresholds 0.6, 3, and 18.

A ball looks like a triangle in a tail tree plot. The bullets are always on the upper side of the wedge, which is shown in Figure 6.10 as dashed lines. For every data set a tail tree plot has the triangle shape at the bottom. The triangle is not filled at every sector if the data is not spherical. For spherical data the triangle is not filled at the upper levels, except in the uniformly distributed case. Tail tree plots visualize the

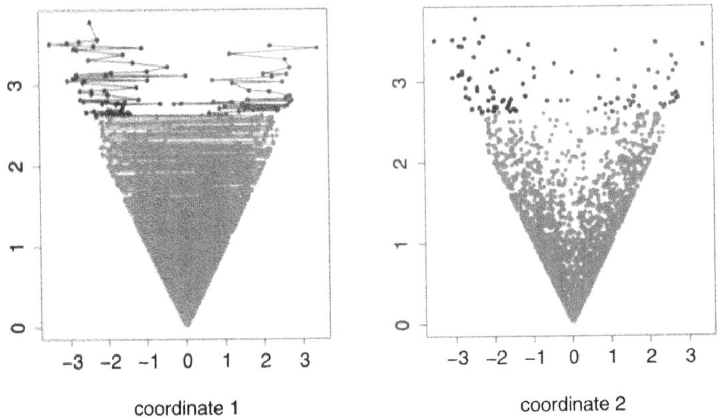

Figure 6.8 *(Overplotting.)* (a) The first coordinate of a tail tree plot where the lines are causing overplotting. (b) The first coordinate without the lines.

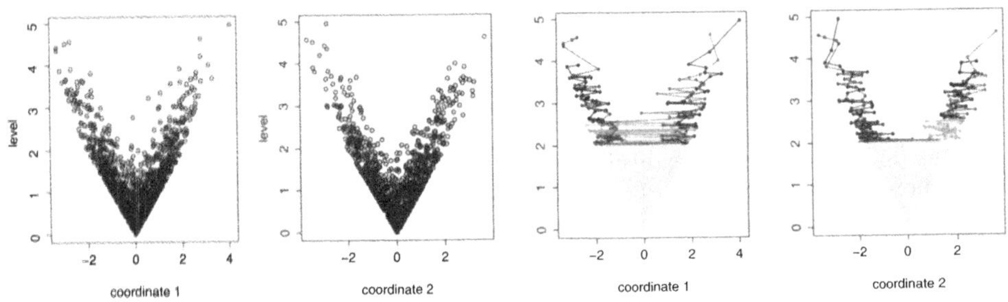

Figure 6.9 *(Parallel level plot versus tail tree plot.)* (a–b) A parallel level plot of the data in Figure 1.19 with the distance from the arithmetic mean as the level. (c–d) A tail tree plot of the data. The resolution threshold is $\rho = 0.65$.

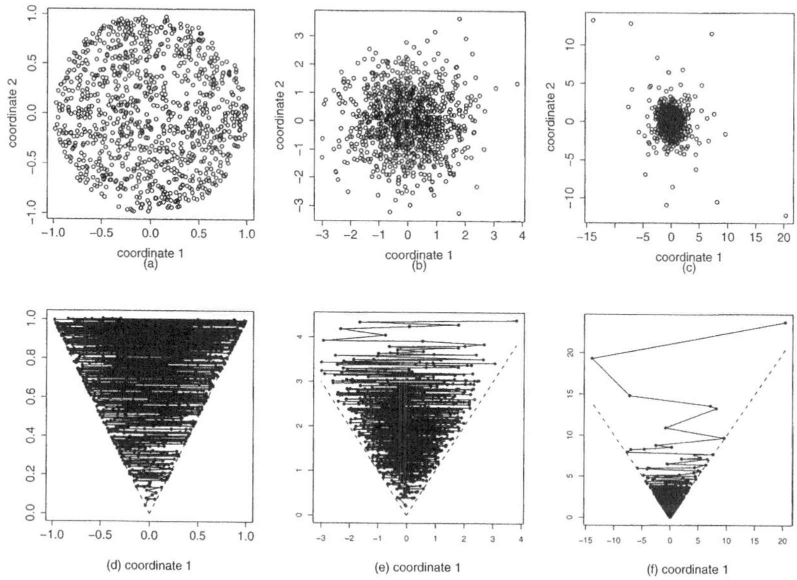

Figure 6.10 *(Spherically symmetric data.)* The upper row shows three scatter plots of 1000 observations. The lower row shows the x-coordinates of tail tree plots. (a) and (d) Uniformly distributed data on the unit ball, (b) and (e) standard Gaussian data, (c) and (f) Student distribution with degrees of freedom 3.

spread by showing how the observations become more sparse in the upper levels of the tail tree plot.

Ellipsoids Ellipsoids have two tails and thus tail trees of ellipsoidal point clouds have typically two branches. Figure 6.11 show a series of rotated 2D ellipses and the corresponding tail tree plots. We have rotated a sample of size $n = 2000$, which was generated from the 2D Gaussian distribution whose marginal standard deviations are 1 and 2, and the correlation is 0.

1. In the upper row of Figure 6.11 the tail tree plot shows that the observations in the blue branch have large values in the first coordinate and small values in the second coordinate, which means that this branch lies in the southeast direction.

2. In the middle row of Figure 6.11 the first window of the tail tree plot shows that the two branches have the same location with respect to the first coordinate (we distinguish the branches only through the coloring of the nodes). The second window of the tail tree plot shows that the locations of the branches differ in the second coordinate, the other lying in the south and the other in the north.

3. In the lowest row of Figure 6.11 the tail tree plot shows that the observations in the blue branch have small values in both coordinates, which means that this branch lies in the southwest direction.

Copulas The definitions of the copulas below can be found in Section 10.1.4.

Elliptical Copulas We show examples of tail tree plots of data generated from elliptical copulas

1. Figure 6.12 shows a sample of size $n = 1000$ from the distribution whose marginals are two independent Student distributions with degrees of freedom 4; that is, the distribution has the Gaussian copula with shape parameter (correlation coefficient) $r = 0$, and Student marginals. We show a scatter plot and a tail tree plot. The data has four tails along each coordinate half-axis. The T1-tail and the T4-tail have the same x-coordinate, and the T2-tail and the T3-tail have the same y-coordinate. The tails are distinguished with the labeling in the tail tree plot.

2. Figure 6.13 shows a sample of size $n = 1000$ from the distribution that has the Gaussian copula with shape parameter $r = 0.5$ and Student marginals with degrees of freedom 3. The data has four tails that are not exactly along the coordinate axis, due to the positive shape parameter of the Gaussian copula.

3. Figure 6.14 shows a sample of size $n = 1000$ from the distribution that has the Student copula with shape parameter $r = 0.6$ and standard Gaussian marginals. The data has four sharp tails, which lie along the diagonals between the coordinate axis. The tails lie close together in the tail tree plot, since the

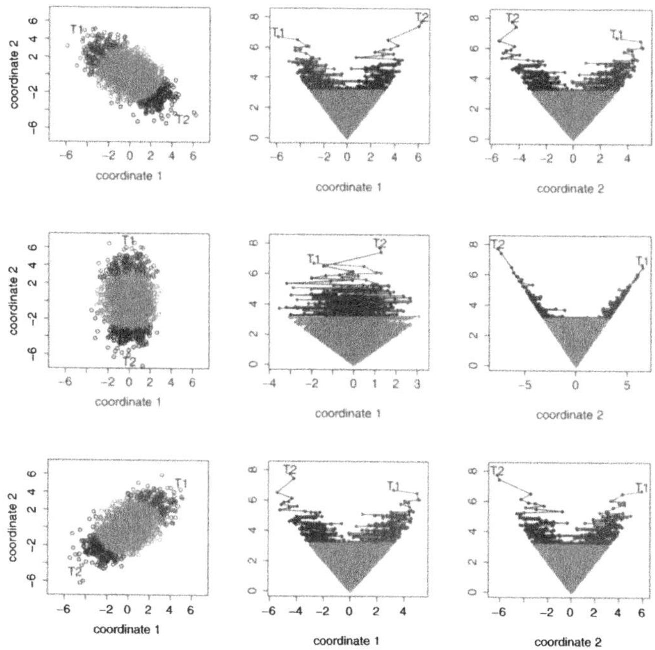

Figure 6.11 *(Tail tree plots of 2D ellipsoidal point clouds.)* In the first row the ellipsoid has the orientation from the southeast to the northwest. In the second row the ellipsoid has the orientation from the south to the north. In the third row the ellipsoid has the orientation from the southwest to the northeast.

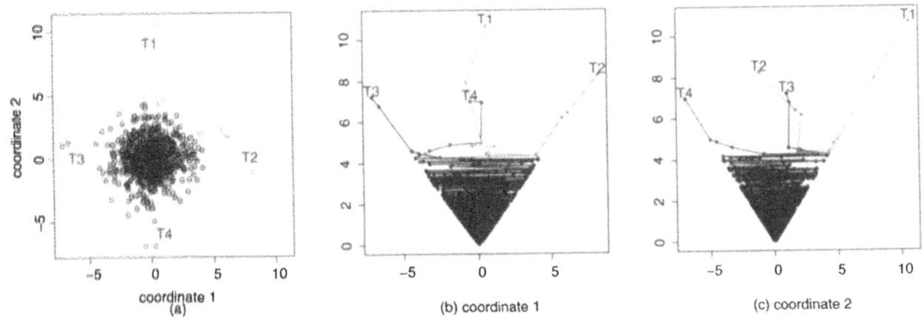

Figure 6.12 *(Gaussian copula with shape parameter* $r = 0$*, Student marginals with degrees of freedom* 4.*)* (a) A scatter plot with $n = 1000$ observations and (b–c) a tail tree plot with the resolution threshold $\rho = 1.5$.

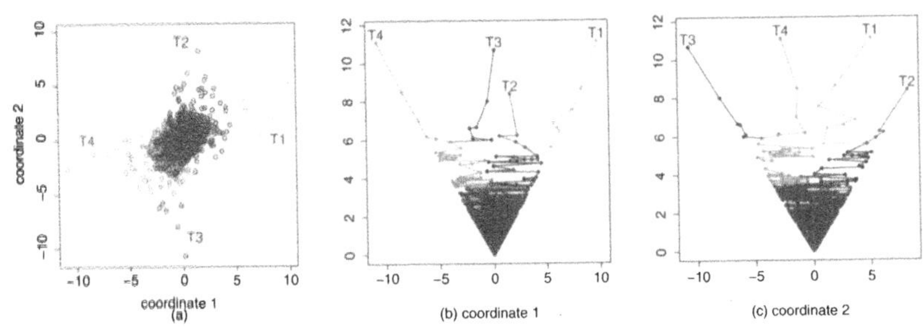

Figure 6.13 *(Gaussian copula with shape parameter* $r = 0.5$*, Student marginals with degrees of freedom* 3.*)* (a) A scatter plot with $n = 1000$ observations and (b–c) a tail tree plot with the resolution threshold $\rho = 1.6$.

T1-tail and the T2-tail have the same y-coordinate, and so have the T3-tail and the T4-tail, whereas the T1-tail and the T3-tail have the same x-coordinate, the same as the T2-tail and the T4-tail. The labeling distinguishes the separate tails in the tail tree plot.

Archimedean Copulas We show three examples of tail tree plots of data generated from Archimedean copulas.

1. Figure 6.15 shows 1000 observations from the distribution with the Clayton copula with parameter 4 and the standard Gaussian marginals. The dependence is greater in the T1-tail, and the tail tree plot shows this tail as a narrower tail than the T2-tail.

2. Figure 6.16 shows 1000 observations from the distribution with the Gumbel copula with parameter 3 and the standard Gaussian marginals. As for the Clayton copula the dependence is greater in the T1-tail. However, the T2-tail

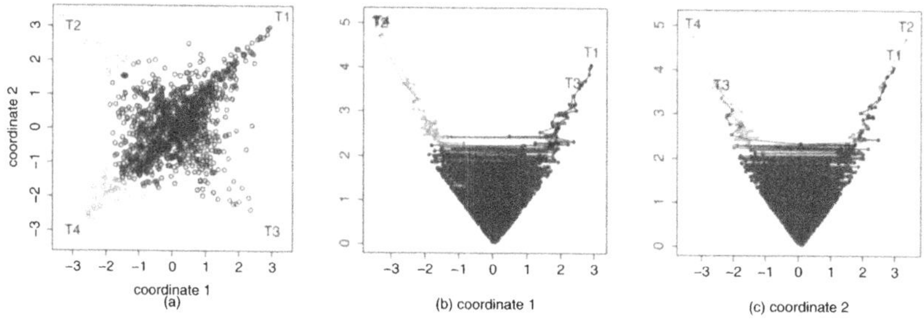

Figure 6.14 *(Student copula with shape parameter $r = 0.6$, standard Gaussian marginals.)* (a) A scatter plot with $n = 1000$ observations and (b–c) a tail tree plot with the resolution threshold $\rho = 1$.

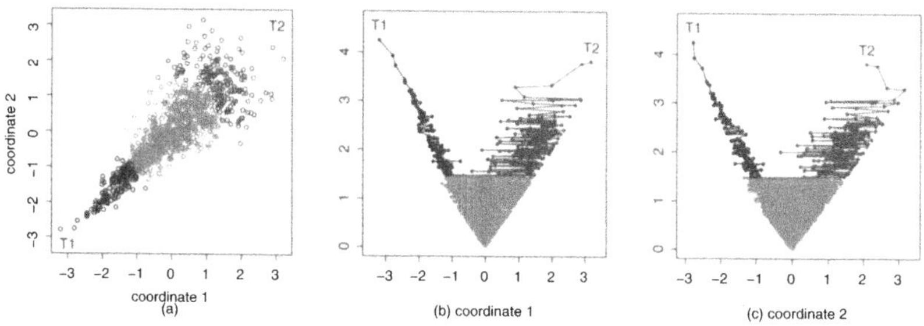

Figure 6.15 *(Clayton copula with parameter 4, standard Gaussian marginals.)* (a) A scatter plot with $n = 1000$ observations and (b–c) a tail tree plot with the resolution threshold $\rho = 1$.

is not spreading as wide at the extreme tail as in the case of the Clayton copula. In the tail tree plot there can be seen how the T2-tail has an arrow-like shape, whereas the T2-tail in the tail tree plot of the Clayton copula has almost uniform wideness.

3. Figure 6.17 shows 2000 observations from the distribution with the Frank copula with parameter 5, and the standard Gaussian marginals. The two tails of the Frank copula are wider than in an ellipse, as can be seen by comparing Figure 6.17 with the third row in Figure 6.11.

6.3 TAIL FREQUENCY PLOT

A tail frequency plot visualizes the heaviness of the tails of the underlying distribution. In the multivariate case the tails of the distribution may have anisotropic heaviness: the tails may decrease at different rates in different directions. For elliptical distributions,

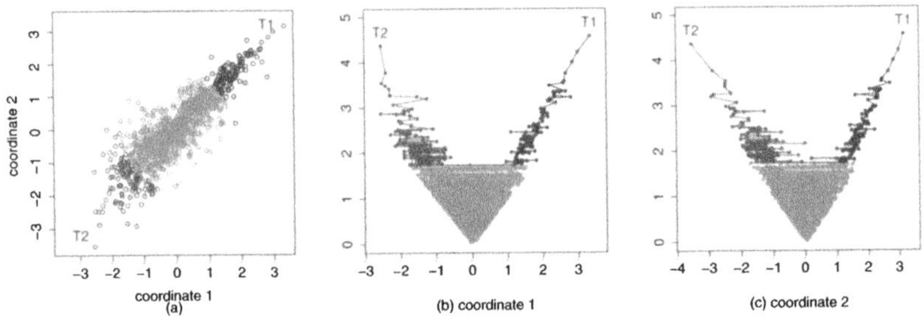

Figure 6.16 *(Gumbel copula with parameter* 3*, standard Gaussian marginals.)* (a) A scatter plot with $n = 1000$ observations and (b–c) a tail tree plot with the resolution threshold $\rho = 1$.

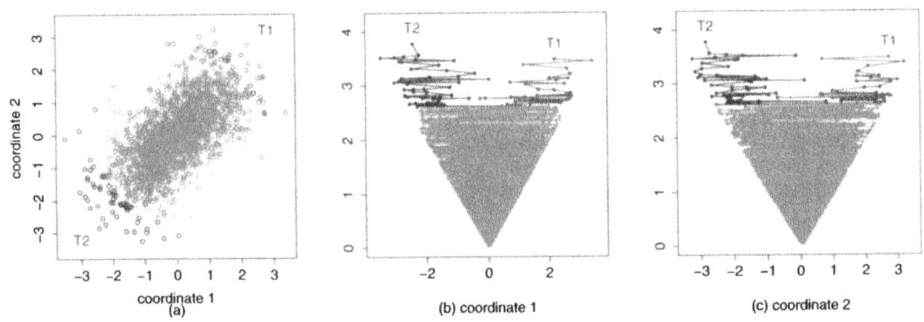

Figure 6.17 *(Frank copula with parameter* 5*, standard Gaussian marginals.)* (a) A scatter plot with $n = 2000$ observations and (b–c) a tail tree plot with the resolution threshold $\rho = 1.1$.

as defined in Section 10.1.3, the density has isotropic tails, determined by the 1D generator function, but in the general case, the tails are anisotropic.

The nodes of a tail tree are associated with subsets of the data. A tail frequency plot visualizes a tail tree so that each node of the tree is drawn as a line whose length is proportional to the number of observations in the set associated with the node. We identify the lines as level sets of a 1D function, and a tail frequency plot is a special case of a function tree plot, as defined in Definition 3.5. A tail frequency plot is a plot of a 1D piecewise constant function, which is defined by associating each node of the tail tree to separated component of a level set: (1) the length of the separated component of a level set is equal to the number of observations in the node, (2) the height of the separated component of a level set is equal to the distance of the closest observation from the center point, among all observations associated with the node, and (3) the separated components of level sets are nested according to the parent–child relations.

Definition 6.6 (Tail frequency function, Tail frequency plot.) *A* tail frequency function $v(A;T) : \mathbf{R} \to \mathbf{R}$ *of data* $A = \{x_1, \ldots, x_n\} \subset \mathbf{R}^d$, *associated with a tail tree* T *of* A, *is a 1D function generated by a function tree as in Definition 3.5. We get the function tree, in the sense of Definition 3.3, from the tail tree* T *by defining the height and the length of the nodes in the following way.*

1. *The height associated to a node of the level set tree is equal to the level associated with the node (the Euclidean distance of the closest observation from the center point, among all observations associated with the node).*

2. *The length associated to node* m *of the tail tree is equal to cardinality* $\#(A_m)$ *where* A_m *is the set associated to node* m.

A tail frequency plot *is a plot of* $v(A;T)$.

Illustrations We illustrate tail frequency plots.

1. Figure 6.18(a) shows a tail frequency plot of the data of size $n = 2000$ in Figure 6.2(a). The tail tree is shown in Figure 6.2(b) and the corresponding tail tree plot is shown in Figure 6.4. The tail frequency function has three modes, and the size of the modes visualizes the number of observations in each tail. Panel (b) zooms into the function showing details of the modes T2 and T3.

2. Figure 6.19 shows a tail frequency plot of the tail tree in Figure 6.3(d) and in Figure 6.5. There are 11 data points. The tail frequency plot shows the four tails of the data.

Remarks We make remarks concerning tail frequency plots.

1. *(Colors.)* By labeling the modes of the function in a tail frequency plot we can indicate the correspondence between the modes of the function and the

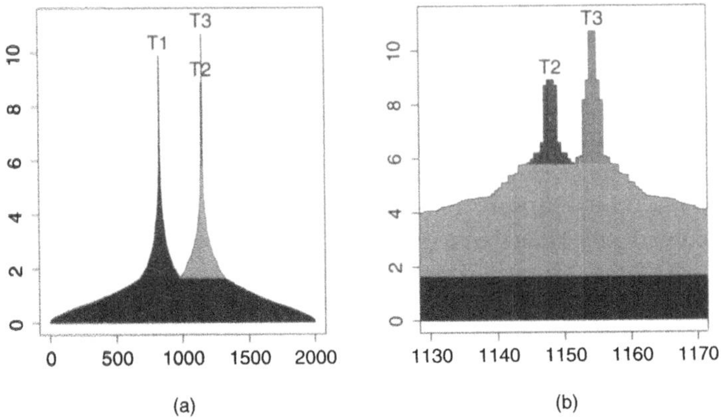

Figure 6.18 *(Tail frequency plot.)* (a) A tail frequency plot of the tail tree in Figure 6.2(b). The corresponding tail tree plot is shown in Figure 6.4. (b) Zooming into details of the plot.

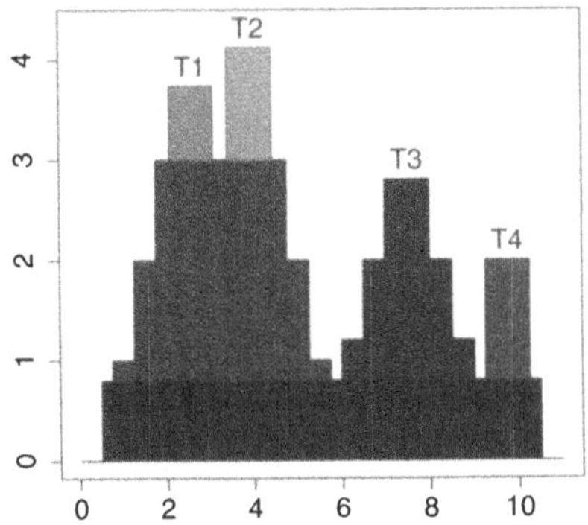

Figure 6.19 *(Tail frequency plot.)* Shown is a tail frequency plot of the tail tree in Figure 6.3.

tails of the data as shown in a tail tree plot. We can use colors to enhance the visualization of the correspondence. In a tail tree plot every node (observation) is colored. A tail frequency plot is a plot of a piecewise constant function, which has a finite number of distinct level sets. These distinct level sets correspond to the nodes of the tail tree, and we color these level sets with the same color with which the corresponding observation was colored. We color the whole area under the graph of the function in such a way that a color is changed when two separated level sets are merging.

2. *(Tail probability function.)* When a shape tree corresponds to a low level set (or the support) of the underlying density, a tail frequency function is an empirical version of the tail probability function, as defined in Definition 5.7. A shape tree makes a recursive tree approximation of a set and we decorate the nodes of the tree with the probability content of the set associated with the node to get a tail probability function.

3. *(Scale curves.)* In analyzing multivariate data it is useful to study 1D spherical distribution function $G_n : [0, \infty) \to [0, 1]$,

$$G_n(r) = n^{-1}\#\{x_i : x_i \in B_r(\mu), i = 1, \ldots, n\},$$

where $B_r(\mu) = \{z \in \mathbf{R}^d : \|z - \mu\| \leq r\}$, and $\mu \in \mathbf{R}^d$ is a given center point. Function G_n gives the empirical probabilities of an increasing sequence of balls, and thus G_n visualizes the heaviness of the tail of the underlying density. Visualization of the spread with 1D functions is discussed more in detail in Section 2.2. For example, Liu et al. (1999) replace the balls $B_r(\mu)$ with depth regions and call the related function a "scale curve" or a "scalar form of dispersion".

When a tail tree does not have branches, a tail frequency function of the tail tree visualizes exactly the same information as function G_n. However, when a tail tree has many leaf nodes, a tail frequency function visualizes more information: it visualizes the heaviness of the tails separately for each disconnected tail of the underlying density. Spherical distribution function may be used for spherically symmetric densities, but with the help of tail frequency function we may visualize anisotropic spread.

Examples We give examples of tail frequency plots.

1. Figure 6.20 shows tail frequency plots for the data sets coming from uniform, Gaussian, and Student distributions. The data are the same as in the upper row of Figure 6.10, and the first coordinates of the corresponding tail tree plots are shown in the lower row of Figure 6.10.

2. Figure 6.21, the upper row, shows tail frequency plots corresponding to the tail tree plots in Figures 6.12 through 6.14 (independent Student marginals, a Gaussian copula with Student marginals, and a Student copula with Gaussian

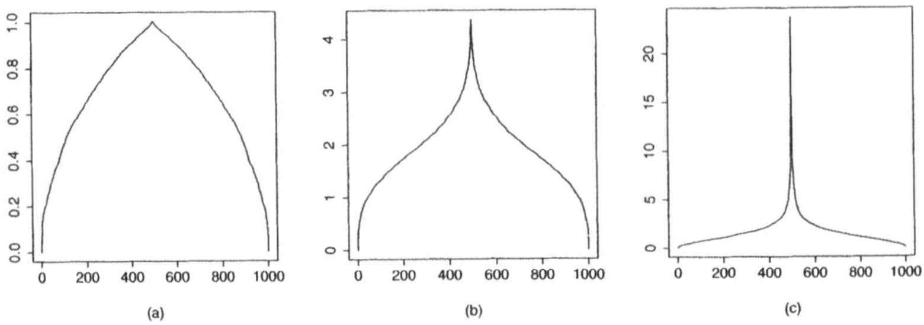

Figure 6.20 *(Tail frequency plots.)* Shown are tail frequency plots of spherically symmetric data; (a) uniform, (b) Gaussian, and (c) Student. The tail frequency plots correspond to the tail tree plots in Figure 6.10.

marginals). The modes have such small size that we need to zoom into the graph of the function to show the modes, and this is done in the lower row.

3. Figure 6.22 shows tail frequency plots corresponding to the tail tree plots in Figures 6.15 through 6.17 (Clayton, Gumbel, and Frank copula).

6.4 SEGMENTATION OF DATA

Tail trees may be used to define a partition of the data. The partition can be applied to enhance parallel coordinate plots, graphical matrices, and the grand tour.

The segmentation of a tree into branches is defined in Definition D.3.1. Since the nodes of a tail tree are associated with observations, a segmentation of the nodes of a tail tree induces a segmentation of the data. A segmentation of the data, corresponding to a tail tree, is such that the observations corresponding to the nodes that are in the same branch belong to the same segment: a child of node m belongs to the same segment as m if it does not have siblings (it is the only child); otherwise, the children of m are the starting nodes of new branches, and thus each is a starting node of a new segment.

Illustrations We illustrate the segmentation of data with tail trees.

1. The segmentation induced by a tail tree is illustrated in Figure 6.2, where the five segments are colored in the tree structure and in the scatter plot.

2. The scatter plots in Figures 6.11 through 6.17 illustrate the segmentation induced by a tail tree: the segments are colored with separate colors in these scatter plots.

Remarks We give remarks concerning the segmentation of data with tail trees.

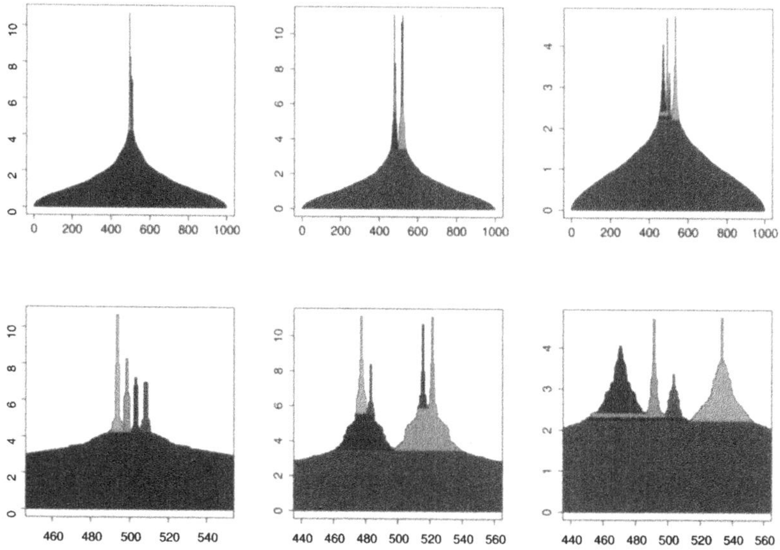

Figure 6.21 *(Independent Student marginals, a Gaussian copula with Student marginals, and a Student copula with Gaussian marginals.)* Shown are tail frequency plots corresponding to the tail tree plots in Figures 6.12 through 6.14. The lower row shows detailed views of the modes in the tail frequency plots of the upper row.

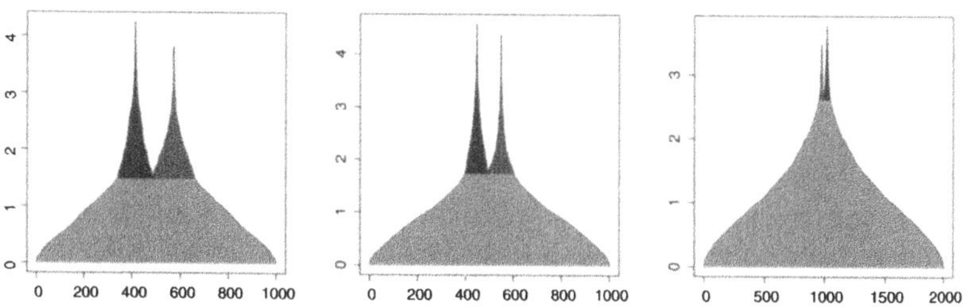

Figure 6.22 *(Clayton, Gumbel, and Frank copula.)* Shown are tail frequency plots corresponding to the tail tree plots in Figures 6.15 through 6.17.

1. *(Segmentation of a level set.)* A population concept may be defined as corresponding to a segmentation of a data set. We take this population concept to be a segmentation of the support or a low level set of the underlying density. Analogously to the definition of the partition of the nodes of a tail tree, we can define a partition of the nodes of a shape tree. A shape tree is defined in Chapter 5. We construct a shape tree from the support or from a low level set of the underlying density and with the help of a partition of the nodes of the shape tree we can define a partition of the support or the level set of the density.

2. *(Projections and slices.)* The segmentation induced by a tail tree may be used to enhance projections of multivariate point clouds: the observations are colored so that the observations in the same segment have the same color. This coloring technique has been used in the scatter plots of Figures 6.11 through 6.17. This kind of coloring can be extremely helpful when one takes the grand tour along the projections of the data The grand tour is discussed in Section 1.1.3.

3. *(Graphical matrices and parallel coordinate plots.)* The enhancement of graphical matrices and parallel coordinate plots with a tail clustering is illustrated in Section 8.4.

6.5 BIBLIOGRAPHIC NOTES

Tail trees were introduced in Klemelä (2007*b*). The approach of moving from connected sets to ρ-connected sets is analogous to the approach of computational topology, see Robins, Abernethy, Rooney and Bradley (2004).

6.5.1 Other Tree Structures

A tail tree is a tree structure among the observations. Other tree structures among the observations include a minimal spanning tree and a shortest path tree. A spanning tree of a graph G is a tree that contains all the nodes (vertices) of G. Spanning trees include minimal spanning trees and shortest path trees. The minimal spanning tree problem is to select a set of edges so that there is a path between each node and the sum of the edge lengths is to be minimized. The shortest path tree problem is to find the set of edges connecting all nodes such that the sum of the edge lengths from the root to each node is minimized. Friedman and Rafsky (1981) propose to visualize a minimal spanning tree by plotting the data in 2D so that the $n - 1$ edge lengths are preserved, and showing in the plot the lines connecting the observations. A tail tree plot (a spatial tree plot) is a visualization tailored for tail trees. We may visualize also a minimal spanning tree and a shortest path tree with spatial tree plots, as defined in Definition 3.2. Michailidis and de Leeuw (2001) visualize multivariate categorical data by representing the data as a graph.

6.5.2 Database Exploration

Database exploration in Keim and Kriegel (1994) has some similarities with tail tree plots. Their center point is determined by the user (as a query to the data base), whereas we have calculated the center point with an algorithm (as the arithmetic mean). They sort the data points according to a distance to the center point. The center point is drawn at the center of the display, and the other data points are arranged on a rectangular spiral originating from the center point. The distances to the center point are visualized with colors. In addition an own display is dedicated for each coordinate. The position of the data points in the coordinate displays is the same as in the main display, but the coloring is determined by the univariate distance in the given coordinate from the observation to the center point. (In information visualization a coordinate is called also a "dimension" or a "predicate".)

A weakness of database exploration might come from the fact that colors are not well suited for expressing quantitative information like distances. In the tail tree plots we have expressed distances by using scatter plots where the y-coordinate gives the overall distance and the x-coordinate gives the univariate distance in the given coordinate, and colors are used to express nominal information (the labeling of the tails).

CHAPTER 7

SCALES OF DENSITY ESTIMATES

Nonparametric density estimators depend often on a real-valued smoothing parameter or even on a vector or a matrix of smoothing parameters. We call a *scale of density estimates* a sequence of density estimates corresponding to a sequence of smoothing parameters. A smoothing parameter has a substantial influence on the quality of the estimates, and there exists a large literature on automatic selection of the smoothing parameters.

Visualization of a scale of density estimates is important in supporting the process of smoothing parameter selection. For example, it is important to know how sensitive the estimates are on the choice of the smoothing parameters. On the other hand, when we visualize the complete scale of density estimates, it might turn out that we do not need to choose a single estimate from the sequence, and the scale alone provides the necessary support for the inference. Indeed in one- and two-dimensional cases the "art of smoothing" has typically consisted of the inspection of the change in the estimates as the smoothing parameter changes. We need visualization tools for the multivariate case that would make it possible conveniently to scan through a scale of smoothing parameters. Such tools may be used in the spirit of *scale space analysis*, as discussed by Chaudhuri and Marron (2000).

A *mode tree* is a useful visualization tool for scanning through estimates. The mode tree for one- and two-dimensional cases was introduced in Minnotte and Scott (1993) and used to visualize the number and the locations of the modes as the smoothing parameter changed.

Section 7.1 defines a *multiframe mode graph* to be used as a road map directing the scanning through a scale of estimates. A multiframe mode graph makes a mode graph separately for each coordinate and uses colors to identify the modes across different windows. We want to apply multiframe mode graphs not only for kernel estimators but for any estimator whose smoothness is controlled with a real-valued smoothing parameter. We use the term "graph" and not the term "tree" because we consider such scales of estimates where the number of modes is not monotonic as the function of the smoothing parameter, unlike in the case of univariate kernel estimates with the standard Gaussian kernel.

A mode graph does not visualize the relative importance of the modes. For this purpose one may use *a branching map*, defined in Section 7.2, A branching map is a perspective plot of a 2D function whose surface is colored. With the colors and with the values of the function we visualize the levels where the level sets of the estimates are decomposing to separate regions and also the excess masses associated with those separate regions, simultaneously for a scale of estimates. The definition of the branching map is based on the concept of a level set tree of a function, defined in Chapter 4.

7.1 MULTIFRAME MODE GRAPH

In a one-dimensional mode tree the locations of the modes of kernel estimates are plotted when the smoothing parameter ranges over an interval. A two-dimensional mode tree was defined to be a three-dimensional plot of the mode locations and bandwidth. A multivariate mode tree may also be defined as a tree that shows how the number of modes of a kernel estimate is increasing as a function of the smoothing parameter, without any spatial information. This kind of multivariate mode tree was considered in Scott and Szewczyk (2000), where it was applied to clustering.

We will define *multivariate mode graphs* as plots where we show a one-dimensional mode graph separately for each coordinate. A tree in the d-dimensional space is a 1D structure that can be visualized with d projections. For the projections to be useful, one needs to identify the same node in the different windows. Only the leaf nodes need to be labeled so as to identify uniquely all the nodes in the different windows. We can, however, considerably ease the identification with a coloring scheme. While it would not be feasible to choose a different color for each node, we get a useful coloring by choosing a different color for each branch of the tree.

A multiframe mode graph is associated to a collection of density estimates

$$\hat{f}_h : \mathbf{R}^d \to \mathbf{R}, \qquad h \in H, \tag{7.1}$$

where $H \subset \mathbf{R}$ is a finite set of smoothing parameters. Denote by

$$M_1^{(h)}, \ldots, M_{m_h}^{(h)} \in \mathbf{R}^d, \qquad h \in H, \tag{7.2}$$

the locations of the modes of the estimates, where m_h is the number of modes of the estimate with smoothing parameter h, $M_j^{(h)} = (M_{j,1}^{(h)}, \ldots, M_{j,d}^{(h)}) \in \mathbf{R}^d$ is the jth mode of the estimate with smoothing parameter h, $j = 1, \ldots, m_h$. We treat h as if it were the smoothing parameter of the kernel estimator: small values of h correspond to undersmoothed estimates and large values of h correspond to oversmoothed estimates.

Definition 7.1 (Multiframe mode graph.) A multiframe mode graph, *associated to a scale of density estimates (7.1) with mode locations (7.2), consists of d windows.*

1. *The x-axis of the ith window corresponds to the ith coordinates of the modes and the y-axis of the windows corresponds to the scale H. That is, the ith window, $i = 1, \ldots, d$, consists of the plot of points*

$$\left(M_{k,i}^{(h)}, h\right), \qquad h \in H, \ k = 1, \ldots, m_h,$$

2. *To identify the same mode between different windows, we use the same color to plot the same mode in different windows, but different colors to plot the different modes with the same h-value. That is, for each $h \in H$, $k = 1, \ldots, m_h$, points*

$$\left(M_{k,i}^{(h)}, h\right), \qquad i = 1, \ldots, d,$$

have the same color, and for each $h \in H$, $i = 1, \ldots, d$, points

$$\left(M_{k,i}^{(h)}, h\right), \qquad k = 1, \ldots, m_h,$$

have different colors.

Remarks We make the following observations concerning the definition of a mode graph:

1. *(Tree or graph structure.)* In Definition 7.1 we did not define a tree or a graph. By adding to the mode graph parent–child connections, we make the plot more readily interpretable. However, there does not seem to exist a distinguished choice for the parent–child relations, so we define these relations separately, instead of defining them in the proper definition of a multiframe mode graph. We give a rule for determining parent–child relations in Appendix C.

2. *(Coloring.)* Definition 7.1 gave a minimal condition for the coloring. We use the parent–child relations to enhance the coloring. The coloring of the nodes is determined so that we choose first distinct colors for the modes at the root level, that is, for the modes corresponding to the largest smoothing parameter. (Typically there is only one root.) For one of the children we choose the same color as that of the parent. For the other children, we choose new distinct colors. This coloring principle is called the root-first coloring in Section 3.1, page 86. The precise rule is given in Appendix C.

Example We give an example of a scale of estimates and a corresponding mode graph.

1. Figure 7.1(a) shows a contour plot of the density that we use for the illustration, and a sample of size 200 generated from this density. We apply Bartlett–Epanechnikov product kernel, defined by $(x_1, \ldots, x_d) \mapsto (3/4)^d \Pi_{i=1}^d \max\{0, 1-x_i^2\}$, We construct a scale of kernel estimates. Figure 7.1(b) through (h) show seven kernel estimates corresponding to smoothing parameter values 2.30, 1.59, 1.20, 1.04, 0.94, 0.89, and 0.85.

2. Figure 7.2 shows a multiframe mode graph, associated to a scale of two-dimensional kernel estimates. The kernel estimates were constructed from a sample of size 200 from the three-modal density shown in Figure 7.1(a). We applied a grid of 100 smoothing parameters in interval $[0.85, 2.3]$. The grid was not equally spaced, but we used a logarithmic spacing and the h-axis has a logarithmic scale.

   Figure 7.2 shows that when we use smoothing parameter $h = 2.3$, then the estimate has one mode. The node corresponding to this mode is labeled with M1. The multiframe mode graph shows that the location of this mode is approximately $(1.8, 2.8)$. When the smoothing parameter is decreased, branches M2 and M3 appear. The branches M1, M2, and M3 correspond to the true modes of the density. The mode M2 is at the location close to $(3.5, 0)$ and the mode M3 is at the location close to $(0, 0)$. The fourth branch appears at a tail region, when $h \approx 1.5$. Finally, when the smoothing parameter is $h = 0.85$, there are seven branches and thus seven modes.

7.2 BRANCHING MAP

A branching map is associated to a scale of level set trees of density estimates.

A function $f : \mathbf{R}^d \to \mathbf{R}$ may be regarded in two ways: as a mapping $x \mapsto f(x)$ that assigns value $f(x)$ to a vector $x \in \mathbf{R}^d$, or as a mapping $\lambda \mapsto \{x \in \mathbf{R}^d : f(x) \geq \lambda\}$ which assigns a level set to a level. A mode graph regards a function in the first way, but a map of branches regards a function in the second way. Since a level is real number, the change of the viewpoint makes it possible to visualize much information about a scale of functions in a concentrated way.

A branching map of level set trees of estimates in a scale of estimates is a perspective plot of a 2D function whose arguments are the level and the smoothing parameter, and the values are the excess masses of the nodes of the level set trees of estimates. As we move to the higher levels, a map of branches shows for each estimate the levels where the level sets of the estimate are decomposing to separated components, and it visualizes also the probability masses associated with these separated components. The probability masses (excess masses) associated to the separated regions of level sets measure the relative importance of modes and bundles of modes.

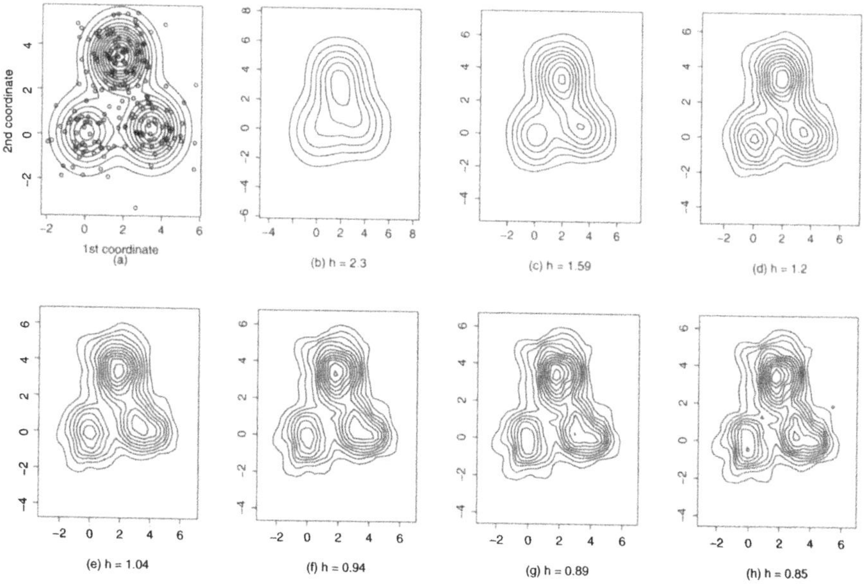

Figure 7.1 *(Scale of kernel estimates.)* (a) The sample together with a contour plot of the density from which we generated the sample. (b–h) Contour plots of seven kernel estimates with decreasing smoothing parameters.

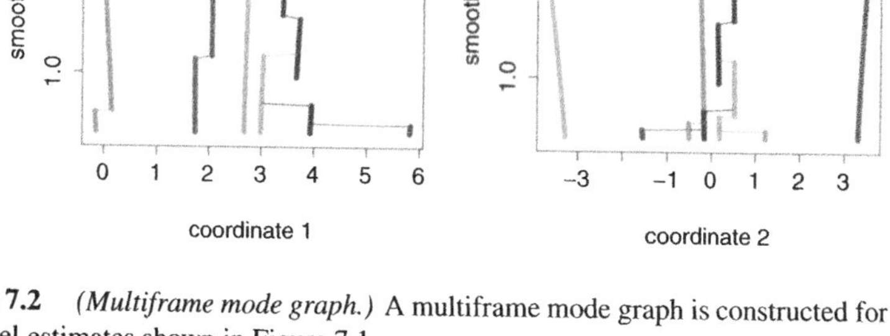

Figure 7.2 *(Multiframe mode graph.)* A multiframe mode graph is constructed for the scale of kernel estimates shown in Figure 7.1.

We will use terms "branching map" and "map of branches" interchangeably. To define a branching map we need four preparatory concepts: (1) a level set tree, (2) excess mass, (3) branching node, and (4) branching profile.

7.2.1 Level Set Tree

Level set trees were defined in Chapter 4. Figure 7.3(a) shows a level set tree of the estimate in Figure 7.1(d). The nodes corresponding to the three modes of the estimate are labeled M1 to M3.

7.2.2 Excess Mass

We want to condense the information contained in a level set tree in order to make it possible to represent information about a scale of multivariate estimates with a single 2D function. The first concept we use for the condensation of the information is the *excess mass associated with a node of a level set tree*, defined in Definition 4.8. The excess mass associated with a given node of a level set tree is the volume of the area that the density delineates over the level of the node, on the separated region of the level set associated with the node. In Figure 7.3(a) we have annotated five of the nodes of the level set tree with their excess masses.

7.2.3 Branching Node

The second concept we need for the condensation of the information contained in a level set tree is the concept of a *branching node of a level set tree*.

Definition 7.2 (Branching node.) *The* branching nodes of a level set tree of a function $f : \mathbf{R}^d \to \mathbf{R}$ *are the nodes that have more than one child.*

Figure 7.3(a) shows the two branching nodes as (red) rectangles. The children of the branching nodes and the root node are shown as (blue) triangles. The tree that consists only of the root nodes of a level set tree, from the branching nodes and from the children of the branching nodes, is called a *reduced level set tree*. A reduced level set tree is closely connected to a cluster tree as defined in Stuetzle (2003) and to a contour tree; see Section 4.6.2.

7.2.4 Branching Profile

Branching profile of a level set tree is a 1D plot that visualizes the number and the levels of branching of a level set tree of the function. It visualizes also the excess masses associated with the branching nodes and with the children of the branching nodes of a level set tree. The values of the plotted function are equal to the excess masses of the branching nodes. We divide the graph of the function between two levels of branching to two bands, which are colored in red and blue. The lengths of the two bands are proportional to the excess masses of the children of the branching

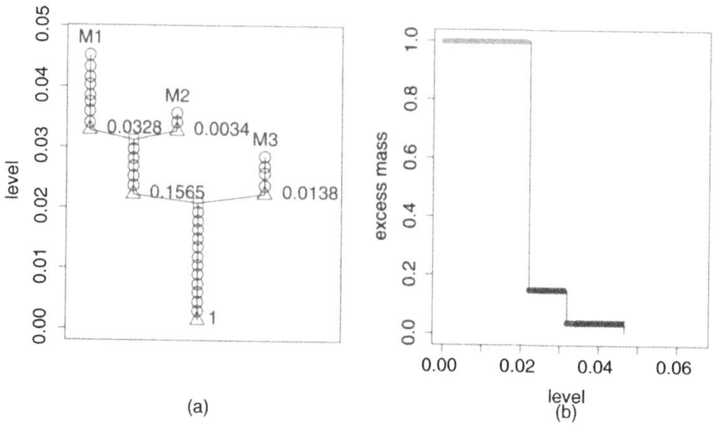

Figure 7.3 (a) A level set tree of the density estimate shown in Figure 7.1(d). (b) The branching profile corresponding to the level set tree in (a).

nodes, that is, to the excess masses of the separated regions that are separating at this level.

Definition 7.3 (Branching profile.) *A* branching profile of a level set tree *of density* $f : \mathbf{R}^d \to \mathbf{R}$ *is a plot of the colored excess mass function, defined in the following:*

Let $b_1, \ldots, b_M$ *be the branching nodes of the level set tree of* f. *Let* λ_m *be the level of* b_m, $m = 1, \ldots, M$. *We assume that the branching nodes are ordered so that* $0 < \lambda_1 < \cdots < \lambda_M < \|f\|_\infty$, *where* $\|f\|_\infty = \sup_{x \in \mathbf{R}^d} f(x)$. *Denote with* $\text{excmas}(b)$ *the excess mass of node* b.

- *Define the* excess mass function $e : [0, \infty) \to [0, 1]$ *to be the function that gives for every level of branching the excess mass of the branching node. The function remains constant until the next level of branching. The function is equal to 1 at the origin. Thus we define the excess mass function by*

$$e(t) = \sum_{m=0}^{M} \text{excmas}(b_m) I_{[\lambda_m, \lambda_{m+1})}(t),$$

where we denote $\lambda_0 = 0$, $\lambda_{M+1} = \|f\|_\infty$, *and* $\text{excmas}(b_0) = 1$.

- *We color each constant segment of the graph of the excess mass function so that the colors give information on the excess masses of the children of the branching nodes. We need to take into account that a level set tree can have several root nodes. Let* $r_1, \ldots, r_L$ *be the root nodes of the level set tree of* f. *Define two vectors of colors:* $\text{rootpaletti} = (seagreen, violet, \ldots)$ *and* $\text{paletti} = (red, blue, green, \ldots)$.

- *Divide interval* $[0, \lambda_1)$ *to* L *subintervals so that* l*th subinterval* I_{0l} *has length* $\text{excmas}(r_l)/\lambda_1$, $l = 1, \dots, L$. *Choose color* $\text{col}(I_{0l}) = \text{rootpaletti}(l)$ *for each interval, assuming that the intervals are ordered so that* $\text{length}(I_{01}) < \cdots < \text{length}(I_{0L})$.
- *Let node* b_m *have* N *children* $c_1, \dots, c_N$. *Divide interval* $[\lambda_m, \lambda_{m+1})$, $m = 1, \dots, M$, *to* N *subintervals so that* i*-th subinterval* I_{mi} *has length*

$$\frac{(\lambda_{m+1} - \lambda_m) \cdot \text{excmas}(c_i)}{\sum_{j=1}^{N} \text{excmas}(c_j)},$$

  $m = 1, \dots, M$, $i = 1, \dots, N$. *Choose colors* $\text{col}(I_{mi}) = \text{paletti}(i)$ *for each interval, assuming that the intervals are ordered so that* $\text{length}(I_{m1}) < \cdots < \text{length}(I_{mN})$.

- *The branching profile is a plot of the graph* $(\lambda, e(\lambda))$, $\lambda \in [0, \|f\|_\infty]$, *where* $e(\lambda)$ *has color* $\text{col}(I)$ *when* $\lambda \in I$.

Figure 7.3(b) shows the branching profile corresponding to the level set tree in Figure 7.3(a). A jump in the branching profile (red color) indicates always a new branch in the level set tree. There are two branching nodes: the first at the level 0.02 and the second at the level 0.03. The branching profile makes jumps at these levels. The red band starting at level 0.02 shows the size of the excess mass 0.0134 of mode M3 in relation to the size 0.01468 of the mode bundle consisting of modes M1 and M2. The red band starting at level 0.03 shows the size of the excess mass 0.0074 of mode M2 in relation to the size 0.0469 of mode M1. Note that the absolute lengths of the color bands do not contain information, but the relative length of a red and blue band tells how the excess mass is distributed over the two branches.

Note a delicate issue in Definition 7.3. We should not define the excess mass function $e : [0, \infty) \to [0, 1]$ so that it gives for a level λ the probability mass of the level set with level λ: $e(\lambda) \neq P_f(\{x \in \mathbf{R}^d : f(x) \geq \lambda\})$, where P_f is the probability measure corresponding to density f. Indeed we have to take into account that the level set tree has several branches corresponding to the various modes and the total excess mass does not give information about the individual branches.

7.2.5 Branching Map

Now we are ready to define a branching map. By combining together the branching profiles of level set trees of estimates in a scale of estimates we get the branching map.

Definition 7.4 (Branching map.) *A branching map of a scale of level set trees, associated with the scale of estimates in (7.1), is the perspective plot of the 2D function* $B : [0, \infty) \times H \to [0, 1]$ *whose slices* $B(\cdot, h) : [0, \infty) \to [0, 1]$ *are the branching profiles of the estimates, defined in Definition 7.3*

Figure 7.4 shows a branching map, constructed from the same scale of kernel estimates as the mode graph of Figure 7.2. The map of branches has color seagreen

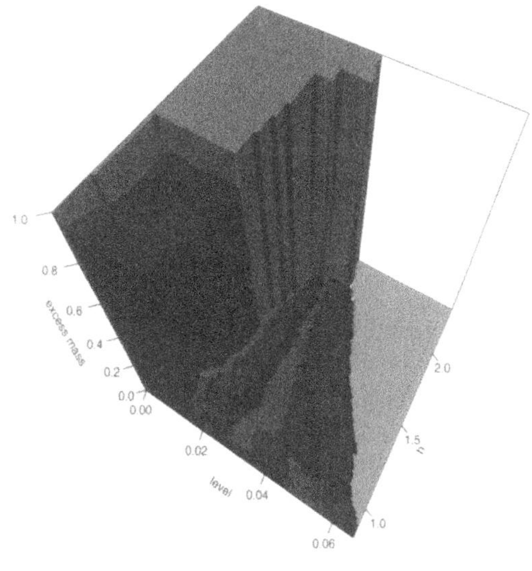

Figure 7.4 *(Map of branches.)* A map of branches of the scale of kernel estimates shown in Figure 7.1

when h is large, and this means that the estimates are unimodal for large h. The appearance of red bands signals the appearance of modes: a red band shows where the level set is decomposing to separate regions. One branching increases the number of modes by one. There appear two red bands almost simultaneously (when h is decreased), and this means that there appear three modes (first branching implies two modes, second branching implies three modes, etc.). When h is further decreased, a third red band appears at a low level, and this means that a fourth mode appears. The relative widths of red and blue bands tell how the excess mass is distributed between the two modes (between the "old" mode and the "new" mode). The height of the surface shows the total remaining excess mass at this level to be distributed between the modes.

One sees that the three modes, which appear when $h \approx 2$, emerge at relative high levels $0.02 - 0.03$, and the roots of the branches leading to these modes have nonnegligible excess masses. The fourth mode, which appears when $h \approx 1.5$, emerges at a low level and it has a small excess mass. The rest of the modes appear at moderate levels, and they have small excess masses.

7.3 BIBLIOGRAPHIC NOTES

7.3.1 Mode Trees

Multiframe mode graphs were introduced in Klemelä (2008*b*). Mode trees in the one- and two-dimensional cases were introduced by Minnotte and Scott (1993). Marchette and Wegman (1997) observe that a mode tree with kernel estimates can be hard to interpret, for example, if the data contains outliers, and they propose a filtered kernel estimator, where the kernel function is a mixture. Minnotte, Marchette and Wegman (1998) note that the basic mode tree can be strongly affected by small changes in the data, and they suggest that the basic mode tree be modified by generating several mode trees using bootstrap resamples or jittering (adding random noise to each data point) and then aggregating the resulting mode trees. Scott and Szewczyk (2000) consider using mode trees in multivariate clustering. Silverman (1981) showed that in the one-dimensional case the number of modes of kernel estimates with the Gaussian kernel is nondecreasing as the smoothing parameter decreases. This property leads to attractive mode trees.

7.3.2 Mode Testing

An approach to mode testing where one is testing zero-crossings of derivatives of the functions is presented in Chaudhuri and Marron (1999). The two-dimensional case is considered by Godtliebsen, Marron and Chaudhuri (2002). Various mode testing procedures are presented in Good and Gaskins (1980), Silverman (1981), Hartigan and Hartigan (1985), Silverman (1986), Donoho (1988), Hartigan and Mohanty (1992), Mammen, Marron and Fisher (1992), Fisher, Mammen and Marron (1994), and Davies and Kovac (2001).

Minnotte and Scott (1993) suggest that the probability mass of the mode be visualized by the widths of the mode traces. Mode testing associated with the mode tree was introduced in Minnotte (1997). Excess masses were applied in cluster analysis and mode testing also by Hartigan (1987), and Müller and Sawitzki (1991). The theoretical excess mass functional for distribution P_f with density f is defined by

$$\text{excmass}(A) = -P_f(A) + \lambda \cdot \text{volume}(A), \qquad A \subset \mathbf{R}^d,$$

where $\lambda > 0$. The level set $\Lambda(f, \lambda)$ with level λ minimizes the theoretical excess mass functional. Indeed we have that

$$\begin{aligned}
&\int_{\mathbf{R}^d} |f(x) - \lambda|\, I_{\Lambda(f,\lambda)\Delta A}(x)\, dx \\
&= \int_{\mathbf{R}^d} (f(x) - \lambda)\, I_{\Lambda(f,\lambda)\setminus A}(x)\, dx + \int_{\mathbf{R}^d} (\lambda - f(x))\, I_{A\setminus\Lambda(f,\lambda)}(x)\, dx \\
&= 2\left[P_f(\Lambda(f,\lambda)) - \lambda \cdot \text{volume}(\Lambda(f,\lambda)) - P_f(A) + \lambda \cdot \text{volume}(A)\right] \\
&= 2\left[\text{excmass}(\Lambda(f,\lambda)) - \text{excmass}(A)\right],
\end{aligned}$$

where $A\Delta B = (A \setminus B) \cup (B \setminus A)$ is the symmetric difference. Thus

$$\begin{aligned}
&\text{excmass}(A) \\
&= \text{excmass}(\Lambda(f,\lambda)) + \frac{1}{2}\int_{\mathbf{R}^d} |f(x) - \lambda|\, I_{\Lambda(f,\lambda)\Delta A}(x)\, dx, \qquad (7.3)
\end{aligned}$$

and we see that excmass(A) is minimized by choosing $A = \Lambda(f, \lambda)$. Compare this fact with Lemma 2.4. Level sets minimize the excess mass functional, and thus the level set with level λ approximately minimizes the empirical excess mass functional

$$\gamma_n(A, \lambda) = -\frac{1}{n}\sum_{i=1}^{n} I_A(X_i) + \lambda \cdot \text{volume}(A), \qquad A \subset \mathbf{R}^d.$$

Müller and Sawitzki (1991) suggest to test multimodality by test statistics

$$\Delta_m = \sup_{\lambda > 0}[E_m(\lambda) - E_{m-1}(\lambda)],$$

where $m \geq 2$, and

$$E_m(\lambda) = -\inf_{A \in \mathcal{A}_m} \gamma_n(A, \lambda),$$

where $m \geq 1$ and $\mathcal{A}_m$ is a collection of sets that consist of m separated components so that $\mathcal{A}_1$ contains only connected sets. If f has no more than $m - 1$ modes, then Δ_m should vanish, where $m \geq 2$.

CHAPTER 8

CLUSTER ANALYSIS

We apply cluster analysis to visualize the data and we apply visualization in validating the results of cluster analysis.

Clustering in the widest sense may be defined as finding such groups that consist of similar observations and that are separate from each other. The collection of observations is partitioned into groups in such a way that similar observations belong to the same group (homogeneous groups) and dissimilar observations belong to different groups (separateness). Clustering has been called classification, unsupervised classification, grouping, or numerical taxonomy (in biology).

There exists many interpretations of cluster analysis: clusters may be defined as connected subsets separated from each other, minimum variance partitions, high-density regions, or tail regions separated from each other. Equally varied are the applications of cluster analysis. Clustering has been used, for example, for the segmentation of images, for data compression (vector quantization), for clustering of Web documents, in the segmentation of customer databases, and to construct a taxonomy for organisms.

We will look at cluster analysis in two ways:

1. We are interested in cluster analysis as a tool for visualization. The observations can be colored so that the observations in the same cluster have the same color

(and the observations in the different groups have different colors). Coloring provides a powerful enhancement for the grand tour and parallel coordinate plots, for example. In addition clustering can be used to find a permutation to be applied in a graphical matrix: the observations are permutated so that the observations in the same group have nearby indexes.

2. Conversely, we can use visualization to validate the results of a cluster analysis. Clustering algorithms partition the data even when there do not exist any natural clusters, and many clustering algorithms do not help in choosing the right number of clusters. While formal tests can be developed for choosing the right number of clusters, visualization is indispensable.

There exists a large collection of clustering algorithms. We concentrate on four algorithms. We suggests for each algorithm a population interpretation. A clustering algorithm defines a partition of the sample $X_1, \ldots, X_n \subset \mathbf{R}^d$, and the population interpretation defines a partition of (a subset) of $\mathbf{R}^d$.

1. *Hierarchical clustering.* The suggested population version is the partition of a low level set of the density to disconnected components. This interpretation holds in the case of agglomerative clustering.

2. *k-Means clustering.* The suggested population version is the minimal variance partition.

3. *Density-based clustering, or mode clustering.* Clusters are defined in terms of the underlying density as regions of high density separated from other such regions by regions of low density.

4. *Tail clustering.* Clusters are defined as regions of low density separated from other such regions after removing the regions of high density and regions of zero density.

Clustering algorithms define directly a partition of the sample, but it is possible to start with a population version and find a partition of the sample space, and then find the induced partition of the sample. For example, in the density-based clustering one may estimate the density, find the regions of high density, and finally find the observations inside the separated high density regions. A similar approach can be applied in the case of tail clustering. The k-means clustering and hierarchical clustering partition directly the sample and the underlying population interpretation is less important.

Everitt (1974), Bock (1974), Hartigan (1975*a*), and Gordon (1981) give expositions of cluster analysis.

Section 8.1 discusses hierarchical clustering, Section 8.2 dicusses k-means clustering, Section 8.3 discusses high-density clustering, and Section 8.4 discusses tail clustering.

8.1 HIERARCHICAL CLUSTERING

8.1.1 Algorithms

Hierarchical clustering algorithms are called *agglomerative algorithms* when they proceed by joining smaller groups to larger groups, and they are called *divisive clustering algorithms* when the observations become separated successively into finer groups.

Agglomerative Clustering An agglomerative hierarchical clustering algorithm starts by taking every observation as a separate cluster. Then the two closest observations are merged to a single cluster according to a closeness criterion. The process continues, at each step joining the two closest clusters. In the end all the observations belong to the same cluster, unless one interrupts the merging at the ith step, and in this case one gets $n - i + 1$ clusters.

The definition of an agglomerative algorithm depends on the definition of the closeness between two subsets of the observations. Let $A, B \subset \mathbf{R}^d$ be disjoint sets of finite cardinality, $A \cap B = \emptyset$, $\#A, \#B < \infty$. There exist several proposals for the choice of the distance D between two clusters A and B. The distance D between clusters is defined in terms of the distance $d : \mathbf{R}^d \times \mathbf{R}^d \to \mathbf{R}$ between individual vectors. Typically the Euclidean distance $d(x, y) = \|x - y\|$ is chosen, but one may also consider a density-based distance $d(x, y) = 1/\hat{f}(z_{x,y})$, where $z_{x,y} = x + (x - y)/2$ is the midpoint between x and y, and $\hat{f} : \mathbf{R}^d \to \mathbf{R}$ is a density estimate. The proposals for the choice of D include the following:

1. Single linkage: $D(A, B) = \min\{d(x, y) : x \in A, y \in B\}$,

2. Average linkage: $D(A, B) = (\#A \cdot \#B)^{-1} \sum_{x \in A, y \in B} d(x, y)$,

3. Complete linkage: $D(A, B) = \max\{d(x, y) : x \in A, y \in B\}$,

4. Average group linkage:

$$D(A, B) = \frac{\sum_{x,y \in A \cup B, x \neq y} d(x, y)}{(\#A + \#B) \cdot (\#A + \#B - 1)}, \tag{8.1}$$

5. Centroid: $D(A, B) = d(\bar{x}_A, \bar{x}_B)$, where $\bar{x}_A = (\#A)^{-1} \sum_{x \in A} x$ is the average of the vectors in A, and $\bar{x}_B$ is defined similarly,

6. Ward (minimum variance method):

$$D(A, B) = \frac{\|\bar{x}_A - \bar{x}_B\|^2}{(\#A)^{-1} + (\#B)^{-1}},$$

where $\bar{x}_A = (\#A)^{-1} \sum_{x \in A} x$ is the average of the vectors in A, and $\bar{x}_B$ is defined similarly.

Single Linkage Clustering Single linkage clustering may be defined and calculated with the help of a minimal spanning tree, defined in Section 6.5, page 180. When one breaks the $k-1$ longest edges in a minimal spanning tree, the disconnected nodes are equal to the k clusters of single linkage clustering. Single linkage clustering is also related to the level sets of a 1-nearest neighbor estimate; see Section 14.4. The longest edges of a minimal spanning tree tend to be in low-density regions, and thus single linkage clustering tends to produce clusters that consist only of outliers.

Divisive Clustering Divisive clustering may be done at least in the following three ways:

1. We find a set A_i to be splitted among the sets in the current partition $A_1, \ldots, A_m$. Then we split set A_i to two sets B_1 and B_2. The process continues until a global criterion for the homogeneity or the dissimilarity of the sets in the clustering is fulfilled.

   For example, Edwards and Cavalli-Sforza (1964) proposed splitting the partition $A_1, \ldots, A_m$ to a finer partition $A_1, \ldots, A_{i-1}, B_1, B_2, A_{i+1}, \ldots, A_m$ by maximizing
   $$\frac{(\#B_1)(\#B_2)}{n} \left\|\bar{x}_{B_1} - \bar{x}_{B_2}\right\|^2 .$$
   This will minimize the variance
   $$\sum_{i=1}^{m+1} \sum_{k \in C_i} \left\|x_k - \bar{x}_{C_i}\right\|^2 .$$

2. We split each set independently until a local homogeneity criterion is fulfilled.

   For example, Macnaughton Smith, Williams, Dale and Mockett (1965) proposed the following algorithm:

   - Start by finding the observation whose average distance from the other observations is the largest, and making this observation the first member of the second cluster. At each step, find the observation whose average distance from the other observations in the first cluster minus the average distance from the observations in the second cluster is the largest, and joins this observation to the second cluster. Continue so long as there are observations for which the difference between the average distances is positive.
   - Continue splitting until each cluster has only one observation, or define a height for each cluster and use the height in formulating a stopping rule for the clustering: do not continue splitting those clusters whose height is larger than a threshold. The height could be the diameter
     $$H_{dia}(A) = \max_{x,y \in A} d(x,y) \tag{8.2}$$

of the cluster, or the average distance

$$H_{ave}(A) = \frac{1}{\#A \cdot (\#A - 1)} \sum_{x \in A} \sum_{y \in A, y \neq x} d(x, y) \qquad (8.3)$$

between the observations in the cluster. Note that the average group linkage-distance in (8.1) may be written as $H_{ave}(A \cup B)$.

3. We construct a symmetric splitting starting with a single set, splitting it to two sets, then constructing 4 sets, 8 sets, 16 sets, and so forth. The process is continued until a global criterion is fulfilled.

   For, example, A dyadic division algorithm would choose a single variable X_j and split the set A_i to two parts $\{X \in A_i : X_j \leq s\}$, $\{X \in A_i : X_j > s\}$.

8.1.2 Visualization

Dendrogram The process of agglomerating or dividing the observations returns a rooted binary tree; the leafs of the tree are the observations (singleton clusters), and the root of the tree is the set of all observations. All nodes of the tree are identified (or annotated) with a cluster.

A height can be attached to each node of the binary tree. In the case of agglomerative clustering the leaf nodes have height 0, and the height of the other nodes may be defined to be equal to the distance between the clusters that are merged to build this node (the height of a node is the distance between its children). For the divisive clustering we mentioned proposals (8.2) and (8.3) for the definition of heights.

A *dendrogram*, or a hierarchical clustering tree, is a graphical display of the binary tree where each node is plotted at its height. We use the term "dendrogram" also to denote the underlying binary tree whose nodes are associated with the heights and the clusters, and not only its graphical display.

A dendrogram gives a transparent visualization of the tree when the height of the parent is greater than the height of its children. When the height of the parent is greater than the height of its children, then one can use the dendrogram in selecting the clustering. This is because the cutting of the dendrogram at a given height is equivalent to the choosing of clustering that consists of those clusters whose height is greater than the given threshold. In the case of agglomerative clustering, condition (8.5) guarantees that the height of the parent is greater than the height of its children.

We draw dendrograms for the various choices of distance D.

1. Figure 8.1 shows dendrograms with the single linkage clustering, average linkage clustering, complete linkage clustering, average group linkage (McQuitty distance), centroid distance, and Ward distance. The data of size 300 is generated from the 2D three-modal density shown in Figure 2.7. The dendrograms were drawn with the R-function "hclust".

2. Figure 8.2 shows scatter plots of the data when we have colored the clusters. Panel (a) shows average linkage clusters and panel (b) shows complete linkage

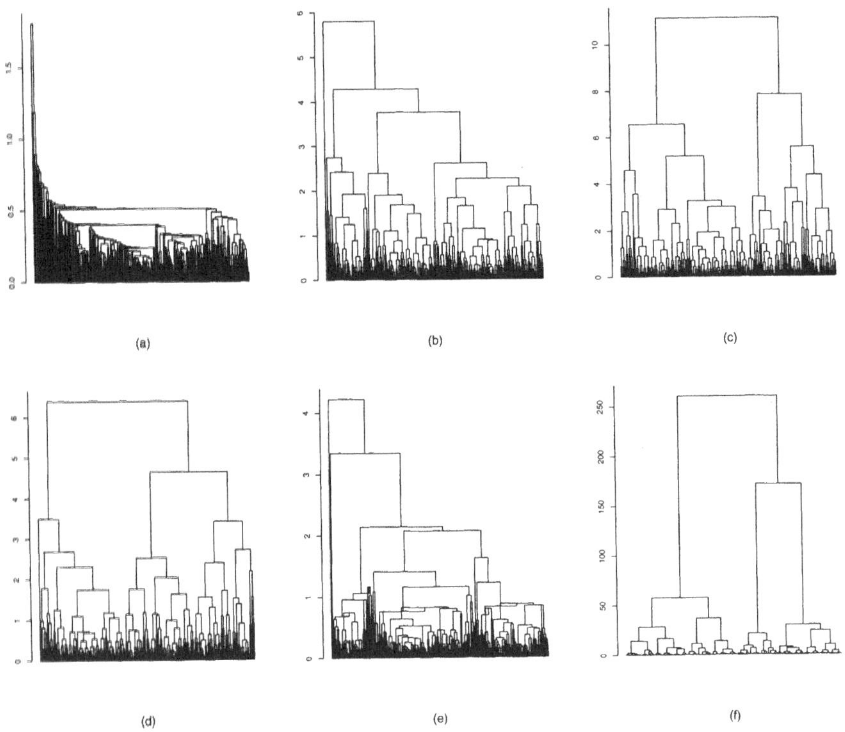

Figure 8.1 *(Dendrograms.)* Dendrograms with the (a) single linkage, (b) average linkage, (c) complete linkage, (d) average group linkage, (e) centroid distance, and (f) Ward distance clustering. The data of size 300 is generated from the 2D three-modal density shown in Figure 2.7.

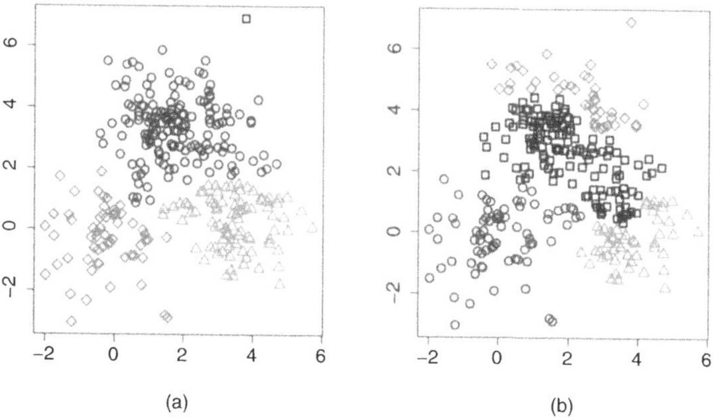

Figure 8.2 *(Scatter plots.)* Scatter plots of clusters generated by the dendrograms in Figure 8.1(b)-(c). (a) Average linkage clusters and (b) complete linkage clusters. The symbols (circle, square, diamond, triangle) indicate the membership in the clusters.

clusters. The clusters are obtained by cutting the dendrograms of Figure 8.1 at certain heights. In panel (a) we have cutted the average linkage dendrogram at height 3. When we cut the dendrogram at height 3, then the sample is divided into four clusters. Three of the clusters correspond to the modes of the underlying density. One observation in the tail forms a single cluster (the square observation). In panel (b) we have cutted the complete linkage dendrogram at the height 6.

Spatial Tree Plot Dendrograms are trees whose nodes are annotated with subsets of the observations. We may summarize the locations of the clusters by calculating the average of the observations in each cluster. Thus we annotate each node with a d-dimensional vector, giving the center of a cluster. This transforms a dendrogram to a spatial tree in the sense of Definition 3.1.

The spatial tree plot was defined in Definition 3.2. We use this plot to visualize dendrograms. This visualization method shows the locations of the clusters, unlike a dendrogram, which shows only the tree structure of the agglomerating process. We apply the root-first coloring in the spatial tree plots of dendrograms. The root-first coloring is discussed in Section 3.1, page 86. In the root-first coloring we start by choosing a color for the root. We travel toward the leafs, and when the tree is branching, we choose new colors for the children (but one of the children has the same color as the parent). Below a certain height we fix the colors and choose always the color of the parent for the children. Fixing the color corresponds to the cutting of the dendrogram at the given height.

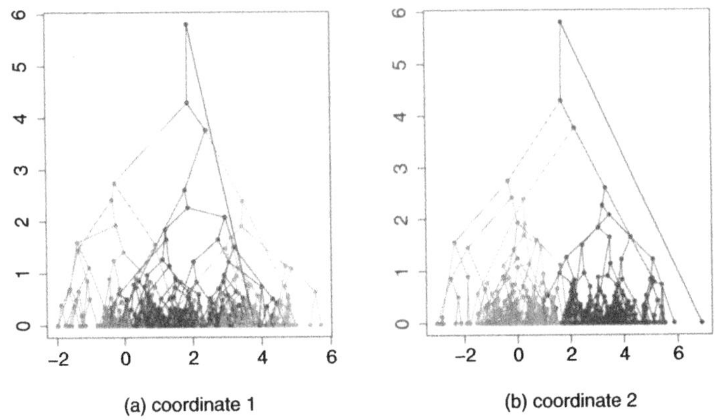

Figure 8.3 *(Spatial tree plot.)* (a–b) A spatial tree plot of the average linkage dendrogram in Figure 8.1(b). The corresponding scatter plot of the clusters is shown in Figure 8.2(a).

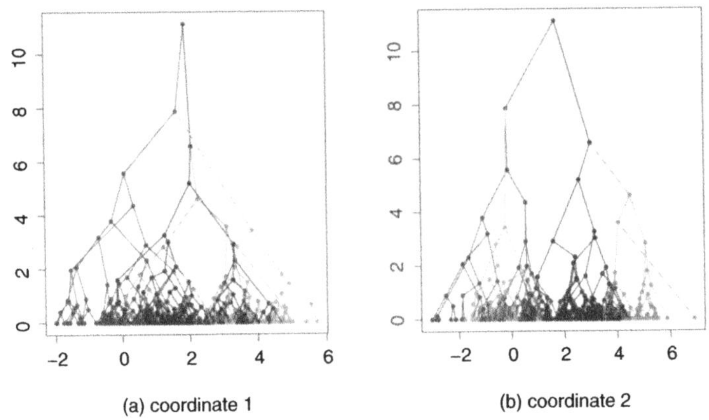

Figure 8.4 *(Spatial tree plot.)* (a–b) A spatial tree plot of the complete linkage dendrogram in Figure 8.1(c). The corresponding scatter plot of the clusters is shown in Figure 8.2(b).

1. Figure 8.3 shows a spatial tree plot of the average linkage dendrogram in Figure 8.1(b). The clusters are obtained by cutting at height 3. The corresponding scatter plot of the clusters is shown in Figure 8.2(a).

2. Figure 8.4 shows a spatial tree plot of the complete linkage dendrogram in Figure 8.1(c). The clusters are obtained by cutting at height 6. The corresponding scatter plot of the clusters is shown in Figure 8.2(b).

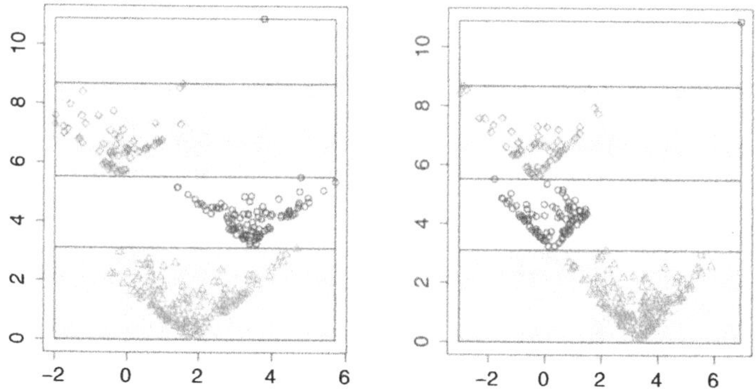

Figure 8.5 *(Parallel level plot.)* Shown is a parallel level plot of average linkage clusters. The dendrogram is shown in Figure 8.1(b), and the scatter plot of the clusters is shown in Figure 8.2(a).

Parallel Level Plot Parallel level plots were discussed in Section 1.3. A parallel level plot consists of d scatter plots, and the observations are identified across the windows with a level function. It was noted that the data may be visualized by parallel level plots well when the level of the observations is well chosen. A useful way to choose the level is to put observations together that belong to the same cluster; the level of the observations is chosen so that the observations that belong to a same cluster have a nearby level. In addition we will define the level function in such a way that inside each cluster the level is chosen according to the Euclidean distance to the center point of the cluster (the arithmetic mean of the cluster).

1. Figure 8.5 shows a parallel level plot where average linkage clusters are applied to define the level function. The corresponding dendrogram is shown in Figure 8.1(b) and the scatter plot of the clusters is shown in Figure 8.2(a). The clusters are obtained by cutting at height 3.

2. Figure 8.6 shows a parallel level plot where complete linkage clusters are applied to define the level function. The corresponding dendrogram is shown in Figure 8.1(c) and the scatter plot of the clusters is shown in Figure 8.2(b). The clusters are obtained by cutting at height 6.

Graphical Matrix Graphical matrices were discussed in Section 1.4. Graphical matrices are closely related to parallel level plots. The usefulness of a graphical matrix depends on the choice of an informative permutation of rows and columns of the matrix (analogously as the usefulness of a parallel level plot depends on an informative choice of the level). It is natural to choose the permutation of a graphical matrix so that the observations belonging to the same group are together.

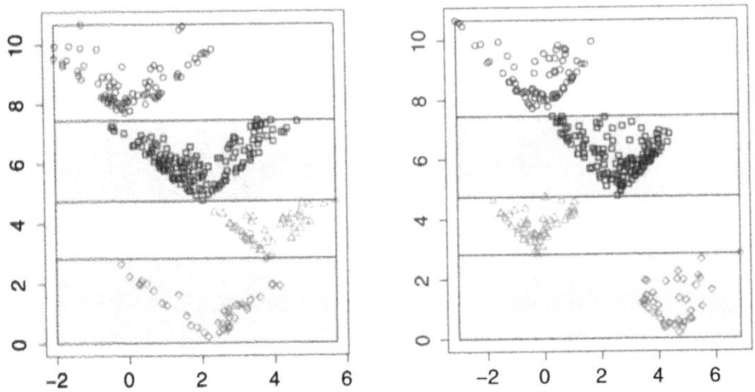

Figure 8.6 *(Parallel level plot.)* Shown is a parallel level plot of complete linkage clusters. The dendrogram is shown in Figure 8.1(c), and the scatter plot of the clusters is shown in Figure 8.2(b).

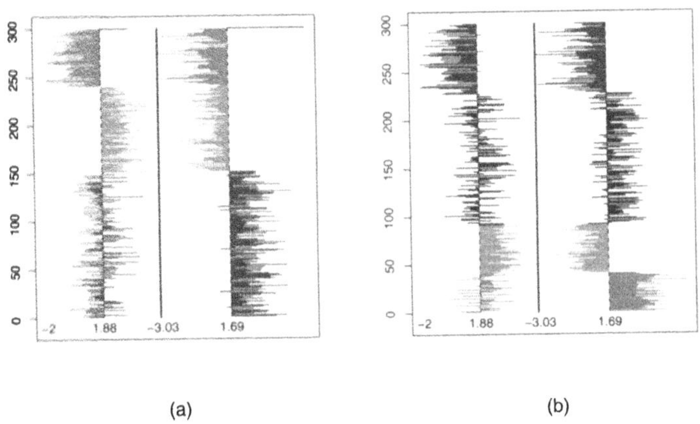

Figure 8.7 *(Graphical matrices.)* (a) A graphical matrix of the average linkage clusters whose scatter plot is shown in Figure 8.2(a). (b) A graphical matrix of the complete linkage clusters whose scatter plot is shown in Figure 8.2(b).

1. Figure 8.7(a) shows a graphical matrix where average linkage clusters are applied to define the permutation. The corresponding dendrogram is shown in Figure 8.1(b) and the scatter plot of the clusters is shown in Figure 8.2(a). The clusters are obtained by cutting at height 3.

2. Figure 8.7(b) shows a graphical matrix where complete linkage clusters are applied to define the permutation. The corresponding dendrogram is shown in Figure 8.1(b), and the scatter plot of the clusters is shown in Figure 8.2(a). The clusters are obtained by cutting at height 3.

8.1.3 Population Interpretation

We can interpret the clusters of agglomerative hierarchical clustering as corresponding to connected and pairwise separated components of the support of the underlying density or corresponding to a low level set of the underlying density. This interpretation can be justified by noting that it is possible to define the concepts of separateness and connectivity in such a way that the clusters of agglomerative clustering are connected and pairwise separated. We will state this in Proposition 8.1 below.

We have defined ρ-separated sets and a ρ-connected set in Definition 6.1. If the sets have finite cardinality, then in fact the single linkage distance is used in Definition 6.1 to define the separateness of sets. We may generalize the definition to cover other clustering distances.

Definition 8.1 ((ρ, D)-separated sets, (ρ, D)-connected set.) *Let D be a distance of agglomerative clustering and let $A, B \subset \mathbf{R}^d$ have finite cardinality.*

1. *Sets $A, B \subset \mathbf{R}^d$ are* separated *for the resolution threshold $\rho > 0$ ((ρ, D)-separated) if $D(A, B) > 2\rho$.*

2. *Set $A \subset \mathbf{R}^d$ is* connected *for the resolution threshold $\rho > 0$ ((ρ, D)-connected) if for every nonempty B, C such that $A = B \cup C$, B and C are not separated for the resolution threshold ρ.*

Proposition 8.1 states that if we have a set $\{x_1, \ldots, x_n\} \subset \mathbf{R}^d$ of cardinality n, then agglomerative hierarchical clustering finds a finite sequence of intervals

$$[0, \rho_1), \ldots, [\rho_{n-1}, \infty)$$

so that when $\rho \in [\rho_{i-1}, \rho_i)$, $i = 1, \ldots, n$, $\rho_0 = 0$, $\rho_n = \infty$, set has $n - i + 1$ connected components that are pairwise separated. We call values $\rho_1, \ldots, \rho_{n-1}$ the critical resolution thresholds. The agglomerative hierarchical clustering is a dynamical programming algorithm for finding the critical resolution thresholds and the corresponding connected and pairwise separated components. The algorithm starts by finding the two closest singletons, and joining them together into a cluster. Let the distance between the two closest singletons be equal to $2\rho_1$. At step k, $k = 1, \ldots, n - 1$, we find the two closest clusters among the set of $n - k + 1$ previously found clusters, and join these clusters together. Let the distance between these two closest clusters be equal to $2\rho_k$. Finally at step $n-1$ we merge together the remaining two clusters, and let the distance between these clusters to be equal to $2\rho_{n-1}$. This process has defined the critical resolution thresholds.

Proposition 8.1 *At step i of the agglomerative hierarchical clustering algorithm, $i = 1, \ldots, n$, the $n-i+1$ clusters are ρ-connected and pairwise ρ-separated, for each $\rho \in [\rho_{i-1}, \rho_i)$, where $2\rho_{i-1}$ is the distance between the clusters that were joined at step $i-1$, and $2\rho_i$ is the closest distance between the current clusters, where we denote $\rho_0 = 0$, $\rho_n = \infty$. We need to assume that distance D satisfies the properties (8.4) and (8.5) below.*

Proof. The proposition is proved by induction.

Step 1: All singletons are ρ-connected for each $\rho \geq 0$ and the singletons are ρ-separated from each other for $0 \leq \rho < \rho_1$, since $2\rho_1$ is the distance between the two closest singletons.

Inductive step: By the induction hypothesis all clusters at step $i-1$ are ρ' connected for $\rho' < \rho_{i-1}$. Thus these clusters are ρ-connected for $\rho \geq \rho_{i-1} > \rho'$. Here we need the following property:

$$\begin{aligned} &\text{When } A, B \subset \mathbf{R}^d \text{ are } \rho'\text{-connected,} \\ &\text{then } A, B \text{ are } \rho\text{-connected for } \rho \geq \rho'. \end{aligned} \tag{8.4}$$

At step i there is also a cluster $C = A \cup B$ that we get by joining two previous clusters A and B. Using property (8.4), it can easily be inferred that cluster C is ρ-connected for $\rho_{i-1} \leq \rho < \rho_i$. The clusters at step i are ρ-separated for $\rho < \rho_i$ because $2\rho_i$ is the closest distance between the current clusters. Here one needs to exclude the case where the closest distant decreases. The closest distant could decrease if the process of joining two clusters together would result in the decreasing distance to the other clusters. This case is excluded by the following condition on the distance D:

$$\begin{aligned} &\text{for each disjoint } A, B, C \subset \{x_1, \ldots, x_n\}, \\ &\min\{D(A, C), D(B, C)\} \leq D(A \cup B, C). \end{aligned} \tag{8.5}$$

We have proved the proposition. □

8.2 THE k-MEANS CLUSTERING

The k-means algorithm belongs to the group of combinatorial clustering algorithms. Combinatorial clustering algorithms search for a partition that minimizes an objective function. In the case of k-means algorithm the objective function is a minimal variance functional. Since the number of all partitions is very large, it is reasonable to use greedy algorithms in solving the optimization problem.

8.2.1 Algorithms

The k-means algorithm starts with an initial choice for the k cluster centers, finds for each cluster center its Voronoi neighborhood, and updates the cluster centers to be the average of the observations in the Voronoi neighborhood. The Voronoi neighborhood of a cluster center is the set of those observations that are closer to this cluster center than to the other cluster centers. We give below the k-means algorithm for clustering set $\mathcal{X} = \{x_1, \ldots, x_n\} \subset \mathbf{R}^d$.

1. Choose k cluster centers $\mathcal{M} = \{m_1, \ldots, m_k\} \subset \mathbf{R}^d$.
2. Go through the cluster centers $m \in \mathcal{M}$ until convergence.
   (a) Find the Voronoi neighborhood $\mathcal{X}_m$ of the cluster center m (the set of those observations that are closer to m than to any other cluster center):
   $$\mathcal{X}_m = \{x \in \mathcal{X} : \|x - m\| = \min\{\|x - m'\| : m' \in \mathcal{M}\}\}.$$
   (b) Update m to be the average of the observations in $\mathcal{X}_m$:
   $$m = \frac{1}{\#\mathcal{X}_m} \sum_{x \in \mathcal{X}_m} x.$$

The algorithm finds the cluster centers, and we define the clusters to be the Voronoi neighborhoods of the cluster centers.

Definition 8.2 (k-Means clusters.) *The* k-means clusters *are the Voronoi neighborhoods $\mathcal{X}_m$, $m \in \mathcal{M}$, where cluster centers $\mathcal{M}$ are found by the k-means algorithm.*

The k-means algorithm may be considered as a greedy algorithm to find the sample version of a minimum variance partition, or to find the sample versions of principal points, as stated below.

1. *(The sample version of minimum variance partition.)* Find a partition $\mathbb{P}$ of $\{x_1, \ldots, x_n\}$ of size k such that
   $$\sum_{i=1}^{n} \|x_i - \mu(x_i; \mathbb{P})\|_2^2$$
   is minimized with respect to all partitions of size k, where $\mu(x_i; \mathbb{P})$ is the arithmetic mean over the cluster where x_i belongs to
   $$\mu(x_i; \mathbb{P}) = \sum_{\mathcal{X} \in \mathbb{P}} \mu_{\mathcal{X}} I_{\mathcal{X}}(x_i), \qquad \mu_{\mathcal{X}} = \frac{1}{\#\mathcal{X}} \sum_{i=1}^{n} x_i I_{\mathcal{X}}(x_i).$$
2. *(The sample version of principal point problem.)* Find points $m_1, \ldots, m_k \in \mathbf{R}^d$ such that
   $$\sum_{i=1}^{n} \|x_i - \mu(x_i; m_1, \ldots, m_k)\|_2^2$$
   is minimized with respect to all sets of k distinct points, where
   $$\mu(x; m_1, \ldots, m_k) = \text{argmin}_{a \in \{m_1, \ldots, m_k\}} \|x - a\|_2.$$

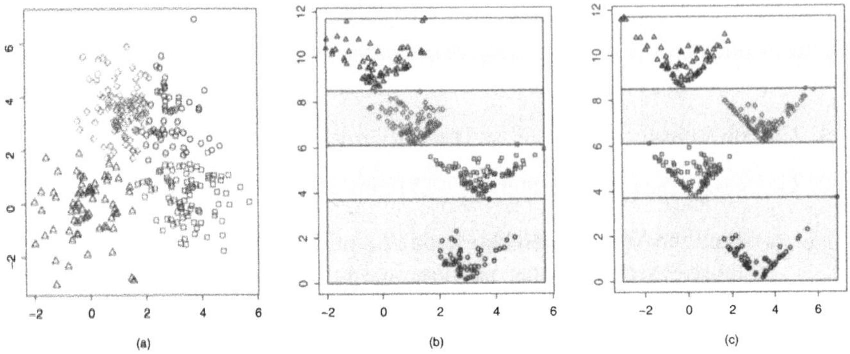

Figure 8.8 *(k-Means clustering with $k = 4$.)* (a) A scatter plot of the data with the four clusters indicated by different symbols (square, triangle, diamond, and circle). (b–c) Parallel level plot of the data, when clustering is applied to determine the level. The data of size 300 is generated from the 2D three-modal density shown in Figure 2.7.

8.2.2 Visualization

As with any clustering, k-means clustering may be visualized using parallel level plots and graphical matrices. Parallel level plots are discussed in Section 1.3, and graphical matrices are discussed in Section 1.4.

1. Figure 8.8(a) shows a scatter plot of k-means clustering with $k = 4$. The cluster membership is indicated by the symbols square, triangle, diamond, and circle. The data has size 300 and is generated from the 2D three-modal density shown in Figure 2.7.

2. Figure 8.8(b) and (c) shows a parallel level plot of the clustering in panel (a). We have defined the level of the observations so that the observations belonging to the same cluster have a nearby level and inside each cluster the level is determined according to the Euclidean distance from the arithmetic mean of the cluster.

3. Figure 8.9 shows graphical matrices of k-means clustering for $k = 2, 3, \ldots, 9$. Graphical matrices give a compact visualization of the clusters and are suitable for the visualization of a sequence of clusterings. Visualizing a sequence of clusterings helps in the choice of an optimal k. The data is the same as in Figure 8.8. The graphical matrices are designed in such a way that the observations that belong to a same cluster are put in neighboring rows.

8.2.3 Population Interpretation

We may consider a k-means clustering as a sample version of the minimum variance partition. The fundamental lemma of cluster analysis states that the minimum

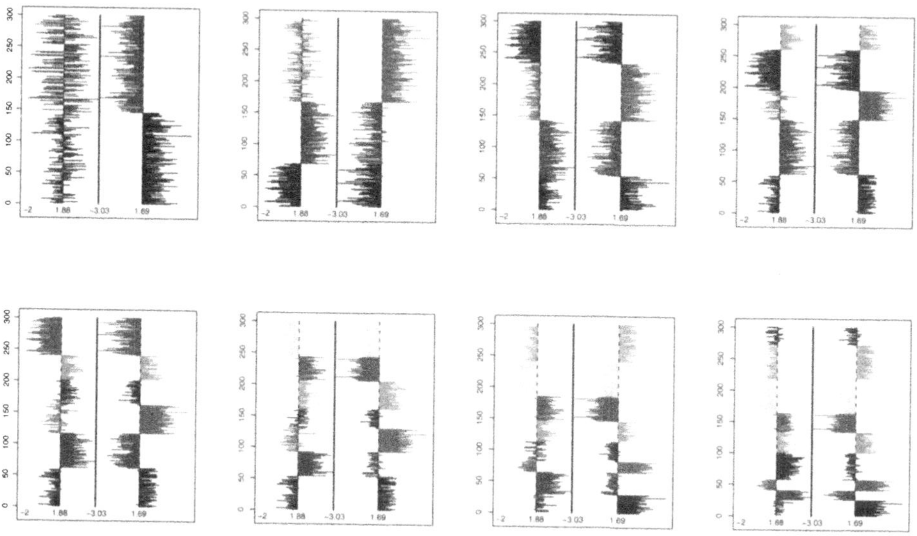

Figure 8.9 *(k-Means clustering with $k = 2 - 9$.)* Shown are graphical matrices corresponding to the number of clusters $k = 2, 3, \ldots, 9$. The data of size 300 is generated from the 2D three-modal density shown in Figure 2.7.

variance partition can be defined in terms of the principal points, and conversely. Before stating the lemma, let us define the minimum variance partition and the principal points. Let $f : \mathbf{R}^d \to \mathbf{R}$ be a density function.

1. *(Minimum variance partition.)* Find a partition $\mathbb{P}$ of $\mathbf{R}^d$ of size k such that

$$\int_{\mathbf{R}^d} \|x - \mu(x; \mathbb{P})\|_2^2 \, f(x) \, dx$$

is minimized with respect to all measurable partitions of size k, where

$$\mu(x; \mathbb{P}) = \sum_{A \in \mathbb{P}} \mu_A I_A(x), \tag{8.6}$$

where μ_A is the conditional expectation of f in A, which may be defined with the help of the conditional density f_A,

$$\mu_A = \int_{\mathbf{R}^d} x f_A(x) \, dx, \qquad f_A(x) = \frac{f(x) I_A(x)}{\int_A f}, \tag{8.7}$$

$x \in \mathbf{R}^d$, $A \in \mathbb{P}$. Note that the conditional expectation of $X \sim f$ with respect to σ-algebra $\mathcal{P}$ generated by $\mathbb{P}$ is equal to

$$E(X \mid \mathcal{P}) = \sum_{A \in \mathbb{P}} \mu_A I_A. \tag{8.8}$$

2. *(Principal point problem.)* Find points $m_1, \ldots, m_k \in \mathbf{R}^d$ such that

$$\int_{\mathbf{R}^d} \|x - \mu(x; m_1, \ldots, m_k)\|_2^2 \, f(x) \, dx$$

is minimized with respect to all sets of k distinct points, where

$$\mu(x; m_1, \ldots, m_k) = \text{argmin}_{a \in \{m_1, \ldots, m_k\}} \|x - a\|_2.$$

We now formulate the fundamental lemma of cluster analysis, which states the equivalence of the problem of finding the minimum variance partition and the principal point problem.

Lemma 8.2 *The minimum variance partition and principal points are connected in the following ways:*

1. *Let $m_1, \ldots, m_k \in \mathbf{R}^d$ solve the principal point problem. Let B_j be the Voronoi neighborhood of m_j,*

$$B_j = \left\{ x : j = \text{argmin}_{l=1,\ldots,k} \|x - m_l\|_2 \right\}.$$

Then $\{B_1, \ldots, B_k\}$ is the minimum variance partition of size k.

2. *Conversely, if $\{B_1, \ldots, B_k\}$ is the minimum variance partition of size k, then points*

$$m_j = \int_{B_j} x f_{B_j}(x) \, dx, \qquad j = 1, \ldots, k,$$

solve the principal point problem, where f_{B_j} is the conditional density defined in (8.7).

8.2.4 Bibliographic Notes

Modifications We may find several modifications of the problem of finding the minimum variance partition. First we note that the minimum variance partition and the principal point problem may be written in a modified form.

1. *(Minimum variance partition.)* Find a partition $\mathbb{P}$ of $\mathbf{R}^d$ of size k such that

$$\int_{\mathbf{R}^d} \|\mu(x;\mathbb{P})\|_2^2\, f(x)\, dx$$

is maximized with respect to all measurable partitions of size k, where $\mu(x;\mathbb{P})$ is defined in (8.6).

2. *(Principal point problem.)* Find points $m_1, \ldots, m_k \in \mathbf{R}^d$ such that

$$\int_{\mathbf{R}^d} \max_{j=1,\ldots,k} \left(m_j^T x - \frac{1}{2}\|m_j\|_2^2 \right) f(x)\, dx$$

is maximized with respect to all sets of k distinct points.

The equivalence of the formulations of the minimum variance partition follows from

$$E\|X - E(X\,|\,\mathcal{P})\|_2^2 = E\|X\|_2^2 + E\|E(X\,|\,\mathcal{P})\|_2^2,$$

where we used the notation of (8.8). The new way of writing the optimization problems suggests natural modifications.

1. Find a partition $\mathbb{P}$ of $\mathbf{R}^d$ of size k such that

$$\int_{\mathbf{R}^d} \|\mu(x;\mathbb{P})\|_2\, f(x)\, dx$$

is maximized with respect to all measurable partitions of size k, where $\mu(x;\mathbb{P})$ is defined in (8.6).

2. Find points $m_1, \ldots, m_k \in \mathbf{R}^d$ such that

$$\int_{\mathbf{R}^d} \max_{j=1,\ldots,k} \left(m_j^T x \right) f(x)\, dx$$

is maximized with respect to all sets of k distinct points and define $\mathbb{P} = \{B_1, \ldots, B_k\}$ by

$$B_j = \{x : j = \mathrm{argmax}_{l=1,\ldots,k} m_l^T x\}, \qquad j = 1, \ldots, k.$$

Problem 1 was proposed in Pötzelberger and Strasser (2001), and problem 2 was called there the "Kohonen problem". The equivalence of these optimization problems was proved in Pötzelberger and Strasser (2001). There exists also a modification called generalized principal point problem:

Find points $m_1, \ldots, m_k \in \mathbf{R}^d$ such that

$$\int_{\mathbf{R}^d} \min_{j=1,\ldots,k} \Phi(\|x - m_j\|_2) f(x)\, dx$$

is minimized with respect to all sets of k distinct points, where $\Phi : [0, \infty) \to \mathbf{R}$, and define $\mathbb{P} = \{B_1, \ldots, B_k\}$ by

$$B_j = \{x : j = \text{argmin}_{l=1,\ldots,k} \Phi(\|x - m_l\|_2)\}, \qquad j = 1, \ldots, k.$$

Minimum Variance Problem Several optimization problems are related to the minimum variance problem. Let us rename the sample points as

$$\{x_1, \ldots, x_n\} = \{y_{ij} : i = 1, \ldots, k, j = 1, \ldots, n_i\},$$

where we have k groups and each group has n_i observations. Then we can write the covariance matrix, defined in (1.5), as

$$\frac{1}{n} \sum_{i=1}^{k} \sum_{j=1}^{n_i} (y_{ij} - \bar{x}_n)^T (y_{ij} - \bar{x}_n) = W + B,$$

where $\bar{x}_n$ is the arithmetic mean, W is the within-group covariance matrix, and B is the between-group covariance matrix,

$$W = \frac{1}{n-k} \sum_{i=1}^{k} \sum_{j=1}^{n_i} (y_{ij} - \bar{y}_i)^T (y_{ij} - \bar{y}_i),$$

$$B = \sum_{i=1}^{k} n_i (\bar{y}_i - \bar{x}_n)^T (\bar{y}_i - \bar{x}_n),$$

with $\bar{y}_i$ as the arithmetic mean of the group i. When $d = 1$, this is the decomposition of the total sum of squares to the within-group and between-group sum of squares, and one could find the clusters as the partition minimizing W, which is equivalent to maximizing B. In the multivariate case we can minimize the trace of W (maximize the trace of B). We can also minimize the determinant of W (maximize the determinant of B); see Marriot (1971). Yet another way to partition is to maximize the trace of BW^{-1}; see Friedman and Rubin (1967). Finding the optimal partition cannot be done with a complete enumeration, since

$$N(n,k) = \frac{1}{k!} \sum_{i=0}^{k} (-1)^{k-i} \binom{k}{i} i^n,$$

where $N(n,k)$ is the number of partitions of n observations to k distinct nonempty groups; see Liu (1968).

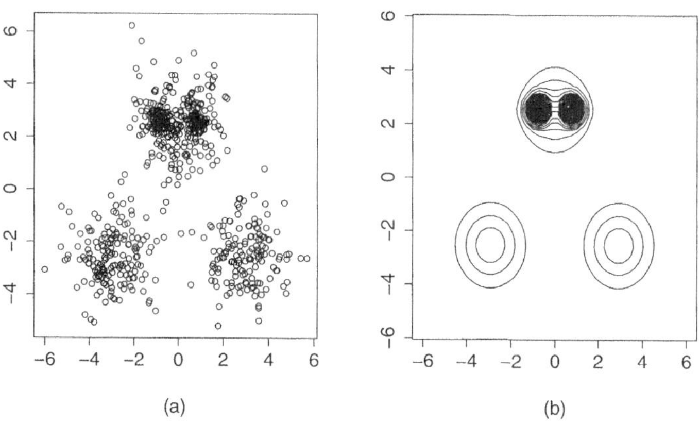

Figure 8.10 *(Nested clusters.)* (a) A scatter plot of data with nested clusters ($n = 700$). (b) A contour plot of the underlying density. The density is a mixture of 5 Gaussian densities.

8.3 HIGH-DENSITY CLUSTERING

In high-density clustering clusters are defined as regions corresponding to the modes of the underlying density. Clusters are defined as regions of high density separated from other such regions by regions of low density (Hartigan (1975*a*, p. 205)). The clustering algorithms are based on density estimation and mode seeking. Finding the right number of clusters is related to mode testing. In the framework of density-based clustering we may find more readily an answer to the question of the right number of clusters, as compared to the hierarchical clustering and k-means clustering.

The most natural conceptual framework for defining the population version of high-density clustering and the high-density clustering algorithms is the framework of level set trees, as discussed in Chapter 4. This because clusters, in general, have a nested structure, like the one shown in Figure 8.10. Panel (a) shows a data of size $n = 700$ composed of three super-clusters and two sub-clusters within the north-cluster. Panel (b) shows a contour plot of the underlying density, which is a mixture of five Gaussian densities.

Level set trees are introduced in Klemelä (2004). Related trees, which are called *cluster trees*, are introduced and applied to cluster analysis in Stuetzle (2003). The approach of Ankerst, Breunig, Kriegel and Sander (1999) aims also at finding nested clusters.

8.3.1 Population Interpretation

The population version of high-density clustering may be formulated with the help of a level set tree, as defined in Definition 4.2. A level set tree is a tree structure where each node is annotated with a subset of the sample space. The roots of the tree are

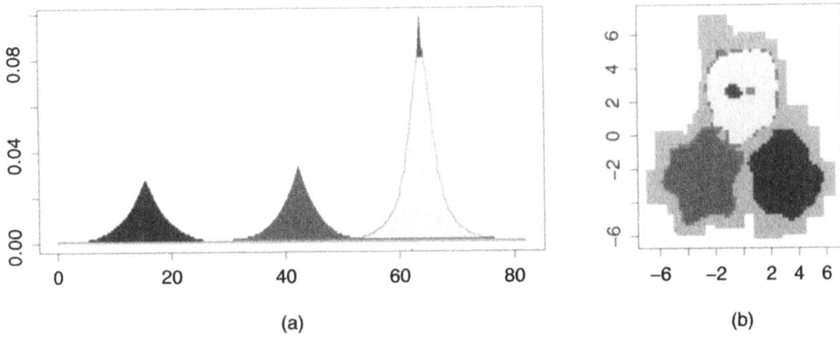

Figure 8.11 *(High-density clustering with a level set tree.)* (a) A volume function of a level set tree of a kernel estimate based on the data in Figure 8.10(a). (b) A part of the support of the kernel estimate with sub-modes shaded.

annotated with the separated components of the support of the underlying density, and the upper levels of the tree are annotated with increasingly small subsets of the sample space. A clustering is obtained by cutting the tree at a given level.

Definition 8.3 (Population version of high-density clustering.) *A population version of* high-density clustering *corresponds to a level set tree of a density function* $\mathbf{R}^d \to \mathbf{R}$*, and to a level* $\lambda \geq 0$*. The clustering consists of the subsets* $A_1, \ldots, A_k \subset \mathbf{R}^d$*, such that (1) each* A_i*,* $i = 1, \ldots, k$*, is annotated with a node of the level set tree whose level is* $\geq \lambda$*, and (2) for every other set* B*, that is annotated with a node of the level set tree whose level is* $\geq \lambda$*, either* $B \subset A_i$ *or* $B \cap A_i = \emptyset$*.*

Thus we find the population clustering by going through all the nodes of the level set tree, taking all the sets that are annotated to nodes whose level is $\geq \lambda$, and then removing all sets that are included in some larger set whose level is $\geq \lambda$ (we take only the largest sets, and these sets can be charcterized as sets that do not have a parent among the sets with level $\geq \lambda$).

A population version of high-density clustering makes only a slice of the level set tree; the tree itself is more interesting and its visualization is discussed in Chapter 4. Figure 8.11(a) shows a volume function of a kernel estimate based on the data in Figure 8.10(a). The volume function shows the three super-modes and the two sub-modes. Panel (b) shows a part of the support of the kernel estimate in such a way that the regions corresponding to the super- and sub-modes are colored.

A population version of high-density clustering does not make a partition of the support of the density, unlike the minimum variance partition, for example. The union of the sets in a high-density clustering is a genuine subset of the support.

8.3.2 Algorithms

We find an estimate of a population version of high-density clustering by estimating the density and constructing a level set tree of the density estimate. If $A_1, \ldots A_k \subset \mathbf{R}^d$ is a high-density clustering, then we use data $X_1, \ldots, X_n$ to construct a density estimate $\hat{f}$, and use the density estimate to construct an estimate $\hat{A}_1, \ldots, \hat{A}_k$ of the clustering.

Note that hierarchical clustering and k-means clustering give partitions of the sample $X_1, \ldots, X_n$. In contrast, high-density clustering is composed of subsets of the sample space. However, high-density clustering induces a partition of a part of the sample. Given a high-density clustering $A_1, \ldots A_k \subset \mathbf{R}^d$, define

$$B_j = \{X_i \in A_j : i = 1 \ldots, n\}, \qquad j = 1, \ldots, k. \tag{8.9}$$

This induced clustering does not typically make a partition of the complete sample, since typically $\{X_1, \ldots, X_n\} \setminus \cup_{j=1}^k B_j \neq \emptyset$. When $A_1, \ldots A_k$ makes a near partition of the support of the density, it can happen that $B_1, \ldots, B_k$ makes a partition of the sample $X_1, \ldots, X_n$.

Any level set tree induces a tree structure among the set of observations, and this tree structure is more useful than any fixed clustering of the observations. We call the tree induced by a level set tree a *likelihood tree* because the level of an observation in the tree is (close) to its likelihood.

Definition 8.4 (Likelihood tree.) *A likelihood tree, corresponding to a level set tree, is a tree structure where each node is annotated with a subset of the observations $X_1, \ldots, X_n \in \mathbf{R}^d$. A likelihood tree has the same tree structure as the level set tree. When a node of the level set tree is annotated with set $A \subset \mathbf{R}^d$, the corresponding node of the likelihood tree is annotated with set $\{X_i \in A : i = 1, \ldots, n\}$.*

We may obtain a clustering of data $X_1, \ldots, X_n$ by cutting the likelihood tree or cutting the level set tree and then using the rule (8.9) to obtain a clustering. The both procedures give the same clustering.

Note that a likelihood tree is a different kind of tree than a dendrogram, although the nodes of both trees are annotated with the subsets of the observations, and the root of the tree is in both cases annotated with the set of all observations. In a dendrogram, a parent is always a union of its children, whereas in a likelihood tree, a parent is typically a genuine subset of its children. In a dendrogram, the leaf nodes contain all the observations, but in a likelihood tree, the leaf nodes contain only few observations.

8.3.3 Visualization

A likelihood tree was defined as a tree whose nodes are annotated with subsets of observations. We may derive a tree from a likelihood tree where each node is annotated with a single observation, and use this tree to visualize the data. We call this tree a likelihood tree of observations, or just a likelihood tree, when there is no

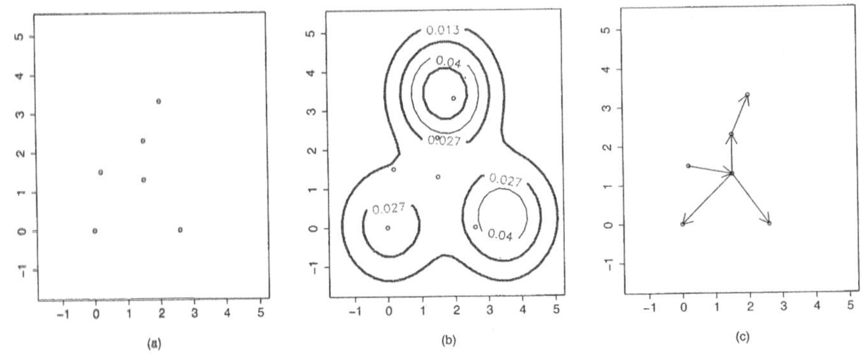

Figure 8.12 *(Likelihood tree of the observations.)* (a) The data; (b) a contour plot of a function, which defines a level set tree and induces a likelihood tree; (c) the corresponding likelihood tree of the observations.

danger of confusion. A likelihood tree of observations is constructed by subtracting those observations from a set annotated to a node of the likelihood tree that belong also to higher level nodes.

Definition 8.5 (Likelihood tree of observations.) *A likelihood tree of observations $X_1, \ldots, X_n$ is a tree among the observations, and this tree is constructed with the help of a likelihood tree T as defined in Definition 8.4.*

1. *An observation X_i is an ancestor of X_j if X_j belongs to some set in the likelihood tree T that has a higher level than any set to which X_i belongs.*

2. *If the highest level of X_i and X_j in the likelihood tree T are equal, then the ancestor relation is determined arbitrarily.*

Illustrations We illustrate the definition of a likelihood tree of observations.

1. Figure 8.12 illustrates Definition 8.5. Panel (a) shows six observations. Panel (b) shows a contour plot of a function (e.g., a density estimate based on the observations). This contour plot defines a level set tree, and the level set tree defines a likelihood tree among the subsets of the observations. The thick contour lines indicate the tree structure of the likelihood tree. Panel (c) shows the likelihood tree of the observations. The order of the first two observations is determined arbitrarily. This is because the resolution of the likelihood tree does not differentiate these observations, but the other observations are uniquely differentiated by the likelihood tree.

2. A likelihood tree of observations is a spatial tree, as in Definition 3.1, since each node is annotated by a vector. Thus we can visualize a likelihood tree of observations with a spatial tree plot, as defined in Definition 3.2.

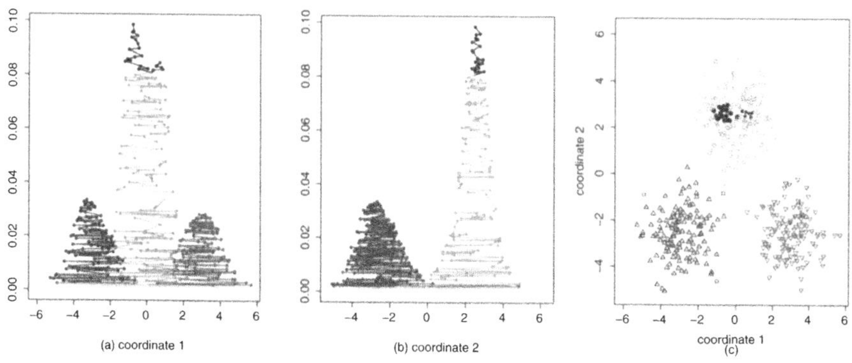

Figure 8.13 *(Spatial tree plot of a likelihood tree.)* Panels a-b) show a spatial tree plot of a likelihood tree of the data in Figure 8.10(a), corresponding to the level set tree in Figure 8.11(a). Panel c) shows a scatter plot of the data so that the observations are labeled according to the clustering structure.

Figure 8.13(a) and (b) shows a spatial tree plot of the data in Figure 8.10(a). The spatial tree is the likelihood tree corresponding to the level set tree in Figure 8.11(a). In figure 8.13 is a scatter plot of the data with the observations labeled (bullet, circle, solid circle, square, diamond, triangle) according to the clustering structure induced by the level set tree of the kernel estimate.

Note that Figure 8.13 uses subsetting in visualizing the observations. Among the observations belonging to the same rectangle of the piecewise constant kernel estimate, only one is shown.

8.4 TAIL CLUSTERING

8.4.1 Population Interpretation

Tail clusters may be defined as regions of low density separated from other such regions after removing the regions of high density and regions of zero density (or regions of extreme low density). Tail clustering may be used in the cases where the underlying distribution is unimodal. For example, one might want to cluster a database of customers in order to concentrate marketing efforts on specific groups that may lie in the tails of the distribution.

8.4.2 Algorithms

Just as a likelihood tree provides a data-based version of high-density clustering, a tail tree as defined in Chapter 6 provides a data-based version of tail clustering. The comprehensive tail clustering would go through all the levels of a likelihood tree and make a tail clustering for each separated component in the given level.

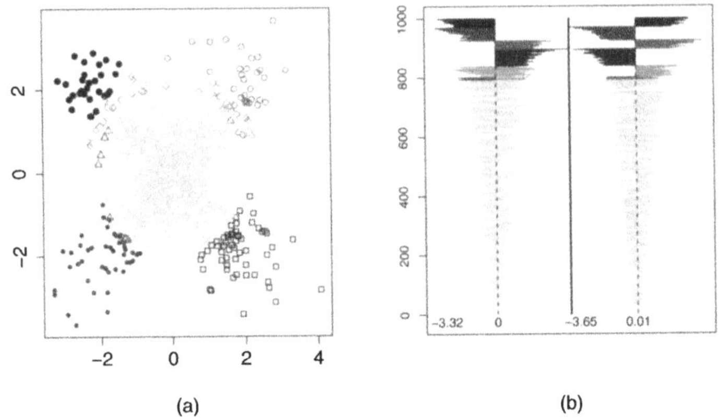

Figure 8.14 *(Tail segments and a graphical matrix.)* (a) Data of size 1000 with the segments given by a tail tree colored. The colors are chosen according to the segmentation induced by the tail tree in Figure 6.9(c) and (d). (b) A graphical matrix where the observations belonging to the same segment are grouped together.

The segmentation of a tree to branches is defined in Definition D.3.1. A segmentation of a tail tree induces a segmentation of the data, since the nodes of a tail tree are associated with observations. A segmentation of the data, corresponding to a tail tree, is such that the observations corresponding to the nodes that are in the same branch belong to the same segment: a child of node m belongs to the same segment as m if it does not have siblings; otherwise, the children of m are the starting nodes of new branches, and thus each is a starting node of a new segment. The *tail clusters* are those segments whose branches contain a leaf node.

Figure 8.14(a) shows a scatter plot of data with four tail clusters. The segments given by the tail tree are colored.

Tail clustering has similarities with divisive hierarchical clustering. Divisive clustering takes as the first cluster the complete data, as does the tail tree. After that divisive clustering makes a recursive partition of the data to smaller clusters. In a tail tree, however, the number of observations is decreasing at each step, whereas in a divisive clustering tree a parent is always a union of its children. Hierarchical clustering tree (dendrogram), in general, makes a recursive partition of the observations so that the union of the sets associated with children is always equal to the set associated with the parent, and thus has a different structure than a tail tree.

8.4.3 Visualization

Graphical Matrices Tail clusters may be visualized with graphical matrices by using the permutation given by the the tail clustering. One finds tail clusters of the

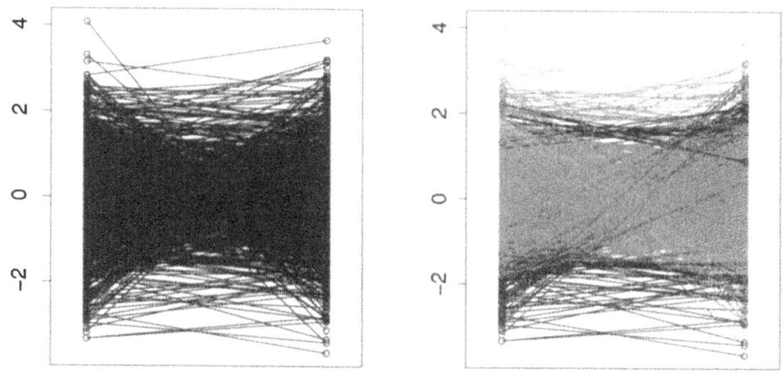

(a) (b)

Figure 8.15 *(Parallel coordinate plots.)* (a) A parallel coordinate plot of the data in Figure 8.14(a). (b) A parallel coordinate plot where the lines and points are colored according to the segmentation.

data, and chooses a permutation where the indexes of the observations that belong to the same cluster are together in the graphical matrix.

Figure 8.14(b) shows a graphical matrix after a permutation corresponding to the segmentation in panel (a). The observations belonging to the same segment are together, and the observations are ordered inside each group according to the distance from the arithmetic mean of the complete data. The four tails are clearly evident. For example, the "flag" at the top of the graphical matrix points to the left in the column of the first coordinate, and to the right in the column of the second coordinate. Thus we know that this tail lies in the northwest direction from the center.

Parallel Coordinate Plots The segmentation of the data induced by a tail tree can be used to enhance parallel coordinate plots. Figure 8.15(a) shows a parallel coordinate plot of the data in Figure 8.14(a). In panel (b) is a colored parallel coordinate plot, where the lines and points are colored according to the segmentation.

PART II

ANALYTICAL AND ALGORITHMIC TOOLS

CHAPTER 9

DENSITY ESTIMATION

In Section 9.1 we define the concepts of a density function and a density estimator. In Section 9.2 we consider preprocessing of data: before starting density estimation it may be useful to preprocess data. We mention two preprocessing methods: data sphering and copula preprocessing.

In Section 9.3 we discuss time series concepts related to density estimation and the setting of inverse problems. In density estimation it is typically assumed that the observations $X_1, \ldots, X_n$ are i.i.d. (independent and identically distributed) with density function $f : \mathbf{R}^d \to \mathbf{R}$. These assumptions can be relaxed. In Section 9.3.1 we consider the relaxation of the assumption of identically distributed observations. In Section 9.3.2 we define numerical measures of dependency that can be used to define relaxations of the assumption of independence. Later, in Section 14.3 of Part III we analyze asymptotic properties of kernel estimators under relaxed assumptions, but the other estimators will be analyzed in Part III under the assumption of i.i.d. observations. Section 9.3.3 defines the concepts of stationarity and approximate stationarity and discusses estimation of the joint distribution of consecutive observations in the setting of serial dependency. In Section 9.3.4 we define the concept of a statistical inverse problem. Two examples will be mentioned: convolution operator and Radon transform.

In Section 9.4 we define the following three statistical estimation problems related to density estimation: regression function estimation, classification function estimation, and signal function estimation in the Gaussian white noise model. The Gaussian white noise model will be heavily utilized as a heuristic tool in the theory of lower bounds in Chapter 11, and in the asymptotic analysis of the minimization estimators in Chapter 15.

9.1 DENSITY FUNCTIONS AND ESTIMATORS

9.1.1 Density Function

A density function is a nonnegative function $f : \mathbf{R}^d \to \mathbf{R}$ that integrates to one:

- $\int_{\mathbf{R}^d} f = 1$,
- $f(x) \geq 0$ for almost all $x \in \mathbf{R}^d$.

The integration is with respect to the Lebesgue measure, and "for almost all $x \in \mathbf{R}^d$" means that the property holds in a set whose complement has Lebesgue measure zero. A density function is a density function of probability measure P if the probabilities of sets may be written as the integrals of the density function:

$$P(A) = \int_A f$$

for every Borel set $A \subset \mathbf{R}^d$. A given probability measure has many densities. The equivalence class of the densities of a given probability measure consists of those densities that differ only in the sets of Lebesgue measure zero.

A density function describes how the probability mass is distributed over the d-dimensional Euclidean space. For example, local maxima of a density function (modes) indicate a concentration of the probability mass, and heavy tails of a density function indicate a high probability of extreme events. A density function gives an intuitive description of the probability distribution, at least in one- and two-dimensional cases.

9.1.2 Density Estimator

Our data are realizations of identically distributed random vectors $X_1, \ldots, X_n \in \mathbf{R}^d$ whose common distribution has density $f : \mathbf{R}^d \to \mathbf{R}$. We want to estimate the density function using these data. It is also possible to estimate a density if we have only approximately identically distributed observations. A density estimator is a measurable map from $(\mathbf{R}^d)^n$ into the set of multivariate functions: $(X_1, \ldots, X_n) \mapsto \hat{f}(\cdot, X_1, \ldots, X_n)$. We denote density estimators with

$$\hat{f} : \mathbf{R}^d \to \mathbf{R}$$

and do not show the dependence of $\hat{f}$ from the sample $X_1, \ldots, X_n$ explicitly. Sometimes a density estimate is not a density: it may take negative values or it might not integrate to unity. We call a *density estimator* a mapping from the observations to the set of multivariate functions (or to the set of densities), and a *density estimate* is a function that is the value of this mapping for a particular sample.

9.2 PREPROCESSING OF DATA

Before starting density estimation, it is useful to preprocess the data. We discuss two methods of preprocessing: data sphering and copula preprocessing.

9.2.1 Data Sphering

We can make the scales of variables compatible by normalizing each column of the data matrix to have unit variance. Data sphering is more extensive transformation; we make such linear transformation of data that the covariance matrix becomes the identity matrix. The sphering is almost the same as the principal component transformation, defined in Section 1.8.1. In the principal component transformation the covariance matrix was diagonalized, but it was not made the identity matrix.

1. Sphering of a random vector $X \in \mathbf{R}^d$ means that we make a linear transform of X so that the new random variable has expectation zero and the identity covariance matrix. Let

$$\Sigma = E(X - EX)(X - EX)^T$$

be the covariance matrix and make the spectral representation of Σ:

$$\Sigma = A\Lambda A^T,$$

where A is orthogonal and Λ is diagonal. Then

$$Y = \Lambda^{-1/2} A^T (X - EX)$$

is the sphered random vector. Indeed

$$\mathrm{Cov}(Y) = \Lambda^{-1/2} A^T \mathrm{Cov}(X) A \Lambda^{-1/2} = \Lambda^{-1/2} A^T \Sigma A \Lambda^{-1/2} = I_d.$$

2. Data sphering of the data means that the data matrix is transformed so that the arithmetic mean of each column is zero and the empirical covariance matrix is the unit matrix. Let Σ_n be the empirical covariance matrix, as defined in (1.5),

$$\Sigma_n = \frac{1}{n} \sum_{i=1}^{n} (X_i - \bar{X}_n)(X_i - \bar{X}_n)^T,$$

where $\bar{X}_n$ is the $d \times 1$ column vector of arithmetic means. We find the spectral representation of Σ_n, as defined in (1.6),

$$\Sigma_n = A_n \Lambda_n A_n^T.$$

The sphered data matrix is

$$\mathbb{Y}_n = \left(\mathbb{X}_n - 1_{n\times 1}\bar{X}_n^T\right) \times A_n \Lambda_n^{-1/2},$$

where $\mathbb{X}_n$ is the original $n \times d$ data matrix, and $1_{n\times 1}$ is the $n \times 1$ column vector of ones.

9.2.2 Copula Preserving Transform

In multivariate density estimation we are interested in estimating the dependence between the variables. Marginal densities are easier to estimate since their estimation is a univariate problem. Thus it makes sense to transform the data in such a way that the influence of the marginal distributions is eliminated and only the information about the dependence is left in the data. Copula preserving preprocessing makes such a transformation. Copula preserving preprocessing is a data version of a transformation that leads to a nonstandard copula. The nonstandard copula is defined in (10.15). Copula preserving preprocessing of data matrix $\mathbb{X}_n = (x_i^j)$, $i = 1, \ldots, n$, $j = 1, \ldots, d$, is defined in two steps.

1. Make each margin approximately uniformly distributed by the level transformation of (1.1). Let z_i^j, $i = 1, \ldots, n$, $j = 1, \ldots, d$, be the number of observations smaller or equal to x_i^j, divided by n (z_i^j is the rank of x_i^j, divided by n):
$$z_i^j = n^{-1}\#\left\{x_l^j : x_l^j \le x_i^j,\ l = 1, \ldots, n\right\}.$$

2. From step 1 each margin is approximately uniformly distributed. When $X \sim \mathrm{Unif}([0,1])$, then $F^{-1}(X) \sim F$, where F is any continuous distribution function. Next make the margins to be approximately normally distributed by defining
$$y_i^j = \Phi^{-1}(z_i^j), \qquad i = 1, \ldots, n,\ j = 1, \ldots, d,$$
where Φ is the distribution function of the standard Gaussian distribution. The copula preprocessed data matrix is $\mathbb{Y}_n = (y_i^j)$, $i = 1, \ldots, n$, $j = 1, \ldots, d$.

We call a copula preserving transform for shortness also a copula transform.

9.2.3 Illustrations

We illustrate data sphering and the copula transform.

1. Figure 9.1(a) shows a scatter plot of a simulated sample of size 500, where the marginals are uniform on $[0, 1]$ and independent from each other. Panel (b)

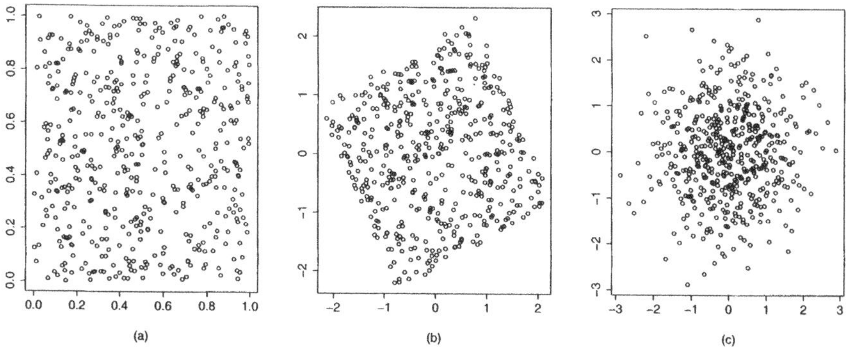

Figure 9.1 *(Independent uniform marginals.)* Data preprocessing with data whose marginals are independent and uniformly distributed. (a) A scatter plot of the data, (b) the sphered data, and (c) the copula transformed data.

shows sphered data and panel (c) shows copula transformed data, where the marginals are approximately standard Gaussian. The data in panel (c) is distributed as standard 2D Gaussian.

2. Figure 9.2(a) shows a scatter plot of exchange rates of Brazilian Real and Mexican new Peso between 1995-01-05 and 2007-09-26. The rates are with respect to one U.S. dollar and transformed to returns (we make the mapping $r_i \mapsto (r_i - r_{i-1})/r_{i-i}$). There are 3197 observations. Panel (b) shows sphered data, and panel (c) shows copula transformed data, where the marginals are approximately standard Gaussian. The data are provided by the Federal Reserve Economic Data (http://research.stlouisfed.org).

3. Figure 9.3(a) shows a scatter plot of the returns of the German stock index DAX and the French stock index CAC between 1990-01-05 and 2008-01-14. There are 4277 observations. Panel (b) shows sphered data, and panel (c) shows copula transformed data, where the marginals are approximately standard Gaussian. The marginals appear to be almost independent. The data are provided by Yahoo.

9.3 SETTINGS OF DENSITY ESTIMATION

In the basic setting that we will consider we are interested in estimating the distribution of the random vector $X = (X^1, \ldots, X^d) \in \mathbf{R}^d$. We may estimate the distribution of X if we have a sequence $X_1, \ldots, X_n$ of identically distributed copies of X. We mention two complications that can arise.

1. The first complication arises when $X_1, \ldots, X_n$ are not identically distributed but only approximately identically distributed. If the observations are not

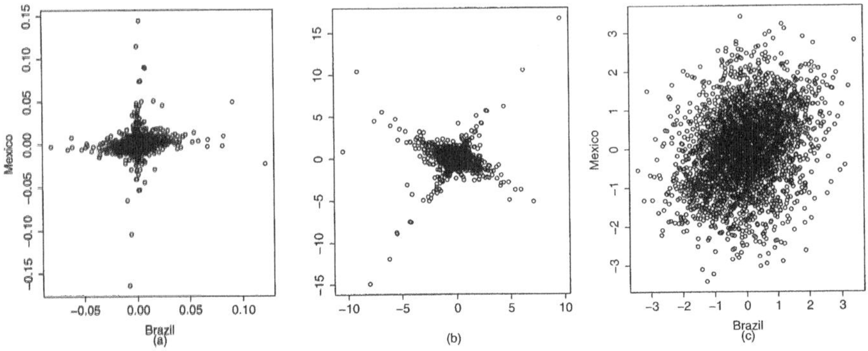

Figure 9.2 *(Exchange rates.)* Data preprocessing with data of exchange rates of Brazilian Real and Mexican Peso ($n = 3197$). (a) A scatter plot of the data, (b) the sphered data, and (c) the copula transformed data.

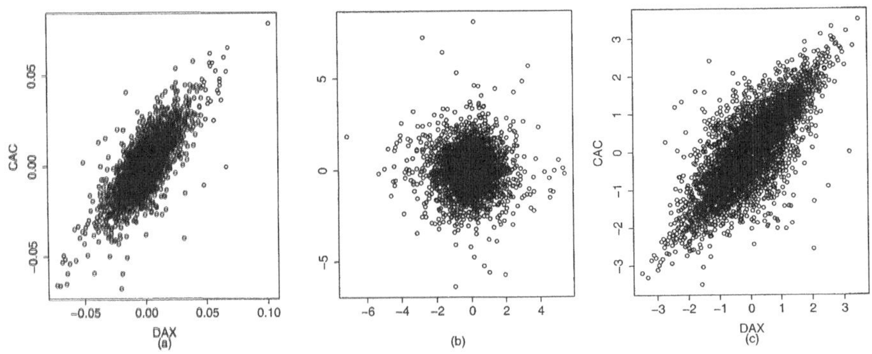

Figure 9.3 *(Stock indexes.)* Data preprocessing with data of the German stock index DAX and the French stock index CAC ($n = 4277$). (a) A scatter plot of the data, (b) the sphered data, and (c) the copula transformed data.

identically distributed, but the distribution is changing continuously as the index of the observations changes, then we may estimate the distribution of X_{i_0}, for a fixed $i_0 \in \{1, \ldots, n\}$, by choosing a window of nearby observations

$$X_{i_0-k}, \ldots, X_{i_0}, \ldots, X_{i_0+k},$$

and using these to estimate the distribution of X_{i_0}.

2. The second complication arises when $X_1, \ldots, X_n$ are not independent. If the later observations depend heavily on the first observation, then it is improbable that one could estimate accurately the distribution of X using the sequence $X_1, \ldots, X_n$; under heavy dependence the information in the sequence would be equal to the information in a single (the first) observation. We will need to assume that the observations are approximately independent, in the sense that only nearby observations depend on each other, and the dependence decays when the distance between the observations grows.

In the *time series* setting we observe a sequence $X_1, \ldots, X_T \in \mathbf{R}^p$, where X_i is observed earlier than X_j if $i < j$. It is important to distinguish between the *cross-sectional dependency* and the *serial dependency.*

1. Estimation of the cross-sectional dependency is identical to our basic setting (when we make the identifications $n = T$ and $d = p$). We want to estimate the joint distribution of the components in X_{t_0} for some $t_0 \in \{1, \ldots, T\}$. In other words, we are interested in estimating the joint distribution of real valued random variables $X_{t_0}^1, \ldots, X_{t_0}^p \in \mathbf{R}$.

2. Serial dependency means the dependency between consecutive observations. The interest in serial dependency arises when the random vectors in the sequence $X_1, \ldots, X_T$ are not independent. For example, we may be interested in the estimation of the distribution of the pair of two consecutive observations (X_1, X_2). In the identically distributed case the estimation may be done with the help of the observations

$$(X_1, X_2), \ldots, (X_{T-1}, X_T).$$

We are in the basic setting when we make the identifications $n = T - 1$ and $d = 2p$. More generally, we may be in the non-identically distributed case, and we may be interested in the estimation of the distribution of the pair (X_{t_0}, X_{t_0+1}) for a fixed $t_0 \in \{1, \ldots, T\}$. The estimation may be done with the help of a window of observations

$$(X_{t_0-k}, X_{t_0+1-k}), \ldots, (X_{t_0}, X_{t_0+1}), \ldots, (X_{t_0+k}, X_{t_0+1+k}),$$

if these pairs are approximately identically distributed, in which case we say that the time series is approximately stationary; see Definition 9.3. The time series setting is closely related to the *spatial* setting where the observations are not indexed with the time index but with the location.

In some cases we are not interested in estimating the density of the observations but some underlying density. In *statistical inverse problems* the observations are available from function Af, where A is an operator, but we want to estimate f. In the case of density estimation this means that we observe $Y_1, \ldots, Y_n$, whose density is Af, but we want to estimate density f.

Section 9.3.1 discusses the concept of approximately identically distributed observations. Section 9.3.2 discusses the measures of dependence. Section 9.3.3 discusses the estimation of serial dependency and the concept of stationarity. Section 9.3.4 discusses statistical inverse problems related to density estimation.

9.3.1 Locally Identically Distributed Observations

In our basic setting the observations $X_1, \ldots, X_n$ are identically distributed, and we want to estimate the distribution of a single observation, say the distribution of X_n. However, in many cases the observations are not exactly identically distributed.

1. We call a *vector time series* a sequence of random vectors $X_1, \ldots, X_n \in \mathbf{R}^d$, where X_i is observed earlier than X_j, when $1 \le i < j \le n$. We might want to estimate the density at some time point, for example, the density of the last observation X_n. In a time series setting the distribution might be changing over time. For example, X_i could be the return of a stock at the ith day, but the distribution of returns is changing over time because of the changes in the characteristics of the company and the economy.

2. We call *spatial data* such data where the observations are made at different geographical locations. For example, X_i might be the amount of radiation emitted at a given location. Then it is reasonable to assume that the distribution depends on the geographical location and is not identically distributed over locations. In some cases we may treat the geographical location as an additional random variable, but in other cases the locations are not chosen randomly.

In order for the estimation of the density of X_i to be possible, we need observations that are at least approximately identically distributed. If the distribution is changing without any continuity, then we would not have enough information to estimate the density of X_i. We need a concept of locally identically distributed observations. When there exists a probability measure P such that a sufficient number of random vectors in the sequence have a distribution close to P, we say that the random vectors are locally identically distributed. Then we may also find an estimator that accurately estimates P. We call probability measure P the true distribution.

We assume that there exists a mapping $t : \{1, \ldots, n\} \to \mathbf{T}$ that gives the location of the observation in time or space. In the time series setting, we may take $\mathbf{T} = [0, 1]$ and $t(i) = i/n$. In the spatial setting, we may have $\mathbf{T} = [0, 1]^2$, $\mathbf{T} = [0, 1]^3$, or $\mathbf{T} = \mathbf{S}_2$, where $\mathbf{S}_2$ is the unit sphere in $\mathbf{R}^3$ and $t(i)$ gives the spatial location of the observation X_i. We use the time series terminology and call $t(i)$ the time point of the observation.

Approximately locally distributed observations are such that when $t(i)$ and $t(j)$ are close, the densities of X_i and X_j are close with respect to some distance between the densities. We assume that $\mathbf{T}$ is a metric space with metric $\delta : \mathbf{T} \times \mathbf{T} \to [0, \infty)$. We use the L_2 distance between densities to measure the distance between probability distributions.

We may suppose that there exists a collection of density functions $\{g_u : u \in \mathbf{T}\}$, and we want to estimate density function g_{t_0} for a fixed $t_0 \in \mathbf{T}$. If the density of X_i is close to g_{t_0} when $t(i)$ is close to t_0, and if there is enough such indexes i that $t(i)$ is close to t_0, then sequence $X_1, \ldots, X_n$ may be used to estimate g_{t_0}.

We define the concept of locally identically distributed observations as an asymptotic concept, as $n \to \infty$. We can define the concept quantitatively according to the rate of convergence. When the rate is slow, we can guarantee only the existence of consistent estimators. With faster rates of distributions approaching closeness we can guarantee the existence of estimators with faster rates of convergence.

Definition 9.1 (Locally identically distributed observations.) *A sequence of random vectors $X_1, \ldots, X_n \in \mathbf{R}^d$ is* locally identically distributed, *with rate function $r : [0, \infty) \to [0, \infty)$, with rate $o(1)$, at time point $t_0 \in \mathbf{T}$, when there exists density $f : \mathbf{R}^d \to \mathbf{R}$ such that*

$$\lim_{n\to\infty} \frac{1}{n} \sum_{i=1}^{n} r[\delta(t(i), t_0)] \, \|f_i - f\|_2 = 0 \tag{9.1}$$

and with rate $O(n^{-1/2})$, when

$$\limsup_{n\to\infty} n^{-1/2} \sum_{i=1}^{n} r[\delta(t(i), t_0)] \, \|f_i - f\|_2 < \infty, \tag{9.2}$$

where $f_i : \mathbf{R}^d \to \mathbf{R}$, $i = 1, \ldots, n$, is the density function of X_i.

In order the setting to be reasonable we need to assume that around the time point t_0 of interest there are asymptotically a nonnegligible amount of observations.

Assumption 9.1 *Let $t_0 \in \mathbf{T}$ be the point of interest, and let $r : [0, \infty) \to [0, \infty)$ be the rate function in Definition 9.1. We assume that*

$$\liminf_{n\to\infty} \frac{1}{n} \sum_{i=1}^{n} r[\delta(t(i), t_0)] > 0.$$

For example, when $\mathbf{T} = [0, 1]$, $\delta(t, s) = |t - s|$, $t(i) = i/n$, and $t_0 = 1$, we have

$$\frac{1}{n} \sum_{i=1}^{n} r[\delta(t(i), t_0)] = \frac{1}{n} \sum_{i=1}^{n} r(|i/n - 1|) \sim \int_0^1 r(|s - 1|)\, ds = \int_0^1 r,$$

as $n \to \infty$. Assumption 9.1 holds when $0 < \int_0^1 r < \infty$.

We make several observations concerning Definition 9.1.

1. To construct a consistent estimate for f, we need to assume (9.1); to construct an estimate with fast rates of convergence, we need to assume (9.2). See Theorem 14.4 for the rates of convergence of a kernel estimator and Theorem 14.5 for the consistency of a kernel estimator.

2. It is natural to assume that the rate function $r : [0, \infty) \to [0, \infty)$ is monotonically decreasing; for example, $r(x) = (1 - x^q)_+$ or $r(x) = \exp(-x^q)$, for $q \geq 1$. Then Definition 9.1 says that when $t(i)$ are close to t_0, then f_i are close to f up to a negligible number of indexes i. Rate function r quantifies the closeness of f_i to f. If Definition 9.1 holds for rate function r, then it holds for "slower" or "smaller" rate functions as well. If r satisfies (9.1) or (9.2), then r^* satisfies (9.1) or (9.2) provided that
$$\limsup_{n \to \infty} \max_{i=1,\ldots,n} \frac{r^*[\delta(t(i), t_0)]}{r[\delta(t(i), t_0)]} < \infty.$$

3. We have defined locally identically distributed random vectors as an asymptotic concept, as $T \to \infty$. The type of asymptotics we used is called time rescaling (Dahlhaus 1997) or in-fill asymptotics. The in-fill asymptotics is often used to analyze estimators in the setting where one assumes that there exists a continuous time stochastic process $Z(t)$, $t \in [0, 1]$, and the observations are sampled from this process: $X_i = Z(i/n)$, $i = 1, \ldots, n$.

4. Two nonstationary models that have been used in statistics include the mean-nonstationary model
$$X_i = \mu_i + Z_i,$$
where $\mu_i \in \mathbf{R}^d$ and Z_i are i.i.d. with $EZ_i = 0$, and the volatility-nonstationary model
$$X_i = \sigma_i Z_i,$$
where $\sigma_i > 0$ and Z_i are i.i.d. This volatility-nonstationary model has been used as an alternative for stationary GARCH modeling; see Stărică and Granger (2005). For the mean-nonstationary model one has $f_i(x) = f_Z(x - \mu_i)$ where f_Z is the density of Z_i, and thus one can write, under smoothness assumptions on f_Z,
$$\|f_i - f_j\|_2 \leq \|\mu_i - \mu_j\| \cdot \|Df_Z\|_2 .$$
Thus X_i are locally identically distributed in the sense of Definition 9.1 if μ_i is sufficiently smooth as function of i. For the volatility-nonstationary model we have $f_i(x) = f_Z(x/\sigma_i)/\sigma_i^d$ and thus we can write, under smoothness assumptions on f_Z,
$$\begin{aligned}
&\|f_i - f_j\|_2 \\
&\quad \leq \left\|[f_Z(x/\sigma_i) - f_Z(x/\sigma_j)]/\sigma_i^d\right\|_2 + \left\|f_Z(x/\sigma_j)/\sigma_i^d - f_Z(x/\sigma_j)/\sigma_j^d\right\|_2 \\
&\quad = \sigma_i^{-d/2} \|f_Z(x) - f_Z(x\sigma_i/\sigma_j)\|_2 + \left|\sigma_i^{-d} - \sigma_j^{-d}\right| \sigma_j^{d/2} \|f_Z\|_2 \\
&\quad = |1 - \sigma_i/\sigma_j| \, \sigma_i^{-d/2} \left\|x^T Df_Z(\xi)\right\|_2 + \left|1 - (\sigma_j/\sigma_i)^d\right| \sigma_j^{-d/2} \|f_Z\|_2 ,
\end{aligned}$$

where $\xi = \xi_{x,\sigma_i,\sigma_j}$ is between x and $x\sigma_i/\sigma_j$. We use the abuse of notation $\|f\|_2 = \|f(x)\|_2$ when x is the integration variable. Thus X_i are locally identically distributed in the sense of Definition 9.1 if the ratio σ_j/σ_i is close to one and σ_i^{-1} are bounded.

9.3.2 Quantifying Dependence

The definition of independence of random vectors X_1 and X_2 is that

$$P(X_1 \in A,\, X_2 \in B) = P(X_1 \in A) \cdot P(X_2 \in B),$$

for all measurable $A, B \subset \mathbf{R}^d$. This is equivalent to $P(X_1 \in A \mid X_2) = P(X_1 \in A)$, so knowledge of X_2 does not affect the probability evaluations of X_1. The complete dependence between random vectors X_1 and X_2 occurs when there is a bijection $G : \mathbf{R}^d \to \mathbf{R}^d$ so that $X_2 = G(X_1)$ holds almost everywhere. Real-valued random vectors V and W are said to be comonotonic if there is a strictly increasing function $g : \mathbf{R} \to \mathbf{R}$ so that $W = g(V)$ almost everywhere, and V and W are said to be countermonotonic if there is a strictly decreasing function $g : \mathbf{R} \to \mathbf{R}$ so that $W = g(V)$ almost everywhere. In the other cases we may try to quantify the dependency between two random vectors. We may also say that two random vectors have the same dependency when they have the same copula, and the copula is defined as in Section 10.1.4.

Dependency measures can be used to define approximate independence, and these measures can also be used in solving some optimization problems. For example, we might want to find a portfolio of random variables that are as independent as possible. In order to formulate the corresponding optimization problem we need to quantify the dependence.

Concordance and Dependence Measures A concordance measure takes values on the interval $[-1, 1]$. Independence implies the value 0, total concordance implies value 1, and total discordance implies value -1. A dependence measure takes values on the interval $[0, 1]$. Independence implies the value 0 and comonotonicity implies the value 1.

Concordance is measured (1) between two real-valued random variables (real vs. real) and (2) between a real-valued random variable and a vector of real-valued random variables (real vs. vector). Dependence is measured (3) between two vectors of real-valued random variables (vector vs. vector).

Real versus Real Concordance measures between two real-valued random variables include the linear correlation coefficient, Spearman's rank correlation, and Kendall's rank correlation.

Linear Correlation The linear correlation coefficient is the normalized covariance

$$\rho(X^1, X^2) = \frac{\mathrm{Cov}(X^1, X^2)}{\mathrm{std}(X^1)\,\mathrm{std}(X^2)}, \tag{9.3}$$

where we denote $\mathrm{Cov}(X^1, X^2) = E[(X^1 - EX^1)(X^2 - EX^2)]$ and $\mathrm{std}(X^j) = \sqrt{\mathrm{Var}(X^j)}$. When $X^1$and X^2 are continuous, we have

$$\begin{aligned}\mathrm{Cov}(X^1, X^2) &= E(X^1X^2) - EX^1EX^2\\ &= \int_{\mathbf{R}^2} \left(F_{X^1,X^2}(u,v) - F_{X^1}(u)F_{X^2}(v)\right) du\, dv,\\ &= \int_{[0,1]^2} [C(u,v) - uv]\, dF_{X^1}^{-1}(u)\, dF_{X^2}^{-1}(v),\end{aligned}$$

where $C(u,v) = F_{X^1,X^2}(F_{X^1}^{-1}(u), F_{X^2}^{-1}(v))$, $u, v \in [0,1]$, is the copula as defined in Section 10.1.4. Note that the linear correlation is not solely a function of the copula, it depends also on the marginal distributions F_{X^1}, F_{X^2}. When $\rho(X^1, X^2) = 0$, this does not imply that X^1 and X^2 are independent. It can happen that $\rho(X^1, X^2) = 0$ even when X^1 and X^2 are deterministically related with each other; see Exercise 9.1 and Exercise 9.2.

The linear correlation can be estimated with the sample correlation when we have samples $X_1^j, \ldots, X_n^j$, $j = 1, 2$, from the distributions of X^1 and X^2, respectively. We define the sample correlation coefficient as

$$\begin{aligned}\rho_n &= \rho_n(\{X_1^1, \ldots, X_n^1\}, \{X_1^2, \ldots, X_n^2\})\\ &= \frac{1}{\sigma_{X^1}\sigma_{X^2}} \sum_{i=1}^{n} (X_i^1 - \bar{X}^1)(X_i^2 - \bar{X}^2),\end{aligned} \tag{9.4}$$

where $\bar{X}^j = n^{-1}\sum_{i=1}^n X_i^j$, $(\sigma_{X^j})^2 = n^{-1}\sum_{i=1}^n (X_i^j - \bar{X}^j)^2$.

Spearman's Rank Correlation Spearman's rank correlation (Spearman's rho) is defined by

$$\rho_S(X^1, X^2) = \rho(F_1(X^1), F_2(X^2)),$$

where F_j is the distribution function of X^j. If X^1 and X^2 have a continuous distribution, then

$$\begin{aligned}\rho_S(X^1, X^2) &= 12 \int_{\mathbf{R}^2} \left[F_{X^1,X^2}(u,v) - F_{X^1}(u)F_{X^2}(v)\right] dF_{X^1}(u)\, dF_{X^2}(v)\\ &= 12 \int_{[0,1]^2} [C(u,v) - uv]\, du\, dv\\ &= 12 \int_{[0,1]^2} C(u,v)\, du\, dv - 3\\ &= 12 \int_{[0,1]^2} u \cdot v\, dC(u,v) - 3,\end{aligned}$$

where $C(u,v) = F_{X^1,X^2}(F_{X^1}^{-1}(u), F_{X^2}^{-1}(v))$, $u, v \in [0,1]$, is the copula as defined in Section 10.1.4. We have still an other way of writing Spearman's rank correlation. Let

Z and Y be distributed as $X = (X^1, X^2)$: $Z = (Z^1, Z^2) \sim X, Y = (Y^1, Y^2) \sim X$, and let X, Y, and Z be independent. Then

$$\rho_S(X^1, X^2) = 2\left(P[(X^1 - Y^1)(X^2 - Z^2) > 0] - P[(X^1 - Y^1)(X^2 - Z^2) < 0]\right).$$

The data version is based on the ranks. Let us denote with $r_j : \{1, \ldots, n\} \to \{1, \ldots, n\}$ the rank of the observations $X_1^j, \ldots, X_n^j$, $j = 1, 2$. The ranks give the order of the observation:

$$r_j(i) = \begin{cases} 1, & \text{when } X_i^j = X_{(1)}^j, \\ 2, & \text{when } X_i^j = X_{(2)}^j, \\ \vdots & \end{cases}$$

where $X_{(1)}^j < \cdots < X_{(n)}^j$ are the ordered observations, $j = 1, 2$. (We may write $X_{r^{-1}(1)}^j < \cdots < X_{r^{-1}(n)}^j$.) Then the data version of the Spearman's rank correlation is defined by

$$\rho_{S,n}(X^1, X^2) = \rho_n(\{r_1(1), \ldots, r_1(n)\}, \{r_2(1), \ldots, r_2(n)\}),$$

where ρ_n is the empirical correlation coefficient defined in (9.4).

Kendall's Rank Correlation Kendall's rank correlation (Kenadall's tau) is defined by

$$\rho_\tau(X^1, X^2) = P[(X^1 - Y^1)(X^2 - Y^2) > 0] - P[(X^1 - Y^1)(X^2 - Y^2) < 0],$$

where $Y = (Y^1, Y^2) \sim X = (X^1, X^2)$, and X and Y are independent. When X^1 and X^2 are continuous, we have

$$\rho_\tau(X^1, X^2) = 2P[(X^1 - Y^1)(X^2 - Y^2) > 0] - 1,$$

and we can write

$$\begin{aligned} \rho_\tau(X^1, X^2) &= 4\int_{\mathbf{R}^2} F_{X^1,X^2}(u,v)\, dF_{X^1,X^2}(u,v) - 1 \\ &= 4\int_{[0,1]^2} C(u,v)\, dC(u,v) - 1, \end{aligned}$$

where $C(u,v) = F_{X^1,X^2}(F_{X^1}^{-1}(u), F_{X^2}^{-1}(v))$, $u, v \in [0,1]$, is the copula as defined in Section 10.1.4.

Let us define an estimator $\rho_{\tau,n}$ for $\rho_\tau(X^1, X^2)$. We call a pair (X_i^1, X_j^1) and (X_i^2, X_j^2) concordant if the ranks have the same order: (1) $r_1(i) < r_1(j)$ and $r_2(i) < r_2(j)$ or (2) $r_1(i) > r_1(j)$ and $r_2(i) > r_2(j)$. Otherwise, the pair is called discordant. Let n_c be the number of concordant pairs over all $n(n-1)/2$ pairs, and let n_d be the number of discordant pairs: $n_d = n(n-1)/2 - n_c$. Define

$$\rho_{\tau,n} = \frac{n_c - n_d}{n_c + n_d}.$$

We have assumed that there are no ties.

Real versus Vector We define the generalized correlation between a real-valued random variable and a vector of real-valued random variables. Let $Y \in \mathbf{R}$ be a random variable and $X = (X^1, \ldots, X^d) \in \mathbf{R}^d$ be a random vector, where $d \geq 2$. Define (the canonical coefficient of d-correlation)

$$\rho_G = \max_{\phi} \rho(Y, \phi^T X),$$

where ρ is the linear correlation defined in (9.3) and the maximum is over d-vectors. We may show that

$$\rho_G = \mathrm{Cov}(Y, X)\,\mathrm{Cov}(X)^{-1}\,\mathrm{Cov}(X, Y)\,\mathrm{Var}(Y)^{-1}, \tag{9.5}$$

where $\mathrm{Cov}(Y, X)$ is the $1 \times d$ vector of pairwise covariances, $\mathrm{Cov}(X)$ is the $d \times d$-covariance matrix defined by (1.4), and $\mathrm{Cov}(X, Y)$ is the $d \times 1$ vector of pairwise covariances. In addition the maximum over ϕ is achieved by the eigenvector corresponding to the largest eigenvalue of the $d \times d$ matrix

$$\mathrm{Cov}(X)^{-1}\,\mathrm{Cov}(X, Y)\,\mathrm{Var}(Y)^{-1}\,\mathrm{Cov}(Y, X),$$

and ρ_G is this largest eigenvalue.

Vector versus Vector Dependence measures between two random vectors include the regular mixing coefficients and the strong mixing coefficients.

Regular Mixing Coefficients Regular mixing coefficients are also called β-mixing coefficients. These coefficients were introduced by Volkonskii and Rozanov (1959) and Kolmogorov and Rozanov (1961). Let $X \in \mathbf{R}^d$, $X' \in \mathbf{R}^{d'}$ be random vectors, let $f_X, f_{X'}$ be their density functions, and let $f_{X,X'}$ be the density function of the joint distribution of X, X'. The β-mixing coefficient is defined as the half of the L_1 distance between the joint density and the product of marginals:

$$\beta(X, X') = \frac{1}{2} \int_{\mathbf{R}^d \times \mathbf{R}^{d'}} |f_{X,X'} - f_X f_{X'}| \,. \tag{9.6}$$

We denote $\beta(X, X) = 1$. We can write the β-mixing coefficient in a more general way, using Scheffé's equality given in (11.5). Let $P_X, P_{X'}$ be the probability measures of random vectors X, X', and let $P_{X,X'}$ be the probability measure of the joint distribution of X, X'. Let $P_X \otimes P_{X'}$ be the product measure of P_X and $P_{X'}$. Then

$$\begin{aligned}
&\beta(X, X') \\
&= \|P_{X,X'} - P_X \otimes P_{X'}\|_{tot} \\
&= \frac{1}{2} \sup_{\mathbb{A},\mathbb{B}} \sum_{A \in \mathbb{A}, B \in \mathbb{B}} [P(X \in A, X' \in B) - P(X \in A) \cdot P(X' \in B)] \quad (9.7) \\
&= \sup_W \int W \, d(P_{X,X'} - P_X \otimes P_{X'}), \quad (9.8)
\end{aligned}$$

where the supremum in (9.7) is over all measurable finite partitions $\mathbb{A}$ of $\mathbf{R}^d$ and $\mathbb{B}$ of $\mathbf{R}^{d'}$, and the supremum in (9.8) is over all random variables $0 \le W \le 1$, measurable with respect to the product sigma-algebra of the sigma-algebras generated by X and X'.

Strong Mixing Coefficients Strong mixing coefficients are also called α-mixing coefficients. Strong mixing coefficients were introduced by Rosenblatt (1956*a*). We have written the definition of the β-mixing coefficient in (9.7) so that it can be modified to get the definition of the α-mixing coefficient. The α-mixing coefficient is defined as

$$\begin{aligned} \alpha(X, X') &= \sup_{A,B} [P(X \in A, X' \in B) - P(X \in A) \cdot P(X' \in B)] \\ &= \sup_{A,B} \int I_{A \times B}\, d(P_{X,X'} - P_X \otimes P_{X'}), \end{aligned} \tag{9.9}$$

where the supremum is taken over Borel measurable subsets of $\mathbf{R}^d$. Strong mixing coefficients are weaker than regular mixing coefficients in the sense that

$$2\alpha(X, X') \le \beta(X, X');$$

see Doukhan (1994). The α-mixing coefficients are used sometimes in probability theory because they provide a natural weakening of the β-mixing coefficients.

Distance between Densities We may generalize the L_1 distance in (9.6) to other L_p distances and consider the dependence measure

$$\int_{\mathbf{R}^d \times \mathbf{R}^{d'}} |f_{X,X'} - f_X f_{X'}|^p .$$

Note, however, that only for the case $p = 1$ the measure of dependence is not influenced by the marginal distributions, in which case we get that

$$\int_{\mathbf{R}^d \times \mathbf{R}^{d'}} |f_{X,X'} - f_X f_{X'}| = \int_{[0,1]^d \times [0,1]^{d'}} |c - 1|, \tag{9.10}$$

where $c : [0,1]^d \times [0,1]^{d'} \to \mathbf{R}$ is the density of the copula C as defined in Section 10.1.4: $C(u,v) = F_{X,X'}(F_X^{-1}(u), F_{X'}^{-1}(v))$, $(u,v) \in [0,1]^d \times [0,1]^{d'}$. Indeed, we use the fact (compare to (10.19))

$$f_{X,X'}(u,v) = c(F_X(u), F_{X'}(v))\, f_X(u) f_{X'}(v)$$

to get (for the notational simplicity we write the equations for the case $X, X' \in \mathbf{R}$)

$$\begin{aligned} &\int_{\mathbf{R}^2} |f_{X,X'} - f_X f_{X'}|^p \\ &= \int_{\mathbf{R}^2} |[c(F_X(u), F_{X'}(v)) - 1]\, f_X(u) f_{X'}(v)|^p \, du\, dv \\ &= \int_{[0,1]^2} |c(u,v) - 1|^p \, \frac{|f_X(F_X^{-1}(u)) f_{X'}(F_{X'}^{-1}(v))|^p}{f_X(F_X^{-1}(u)) f_{X'}(F_{X'}^{-1}(v))} \, du\, dv. \end{aligned}$$

Also in the case of the Hellinger distance the dependence measure does not depend on the marginal distributions:

$$\frac{1}{2}\int_{\mathbf{R}^d\times\mathbf{R}^{d'}}\left(\sqrt{f_{X,X'}}-\sqrt{f_X f_{X'}}\right)^2=\int_{[0,1]^d\times[0,1]^{d'}}\left(1-\sqrt{c}\right)\,du_1du_2. \quad (9.11)$$

The Hellinger distance was considered by Granger, Maasoumi and Racine (2004).

Distance between Copulas We have measured dependence in (9.10) and (9.11) using the copula densities. However, it is also possible to use the copulas (distribution functions) to measure the dependence. Dependence may be measured by the L_p distance between the copula and the product copula. We give the definition for the case of two real-valued random variables. The measure of dependence is defined by

$$k_p\left(\int_{[0,1]^2}|C(u,v)-uv|^p\,du\,dv\right)^{1/p},$$

where k_p is the normalizing constant

$$k_p^{-1}=\|B(u,v)-uv\|_p=\|T(u,v-uv\|_p,$$

with $B(u,v)=\max\{u,v-1,0\}$ and $T(u,v)=\min(u,v)$. Fréchet-Hoeffding bounds state that for every copula C,

$$B(u,v)\le C(u,v)\le T(u,v),\qquad (u,v)\in[0,1]^2.$$

This distance was introduced by Schweizer and Wolff (1981).

Approximately Independent Observations When the dependence between the observations is weak, we may reduce the asymptotic theory of estimation to the case of independent observations. We define a concept of dependency in terms of β- and α-coefficients. We mention also the case of a spatial Poisson process.

β-Mixing Coefficients Dependence may be quantified in terms of the largeness of the regular mixing coefficients $\beta(X_i, X_j)$, defined in (9.8).

Assumption 9.2 *Let $r:[0,\infty)\to[0,\infty)$ be the rate function of Definition 9.1, and let $t_0\in\mathbf{T}$. Assume that*

$$\limsup_{n\to\infty}\frac{1}{n}\sum_{i,j=1}^{n}r[\delta(t(i),t_0)]\cdot r[\delta(t(j),t_0)]\,\beta(X_i,X_j)<\infty.$$

We make remarks concerning the Assumption 9.2.

1. Assumption 9.2 defines a localized dependency property, since it depends on $t_0\in\mathbf{T}$. We will take a rate function r the same as in Definition 9.1. Thus r is typically monotonically decreasing. We have multiplied with $r[\delta(t(i),t_0)]$ and

$r[\delta(t(j), t_0)]$, which are small when $t(i)$ or $t(j)$ are far away from t_0, when r is monotonically decreasing. Thus the assumption of the smallness of $\beta(X_i, X_j)$ is not so restrictive as in the case where we take $r \equiv 1$.

2. Let us consider the case where $X_1, \ldots, X_n$ is strictly stationary, as in Definition 9.2. We define $b_k = \beta(X_i, X_{i+k})$, for $i = 1, \ldots, n$, $k = 1 - i, \ldots n - i$. Numbers b_k do not depend on i, due to the strict stationarity. We have

$$\frac{1}{n} \sum_{i,j=1}^{n} \beta(X_i, X_j) = \frac{1}{n} \sum_{i=1}^{n} \sum_{k=1-i}^{n-i} \beta(X_i, X_{i+k}) \le 2 \sum_{k=0}^{\infty} b_k. \tag{9.12}$$

Thus the assumption $\sum_{k=0}^{\infty} b_k < \infty$ implies Assumption 9.2 with the rate function r as identically 1: $r \equiv 1$.

α-Mixing Coefficients Dependence may be quantified in terms of the largeness of the strong mixing coefficients $\alpha(X_i, X_j)$, defined in (9.9). We have to pose more stringent conditions on the α-mixing coefficients than we did on the β-mixing coefficients.

Assumption 9.3 *Let $r : [0, \infty) \to [0, \infty)$ be the rate function of Definition 9.1, and let $t_0 \in \mathbf{T}$. Assume that*

$$n^{-1} \sum_{i=1}^{n} \sum_{j=1, |j-i|>m}^{n} r[\delta(t(i), t_0)] \cdot r[\delta(t(j), t_0)]\ \alpha(X_i, X_j) \le C \cdot m^{1-a}, \tag{9.13}$$

and

$$n^{-1} \sum_{i=1}^{n} \sum_{j=1, |j-i|\le m}^{n} r[\delta(t(i), t_0)] \cdot r[\delta(t(j), t_0)] \le C \cdot m, \tag{9.14}$$

for integers $1 \le m < n$, where C and a are positive constants.

We make some observations concerning the Assumption 9.3.

1. Assumption (9.14) holds when r is bounded.

2. Let us consider the case where $X_1, \ldots, X_n$ is strictly stationary, as in Definition 9.2. Define $a_k = \alpha(X_i, X_{i+k})$, for $i = 1, \ldots, n$, $k = 1 - i, \ldots, n - i$. We have, the same as in (9.12),

$$\begin{aligned} \frac{1}{n} \sum_{i=1}^{n} \sum_{j=1, |j-i|>m}^{n} \alpha(X_i, X^j) &= \frac{1}{n} \sum_{i=1}^{n} \sum_{k=1-i, |k|>m}^{n-i} \alpha(X_i, X_{i+k}) \\ &\le 2 \sum_{k=m+1}^{\infty} a_k. \end{aligned}$$

When the coefficients a_k have algebraic decrease $a_k \le C'k^{-a}$, where $a > 1$ and C' is a positive constant, we have

$$\sum_{k=m+1}^{\infty} a_k \le C' \frac{(m+1)^{1-a}}{a-1}.$$

Thus the assumption $a_k \le C'k^{-a}$ implies assumption (9.13) with the rate function r as identically 1: $r \equiv 1$.

9.3.3 Serial Dependency

When $X_1, X_2, \ldots$ is a time series, we may be interested in the prediction of, for example, one step ahead with one explanatory variable, so consider a new time series $Y_1 = (X_1, X_2), Y_2 = (X_2, X_3), \ldots$. If the original time series is stationary, then the new time series consists of identically distributed random vectors. We say that the original time series is locally stationary if the new time series is locally identically distributed. Local stationarity seems more natural in the time series setting than in the spatial setting, and thus we restrict ourselves to the time series setting.

Definition 9.2 (Stationary time series.) *We call time series $X_1, \ldots, X_n \in \mathbf{R}^d$ (strongly, strictly)* stationary *when for all $1 \le i_1 < \cdots < i_l \le n-1$, $k = 1, \ldots, n - i_l$,*

$$(X_{i_1}, \ldots, X_{i_l}) \textit{ and } (X_{i_1+k}, \ldots, X_{i_l+k}) \textit{ are identically distributed.}$$

The assumption of identically distributed observations is weaker than the assumption of stationarity. The assumption of identically distributed observations may be written as: for all $i = 1, \ldots, n-1$, $k = 1, \ldots, n-i$,

$$X_i \text{ and } X_{i+k} \text{ are identically distributed.} \tag{9.15}$$

When the observations are independent and (9.15) holds, the observations are stationary.

We want to weaken the assumption of stationarity to the assumption of *local stationarity*. Let $l \ge 1$, and let $k_i \ge 1$ be such that $1 + k_1 + \cdots + k_l \le n$. Denote $|k| = k_1 + \cdots + k_l$. We create data matrix

$$A = \begin{bmatrix} X_1 & X_{1+k_1} & \cdots & X_{1+|k|} \\ X_2 & X_{2+k_1} & \cdots & X_{2+|k|} \\ \vdots & \vdots & & \vdots \\ X_{n-|k|} & X_{n-|k|+k_1} & \cdots & X_n \end{bmatrix}.$$

Data matrix A is of dimension $m \times p$, where

$$m = n - |k|, \qquad p = (l+1) \cdot d,$$

when X_i are interpreted as row vectors of length d. Denote with

$$Y_i = (X_i, X_{i+k_1}, \ldots, X_{i+|k|}) \in \mathbf{R}^p, \qquad i = 1, \ldots, m, \tag{9.16}$$

the ith row of matrix A. Definition 9.2 of stationarity is equivalent to saying that sequence $X_1, \ldots, X_n \in \mathbf{R}^d$ is stationary when for all choices of $l \geq 1$ and $k_i \geq 1$ such that $1 + k_1 + \cdots + k_l \leq n$,

$$Y_1, \ldots, Y_m \text{ are identically distributed.}$$

The local stationarity will be defined to mean that time series $Y_1, \ldots, Y_m$ is locally identically distributed. In addition the definition of local stationarity will be made to depend on a given choice of $k_1, \ldots, k_l$. This is natural since time series $Y_1 = (X_1, X_2), Y_2 = (X_2, X_3), \ldots$ might be locally identically distributed but the time series $Y_1 = (X_1, X_{101}), Y_2 = (X_2, X_{102}), \ldots$ might not.

Definition 9.3 (Locally stationary time series.) *Time series $X_1, \ldots, X_n \in \mathbf{R}^d$ is* locally stationary, *with shifting steps $k_1, \ldots, k_l$, with rate function $r : [0, \infty) \to [0, \infty)$, with rate $o(1)$ or $O(n^{-1/2})$, at time point $i_0 \in \{1, \ldots, n - |k|\}$, when time series $(Y_1, \ldots, Y_m)$, defined with (9.16), is locally identically distributed with rate function r, rate $o(1)$ or $O(n^{-1/2})$, and time point i_0, as defined in Definition 9.1.*

When the time series is stationary, we may apply a multivariate ($(l + 1) \cdot d$)-dimensional) density estimator to estimate the joint density of

$$Y_1 = (X_1, X_{1+k_1}, \ldots, X_{1+k_1+\cdots+k_l}),$$

where $k_i \geq 1$, $1 + k_1 + \cdots + k_l \leq n$ (and $X_i \in \mathbf{R}^d$ is a row vector). When the time series is locally stationary we may apply the time localized kernel estimator, defined in (14.7), to estimate the density of Y_{i_0} for a fixed $i_0 \in \{1, \ldots, m\}$.

The definition of locally identically and locally stationary distributed observations is taken from Klemelä (2008*a*). A different concept of a nonstationary time series was introduced by Priestley (1965), in terms of the spectral density. Dahlhaus (1997) studied local stationarity in the context of covariance evolution and developed an asymptotic theory with the help of time rescaling.

9.3.4 Inverse Problems

In an inverse density estimation problem we observe realizations of random variables

$$Y_1, \ldots, Y_n \in \mathbf{Y}, \tag{9.17}$$

whose common density function is

$$Af : \mathbf{Y} \to \mathbf{R}, \tag{9.18}$$

where $f : \mathbf{R}^d \to \mathbf{R}$ is a function that we want to estimate, $(\mathbf{Y}, \mathcal{Y}, \nu)$ is a Borel space, and A is an operator that maps functions $f : \mathbf{R}^d \to \mathbf{R}$ to densities $g : \mathbf{Y} \to \mathbf{R}$. The Gaussian white noise model for inverse problems will be defined in (9.47).

In applied mathematics an inverse problem is said to be the problem of finding f given y, where $y = Af$. An inverse problem is well posed when (1) the solution exists, (2) it is unique, and (3) it is stable, so it depends continuously on the data: A^{-1} has to be continuous. An inverse problem with deterministic noise is a problem where we have observation $y = Af + \epsilon$, where ϵ is deterministic noise, and we want to estimate f or some of its functional. For example, A could be a sampling operator: $Af = (f(x_1), \ldots, f(x_n)) \in \mathbf{R}^n$, where $x_1, \ldots x_n \in \mathbf{R}^d$ are fixed points. Thus all statistical problems are inverse problems in the terminology of applied mathematics, with a difference that in statistics one considers stochastic noise and not deterministic noise. Indeed in statistics the objective is to recover a function or some functionals of a function based on a finite number of noisy measurements. In statistics the term "inverse problem" is reserved only to problems where there is an additional operator involved in addition to the sampling operator. The sampling operator does not have a unique inverse. In the statistical setting we assume that the operator does have a unique inverse, but we allow that it is unstable, in the sense that small perturbations of the argument f lead to big changes in the value Af.

Convolution and Measurement Errors Density estimation involves a convolution operator when the observations are contaminated with measurement errors.

Stefanski and Carroll (1990) proposed a kernel estimator for the problem of deconvolving a density. Carroll and Hall (1988) showed that the rates of convergence are logarithmic when the error density is Gaussian. Consistency was studied by Devroye (1989). Optimal rates of convergence were obtained for the convolution problem in various settings by Ermakov (1989), Fan (1991), Donoho and Low (1992), Koo (1993), and Korostelev and Tsybakov (1993). We show in Corollary 15.4 that a δ-net estimator can achieve the optimal rates of convergence for the problem of deconvolution and in Theorem 14.6 that a kernel estimator can achieve the optimal rates of convergence.

Convolution The convolution operator A is defined by

$$Af = a * f, \qquad f : \mathbf{R}^d \to \mathbf{R},$$

where $a : \mathbf{R}^d \to \mathbf{R}$ is a known function and the convolution is defined by

$$a * f(x) = \int_{\mathbf{R}^d} f(x - y) a(y) \, dy, \qquad x \in \mathbf{R}^d.$$

Measurements Errors In the setting of deconvolution we observe realizations of random vectors $Y_1, \ldots, Y_n \in \mathbf{R}^d$ that are contaminated by measurement errors:

$$Y_i = X_i + \epsilon_i, \tag{9.19}$$

where ϵ_i are measurement errors. We assume that X_i are identically distributed and that ϵ_i are identically distributed, and we want to estimate the density of X_i, denoted by $f : \mathbf{R}^d \to \mathbf{R}$. When ϵ_i and X_i are independent, the density of Y_i is the convolution of f and a, where a the density of ϵ_i. Thus in deconvolution the operator in (9.18) is defined by

$$Af = a * f \tag{9.20}$$

and $\mathbf{Y} = \mathbf{R}^d$.

Inverse of a Convolution We can find the inverse of a convolution operator with the help of Fourier transform.

Lemma 9.1 *The inverse of A is defined for $g \in L_1(\mathbf{R}^d)$ by*

$$A^{-1}g = F^{-1}\left(\frac{Fg}{Fa}\right),$$

where F denotes the Fourier transform, and we assume that $(Fa)(\omega) > 0$ for all $\omega \in \mathbf{R}^d$.

Proof. For $g : \mathbf{R}^d \to \mathbf{R}$, $f : \mathbf{R}^d \to \mathbf{R}$,

$$\begin{aligned} g = Af \quad &\Leftrightarrow \quad Fg = F(Af) \\ &\Leftrightarrow \quad Fg = (Fa)(Ff), \end{aligned} \tag{9.21}$$

where in (9.21) we used the property (B.7) of the convolution. □

Adjoint of the Inverse We will need in Chapter 15, equation (15.6), the adjoint of the inverse of a convolution operator. The adjoint of the inverse is denoted by $(A^{-1})^*$

Lemma 9.2 *We have $(A^{-1})^* = Q$, where Q is defined for $g : \mathbf{R}^d \to \mathbf{R}$ by*

$$Qg = F^{-1}\left(\frac{Fg}{Fa}\right),$$

where F denotes the Fourier transform.

Proof. Lemma 9.1 implies that

$$FA^{-1} = G, \tag{9.22}$$

where G is defined, for $h : \mathbf{R}^d \to \mathbf{R}$, by

$$Gh = \frac{Fh}{Fa}.$$

Thus, for $h : \mathbf{R}^d \to \mathbf{R}$, $g : \mathbf{R}^d \to \mathbf{R}$, by applying (9.22) and twice Parseval's theorem given in (B.6), we have

$$\begin{aligned} \int_{\mathbf{R}^d} (A^{-1}h)g &= (2\pi)^{-d} \int_{\mathbf{R}^d} (Gh)(Fg) \\ &= (2\pi)^{-d} \int_{\mathbf{R}^d} \frac{(Fh)(Fg)}{Fa} \\ &= \int_{\mathbf{R}^d} h(Qg). \end{aligned}$$

We have proved the lemma. This proof was written in a lengthy way to make it similar to the more difficult proof of Lemma 9.4. □

Singular Value Decomposition Let the collection (ϕ_{jk}), $(j,k) \in \{0,1,\ldots\}^d \times K_j$, of trigonometric polynomials be defined in (10.29). This collection is a basis for 1-periodic functions on $L_2([0,1]^d)$. When the convolution kernel a is an 1-periodic function in $L_2([0,1]^d)$, we can write

$$a(x) = \sum_{j_1=0,\ldots,j_d=0}^{\infty} \sum_{k \in K_j} b_{jk}\phi_{jk}(x). \tag{9.23}$$

The functions ϕ_{jk} are the eigenfunctions of the operator A^*A and the values b_{jk}^2 are the corresponding eigenvalues.

Radon Transform and Tomography The Radon transform of a function $f : \mathbf{R}^d \to \mathbf{R}$ is the collection of the integrals of f over the $(d-1)$-dimensional hyperplanes. In emission tomography we observe a collection of planes. The density of the planes is determined from the underlying density through the Radon transform. The Radon transform has been discussed in Deans (1983) and Natterer (2001).

Optimal rates of convergence for the problem of tomography were obtained by Johnstone and Silverman (1990), Korostelev and Tsybakov (1991), and Donoho and Low (1992). Theorem 14.7 shows that a kernel estimator can achieve the optimal rates of convergence in the problem of tomography. The exposition of kernel estimation follows Korostelev and Tsybakov (1993). We show in Corollary 15.5 that a δ-net estimator can achieve the optimal rates of convergence in inverse problems with the Radon transform.

Radon Transform We define the Radon transform as the integral of a d-dimensional function over the $(d-1)$-dimensional hyperplanes. We may parametrize the $(d-1)$-dimensional hyperplanes in the d-dimensional Euclidean space with the help of a direction vector $\xi \in \mathbf{S}_{d-1} = \{x \in \mathbf{R}^d : \|x\| = 1\}$ and a distance from the origin $u \in [0,\infty)$:

$$P_{\xi,u} = \{z \in \mathbf{R}^d : z^T\xi = u\}, \qquad \xi \in \mathbf{S}_{d-1}, \; u \in [0,\infty). \tag{9.24}$$

We can write

$$P_{\xi,u} = \left\{u\xi + x : x \in \xi^\perp\right\}, \tag{9.25}$$

where $\xi^\perp = \{x \in \mathbf{R}^d : x^T\xi = 0\}$ is the subspace orthogonal to ξ. In the 2D case we can write $P_{\xi,u} = \{u\xi + t\xi^\perp : t \in \mathbf{R}\}$. Figure 9.4 illustrates this parametrization of hyperplanes in the 2D case. The lines are parametrized by the length u of the perpendicular from the origin to the line and by the orientation of this perpendicular.

We define the Radon transform of $f : \mathbf{R}^d \to \mathbf{R}$ as

$$(Rf)(\xi,u) = \int_{P_{\xi,u}} f, \qquad \xi \in \mathbf{S}_{d-1}, \; u \in [0,\infty), \tag{9.26}$$

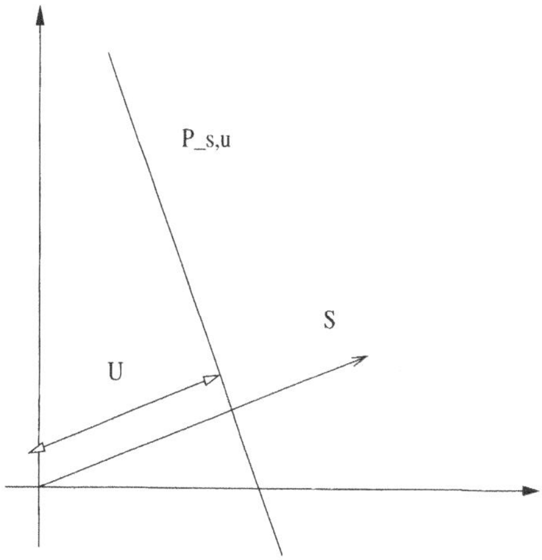

Figure 9.4 *(Parametrization of hyperplanes.)* Plane $P_{s,u}$ is parametrized with direction vector s and distance u.

where the integration is with respect to the $(d-1)$-dimensional Lebesgue measure. Using (9.25), we can write

$$(Rf)(\xi, u) = \int_{\xi^\perp} f(u\xi + x)\, dx.$$

Tomography The positron emission tomography is a density estimation problem, whereas the X-ray tomography is a regression problem. In the setting of emission tomography, events occur at points $X_1, \ldots, X_n \in \mathbf{R}^d$. These points are i.i.d. random vectors with density f. We do not observe the location of the points but only that an event has occurred on a hyperplane P_i containing the point; at every sample point an event happens that causes an emission of two particles at the opposite directions. The detectors on the boundary of the region register the emissions, and this is equivalent to observing the hyperplanes. When $d = 2$, we observe lines and when $d = 3$, we observe two-dimensional planes. Figure 9.5 illustrates the observations of tomography. The left panrl shows a contour plot of a density and a sample of size 60 from the density. The right panel shows the lines going through the sample points. The orientation of the lines is uniform. We want to estimate the density whose contour plot is shown in the left panel.

When we use the parametrization of planes given in (9.24), the observation of hyperplanes is equivalent to observing the sequence

$$((\xi_1, U_1), \ldots, (\xi_n, U_n)) \in \mathbf{S}_{d-1} \times [0, \infty). \tag{9.27}$$

We assume that the hyperplanes are uniformly oriented. The fundamental observation is that the distance of the hyperplane from the origin, conditional on the direction of

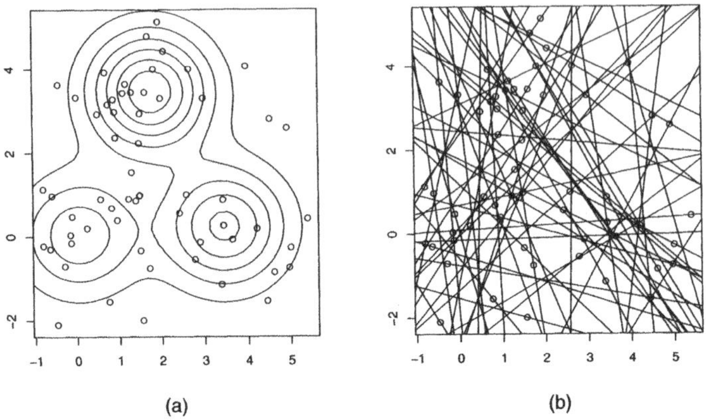

Figure 9.5 *(Observations of tomography.)* (a) A contour plot of a density and a sample of size 60 from the density. The sample points are not observed in tomography but instead the lines shown in (b).

the hyperplane, is given by the slice of the Radon transformed density. That is,

$$\xi \sim \mathrm{Unif}(\mathbf{S}_{d-1}), \qquad U \mid \xi = s \sim (Rf)(s, \cdot). \tag{9.28}$$

Thus the operator in (9.18) is defined by

$$(Af)(\xi, u) = \frac{1}{\mu(\mathbf{S}_{d-1})} (Rf)(\xi, u), \qquad \xi \in \mathbf{S}_{d-1}, \ u \in [0, \infty), \tag{9.29}$$

and $\mathbf{Y} = \mathbf{S}_{d-1} \times [0, \infty)$.

Inverse of the Radon Transform We find the inverse of the Radon transform in the following lemma:

Lemma 9.3 *We have that*

$$R^{-1} = F^{-1} G, \tag{9.30}$$

where F denotes Fourier transform and G is defined, for $h : \mathbf{S}_{d-1} \times [0, \infty) \to \mathbf{R}$, by

$$(Gh)(\omega) = (\mathcal{H}_{\omega/\|\omega\|} h)(\|\omega\|), \qquad \omega \in \mathbf{R}^d, \tag{9.31}$$

where $\mathcal{H}_\xi$ is the Fourier transform of $h(\xi, \cdot\,)$ for fixed $\xi \in \mathbf{S}_{d-1}$:

$$\mathcal{H}_\xi h = F_1(h(\xi, \cdot\,)), \qquad \xi \in \mathbf{S}_{d-1}.$$

Proof. The projection theorem in Lemma B.2 states that for $f : \mathbf{R}^d \to \mathbf{R}$, for fixed $\xi \in \mathbf{S}_{d-1}$, and for the Radon transform R, the Fourier transform of $u \mapsto (Rf)(\xi, u)$ is equal to $(Ff)(\xi u)$:

$$\mathcal{H}_\xi R f = (Ff)(\xi\, \cdot), \qquad \xi \in \mathbf{S}_{d-1}. \tag{9.32}$$

Then, for $h : \mathbf{S}_{d-1} \times [0, \infty) \to \mathbf{R}$, $f : \mathbf{R}^d \to \mathbf{R}$,

$$
\begin{aligned}
h = Rf \quad &\Leftrightarrow \quad \mathcal{H}_\xi h = \mathcal{H}_\xi Rf, \qquad \xi \in \mathbf{S}_{d-1} \\
&\Leftrightarrow \quad \mathcal{H}_\xi h = (Ff)(\xi \cdot), \qquad \xi \in \mathbf{S}_{d-1} && (9.33) \\
&\Leftrightarrow \quad (\mathcal{H}_{\omega/\|\omega\|} h)(\|\omega\|) = (Ff)(\omega), \qquad \omega \in \mathbf{R}^d \\
&\Leftrightarrow \quad (Gh)(\omega) = (Ff)(\omega), \qquad \omega \in \mathbf{R}^d && (9.34) \\
&\Leftrightarrow \quad F^{-1}Gh = f,
\end{aligned}
$$

where in (9.33) we applied the projection theorem as given in (9.32), and in (9.34) we applied the definition of G as given in (9.31). □

Adjoint of the Inverse We will need in Chapter 15, equation (15.6), the adjoint of the inverse of a convolution operator. The adjoint of the inverse is denoted by $(R^{-1})^*$.

Lemma 9.4 *We have $(R^{-1})^* = Q$ where Q is defined for $g : \mathbf{R}^d \to \mathbf{R}$ by*

$$(Qg)(\xi, u) = (2\pi)^{d-1} \cdot (F_1^{-1} \mathcal{I}_\xi g)(u), \qquad \xi \in S_d, \ u \in [0, \infty),$$

where

$$(\mathcal{I}_\xi g)(t) = (Fg)(t\xi), \qquad \xi \in S_d, \ t \in [0, \infty). \tag{9.35}$$

Proof. Lemma 9.3 implies that

$$FR^{-1} = G, \tag{9.36}$$

where G is defined in (9.31). Thus, for $h : \mathbf{S}_{d-1} \times [0, \infty) \to \mathbf{R}$, $g : \mathbf{R}^d \to \mathbf{R}$,

$$
\begin{aligned}
&\int_{\mathbf{R}^d} (R^{-1}h) g \\
&= (2\pi)^d \int_{\mathbf{R}^d} (Gh)(Fg) && (9.37) \\
&= (2\pi)^d \int_{\mathbf{R}^d} (\mathcal{H}_{\omega/\|\omega\|} h)(\|\omega\|)(Fg)(\omega)\, d\omega && (9.38) \\
&= (2\pi)^d \int_{\mathbf{S}_{d-1}} \int_0^\infty t^{d-1} (\mathcal{H}_\xi h)(t)(Fg)(t\xi)\, dt\, d\mu(\xi) && (9.39) \\
&= (2\pi)^{d-1} \int_{\mathbf{S}_{d-1}} \int_0^\infty u^{d-1} h(\xi, u)(F_1^{-1} I_\xi g)(u)\, du\, d\mu(\xi) && (9.40) \\
&= \int_{\mathbf{Y}} h(Qg). && (9.41)
\end{aligned}
$$

In (9.37) we applied Parseval's theorem from (B.6) and (9.36), in (9.38) we applied the definition of G given in (9.31), in (9.39) we applied the change to the polar coordinates, in (9.40) we applied Parseval's theorem in $L_2([0, \infty))$ with the measure $t^{d-1}\, dt$, and in (9.41) we applied the definition of Q. We have proved the lemma. □

2D Radon Transform In the 2D case we consider reconstructing a two-dimensional function from observations of its integrals over lines. Let $D = \{x \in \mathbf{R}^2 : \|x\| \le 1\}$ be the unit disk in $\mathbf{R}^2$. The plane in (9.25) can be written as $P_{\xi,u} = \{u\xi + t\xi^\perp : t \in \mathbf{R}\}$, where $\xi^\perp$ is a vector orthogonal to ξ. We can write $\xi = (\cos\phi, \sin\phi)$ and $\xi^\perp = (-\sin\phi, \cos\phi)$. Thus we parametrize the lines by the length $u \in [0,1]$ of the perpendicular from the origin to the line and by the orientation $\phi \in [0, 2\pi)$ of this perpendicular. A common way to define the 2D Radon transform is

$$Rf(u,\phi) = \frac{\pi}{2\sqrt{1-u^2}} \int_{\sqrt{1-u^2}}^{\sqrt{1-u^2}} f(u\cos\phi - t\sin\phi, u\sin\phi + t\cos\phi)\, dt, \quad (9.42)$$

where $(u,\phi) \in \mathbf{Y} = [0,1] \times [0, 2\pi]$, and we suppose that $f \in L_1(D) \cap L_2(D)$. Now the Radon transform is π times the average of f over the line segment that intersects D. We consider Rf as the element of $L_2(\mathbf{Y}, \nu)$, where ν is the measure defined by $d\nu(u,\phi) = 2\pi^{-1}\sqrt{1-u^2}\, du\, d\phi$.

Singular Value Decomposition We consider the 2D Radon transform as defined in (9.42). The singular value decomposition of the Radon transform is given in Deans (1983). Let

$$\tilde{\phi}_{jk}(r,\theta) = \pi^{-1/2}(j+k+1)^{1/2} Z_{j+k}^{|j-k|}(r) e^{i(j-k)\theta}, \qquad (r,\theta) \in D = [0,1] \times [0, 2\pi),$$

where Z_a^b denotes the Zernike polynomial of degree a and order b. Functions $\tilde{\phi}_{jk}$, $j,k = 0,1,\ldots, (j,k) \ne (0,0)$, constitute an orthonormal complex-valued basis for $L_2(D)$. The corresponding orthonormal functions in $L_2(\mathbf{Y}, \nu)$ are

$$\tilde{\psi}_{jk}(u,\phi) = \pi^{-1/2} U_{j+k}(u) e^{i(j-k)\phi}, \qquad (u,\phi) \in \mathbf{Y} = [0,1] \times [0, 2\pi),$$

where $U_m(\cos\theta) = \sin((m+1)\theta)/\sin\theta$ are the Chebyshev polynomials of the second kind. We have

$$A\tilde{\phi}_{jk} = b_{jk}\tilde{\psi}_{jk},$$

where the singular values are

$$b_{jk} = \pi^{-1}(j+k+1)^{-1/2}. \quad (9.43)$$

We will identify the complex bases with the equivalent real orthonormal bases by

$$\phi_{jk} = \begin{cases} \sqrt{2}\,\mathrm{Re}(\tilde{\phi}_{jk}) & \text{if } j > k \\ \tilde{\phi}_{jk} & \text{if } j = k \\ \sqrt{2}\,\mathrm{Im}(\tilde{\phi}_{jk}) & \text{if } j < k. \end{cases}$$

9.4 RELATED TOPICS

Density estimation provides a general approach for regression function estimation and classification. A regression function may be found as the mean of a conditional distribution, and classification may be based on the class probability densities. Density estimation provides also a general approach for cluster analysis if we define clusters as regions of high density. However, there are two different types of caveates on using density estimation to solve other problems. (1) We should never try to solve a problem that is more general than is actually needed (Vapnik (1998)). (2) The density may not be well defined. For example, the distribution may have singular components.

Further keep in mind the counterpart of Vapnik's principle: never be satisfied in solving a problem that is less general than we actually need to solve. For example, when we are interested in predicting a value of a response variable Y using explanatory variables X, the conditional mean is not the only relevant characteristic of the distribution but we might need to estimate the variability of Y given X (the conditional distribution), and we might need to estimate the variability of X (the marginal distribution). This brings us back to the density estimation problem. However, there are cases where we really need to estimate only some simple characteristics of the distribution or where we can be satisfied by automatic classification and prediction procedures.

Section 9.4.1 gives an overview of regression function estimation, and Section 9.4.2 gives an overview of classification. (An overview of cluster analysis is given in Chapter 8.)

9.4.1 Regression Function Estimation

In regression function estimation we have identically distributed observations (X_i, Y_i), $i = 1, \ldots, n$, where $X_i \in \mathbf{R}^d$ and $Y_i \in \mathbf{R}$. Let $(X, Y) \sim (X_1, Y_1)$. The variable Y has a special interpretation: it is the response variable, and we are interested in the dependence of Y from X, that is, we are interested in the conditional distribution $Y|X$. Components of random vector X are called explanatory variables. Estimating the conditional distribution $Y|X$ is not very far from estimating the joint distribution. Indeed we can write

$$f_{Y|X=x}(y) = \frac{f_{X,Y}(y, x)}{f_X(x)},$$

where $f_{X,Y}$ is the joint density of (X, Y) and f_X is the density of X. Note, however, that Y and some components of X may be discrete random variables, in which case the joint distribution (X, Y) does not have a density with respect to the Lebesgue measure. In the *fixed design regression* the values of the explanatory variables are fixed, and not random. Also in this case we do not have a joint density.

Regression Function Often one is satisfied with estimating the regression function

$$g(x) = E(Y \mid X = x), \qquad x \in \mathbf{R}^d.$$

Then we may write the regression model as

$$Y_i = g(X_i) + \epsilon \, \xi_i, \qquad i = 1, \ldots, n,$$

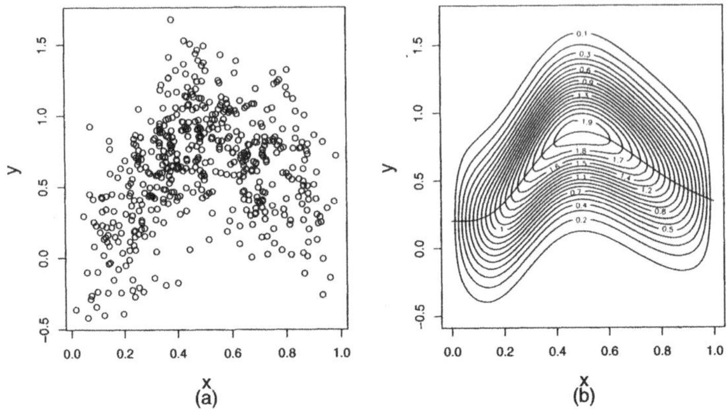

Figure 9.6 *(Scatter plot of regression data.)* (a) A scatter plot of simulated data. (b) A contour plot of the joint density of an explanatory variable and a response variable. The regression function is shown as the solid line.

where $\xi_i \in \mathbf{R}$ are identically distributed noise variables. and we assume that $E\xi_i = 0$. The amount of noise is given by $\epsilon > 0$, and often $\text{Var}(\xi_i) = 1$ is assumed. We can write

$$f_{Y|X=x}(y) = f_\xi(y - g(x)),$$

where f_X is the density of the explanatory variable and f_ξ is the density of the noise.

Illustration Figure 9.6 shows a simulated data of size $n = 500$ and a contour plot of the density function of $(X, Y) \in [0, 1]^2$. The density of X is a scaled and translated Bartlett density: $f_X(x) = 3(1 - [(x - 0.5)/h)]^2)_+/(4h)$, where $h = 0.5$. The distribution of the errors ξ is the standard Gaussian and $\epsilon = 0.3$. The regression function is $g(x) = 0.2 + \exp\{-2(x/h)\}(x/h)^5$, where $h = 0.2$. The regression function is shown in the figure as a solid line.

Regression Function Estimation versus Density Estimation In density estimation we are interested in the distribution of the probability mass: we want to find the shape of the density function. In regression function estimation we may be interested in finding the nature of the dependence of Y from X: for example, the shape of the regression function. On the other hand, we may be interested in predicting the value of Y after we have observed $X = x$. To make a prediction, we need to only evaluate regression function: the prediction will be $g(x)$. In density estimation the mere evaluation of the density function at individual points does not bring us much interesting information. To make density estimation interesting we are even more dependent on visualization than in the case of regression function estimation.

Reduction of Density Estimation to Regression Function Estimation It is possible to estimate densities by the methods of regression function estimation.

1. *(Empirical distribution function as a response.)* We may calculate the empirical distribution function F_n, defined in (11.12), based on the observations $X_1, \ldots, X_n$. Then we may define a response variable $Y_i = F_n(X_i)$ and consider the regression data $(X_1, Y_1), \ldots, (X_n, Y_n)$. This approach was applied in Weston, Gammerman, Stitson, Vapnik, Vovk and Watkins (1998).

2. *(Presmoothing.)* In presmoothing we make a fine binning $\mathcal{P}$ of a rectangle containing the observations and calculate the number of observations in each bin. Then we consider the problem of fixed design regression where the response variable is a function of the frequency in each bin. Let

$$Y_R = \frac{n_R}{n \cdot \text{volume}(R)},$$

where

$$n_R = \#\{X_i \in R : i = 1, \ldots, n\}, \qquad R \in \mathcal{P}.$$

The regression data are (x_R, Y_R), $R \in \mathcal{P}$, where x_R is the center point of rectangle R.

Note that frequency n_R is binomially distributed: $n_R \sim \text{Bin}(n, \int_R f)$. Thus $EY_R = \int_R f/\text{volume}(R)$. If we choose a sequence of partitions $\mathcal{P}_l$, $l = 1, 2, \ldots$, so that $\lim_{l \to \infty} \max_{R \in \mathcal{P}_l} \text{volume}(R) = 0$, then for every sequence $R_l \in \mathcal{P}_l$, $l = 1, 2, \ldots$, $EY_{R_l} \sim f(x_{R_l})$, as $l \to \infty$, for which we assume continuity of density f. We can write

$$Y_{R_l} = f(x_{R_l}) + n^{-1/2}\xi_{R_l} + \delta_{n,l}, \tag{9.44}$$

where $\xi_{R_l} = n^{1/2}(Y_{R_l} - EY_{R_l})$, $\delta_{n,l} = EY_{R_l} - f(x_{R_l}) \to 0$, as $l \to \infty$. Random variable ξ_R has the expectation zero: $E\xi_R = 0$ and variance $\text{Var}(\xi_R) = \int_R f(1 - \int_R f)/\text{volume}(R)^2$. We have an approximation $\text{Var}(\xi_{R_l}) \sim f(x_{R_l})/\text{volume}(R_l)$, as $l \to \infty$, and ξ_R is converging weakly to a Gaussian distribution. We have written an approximate fixed design regression model in (9.44). However, this is a heteroscedastic regression model, in the sense that the variance of the error distribution is different in various regions of the design space. Furthermore error variables are correlated, but this correlation is negligible for ξ_{R_1} and ξ_{R_2} when sets R_1 and R_2 are distant from each other.

3. *(Deviation from a reference density.)* We can reduce the density estimation to 0-1-regression estimation in the following way: Let f_0 be a specified probability density function. Generate a random sample $X_{n+1}, \ldots, X_{2n}$ from this density. Let

$$Y_i = \begin{cases} 1, & \text{if } 1 \le i \le n, \\ 0, & \text{if } n+1 \le i \le 2n. \end{cases}$$

We estimate

$$\mu(x) = E(Y \mid X = x) = \frac{f(x)}{f(x) + f_0(x)}$$

with the data $(Y_1, X_1), \ldots, (Y_{2n}, X_{2n})$. Then we set

$$\hat{f}(x) = f_0(x) \frac{\hat{\mu}(x)}{1 - \hat{\mu}(x)}.$$

9.4.2 Supervised Classification

Supervised classification is called also discriminant analysis and pattern recognition. In some sense the relation of regression analysis and supervised classification is like the relation of density estimation and clustering. Like in the case of regression function estimation, it is sometimes enough to find a classification rule, and we are not interested in the interpretation and visualization of the results.

Basic Setting In the supervised classification we observe identically distributed random variables (X_i, Y_i), $i = 1, \ldots, n$, where $X_i \in \mathbf{R}^d$, $Y_i \in \{1, \ldots, k\}$. Thus the setting is similar to the setting of regression function estimation, but now the response variable takes values on a finite set. We construct classification function $\hat{g} : \mathbf{R}^d \to \{1, \ldots, k\}$ with the help of observations (X_i, Y_i) and we measure the quality of the estimator $\hat{g}$ with the loss function

$$L(P, \hat{g}) = P\left(\hat{g}(X) \neq Y\right),$$

where P is the joint distribution of (X, Y) where $(X, Y) \sim (X_1, Y_1)$. The loss is equal to the probability of misclassification.

Fixed Class Size In the supervised classification we may also have a fixed design, as in the case of regression function estimation. Even when the design is random, we may have a setting where the number of observations is fixed in each class; that is, we observe

$$X_1^{(j)}, \ldots, X_{n_j}^{(j)} \sim X|Y = j$$

for $j = 1, \ldots, k$. Then the total sample size is $\sum_{j=1}^k n_j$. For example, when we were to classify handwritten digits, we would choose the sample size of the learning sample to be equal for each of the digits. The loss function for the case of the fixed class size is

$$L(P, \hat{g}) = \sum_{j=1}^{k} \pi_j P\left(\hat{g}(X) \neq i\right),$$

where $\pi_j \geq 0$, $\sum_{j=1}^k \pi_j = 1$. Typically $\pi_j = 1/k$, $j = 1, \ldots, k$.

9.4.3 The Gaussian White Noise Model

The Gaussian white noise model is extremely useful as a heuristic tool. Many density estimators have a counterpart in the Gaussian white noise model. We can gain insight into the (asymptotical) properties of a density estimator when we analyze its counterpart in the Gaussian white noise model. In the Gaussian white noise model the observation is a realization of the process

$$dX_n(x) = dX_{n,f}(x) = f(x)\,dx + n^{-1/2}\,dW(x), \qquad x \in \mathbf{R}^d, \tag{9.45}$$

where $f \in L_2(\mathbf{R}^d)$ and $W(x)$ is the Brownian sheet in $\mathbf{R}^d$.

Brownian Sheet The Brownian sheet may be thought of as a Gaussian stochastic process indexed with functions $h \in L_2(\mathbf{R}^d)$. The process

$$\left(\int_{\mathbf{R}^d} h\,dW\right)_{h\in L_2(\mathbf{R}^d)}$$

is such that its finite dimensional marginals

$$\left(\int_{\mathbf{R}^d} h_1\,dW, \ldots, \int_{\mathbf{R}^d} h_k\,dW\right)$$

are Gaussian with zero mean and the covariance matrix $[\int_{\mathbf{R}^d} h_i h_j]_{i,j=1,\ldots,k}$. For the process X_n defined in (9.45), this implies that for each $g, h \in L_2(\mathbf{R}^d)$,

$$\left(\int_{\mathbf{R}^d} g\,dX_n, \int_{\mathbf{R}^d} h\,dX_n\right) \sim N\left(\begin{bmatrix} \int_{\mathbf{R}^d} fg \\ \int_{\mathbf{R}^d} fh \end{bmatrix}, n^{-1}\begin{bmatrix} \|g\|_2^2 & \int_{\mathbf{R}^d} gh \\ \int_{\mathbf{R}^d} gh & \|h\|_2^2 \end{bmatrix}\right).$$

Brownian Motion In the one-dimensional case the Brownian sheet may be called the Brownian motion. The Brownian motion $W(t)$, $t \in \mathbf{R}$, is a stochastic process that has independent Gaussian increments:

- for $t < u$, $W(u-t) \sim N(0, u-t)$,
- $W(u-t)$ and $W(t)$ are independent,

where $t, u \in \mathbf{R}$. In the one-dimensional case the Gaussian white noise model can be defined by

$$X_n(t) = X_{n,f}(t) = \int_{-\infty}^{t} f(t)\,dt + n^{-1/2}W(t), \qquad t \in \mathbf{R}.$$

Brownian Process We may generalize the Brownian sheet from the Euclidean space to more general Borel spaces. Process W is the Brownian process in Borel space $(\mathbf{Y}, \mathcal{Y}, \nu)$, if for $h_1, \ldots, h_k \in L_2(\mathbf{Y})$ the finite dimensional marginal

$$\left(\int_{\mathbf{Y}} h_1\,dW, \ldots, \int_{\mathbf{Y}} h_k\,dW\right)$$

is Gaussian with mean zero and covariance matrix $[\int_{\mathbf{Y}} h_i h_j\,d\nu]_{i,j=1,\ldots,k}$. Analogously to the one-dimensional Brownian motion, we may characterize the general Brownian process with the two properties: for measurable sets $B, B_1, B_2 \in \mathcal{Y}$,

- $\int_B dW \sim N(0, \nu(B))$,
- if $B_1 \cap B_2 = \emptyset$, then $\int_{B_1} dW$ and $\int_{B_2} dW$ are independent.

We may generalize the Gaussian white noise model in (9.45) by

$$dY_n(y) = dY_{n,f}(y) = f(y)\, dy + n^{-1/2}\, dW(y), \qquad y \in \mathbf{Y}, \tag{9.46}$$

where $f \in L_2(\mathbf{Y})$. This process has Gaussian marginals: $(\int_{\mathbf{Y}} h_1\, dY_n, \int_{\mathbf{Y}} h_2\, dY_n)$ has Gaussian distribution with mean $(\int_{\mathbf{Y}} h_1 f\, d\nu, \int_{\mathbf{Y}} h_2 f\, d\nu)$ and covariance matrix $n^{-1}[\int_{\mathbf{Y}} h_i h_j\, d\nu]_{i,j=1,2}$. We use the generalization (9.46) in the connection of inverse problems. In inverse problems we observe a transformation Af of function $f \in L_2(\mathbf{R}^d)$:

$$dY_n(y) = dY_{n,Af}(y) = Af(y)\, dy + n^{-1/2}\, dW(y), \qquad y \in \mathbf{Y}, \tag{9.47}$$

where $f \in L_2(\mathbf{R}^d)$ and $A : L_2(\mathbf{R}^d) \to L_2(\mathbf{Y})$. For example, when A is the Radon transform, $\mathbf{Y} = \mathbf{S}_{d-1} \times [0, \infty)$; see Section 9.3.4.

Likelihood To prove lower bounds in Chapter 11 we need the Radon–Nikodym density for the processes in (9.46). Let P_i, $i = 0, 1$, be the probability measures corresponding to processes Y_{n,g_0} and Y_{n,g_1}. Then measures P_0 and P_1 are equivalent: P_0 dominates P_1 and P_1 dominates P_0. We say that measure P_0 dominates P_1 if for all measurable A, $P_0(A) = 0 \Rightarrow P_1(A) = 0$. Then

$$\frac{dP_1}{dP_0}(Y_{n,g_0}) = \exp\left\{ \int_{\mathbf{Y}} (g_1 - g_0)\, dY_{n,g_0} - \frac{n}{2} \left(\|g_1\|_{2,\mathbf{Y}}^2 - \|g_0\|_{2,\mathbf{Y}}^2 \right) \right\}. \tag{9.48}$$

We can write:

$$\mathcal{L}\left(\frac{dP_1}{dP_0} \,\Big|\, P_0 \right) = \exp\left\{ n^{1/2} \|g_1 - g_0\|_{2,\mathbf{Y}} \cdot \xi - \frac{n}{2} \|g_1 - g_0\|_{2,\mathbf{Y}}^2 \right\}, \tag{9.49}$$

where $\xi \sim N(0, 1)$, since

$$\int_{\mathbf{Y}} (g_1 - g_0)\, dY_{n,g_0} \sim N\left(\int_{\mathbf{Y}} g_1 g_0\, d\nu - \|g_0\|_{2,\mathbf{Y}}^2, n^{-1} \|g_1 - g_0\|_{2,\mathbf{Y}}^2 \right).$$

Note the analogous finite dimensional formula: for densities $f_0, f_1 : \mathbf{R}^d \to \mathbf{R}$, $f_0 \sim N(\mu_0, \sigma I_d)$, $f_1 \sim N(\mu_1, \sigma I_d)$,

$$\frac{f_1(x)}{f_0(x)} = \exp\left\{ \frac{1}{\sigma^2} x^T (\mu_1 - \mu_2) - \frac{1}{2\sigma^2} \left(\|\mu_1\|^2 - \|\mu_0\|^2 \right) \right\}, \qquad x \in \mathbf{R}^d.$$

Bibliographic Notes The Gaussian white noise model was made popular by Ibragimov and Hasminskii (1981*b*) as a heuristic tool in statistics. A program initiated by Le Cam, Brown and Low (1996), and Nussbaum (1996) analyzes estimators in models related to the Gaussian white noise model and then derives asymptotic properties of density estimators through reduction.

EXERCISES

9.1 *(Dependence.)* Let U be uniformly distributed on $[0, 1]$, and let $V = U/\theta$, when $U \in [0, \theta]$ and $V = (1-U)/(1-\theta)$, when $U \in [\theta, 1]$, where $\theta \in (0, 1)$. Show that $\rho(U, V) = 1 - 2\theta$. Thus for $\theta = 1/2$, $\rho(U, V) = 0$, although U determines V. This example is from Malevergne and Sornette (2005).

9.2 *(Dependence.)* Let Z be a standard Gaussian random variable. Show that

$$\rho\left(e^Z, e^{-\sigma Z}\right) = \frac{e^{-\sigma} - 1}{\sqrt{(e-1)(e^{\sigma^2} - 1)}}, \quad \rho\left(e^Z, e^{\sigma Z}\right) = \frac{e^{\sigma} - 1}{\sqrt{(e-1)(e^{\sigma^2} - 1)}},$$

where $\sigma > 0$. Study functions $\sigma \mapsto \rho\left(e^Z, e^{-\sigma Z}\right)$ and $\sigma \mapsto \rho\left(e^Z, e^{\sigma Z}\right)$, as σ grows. Hint: Use the fact that $Ee^{\sigma Z} = e^{\sigma^2/2}$. This example is from Embrechts, McNeil and Straumann (2002).

9.3 *(Dependence.)* Prove Equation (9.11).

CHAPTER 10

DENSITY CLASSES

We discuss two types of density classes: (1) classes defined by structural restrictions and (2) smoothness classes. Parametric and semiparametric classes can be considered as defined by structural restrictions.

It is often reasonable to assume that the estimated function satisfies various structural restrictions. For example, it may often be assumed that the multivariate density is the product of univariate densities so that the individual variables are independent. Under such structural assumptions the accuracy of estimation that is reached is close to the one-dimensional case. We give also examples of parametric classes and semiparametric classes. In semiparametric classes functions are defined in terms of a finite dimensional component and an infinite dimensional component.

One may construct a finite dimensional model for the data-generating process and then design an estimator that is efficient in this model. A second possibility is to consider completely model-free estimators, like kernel estimators and histogram estimators, which do not utilize any model assumptions. A third possibility is to consider structured nonparametric methods that utilize certain more flexible assumptions than parametric assumptions.

We discuss also smoothness classes. For example, a class of densities with bounded second derivatives is a smoothness class. We view such smoothness classes

Smoothing of Multivariate Data: Density Estimation and Visualization. By Jussi Klemelä

as *test beds*. Test beds are collections of densities that describe various possibilities that might arise in reality. With the help of a test bed we may understand what kind of densities we are able to estimate efficiently with our estimators. Given a test bed, we are interested in two aspects: (1) how our estimator works under simulated data generated from a given density in the test bed and (2) what are the asymptotical properties of the estimator, when we suppose that the density lies in a given test bed, and the number of observations tends to infinity. The same as smoothness classes, parametric density classes may also be used as test beds.

After defining several density classes, we want to calculate the order of entropy and capacity numbers (covering and packing numbers) of these classes. The entropy and capacity numbers were introduced in Kolmogorov and Tikhomirov (1961). The entropy number of a bounded set in a metric space is the smallest number of balls needed to cover the set, and the capacity number of a bounded set is the largest number of disjoint balls that can be packed in to the set. The entropy and capacity numbers characterize the largeness of a function class. They have a twofold use in statistical estimation:

1. If the entropy and capacity are not too large, then it can be proved that there exists estimators with fast rate of convergence. In particular, the minimization estimators analyzed in Chapter 15 achieve fast rates of convergence if the entropy and capacity are not too large.

2. If the entropy and capacity are large, this implies that there do not exist estimators with fast rates of convergence. This issue of lower bounds is covered in Chapter 11.

In Section 10.1 we consider density classes that are defined through structural restrictions, and we consider also parametric and semiparametric models. We define classes of elliptical densities, copula models, and parametric families of skewed densities. In Section 10.2 we define various smoothness classes. We define classes of anisotropic Besov classes, spaces of dominating mixed derivatives, and infinite mixture classes. In Section 10.3 we define entropy and capacity numbers and calculate the order of these numbers for some function classes.

10.1 STRUCTURAL AND PARAMETRIC RESTRICTIONS

10.1.1 1D Parametric Families

Symmetric Families Symmetric densities with different tail behavior include the standard Gaussian, double exponential, Student, and polynomial densities:

1. Standard Gaussian,

$$f(t) = (2\pi)^{-1/2} \exp\{-t^2/2\}, \qquad t \in \mathbf{R}.$$

2. Double exponential,

$$f(t) = \exp\{-|t|\}/2, \qquad t \in \mathbf{R}.$$

3. Student,

$$f(t) = c\,(1 + t^2/\nu)^{-(\nu+1)/2}, \qquad t \in \mathbf{R}, \qquad (10.1)$$

where

$$c = \frac{\Gamma((\nu+1)/2)}{(\nu\pi)^{1/2}\Gamma(\nu/2)}$$

and $\nu > 0$ (degrees of freedom).

4. Polynomial:

$$f(t) = c\,(1 - |t|^s)_+, \qquad t \in \mathbf{R},$$

where

$$c = [2(1 - 1/(s+1)]^{-1},$$

and $s > 0$, $(t)_+ = \max\{0, t\}$.

Figure 2.1 shows graphs of these densities.

Asymmetric Families Families that include asymmetric (skew) densities are the log-normal, gamma, and beta family.

1. The log-normal density is

$$f(t) = \frac{1}{t\sigma\sqrt{2\pi}} \exp\left\{\frac{-[(\log_e t) - \mu]^2}{2\sigma^2}\right\}, \qquad t > 0,$$

where $\mu \in \mathbf{R}$ and $\sigma > 0$ are the parameters. A variable X is log-normally distributed with parameters μ and σ, if $\log_e X \sim N(\mu, \sigma^2)$; see (B.22). When $\mu = 0$, the expectation of a log-normally distributed random variable is $e^{\sigma^2/2}$, the variance is $e^{\sigma^2}(e^{\sigma^2} - 1)$, and the mode of the density is $e^{-\sigma^2}$.

2. The gamma density is

$$f(t) = \frac{1}{\Gamma(a)\,b^a}\, t^{a-1} e^{-t/b}, \qquad t > 0.$$

The expectation of a gamma distributed random variable is is ab, the variance is ab^2, and the mode of the density is $(a-1)b$.

3. The beta density is

$$f(t) = \frac{\Gamma(a+b)}{\Gamma(a)\Gamma(b)}\, t^{a-1}(1-t)^{b-1}, \qquad t \in [0, 1],$$

where $a, b > 1$. The expectation of a beta distributed random variable is $a/(a+b)$, the variance is $ab/[(a+b)^2(a+b+1)]$, and the mode of the density is $(a-1)/(a+b-2)$.

10.1.2 Structural Restrictions

Product Densities In multiplicative models the density function is a product of low-dimensional densities. Consider, for example, the model

$$f(x) = g(x_1, \ldots, x_k) \prod_{i=k+1}^{d} f_i(x_i), \qquad x \in \mathbf{R}^d, \tag{10.2}$$

where $g : \mathbf{R}^k \to \mathbf{R}$ is a k-dimensional density and f_i, $i = k+1, \ldots, d$ are univariate densities. The density in (10.2) is effectively k-dimensional. When $k = 1$, we have a density of a random vector whose components are independent, and this would be effectively a one-dimensional model.

Model of Independent Components A more general model of independent components assumes that we observe realizations of random vector $X \in \mathbf{R}^d$,

$$X = A^T Z,$$

where $Z \in \mathbf{R}^d$ is a random vector whose components are independent, and A is an $d \times d$ invertible matrix. For $B = A^{-1}$, we can write

$$Z = B^T X,$$

and the density of X is equal to

$$f_X(x) = \det(B)\, f_Z(B^T x),$$

where f_Z is the density of Z. We can write the density of X as

$$f_X(x) = \det(B) \prod_{i=1}^{d} f_i\left(x^T \beta_i\right), \qquad x \in \mathbf{R}^d, \tag{10.3}$$

where f_i are (unknown) univariate densities, $\beta_i \in \mathbf{R}^d$, $i = 1, \ldots, d$, are (unknown) linearly independent vectors, and $\det(B)$ is the determinant of the matrix $B = (\beta_1, \ldots, \beta_d)$ whose columns are equal to the vectors β_i. Here we have a product of univariate densities whose "principal directions" are unknown. The density in (10.3) is effectively one-dimensional. The problem of estimating density f in (10.3) is called a semiparametric problem, since we have to estimate both the densities f_i and the parameters β_i.

Examples of Product Densities Figure 10.1 shows product densities. In the upper row the marginals are polynomial densities with the exponents $s = 10$, $s = 2$, $s = 1$, and $s = 0.5$. In the lower row the first panel shows a product density with the standard exponential marginals and after that the marginals are Student densities with degrees of freedom $\nu = 3$, $\nu = 1$, and $\nu = 0.5$.

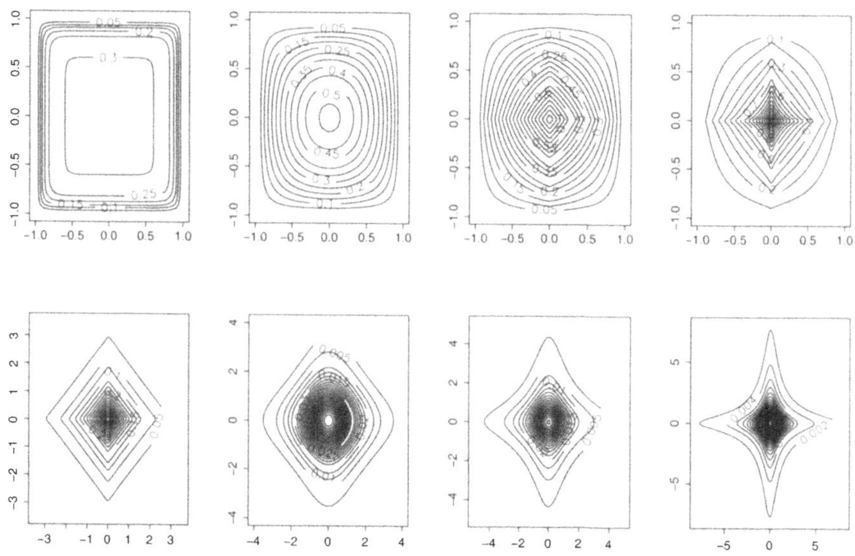

Figure 10.1 *(Product densities.)* The marginals in the upper row are polynomial with the exponents $s = 10$, $s = 2$, $s = 1$, and $s = 0.5$. The marginals in the lower row are the standard exponential and Student densities with degrees of freedom $\nu = 3$, $\nu = 1$, and $\nu = 0.5$.

Restrictions in Terms of Level Sets We may pose structural restrictions to the density in terms of its level sets. For example, the assumption of unimodality translates to the assumption of the connectedness of the level sets. In many applications we may assume that the level sets are star shaped.

Mixture Densities An M component mixture class is

$$\left\{ \sum_{i=1}^{M} \lambda_i f_i : \Lambda \in \mathcal{W}_M, \; f_i \in \mathcal{G}_i \right\},$$

where $\Lambda = (\lambda_i)_{i=1,\ldots,M}$, and $\mathcal{W}_M$ is the set of probability distributions on a set of cardinality M:

$$\mathcal{W}_M = \left\{ (\lambda_i)_{i=1,\ldots,M} \in [0,1]^M : \sum_{i=1}^{M} \lambda_i = 1, \; \lambda_i \geq 0 \right\}.$$

Note the fundamental difference to the convex hull defined in (10.50). Here M is fixed and $\Lambda \in \mathbf{R}^M$ is a finite dimensional parameter. The classes $\mathcal{G}_i$ may be parametric or infinite dimensional.

10.1.3 Elliptical Densities

A density of an elliptical distribution, or an elliptical density, has the form

$$f(x) = |\det(\Sigma)|^{-1/2} g\left\{ (x-\mu)^T \Sigma^{-1} (x-\mu) \right\}, \qquad x \in \mathbf{R}^d, \tag{10.4}$$

where $g : [0,\infty) \to [0,\infty)$ is called the density generator, dispersion matrix Σ is a symmetric positive definite $d \times d$ matrix, and $\mu \in \mathbf{R}^d$ is a location vector. Since Σ is positive definite, it has inverse Σ^{-1} which is positive definite, and this implies that for all $z \in \mathbf{R}^d$, $z^T \Sigma^{-1} z \geq 0$. Thus g need to be defined only on the nonnegative real axis.

For a fixed density generator g, the corresponding elliptical family has $2d + d(d-1)/2$ parameters (d parameters in μ, d parameters in the diagonal of Σ, and $d(d-1)/2$ parameters in the off-diagonals of symmetric matrix Σ). We may get tractable models by restricting the form of Σ.

Two-dimensional Case Let $d = 2$, and let the diagonal elements of Σ be σ_1^2 and σ_2^2, and the non-diagonal elements be $\rho\sigma_1\sigma_2$, where $\sigma_1, \sigma_2 > 0$, $\rho \in (-1,1)$. Then we can write

$$\begin{aligned} f(x) &= (\sigma_1\sigma_2)^{-1}(1-\rho^2)^{-1/2} \\ &\quad \times g\left(\frac{1}{(1-\rho^2)} \left[\left(\frac{x_1}{\sigma_1}\right)^2 + \left(\frac{x_2}{\sigma_2}\right)^2 - 2\rho \frac{x_1 x_2}{\sigma_1\sigma_2} \right] \right), \end{aligned}$$

$x = (x_1, x_2) \in \mathbf{R}^2$.

Examples of Elliptical Densities We give the density generators for the Gaussian, Student, and Bartlett cases.

1. The Gaussian density generator is defined by

$$g(t) = c \cdot \exp\{-t/2\}, \qquad t \in \mathbf{R}, \tag{10.5}$$

   where $c = (2\pi)^{-d/2}$. We say that a Gaussian density is the *standard Gaussian density* when $\mu = 0$ and $\Sigma = I_d$ in (10.4). When Σ is a diagonal matrix, we get a product density.

2. The Student density generator is defined by

$$g(t) = c \cdot (1 + t/\nu)^{-(d+\nu)/2}, \qquad t \in \mathbf{R}, \tag{10.6}$$

   where $\nu > 0$ is the parameter (degrees of freedom) and

$$c = \frac{\Gamma((\nu + d)/2)}{(\pi\nu)^{d/2}\Gamma(\nu/2)},$$

   with Γ denoting the gamma function. When $\nu \to \infty$, the Student density approaches the Gaussian density. Indeed $(1 + t/\nu)^{-(d+\nu)/2} \to \exp\{-t/2\}$, as $\nu \to \infty$, since $(1 + a/\nu)^\nu \to e^a$ when $\nu \to \infty$. The Student density has tails $f(x) \asymp \|x\|^{-(d+\nu)}$, as $\|x\| \to \infty$. The Student density generator has tails $g(t) \asymp t^{-(d+\nu)/2}$, as $t \to \infty$, and thus the density function is integrable when $\nu > 0$, according to (10.9). Note that the multivariate Student density has covariance matrix $\Sigma \cdot \nu/(\nu - 2)$, for $\nu > 2$.

3. The Bartlett density generator is defined by

$$g(t) = c \cdot (1 - t)_+, \qquad t \in \mathbf{R}, \tag{10.7}$$

   where $(a)_+ = \max\{0, a\}$, and

$$c = \frac{d(d+2)}{2\text{volume}(\mathbf{S}_{d-1})}, \tag{10.8}$$

   with volume$(\mathbf{S}_{d-1})$ as defined in (B.12).

4. In general, let $g_1 : [0, \infty) \to [0, \infty)$ be such that $\int_0^\infty t^{d/2-1} g_1(t)\, dt < \infty$. Then $g = c \cdot g_1$ is a density generator where c is the normalization constant

$$\begin{aligned} c^{-1} &= \int_{\mathbf{R}^d} g_1\left(\|x\|^2\right)\, dx \\ &= \text{volume}(\mathbf{S}_{d-1}) \cdot 2^{-1} \int_0^\infty t^{d/2-1} g_1(t)\, dt, \end{aligned} \tag{10.9}$$

   where we used the integration formula (B.2).

When g is unimodal, the densities in the corresponding elliptical family are unimodal. We get a rich family of multimodal densities by considering mixture densities whose components belong to an elliptical family with diagonal covariance matrices.

10.1.4 Copulas

Distribution function F may be decomposed into a part that describes the dependency and into a part that describes the marginal distributions. We call a *copula* the part that describes the dependency. The basic idea is that any distribution function $F : \mathbf{R}^d \to \mathbf{R}$ of a random vector $(X_1, \ldots, X_d)$ may be written as $F(x_1, \ldots, x_d) = C(F_1(x_1), \ldots, F_d(x_d))$, where C is the copula and F_i, $i = 1 \ldots, d$, are the marginal distribution functions: $F_i(x_i) = P(X_i \le x_i)$. A copula is a function $C : [0,1]^d \to [0,1]$ that describes the dependency between components of a random vector; it links univariate marginals to their multivariate distribution. We may construct parametric and semiparametric families of multivariate distributions by giving separately a model for the dependency and separately a model for the marginal distributions.

Copulas were defined by Sklar (1959), where Theorem 10.1 was proved. An exposition of copulas is given by Nelsen (1999). Malevergne and Sornette (2005, ch. 5) give a detailed discussion of the estimation of copulas. An application for the calculation of value-at-risk is given in Franke, Härdle and Hafner (2004).

Definition of a Copula Copulas are multivariate distribution functions whose one-dimensional marginal distributions are uniform on the interval $[0, 1]$. For simplicity, we write the definition first for the two-dimensional case.

Definition 10.1 (Copula, two-dimensional.) A two dimensional copula *is a function* $C : [0,1]^2 \to [0,1]$ *satisfying the following properties:*

1. *For each* $u \in [0,1]$, $C(0,u) = C(u,0) = 0$.

2. *For each* $u \in [0,1]$, $C(1,u) = C(u,1) = u$.

3. *For each* $(u_1, u_2), (v_1, v_2) \in [0,1] \times [0,1]$ *with* $u_1 \le v_1$ *and* $u_2 \le v_2$,

$$C(v_1, v_2) - C(v_1, u_2) - C(u_1, v_2) + C(u_1, u_2) \ge 0. \tag{10.10}$$

We generalize Definition 10.1 to the multivariate case.

Definition 10.2 (Copula.) A copula *is a function* $C : [0,1]^d \to [0,1]$ *satisfying the following properties:*

1. *For each* $x \in [0,1]^d$, $C(x) = 0$ *if* $x_i = 0$ *for some* $i \in \{1, \ldots, d\}$.

2. *For* $0 \le u \le 1$, $C(1, \ldots, 1, u, 1, \ldots, 1) = u$.

3. *For each* $x, y \in [0,1]^d$, $x_i \le y_i$, $i = 1, \ldots, d$,

$$\Delta^{(d)}_{x_d, y_d} \cdots \Delta^{(1)}_{x_1, y_1} C \ge 0, \tag{10.11}$$

where for $G : \mathbf{R}^k \to \mathbf{R}$, $a \le b$, $\Delta^{(i)}_{a,b} G : \mathbf{R}^{k-1} \to \mathbf{R}$ *is defined by*

$$\begin{aligned} &\Delta^{(i)}_{a,b} G(x_1, \ldots, x_{k-1}) \\ &= G(x_1, \ldots, x_{i-1}, b, x_{i+1}, \ldots, x_k) - G(x_1, \ldots, x_{i-1}, a, x_{i+1}, \ldots, x_k), \end{aligned}$$

$k \in \{1, \ldots, d\}$, $i \in \{1, \ldots, k\}$.

For distribution functions $F : \mathbf{R}^d \to \mathbf{R}$, for $x, y \in \mathbf{R}^d$, $x_i \le y_i$,

$$\Delta^{(d)}_{x_d, y_d} \cdots \Delta^{(1)}_{x_1, y_1} F = \int_{x_d}^{y_d} \cdots \int_{x_1}^{y_d} dF.$$

Thus $\Delta^{(d)}_{x_d, y_d} \cdots \Delta^{(1)}_{x_1, y_1} F$ is equal to the probability of the hyper-rectangle $[x_1, y_1] \times \cdots \times [x_d, y_d]$. A function with the property (10.10) or (10.11) is called 2-increasing. Theorem 10.1 motivates the definition of a copula.

Theorem 10.1 (Sklar's theorem.) *Let* $F : [0,1]^2 \to [0,1]$ *be a distribution function with marginal distribution functions* $F_1, F_2 : [0,1] \to [0,1]$. *There exists a copula C such that*

$$F(x_1, x_2) = C(F_1(x_1), F_2(x_2)). \tag{10.12}$$

Copula C *is unique when* F_1 *and* F_2 *are continuous.*

Partial Proof. We prove the existence. The uniqueness is the difficult part. Let X_1 and X_2 be random variables with distribution functions F_1 and F_2. We have

$$\begin{aligned} F(x_1, x_2) &= P(X_1 \le x_1, X_2 \le x_2) \\ &= P(F_1(X_1) \le F_1(x_1), F_2(X_2) \le F_2(x_2)) \\ &= C(F_1(x_1), F_2(x_2)), \end{aligned} \tag{10.13}$$

where

$$C(u, v) = P(F_1(X_1) \le u, F_2(X_2) \le v), \qquad u, v \in [0,1]. \tag{10.14}$$

Put in words, copula C is the joint distribution function of uniformly distributed random vectors $F_1(X_1), F_2(X_2)$. □

Nonstandard Copulas A copula is a distribution function with uniform marginals. We call this kind of copula a *standard copula*. However, we can choose the marginal distributions of a copula to be some other continuous distribution than the uniform distribution on $[0,1]$. It turns out that we get simpler copulas by choosing the marginal distributions of a copula to be the standard Gaussian distribution; see Section 10.1.4. As in (10.13) we can write F as

$$F(x_1, x_2) = C\left(\Phi^{-1}(F_1(x_1)), \Phi^{-1}(F_2(x_2))\right)$$

where $\Phi : \mathbf{R} \to \mathbf{R}$ is the distribution function of the standard Gaussian distribution and

$$C(u, v) = P\left(\Phi^{-1}(F_1(X_1)) \le u, \Phi^{-1}(F_2(X_2)) \le v\right), \qquad u, v \in \mathbf{R}. \tag{10.15}$$

Now C is a distribution function whose marginals are standard Gaussian.

Construction of Copulas Sklar's theorem gives a method to construct copulas. Let $F : \mathbf{R}^2 \to \mathbf{R}$ be a distribution function with marginal distribution functions $F_i : \mathbf{R} \to \mathbf{R}$.

1. *(Standard copula.)* Function C defined by (10.14) is a copula. When F_1 and F_2 are continuous,

$$C(u,v) = F\left(F_1^{-1}(u), F_2^{-1}(v)\right), \qquad u,v \in [0,1], \tag{10.16}$$

is a copula. When copula C is defined by (10.16), the copula density is

$$c(u,v) = \frac{f\left(F_1^{-1}(u), F_2^{-1}(v)\right)}{f_1(F_1^{-1}(u)) \cdot f_2(F_2^{-1}(v))} \tag{10.17}$$

because $(\partial/\partial u)F_i^{-1}(u) = 1/f_i(F_i^{-1}(u))$, where f is the density of F and f_1, f_2 are the densities of F_1, F_2.

2. *(Nonstandard copula.)* Function C defined by (10.15) is a nonstandard copula with standard Gaussian marginals. When F_1 and F_2 are continuous,

$$C(u,v) = F\left(F_1^{-1}\left(\Phi(u)\right), F_2^{-1}\left(\Phi(v)\right)\right), \qquad u,v \in \mathbf{R}, \tag{10.18}$$

is a nonstandard copula with standard Gaussian marginals. When copula C is defined by (10.18), the copula density is

$$\begin{aligned} c(u,v) &= f\left(F_1^{-1}(\Phi(u)), F_2^{-1}(\Phi(v))\right) \\ &\quad \times \frac{\phi(u)\phi(v)}{f_1(F_1^{-1}(\Phi(u))) \cdot f_2(F_2^{-1}(\Phi(v)))}, \end{aligned}$$

where f is the density of F, f_1, f_2 are the densities of F_1, F_2, and ϕ is the density of the standard Gaussian distribution.

Construction of a Distribution with a Given Copula We can construct a distribution with a given copula C and given marginals G_1 and G_2.

1. *(Standard case.)* When C is a standard copula, then distribution function

$$F(x_1, x_2) = C(G_1(x_1), G_2(x_2))$$

has copula C and marginals G_i, and the density function of this distribution is

$$f(x_1, x_2) = c(G_1(x_1), G_2(x_2)) \cdot g_1(x_1) g_2(x_2), \tag{10.19}$$

where

$$c(u,v) = \frac{\partial^2}{\partial u \partial v} C(u,v),$$

and g_1, g_2 are the densities corresponding to the distribution functions G_1, G_2.

2. *(Nonstandard case.)* When C is a nonstandard copula with the standard Gaussian marginals, distribution function

$$F(x_1, x_2) = C\left(\Phi^{-1}(G_1(x_1)), \Phi^{-1}(G_2(x_2))\right)$$

has copula C and marginals G_i, and the density function of this distribution is

$$f(x_1, x_2) = c(G_1(x_1), G_2(x_2)) \cdot \frac{g_1(x_1) g_2(x_2)}{\phi(G_1(x_1))\phi(G_2(x_2))}.$$

Copula Families We give examples of parametric families of copulas. The copulas are given in the standard form, where the marginals are uniform.

The Minimum Copula The minimum copula is given by

$$C_{min}(u) = \min\{u_1, \ldots, u_d\}, \qquad u \in [0,1]^d.$$

The Product Copula The product copula is given by

$$C_{prod}(u) = u_1 \cdots u_d, \qquad u \in [0,1]^d.$$

Distributions with independent marginals have the product copula. The density of the product copula is the uniform density $I_{[0,1]^d}$.

Gaussian Copulas The Gaussian copula, with $d \times d$ correlation matrix ρ, is defined by

$$C_{gauss}(u; \rho) = \Phi_\rho\left(\Phi^{-1}(u)\right), \qquad u \in [0,1]^d,$$

where Φ is the distribution function of the standard Gaussian distribution, $\Phi^{-1}(u) = (\Phi^{-1}(u_1), \ldots, \Phi^{-1}(u_d))$, and Φ_ρ is the Gaussian distribution function with covariance matrix ρ (we take $\Sigma = \rho$ and $\mu = 0$ in (10.4) and (10.5)). The diagonal elements of the correlation matrix ρ are 1, and the nondiagonal elements are between -1 and 1. Here we used the construction (10.16), and the density is given by (10.17):

$$c_{gauss}(u; \rho) = \frac{1}{\sqrt{\det(\rho)}} \exp\left\{-\frac{1}{2}\Phi^{-1}(u)^T(\rho^{-1} - I_d)\Phi^{-1}(u)\right\}, \qquad u \in [0,1]^d.$$

When $\rho = I_d$, then $C_{gauss} = C_{prod}$, where I_d is the $d \times d$ identity matrix.

Student Copulas The Student copula, with the $d \times d$ correlation matrix ρ and degrees of freedom $\nu > 0$, is defined by

$$C_t(u; \rho, \nu) = T_{\rho,\nu}\left(T_\nu^{-1}(u)\right), \qquad u \in [0,1]^d, \qquad (10.20)$$

where T_ν is the one-dimensional Student distribution function with degrees of freedom ν, defined by (10.1), $T_\nu^{-1}(u) = (T_\nu^{-1}(u_1), \ldots, T_\nu^{-1}(u_d))$, and $T_{\rho,\nu}$ is the d-dimensional Student distribution function with the dispersion matrix ρ and with degrees of freedom ν (we take $\Sigma = \rho$ and $\mu = 0$ in (10.4) and (10.6)). Here we used again the construction (10.16). The density is given by (10.17), which can be written, using (10.6),

$$c_t(u;\rho,\nu) = \frac{c}{\sqrt{\det(\rho)}} \prod_{k=1}^{d} \left(1 + \frac{T_\nu^{-1}(u_k)^2}{\nu}\right)^{(\nu+1)/2} \times \left(1 + \frac{T_\nu^{-1}(u)^T \rho^{-1} T_\nu^{-1}(u)}{\nu}\right)^{-(\nu+d)/2},$$

where $u \in [0,1]^d$,

$$c = \frac{\Gamma((\nu+d)/2)[\Gamma(\nu/2)]^{d-1}}{[\Gamma((\nu+1)/2)]^d}.$$

When $\nu \to \infty$, the Student copula approaches the Gaussian copula.

Gumbel–Hougaard Copulas The Gumbel-Hougaard family of copulas is defined by

$$C_{gh}(u;\theta) = \exp\left\{-\left[\sum_{i=1}^{d}(-\log_e u_i)^\theta\right]^{1/\theta}\right\}, \qquad u \in [0,1]^d,$$

where $\theta \in [1,\infty)$ is the parameter. When $\theta = 1$, $C_{gh} = C_{prod}$, and when $\theta \to \infty$, $C_{gh}(u;\theta) \to C_{min}(u)$.

Clayton Copulas Clayton's family of copulas is defined by

$$C_{cl}(u;\theta) = \left(1 - d + \sum_{i=1}^{d} u_i^{-\theta}\right)^{-1/\theta}, \qquad \theta > 0, \tag{10.21}$$

$C_{cl}(u;\theta) = \Pi_{i=1}^d u_i$, when $\theta = 0$, where $u \in [0,1]^d$. As the parameter θ increases, the dependence between coordinate variables increases. The dependence is larger in the negative orthant. The Clayton family was discussed in Clayton (1978).

Frank Copulas Franks's family of copulas is defined by

$$C_f(u;\theta) = -\frac{1}{\theta}\log_e\left\{1 + \frac{\Pi_{i=1}^d(e^{-\theta u_i}-1)}{(e^{-\theta}-1)^{d-1}}\right\}, \qquad u \in [0,1]^d,$$

where $\theta \in \mathbf{R} \setminus \{0\}$.

Elliptical and Archimedean Copulas Elliptical distributions are defined in Section 10.1.3. An elliptical copula is obtained from an elliptical distribution F by the construction (10.16). The Gaussian copula and the Student copula are elliptical copulas.

Archimedean copulas have the form

$$C(u) = \phi^{-1}\left(\phi(u_1) + \cdots + \phi(u_d)\right), \qquad u \in [0,1]^d,$$

where $\phi : [0, 1] \longrightarrow [0, \infty)$ is strictly decreasing, continuous, convex, and $\phi(1) = 0$. For C to be a copula, we need that $(-1)^i \partial^i \phi^{-1}(t)/\partial t^i \geq 0$, $i = 1, \ldots, d$. The function ϕ is called the generator. The product copula, Gumbel copula, Clayton copula, and Frank copula are all Archimedean copulas and we have:

- Product copula: $\phi(t) = \log_e t$.
- Gumbel copula: $\phi(t) = (-\log_e t)^\theta$.
- Clayton copula: $\phi(t) = t^{-\theta} - 1$.
- Frank copula: $\phi(t) = -\log_e[(e^{-\theta t} - 1)/(e^{-\theta} - 1)]$.

The density of an Archimedean copula is

$$c(u) = \psi\left(\phi(u_1) + \cdots + \phi(u_d)\right) \phi'(u_1) \cdots \phi'(u_d),$$

where ψ is the second derivative of ϕ^{-1}: $\psi(y) = -\phi''(\phi^{-1}(y))(\phi'(\phi^{-1}(y))^{-3}$ (because $(\partial/\partial x)\phi^{-1}(x) = 1/\phi'(\phi^{-1}(x))$). We make the list of the common examples:

- Gumbel copula: $\phi'(t) = -\theta t^{-1}(-\log_e t)^{\theta-1}$, $\phi''(t) = \theta t^{-2}(-\log_e t)^{\theta-2}(\theta - 1 - \log_e t)$, $\phi^{-1}(t) = \exp(-t^{1/\theta})$.
- Clayton copula: $\phi'(t) = -\theta t^{-\theta-1}$, $\phi''(t) = \theta(\theta+1)t^{-\theta-2}$, $\phi^{-1}(t) = (t + 1)^{-1/\theta}$.
- Frank copula: $\phi'(t) = \theta e^{-\theta t}/(e^{-\theta t} - 1)$, $\phi''(t) = \theta^2 e^{-\theta t}/(e^{-\theta t} - 1)^2$, $\phi^{-1}(t) = -\theta^{-1} \log_e[(e^{-\theta} - 1)e^{-t} + 1]$.

Figures of Copula Densities In addition to R-package "denpro" we have used R-package "copula" to draw the figures; see Yan (2007).

Gaussian Copula We give examples of Gaussian copulas both with the uniform marginals and Student marginals. Note that when $\rho = 0$, the Gaussian copula is the product copula, and thus Figure 10.1 shows densities with the Gaussian copula for the case $\rho = 0$.

Uniform Marginals

1. Figure 10.2 shows Gaussian copula densities when $\rho = 0.1$, $\rho = 0.4$, and $\rho = 0.8$. The upper row shows contour plots, and the lower row shows perspective plots.
2. Figure 2.9 shows the density of the Gaussian copula when $\rho = 0.2$.

Gaussian copulas with uniform marginals are quite complex. For $0 < \rho < 1$ there are five local maxima: three local maxima at $(0, 0)$, at $(1, 1)$, and at $(0.5, 0.5)$; two local maxima at $(0, 1)$ and at $(1, 0)$. That is why we prefer to use the standard Gaussian marginals to define copulas.

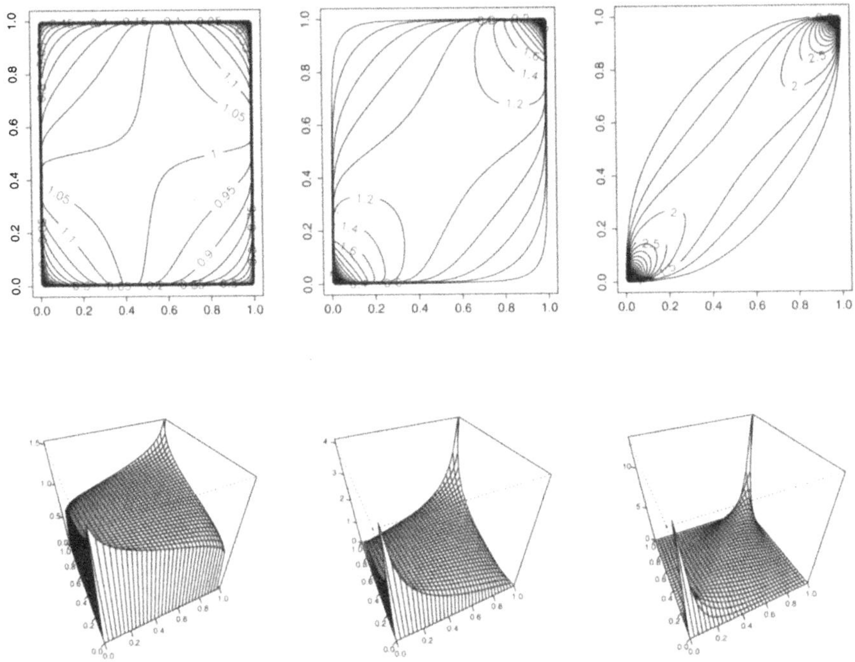

Figure 10.2 *(Gaussian copula, uniform marginals.)* Densities of the Gaussian copula with $\rho = 0.1$, $\rho = 0.4$, and $\rho = 0.8$. In the upper row are the contour plots, and in the lower row are the perspective plots.

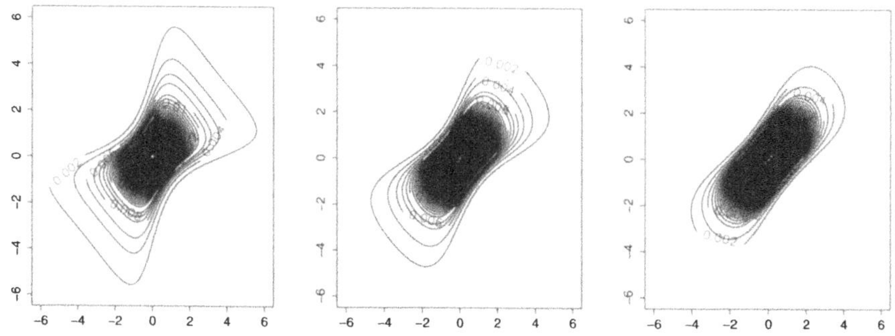

Figure 10.3 *(Gaussian copula, Student marginals.)* Contour plots of the densities of the Gaussian copula with $\rho = 0.8$, with Student marginals when the degrees of freedom are $\nu = 1$, $\nu = 2$, and $\nu = 4$.

Student Marginals

1. Figure 10.3 shows Gaussian copula densities when the correlation parameter is $\rho = 0.8$. The degrees of freedom of the Student marginals are $\nu = 1$, $\nu = 2$, and $\nu = 4$. See Figure 2.8 for a warning on the interpretation of these contour plots.

2. Figure 10.4 visualizes the 0.5% level set of the density in Figure 10.3(c) (Gaussian copula with $\rho = 0.8$, Student marginals with $\nu = 4$). Panel (a) shows the level set, panel (b) the tail probability plot of this level set, and panels (c) and (d) show the corresponding location plot.

3. Figure 10.5 shows data of size $n = 1000$ generated from the distribution in Figure 10.3(c) ($\rho = 0.8, \nu = 4$). Panel (a) shows a scatter plot, panel (b) shows a tail frequency plot with the resolution threshold 1.4, and panels (c) and (d) show the corresponding tail tree plots. The tail frequency plot shows the one separate observation as a vertical line whose length is equal to the distance of the observation from the reference point (the origin).

4. Figure 10.6 shows shape plots of the 0.5% level set of the density in Figure 10.3(a). Panel (a) shows the level set, panel (b) shows a radius plot, panel (c) shows a tail probability plot, and panel (d) shows a probability content plot.

Student Copula We give examples of Student copulas both for uniform marginals and Gaussian marginals.

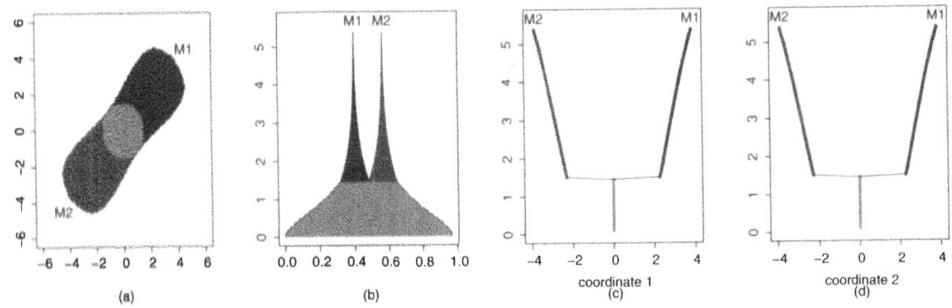

Figure 10.4 *(Gaussian copula, Student marginals, $\rho = 0.8$, $\nu = 4$, 0.5% level set.)* (a) The level set, (b) a tail probability plot, and (c–d) the location plot. The level set is a level set of the density in Figure 10.3(c).

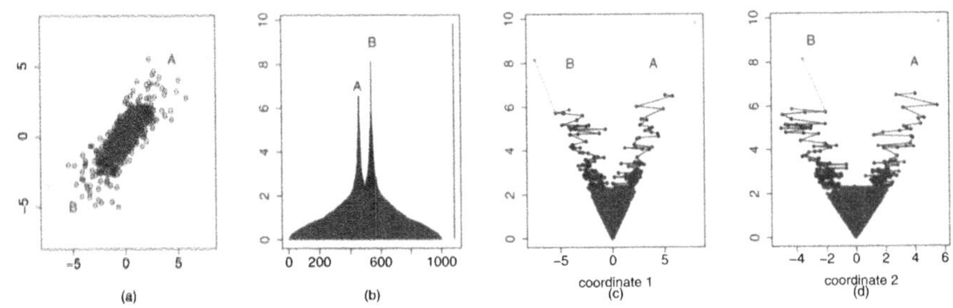

Figure 10.5 *(Gaussian copula, Student marginals, $\rho = 0.8$, $\nu = 4$.)* (a) A scatter plot of data of size $n = 1000$ generated from the distribution in Figure 10.3(c), (b) a tail frequency plot, with the resolution threshold 1.4, and (c–d) the corresponding tail tree plot.

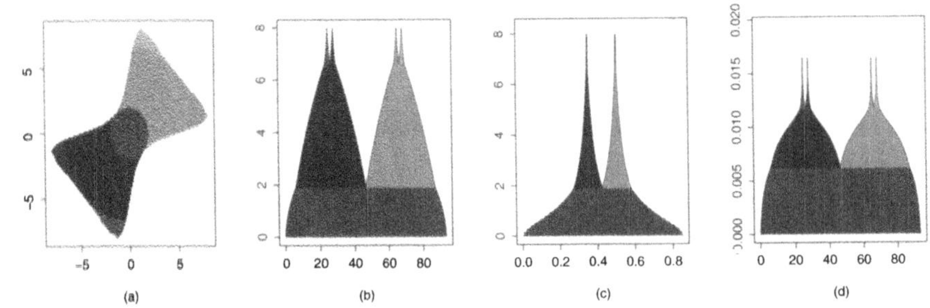

Figure 10.6 *(Gaussian copula, Student marginals, $\rho = 0.8$, $\nu = 1$, 0.5% level set.)* (a) The level set, (b) a radius plot, (c) a tail probability plot, and (d) a probability content plot. The level set is a level set of the density in Figure 10.3(a).

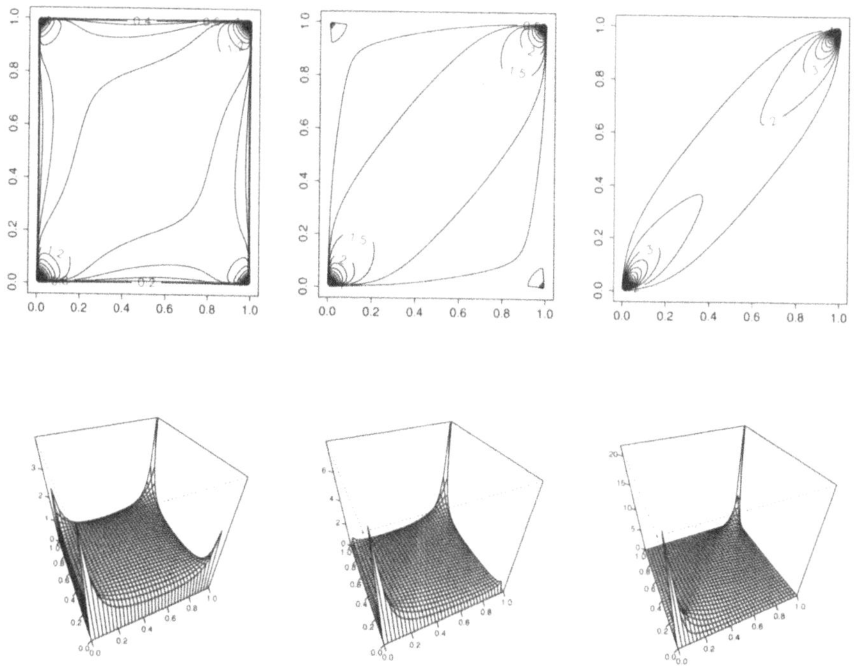

Figure 10.7 *(Student copula, uniform marginals.)* Shown are densities of the Student copulas with degrees of freedom 4 and with $\rho = 0.1$, $\rho = 0.4$, and $\rho = 0.8$. The upper row shows contour plots, and the lower row shows perspective plots.

Uniform Marginals

Figure 10.7 shows Student copula densities, with degrees of freedom $\nu = 4$ when $\rho = 0.1$, $\rho = 0.4$, and $\rho = 0.8$. The upper row shows contour plots, and the lower row shows perspective plots.

Gaussian Marginals

1. Figure 10.8 shows Student copula densities with the correlation parameter $\rho = 0.6$. The degrees of freedom of the copula are $\nu = 1, 2, 4$. The marginals are standard Gaussian.

2. Figure 10.9 visualizes the 0.5% level set of the density in Figure 10.8(a) (Student copula with $\rho = 0.6$ and $\nu = 1$, standard Gaussian marginals).

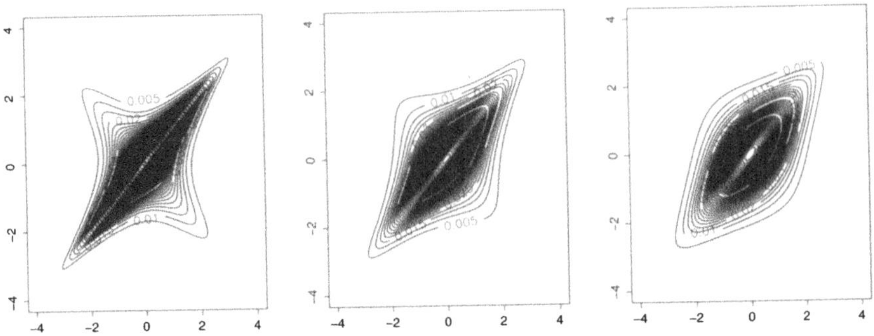

Figure 10.8 *(Student copula, Gaussian marginals.)* Shown are contour plots of the densities of the Student copula with $\rho = 0.6$. The degrees of freedom are $\nu = 1, 2, 4$. The marginals are standard Gaussian.

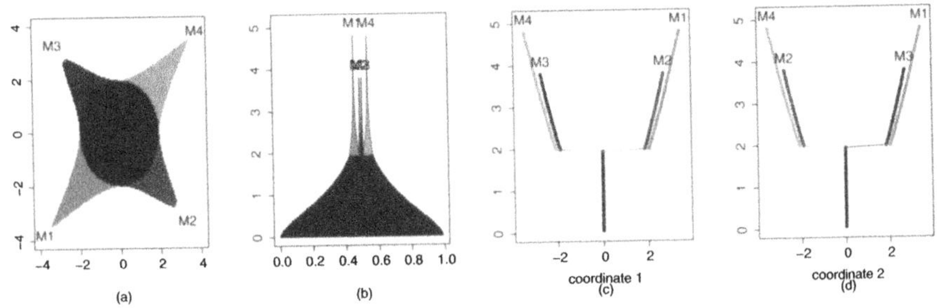

Figure 10.9 *(Student copula, Gaussian marginals, $\rho = 0.6$, $\nu = 1$, 0.5% level set.)* (a) The level set, (b) a tail probability plot, and (c–d) the location plot. The level set is a level set of the density in Figure 10.8(a).

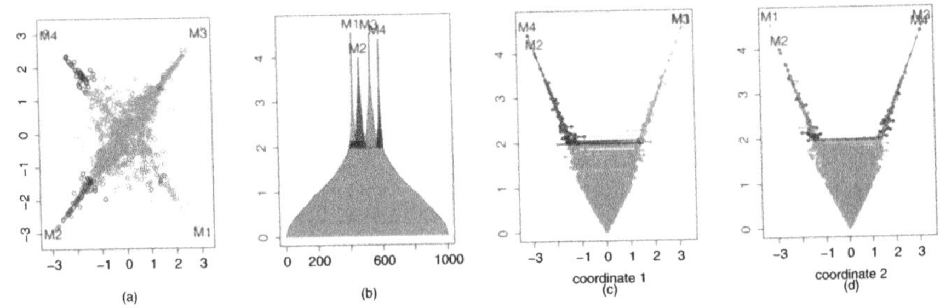

Figure 10.10 *(Student copula, Gaussian marginals, $\rho = 0.6$, $\nu = 1$.)* (a) A scatter plot of data of size $n = 1000$ generated from the distribution in Figure 10.8(a), (b) a tail frequency plot, with the resolution threshold 1, and (c–d) the corresponding tail tree plot.

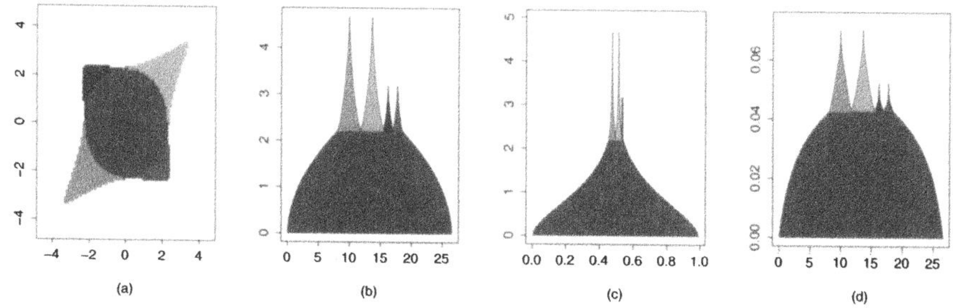

Figure 10.11 *(Student copula, Gaussian marginals, $\rho = 0.6$, $\nu = 2$, 0.5% level set.)* (a) The level set, (b) a radius plot, (c) a tail probability plot, and (d) a probability content plot. The level set is a level set of the density in Figure 10.8(b).

Panel (a) shows the level set, panel (b) shows the tail probability plot of this level set, and panels (c) and (d) show the corresponding location plot.

3. Figure 10.10 shows a data of size $n = 1000$ generated from the distribution in Figure 10.8(a) ($\rho = 0.6$, $\nu = 1$). Panel (a) shows a scatter plot, panel (b) shows a tail frequency plot with the resolution threshold 1, and panels (c) and (d) show the corresponding tail tree plot.

4. Figure 10.11 shows shape plots of the 0.5% level set of the density in Figure 10.8(b). Panel (a) shows the level set, panel (b) shows a radius plot, panel (c) shows a tail probability plot, and panel (d) shows a probability content plot.

The Gumbel–Hougaard Copula We give examples of Gumbel copula densities both for the uniform marginals and the standard Gaussian marginals.

Uniform Marginals

Figure 10.12 shows Gumbel copula densities with $\theta = 1.5$, $\theta = 2$, and $\theta = 4$. The upper row shows contour plots and the lower row shows perspective plots.

Gaussian Marginals

1. Figure 10.13 shows contour plots of densities with the Gumbel copula when $\theta = 1, 1.5$, and 4, with the standard Gaussian marginals.

2. Figure 10.14 visualizes the 0.5% level set of the density in Figure 10.13(a) (Gumbel copula with $\theta = 4$, standard Gaussian marginals). Panel (a) shows the level set, panel (b) shows the tail probability plot of this level set, and panels (c) and (d) show the corresponding location plot.

3. Figure 10.15 shows a data of size $n = 1000$ generated from the distribution in Figure 10.13(c) ($\theta = 4$). Panel (a) shows a scatter plot, panel (b) shows a tail

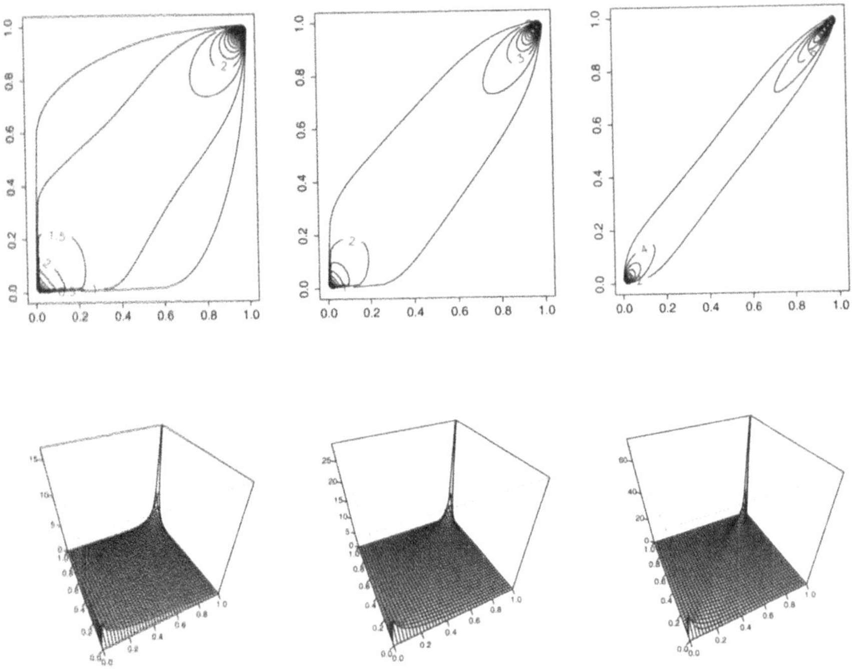

Figure 10.12 *(Gumbel copula, uniform marginals.)* Shown are the densities of the Gumbel copula with $\theta = 1.5, 2$, and 4. The upper row shows contour plots, and the lower row shows perspective plots.

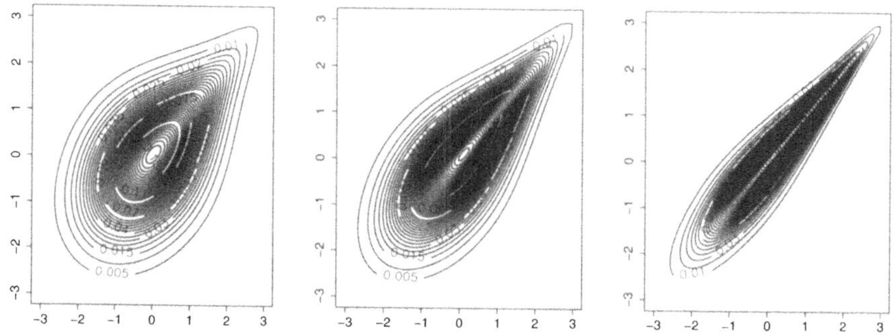

Figure 10.13 *(Gumbel copula, Gaussian marginals.)* Shown are contour plots of the densities of the Gumbel copula with $\theta = 1.5, 2$, and 4, with the standard Gaussian marginals.

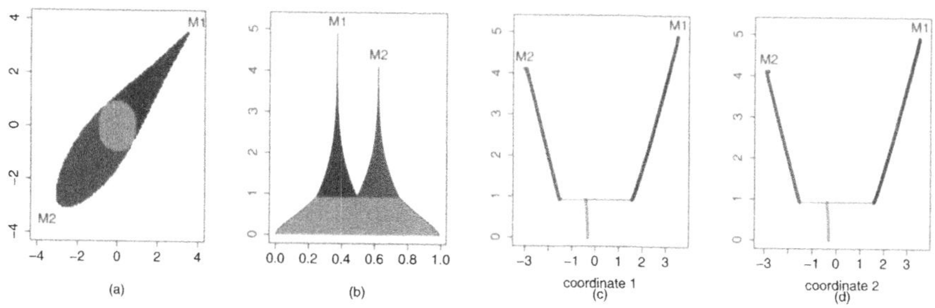

Figure 10.14 *(Gumbel copula, Gaussian marginals, $\theta = 4$, 0.5% level set.)* (a) The level set, (b) a tail probability plot, and (c–d) the location plot. The level set is a level set of the density in Figure 10.13(c).

frequency plot with the resolution threshold 1, and panels (c) and (d) show the corresponding tail tree plot.

4. Figure 10.16 shows shape plots of the 0.5% level set of the density in Figure 10.13(b). Panel (a) shows the level set, panel (b) shows a radius plot, panel (c) shows a tail probability plot, and panel (d) shows a probability content plot.

The Clayton Copula We give examples of Clayton copula densities both for the uniform marginals and the standard Gaussian marginals.

Uniform Marginals

Figure 10.17 shows Clayton copula densities with $\theta = 1$, $\theta = 2$, and $\theta = 4$. The upper row shows contour plots, and the lower row shows perspective plots.

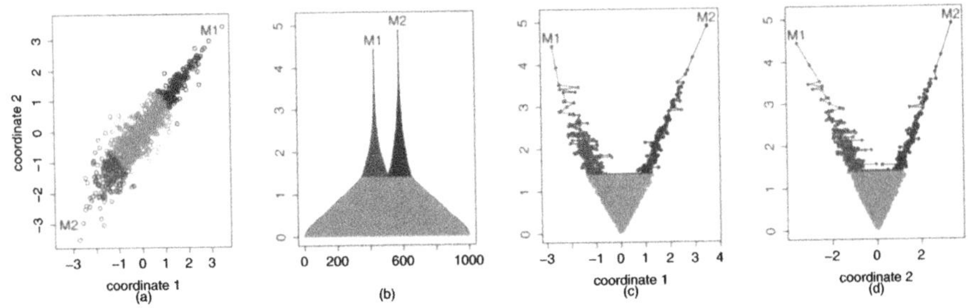

Figure 10.15 *(Gumbel copula, Gaussian marginals, $\theta = 4$.)* (a) A scatter plot of data of size $n = 1000$ generated from the distribution in Figure 10.13(c), (b) a tail frequency plot, with the resolution threshold 1, and (c–d) the corresponding tail tree plot.

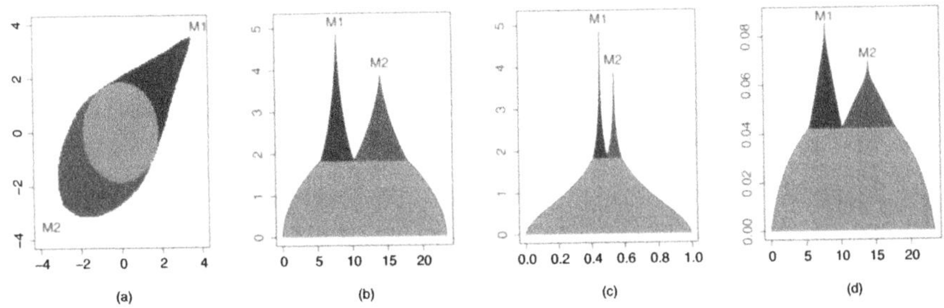

Figure 10.16 *(Gumbel copula, Gaussian marginals, $\theta = 2$, 0.5% level set.)* (a) The level set, (b) a radius plot, (c) a tail probability plot, and (d) a probability content plot. The level set is a level set of the density in Figure 10.13(b).

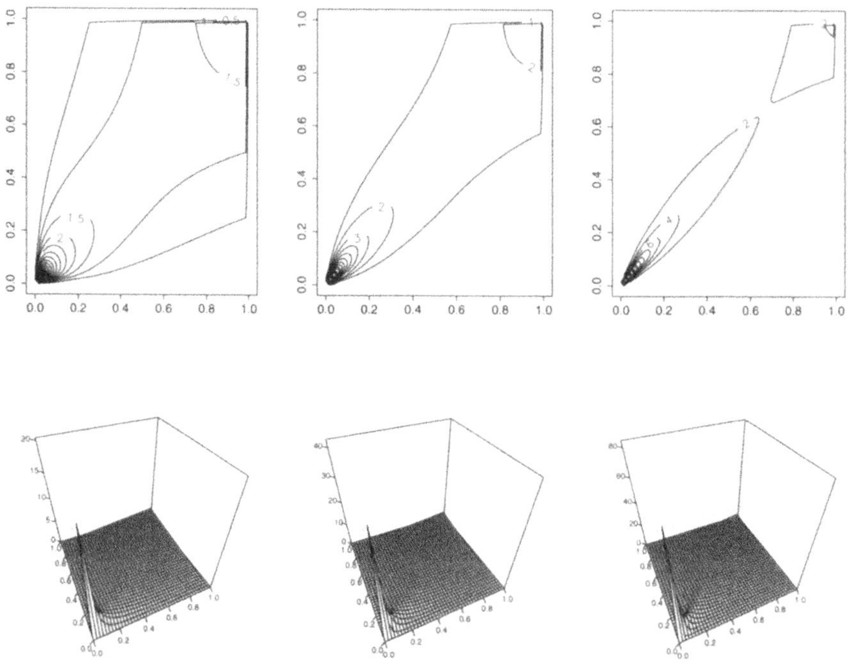

Figure 10.17 *(Clayton copula, uniform marginals.)* Shown are the densities of the Clayton copula with $\theta = 1, 2$, and 4. The upper row shows contour plots and the lower row shows perspective plots.

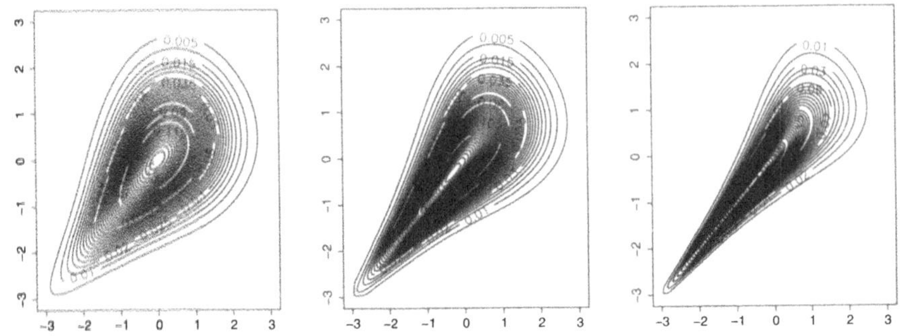

Figure 10.18 *(Clayton copula, Gaussian marginals.)* Shown are contour plots of the densities of the Clayton copula with $\theta = 1, 2$, and 4, with the standard Gaussian marginals.

Gaussian Marginals

1. Figure 10.18 shows contour plots of densities with the Clayton copula when $\theta = 1, 2$, and 4, with the standard Gaussian marginals.

2. Figure 10.19 visualizes the 0.5% level set of the density in Figure 10.13(a) (Clayton copula with $\theta = 4$, standard Gaussian marginals). Panel (a) shows the level set, panel (b) shows the tail probability plot of this level set, and panels (c) and (d) show the corresponding location plot.

3. Figure 10.20 shows a data of size $n = 1000$ generated from the distribution in Figure 10.18(c) ($\theta = 4$). Panel (a) shows a scatter plot, panel (b) shows a tail frequency plot with the resolution threshold 1, and panels (c) and (d) show the corresponding tail tree plot.

4. Figure 10.21 shows shape plots of the 0.5% level set of the density in Figure 10.18(b). Panel (a) shows the level set, panel (b) shows a radius plot, panel (c) shows a tail probability plot, and panel (d) shows a probability content plot.

The Frank Copula We give examples of Frank copula densities both for the uniform marginals and the standard Gaussian marginals.

Uniform Marginals

Figure 10.17 shows Frank copula densities with $\theta = -1$, $\theta = 3$, and $\theta = 5$. The upper row shows contour plots, and the lower row shows perspective plots.

Gaussian Marginals

1. Figure 10.23 shows contour plots of densities with the Frank copula when $\theta = -1, 3$, and 5, with the standard Gaussian marginals.

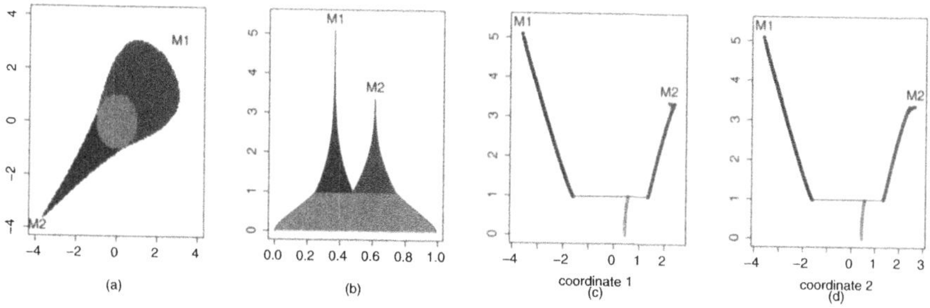

Figure 10.19 *(Clayton copula, Gaussian marginals, $\theta = 4$, 0.5% level set.)* (a) The level set, (b) a tail probability plot, and (c–d) the location plots. The level set is a level set of the density in Figure 10.18(c).

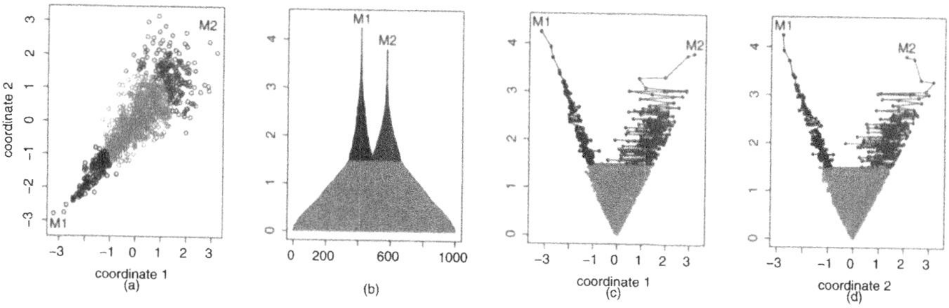

Figure 10.20 *(Clayton copula, Gaussian marginals, $\theta = 4$.)* (a) A scatter plot of data of size $n = 1000$ generated from the distribution in Figure 10.18(c), (b) a tail frequency plot, with the resolution threshold 1, and (c–d) the corresponding tail tree plots.

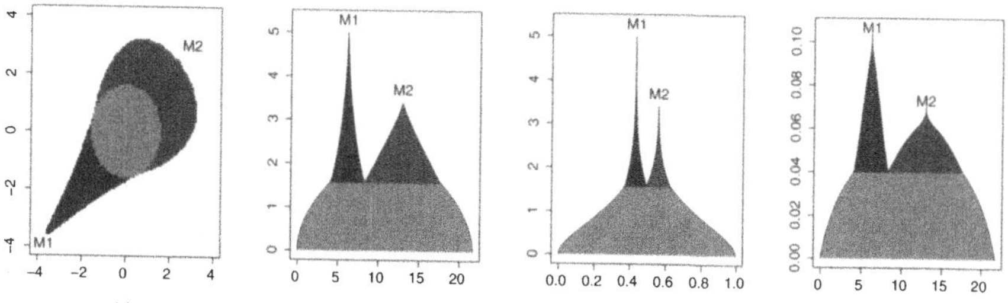

Figure 10.21 *(Clayton copula, Gaussian marginals, $\theta = 2$, 0.5% level set.)* (a) The level set, (b) a radius plot, (c) a tail probability plot, and (d) a probability content plot. The level set is a level set of the density in Figure 10.18(b).

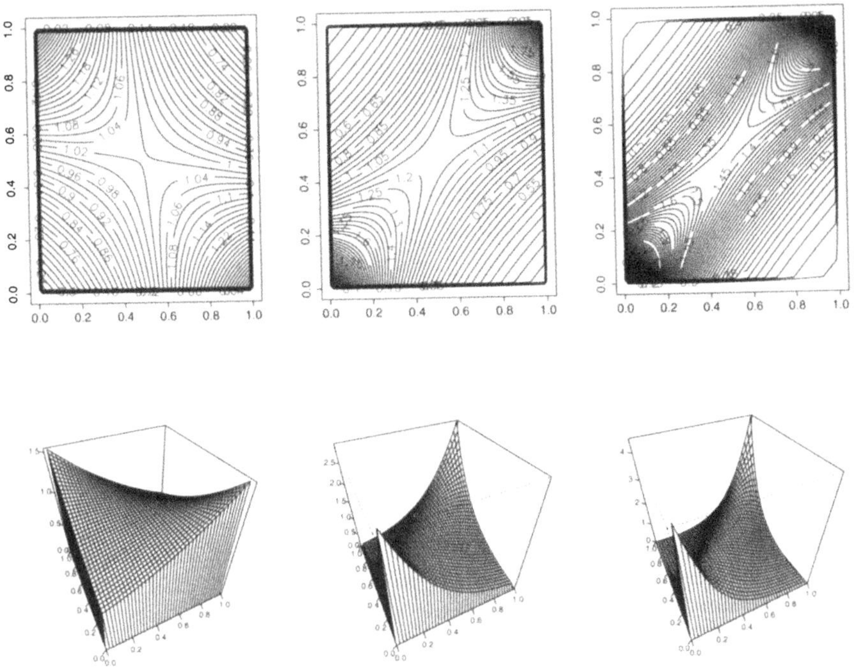

Figure 10.22 *(Frank copula, uniform marginals.)* Shown are the densities of the Frank copula with $\theta = -1, 3$, and 5. The upper row shows contour plots, and the lower row shows perspective plots.

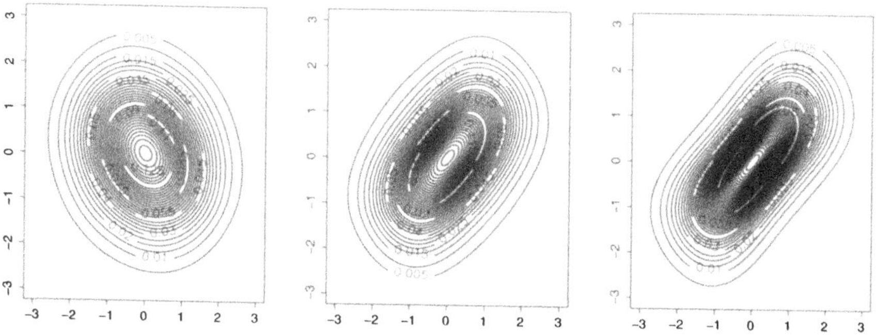

Figure 10.23 *(Frank copula, Gaussian marginals.)* Shown are contour plots of the densities of the Frank copula with $\theta = -1, 3$, and 5, with the standard Gaussian marginals.

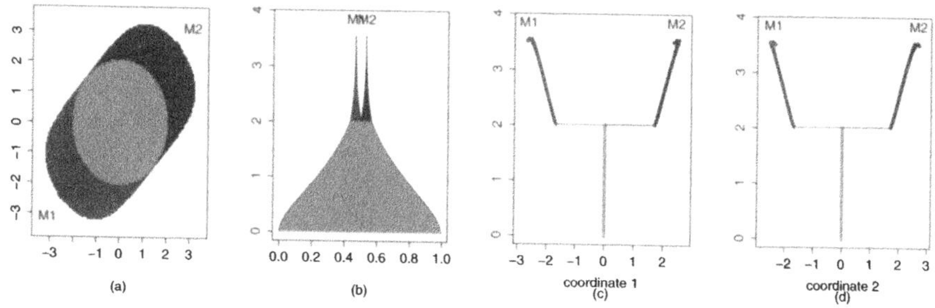

Figure 10.24 *(Frank copula, Gaussian marginals, $\theta = 5$, 0.5% level set.)* (a) The level set, (b) a tail probability plot, and (c–d) the location plots. The level set is a level set of the density in Figure 10.23(c).

2. Figure 10.24 visualizes the 0.5% level set of the density in Figure 10.23(c) (Frank copula with $\theta = 5$, standard Gaussian marginals). Panel (a) shows the level set, panel (b) shows the tail probability plot of this level set, and panels (c) and (d) show the corresponding location plots.

3. Figure 10.25 shows a data of size $n = 1000$ generated from the distribution in Figure 10.23(c) ($\theta = 5$). Panel (a) shows a scatter plot, panel (b) shows a tail frequency plot with the resolution threshold 1.3, and panels (c) and (d) show the corresponding tail tree plots.

4. Figure 10.26 shows shape plots of the 0.5% level set of the density in Figure 10.23(b). Panel (a) shows the level set, panel (b) shows a radius plot, panel (c) shows a tail probability plot, and panel (d) shows a probability content plot.

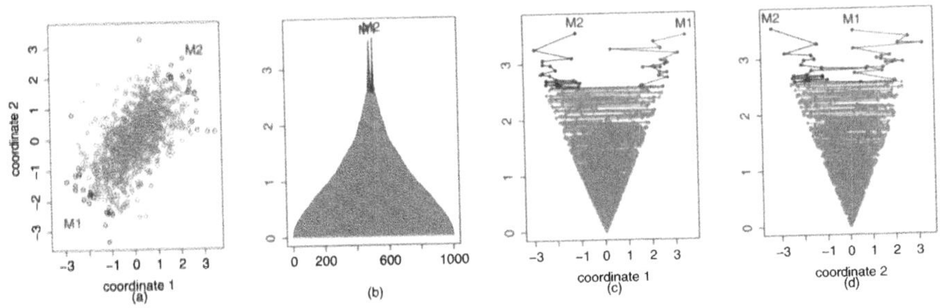

Figure 10.25 *(Frank copula, Gaussian marginals, $\theta = 5$.)* (a) A scatter plot of data of size $n = 1000$ generated from the distribution in Figure 10.23(c), (b) a tail frequency plot, with the resolution threshold 1.3, and (c–d) the corresponding tail tree plots.

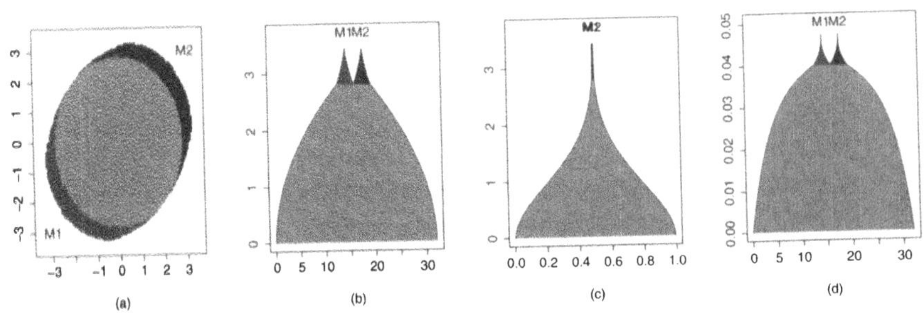

Figure 10.26 *(Frank copula, Gaussian marginals, $\theta = 3$, 0.5% level set.)* (a) The level set, (b) a radius plot, (c) a tail probability plot, and (d) a probability content plot. The level set is a level set of the density in Figure 10.23(b).

Estimation A copula C or a copula density c can be estimated when we estimate first the marginal distribution functions with $\hat{F}_1, \ldots, \hat{F}_d$. Then we make the transform $Z_i = \hat{F}(X_i)$, and use the data Z_i, $i = 1, \ldots, n$, to estimate C or c. This approach was indeed proposed in Section 9.2, where the copula preprocessing gives variables Z_i with $\hat{F}_i$ as the empirical distribution function, and the additional standardization to standard Gaussian marginals was proposed.

In the semiparametric setting it is assumed that the copula belongs to a parametrized family. We can estimate marginals nonparametrically, and then we can estimate the parameter of the copula with the maximum likelihood method. This is called pseudo maximum likelihood estimation because we estimate first the marginal distributions and then plug these estimates in to the likelihood. Let $\{C(\cdot\,;\theta) : \theta \in \Theta\}$ be a parametrized class of copulas. The likelihood is given, using (10.19), by

$$L_n(\theta) = \prod_{i=1}^{n} c\left(\hat{F}_1(X_i^1), \ldots, \hat{F}_d(X_i^d); \theta\right) \cdot \hat{f}_1(X_i^1) \cdots \hat{f}_d(X_i^d),$$

where

$$c(u, v; \theta) = \frac{\partial^2}{\partial u \partial v} C(u, v; \theta)$$

is the density of the copula, $\hat{F}_1, \ldots, \hat{F}_d$ are estimates of the marginal distribution functions, and $\hat{f}_1, \ldots, \hat{f}_d$ are estimates of the marginal densities. Thus the pseudo maximum likelihood estimate is

$$\hat{\theta} = \operatorname{argmax}_\theta \sum_{i=1}^{n} \log c\left(\hat{F}_1(X_i^1), \ldots, \hat{F}_d(X_i^d); \theta\right).$$

10.1.5 Skewed Densities

We consider classes of skewed densities. We follow Azzalini (2005) in this section. Skewed densities are constructed by perturbing a symmetric density f_0. The following lemma gives a general way to perturb a symmetric density.

Lemma 10.2 *Let $f_0 : \mathbf{R}^d \to \mathbf{R}$ be a density such that $f_0(x) = f_0(-x)$, let $G : \mathbf{R} \to \mathbf{R}$ be a distribution function such that G' is a symmetric density about 0, and let $w : \mathbf{R}^d \to \mathbf{R}$ be such that $w(-x) = -w(x)$. Then*

$$f(x) = f_0(x)\, 2G(w(x)), \qquad x \in \mathbf{R}^d,$$

is a density.

Proof. Let $Y \sim f_0$ and $X \sim G'$. Then $w(Y)$ has symmetric distribution, and thus also $X - w(Y)$ has symmetric distribution. Thus

$$\frac{1}{2} = P(X - w(Y) \le 0) = E\, E[P(X - w(Y) \mid Y] = \int_{-\infty}^{\infty} G(w(x)) f_0(x)\, dx.$$

Also $f_0 \ge 0$. □

The factor $2G(w(x))$ is a perturbation factor. When we choose $w \equiv 0$, we have $f = f_0$.

We give two examples of skewed densities.

1. *(Skewed Gaussian density.)* The skewed Gaussian density may be defined by

$$f_{\mu,\Sigma}(x)\, 2\Phi(((x_1 - \mu_1)/\sigma_1, \ldots, (x_d - \mu_d)/\sigma_d) \cdot \alpha),$$

   where $f_{\mu,\Sigma}$ is the Gaussian density with the expectation $\mu = (\mu_1, \ldots \mu_d)$ and covariance matrix Σ, whose diagonal is $(\sigma_1^2, \ldots, \sigma_d^2)$, Φ is the distribution function of the standard Gaussian density, and $\alpha \in \mathbf{R}^d$ is the skewness parameter. Figure 5.13(a) shows a skewed Gaussian density.

2. *(Skewed Student density.)* The skewed Student density may be obtained by choosing f_0 to be the Student density with degrees of freedom ν, G to be the distribution function of the 1D Student distribution with degrees of freedom $\nu + d$, and $w(x) = \alpha^T \omega x / x^T \Sigma^{-1} x$, where Σ is a symmetric positive semidefinite matrix, ω is the $d \times d$ diagonal matrix whose diagonal consists of the squared roots of the diagonal elements of Σ (standard deviations), and $\alpha \in \mathbf{R}^d$.

10.2 SMOOTHNESS CLASSES

Seminorms and Function Classes We define smoothness classes with the help of seminorms. With the help of a seminorm ρ we may define the smoothness class

$$\{f : \rho(f) \le L\},$$

$0 < L \le \infty$. A seminorm does not necessarily satisfy the uniqueness condition: $\rho(f) = 0 \Rightarrow f \equiv 0$. If a seminorm satisfies the uniqueness condition, then it is a norm. See Appendix B for the definition of a norm.

Densities Smoothness classes defined with the help of seminorms contain functions that are not densities. Given a smoothness class $\mathcal{G}$, we define the corresponding class of densities:

$$\mathcal{F} = \mathcal{G} \cap \mathcal{F}_{den}, \qquad \mathcal{F}_{den} = \left\{ f : \int_{\mathbf{R}^d} f = 1, f \ge 0 \right\}. \tag{10.22}$$

10.2.1 Sobolev Classes

We define the seminorm, for $s \ge 1$ an integer, as

$$\rho_s^2(f) = \sum_{\{\alpha : |\alpha| = s\}} \int_{\mathbf{R}^d} (D^\alpha f)^2, \tag{10.23}$$

where multi-index $\alpha = (\alpha_1, \ldots, \alpha_d)$ is a vector of nonnegative integers, $D^\alpha f = D_1^{\alpha_1} \cdots D_d^{\alpha_d} f$ is the weak partial derivative (see (A.1)), and we denote $|\alpha| = \alpha_1 + \cdots + \alpha_d$. We denote the Sobolev ball by

$$S_s(L) = \{f : \rho_s(f) \le L\}. \tag{10.24}$$

When $f : [0,1]^d \to \mathbf{R}$, we assume that f is a periodic function. This way the differentiation is well defined, and we can define the seminorm as

$$\rho_s^2(f) = \sum_{\{\alpha:|\alpha|=s\}} \int_{[0,1]^d} (D^\alpha f)^2 . \tag{10.25}$$

Note that the L_p-Sobolev space is defined as

$$S_{s,p} = \left\{ f : \mathbf{R}^d \to \mathbf{R} : \sum_{|\alpha| \le s} \|D^\alpha f\|_p < \infty \right\}, \tag{10.26}$$

for $s \ge 1$ integer and $p \ge 1$, see for example Stein (1970, p. 122).

Fourier Transform We may define the Sobolev seminorm in terms of the Fourier transform. We write

$$F(D^\alpha f)(\omega) = i^{|\alpha|}\omega^\alpha (Ff)(\omega), \qquad \omega \in \mathbf{R}^d,$$

where i is the imaginary unit. Thus, from (B.6), we have

$$\begin{aligned} \rho_s^2(f) &= (2\pi)^{-d} \sum_{\{\alpha:|\alpha|=s\}} \int_{\mathbf{R}^d} |F(D^\alpha f)|^2 \\ &= (2\pi)^{-d} \sum_{\{\alpha:|\alpha|=s\}} \int_{\mathbf{R}^d} |\omega^\alpha F(f)(\omega)|^2 \, d\omega \\ &= (2\pi)^{-d} \int_{\mathbf{R}^d} \|\omega\|^{2s} |F(f)(\omega)|^2 \, d\omega, \end{aligned} \tag{10.27}$$

since

$$\left(\sum_{i=1}^d \omega_i^2 \right)^s = \sum_{|\alpha|=s} \omega_i^{2\alpha}. \tag{10.28}$$

Ellipsoid A Sobolev ball can be written as an ellipsoid. We denote the trigonometric polynomials by

$$\phi_{jk}(x) = \prod_{i=1}^d \sqrt{2} \left[(1-k_i)\cos(2\pi j_i x_i) + k_i \sin(2\pi j_i x_i) \right], \qquad x \in [0,1]^d, \tag{10.29}$$

where $j \in \{0,1,\ldots\}^d$, $k \in K_j$, and

$$K_j = \left\{ k \in \{0,1\}^d : k_i = 0, \text{ when } j_i = 0 \right\}.$$

The cardinality of K_j is $2^{d-\alpha(j)}$, where $\alpha(j) = \#\{j_i : j_i = 0\}$. The collection (ϕ_{jk}), $(j,k) \in \{0,1,\ldots\}^d \times K_j$, is a basis for 1-periodic functions on $L_2([0,1]^d)$. When $s \geq 1$ is an integer, for $|\alpha| = s$ we have

$$\begin{aligned} & D^\alpha \phi_{jk}(x) \\ = \; & (2\pi j)^\alpha \times \prod_{i=1}^d \sqrt{2}\left[(-1)^{l_i}(1-k_i')\cos(2\pi j_i x_i) + (-1)^{m_i} k_i' \sin(2\pi j_i x_i)\right], \end{aligned}$$

for some k', l, m. Thus for

$$f(x) = \sum_{j_1=0,\ldots,j_d=0}^{\infty} \sum_{k \in K_j} \theta_{jk}\phi_{jk}(x), \qquad x \in [0,1]^d,$$

we have

$$\int_{[0,1]^d} (D^\alpha f)^2 = \sum_{j_1=0,\ldots,j_d=0}^{\infty} \sum_{k \in K_j} (2\pi j)^{2\alpha} \theta_{jk}^2.$$

As in (10.28), $\sum_{|\alpha|=s} j^{2\alpha} = |j|^{2s}$. So in the collection of 1-periodic functions on $[0,1]^d$ we have

$$\{f : \rho_s(f) \leq L\} = \left\{ \sum_{j_1=0,\ldots,j_d=0}^{\infty} \sum_{k \in K_j} \theta_{jk}\phi_{jk}(x) : (\theta_{jk}) \in \Theta \right\},$$

where ρ_s is defined in (10.25) and

$$\Theta = \left\{ \theta : \sum_{j_1=0,\ldots,j_d=0}^{\infty} \sum_{k \in K_j} a_{jk}^2 \theta_{jk}^2 \leq L^2 \right\}, \tag{10.30}$$

with

$$a_{jk} = |2\pi j|^s.$$

We can define the ellipsoid in (10.30) for any $s > 0$.

L_p-Sobolev Seminorm We have written in (10.27) the Sobolev seminorm using the Fourier transform. There the smoothness index s does not have to be an integer although in (10.23) we had to assume that s is an integer. Definition (10.27) may also be generalized to cover cases $p \neq 2$. Denote

$$\rho_{s,p}^p(f) = (2\pi)^{-d} \int_{\mathbf{R}^d} \|\omega\|^q |F(f)(\omega)|^p \, d\omega, \tag{10.31}$$

where $q = p(s + d/2 - d/p)$, $p \geq 1$, $s > d/2$. When $p = 1$, we have $q = s - d/2$. When $p = 2$, we have $q = 2s$. Note that for $s' = s - d/2$, we have $\rho_{s,1} = \rho_{s'}$, where $\rho_{s'}$ is defined in (10.57).

10.2.2 Hölder Classes

Isotropic Hölder Seminorm For $0 < s \le 1$ and $f : \mathbf{R}^d \to \mathbf{R}$ we may define the Hölder seminorm as

$$\rho_s(f) = \sup_{x,y \in \mathbf{R}^d, x \neq y} \|x - y\|^{-s} |f(x) - f(y)|. \tag{10.32}$$

When $f \in \{g : \rho_s(g) \le L\}$, we have $|f(x) - f(y)| \le L\|x - y\|^s$. That is, function f is not only continuous, but the variation of f may be bounded by $L\|x - y\|^s$, which becomes smaller as the smoothness index s grows. When $s > 1$, let

$$\rho_s(f) = \max_{|\alpha| = k} \rho_{s-k}(D^\alpha f),$$

where k is the largest integer strictly less than s.

Anisotropic Hölder Seminorm Anisotropic smoothness classes are interesting because functions that are not smooth in few directions can be estimated efficiently. We define next the anisotropic Hölder seminorm. Let $s = (s_1, \ldots, s_d)$, $s_i > 0$. We let k_i be the largest integer strictly less than s_i. We denote $\alpha(i) = (0, \ldots, 0, k_i, 0, \ldots, 0)$. For $f : \mathbf{R}^d \to \mathbf{R}$, we define

$$\rho_{s_i}^{(i)}(f) = \sup_{x,y \in \mathbf{R}^d, x \neq y} |x_i - y_i|^{k_i - s_i} \left| D^{\alpha(i)} f(x) - D^{\alpha(i)} f(y) \right|$$

and

$$\rho_s(f) = \max_{i=1,\ldots,d} \rho_{s_i}^{(i)}(f).$$

10.2.3 Besov Classes

We define first the univariate Besov seminorms, then we generalize these to the multivariate Besov seminorms, and finally we define the anisotropic Besov seminorms. Besov spaces are covered in more detail in Nikol'skii (1975). For alternative definitions, see Bergh and Löfström (1976), Peetre (1976), Triebel (1992), and Härdle, Kerkyacharian, Picard and Tsybakov (1998).

Univariate Besov Classes We can define the Hölder seminorm for $f : [0,1] \to \mathbf{R}$, $0 < s \le 1$, as

$$\rho_s(f) = \sup_{x,y \in [0,1]} |x - y|^{-s} |f(x) - f(y)|.$$

To generalize ρ_s it is helpful to write the Hölder seminorm as

$$\rho_s(f) = \sup_{|h| \le 1} \sup_{x \in A_h} |h|^{-s} |f(x + h) - f(x)|,$$

where

$$A_h = [\max\{0, -h\}, \min\{1, 1 - h\}].$$

First, we generalize ρ_s by replacing the inner sup-norm with the L_p norm:

$$\rho_{sp}(f) = \sup_{|h|\le 1} |h|^{-s} \|\Delta_h f\|_{L_p(A_h)}, \qquad 1 \le p \le \infty, \tag{10.33}$$

where

$$\Delta_h f(x) = f(x+h) - f(x)$$

is the difference of f. Second, we continue the generalization by replacing the supremum over h with the L_q norm:

$$\rho_{spq}(f) = \left[\int_{-1}^{1} |h|^{-sq-1} \|\Delta_h f\|_{L_p(A_h)}^q \, dh \right]^{1/q}, \qquad 1 \le p \le \infty,\ 1 \le q < \infty.$$

When $q = \infty$, then ρ_{spq} is defined by (10.33). We have so far defined the Besov seminorm for $0 < s \le 1$. We want to generalize the seminorm to the cases $s > 1$. So we define the lth difference $\Delta_y^l f$,

$$\Delta_y^l f(x) = \Delta_y \Delta_y^{l-1} f(x), \qquad \Delta_y^1 = \Delta_y \tag{10.34}$$

for $l = 2, 3, \ldots$. The lth-order difference of a function satisfies

$$\Delta_y^l f(x) = \sum_{j=0}^{l} (-1)^{j+l} \binom{l}{j} f(x + jy). \tag{10.35}$$

Exercise 10.4 requires us to show this. We are ready to define the one-dimensional Besov seminorm. Let $s > 0$, and $p, q \ge 1$. Besov seminorm for $f : [0,1] \to \mathbf{R}$ is defined by

$$\rho_{spq}(f) = \begin{cases} \left[\int_{-1}^{1} |h|^{-sq-1} \|\Delta_h^l f\|_{L_p(A_h)}^q \, dh \right]^{1/q}, & p \ge 1,\ 1 \le q < \infty, \\ \sup_{|h|\le 1} |h|^{-s} \|\Delta_h^l f\|_{L_p(A_h)}, & p \ge 1,\ q = \infty, \end{cases} \tag{10.36}$$

where $l \ge s$ is an integer.

Remarks Several smoothness classes are special cases of Besov classes.

1. When $p = q = \infty$, the Hölder seminorm is equal to the Besov seminorm.
2. When $p = q = 2$, the L_2 Sobolev seminorm is equal to the Besov seminorm.
3. We have that $B_{sp,1} \subset S_{sp} \subset B_{sp,\infty}$, where S_{sp} is the L_p Sobolev class.
4. We have that $B_{1,1,1} \subset TV \subset B_{1,1,\infty}$ where TV is the total variation class.
5. We have that $B_{1,1,1}$ is equal to the bump algebra. The seminorm of the bump algebra is $\|f\|_B = \inf \sum_{i=0}^{\infty} |\alpha_i|$, where the infimum is over representations $f = \sum_{i=0}^{\infty} \alpha_i g_{t_i,s_i}$ where $\sum_{i=0}^{\infty} |\alpha_i| < \infty$ and $g_{t,s}(x) = \exp\{-(x-t)^2/(2s^2)\}$. Bump algebra was defined in Meyer (1990, ch. VI,6).

Isotropic Besov Classes We can generalize (10.36) to the multivariate case and define the isotropic Besov seminorm for $f : [0,1]^d \to \mathbf{R}$ by

$$\rho_{spq}(f) = \begin{cases} \left[\int_{-1}^{1} \|h\|^{-sq-d} \|\Delta_h^l f\|_{L_p(A_h)}^q \, dh\right]^{1/q}, & p \ge 1,\ 1 \le q < \infty, \\ \sup_{|h| \le 1} \|h\|^{-s} \|\Delta_h^l f\|_{L_p(A_h)}, & p \ge 1,\ q = \infty, \end{cases}$$

where $l \ge s$ is an integer.

Sequence Space Let $\theta = (\theta_{mk})$ be a sequence of real values, where $m \in \{0, 1, \ldots\}$, $k = (k_1, \ldots, k_d) \in K_m = \{0, \ldots, 2^m - 1\}^d$. Let us define a seminorm in the space of sequences by

$$\rho_{spq}(\theta) = \begin{cases} \left(\sum_{m=0}^{\infty} \left[2^{ms} \|\theta_{m\cdot}\|_{l_p}\right]^q\right)^{1/q}, & 1 \le q < \infty, \\ \sup_{m=0,1,\ldots} 2^{ms} \|\theta_{m\cdot}\|_{l_p}, & q = \infty, \end{cases} \tag{10.37}$$

where $s = \sigma - (d/2 - d/p)_+$, $\sigma > (d/p - d/2)_+$, $1 \le p \le \infty$, $1 \le q \le \infty$,

$$\|\theta_{m\cdot}\|_{l_p} = \left(\sum_{k \in K_j} |\theta_{mk}|^p\right)^{1/p}$$

when $p < \infty$ and is defined for $p = \infty$ in the usual way. Let us denote $\Theta_{spq}(L) = \{\theta : \rho_{spq}(\theta) \le L\}$. When $f \in B_{spq}(L')$, the sequence of wavelet coefficients of f belongs to $\Theta_{spq}(L)$, and conversely, for some $L', L > 0$. We define the wavelet coefficients of function $f \in L_2([0,1]^d)$ by

$$\theta_{mk}^i = \int_{[0,1]^d} h_{j_m k}^i f,$$

where h_{jk}^i is a product of univariate scaling functions or wavelets and $j_m = (m, \ldots, m) \in \{0, 1, \ldots\}^d$. The definition is given in (16.16) for the Haar wavelets. Functions h_{jk}^i are considered to belong to the multiresolution basis as defined in Definition 16.2. We have to assume that the wavelet regularity is $r > s$. Thus the Haar wavelets can be used only when $s < 1$. In the case where $s \ge 1$ we have to use smoother wavelets. The sequence space representation is derived in Meyer (1990) for the univariate case.

Anisotropic Besov Classes Let $0 < s_i < \infty$ and $1 \le p_i \le \infty$ for $i = 1, \ldots, d$, and let $1 \le q \le \infty$. Define the directional Besov seminorm for $f : [0,1]^d \to \mathbf{R}$ as

$$\rho_{s_i p_i q}^{(i)}(f) = \begin{cases} \left[\int_{-1}^{1} |h|^{-s_i q} \|\Delta_{he_i}^{l_i} f\|_{L_{p_i}(A_{i,hl_i})}^q \frac{dh}{h}\right]^{1/q}, & 1 \le q < \infty, \\ \sup_{|h| \le 1} |h|^{-s_i} \|\Delta_{he_i}^{l_i} f\|_{L_{p_i}(A_{i,hl_i})}, & q = \infty, \end{cases} \tag{10.38}$$

where $l_i \ge s_i$, $i = 1, \ldots, d$, are integers, $e_i = (e_{i1}, \ldots, e_{id}) \in \mathbf{R}^d$ is the direction vector with

$$e_{ij} = \begin{cases} 1, & \text{when } i = j \\ 0, & \text{when } i \neq j, \end{cases} \tag{10.39}$$

and

$$A_{i,h} = \{(x_1, \ldots, x_d) : 0 \le x_j \le 1, j \neq i, \ \max\{0, -h\} \le x_i \le \min\{1, 1-h\}\} \tag{10.40}$$

for $i = 1, \ldots, d$. We are ready to define the anisotropic Besov seminorm.

Definition 10.3 (Besov seminorm.) *Let* $s = (s_1, \ldots, s_d)$, $p = (p_1, \ldots, p_d)$, $0 < s_i < \infty$, $1 \le p_i \le \infty$, *and* $1 \le q \le \infty$. *The* anisotropic Besov seminorm *is defined by*

$$\rho_{spq}(f) = \sum_{i=1}^{d} \rho^{(i)}_{s_i p_i q}(f).$$

We denote a Besov ball by

$$B_{spq}(L) = \{f : \rho_{spq}(f) \le L\}. \tag{10.41}$$

Sequence Space We will derive a bound for the wavelet coefficients of a function in an anisotropic Besov ball. The bound is proved only for smoothness indexes $s_i \le 1$ and for $q = \infty$. We define the wavelet coefficients of function $f \in L_2([0,1]^d)$ by

$$\theta^i_{jk} = \int_{[0,1]^d} h^i_{jk} f, \tag{10.42}$$

where h^i_{jk} is a product of univariate scaling functions or wavelets, defined in (10.42) and $j \in \{0, 1, \ldots\}^d$, $k = (k_1, \ldots, k_d) \in K_j = \{k : 0 \le k_l \le 2^{j_l} - 1\}$, $i \in \{0,1\}^d$. We take functions h^i_{jk} as basis functions in the anisotropic basis as defined in Definition 16.3.

Lemma 10.3 *Let* $j = (j_1, \ldots, j_d) \in \{0, 1, \ldots\}^d$ *be a resolution index. Let* $f \in B_{spq}(L)$, *where* $B_{spq}(L)$ *is the anisotropic Besov ball defined in (10.41), and* $s = (s_1, \ldots, s_d) \in (0,1]^d$, $p = (p_1, \ldots, p_d) \in [1, \infty]^d$, $q = \infty$. *Then, for* $l \in \{1, \ldots, d\}$, *when* $i = (i_1, \ldots, i_d)$ *with* $i_l = 1$, $1 \le p_l < \infty$,

$$\left(\sum_{k \in K_j} |\theta^i_{jk}|^{p_l} \right)^{1/p_l} \le L 2^{-j_l s_l} 2^{-|j|(1/2 - 1/p_l)} \tag{10.43}$$

and when $p_l = \infty$,

$$\max_{k \in K_j} |\theta^i_{jk}| \le L 2^{-j_l s_l} 2^{-|j|/2}, \tag{10.44}$$

where $|j| = j_1 + \cdots + j_d$. *Denote*

$$\tilde{p}_l = \min\{p_l, 2\}, \qquad l = 1, \ldots, d. \tag{10.45}$$

Then, for $l \in \{1, \ldots, d\}$, *when* $i = (i_1, \ldots, i_d)$ *with* $i_l = 1$,

$$\left(\sum_{k \in K_j} |\theta^i_{jk}|^{\tilde{p}_l} \right)^{1/\tilde{p}_l} \leq L 2^{-j_l s_l} 2^{-|j|(1/2 - 1/\tilde{p}_l)}, \tag{10.46}$$

A proof of Lemma 10.3 is given in the Appendix E.1.

10.2.4 Spaces of Dominating Mixed Derivatives

Let us define the seminorm of the space of dominating mixed derivatives by

$$\rho_s^2(f) = \sum_{\{\alpha : |\alpha|_\infty \leq s\}} \int_{\mathbf{R}^d} (D^\alpha f)^2, \tag{10.47}$$

where $f : \mathbf{R}^d \to \mathbf{R}$, $s \geq 1$ is an integer, multi-index $\alpha = (\alpha_1, \ldots, \alpha_d)$ is a vector of nonnegative integers, $D^\alpha f = D_1^{\alpha_1} \cdots D_d^{\alpha_d} f$ is the weak partial derivative (see (A.1)), and $|\alpha|_\infty = \max\{\alpha_i : i = 1, \ldots, d\}$. We denote the corresponding ball by

$$D_s(L) = \{f : \rho_s(f) \leq L\}. \tag{10.48}$$

The ball $D_s(L)$ is much smaller than the Sobolev ball $S_s(L)$ defined in (10.24) because now we make a stronger restriction on the mixed derivatives: the squared L_2 norm $\|D^{(s,\ldots,s)} f\|_2^2$ is included in the sum in (10.47).

The spaces of dominating mixed derivatives were considered by Korobov (1963) and by Schmeisser and Triebel (1987, ch. 2).

Sequence Space Let $\theta = (\theta_{jk})$ be a sequence of real values, where $j \in \{0, 1, \ldots\}^d$, $k = (k_1, \ldots, k_d) \in K_j = \{k : 0 \leq k_l \leq 2^{j_l} - 1\}$. We define a seminorm in the space of sequences by

$$\rho_s^2(\theta) = \sum_{j \in \{0,1,\ldots\}^d} 2^{2|j|s} \sum_{k \in K_j} |\theta_{jk}|^2, \tag{10.49}$$

where $s > 0$. Let us write $\Theta_{spq}(L) = \{\theta : \rho_{spq}(\theta) \leq L\}$. When $f \in D_s(L')$, the sequence of wavelet coefficients of f belongs to $\Theta_{spq}(L)$, for some $L', L > 0$. The wavelet coefficients are defined as $\theta^i_{jk} = \int_{[0,1]^d} h^i_{jk} f$, where h^i_{jk} is a product of univariate wavelets and scaling functions and belongs to the anisotropic basis as defined in Definition 16.3. We have to assume that the wavelet regularity is $r > s$. Definition 16.3 is written for the Haar wavelets but the Haar wavelets can be used only when $s < 1$, and in the general case we have to use smoother wavelets.

Remarks We make some remarks concerning spaces of dominating mixed derivatives.

1. We can define the seminorm by taking only the largest value for the components of the multi-index:

$$\rho_s(f) = \left\| D^{(s,\ldots,s)} f \right\|_2 .$$

Then, as in (10.27), we have

$$\rho_s^2(f) = (2\pi)^{-d} \int |\omega_1|^{2s} \cdots |\omega_d|^{2s} |F(f)(\omega)|^2 \, d\omega.$$

Note that the L_2 Sobolev seminorm as written in (10.27) involves the multiplier $(\omega_1^2 + \cdots + \omega_d^2)^s = \sum_{|\alpha|=s} \omega_1^{2\alpha_1} \cdots \omega_d^{2\alpha_d}$.

2. We can define an anisotropic seminorm by

$$\rho_s^2(f) = \sum_{\{\alpha: 0 \le \alpha_i \le s_i\}} \|D^\alpha f\|_2^2,$$

where $s = (s_1, \ldots, s_d)$. This is not an essential generalization, since the estimation and the approximation rate is determined by $\min_{i=1,\ldots,d} s_i$.

3. The product density (10.3) (for $B = I_d$) belongs to $D_s(L)$ when the component densities are in the one-dimensional Sobolev space with the smoothness index s.

10.2.5 Convex Hulls and Infinite Mixtures

Convex Hull The convex hull of a base class $\mathbb{G}$ of functions $\mathbf{R}^d \to \mathbf{R}$ is the class of finite convex combinations of the functions in $\mathbb{G}$. We denote the convex hull by $\text{co}(\mathbb{G})$. We write

$$\text{co}(\mathbb{G}) = \left\{ \sum_{\phi \in \mathbb{G}} \lambda_g g(x) : \Lambda \in \mathcal{W}(\mathbb{G}) \right\}, \tag{10.50}$$

where $\Lambda = (\lambda_g)_{g \in \mathbb{G}}$, and $\mathcal{W}(\mathbb{G})$ is the set of vectors of coefficients of finite convex combinations, that is, the set finite discrete probability measures on $\mathbb{G}$:

$$\mathcal{W}(\mathbb{G}) = \left\{ (\lambda_g)_{g \in \mathbb{G}} \in \mathbf{R}^{\mathbb{G}} : \lambda_g \ge 0, \ \sum_{g \in \mathbb{G}} \lambda_g = 1, \ \#\{\lambda_g > 0\} < \infty \right\}. \tag{10.51}$$

The L_2 closure of the convex hull which is denoted by

$$\bar{\text{co}}(\mathbb{G}).$$

Infinite Mixture Family Assume that the base class $\mathbb{G}$ is a parametrized class:

$$\mathbb{G} = \{g(\cdot, \theta) : \theta \in \Theta\} \tag{10.52}$$

with $\Theta \subset \mathbf{R}^k$. The infinite mixture class is

$$\mathcal{G}(\mathbb{G}) = \left\{ \int_\Theta g(\cdot, \theta) \, dQ(\theta) : Q \in \mathbb{Q}(\Theta) \right\}, \tag{10.53}$$

where $\mathbb{Q}(\Theta)$ is the set of probability measures on Θ. If the set of probability measures $\mathbb{Q}(\Theta)$, applied in (10.53) to define $\mathcal{G}(\mathbb{G})$, is the set of point measures (for each $Q \in \mathbb{Q}(\Theta)$ there is $\theta \in \Theta$ so that $Q(\{\theta\}) = 1$), then $\mathcal{G}(\mathbb{G}) \subset \mathbb{G}$. However, $\mathbb{Q}(\Theta)$ is the set of all probability measures on Θ, and thus $\mathcal{G}(\mathbb{G})$ is a rich nonparametric class. See Exercise 10.5 for a connection between $\bar{\text{co}}(\mathbb{G})$ and $\mathcal{G}(\mathbb{G})$.

Remarks We make some remarks concerning mixture classes.

1. *(Gaussian mixture family.)* We can choose the base class to consist of Gaussian densities. Let

$$\mathbb{G} = \{g(\cdot, \mu, \sigma) : (\mu, \sigma) \in M \times S\}, \tag{10.54}$$

where $M \subset \mathbf{R}^d$ and $S \subset (0, \infty)$, and

$$g(x, \mu, \sigma) = (2\pi)^{-d/2} \sigma^{-d} \exp\left\{ \frac{-\|x - \mu\|^2}{2\sigma^2} \right\}, \qquad x \in \mathbf{R}^d. \tag{10.55}$$

Isotropic Liouville spaces, or Bessel potential spaces, are contained in the space of Gaussian mixtures, under some conditions; see Exercise 10.6. Liouville space is defined in Stein (1970, p. 134), as

$$L_{s,p} = \left\{ G_s * \phi \mid \phi : \mathbf{R}^d \to \mathbf{R}, \ \|\phi\|_p < \infty \right\}, \tag{10.56}$$

where G_s is the kernel of the Bessel potential (Bessel-Macdonald function)

$$G_s(x) = \frac{1}{(4\pi)^{s/2}} \frac{1}{\Gamma(s/2)} \int_0^\infty e^{-\pi\|x\|^2} e^{-\sigma/(4\pi)} \sigma^{(s-d)/2} \sigma^{-1} \, d\sigma,$$

$x \in \mathbf{R}^d$. We can define the norm of $f \in L_{s,p}$ as $\rho_{s,p}(f) = \|\phi\|_p$, when $f = G_s * \phi$. Theorem 3 in Stein (1970), page 135, states that $L_{s,p} = S_{s,p}$ for integers $s \geq 1$ and $1 < p < \infty$, where $S_{s.p}$ is the Sobolev space defined in (10.26). Note also that the Fourier transform of G_s is $F(G_s)(\omega) = \left(1 + 4\pi^2\|\omega\|^2\right)^{s/2}$.

2. *(L_1 Fourier class.)* The L_1 Fourier seminorm is defined by

$$\rho_s(f) = (2\pi)^{-d} \int_{\mathbf{R}^d} \|\omega\|^s \, |F(f)(\omega)| \, d\omega, \tag{10.57}$$

where $s \geq 0$ and the Fourier transform of $f : \mathbf{R}^d \to \mathbf{R}$ is defined in (B.4). The L_1 Fourier seminorm is a special case of the L_p Fourier seminorm defined in (10.31). The approximation of functions in a L_1 Fourier class was studied in Barron (1993) and statistical estimation was studied in Juditsky and Nemirovski (2000), for example. The L_1 Fourier ball is included in a mixture class, under some conditions; see Exercise 10.7.

10.3 COVERING AND PACKING NUMBERS

We define covering and packing numbers and calculate the order of these numbers for some function classes.

10.3.1 Definitions

Let Θ be a subset of a metric space $\mathcal{R}$ with metric $D : \mathcal{R} \times \mathcal{R} \to [0, \infty)$. See Appendix B.8 for the definition of a metric. We characterize the largeness of a set with the δ-covering number, which is the smallest number of balls needed to cover a set.

Definition 10.4 (δ-net, δ-covering number, δ-entropy function.)

1. *Set* $\Theta_0 \subset \mathcal{R}$ *is a* δ-net *of* Θ, *if for each* $\theta \in \Theta$ *there is* $\theta' \in \Theta_0$ *such that* $D(\theta, \theta') \leq \delta$.

2. *Number* $N_\delta = N(\delta, \Theta, D)$ *is the* δ covering number, *if* N_δ *is the smallest cardinality of any* δ*-net for* Θ. *We say that* $\mathcal{N}_\delta = \log_e N_\delta$ *is the* entropy function. *Function* $\mathcal{N}_\delta$ *is also called the covering* δ*-entropy.*

We define the closely related concepts of δ-packing in Definition 10.5

Definition 10.5 (δ-packing set, δ-packing number, δ-capacity function.)

1. *Set* $\Theta_0 \subset \Theta$ *is a* δ-packing set *of* Θ, *if* $D(\theta, \theta') > \delta$ *for all* $\theta, \theta' \in \Theta_0$, $\theta \neq \theta'$.

2. *Number* $M_\delta = M(\delta, \Theta, D)$ *is the* δ-packing number, *if* M_δ *is the largest cardinality of any* δ*-packing set for* Θ. *We say that* $\mathcal{M}_\delta = \log_e M_\delta$ *is the* capacity function. *Function* $\mathcal{M}_\delta$ *is also called the packing* δ*-entropy.*

We may show that the δ-packing number and the δ-covering number have the same order, as $\delta \to 0$.

Lemma 10.4 *Let* M_δ *be the* δ*-packing number of set* Θ, *for distance* D, *as in Definition 10.5. Let* N_δ *be the corresponding* δ*-covering number as in Definition 10.4. Then*

$$N_\delta \leq M_\delta \leq N_{\delta/2}.$$

Proof. Let $x_1, \ldots, x_M$ be a δ-packing set of largest cardinality. Then the balls $B_\delta(x_i)$ cover Θ (they form a δ-net for Θ); otherwise, there would exist a larger set whose members are δ apart. We have proved the first inequality. Let $y_1, \ldots, y_N$ be a $\delta/2$-net. Then $M \leq N$ because each ball $B_{\delta/2}(y_j)$ can cover at most one point x_i; a set of diameter δ can contain at most one of a set of points more than δ apart. We have proved the second inequality. □

We will need the concept of a bracketing number in the theory of empirical processes.

Definition 10.6 (δ-bracket, δ-bracketing net, δ-bracketing number.) *Let* Θ *be a set of functions.*

1. *Let* f_l *and* f_u *be two functions. The* bracket $[f_l, f_u]$ *is the set of all functions* f *with* $f_l \leq f \leq f_u$.

2. *Set* $\Theta_0 \subset \mathcal{R} \times \mathcal{R}$ *is a* δ-bracketing net *of* Θ *if for each* $f \in \Theta$ *there is* $(g_l, g_u) \in \Theta_0$ *such that* $g_l \le f \le g_u$ *and* $D(g_l, g_u) \le \delta$.

3. *We say that* $N_{\delta,[]} = N_{[]}(\delta, \Theta, D)$ *is the* bracketing number *if* $N_{\delta,[]}$ *is the smallest cardinality of any* δ*-bracketing net of* Θ.

Lemma 10.5 *Let* $\|f\|_D = D(0, f)$ *satisfy the Riesz property: if* $|f| \le |g|$*, then* $\|f\|_D \le \|g\|_D$*. Then*

$$N(\delta, \Theta, D) \le N_{[]}(2\delta, \Theta, D). \tag{10.58}$$

Proof. If f is in the 2δ bracket $[g_l, g_u]$, then it is in the ball of radius δ whose center is $(g_l + g_u)/2$. □

When $\|\cdot\|_D$ is the uniform norm $\|\cdot\|_\infty$, there is equality in (10.58), but in general one cannot bound the bracketing number in terms of the covering number.

Bibliographic Notes We have followed Kolmogorov and Tikhomirov (1961), van der Vaart and Wellner (1996, secs. 2.6, 2.7), Dudley (1999, sec. 1.2), and van de Geer (2000). Kolmogorov and Tikhomirov (1961) assume that A is a subset of a metric space $\mathcal{R}$, and call an ϵ-covering of A a collection γ of sets $U \subset \mathcal{R}$ such that the diameter of sets U is less that 2ϵ and $A \subset \cup_{U \in \gamma} U$. An ϵ-net $A_0 \subset \mathcal{R}$ of A is such that for every $x \in A$ there is $y \in A_0$ with $D(x, y) \le \epsilon$. They define $\mathcal{N}_\epsilon(A)$ to be the minimal number of sets in an ϵ-covering and $\mathcal{N}_\epsilon^{\mathcal{R}}(A)$ to be the minimal number of points in an ϵ-net. Note that $\mathcal{N}_\epsilon(A)$ does not depend on $\mathcal{R}$ but $\mathcal{N}_\epsilon^{\mathcal{R}}(A)$ depends on $\mathcal{R}$. Entropy numbers are defined using logarithms to the base 2.

10.3.2 Finite Dimensional Sets

We derive an upper bound for the δ-packing number and for the δ-covering number of a bounded set in the Euclidean space.

Lemma 10.6 *Let* $A \subset \mathbf{R}^d$ *and let* D *be a metric in* $\mathbf{R}^d$*. Then*

$$N(\delta, A, D) \le M(\delta, A, D) \le \frac{\text{volume}(A_{\delta/2})}{\text{volume}(B_{\delta/2}(0))},$$

where A_r *is the* r*-corona of* A*:* $A_r = \{x \in \mathbf{R}^d : \inf_{y \in A} D(x, y) \le r\}$ *and* $B_r(y) = \{x \in \mathbf{R}^d : D(x, y) \le r\}$.

Proof. Let $x_1, \ldots, x_M$ be a δ-packing set of A. Sets $B_{\delta/2}(x_i)$ are disjoint by the definition of a δ-packing set. Also $\bigcup_{i=1}^M B_{\delta/2}(x_i) \subset A_{\delta/2}$ by the triangle inequality, since all $x_i \in A$. Thus

$$M \cdot \text{volume}(B_{\delta/2}(0)) = \text{volume}\left(\bigcup_{i=1}^M B_{\delta/2}(x_i)\right) \le \text{volume}(A_{\delta/2}),$$

which implies the bound for $M(\delta, A, D)$. By Lemma 10.4, this implies a bound also for the covering number $N(\delta, A, D)$. □

As an application of Lemma 10.6 we have the following bounds for the packing number of a ball $B_R(0) = \{x \in \mathbf{R}^d : \|x\| \le R\}$:

Lemma 10.7 *When* $R/\delta \ge 1$,

$$2^{-d/2}(R/\delta)^d \le M(\delta, B_R(0), \|\cdot\|) \le 3^d(R/\delta)^d.$$

Proof. Lemma 10.6 implies that

$$M(\delta, B_R(0), \|\cdot\|) \le \frac{\text{volume}(B_{R+\delta/2}(0))}{\text{volume}(B_{\delta/2}(0))} \le (2R/\delta + 1)^d,$$

where we applied the fact that for each $r > 0$, volume$(B_r(0)) = C_d r^d$, where $C_d =$ volume$(B_1(0))$ is defined in (B.10). This implies the upper bound. Let us denote $P_r(0) = \{x : \max_{i=1,\ldots,d} |x_i| \le r\}$ the rectangle with radius r. The lower bound follows because $P_{R/\sqrt{2}}(0) \subset B_R(0)$ and $B_\delta(0) \subset P_\delta(0)$. □

10.3.3 Ellipsoids

We define the ellipsoid of sequences $\theta = (\theta_1, \theta_2, \ldots)$ as

$$\Theta = \left\{ \theta \in \mathbf{R}^\infty : \sum_{k=1}^{\infty} a_k^2 \theta_k^2 \le L^2 \right\}. \tag{10.59}$$

We assume that there exists positive constants C_1, C_2 such that for all $k \in \{1, 2, \ldots\}$

$$C_1 \cdot k^s \le a_k \le C_2 \cdot k^s. \tag{10.60}$$

Let

$$M = \lceil (C_1^{-1} 2^{1/2} L \delta^{-1})^{1/s} \rceil. \tag{10.61}$$

Let $\Theta_\delta(M)$ be a $\delta/2$-net of

$$E_M = \left\{ (\theta_k)_{k \in \{1,\ldots,M\}} : \sum_{k=1}^{M} a_k^2 \theta_k^2 \le L^2 \right\},$$

and let $\Theta_\delta^*(M)$ be a δ-packing set of E_M. We can choose $\Theta_\delta(M)$ in such a way that its cardinality satisfies

$$\#\Theta_\delta(M) \le C \, \frac{\text{volume}(E_M)}{\text{volume}(B_\delta^{(M)})},$$

where $B_\delta^{(M)}$ is a ball of radius δ in the M-dimensional Euclidean space. We can choose $\Theta_\delta^*(M)$ in such a way that its cardinality satisfies

$$\#\Theta_\delta^*(M) \ge C' \, \frac{\text{volume}(E_M)}{\text{volume}(B_\delta^{(M)})}. \tag{10.62}$$

Define the δ-net by

$$\Theta_\delta = \big\{(\theta_k)_{k\in\{1,\ldots,\infty\}} \;:\; (\theta_k)_{k\in\{1,\ldots,M\}} \in \Theta_\delta(M),\; \theta_k = 0,\ \text{for } k \notin \{1,\ldots,M\}\big\} \tag{10.63}$$

and the δ-packing set by

$$\Theta_\delta^* = \big\{(\theta_k)_{k\in\{1,\ldots,\infty\}} \;:\; (\theta_k)_{k\in\{1,\ldots,M\}} \in \Theta_\delta^*(M),\; \theta_k = 0,\ \text{for } k \notin \{1,\ldots,M\}\big\}. \tag{10.64}$$

Lemma 10.8 *When the lower bound holds in (10.60), collection Θ_δ is a δ-net of Θ and*

$$\log(\#\Theta_\delta) \le C\delta^{-1/s}. \tag{10.65}$$

Collection Θ_δ^* is a δ-packing set of Θ and when the upper bound holds in (10.60),

$$\log(\#\Theta_\delta^*) \ge C'\delta^{-1/s}, \tag{10.66}$$

for positive constants C, C'.

Proof. The proof is similar to the proof in Kolmogorov and Tikhomirov (1961).

δ-capacity property It is obvious by construction that the sequences in Θ_δ^* are δ-apart.

δ-net property We need to prove that Θ_δ, defined in (10.63), is a δ-net of the ellipsoid Θ. For each $\theta \in \Theta$ there is $\theta_\delta \in \Theta_\delta$ such that $\|\theta - \theta_\delta\|_{l_2} \le \delta$. Indeed, let $\theta \in \Theta$ and let $\theta_\delta \in \Theta_\delta$ be such that

$$\sum_{k=1}^{M} (\theta_k - \theta_{\delta,k})^2 \le \frac{\delta^2}{2}.$$

Then

$$\|\theta - \theta_\delta\|_{l_2}^2 = \sum_{k=1}^{M} (\theta_k - \theta_{\delta,k})^2 + \sum_{k=M+1}^{\infty} \theta_k^2 \le \delta^2,$$

where we used the fact that

$$\sum_{k=M+1}^{\infty} \theta_k^2 \le C_1^{-2} \cdot M^{-2s} \sum_{k=M+1}^{\infty} a_k^2\theta_k^2 \le C_1^{-2} M^{-2s} L^2 \le \frac{\delta^2}{2}, \tag{10.67}$$

because, when $k \notin \{1,\ldots,M\}$,

$$a_k^{-2} \le C_1^{-2} \cdot k^{-2s} \le C_1^{-2} \cdot M^{-2s} \le \frac{\delta^2}{2L^2}.$$

Cardinality We prove the bounds (10.65) and (10.66). We have

$$\text{volume}(E^{(M)}) = C_M \cdot L^M \prod_{k=1}^{M} a_k^{-1}$$

and

$$\text{volume}(B_\delta^{(M)}) = C_M \cdot \delta^M,$$

where C_M is the volume of the unit ball in the M-dimensional Euclidean space. Thus the cardinality of Θ_δ satisfies

$$\#\Theta_\delta = \#\Theta_\delta(M) \le C\,\frac{L^M \prod_{k=1}^M a_k^{-1}}{\delta^M},$$

and the cardinality of Θ_δ^* satisfies

$$\#\Theta_\delta^* = \#\Theta_\delta^*(M) \ge C'\,\frac{L^M \prod_{k=1}^M a_k^{-1}}{\delta^M}.$$

We have

$$C'(M!)^{-s} = C' \prod_{k=1}^M k^{-s} \le \prod_{k=1}^M a_k^{-1} \le C \prod_{k=1}^M k^{-s} = C(M!)^{-s}.$$

Applying the inequality (B.19), we get

$$M^{M+1/2}e^{-M} < M! < 2M^{M+1/2}e^{-M}.$$

Thus

$$\begin{aligned}
&\log(\#\Theta_\delta)\\
&\quad\le\ M\log(L) - s\log(M!) + M\log(\delta^{-1}) + C\\
&\quad\le\ M\log(L) - s(M+1/2)\log M + sM + M\log(\delta^{-1}) + C\\
&\quad\le\ M(\log(L)+s) - sM\log M + M\log(\delta^{-1}) + C\\
&\quad\le\ M(\log(L)+s+C') + C\\
&\quad\le\ \delta^{-1/s}C'' + C,
\end{aligned} \tag{10.68}$$

since $M = C'''\delta^{-1/s}$. Also

$$\begin{aligned}
&\log(\#\Theta_\delta^*)\\
&\quad\ge\ M\log(L) - s\log(M!) + M\log(\delta^{-1}) + C'\\
&\quad\ge\ M\log(L) - s(M+1/2)\log M + sM + M\log(\delta^{-1}) + C''\\
&\quad=\ M(\log(L)+s) - sM\log M + M\log(\delta^{-1}) - (s/2)\log M + C''\\
&\quad\le\ M(\log(L)+s)C''' + C''\\
&\quad\ge\ \delta^{-1/s}C'''' + C''.
\end{aligned}$$

We have proved the lemma. □

Multivariate Case In the multivariate case the ellipsoid has the form

$$\Theta = \left\{ \theta : \sum_{k_1=1,\ldots,k_d=1}^{\infty} a_k^2 \theta_k^2 \le L^2 \right\}, \tag{10.69}$$

where $k = (k_1, \ldots, k_d) \in \{1, 2, \ldots\}^d$ is a multi-index. We assume that there exists positive constants C_1, C_2 such that for all $k \in \{1, 2, \ldots\}^d$

$$C_1 \cdot |k|^s \le a_k \le C_2 \cdot |k|^s,$$

where $|k| = k_1 + \cdots + k_d$. The multivariate case can be reduced to the ellipsoid in (10.59) with a suitable enumeration. In the new enumeration we replace M in (10.61) by M^d, and we get

$$\log(\#\Theta_\delta) \le C\delta^{-d/s} \tag{10.70}$$

and

$$\log(\#\Theta_\delta^*) \ge C''\delta^{-d/s}. \tag{10.71}$$

A Local δ-Packing Set It turns out to be useful to have available a local δ-packing set of ellipsoid (10.59). We will need this local δ-packing set to prove a lower bound to the minimax rate of convergence, in the case of inverse problems; see Corollary 11.13 and Corollary 11.14. Let θ^* be a fixed sequence with $\sum_{j=0}^{\infty} a_j^2 \theta_j^{*2} = L^* < L$, and let $\Theta_\delta^*(M)$ be a δ-packing set of

$$E_M^* = \left\{ (\theta_j)_{j\in\{M^*,\ldots,M\}} : \sum_{j=M^*}^{M} a_j^2 \theta_j^2 \le (L - L^*)^2 \right\},$$

where

$$M^* = [M/2], \qquad M \asymp \delta^{-1/s}.$$

We can choose the δ-packing set $\Theta_\delta^*(M)$ analogously to (10.62) so that its cardinality satisfies

$$\log(\#\Theta_\delta^*(M)) \ge C^*\delta^{-1/s}. \tag{10.72}$$

Define

$$\begin{aligned}\Theta_\delta^* = \{(\theta_j)_{j\in\{0,\ldots,\infty\}} \quad &: \quad (\theta_j - \theta_j^*)_{j\in\{M^*,\ldots,M\}} \in \Theta_\delta^*(M), \\ &\theta_j = \theta_j^*, \ \text{for } j \notin \{M^*, \ldots, M\}\}. \end{aligned} \tag{10.73}$$

The bound (10.72) follows similarly as the upper bound (10.68). In the case of Gaussian white noise, we can use this construction with $\theta^* = 0$ and $L^* = 0$. In the case of density estimation, another choice of θ^* may be appropriate to ensure that the corresponding functions are bounded from above and from below away from zero. This would allow us to use the bound of Lemma 11.6 to carry over bounds on the L_2

norm to corresponding bounds on the Kullback-Leibler distance. The δ-packing set that we have constructed is a local because for $\theta, \theta' \in \Theta_\delta^*$,

$$\|\theta - \theta'\|_{l_2}^2 = \sum_{i=M^*}^{M} (\theta_i - \theta_i')^2 = \sum_{i=M^*}^{\infty} (\theta_i - \theta_i')^2 \le C\delta^2, \tag{10.74}$$

which follows after using a similar calculation as in (10.67). Now we have needed both the upper bound and the lower bound in (10.60).

Parametric Families When $\Theta \subset \mathbf{R}^k$ or $\Theta \subset \mathbf{R}^\infty$,

$$\mathcal{F} = \{g(\cdot, \theta) : \theta \in \Theta\}$$

and for all $\theta, \theta' \in \Theta$,

$$C' D(\theta, \theta') \le \|g(\cdot, \theta) - g(\cdot, \theta')\|_2 \le C D(\theta, \theta') \tag{10.75}$$

for positive constants C, C', we have

$$M(\delta, \mathcal{F}, \|\cdot\|_2) \asymp M(\delta, \Theta, D). \tag{10.76}$$

In particular, if $\phi_i : \mathbf{R}^d \to \mathbf{R}$, $i = 1, 2, \ldots$, is an orthonormal system in $L_2(\mathbf{R}^d)$ and

$$\mathcal{F} = \left\{ \sum_{i=1}^{\infty} \theta_i \phi_i : \theta \in \Theta \right\},$$

then

$$M(\delta, \mathcal{F}, \|\cdot\|_2) \asymp M(\delta, \Theta, \|\cdot\|_{l_2}),$$

where $\|\theta\|_{l_2}^2 = \sum_{i=1}^\infty \theta_i^2$.

10.3.4 Global and Local δ-Nets

We can apply in some cases a general scheme, where a global δ-net is constructed with the help of local δ-nets. Sometimes it is easy to construct a local δ-net and if we can do this, then we know how to construct a global δ-net, as defined in Definition 10.4.

Definition 10.7 (Local δ-net.) *A* local δ-net *at $\theta_0 \in \Theta$, with parameter $\kappa > 1$, for $\delta > 0$, is a δ-net of $\{\theta : D(\theta, \theta_0) \le \kappa\delta\}$.*

Let us construct a global δ-net $\mathcal{G}_\delta$ of $\mathcal{F}$. The global δ-net $\mathcal{G}_\delta$ of $\mathcal{F}$ will be constructed recursively. We assume that $0 \in \mathcal{F}$ and $\sup_{f \in \mathcal{F}} D(0, f) \le R$. Choose

$$B_\delta = \lceil \log_\kappa(R/\delta) \rceil,$$

so that $R/\kappa^{B_\delta} \le \delta$. Denote

$$\delta_i = R/\kappa^i, \qquad i = 0, \ldots, B_\delta.$$

1. The class $\mathcal{F}$ is contained in the ball $\{f : D(0, f) \leq R\}$. Denote $\mathcal{G}^{(0)} = \{0\}$. This singleton is an R-net of $\mathcal{F}$.

2. At stage i, $i = 1, \ldots, B_\delta$, we construct for each $f_0 \in \mathcal{G}^{(i-1)}$ a local δ_i-net of $\{f \in \mathcal{F} : D(f_0, f) \leq \delta_{i-1}\}$. This is denoted by $\mathcal{G}_{\delta_i}(f_0)$. Denote [2]

$$\mathcal{G}^{(i)} = \bigcup \left\{ \mathcal{G}_{\delta_i}(f_0) : f_0 \in \mathcal{G}^{(i-1)} \right\}.$$

3. The global δ-net is defined as

$$\mathcal{G}_\delta = \mathcal{G}^{(B_\delta)}. \tag{10.77}$$

Now we can proof that $\mathcal{G}_\delta$ is a δ-net of $\mathcal{F}$.

Lemma 10.9 *For each $\delta > 0$, $\mathcal{G}_\delta$ is a δ-net of $\mathcal{F}$.*

Proof. Let $f \in \mathcal{F}$. We can find $f_0 \in \mathcal{G}^{(0)}$ such that $D(f, f_0) \leq \delta_0 = R$. As the induction hypothesis assume that we can find $f_0 \in \mathcal{G}^{(i-1)}$ such that $D(f, f_0) \leq \delta_{i-1}$. Then $f \in \{g : D(g, f_0) \leq \delta_{i-1}\}$, and $\mathcal{G}_{\delta_i}(f_0)$ is a δ_i-net of this set. Thus there is $g_f \in \mathcal{G}_{\delta_i}(f_0)$ such that $D(f, g_f) \leq \delta_i$. Thus there is $f_0 = g_f \in \mathcal{G}^{(i)}$ such that $D(f, f_0) \leq \delta_i$. We have proved inductively that for $i = 0, \ldots, B_\delta$, there is $f_0 \in \mathcal{G}^{(i)}$ such that $D(f, f_0) \leq \delta_i$. In particular, there is $f_0 \in \mathcal{G}^{(B_\delta)} = \mathcal{G}_\delta$ such that $D(f, f_0) \leq \delta_{B_\delta} = R/\kappa^{B_\delta} \leq \delta$. □

Cardinality of the Global δ-Net The cardinality of the δ-net $\mathcal{G}_\delta$ satisfies

$$\#\mathcal{G}_\delta \leq \prod_{i=0}^{B_\delta} \#\mathcal{G}^{(i)}.$$

Assume that the cardinalities of the local δ-nets satisfy

$$\log(\#\mathcal{G}^{(i)}) \asymp \delta_i^{-a},$$

where $a > 0$. The logarithm of the cardinality of the global δ-net satisfies

$$\begin{aligned}
\log(\#\mathcal{G}_\delta) &\leq \sum_{i=0}^{B_\delta} \log\left(\#\mathcal{G}^{(i)}\right) \asymp \sum_{i=0}^{B_\delta} \delta_i^{-a} = \sum_{i=0}^{B_\delta} \left(\frac{\kappa^i}{R}\right)^a \\
&= \left(\frac{\kappa^{B_\delta}}{R}\right)^a \sum_{i=0}^{B_\delta} \left(\frac{\kappa^i}{\kappa^{B_\delta}}\right)^a \asymp \delta^{-a} \int_0^1 t^{a-1}\, dt \\
&\asymp \delta^{-a},
\end{aligned}$$

as $\delta \to 0$.

[2] Note that $\mathcal{G}^{(1)} = \mathcal{G}_{\delta_1}(0)$ and $\mathcal{G}^{(1)}$ is in general not equal to $\mathcal{G}^{(0)} = \{0\}$ since $\delta_1 = R/\kappa$ and $\kappa > 1$.

Local δ-Net for a Hölder Class As an example we consider the Hölder class of functions $f : [0,1]^d \to \mathbf{R}$, defined through the seminorm ρ_s in (10.32), for $0 < s \le 1$. We will denote $\mathcal{F}_s(L) = \{f : [0,1]^d \to \mathbf{R} \mid \rho_s(f) \le L\}$. We apply the supremum norm. Note that the δ-net property in the L_2-norm follows because $\|f\|_2^2 \le \|f\|_\infty^2 \cdot \text{volume}([0,1]^d) = \|f\|_\infty^2$. We will construct a local δ-net of $\mathcal{F}_s(L)$. This construction is instructive, since we will later use a similar construction to prove lower bounds to the packing numbers of function classes.

Let $G \in \mathcal{F}_s(L')$, $G : [0,1]^d \to \mathbf{R}$. Define

$$G_k(x) = m^{d/2} G(mx - k), \qquad x \in \mathbf{R}^d,$$

where $k \in I$, and I is the set of d-dimensional multi-indexes:

$$I = \{(k_1, \ldots, k_d) : k_i = 0, \ldots, m-1\},$$

where

$$m = \left\lceil \left(\frac{\delta}{C_m L} \right)^{-1/s} \right\rceil,$$

and $C_m > 0$ is a constant. Note that function G_k has the support on the cell S_k where cells S_k constitute an almost sure partition of $[0,1]^d$: $[0,1]^d = \cup_{k \in I} S_k$,

$$S_k = \prod_{i=1}^d \left[k_i m^{-1}, (k_i + 1) m^{-1} \right].$$

Define

$$f_a = f_0 + \delta^{(2s+d)/(2s)} \sum_{k \in I} a_k G_k,$$

where $a = (a_k)_{k \in I} \in \{-1, 1\}^I$, and $f_0 \in \mathcal{F}_s(L)$. Define

$$\mathcal{G}_\delta(f_0) = \left\{ f_a : a \in \{-1,1\}^I \right\}.$$

The cardinality of the local δ-net satisfies

$$\log_2(\#\mathcal{G}_\delta(f_0)) = \#I = m^d \asymp \delta^{-d/s}.$$

We prove that $\mathcal{G}_\delta(f_0)$ is a δ-net of

$$\{f' \in \mathcal{F}_s(L) : \|f' - f_0\|_2 \le \kappa\delta\}$$

with parameter

$$\kappa = 2\lceil (C_m L)^{1/s} \rceil^{d/2} G(1/2, \ldots, 1/2) > 1. \tag{10.78}$$

In addition we need to choose the parameters so that

$$2/C_m + \lceil (C_m L)^{1/s} \rceil^{d/2} \cdot L' + \kappa/2 \le 1.$$

Let $f \in \{f' \in \mathcal{F}_s(L) : \|f' - f_0\|_2 \le \kappa\delta\}$. We choose $a \in \{-1,1\}^I$ so that

$$a_k = \begin{cases} +1, & \text{if } f(x_k) > f_0(x_k), \\ -1, & \text{if } f(x_k) \le f_0(x_k), \end{cases}$$

where x_k is the center of the cell S_k, $k \in I$. Let $x \in [0,1]^d$. We prove that $|f(x) - f_a(x)| \le \delta$. Assume that $x \in S_k$, where $k \in I$. We have

$$\begin{aligned} |f(x) - f_a(x)| &= |f(x) - f_0(x) - a_k G_k(x)| \\ &\le |f(x) - f(x_k)| + |f_0(x) - f_0(x_k)| + |G_k(x) - G_k(x_k)| \\ &\quad + |f(x_k) - f_0(x_k) - a_k G_k(x_k)| \\ &\le 2\delta/C_m + \delta\,\lceil (C_m L)^{1/s}\rceil^{d/2} \cdot L' + \kappa\delta/2 \\ &\le \delta \end{aligned}$$

because

$$|f(x) - f(x_k)| \le L\|x - x_k\|^s \le Lm^{-s} \le \delta/C_m.$$

Similarly, $|f_0(x) - f_0(x_k)| \le \delta/C_m$. Also

$$\begin{aligned} |G_k(x) - G_k(x_k)| &= \delta^{(2s+d)/2s} m^{d/2} \cdot |G(mx - k) - G(mx_k - k)| \\ &\le \delta^{(2s+d)/2s} m^{d/2} \cdot L' m^s \|x - x_k\|^s \\ &\le \delta\,\lceil (C_m L)^{1/s}\rceil^{d/2} \cdot L' \end{aligned}$$

because $\delta^{(2s+d)/2s}(\delta^{-1/s})^{d/2} = \delta$. Finally,

$$\begin{aligned} &|f(x_k) - f_0(x_k) - a_k G_k(x_k)| \\ &= |f(x_k) - f_0(x_k) - \delta^{(2s+d)/2s} a_k m^{d/2} G(mx_k - k)| \\ &= |f(x_k) - f_0(x_k) - a_k \kappa\delta/2| \\ &\le \kappa\delta/2, \end{aligned}$$

where we used the choice of κ is (10.78), the facts $mx_k - k = (1/2, \ldots, 1/2)$, and $|f(x_k) - f_0(x_k)| \le \kappa\delta$.

10.3.5 Varshamov–Gilbert Bound

The Varshamov–Gilbert bound gives a lower bound to the packing number of a collection of finite 0–1 sequences when the Hamming distance is used to measure a distance between sequences. We will apply the Varshamov–Gilbert bound to derive lower bounds for packing numbers of function spaces. Let us define the Hamming distance H for sequencies in $\{0,1\}^N$, $N \ge 1$, as

$$H(a, b) = \sum_{i=1}^{N} |a_i - b_i|, \tag{10.79}$$

where $a = (a_1, \ldots, a_N) \in \{0,1\}^N$, $b = (b_1, \ldots, b_N) \in \{0,1\}^N$. The Hamming distance is equal to the number of positions where vectors a and b differ.

As an example from information theory, consider the case where we want to transmit a message through a medium that can transmit only sequencies of zeros and ones. We choose number N to be sufficiently large so that every word in our dictionary can be coded by a 0–1 sequence in $\{0,1\}^N$. These sequencies are the code words. When the code words are centers of disjoint balls with radius $\delta = 0, 1, 2, \ldots$, in the Hamming distance, we may restore the code words without errors, provided that the errors occur in δ-bits or less (in δ positions or less). Thus it is of interest to have a large number of code words, which are nevertheless far away in the Hamming distance. When we construct a δ-packing set, we get a set of disjoint balls with radii $\delta/2$. Lemma 10.10 gives an existence result that guarantees the existence of a certain number of code words that are distant from each other.

Lemma 10.10 *We may choose $A \subset \{0,1\}^N$ so that for all $a, b \in A$, $a \neq b$,*

$$H(a,b) > N/3$$

and

$$\#A \geq e^{N/25}.$$

Proof. We start the proof using the same reasoning as in the proof of Lemma 10.6: we construct a covering of $\{0,1\}^N$ whose centers are δ-apart. Note that now $\delta \geq 1$ is an integer. Let $a_1, \ldots, a_M \in \{0,1\}^N$ be a largest set so that distance between $a_i \neq a_j$ is larger than δ. Then the balls $B_\delta(a_j)$ cover $\{0,1\}^N$: they form a δ-net for $\{0,1\}^N$, in the H-metric. (Otherwise there would exist a larger set whose members are δ apart.) Then

$$M \cdot \#B_\delta \geq \#\{0,1\}^N.$$

Clearly, $\#\{0,1\}^N = 2^N$. We need to calculate $\#B_\delta$. We have

$$\#B_\delta = 1 + \binom{N}{1} + \binom{N}{2} + \cdots + \binom{N}{\delta}.$$

In fact B_δ contains the center of the ball, it contains all those sequencies that differ from the center in exactly one position, and there are $\binom{N}{1}$ of such sequencies; it contains all sequencies that differ from the center in exactly two positions, and there are $\binom{N}{2}$ of those sequencies, and so on, up to the sequencies that differ from the center in exactly δ positions. Sometimes the Varshamov–Gilbert bound is stated as

$$\#A = M \geq \frac{2^N}{\sum_{j=0}^{\delta} \binom{N}{j}}.$$

However, we have that

$$2^{-N} \sum_{j=0}^{\delta} \binom{N}{j} = P(S_n \leq \delta),$$

where $S_n \sim \text{Bin}(N, 1/2) = \sum_{i=1}^N Z_i$, where $Z_i \sim \text{Bernoulli}(1/2)$ and Z_i are i.i.d. We have that $EZ_i = 1/2$, $\text{Var}(Z_i) = 1/4$. We apply Bernstein's inequality (see Theorem 12.1) with $\xi = \delta - N/2$, $\sigma = 1/2$, $b = 1/2$, to get

$$\begin{aligned} P(S_n \le \delta) &= P(S_n - N/2 \ge N/2 - \delta) \\ &\le \exp\left\{ -\frac{1}{2} \frac{(N/2 - \delta)^2}{N/4 + (N/2 - \delta)/3} \right\} \\ &\le \exp\{-C_r N\}, \end{aligned}$$

where we chose $\delta = [N/r]$ with $r > 2$ and C_r is defined by

$$C_r = \frac{1}{2} \frac{(1/2 - 1/r)^2}{1/4 + (1/2 - 1/r)/3}. \tag{10.80}$$

Choosing $r = 3$, we get that $C_r \ge 1/25$. □

Packing Number The proof of Lemma 10.10 implies a lower bound to the packing number. Let $\delta < N/2$. Then the δ-packing number $M(\delta, \{0,1\}^N, H)$ satisfies

$$M(\delta, \{0,1\}^N, H) \ge \exp\left\{ \frac{1}{2} \frac{(N/2 - \delta)^2}{N/4 + (N/2 - \delta)/3} \right\}.$$

Note also that a result from Haussler (1991) can be applied:

$$\sum_{j=0}^{\delta} \binom{N}{j} < \left(\frac{eN}{\delta} \right)^{\delta},$$

where $N \ge 1$ and $1 \le \delta \le N$, and whose proof can be found in Barron, Birgé and Massart (1999, lmm. 6).

10.3.6 δ-Packing Sets: Sobolev and Besov

We will construct δ-packing sets for Sobolev and Besov classes. These δ-packing sets can be used to construct lower bounds for the minimax rate of convergence. We prove only lower bounds to the cardinality of the δ-packing sets.

Theorem 10.11 *Let the Sobolev ball $S_s(L)$ be as defined in (10.24). Let the anisotropic Besov ball $B_{sp\infty}(L)$ be as defined in (10.41).*

1. *Let $s > 0$ and $0 < L < \infty$. We have that*

$$\log M(\delta, S_s(L), \|\cdot\|_2) \succeq \delta^{-d/s},$$

   as $\delta \to 0$.

2. *Let $s = (s_1, \ldots, s_d)$, $s_i > 0$, $p = (p_1, \ldots, p_d)$, $1 \le p_i \le \infty$, and $0 < L < \infty$. We have that*

$$\log M(\delta, B_{sp\infty}(L), \|\cdot\|_2) \succeq \delta^{-1/\sigma},$$

as $\delta \to 0$, where

$$\sigma = \left(\sum_{l=1}^{d} s_l^{-1} \right)^{-1} . \tag{10.81}$$

The packing sets may be chosen to be subsets of densities.

Proof. In the anisotropic case, the smoothness index is $s = (s_1, \ldots, s_d)$. In the isotropic case, we misuse the notation and denote $s = s_1 = \cdots = s_d$. Let $\delta > 0$. Let $g_i : \mathbf{R} \to \mathbf{R}$, $i = 1, \ldots, d$, $\int_{-\infty}^{\infty} g_i^2 < \infty$. We assume that the support of g_i is $[0, 1]$, and $g_i(0) = g_i(1) = 0$. Let $G(x) = \prod_{i=1}^{d} g_i(x_i)$ and

$$G_k(x) = m_1^{1/2} \cdots m_d^{1/2} G(m^T x - k), \qquad x \in \mathbf{R}^d,$$

where $k \in I$, and I is the set of d-dimensional multi-indexes:

$$I = \{(k_1, \ldots, k_d) : k_i = 0, \ldots, m_i - 1\},$$

where

$$m = (m_1, \ldots, m_d), \qquad m_i = [\delta^{-1/s_i}], \qquad i = 1, \ldots, d.$$

Note that function G_k has the support on the cell S_k where cells S_k constitute an almost sure partition of $[0, 1]^d$: $[0, 1]^d = \cup_{k \in I} S_k$,

$$S_k = \prod_{i=1}^{d} \left[k_i m_i^{-1}, (k_i + 1) m_i^{-1} \right] . \tag{10.82}$$

Define

$$f_a = f_0 + \delta^{(2\sigma+1)/(2\sigma)} \sum_{k \in I} a_k G_k,$$

where $a = (a_k)_{k \in I} \in \{-1, 1\}^I$, and $f_0 \in \mathcal{F}_s(L)$. Define

$$\mathcal{G}_\delta(f_0) = \left\{ f_a : a \in \{-1, 1\}^I \right\} . \tag{10.83}$$

The functions f_a are densities if f_0 is a density, $\int_0^1 g_i = 0$ (which guarantees that $\int_{\mathbf{R}^d} f_a = 1$), and g_i are bounded (which guarantees that $f_a \geq 0$ for sufficiently small δ). The cardinality of $\mathcal{G}_\delta(f_0)$ satisfies

$$\log_2 \left(\#\mathcal{G}_\delta(f_0) \right) = \#I = m_1 \cdots m_d \sim \delta^{-1/\sigma} . \tag{10.84}$$

Let $A_\delta \subset \{0, 1\}^I$ be such that for all $a, a' \in A_\delta$, $a \neq a'$,

$$\sum_{l \in I} |a_l - a_l'| \geq C \# I,$$

for a positive constant C, and

$$\log(\# A_\delta) \geq C' \# I,$$

for a positive constant C'. Such set A_δ exists by Varshamov–Gilbert bound, which is stated as Lemma 10.10. Since we consider now -1–1 sequences, we have to modify the Hamming distance H in (10.79) to $\frac{1}{2}H$. Let us define the δ-packing set as

$$\mathcal{H}_\delta(f_0) = \{f_a : a \in A_\delta\}. \tag{10.85}$$

We have for $f_a, f_{a'} \in \mathcal{H}_\delta(f_0)$, $a \neq a'$,

$$\begin{aligned}
\|f_a - f_{a'}\|_2^2 &= \delta^{(2\sigma+1)/\sigma} \sum_{k\in I} (a_k - a_k')^2 \|G_k\|_2^2 \\
&= \delta^{(2\sigma+1)/\sigma} 2\|G\|_2^2 \sum_{k\in I} |a_k - a_k'| \\
&\geq \delta^{(2\sigma+1)/\sigma} 2\|G\|_2^2 C \# I \\
&\geq 2\|G\|_2^2 C'' \delta^2 \\
&\geq \delta^2,
\end{aligned}$$

since $\|G_k\|_2^2 = \|G\|_2^2$, and we have chosen g_i so that $2\|G\|_2^2 C'' \geq 1$. We have constructed a set whose members are δ-apart, and the cardinality of the set satisfies

$$\log(\#\mathcal{H}_\delta(f_0)) = \log(\#A_\delta) \succeq \delta^{-1/\sigma}.$$

In the isotropic case, when $s = s_1 = \cdots = s_d$, we have

$$\delta^{-1/\sigma} = \delta^{-d/s}.$$

To prove that that $\mathcal{H}_\delta(f_0)$ is a δ-packing set of $\mathcal{F}$, we have to show that $\mathcal{H}_\delta(f_0) \subset \mathcal{F}$, where $\mathcal{F}$ is either $S_s(L)$ or $B_{sp\infty}(L)$.

Proof of $\mathcal{H}_\delta(f_0) \subset S_s(L)$ The proof of $\mathcal{H}_\delta(f_0) \subset S_s(L)$ is Exercise 10.9.

Proof of $\mathcal{H}_\delta(f_0) \subset B_{sp\infty}(L)$ We consider the case $p_i < \infty$. The case $p_i = \infty$ is Exercise 10.10. Let $f_a \in \mathcal{H}_\delta(f_0)$. Let l_i be the smallest integer $\geq s_i$. We have

$$\rho_s\left(\sum_{k\in I} a_k G_k\right) = \sum_{i=1}^d \sup_{|h|\leq 1} |h|^{-s_i} \left\| \sum_{k\in I} a_k \Delta_{he_i}^{l_i} G_k \right\|_{L_{p_i}(A_{i,hl_i})}$$

Now, for $i \in \{1, \ldots, d\}$, $|h| < 1$, $k \in I$,

$$\Delta_{he_i}^{l_i} G_k(x) = m_1^{1/2} \cdots m_d^{1/2} (\Delta_{hm_ie_i}^{l_i} G)(m^T x - k)$$

as can be seen from (10.35). Then, for $i \in \{1, \ldots, d\}$, $|h| < 1$,

$$\left\| \sum_{k\in I} a_k \Delta_{he_i}^{l_i} G_k \right\|_{L_{p_i}(A_{i,hl_i})}^{p_i}$$

$$
\begin{aligned}
&= \sum_{k\in I} \left\| \Delta_{he_i}^{l_i} G_k(x) \right\|_{L_{p_i}(A_{i,hl_i}\cap S_k)}^{p_i} \\
&= m_1^{p_i/2}\cdots m_d^{p_i/2} \sum_{k\in I} \int_{A_{i,hl_i}\cap S_k} \left|(\Delta_{hm_ie_i}^{l_i} G)(m^T x - k)\right|^{p_i}\, dx \\
&= m_1^{p_i/2}\cdots m_d^{p_i/2} m_1^{-1}\cdots m_d^{-1} \#I \int_{A_{i,hl_i}} \left|(\Delta_{hm_ie_i}^{l_i} G)(x)\right|^{p_i}\, dx \\
&= \left[m_1^{1/2}\cdots m_d^{1/2} \cdot L_i |hm_i|^{s_i} \right]^{p_i} \\
&\le \left[\delta^{-(2\sigma+1)/(2\sigma)} \cdot L_i |h|^{s_i} \right]^{p_i},
\end{aligned}
$$

since $\#I = m_1 \cdots m_d$, $m_1^{1/2} \cdots m_d^{1/2} \le \delta^{-1/(2\sigma)}$ and $m_i^{s_i} \le \delta^{-1}$, and we have assumed that g_i belongs to the univariate Besov class $B_{s_i p_i \infty}(L_i)$. Thus

$$
\rho_s\left(\sum_{k\in I} a_k G_k \right) \le \delta^{-(2\sigma+1)/(2\sigma)} \sum_{i=1}^d L_i
$$

and

$$
\rho_s(f_a) \le \rho_s(f_0) + \sum_{i=1}^d L_i.
$$

To obtain that $f_a \in \mathcal{B}_{sp\infty}(L)$, we need that $\rho_s(f_0) + \rho_s(G) \le L$. For example, we can choose f_0 so that $\rho_s(f_0) = 0$ ($f_0 = I_{[0,1]^d}$) and then let $\sum_{i=1}^d L_i = L$. □

10.3.7 δ-Packing Set: Dominating Mixed Derivatives

We will construct a δ-packing set for the space $D_s(L)$ of dominating mixed derivatives defined in Section 10.2.4. The δ-packing set will be applied in Corollary 11.12 to prove a lower to the minimax rate of convergence.

Theorem 10.12 *We have for $s \ge 1$ integer, $L > 0$, that*

$$
\log M(\delta, D_s(L), \|\cdot\|_2) \succeq \delta^{-1/s} \cdot \left(\log \delta^{-1}\right)^{d-1},
$$

as $\delta \to 0$. The packing sets may be chosen to be subsets of densities.

Proof. The proof follows the same steps as the proof of Theorem 10.11. Let $\delta > 0$. Let

$$
G_{j,k}(x) = 2^{j_1/2} \cdots 2^{j_d/2} \psi(2^{j_1} x_1 - k_1) \cdots \psi(2^{j_d} x_d - k_d), \qquad x \in \mathbf{R}^d,
$$

where $\psi : \mathbf{R} \to \mathbf{R}$ is an s times continuously differentiable wavelet, the support of ψ is contained in $[0,1]$, $(j,k) \in I$, and I is a set of pairs of d-dimensional multi indexes:

$$
I = \{(j,k) : 0 \le |j| \le J_\delta,\ j_i = 0, 1, \ldots, J_\delta,\ k_i = 0, \ldots, 2^{j_i} - 1\},
$$

where $|j| = j_1 + \cdots + j_d$,

$$J_\delta = [\log_2 \delta^{-1/s}].$$

Define

$$f_a = f_0 + \gamma_\delta \sum_{(j,k)\in I} a_{j,k} G_{j,k},$$

where f_0 is chosen later, $a = (a_{j,k})_{(j,k)\in I} \in \{-1,1\}^I$, and

$$\gamma_\delta = C\,\delta^{(2s+1)/(2s)} \left(\log \delta^{-1}\right)^{-(d-1)/2}.$$

Define

$$\mathcal{G}_\delta(f_0) = \{f_a : a \in \{-1,1\}^I\}.$$

The cardinality of $\mathcal{G}_\delta$ satisfies

$$\begin{aligned}
\log_2(\#\mathcal{G}_\delta) &= \#I \\
&= \sum_{j_1=0}^{J_\delta} \sum_{j_2=J_\delta-j_1}^{J_\delta} \cdots \sum_{j_d=J_\delta-j_1-\cdots-j_{d-1}}^{J_\delta} 2^{|j|} \\
&\asymp 2^{J_\delta} J_\delta^{d-1} \\
&\asymp \delta^{-1/s} \cdot \left(\log \delta^{-1}\right)^{d-1}.
\end{aligned}$$

Let $A_\delta \subset \{0,1\}^I$ be such that for all $a, a' \in A_\delta$, $a \neq a'$,

$$\sum_{(j,k)\in I} |a_{j,k} - a_{j',k'}| \geq C\#I, \qquad \log(\#A_\delta) \geq C'\#I.$$

Such set A_δ exists by Varshamov–Gilbert theorem, which is stated as Lemma 10.10. Let us define the δ-packing set as

$$\mathcal{H}_\delta(f_0) = \{f_a : a \in A_\delta\}. \tag{10.86}$$

We have for $f_a, f_{a'} \in \mathcal{H}_\delta(f_0)$, $a \neq a'$,

$$\begin{aligned}
\|f_a - f_{a'}\|_2^2 &= \gamma_\delta^2 \sum_{(j,k)\in I} (a_{j,k} - a'_{j,k})^2 \|G_{j,k}\|_2^2 \\
&= \gamma_\delta^2 2 \sum_{(j,k)\in I} |a_{j,k} - a'_{j,k}| \\
&\geq \gamma_\delta^2 2C\#I \\
&\geq 2C'\delta^2 \\
&\geq \delta^2,
\end{aligned}$$

since $\|G_{j,k}\|_2 = \|G\|_2 = 1$, $G_{j,k}$, $G_{j',k'}$ are orthogonal for $(j,k) \neq (j',k')$, and taking $2C' \geq 1$. We have constructed a set $\mathcal{H}_\delta(f_0)$ whose members are δ-apart, and the cardinality of the set satisfies

$$\log\left(\#\mathcal{H}_\delta(f_0)\right) = \log(\#A_\delta) \succeq \delta^{-1/s} \cdot \left(\log \delta^{-1}\right)^{d-1}. \tag{10.87}$$

Proof of $\mathcal{H}_\delta(f_0) \subset D_s(L)$ Let $f_a \in \mathcal{H}_\delta(f_0)$. Let ρ_s be defined as in (10.47). We have

$$\rho_s^2\left(\sum_{(j,k)\in I} a_{j,k}G_{j,k}\right) = \sum_{|\alpha|_\infty \le s} \left\| \sum_{(j,k)\in I} a_{j,k}D^\alpha G_{j,k}\right\|_2^2 .$$

For $|\alpha|_\infty \le s$, $(j,k) \in I$,

$$D^\alpha G_{j,k}(x) = 2^{\alpha_1 j_1} \cdots 2^{\alpha_d j_d}\, 2^{j_1/2} \cdots 2^{j_d/2} (D^\alpha G)((2^j)^T x - k).$$

Function $G_{j,k}$ has the support on the cell $S_{j,k}$, where

$$S_{j,k} = \prod_{i=1}^d \left[k_i 2^{-j_i}, (k_i+1)2^{-j_i}\right].$$

For a fixed j the cells $S_{j,k}$ constitute an almost sure partition of $[0,1]^d$: $[0,1]^d = \cup_{(j',k)\in I, j'=j} S_{j',k}$. Thus for $|\alpha|_\infty \le s$,

$$\begin{aligned}
\sum_{(j,k)\in I} \int_{[0,1]^d} |D^\alpha G_{j,k}|^2 &= \sum_{(j,k)\in I} 2^{2\alpha_1 j_1} \cdots 2^{2\alpha_d j_d} \int_{[0,1]^d} |D^\alpha G|^2 \\
&\le \#I \cdot 2^{2sJ_\delta} \int_{[0,1]^d} |D^\alpha G|^2 \, dx,
\end{aligned}$$

since $\alpha_1 j_1 + \cdots + \alpha_d j_d \le |\alpha|_\infty |j|$. Also

$$\begin{aligned}
&\sum_{(j,k),(j',k')\in I, j'>j} \int_{[0,1]^d} D^\alpha G_{j,k} D^\alpha G_{j',k'} \\
&\le \sum_{(j,k)\in I} \max_{(j',k')\in I, j'>j} \left| \int_{[0,1]^d} D^\alpha G_{j,k} D^\alpha G_{j',k'} \right| \\
&= O\left(2^{2sJ_\delta}\right).
\end{aligned}$$

Thus for $|\alpha|_\infty \le s$,

$$\int_{[0,1]^d} \left| \sum_{(j,k)\in I} a_{j,k} D^\alpha G_{j,k} \right|^2 \le \gamma_\delta^{-2} \int_{[0,1]^d} |D^\alpha G|^2 \, dx.$$

Hence

$$\rho_s^2\left(\sum_{(j,k)\in I} a_{j,k}G_{j,k}\right) \le \gamma_\delta^{-2} \rho_s^2(G)$$

and

$$\rho_s(f_a) \le \rho_s(f_0) + \rho_s(G).$$

To obtain that $f_a \in D_s(L)$, we need that $\rho_s(f_0) + \rho_s(G) \le L$. □

10.3.8 Convex Hull

Let $\mathbb{G} \subset L_2(Q)$, where Q is a probability measure. Let us denote with $\text{co}(\mathbb{G})$ the collection of convex combinations $\sum_{i=1}^k \lambda_i \phi_i$, where $\phi_i \in \mathbb{G}$, $\sum_{i=1}^k \lambda_i = 1$, $\lambda_i \geq 0$, and $k = 1, 2, \ldots$; see definition (10.50). We denote with $\bar{\text{co}}(\mathbb{G})$ the $L_2(Q)$-closure of $\text{co}(\mathbb{G})$.

Theorem 10.13 *If there exists a δ-net of $\mathbb{G}$ of cardinality $C(1/\delta)^V$, then there exists a δ-net of $\bar{\text{co}}(\mathbb{G})$ of cardinality $C'(1/\delta)^{2V/(V+2)}$, where C' depends only on the envelope of $\mathbb{G}$, on C and V. Here the δ-nets are with respect to the $L_2(Q)$ metric.*

The theorem was proved by van der Vaart and Wellner (1996) and Carl (1997). Generalizations and better constants have been given by Carl, Kyrezi and Pajor (1999) and Mendelson (2002).

EXERCISES

10.1 *(Copulas.)* Let the random vector $(X_1, \ldots, X_d)$ have a continuous distribution function and copula C. If h_i is strictly increasing (in the range of X_i), for $i = 1, \ldots, d$, then $(h_1(X_1), \ldots, h_d(X_d))$ has the same copula C.

10.2 *(Copulas.)* For any copula C,

$$|C(u) - C(v)| \leq \sum_{i=1}^d |u_i - v_i|, \qquad u, v \in [0,1]^d.$$

10.3 *(Copulas; Fréchet–Hoeffding bounds.)* For any copula C,

$$\max\{u_1 + \cdots + u_d - d + 1, 0\} \leq C(u) \leq \min\{u_1, \ldots, u_d\}.$$

The lower bound is a copula only for $d = 2$.

10.4 *(Smoothness classes.)* Prove equation (10.35).

10.5 *(Smoothness classes.)* We may generalize (10.53) and define a partially degenerate mixture class. We assume that $\Theta = \Theta^{(1)} \times \Theta^{(2)}$, where $\Theta^{(i)} \subset \mathbf{R}^{k_i}$, $0 \leq k_i \leq k$, $i = 1, 2$, $k_1 + k_2 = k$. We define the infinite mixture class, with the base class $\mathbb{G}$ and with the degenerate component $\Theta^{(1)}$, to be

$$\mathcal{G}(\mathbb{G}) = \left\{ \int_{\Theta^{(2)}} g\left(\cdot, \theta^{(1)}, \theta^{(2)}\right) dQ\left(\theta^{(2)}\right) : \theta^{(1)} \in \Theta^{(1)}, Q \in \mathbb{Q}\left(\Theta^{(2)}\right) \right\}, \tag{10.88}$$

where $g(\cdot, \theta) = g(\cdot, \theta^{(1)}, \theta^{(2)}) \in \mathbb{G}$. Prove Lemma 10.14 below. (Use the fact that the integral in the definition of $g \in \mathcal{G}(\mathbb{G})$ may be approximated by a Riemann sum.)

Lemma 10.14 *Let $\mathbb{G}$ be defined in (10.52). Let $g(x, \theta^{(1)}, \cdot)$ be Riemann integrable for all $x \in \mathbf{R}^d$, $\theta^{(1)} \in \Theta^{(1)}$. Let*

$$\sup_{\theta^{(2)} \in \Theta^{(2)}} \left\| g\left(\cdot, \theta^{(1)}, \theta^{(2)}\right) \right\|_2 < \infty, \qquad \text{for all } \theta^{(1)} \in \Theta^{(1)}. \tag{10.89}$$

Then

$$\mathcal{G}(\mathbb{G}) \subset \bar{\text{co}}(\mathbb{G}). \tag{10.90}$$

10.6 *(Smoothness classes.)* Prove that when $s > d$,

$$L_{s,1} \subset \mathcal{G}(\mathbb{G}),$$

where $L_{s,1}$ is defined in (10.56), $\mathcal{G}(\mathbb{G})$ is defined in (10.53) with $\mathbb{G}$ defined in (10.54) with $M = \mathbf{R}^d$ and $S = (0, \infty)$.

10.7 *(Smoothness classes.)* Prove Lemma 10.15 below.

Lemma 10.15 *Let $\mathcal{F} = \{f : \rho_s(f) \le L\}$, where $\rho_s(f)$ is defined in (10.57). We have that*

$$\mathcal{F} \subset \mathcal{G}(\mathbb{G}),$$

where $\mathcal{G}(\mathbb{G})$ is defined in (10.88) with $\mathbb{G} = \{g(\cdot, \theta) : \theta \in \Theta\}$,

$$g(x, \theta) = \theta_1 \|\beta\|^{-s} \cos(\theta_2^T x),$$

$\Theta = \Theta^{(1)} \times \Theta^{(2)}$, $\Theta^{(1)} = [0, L]$, $\Theta^{(2)} = \mathbf{R}^d$.

10.8 *(Smoothness classes.)* Prove Lemma 10.16 below.

Lemma 10.16 *Denote $\mathcal{F}_{\beta,p}(L) = \{f : \rho_{\beta,p}(f) \le L\}$, where $\rho_{\beta,p}(f)$ is defined in (10.31). We have that*

$$\mathcal{F}_{\beta_1,1}(L') \cap \{f : \|Ff\|_\infty \le B_\infty\} \subset \mathcal{F}_{\beta_2,2}(L'') \tag{10.91}$$

when $\beta_1 \ge 2\beta_2 + d/2$, for suitable L', L'', where $0 < B_\infty < \infty$. An intersection of two L_2 balls is a subset of a L_1 ball:

$$\mathcal{F}_{\beta_2,2}(L'') \cap \mathcal{F}_{\beta_1 - d/2,2}(L''') \subset \mathcal{F}_{\beta_1,1}(L') \tag{10.92}$$

when $\beta_2 > \beta_1 > d/2$, for suitable L', L'', L'''.

Lemma 10.16 says that to get a L_1 ball inside a L_2 ball, we need to choose the smoothness index of the L_1 ball essentially larger than the smoothness index of the L_2 ball. To get a L_2 ball inside a L_1 ball, we need to choose the smoothness index of the L_2 ball only slightly larger than the smoothness index of the L_1 ball.

10.9 *(Packing numbers.)* Let $\mathcal{H}_\delta(f_0)$ be defined in (10.85). Show that $\mathcal{H}_\delta(f_0) \subset S_s(L)$.

10.10 *(Packing numbers.)* Prove $\mathcal{H}_\delta(f_0) \subset B_{s\infty\infty}(L)$, similarly as the case $\mathcal{H}_\delta(f_0) \subset B_{sp\infty}(L)$ was proved in page 309 for $p_i < \infty$.

CHAPTER 11

LOWER BOUNDS

In statistics one tries to construct as good estimators as possible. One of the most fascinating topics in mathematical statistics is to find out what is possible. There are limits to the accuracy of estimation, and we should find out those limits. We will analyze the accuracy of the estimators in terms of their rate of convergence. The principal epistemological value of lower bounds can be revealed by asymptotic statements. Gnedenko and Kolmogorov (1954): The epistemological value of probability theory is based on the fact that chance phenomena, considered collectively and on a grand scale, create non-random regularity.

An important theme of this chapter is to show how the packing numbers of function classes determine lower bounds for the rate of convergence of any estimator. In Section 10.3 we have defined covering and packing numbers and calculated the order of these numbers for some function classes. We can now apply some of those results.

At least two approaches to the theory of statistical estimation can be considered:

1. Postulate a density class and ask what would be the best possible estimator, when it is known that the true density belongs to this class. This approach has been common in model based statistics, and it leads to decision theoretic concepts like admissibility, minimax risk, and Bayes risk.

2. Construct an estimator and then ask what are the statistical properties of the estimator, when the true density belongs to a given test bed (a given class of densities).

Approaches 1 and 2 have different philosophy, but their realizations can be similar. Approach 1 can be realized in the following way: construct an estimator and calculate its minimax rate of convergence, then derive a lower bound to the minimax rate of convergence. If the rate of convergence of the estimator achieves the lower bound, then we know that the estimator has the best possible rate of convergence in the postulated density class. When we use approach 2, it is of interest to ask whether the estimator that we have constructed is the best possible estimator in a given test bed. If the estimator is not the best possible estimator, or is not sufficiently close to the best possible estimator, then it is of interest to search for a better estimator. When we prove a lower bound to the minimax rate of convergence, and note that the estimator achieves this lower bound, then we know that one cannot find an estimator with a better rate of convergence, in the considered test bed. Thus both approaches 1 and 2 require the calculation of lower bounds.

Section 11.1 defines the basic concepts needed in the theory of lower bounds. Section 11.2 contains the abstract theorems that can be applied to derive lower bounds. Section 11.3 applies the general theorems for some of the smoothness classes for which the capacity numbers are known. The corresponding upper bounds will be proved in Part III.

11.1 RATE OPTIMAL ESTIMATORS

11.1.1 Minimax Risk

The risk of density estimator $\hat{f}$ is the expectation of the loss:

$$E_f L(\hat{f}, f),$$

where f is the true density. We will choose the L_2 loss

$$L(\hat{f}, f) \stackrel{def}{=} \|\hat{f} - f\|_2^2. \tag{11.1}$$

We discuss other loss functions in Section 11.1.2. Finding an estimator minimizing the risk is trivial: we choose the estimator that is equal to the true density for all samples: $\hat{f} = f$ for all realizations of $X_1, \ldots, X_n$. In order to exclude such pathological estimator, we require that estimators perform uniformly well over a class of densities. The minimax risk is defined as

$$\mathcal{R}_n(\mathcal{F}) = \inf_{\hat{g}} \sup_{f \in \mathcal{F}} E_f L(\hat{g}, f), \tag{11.2}$$

where the infimum is taken over all density estimators, and $\mathcal{F}$ is a class of densities. A minimax estimator over class $\mathcal{F}$ is an estimator $\hat{f}$ that achieves the infimum in the definition of the minimax risk: $\hat{f}$ is minimax estimator if

$$\sup_{f \in \mathcal{F}} E_f L(\hat{f}, f) = \mathcal{R}_n(\mathcal{F}).$$

Typically we may hope to find only asymptotically minimax estimators. Estimator $\hat{f}$ is an asymptotically minimax estimator over class $\mathcal{F}$ when

$$\sup_{f \in \mathcal{F}} E_f L(\hat{f}, f) \sim \mathcal{R}_n(\mathcal{F}),$$

where $a_n \sim b_n$ means that $\lim_{n \to \infty}(a_n/b_n) = 1$. Even this kind of exact asymptotic optimality is a strong requirement for an estimator, and we will in most cases construct only rate optimal estimators. Let us define the minimax rate of convergence and a rate optimal estimator. Note that all the notations can be used as well in the Gaussian white noise model as in the density estimation model.

Definition 11.1 (Minimax rate, Rate optimal estimator.) *We say that* $\psi_n > 0$ *is the* minimax rate of convergence, *for the* L_2 *loss* L *defined in (11.1), and for the class* $\mathcal{F}$ *of densities, when for some estimator* $\hat{f}$,

$$\limsup_{n \to \infty} \psi_n^{-2} \sup_{f \in \mathcal{F}} E_f L(\hat{f}, f) < \infty, \tag{11.3}$$

and

$$\liminf_{n \to \infty} \psi_n^{-2} \mathcal{R}_n(\mathcal{F}) > 0. \tag{11.4}$$

When ψ_n *is the minimax rate of convergence, estimator* $\hat{f}$ *satisfying (11.3) is called* a rate optimal estimator. *The same definitions apply in the Gaussian white noise model.*

Thus estimator $\hat{f}$ is rate optimal over class $\mathcal{F}$ when

$$\sup_{f \in \mathcal{F}} E_f L(\hat{f}, f) \asymp \mathcal{R}_n(\mathcal{F}),$$

where $a_n \asymp b_n$ means that $c \le \liminf_{n \to \infty}(a_n/b_n) \le \limsup_{n \to \infty}(a_n/b_n) \le C$ for positive constants c, C. When ψ_n is the minimax rate of convergence and we have a sequence γ_n that satisfies $\gamma_n \asymp 1$, then $\gamma_n \psi_n$ is also a minimax rate of convergence.

Typically $\mathcal{F}$ is a smoothness class indexed with a smoothness parameter. For example, $\mathcal{F} = \mathcal{F}_{s,L}$ could be a class of densities whose sth order derivatives are bounded by L. A rate optimal estimator over $\mathcal{F}_{s,L}$ could depend on parameters s and L which are rather arbitrary and unknown. It is important to construct an estimator that is rate optimal over a wide range of smoothness parameter values. Let $\mathcal{F}_\sigma$, $\sigma \in \Sigma$, be a collection of smoothness classes of densities. We say that estimator $\hat{f}$ is adaptive rate optimal estimator if it has convergence rate $\psi_{\sigma,n}$ simultaneously over this collection, where $\psi_{\sigma,n}$ is the minimax rate of convergence for $\mathcal{F}_\sigma$. A weaker form of adaptive rate optimality requires that an estimator achieve minimax rates simultaneously only up to a logarithmic factor.

Definition 11.2 (Adaptive rate optimal estimator, Adaptive rate optimal estimator up to a factor.) *Estimator $\hat{f}$ is an* adaptive rate optimal estimator, *for the L_2 loss L defined in (11.1), and for the collection of classes $\mathcal{F}_\sigma$, $\sigma \in \Sigma$, when for all $\sigma \in \Sigma$,*

1. $\limsup_{n\to\infty} \psi_{\sigma,n}^{-2} \sup_{f\in\mathcal{F}_\sigma} E_f L(\hat{f}, f) < \infty$,
2. $\psi_{\sigma,n}$ *is the optimal rate of convergence over $\mathcal{F}_\sigma$, as in Definition 11.1.*

Estimator $\hat{f}$ is an adaptive rate optimal estimator, up to the factor $\gamma_{\sigma,n}$, *when for all $\sigma \in \Sigma$,*

1. $\limsup_{n\to\infty} (\gamma_{\sigma,n}/\psi_{\sigma,n})^2 \sup_{f\in\mathcal{F}_\sigma} E_f L(\hat{f}, f) < \infty$, *where $\gamma_{\sigma,n} > 0$,*
2. $\psi_{\sigma,n}$ *is the optimal rate of convergence over $\mathcal{F}_\sigma$, as in Definition 11.1.*

In a typical case the optimal rate of convergence is algebraic with respect to the sample size: $\psi_{\sigma,n} = n^{-h(\sigma)}$ for some $h(\sigma) > 0$; and $\gamma_{\sigma,n}$ is logarithmic: $\gamma_{\sigma,n} = (\log n)^{g(\sigma)}$ for some $g(\sigma) > 0$. Constructing a rate optimal estimator up to a logarithmic factor can be claimed to be well enough because the rate is so close to the optimal algebraic rate.

11.1.2 Loss Functions

L_1 distance A natural measure for the accuracy of a density estimator $\hat{f}$ is the L_1 distance

$$\int_{\mathbf{R}^d} \left| f - \hat{f} \right|,$$

where f is the true (unknown) density. The L_1 distance is natural for three reasons:

1. The L_1 distance is always defined when f and $\hat{f}$ are densities.
2. The L_1 distance is invariant with respect to scale changes; that is, if f_X is the density of X and $f_{X'}$ is the density of X', then

$$\begin{aligned} \int_{\mathbf{R}^d} |f_X - f_{X'}| &= |\det(A)|^{-1} \int_{\mathbf{R}^d} \left| f_X(A^{-1}y) - f_{X'}(A^{-1}y) \right| dy \\ &= \int_{\mathbf{R}^d} |f_{AX} - f_{AX'}|, \end{aligned}$$

   where f_{AX} is the density of AX and $f_{AX'}$ is the density of AX'.

3. We may write the L_1 distance in terms of probabilities

$$\int_{\mathbf{R}^d} \left| f - \hat{f} \right| = 2 \sup_A \left| \int_A f - \int_A \hat{f} \right|, \tag{11.5}$$

   where the supremum is taken over the Borel sets of $\mathbf{R}^d$. This equation is called Scheffé's equality. Exercise 11.1 requires you to prove Scheffé's equality.

Note that the total variation distance between probability measures P and Q is defined as

$$D_{tot}(P,Q) = \sup_A |P(A) - Q(A)|,$$

where the supremum is taken over the Borel sets of $\mathbf{R}^d$.

L_p*, *Hellinger, and Kullback–Leibler Distances We may extend the collection of loss functions by considering other L_p metrics than the L_1 metric:

$$\left\|f - \hat{f}\right\|_p^p = \int_{\mathbf{R}^d} \left|f - \hat{f}\right|^p, \qquad \left\|f - \hat{f}\right\|_\infty = \sup_{x\in\mathbf{R}^d} \left|f(x) - \hat{f}(x)\right|.$$

The squared Hellinger distance is defined as

$$D_H^2(f,g) = \int_{\mathbf{R}^d} \left(\sqrt{f} - \sqrt{g}\right)^2. \tag{11.6}$$

The Hellinger distance is defined for all densities since $\sqrt{f} \in L_2$ for all densities f. The Kullback–Leibler distance is defined as

$$D_K^2(P,Q) = E_P \log_e \left(\frac{dP}{dQ}\right), \tag{11.7}$$

where we assume that probability measure Q dominates probability measure P, that is, for all measurable A, $Q(A) = 0 \Rightarrow P(A) = 0$. If Q does not dominate P, then we may define $D_K^2(P,Q) = \infty$. If f and g are densities of P and Q with respect to the Lebesgue measure, then we may write

$$D_K^2(f,g) = \int_{\mathbf{R}^d\cap\{x:g(x)>0\}} f \log_e \left(\frac{f}{g}\right). \tag{11.8}$$

The Kullback–Leibler distance is not a metric, since it is not symmetric, but we call it a distance nevertheless. See Appendix B.8 for the definition of a metric, or a distance. Exercise 11.2 requires you to prove that the L_1 distance and the Hellinger distance are equivalent. Exercise 11.3 requires you to prove that when the density is bounded and bounded away from zero, the L_2 distance is equivalent to the Hellinger distance, and thus to the L_1 distance.

L_2* *Distance The L_2 distance, defined in (11.1), is convenient for the mathematical analysis because we have the bias-variance decomposition of the expected L_2 error:

$$E_f \int_{\mathbf{R}^d} \left(f - \hat{f}\right)^2 = \int_{\mathbf{R}^d} \left(f - E_f\hat{f}\right)^2 + \int_{\mathbf{R}^d} \mathrm{Var}(\hat{f}), \tag{11.9}$$

where we applied Fubini's theorem to change the order of the expectation and the integration. Exercises 11.2 and 11.3 imply that when the density is bounded and bounded away from zero, the L_2 distance is equivalent to the L_1 distance and to the Hellinger distance. When a density is bounded away from zero, it has to have a compact support. Thus for heavy tailed distributions there might be a difference between these losses. Intuitively the L_2 loss does not get as much contribution from tail regions (from regions where the density takes small values) as do the L_1 loss and the Hellinger loss. On the other hand, when we are interested in the high-density areas, the choice of the loss function from this group does not seem to be crucial.

11.1.3 Historical Notes

Distribution Function Estimation We have generalized the natural L_1 loss function to a larger class of L_p loss functions and to the Hellinger and the Kullback–Leibler loss functions. Another way to extend the class of loss functions is to use Scheffé's characterization of the L_1 distance in (11.5). We may take the supremum in this equation over a more restricted class of sets. For example, we might consider the loss

$$\sup_{x \in \mathbf{R}^d} \left| \int_{(-\infty,x]} f - \int_{(-\infty,x]} \hat{f} \right|, \tag{11.10}$$

where $(-\infty, x] = (-\infty, x_1] \times \cdots \times (-\infty, x_d]$. Using this loss function amounts to the fact that we are interested in estimating the distribution function, since the distribution function function corresponding to density function f is defined as

$$F(x) = \int_{(-\infty,x]} f, \qquad x \in \mathbf{R}^d. \tag{11.11}$$

We can estimate the distribution function with the empirical distribution function

$$F_n(x) = \frac{1}{n} \sum_{i=1}^{n} I_{(-\infty,x]}(X_i). \tag{11.12}$$

The empirical distribution function converges to the true distribution function with the $\sqrt{n}$-rate:

$$\limsup_{n \to \infty} \sqrt{n}\, E_F \|F_n - F\|_\infty < \infty,$$

when the $d = 1$ and F is continuous; see Breiman (1993*b*, thm. 13.16). The result is related to the Kolmogorov–Smirnov test. The rate $\sqrt{n}$ is the optimal rate of convergence in the sense of local asymptotic minimax theory; see Bickel, Klaassen, Ritov and Wellner (1993). The probability measure corresponding to the empirical distribution function is the empirical measure

$$P_n(A) = \frac{1}{n} \sum_{i=1}^{n} I_A(X_i), \qquad A \subset \mathbf{R}^d.$$

In the total variation distance the empirical distribution does not converge to the true probability measure: for all $n \geq 1$,

$$\sup_A |P_n(A) - P(A)| \geq 2$$

with probability one when P is a probability measure with a continuous distribution function. We have shown that estimation of the distribution function with the supremum loss is much easier than the estimation of the density function with the total variation distance (L_1 distance), and in particular, the loss function in (11.10) is too weak to be useful in density estimation.

Local Asymptotic Minimax Risk When the density class $\mathcal{F}$ is finite dimensional, the local asymptotic minimax theory provides a powerful framework for defining asymptotic optimality. Let $\mathcal{F} = \{f_\theta : \theta \in \Theta\}$, where $\Theta \subset \mathbf{R}$ is an open set. An estimator $\hat{\theta}$ of the parameter θ is called asymptotically optimal when for all $\theta_0 \in \Theta$ and for all $0 < R < \infty$,

$$\lim_{n\to\infty} n \sup_{\theta \in \mathcal{N}_n(\theta_0, R)} E_\theta |\theta - \hat{\theta}|^2 = I(\theta_0),$$

where $\mathcal{N}_n(\theta_0, R) = \{\theta_0 + hn^{-1/2} : \|h\| \le R\}$, and $0 < I(\theta_0) < \infty$ is such that for all $0 < R < \infty$,

$$\liminf_{n\to\infty} n \inf_{\theta_n} \sup_{\theta \in \mathcal{N}_n(\theta_0, R)} E_\theta |\theta - \theta_n|^2 \ge I(\theta_0),$$

where the infimum is taken over all estimators of θ. Local asymptotic minimax theory can be generalized to the semiparametric setting; see Bickel et al. (1993).

11.2 METHODS TO PROVE LOWER BOUNDS

By a lower bound we mean a similar type of inequality as in (11.4). A lower bound of this type implies that rate ψ_n vanishes so fast as $n \to \infty$ that there cannot be estimators whose convergence rate is faster. We prove lower bounds both for the density estimation model and for the Gaussian white noise model, defined in Section 9.4.3. This is extremely useful in sketching the general line of reasoning and pointing out the particular technical features of the density estimation model.

11.2.1 The Main Idea

We follow the approach of Hasminskii and Ibragimov (1990). The main idea is to prove that a classification error gives a lower bound to the estimation error. For any finite collection $\mathcal{P} = \{P_1, \ldots, P_N\}$ of probability measures the minimal classification error is defined as

$$\mathrm{class}(\mathcal{P}) = \inf_{\hat{\theta}} \max_{i=1,\ldots,N} P_i(\hat{\theta} \ne P_i), \tag{11.13}$$

where the infimum is taken over all random variables taking values in the set $\mathcal{P}$. In our case we consider a smoothness class $\mathcal{F}$, choose $\mathcal{H}_\delta \subset \mathcal{F}$, and the collection of the probability measures in the classification problem is equal to

$$\mathcal{P} = \left\{ P_{Af}^{(n)} : f \in \mathcal{H}_\delta \right\}, \tag{11.14}$$

where in the density estimation model $P_{Af}^{(n)}$ is the product measure corresponding to density Af, and in the Gaussian white noise model $P_{Af}^{(n)}$ is the measure of process $Y_{n,Af}$ in (9.47).

The derivation of the lower bound proceeds by choosing a finite set $\mathcal{H}_\delta \subset \mathcal{F}$ and considering the classification problem of deciding which density or signal function in $\mathcal{H}_\delta$ has generated the observation. It is assumed that the functions in $\mathcal{H}_\delta$ are δ apart in the L_2 distance. A classification procedure can be defined with the help of any estimator $\hat{f}$: choose the function from $\mathcal{H}_\delta$ which is closest to $\hat{f}$, in the L_2 distance. This leads to a lower bound for the L_2 error in terms of the probability of a classification error.

Lemma 11.1 *Let $\mathcal{H}_\delta \subset \mathcal{F}$ be a finite set for which*

$$\min\{\|f-g\|_2 : f, g \in \mathcal{H}_\delta,\ \ f \neq g\} \geq \delta, \tag{11.15}$$

where $\delta > 0$. Then

$$\inf_{\hat{f}} \sup_{f\in\mathcal{F}} E_f\|\hat{f}-f\|_2^2 \geq \frac{\delta^2}{4}\,\mathrm{class}(\mathcal{P}),$$

where $\mathrm{class}(\mathcal{P})$ *is defined in (11.13), $\mathcal{P}$ is defined in (11.14), and the infimum is taken over all estimators in the density estimation model or in the Gaussian white noise model.*

Proof. Let $\hat{f} : \mathbf{R}^d \to \mathbf{R}$ be an estimator of f. Define a random variable $\hat{\theta}$ taking values in $\mathcal{H}_\delta$:

$$\hat{\theta} = \mathrm{argmin}_{f\in\mathcal{H}_\delta}\|\hat{f}-f\|_2.$$

Note that by (11.15),

$$\hat{\theta} \neq f \in \mathcal{H}_\delta \Rightarrow \|\hat{f}-f\|_2^2 \geq \delta/2,$$

since $\hat{\theta} \neq f \in \mathcal{H}_\delta$ implies that $\hat{f}$ is closer to some other $g \in \mathcal{H}_\delta$ than to f. Then applying Markov's inequality obtains

$$\begin{aligned}\sup_{f\in\mathcal{F}} E_{Af}\|\hat{f}-f\|_2^2 &\geq \max_{f\in\mathcal{H}_\delta} E_{Af}\|\hat{f}-f\|_2^2 \\ &\geq \frac{\delta^2}{4}\max_{f\in\mathcal{H}_\delta} P_{Af}^{(n)}\left(\|\hat{f}-f\|_2^2 \geq \delta^2/4\right) \\ &\geq \frac{\delta^2}{4}\max_{f\in\mathcal{H}_\delta} P_{Af}^{(n)}(\hat{\theta}\neq f).\end{aligned}$$

We have proved the lemma. □

11.2.2 Lower Bounds for the Classification Error

In order to apply Lemma 11.1, we give two methods to obtain a lower bound for the classification error.

Tsybakov's Lemma The classification is the problem of deciding which of the probability measures in a finite set $\mathcal{P}$ of probability measures has generated the observation. We give a lower bound for the maximal error probability $\max_{P\in\mathcal{P}} P(\hat{\theta} \neq P)$ of an arbitrary estimator $\hat{\theta}$ taking values in $\mathcal{P}$.

Lemma 11.2 (Tsybakov (1998), Theorem 6.) *Let $\hat{\theta}$ be a random variable taking values on a finite set $\mathcal{P}$ of probability measures. Denote* $\mathrm{Card}(\mathcal{P}) = N$, *and assume that $N \geq 2$. Let $\tau > 0$ and $0 < \alpha < 1$. Let for some $P_0 \in \mathcal{P}$ and for all $P \in \mathcal{P} \setminus \{P_0\}$,*

$$P\left(\frac{dP_0}{dP} \leq \tau\right) \leq \alpha. \tag{11.16}$$

Then

$$\max_{P\in\mathcal{P}} P(\hat{\theta} \neq P) \geq (1-\alpha)\,\frac{\tau(N-1)}{1+\tau(N-1)}.$$

Proof. Let us denote $\Omega_P = (\hat{\theta} \neq P)$ and $\mathcal{P}_- = \mathcal{P} \setminus \{P_0\}$. Then

$$\begin{aligned}
P_0(\Omega_{P_0}) &\geq P_0\left(\bigcup_{P\in\mathcal{P}_-} \Omega_P^c\right) &(11.17)\\
&= \sum_{P\in\mathcal{P}_-} P_0\left(\Omega_P^c\right)\\
&= \sum_{P\in\mathcal{P}_-} \int_{\Omega_P^c} \frac{dP_0}{dP}\, dP\\
&\geq \tau \sum_{P\in\mathcal{P}_-} P\left(\Omega_P^c \cap \left(\frac{dP_0}{dP} \geq \tau\right)\right)\\
&\geq \tau \sum_{P\in\mathcal{P}_-} \left[P\left(\Omega_P^c\right) - P\left(\frac{dP_0}{dP} \leq \tau\right)\right] &(11.18)\\
&\geq \tau \sum_{P\in\mathcal{P}_-} \left[P\left(\Omega_P^c\right) - \alpha\right] &(11.19)\\
&\geq \tau(N-1)\left[1 - \max_{P\in\mathcal{P}_-} P\left(\Omega_P\right) - \alpha\right].
\end{aligned}$$

At step (11.17), we applied the fact that if $\hat{\theta} = P$ for some $P \in \mathcal{P}_-$, then $\hat{\theta} \neq P_0$; at step (11.18), we applied the inequality $P(A) \leq P(A \cap B) + P(B^c)$; and at step (11.19), we applied assumption (11.16). Now we have

$$\begin{aligned}
\max_{P\in\mathcal{P}} P\left(\Omega_P\right) &= \max\left\{P_0\left(\Omega_{P_0}\right), \max_{P\in\mathcal{P}_-} P\left(\Omega_P\right)\right\}\\
&\geq \min_{0\leq p\leq 1} \max\left\{p, \tau(N-1)(1-p-\alpha)\right\}.
\end{aligned}$$

The minimum is achieved by taking $p = (1-\alpha)\tau(N-1)/[1+\tau(N-1)]$, which proves the lemma. □

A Lower Bound to L_2 Error We combine Lemma 11.1 and Lemma 11.2 to get the following lower bound to the estimation error:

Lemma 11.3 *Let $\mathcal{H}_\delta \subset \mathcal{F}$ be a finite set for which*

$$\min\{\|f-g\|_2 : f,g \in \mathcal{H}_\delta,\ \ f \neq g\} \geq \delta, \tag{11.20}$$

where $\delta > 0$. Assume that for some $f_0 \in \mathcal{H}_\delta$, and for all $f \in \mathcal{H}_\delta \setminus \{f_0\}$,

$$P_{Af}^{(n)}\left(\frac{dP_{Af_0}^{(n)}}{dP_{Af}^{(n)}} \leq \tau\right) \leq \alpha, \tag{11.21}$$

where $0 < \alpha < 1$, $\tau > 0$. In the density estimation model $P_{Af}^{(n)}$ is the product measure corresponding to density Af, and in the Gaussian white noise model $P_{Af}^{(n)}$ is the measure of process $Y_{n,Af}$ in (9.47). Then

$$\inf_{\hat{f}} \sup_{f\in\mathcal{F}} E_f\|\hat{f}-f\|_2^2 \geq \frac{\delta^2}{4}(1-\alpha)\frac{\tau(N_\delta-1)}{1+\tau(N_\delta-1)},$$

where $N_\delta = \#\mathcal{H}_\delta \geq 2$, and the infimum is taken over all estimators (either in the density estimation model or in the Gaussian white noise model).

Distribution Function of the Likelihood Ratio Lemma 11.2 makes an assumption in (11.16) concerning the distribution function of the likelihood ratio. We apply Lemma 11.2 by bounding this distribution function in terms of the Kullback–Leibler distance between the distributions.

Lemma 11.4 *Let D_K be the Kullback–Leibler distance defined in (11.7). Let P be dominated by P_0. We have for $\tau > 0$,*

$$P\left(\frac{dP_0}{dP} \leq \tau\right) \leq \left(\log_e \tau^{-1}\right)^{-1} D_K^2(P,P_0).$$

Proof. We have

$$P\left(\frac{dP_0}{dP} \leq \tau\right) = P\left(\frac{dP}{dP_0} \geq \tau^{-1}\right).$$

By Markov's inequality,

$$P\left(\log_e \frac{dP}{dP_0} \geq \log_e \tau^{-1}\right) \leq \left(\log_e \tau^{-1}\right)^{-1} E_P \log_e\left(\frac{dP}{dP_0}\right).$$

We have proved the lemma. □

Density Estimation We give further bounds for the Kullback–Leibler distance. We start with the density estimation. The first lemma states that the Kullback–Leibler distance between product measures is equal to the sum of the Kullback–Leibler distances between the individual measures. The second lemma states that the Kullback–Leibler distance is equivalent to the L_2 distance when the densities are bounded and bounded away fro zero. The proof of Lemma 11.5 is Exercise 11.4, and the proof of Lemma 11.6 is Exercise 11.5.

Lemma 11.5 *Let P be dominated by P_0. We have that*

$$D_K^2\left(P^{(n)}, P_0^{(n)}\right) = nD_K^2(P, P_0)$$

for product measures $P^{(n)}$ and $P_0^{(n)}$.

Lemma 11.6 *Let D_K be the Kullback–Leibler distance defined in (11.8). We have*

$$D_K^2(f, f_0) \leq \int_{\{x \in \mathbf{R}^d : f_0(x) > 0\}} \frac{(f - f_0)^2}{f_0}.$$

*The upper bound is called the χ^2-*divergence *between f and f_0. In particular, if $\inf_{x \in \mathbf{R}^d} f_0(x) > 0$, then*

$$D_K^2(f, f_0) \leq \frac{\|f - f_0\|_2^2}{\inf_{x \in \mathbf{R}^d} f_0(x)}. \tag{11.22}$$

Also, if f and f_0 are both bounded and bounded away from zero, then

$$D_K^2(f, f_0) \geq \int_{\{x \in \mathbf{R}^d : f_0(x) > 0\}} (f - f_0) + C\,\|f - f_0\|_2^2, \tag{11.23}$$

for a positive constant C.

Gaussian White Noise Model We give a formula for the Kullback–Leibler distance in the case of the Gaussian white noise model. Let $P_f^{(n)}$ and $P_{f_0}^{(n)}$ be the probability measures of the processes $X_{n,f}$ and X_{n,f_0}, defined in (9.45). The distribution of the likelihood ratio was given in (9.49): we have that

$$\frac{dP_f^{(n)}}{dP_{f_0}^{(n)}} = \exp\left\{n^{1/2}\sigma Z + \frac{n\sigma^2}{2}\right\},$$

where $Z \sim N(0, 1)$ under $P_f^{(n)}$, and $\sigma = \|f - f_0\|_2$. Thus

$$D_K^2\left(P_f^{(n)}, P_{f_0}^{(n)}\right) = E_{P_f^{(n)}} \log \frac{dP_f^{(n)}}{dP_{f_0}^{(n)}} = \frac{n}{2}\,\|f - f_0\|_2^2. \tag{11.24}$$

Fano's Lemma As an alternative to Lemma 11.2, lower bounds for the classification error may be given in terms of Shannon's mutual information. These lower bounds are modifications of Fano's (1952) lemma. Alternative proofs are given by Ibragimov and Hasminskii (1981*b*) and Devroye (1987). Han and Verdú (1994) and Birgé (2001) give improved versions of Fano's lemma. Shannon's mutual information is defined for the set of densities $\{f_1, \ldots, f_N\}$ as

$$\begin{aligned} S &= \frac{1}{N}\sum_{i=1}^{N} D_K^2\left(P_{f_i}, \frac{1}{N}\sum_{j=1}^{N} P_{f_j}\right) \\ &= \frac{1}{N}\sum_{i=1}^{N} \int_{\mathbf{R}^d} f_i \log_e\left(\frac{f_i}{\frac{1}{N}\sum_{j=1}^{N} f_j}\right). \end{aligned} \qquad (11.25)$$

Lemma 11.7 (Fano's lemma.) *Let $\hat{\theta}$ be a random variable taking values on the collection of densities $\{f_1, \ldots, f_N\}$. Then*

$$\max_{i=1,\ldots,N} P_{f_i}(\hat{\theta} \neq f_i) \geq 1 - \frac{S + \log_e 2}{\log_e N}.$$

A Lower Bound to L_2 Error We combine Lemma 11.1 and Lemma 11.7 to get the following lower bound to the estimation error:

Lemma 11.8 *Let $\mathcal{H}_\delta \subset \mathcal{F}$ be a finite set for which $\#\mathcal{H}_\delta \geq 2$ and*

$$\min\{\|f - g\|_2 : f, g \in \mathcal{H}_\delta,\ f \neq g\} \geq \delta, \qquad (11.26)$$

where $\delta > 0$. Then

$$\inf_{\hat{f}} \sup_{f \in \mathcal{F}} E_f\|\hat{f} - f\|_2^2 \geq \frac{\delta^2}{4}\left(1 - \frac{S + \log_e 2}{\log_e(\#\mathcal{H}_\delta)}\right),$$

where S is the Shannon's mutual information for the collection $\{P_{Af}^{(n)} : f \in \mathcal{H}_\delta\}$, and the infimum is taken over all estimators (either in the density estimation model or in the Gaussian white noise model). In the density estimation model $P_{Af}^{(n)}$ is the product measure corresponding to density Af, and in the Gaussian white noise model $P_{Af}^{(n)}$ is the measure of process $Y_{n,Af}$ in (9.47).

Bounds to Shannon's Mutual Information To apply Fano's lemma, we have to give a bound to Shannon's mutual information. We have for any distribution Q that

$$S \leq \frac{1}{N}\sum_{i=1}^{N} D_K^2\left(P_{f_i}, Q\right) \leq \max_{i=1,\ldots,N} D_K^2\left(P_{f_i}, Q\right). \qquad (11.27)$$

We have also that

$$S \leq \max_{i,j=1,\ldots,N, i\neq j} D_K^2(f_i, f_j),$$

which has been applied by Birgé (1983) and Devroye (1987, p. 78).

11.2.3 Lower Bounds for the Rate of Convergence

We will give two lower bounds. The first lower bound is based on Lemma 11.2, and the second lower bound is based on Lemma 11.7. The lower bounds are given in the setting of inverse problems, where the density of the observations is Af, as in (9.18), or the signal function of the Gaussian white noise model is Af, as in (9.47). We do not have to assume that A is a linear operator.

A Lower Bound Based on Tsybakov's Lemma Theorem 11.9 below states roughly that we may construct a lower bound for the minimax rate of convergence over class $\mathcal{F}$, in the density estimation model, by finding a finite set $\mathcal{H}_\delta \subset \mathcal{F}$ that is a δ-packing set in the L_2 distance and whose elements are at most δ-apart in the Kullback–Leibler distance, up to constants. For the Gaussian white noise model we use the L_2 distance instead of the Kullback–Leibler distance. Theorem 11.9 implies that a lower bound δ_n for the minimax rate solves the equation

$$\delta_n^2 \varrho_K^2(A, \mathcal{H}_{\delta_n}) = n^{-1} \log(\#\mathcal{H}_{\delta_n}),$$

where $\varrho_K^2(A, \mathcal{H}_{\delta_n})$ is an operator norm defined in the statement of the theorem.

Theorem 11.9 *Assume that for each sufficiently small $\delta > 0$ we find a finite set $\mathcal{H}_\delta \subset \mathcal{F}$ for which*

$$\min\{\|f-g\|_2 : f,g \in \mathcal{H}_\delta,\ f \neq g\} \geq C_0\delta, \tag{11.28}$$

and

$$\begin{cases} \max\{\|f-g\|_2 : f,g \in \mathcal{H}_\delta\} \leq C_1\delta, & \textit{white noise,} \\ \max\{D_K(f,g) : f,g \in \mathcal{H}_\delta\} \leq C_1\delta, & \textit{density estimation,} \end{cases} \tag{11.29}$$

where C_0, C_1 are positive constants. Denote

$$\varrho_K(A, \mathcal{H}_\delta) = \begin{cases} \frac{1}{\sqrt{2}} \max_{f,g\in\mathcal{H}_\delta, f\neq g} \frac{\|A(f-g)\|_2}{\|f-g\|_2}, & \textit{white noise,} \\ \max_{f,g\in\mathcal{H}_\delta, f\neq g} \frac{D_K(Af,Ag)}{\|f-g\|_2}, & \textit{density estimation}\,. \end{cases} \tag{11.30}$$

Let δ_n satisfy

$$\log(\#\mathcal{H}_{\delta_n}) \not\succeq n\delta_n^2 \varrho_K^2(A, \mathcal{H}_{\delta_n}), \tag{11.31}$$

where $a_n \succeq b_n$ means that $\liminf_{n\to\infty} a_n/b_n > 0$. *Assume that*

$$\lim_{n\to\infty} n\delta_n^2 \varrho_K^2(A, \mathcal{H}_{\delta_n}) = \infty. \tag{11.32}$$

Then δ_n is a lower bound for the minimax rate of convergence:

$$\liminf_{n\to\infty} \delta_n^{-2} \inf_{\hat f} \sup_{f\in\mathcal{F}} E_f\|\hat f - f\|_2^2 > 0,$$

where the infimum is taken over all estimators (either in the density estimation model or in the Gaussian white noise model).

Proof. For $f, f_0 \in \mathcal{H}_{\delta_n}$, $f \neq f_0$,

$$\begin{aligned} P_{Af}^{(n)}\left(\frac{dP_{Af_0}^{(n)}}{dP_{Af}^{(n)}} \leq \tau\right) & \\ &\leq \left(\log \tau^{-1}\right)^{-1} D_K^2(P_{Af}^{(n)}, P_{Af_0}^{(n)}) \qquad (11.33)\\ &= \begin{cases} \left(\log \tau^{-1}\right)^{-1} nD_K^2(Af, Af_0), & \text{density estimation,} \\ \left(\log \tau^{-1}\right)^{-1} \frac{n}{2} \|Af - Af_0\|_2^2, & \text{Gaussian white noise,} \end{cases} \qquad (11.34) \end{aligned}$$

where in (11.33) we applied Lemma 11.4, and in (11.34) we applied Lemma 11.5 in the case of density estimation and (11.24) in the case of the Gaussian white noise model. When we choose

$$\tau = \tau_n = \exp\left\{-\alpha^{-1} n[C_1 \varrho_K(A, \mathcal{H}_{\delta_n})\delta_n]^2\right\}$$

for $0 < \alpha < 1$, and apply assumption (11.29), we obtain

$$\begin{aligned} P_{Af}^{(n)}\left(\frac{dP_{Af_0}^{(n)}}{dP_{Af}^{(n)}} \leq \tau\right) &\leq \left(\log \tau^{-1}\right)^{-1} n\varrho_K^2(A, \mathcal{H}_{\delta_n})\|f - f_0\|_2^2 \\ &\leq \left(\log \tau^{-1}\right)^{-1} n\left[\varrho_K(A, \mathcal{H}_{\delta_n})C_1\delta_n\right]^2 \\ &\leq \alpha. \qquad (11.35) \end{aligned}$$

Applying Lemma 11.3, assumption (11.28), and (11.35), we get the lower bound

$$\inf_{\hat{f}} \sup_{f \in \mathcal{F}} E_f\|\hat{f} - f\|_2^2 \geq \frac{(C_0\delta_n)^2}{4}(1 - \alpha)\frac{\tau_n(N_{\delta_n} - 1)}{1 + \tau_n(N_{\delta_n} - 1)}, \qquad (11.36)$$

where $N_{\delta_n} = \#\mathcal{H}_{\delta_n}$. Let n be so large that $\log_e N_{\delta_n} \geq C_2^2 n\varrho_K^2(A, \mathcal{H}_{\delta_n})\delta_n^2$, where $C_2 > C_1$. This is possible by (11.31). Then

$$\begin{aligned} \tau_n N_{\delta_n} &= \exp\left\{\log_e N_{\delta_n} - \alpha^{-1}n[C_1\varrho_K(A, \mathcal{H}_{\delta_n})\delta_n]^2\right\} \\ &\geq \exp\left\{n\varrho_K^2(A, \mathcal{H}_{\delta_n})\delta_n^2(C_2^2 - \alpha^{-1}C_1^2)\right\} \to \infty \end{aligned}$$

as $n \to \infty$, where we have applied (11.32) and chosen α so that $C_2^2 - \alpha^{-1}C_1^2 > 0$, that is, $(C_1/C_2)^2 < \alpha < 1$. Then

$$\lim_{n\to\infty} \frac{\tau_n(N_{\delta_n} - 1)}{1 + \tau_n(N_{\delta_n} - 1)} = 1,$$

and the theorem follows from (11.36). □

A Lower Bound Based on Fano's Lemma Theorem 11.10 below states roughly that we may construct a lower bound for the minimax rate of convergence over class $\mathcal{F}$, in the density estimation model, by finding a δ-packing set set $\mathcal{H}_\delta \subset \mathcal{F}$ in the L_2 distance and a δ-net $\mathcal{G}_\delta \subset \mathcal{F}$ in the Kullback–Leibler distance, up to constants. For the Gaussian white noise model we may use the L_2 distance both in choosing the δ-packing set and the δ-net. We have defined a δ-net in Definition 10.4 and a δ-packing set in Definition 10.5. Theorem 11.10 indicates that a lower bound δ_n for the minimax rate solves the equation

$$\delta_n^2 \varrho_K^2(A, \mathcal{H}_{\delta_n}, \mathcal{G}_{\delta_n}) \asymp n^{-1} \log(\#\mathcal{H}_{\delta_n}),$$

where $\mathcal{H}_\delta \subset \mathcal{F}$ is a δ-packing set of maximal cardinality and $\varrho_K^2(A, \mathcal{H}_{\delta_n}, \mathcal{G}_{\delta_n})$ is an operator norm defined below. The proof is similar to that in Yang and Barron (1999).

The interest in the theorem lies in the fact that the lower bound is stated in terms of standard δ-packing sets and δ-nets, whereas in Theorem 11.9 the lower bound is stated in terms of a local δ-packing set. However, especially in inverse problems the application of Theorem 11.9 may be easier.

Theorem 11.10 *Assume that for each sufficiently small $\delta > 0$ we find a finite set $\mathcal{H}_\delta \subset \mathcal{F}$ for which*

$$\min\{\|f - g\|_2 : f, g \in \mathcal{H}_\delta,\ \ f \neq g\} \geq C_0\delta. \tag{11.37}$$

Additionally we find a finite set $\mathcal{G}_\delta \subset \mathcal{F}$ for which

$$\begin{cases} \max_{h\in\mathcal{H}_\delta} \min_{g\in\mathcal{G}_\delta} \|f - g\|_2 \leq C_1\delta, & \textit{white noise,} \\ \max_{h\in\mathcal{H}_\delta} \min_{g\in\mathcal{G}_\delta} D_K(f, g) \leq C_1\delta, & \textit{density estimation,} \end{cases} \tag{11.38}$$

where C_0, C_1 are positive constants. Denote

$$\varrho_K(A, \mathcal{H}_{\delta_0}, \mathcal{G}_{\delta_1}) = \begin{cases} \frac{1}{\sqrt{2}} \max_{h\in\mathcal{H}_{\delta_0}} \max_{g\in\mathcal{G}_{\delta_1}} \frac{\|A(f-g)\|_2}{\|f-g\|_2}, & \textit{white noise,} \\ \max_{h\in\mathcal{H}_{\delta_0}} \max_{g\in\mathcal{G}_{\delta_1}} \frac{D_K(Af, Ag)}{\|f-g\|_2}, & \textit{density estimation}\,. \end{cases}$$

Let δ_0, δ_1, $\mathcal{H}_{\delta_0}$, and $\mathcal{G}_{\delta_1}$ be such that

$$\frac{n\left[\varrho_K(A, \mathcal{H}_{\delta_0}, \mathcal{G}_{\delta_1}) C_1 \delta_1\right]^2 + \log_e(\#\mathcal{G}_{\delta_1}) + \log_e 2}{\log_e(\#\mathcal{H}_{\delta_0})} \leq \frac{1}{2}. \tag{11.39}$$

Then δ_0 is a lower bound for the minimax rate of convergence:

$$\inf_{\hat f} \sup_{f\in\mathcal{F}} E_f \|\hat f - f\|_2^2 \geq \frac{C_0^2 \delta_0^2}{8},$$

where the infimum is taken over all estimators (either in the density estimation model or in the Gaussian white noise model).

Proof. Let

$$Q = \frac{1}{\#\mathcal{G}_{\delta_1}} \sum_{g\in\mathcal{G}_{\delta_1}} P_{Ag}^{(n)}.$$

For $h \in \mathcal{H}_{\delta_0}$, let $g_h \in \mathcal{G}_{\delta_1}$ be such that

$$\begin{cases} \|h - g_h\|_2 = \min_{g\in\mathcal{G}_{\delta_1}} \|h - g\|_2, & \text{white noise,} \\ D_K(h, g_h) = \min_{g\in\mathcal{G}_{\delta_1}} D_K(h, g), & \text{density estimation.} \end{cases}$$

For $h \in \mathcal{H}_{\delta_0}$,

$$D_K^2\left(P_{Ah}^{(n)}, Q\right) \le D_K^2\left(P_{Ah}^{(n)}, dP_{Ag_h}^{(n)}\right) + \log_e(\#\mathcal{G}_{\delta_1})$$

and

$$D_K^2\left(P_{Ah}^{(n)}, dP_{Ag_h}^{(n)}\right) = \begin{cases} \frac{n}{2}\left\|Ah - Ag_h\right\|_2^2, & \text{white noise,} \\ nD_K^2(Ah, Ag_h), & \text{density estimation,} \end{cases}$$

where we have applied Lemma 11.5 in the case of density estimation and (11.24) in the case of the Gaussian white noise model. Thus, using (11.27), we have

$$\begin{aligned} S &\le \max_{h \in \mathcal{H}_{\delta_0}} D_K^2\left(P_{Ah}^{(n)}, Q\right) \\ &\le n\left[\varrho_K(A, \mathcal{H}_{\delta_0}, \mathcal{G}_{\delta_1})C_1\delta_1\right]^2 + \log_e(\#\mathcal{G}_{\delta_1}). \end{aligned}$$

The theorem follows by (11.39) and Lemma 11.8. □

11.3 LOWER BOUNDS FOR SMOOTHNESS CLASSES

We prove lower bounds for the rate of convergence in the sense of equation (11.4).

11.3.1 Sobolev Spaces and Anisotropic Besov Spaces

We consider here both the Sobolev ball $S_s(L)$ and the anisotropic Besov ball $B_{sp\infty}(L)$. We state lower bounds for the density estimation model and thus consider the intersection of the smoothness classes with the class of densities

$$\mathcal{F}_{den} = \left\{f : \int_{\mathbf{R}^d} f = 1, f \ge 0\right\}.$$

The lower bounds hold obviously also in the Gaussian white noise model.

Corollary 11.11 *Let the Sobolev ball $S_s(L)$ be defined in (10.24) for $s, L > 0$. Let the anisotropic Besov ball $B_{sp\infty}(L)$ be defined in (10.41) for $s = (s_1, \ldots, s_d)$, $s_i > 0$, $p = (p_1, \ldots, p_d)$, $1 \le p_i \le \infty$, and $L > 0$. Then*

$$\liminf_{n \to \infty} \psi_n^{-2} \inf_{\hat{f}} \sup_{f \in \mathcal{F}} E\left\|f - \hat{f}\right\|_2^2 > 0,$$

where the infimum is taken over all estimators in the density estimation model or in the Gaussian white noise model and

1. *when $\mathcal{F} = S_s(L) \cap \mathcal{F}_{den}$, then the rate is*

$$\psi_n = n^{-s/(2s+d)},$$

2. *when* $\mathcal{F} = B_{sp\infty}(L) \cap \mathcal{F}_{den}$, *then the rate is*

$$\psi_n = n^{-\sigma/(2\sigma+1)}, \qquad (11.40)$$

where σ *is defined in (10.81).*

Proof. We apply Theorem 11.9. Let the packing set $\mathcal{H}_\delta$ be defined by (10.85). Since $\mathcal{H}_\delta$ is a packing set, condition (11.28) is satisfied. We have for $f_a, f_{a'} \in \mathcal{H}_\delta$,

$$\begin{aligned} \|f_a - f_{a'}\|_2^2 &= \delta^{(2\sigma+1)/\sigma} \sum_{k \in I} (a_k - a'_k)^2 \|G_k\|_2^2 \\ &\le \delta^{(2\sigma+1)/\sigma} \, \|G\|_2^2 \, 2\#I \\ &\le \delta^2 \|G\|_2^2 2, \end{aligned}$$

since $\|G_k\|_2^2 = \|G\|_2^2$, an upper bound for $\sum_{k\in I}(a_k - a'_k)^2$ is the case where a and a' differ in all I locations, and

$$\#I = \log(\#\mathcal{H}_\delta) \sim \delta^{-1/\sigma}$$

(see (10.84)). Thus condition (11.29) is satisfied. Conditions (11.31) and (11.32) are satisfied because ψ_n satisfies the equation

$$\log(\#\mathcal{H}_{\psi_n}) \asymp \psi_n^{-1/\sigma} \asymp n\psi_n^2.$$

We have proved the corollary. □

Upper Bounds We give a list of upper bounds complementing the lower bounds of this section.

1. An asymptotically exact minimax estimator for the anisotropic periodic Sobolev class was constructed by Nussbaum (1986). The asymptotically minimax estimator was an orthogonal series estimators with the Fourier basis.

2. Theorem 14.4 shows that a kernel estimator can achieve the rate in the lower bound for anisotropic Besov spaces.

3. Corollary 15.4 and Corollary 15.5 give a rate-optimal upper bound for the case of Sobolev spaces using a δ-net estimator.

4. Theorem 16.6 and Theorem 16.7 show that wavelet estimators can achieve the lower bound for anisotropic Besov spaces.

5. In Theorem 18.2 we show that a series estimator with Haar basis functions achieves the lower bound, up to a logarithmic factor, over a scale of anisotropic Besov classes.

11.3.2 Functions with Dominating Mixed Derivatives

We consider the space $D_s(L)$ of functions with dominating mixed derivatives. We prove a lower bound complementing the upper bound of Theorem 16.8, where it is shown that a wavelet estimator can achieve the rate in the lower bound.

Corollary 11.12 *Let $\mathcal{F} = D_s(L) \cap \mathcal{F}_{den}$, where $D_s(L)$ is defined in (10.48) for $s \geq 1$ integer, $L > 0$. Let*

$$\psi_n = \left(\frac{n}{(\log n)^{d-1}} \right)^{-s/(2s+1)} . \tag{11.41}$$

Then

$$\liminf_{n\to\infty} \psi_n^{-2} \inf_{\hat{f}} \sup_{f\in\mathcal{F}} E \left\| f - \hat{f} \right\|_2^2 > 0,$$

where the infimum is taken over all estimators in the density estimation model or in the Gaussian white noise model.

Proof. We apply Theorem 11.9 as in the proof of Corollary 11.11. Let the packing set $\mathcal{H}_\delta$ be defined by (10.86). The satisfaction of conditions (11.28) and (11.29) is checked as in the proof of Corollary 11.11. Conditions (11.31) and (11.32) are satisfied because ψ_n in (11.41) satisfies the equation

$$\log(\#\mathcal{H}_{\psi_n}) \asymp \psi_n^{-1/s} (\log \psi_n^{-1})^{d-1} \asymp n\psi_n^2,$$

where we applied Theorem 10.12. □

11.3.3 Inverse Problems

To prove a lower bound in the case of inverse problems, we need to have bounds for the operator norm $\varrho_K(A, \mathcal{H}_\delta)$, as defined in (11.30). We will calculate the operator norm with the L_2 norm. Lemma 11.6 can be used to derive results for the Kullback–Leibler distance, and thus to derive results for the case of density estimation, assuming that the densities are bounded above and bounded away from zero.

Operator Norms We calculate the operator norm $\varrho_K(A, \mathcal{H}_\delta)$ when the function class $\mathcal{F}$ and its local δ-packing $\mathcal{H}_\delta$ are given by

$$\mathcal{F} = \left\{ \sum_{j=1}^{\infty} \theta_j \phi_j : \theta \in \Theta \right\}, \qquad \mathcal{H}_\delta = \left\{ \sum_{j=1}^{\infty} \theta_j \phi_j : \theta \in \Theta_\delta^* \right\}, \tag{11.42}$$

where Θ is the ellipsoid defined in (10.59) and Θ_δ^* is given in (10.73). We consider two cases.

1. Functions ϕ_j are the singular functions of the operator A, as given in (B.13). Let b_j be the singular values in (B.15). We assume that there exists positive constants q and C, C' such that for all $j \in \{1, 2, \ldots\}$,

$$C j^{-q} \leq b_j \leq C' j^{-q}. \tag{11.43}$$

The functions $A\phi_j$ are orthogonal and $\|A\phi_j\|_2 = b_j$. Indeed

$$\langle A\phi_j, A\phi_l\rangle = \langle \phi_j, A^* A\phi_l\rangle = b_l^2 \langle \phi_j, \phi_l\rangle.$$

Thus, when $g, g' \in \mathcal{H}_\delta$ and we write $g - g' = \sum_{j=M^*}^{M} (\theta_j - \theta_j')\phi_j$, we have

$$\|A(g - g')\|_2^2 = \left\| \sum_{j=M^*}^{M} (\theta_j - \theta_j')^2 A\phi_j \right\|_2^2 = \sum_{j=M^*}^{M} (\theta_j - \theta_j')^2 b_j^2, \qquad (11.44)$$

since in this case for all $\theta \in \Theta_\delta^*$, $\theta_j = \theta_j^*$ when $j \notin \{M^*, \dots, M\}$, where θ^* is a fixed sequence with $\sum_{j=0}^{\infty} a_j^2 \theta_j^{*2} = L^* < L$,

$$M^* = [M/2], \qquad M \asymp \delta^{-1/s}. \qquad (11.45)$$

On the other hand, $\|g - g'\|_2 = \|\theta - \theta\|_{l_2}$. Combining (11.43), (11.44), and (11.45) gives bounds for the operator norm:

$$C\delta^{q/s} \le C' M^{-q} \le \varrho_K(A, \mathcal{H}_\delta) \le C''(M^*)^{-q} \le C''' \delta^{q/s}. \qquad (11.46)$$

2. Functions $\phi_j = \phi_{jk}$ are wavelet basis functions, and we use the wavelet-vaguelette system as given in (B.17). The ellipsoid is defined in (10.37) taking $p = q = 2$. Let b_j be the quasi-singular values in (B.16). We assume that there exists positive constants q and C, C' such that for all $j \in \{1, 2, \dots\}$

$$C 2^{-jq} \le b_j \le C' 2^{-jq}. \qquad (11.47)$$

From (B.16) we get $A\phi_{jk} = b_j v_{jk}$, where (v_{jk}) is a near-orthogonal system. Thus

$$\langle A\phi_{jk}, A\phi_{j'k'}\rangle = b_j b_{j'} \langle v_{jk}, v_{j'k'}\rangle.$$

Then, when $g, g' \in \mathcal{H}_\delta$ and we write $g - g' = \sum_{j=J^*}^{J} \sum_k (\theta_{jk} - \theta_{jk}')\phi_{jk}$,

$$\|A(g - g')\|_2^2 \asymp \sum_{j=J^*}^{J} b_j^{-2} \sum_k (\theta_{jk} - \theta_{jk}')^2. \qquad (11.48)$$

This is because in this case for $\theta \in \Theta_\delta^*$, $\theta_{jk} = \theta_{jk}^*$ when $j \le J^*$ or $j \ge J + 1$; here θ^* is a fixed sequence with $\sum_{j=0}^{\infty} \sum_k a_{jk}^2 {\theta_{jk}^*}^2 = L^* < L$,

$$J^* = J - 1, \qquad 2^J \asymp \delta^{-1/s}. \qquad (11.49)$$

On the other hand, $\|g - g'\|_2 = \|\theta - \theta\|_{l_2}$. Combining (11.47), (11.48), and (11.49) gives bounds for the operator norm:

$$C\delta^{q/s} \le C' 2^{-Jq} \le \varrho_K(A, \mathcal{H}_\delta) \le C'' 2^{-J^* q} \le C''' \delta^{q/s}.$$

Convolution Let A be a convolution operator: $Af = a * f$ where $a : \mathbf{R}^d \to \mathbf{R}$ is a known function. Let the collection (ϕ_{jk}), $(j,k) \in \{0,1,\ldots\}^d \times K_j$, of trigonometric polynomials be defined in (10.29). When the convolution kernel a is a 1-periodic function in $L_2([0,1]^d)$, we can write the Fourier series of a with the Fourier coefficients b_{jk} as in (9.23). The functions ϕ_{jk} are the singular functions of the operator A, and the values b_{jk} are the corresponding singular values. We assume that the underlying function space is given by

$$\mathcal{F} = \left\{ \sum_{j_1=0,\ldots,j_d=0}^{\infty} \sum_{k \in K_j} \theta_{jk}\phi_{jk}(x) : (\theta_{jk}) \in \Theta \right\}, \tag{11.50}$$

where

$$\Theta = \left\{ \theta : \sum_{j_1=0,\ldots,j_d=0}^{\infty} \sum_{k \in K_j} a_{jk}^2 \theta_{jk}^2 \le L^2 \right\}. \tag{11.51}$$

We give the rate of convergence of the δ-net estimator. The corresponding upper bound is given in Theorem 14.6, where a kernel estimator is used and in Corollary 15.4, where a δ-net estimator is used.

Corollary 11.13 *Let $\mathcal{F}$ be defined in (11.50). We assume that the coefficients of the ellipsoid (11.51) satisfy*

$$C|j|^s \le a_{jk} \le C'|j|^s$$

for some $s > 0$ and $C, C' > 0$. We assume that the convolution filter a is 1-periodic function in $L_2([0,1]^d)$ and that the Fourier coefficients of filter a satisfy

$$C|j|^{-q} \le b_{jk} \le C'|j|^{-q}$$

for some $q \ge 0$, $C, C' > 0$. Then

$$\liminf_{n\to\infty} n^{2s/(2s+2q+d)} \inf_{\hat{f}} \sup_{f\in\mathcal{F}} E\left\|\hat{f} - f\right\|_2^2 > 0,$$

where the infimum is taken over all estimators, either in the model of inverse density estimation defined in (9.17) or in the inverse Gaussian white noise model defined in (9.47).

Proof. We apply Theorem 11.9. Assumption (11.28) holds because $\mathcal{H}_\delta$ in (11.42) is a δ-packing set. Assumption (11.29) holds by the construction; see (10.74). Satisfaction of (11.31) and (11.32) follows from (10.71) and (11.46). □

Radon Transform We consider the 2D Radon transform as defined in (9.42). The singular value decomposition is given in Section 9.3.4, page 248. The singular functions are denoted with $\phi_{j_1 j_2}$, and the singular values are given in (9.43). We assume that the underlying function space is given by

$$\mathcal{F} = \left\{ \sum_{j_1=0,j_2=0,(j_1,j_2)\neq(0,0)}^{\infty} \theta_{j_1 j_2}\phi_{j_1 j_2}(x) : (\theta_{j_1 j_2}) \in \Theta \right\}, \tag{11.52}$$

where

$$\Theta = \left\{ \theta : \sum_{j_1=0, j_2=0, (j_1,j_2)\neq(0,0)}^{\infty} a_{j_1 j_2}^2 \theta_{j_1 j_2}^2 \leq L^2 \right\}. \tag{11.53}$$

We give a lower bound to the rate of convergence of any estimator. The corresponding upper bound is given in Theorem 14.7 for a kernel estimator and in Corollary 15.5 for a δ-net estimator. Since the singular functions are given only for the two-dimensional case, the corollary holds only in the two-dimensional case.

Corollary 11.14 *Let $\mathcal{F}$ be the function class as defined in (11.52). We assume that the coefficients of the ellipsoid (11.53) satisfy*

$$C|j| \leq a_{jk} \leq C'|j|^s$$

for some $s > 0$ and $C > 0$. Then, for $d = 2$,

$$\liminf_{n\to\infty} n^{2s/(2s+2d-1)} \inf_{\hat{f}} \sup_{f\in\mathcal{F}} E\left\|\hat{f} - f\right\|_2^2 > 0,$$

where the infimum is taken over all estimators, either in the model of inverse density estimation defined in (9.17) or in the inverse Gaussian white noise model defined in (9.47).

Proof. The proof is similar to the proof of Corollary 11.13, since the singular values in (9.43) satisfy $b_j \asymp |j|^{-1/2}$. □

11.4 BIBLIOGRAPHIC NOTES

We have proved lower bounds as we did in Klemelä and Mammen (2008), where the approach of Hasminskii and Ibragimov (1990) is combined with that of Tsybakov (1998, th. 6), and the lower bounds are generalized to the case of inverse problems.

Fano's lemma was used to derive lower bounds by Samarov (1976), Ibragimov and Hasminskii (1977), Ibragimov and Hasminskii (1978), and Hasminskii (1978). The connection with the minimax rates and the concepts of capacity and entropy was developed by Cencov (1972), Le Cam (1973), Ibragimov and Hasminskii (1980), Ibragimov and Hasminskii (1981*a*), Birgé (1983), Hasminskii and Ibragimov (1990), Yang and Barron (1999), and Ibragimov (2004).

An alternative approach, based on Assouad's lemma (hypercube approach), and less general than the approach based on Fano's lemma, was developed in Bretagnolle and Huber (1979), Assouad (1983). Golubev and Nussbaum (1990), Golubev (1991), Low (1993), and Klemelä (2003) apply the theory of convergence of experiments in the connection of the hypercube approach.

The previous approaches are suitable when deriving lower bounds for the estimation of the whole function, for example, when the L_p losses are applied. The problem of deriving a lower bound when estimating a real-valued functional of a density is a simpler one. Farrell (1972) considered the problem when estimating density at one point. Donoho and Low (1992) discuss general principles in determining the rates of convergence of linear functionals, for the Gaussian white noise model. Donoho and Liu (1991) discuss special issues when deriving lower bounds for estimating a density function and its derivatives at one point. They derive lower bounds by considering convergence of one-dimensional subexperiments. Klemelä (2003) gives an alternative approach based on parameter-local Gaussian limit experiments.

EXERCISES

11.1 *(Loss functions.)* Prove equation (11.5).

11.2 *(Loss functions.)* Prove that for any densities f, g,

$$D_H^2(f,g) \le \int_{\mathbf{R}^d} |f-g| \le 2D_H(f,g),$$

where D_H is the Hellinger distance defined in (11.6).

11.3 *(Loss functions.)* Prove that when the densities f, g are bounded and bounded away zero, then

$$C \cdot D_H^2(f,g) \le \int_{\mathbf{R}^d} (f-g)^2 \le C' \cdot D_H^2(f,g),$$

for positive constants C, C'.

11.4 *(Methods to prove lower bounds.)* Prove Lemma 11.5.

11.5 *(Methods to prove lower bounds.)* Prove Lemma 11.6.

CHAPTER 12

EMPIRICAL PROCESSES

The theory of empirical processes will be applied in Chapter 15 to analyze minimization estimators. It turns out to be extremely useful to study simultaneously density estimation and the Gaussian white noise model, introduced in Section 9.4.3. When we write the proofs for the Gaussian white noise model, the arguments become more transparent, and we can isolate the technical parts of the theory of density estimation. We need to analyze the centered empirical operator, defined by

$$\nu_n(g) = \begin{cases} n^{-1}\sum_{i=1}^n g(X_i) - \int_{\mathbf{R}^d} gf, & \text{density estimation,} \\ \int g\, dX_n - \int_{\mathbf{R}^d} gf, & \text{white noise model,} \end{cases} \qquad (12.1)$$

where $g : \mathbf{R}^d \to \mathbf{R}$. In the model of density estimation $X_1, \ldots, X_n$ are i.i.d. with density $f : \mathbf{R}^d \to \mathbf{R}$ and $\int_{\mathbf{R}^d} gf = E_f g(X_1)$. The Gaussian white noise model is defined in (9.45). We study the quantities

$$Z = \sup_{g\in\mathcal{G}} \nu_n(g), \qquad Z = \sup_{g\in\mathcal{G}} |\nu_n(g)|, \qquad (12.2)$$

where $\mathcal{G}$ is a collection of functions $\mathbf{R}^d \to \mathbf{R}$. We give bounds for the tail probabilities $P(Z > \xi)$ and for the expectation EZ.

The setting of inverse problems was introduced in Section 9.3.4. In the inverse problems one needs to study the random variable $\nu_n(Qg)$, where $Q = (A^{-1})^*$ is the adjoint of the inverse of the original operator A, and the centered empirical operator is defined by

$$\nu_n(h) = \begin{cases} n^{-1}\sum_{i=1}^n h(Y_i) - \int_{\mathbf{Y}} h(Af), & \text{density estimation,} \\ \int h\,dY_n - \int_{\mathbf{Y}} h(Af), & \text{white noise model,} \end{cases} \tag{12.3}$$

where $h : \mathbf{Y} \to \mathbf{R}$. In the model of density estimation $Y_1, \ldots, Y_n$ are i.i.d. with density $Af : \mathbf{Y} \to \mathbf{R}$ and $\int_{\mathbf{Y}} h(Af) = E_{Af}h(Y_1)$. The Gaussian white noise model for inverse problems is defined in (9.47).

12.1 EXPONENTIAL INEQUALITIES

12.1.1 Bernstein's Inequality

For a Gaussian random variable $W \sim N(0, \sigma^2)$ we have the tail bound, for $\xi > 0$,

$$P(W > \xi) \le 2^{-1} \exp\left\{-\frac{\xi^2}{2\sigma^2}\right\}. \tag{12.4}$$

For a proof of (12.4), see Exercise 12.1 or Dudley (1999, prop. 2.2.1). Let $X_1, \ldots, X_n \in \mathbf{R}$ be i.i.d. random variables with variance $\mathrm{Var}(X_1) = \sigma^2$ and denote $\nu_n = \frac{1}{n}\sum_{i=1}^n X_i - EX_1$. We have by the central limit theorem that $P(n^{1/2}\nu_n > \xi) \to P(W > \xi)$, as $n \to \infty$. Thus

$$P(\nu_n > \xi) \approx P(W > n^{1/2}\xi) \le 2^{-1}\exp\{-n\xi^2/(2\sigma^2)\}. \tag{12.5}$$

Bernstein's inequality states an analogous nonasymptotical inequality. In Bernstein's inequality the bound depends both on the variance and on the range of the random variables.

Theorem 12.1 (Bernstein's inequality.) *Let $X_1, \ldots, X_n \in \mathbf{R}$ be i.i.d. random variables with variance* $\mathrm{Var}(X_1) = \sigma^2$ *and with bounded range: $X_i - EX_1 \in [-b, b]$. Then for $\xi > 0$,*

$$P\left(\frac{1}{n}\sum_{i=1}^n X_i - EX_1 \ge \xi\right) \le \exp\left\{-\frac{1}{2}\frac{n\xi^2}{\sigma^2 + b\xi/3}\right\}.$$

A proof of Theorem 12.1 is given in Appendix E.2.1.

Unlike in the Gaussian case of (12.4), Bernstein's inequality contains the additional term $2b\xi/3$. For small ξ we get an upper bound equal to $\exp\{-n\xi^2/(4\sigma^2)\}$, and for large ξ we get an upper bound equal to $\exp\{-n\xi/(4b/3)\}$. Indeed for $\xi > 0$,

$$\exp\left\{-\frac{1}{2}\frac{n\xi^2}{\sigma^2 + b\xi/3}\right\} \le \begin{cases} \exp\left\{-\frac{1}{2}\frac{n\xi^2}{2\sigma^2}\right\}, & \text{when } \xi \le \sigma^2/(b/3) \\ \exp\left\{-\frac{1}{2}\frac{n\xi}{2b/3}\right\}, & \text{when } \xi \ge \sigma^2/(b/3). \end{cases}$$

12.1.2 Borell's and Talagrand's Inequality

Bernstein's inequality is an exponential inequality for a single average of i.i.d. random variables. However, we need also uniform bounds, which give an exponential inequality for a supremum of averages. It is helpful again to start with the Gaussian case: we state the Borell's inequality. Borell's inequality gives a deviation bound for the supremum of a Gaussian process. The deviation is measured from the mean of the supremum.

Theorem 12.2 (Borell's inequality.) *Let X_t be a separable Gaussian stochastic process, indexed by a semimetric space T. Assume that $EX_t = 0$ for all $t \in T$. Denote $Z = \sup_{t\in T} |X_t|$ and $v = \sup_{t\in T} \mathrm{Var}(X_t)$. Then for $\xi > 0$,*

$$P(Z \geq EZ + \xi) \leq 2\exp\left\{-\frac{\xi^2}{2v}\right\}.$$

A proof of Borell's inequality may be found, for example, in van der Vaart and Wellner (1996, prop. A.2.1).

Talagrand's inequality is analogous to the Borell's inequality but it considers a sum of i.i.d. random vectors.

Theorem 12.3 (Talagrand's inequality.) *Let $X_1, \ldots, X_n \in \mathbf{R}^d$ be i.i.d. random vectors. Let $\mathcal{G}$ be a countable set of functions $\mathbf{R}^d \to \mathbf{R}$. Let v and b satisfy*

$$\sup_{g\in\mathcal{G}} \mathrm{Var}(g(X_1)) \leq v, \qquad \sup_{g\in\mathcal{G}} \|g\|_\infty \leq b.$$

Then for $\xi > 0$,

$$P(Z \geq EZ + \xi) \leq \exp\left\{-\frac{1}{2}\frac{n\xi^2}{v + 2bEZ + b\xi/3}\right\}, \tag{12.6}$$

where $Z = \sup_{g\in\mathcal{G}} |\nu_n(g)|$.

Theorem 12.3 follows from Bousquet (2002, thm. 2.3), by applying the inequality $h(t) \geq t^2/(2 + 2t/3)$, $t > 0$, for $h(t) = (1+t)\log(1+t) - t$.

In Talagrand's inequality we do not need entropy conditions, unlike in the inequalities presented in Section 12.1.3 where chaining is applied. However, entropy conditions can be useful in obtaining a bound for the expectation EZ. Bounds for the expectation of the supremum are given in Section 12.2.

12.1.3 Chaining

When $\mathcal{D} \subset \mathcal{G}$ is a finite set, we get from Bernstein's inequality

$$\begin{aligned} P\left(\max_{g\in\mathcal{D}} \nu_n(g) \geq \xi\right) &\leq \#\mathcal{D} \cdot \max_{g\in\mathcal{D}} P(\nu_n(g) \geq \xi) \\ &\leq \exp\left\{-\frac{1}{2}\frac{n\xi^2}{v + b\xi/3} + \log_e(\#\mathcal{D})\right\}, \end{aligned}$$

where $v = \sup_{g\in\mathcal{D}} \mathrm{Var}(g(X_1))$, $b = \sup_{g\in\mathcal{G}} \|g\|_\infty$. This suggests that we can get useful uniform exponential bounds when $\mathcal{G}$ is approximated by a finite set $\mathcal{D}$ of sufficiently small cardinality. These arguments use a chaining technique, and they rely on entropy conditions. The chaining technique was developed by Kolmogorov. The modern theory of chaining can be found in Talagrand (2005)

The Gaussian Case Borell's inequality was stated for a Gaussian stochastic process X_t, where t belongs to a semi-metric space. As a special case we study the stochastic process $\nu_n(g)$, $g \in \mathcal{G}$, defined in (12.1). Borell's inequality gives a deviation bound from the expectation of the supremum, and its application requires bounds for this expectation. In particular, a chaining argument can be applied to bound the expectation of the supremum. We apply a chaining argument to give directly an exponential inequality. We cover the inverse case.

Let $\mathcal{G} \subset L_2(\mathbf{R}^d)$ be such that $\sup_{g\in\mathcal{G}} \|g\|_2 \le R$, and denote by $\mathcal{G}_\delta$ a δ-net of $\mathcal{G}$, $\delta > 0$. A δ-net was defined in Definition 10.4. Define

$$\varrho(Q, \mathcal{G}_\delta) = \max\left\{ \frac{\|Q(f-g)\|_2}{\|f-g\|_2} : f \in \mathcal{G}_\delta, g \in \mathcal{G}_{2\delta}, f \ne g \right\}, \qquad \delta > 0, \quad (12.7)$$

where $Q = (A^{-1})^*$ is the adjoint of the inverse of A, defined through (15.5). Define the entropy integral

$$G(\delta) \stackrel{def}{=} \int_0^\delta \varrho(Q, \mathcal{G}_u)\sqrt{\log_e(\#\mathcal{G}_u)}\, du, \qquad \delta \in (0, R]. \qquad (12.8)$$

Theorem 12.4 *We consider the Gaussian white noise model (9.47) with linear operator A. Assume that*

1. $\delta \mapsto \varrho(Q, \mathcal{G}_\delta)\sqrt{\log_e(\#\mathcal{G}_\delta)}$ *is decreasing on* $(0, R]$,
2. *the entropy integral $G(R)$ is finite,*
3. $\varrho(Q, \mathcal{G}_\delta) = c\delta^{-a}$, *where $0 \le a < 1$ and $c > 0$.*

Then for all

$$\xi \ge n^{-1/2}\,\tilde{G}(R), \qquad \tilde{G}(R) = \max\left\{24\sqrt{2}G(R), cR^{1-a}\sqrt{\log_e 2/C'}\right\}, \quad (12.9)$$

where

$$C' = 12^{-2}(C'')^{-2}, \qquad C'' = (1-a)^{-3/2}\Gamma(3/2)(\log_e 2)^{-3/2}, \qquad (12.10)$$

we have

$$P\left(\sup_{g\in\mathcal{G}} \nu_n(Qg) \ge \xi\right) \le \exp\left\{-\frac{n\xi^2 C'}{c^2 R^{2-2a}}\right\},$$

where ν_n is the centered empirical operator of the Gaussian white noise process, defined in (12.3).

A proof of Theorem 12.4 is given in Appendix E.2.2. Lemma 12.4 is analogous to Lemma 3.2 in van de Geer (2000), which considers the direct case.

Remark We may write $\varrho(Q, \mathcal{F}_\delta)$ in a more transparent way when there exist minimal δ-nets $\mathcal{G}_\delta$ that are nested:

$$\mathcal{G}_{2\delta} \subset \mathcal{G}_\delta .$$

Then we may define alternatively

$$\varrho(Q, \mathcal{G}_\delta) = \max_{f,g \in \mathcal{G}_\delta, f \neq g} \frac{\|Q(f-g)\|_2}{\|f-g\|_2} .$$

The i.i.d. Case The basic chaining argument in the i.i.d. case is similar to the Gaussian case. Now we need to apply Bernstein's inequality and this leads to a chaining technique with truncation.

Let $\mathcal{G} \subset L_2(\mathbf{R}^d)$ be such that $\sup_{g \in \mathcal{G}} \|g\|_2 \le R$. Let $\mathcal{G}_\delta$ be a δ-bracketing net of $\mathcal{G}$, $\delta > 0$. the δ-bracketing net was defined in Definition 10.6. Let us recall that a δ-bracketing net of $\mathcal{G}$ with respect to the L_2 norm is a set of pairs of functions $\mathcal{G}_\delta = \{(g_j^L, g_j^U) : j = 1, \ldots, N_\delta\}$ such that

1. $\|g_j^L - g_j^U\|_2 \le \delta$, $j = 1, \ldots, N_\delta$,
2. for each $g \in \mathcal{G}$ there is $j = j(g) \in \{1, \ldots, N_\delta\}$ such that $g_j^L \le g \le g_j^U$.

Let us denote $\mathcal{G}_\delta^L = \{g_j^L : j = 1, \ldots, N_\delta\}$. Define

$$\varrho(Q, \mathcal{G}_\delta) = \max\left\{\varrho^{(1)}(Q, \mathcal{G}_\delta), \varrho^{(2)}(Q, \mathcal{G}_\delta)\right\}, \tag{12.11}$$

where $Q = (A^{-1})^*$ is the adjoint of the inverse of A, defined by (15.5),

$$\varrho^{(1)}(Q, \mathcal{G}_\delta) = \max\left\{\frac{\|Q(g^U - g^L)\|_2}{\|g^U - g^L\|_2} : (g^L, g^U) \in \mathcal{G}_\delta\right\}, \qquad \delta > 0$$

and

$$\varrho^{(2)}(Q, \mathcal{G}_\delta) = \max\left\{\frac{\|Q(f-g)\|_2}{\|f-g\|_2} : f \in \mathcal{G}_\delta^L, g \in \mathcal{G}_{2\delta}^L, f \neq g\right\}, \qquad \delta > 0.$$

The entropy integral takes the form

$$G(\delta) \stackrel{def}{=} \int_0^\delta \varrho(Q, \mathcal{G}_u)\sqrt{\log_e(\#\mathcal{G}_u)}\, du, \qquad \delta \in (0, R]. \tag{12.12}$$

Theorem 12.5 *Let $Y_1, \ldots, Y_n \in \mathbf{Y}$ be i.i.d. with density Af, where A is linear. Assume that*

1. $\|Af\|_\infty \le B_\infty$, *where* $0 < B_\infty < \infty$,
2. *denote* $\mathcal{G}_\delta^L = \{g^L : (g^L, g^U) \in \mathcal{G}_\delta\}$ *and* $\mathcal{G}_\delta^U = \{g^U : (g^L, g^U) \in \mathcal{G}_\delta\}$ *and assume that* $\sup_{g \in \mathcal{G}_R^L \cup \mathcal{G}_R^U} \|Qg\|_\infty \le B'_\infty$, *where* $0 < B'_\infty < \infty$,

3. $\delta \mapsto \varrho(Q, \mathcal{G}_\delta)\sqrt{\log_e(\#\mathcal{G}_\delta)}$ *is decreasing on* $(0, R]$,

4. *entropy integral* $G(R)$ *is finite,*

5. $\varrho(Q, \mathcal{G}_\delta) = c\delta^{-a}$, *where* $0 \le a < 1$ *and* $c > 0$,

6. *operator* Q *preserves positivity* ($g \ge 0$ *implies* $Qg \ge 0$).

Then for all

$$\xi \ge n^{-1/2}\,\tilde{G}(R), \tag{12.13}$$

where

$$\begin{aligned}\tilde{G}(R) &= B_\infty^{1/2}(9^2 + 96 \cdot 2^{-2a})^{1/2} \\ &\quad \times \max\left\{24\sqrt{2}G(R),\ 4(\log_e(2))^{-1}(1-a)^{-3/2}\Gamma(3/2)cR^{1-a}\right\},\end{aligned} \tag{12.14}$$

we have

$$\begin{aligned}&P\left(\sup_{g\in\mathcal{G}} \nu_n(Qg) \ge \xi\right) \\ &\le\ 4\exp\left\{-\frac{n\xi^2 C'}{B_\infty c^2 R^{2-2a}}\right\} + 2\#\mathcal{G}_R \exp\left\{-\frac{1}{12}\,\frac{n\xi^2}{B_\infty c^2 R^{2(1-a)} + 2\xi B'_\infty/9}\right\},\end{aligned}$$

where ν_n *is the centered empirical operator of the density estimation, defined in (12.3).*

A proof of Theorem 12.5 is given in Appendix E.2.3. The proof uses techniques developed in Bass (1985), Ossiander (1987), Birgé and Massart (1993, prop. 3), van de Geer (2000, thm. 8.13).

We note finally that the second term in the upper bound of Theorem 12.5 may be bounded by a similar term as the first term in the upper bound of Theorem 12.5.

Lemma 12.6 *Under the assumptions of Theorem 12.5, when in addition* ξ *satisfies*

$$2\sqrt{44\log_e(\#\mathcal{G}_R)}B_\infty^{1/2}cR^{1-a}n^{-1/2} \le \xi \le B_\infty c^2 R^{2(1-a)}/B'_\infty, \tag{12.15}$$

we have

$$\begin{aligned}&\#\mathcal{G}_R \exp\left\{-\frac{1}{12}\,\frac{n\xi^2}{B_\infty c^2 R^{2(1-a)} + 2\xi B'_\infty/9}\right\} \\ &\qquad\qquad \le \exp\left\{-\frac{n\xi^2 C'}{B_\infty c^2 R^{2(1-a)}}\right\}.\end{aligned} \tag{12.16}$$

A proof of Lemma 12.6 is given in Appendix E.2.4.

12.2 BOUNDS FOR THE EXPECTATION

Theorem 12.3 involves term EZ, where $Z = \sup_{g \in \mathcal{G}} |\nu_n(g)|$. To apply Theorem 12.3, we need to bound term EZ. We also need to bound term EZ when we analyze the mean integrated squared error of a stagewise minimization estimator in Section 19.2.3.

We start with bounds for the expectation of a maximum over a finite set and over an L_2 ball. Then we study the case of an infinite set and use chaining arguments for the Gaussian case and reduce the i.i.d. case to Gaussian by symmetrization. The bounds hold under entropy conditions or under conditions on VC dimension. Finally, we use the exponential inequalities to bound the expectation of a supremum.

12.2.1 Finite Set

We give a bound for the expectation of a maximum over a finite set.

Lemma 12.7 *Let $Y_1, \ldots, Y_k \in \mathbf{R}$, $k \geq 2$, be random variables such that for all $s > 0$, $i = 1, \ldots, k$,*

$$E\exp\{sY_i\} \leq \exp\left\{\frac{s^2\sigma^2}{2}\right\} \qquad E\exp\{s(-Y_i)\} \leq \exp\left\{\frac{s^2\sigma^2}{2}\right\}, \tag{12.17}$$

where $\sigma > 0$. Then

$$E \max_{i=1,\ldots,k} |Y_i| \leq \sigma\sqrt{2\log_e(2k)}.$$

A proof of Theorem 12.7 is given in Appendix E.2.

Lemma 12.7 makes an assumption concerning the moment-generating function in (12.17). The assumption holds in the Gaussian case and in the case of bounded random variables.

Lemma 12.8 *When $W \sim N(0, \sigma^2)$, then for $s > 0$,*

$$Ee^{sW} \leq e^{s^2\sigma^2/2}, \tag{12.18}$$

Let $EY = 0$ and $a \leq Y \leq b$. Then for $s > 0$,

$$Ee^{sY} \leq e^{s^2(b-a)^2/8}. \tag{12.19}$$

Equation (12.19) was proved by Hoeffding (1963). A further bound for the moment generating function is given in Lemma E.1.

12.2.2 L_2-ball

Let $\mathcal{G}$ be a set of linear combinations:

$$\mathcal{G} = \left\{\sum_{j=1}^{k} \theta_j \phi_j : \sum_{j=1}^{k} \theta_j^2 \leq B_2^2\right\}, \tag{12.20}$$

where $\{\phi_1, \ldots, \phi_k\}$ is an orthonormal system and $0 < B_2 < \infty$. We have a bound for $E \sup_{g \in \mathcal{G}} |\nu_n(g)|$ that depends essentially from $\sqrt{k/n}$.

Lemma 12.9 *Let $\mathcal{G}$ be defined in (12.20). We have for $Z = \sup_{g \in \mathcal{G}} |\nu_n(g)|$ that*

$$EZ \le \begin{cases} B_2 \, (k/n)^{1/2}, & \text{Gaussian white noise,} \\ B_2 \|f\|_\infty^{1/2} \, (k/n)^{1/2}, & \text{density estimation.} \end{cases}$$

Proof. By the Cauchy–Schwartz inequality, for $g = \sum_{j=1}^k \theta_j \phi_j \in \mathcal{G}$,

$$\nu_n(g) = \sum_{j=1}^k \theta_j \nu_n(\phi_j) \le \left(\sum_{j=1}^k \theta_j^2 \sum_{j=1}^k \nu_n(\phi_j)^2 \right)^{1/2}.$$

We have $E|X|^{1/2} \le (E|X|)^{1/2}$. Thus

$$EZ \le B_2 \left(\sum_{j=1}^k E\nu_n(\phi_j)^2 \right)^{1/2}.$$

We have

$$E\nu_n(\phi_j)^2 \le \begin{cases} n^{-1}, & \text{Gaussian white noise,} \\ \|f\|_\infty n^{-1}, & \text{density estimation,} \end{cases}$$

which implies the lemma. □

12.2.3 Chaining

Gaussian White Noise Model We can bound the expectation of a supremum in the Gaussian case by chaining. Let $\mathcal{G} \subset L_2(\mathbf{R}^d)$ be such that $\sup_{g \in \mathcal{G}} \|g\|_2 \le R$, and denote with $\mathcal{G}_\delta$ a δ-net of $\mathcal{G}$, $\delta > 0$. Let the operator norm $\varrho(Q, \mathcal{G}_\delta)$ be defined in (12.7).

Lemma 12.10 *Let the observation come from the Gaussian white noise model. Assume that $\delta \mapsto \varrho(Q, \mathcal{G}_\delta)\sqrt{\log_e(\#\mathcal{G}_\delta)}$ is decreasing on $(0, R]$ and that the entropy integral $G(R)$ defined in (12.8) is finite. Then*

$$E\left(\sup_{g \in \mathcal{G}} |\nu_n(Qg)| \right) \le n^{-1/2} 4\sqrt{2} \int_0^R \varrho(Q, \mathcal{G}_\epsilon) \sqrt{\log_e(2\#\mathcal{G}_\epsilon)} \, d\epsilon.$$

A proof of Theorem 12.10 is given in Appendix E.2.

Density Estimation In the case of density estimation we combine symmetrization and chaining. Let $\mathcal{G} \subset L_2(\mathbf{R}^d)$ be such that $\sup_{g \in \mathcal{G}} \|g\|_\infty \le B_\infty$, and denote with $\mathcal{G}_\delta$ a δ-net of $\mathcal{G}$, $\delta > 0$, with respect to the empirical metric $\|g\|_{2,n}^2 = \sum_{i=1}^n g(X_i)^2$. Denote

$$\varrho(Q, \mathcal{G}_\delta) = \max_{f \in \mathcal{G}_\delta, g \in \mathcal{G}_{2\delta}} \frac{\|Q(f-g)\|_{2,n}}{\|f-g\|_{2,n}}.$$

Lemma 12.11 *Let $X_1, \ldots, X_n$ be i.i.d. Assume that $\delta \mapsto \varrho(Q, \mathcal{G}_\delta)\sqrt{\log_e(\#\mathcal{G}_\delta)}$ is decreasing on $(0, 2B_\infty]$ and that the entropy integral $G(2B_\infty)$ defined in (12.8) is finite. Then*

$$E\left(\sup_{g \in \mathbb{G}} |\nu_n(Qg)|\right) \leq \frac{C}{n^{1/2}} E \int_0^{2B_\infty} \varrho(Q, \mathcal{G}_\epsilon)\sqrt{\log \#\mathcal{G}_\epsilon}\, d\epsilon,$$

where C is a positive constant.

A proof of Theorem 12.11 is given in Appendix E.2.

The expectation in the upper bound can further be bounded by

$$\log \#\mathcal{G}_\epsilon \leq C'(1/\epsilon)^{2(V(\mathcal{G})-1)},$$

where $V(\mathcal{G})$ is the VC-dimension of $\mathcal{G}$, and C' is a positive constant depending on $V(\mathcal{G})$. See van der Vaart and Wellner (1996, thm. 2.6.4).

12.2.4 Application of Exponential Inequalities

It may be convenient to apply exponential inequalities to derive bounds for the expectation of the supremum.

Gaussian White Noise Model In the Gaussian case we apply Lemma 12.4 to derive a bound for the expectation of the supremum.

Lemma 12.12 *Let the observation come from the Gaussian white noise model. Assume that the assumptions of Lemma 12.4 hold. We have that*

$$E_f \sup_{g \in \mathcal{G}} |\nu_n(g)| \leq \frac{C}{n^{1/2}},$$

for a positive constant C.

A proof of Theorem 12.12 is given in Appendix E.2.

Density Estimation In the case of density estimation we apply Lemma 12.5 to derive a bound for the expectation of the supremum.

Lemma 12.13 *Let $X_1, \ldots, X_n$ be i.i.d. Let the assumptions of Lemma 12.5 hold. We have*

$$E_f \sup_{g \in \mathcal{G}} |\nu_n(g)| \leq \frac{C}{n^{1/2}}$$

for a positive constant C.

A proof of Theorem 12.13 is given in Appendix E.2.

EXERCISES

12.1 Prove that when $\psi : \mathbf{R} \to \mathbf{R}$ satisfies $\psi \geq 0$, $\psi' \geq 0$, $\psi \uparrow$, $\psi' \uparrow$,

$$\int_t^\infty \exp\{-\psi(u)\}\,du \leq \frac{\exp\{-\psi(t)\}}{\psi'(t)}$$

for $t > 0$.

12.2 Apply Bernstein's inequality to prove the following lemma, which gives an exponential bound for the deviation of a histogram estimate from a piecewise constant approximation of the density, in the uniform norm.

Lemma 12.14 *Let $\mathcal{P}$ be a finite partition of $[0,1]^d$ to rectangles. Let $\hat{f}(\cdot,\mathcal{P})$ be the histogram defined by (14.2), based on the i.i.d. sample $X_1,\ldots,X_n$ from the distribution of density $f : \mathbf{R}^d \to \mathbf{R}$. Denote a piecewise constant approximation of f by*

$$\bar{f}(x,\mathcal{P}) = \sum_{P\in\mathcal{P}} \frac{\int_R f}{\text{vol}(R)} I_R(x), \qquad x \in \mathbf{R}^d. \tag{12.21}$$

Then for $\xi > 0$,

$$P\left(\left\|\hat{f}(\cdot,\mathcal{P}) - \bar{f}(\cdot,\mathcal{P})\right\|_\infty > \xi\right)$$
$$\leq \#\mathcal{P}\cdot\exp\left\{-\frac{n\xi^2}{2}\left[\min_{R\in\mathcal{P}} \text{vol}(R)\cdot(\|f\|_\infty + \xi/3)\right]^{-1}\right\}.$$

CHAPTER 13

MANIPULATION OF DENSITY ESTIMATES

This chapter discusses algorithms that can be used in connection of density estimation and visualization. Section 13.1 presents two data structures that can be applied to evaluate multidimensional functions and store the values of these functions: evaluation trees and range trees are defined. Evaluation trees are applied in Section 14.3.4 to present an algorithm for evaluating kernel estimates. Section 13.2 contains algorithms for calculating level set trees, shape trees, and tail trees (we use the name "visualization tree" as a generic name for these trees). Two algorithms are described. The first algorithm in Section 13.2.1 is simpler and easier to implement. The second algorithm in Section 13.2.1 illustrates dynamic programming.

13.1 DATA STRUCTURES

An evaluation tree is a useful data structure that can be used to represent a function. We include in our discussion also range trees.

13.1.1 Evaluation Trees

An *evaluation tree* is a binary search tree that represents a setwise constant function. The setwise constant function may be a density estimate, or it may be an approximation of some continuous function. Evaluation trees were discussed in Klemelä (2005). We define first a *partition-generating tree*. Partition-generating trees generate a partition on a rectangle. The partition is constructed by a recursive splitting of the rectangle. Evaluation trees are partition-generating trees whose leaf nodes are annotated with real values.

Definition 13.1 (Set of partition-generating trees.) *The* set of partition-generating trees $\mathbb{T} = \mathbb{T}(R_0)$, *associated with a rectangle* $R_0 \subset \mathbf{R}^d$ *whose sides are parallel to the coordinate axis, consists of trees with a single root node such that each node has* 0, 1, *or* 2 *children, and whose nodes are annotated with rectangles, directions, and split points in the following way:*

1. *The root node is annotated with* R_0.

2. *The non–leaf nodes are annotated with a direction* $l \in \{1, \ldots, d\}$ *and a split point* $s \in \mathbf{R}$. *Let the node be annotated with rectangle* $R = \Pi_{m=1}^d [c_m, d_m]$ *and direction* $l \in \{1, \ldots, d\}$. *Then the annotated split point* s *satisfies*

$$s \in S_l(R) \stackrel{def}{=} (c_l, d_l).$$

3. *Given a non–leaf node that is annotated with rectangle* $R \subset \mathbf{R}^d$, *splitting direction* $l \in \{1, \ldots, d\}$, *and splitting point* $s \in \mathbf{R}$, *denote*

$$R_{l,s}^{(0)}(R) = \{x \in R : x_k \leq s\}, \qquad R_{l,s}^{(1)}(R) = \{x \in R : x_k \geq s\}.$$

If the node has a left child, then this child is annotated with $R_{l,s}^{(0)}(R)$ *and if the node has a right child, then this child is annotated with* $R_{l,s}^{(1)}(R)$.

Definition 13.2 (Evaluation tree.) An evaluation tree *is a partition-generating tree whose terminal nodes are additionally annotated with real values. These values represent the values of the function.*

Remarks We make several observations concerning evaluation trees.

1. *(Set associated with a partition-generating tree.)* There is always a set associated with a partition-generating tree, and this tree is a union of those rectangles that are associated with the leafs of the tree. Let $\mathcal{A}$ be the collection of the annotations of the leaf nodes of T with rectangles:

$$\mathcal{A} = \{A \subset \mathbf{R}^d : A \text{ is a rectangle associated with some leaf node of } T\}. \tag{13.1}$$

The set associated with T is

$$\cup_{A \in \mathcal{A}} A. \tag{13.2}$$

2. *(Function associated with an evaluation tree.)* An evaluation tree represents a rectangularwise constant function. Evaluation tree T represents the function

$$\bar{f}(x) = \sum_{A \in \mathcal{A}} f_A I_A(x), \tag{13.3}$$

where $\mathcal{A}$ is defined in (13.1) and $f_A \in \mathbf{R}$ are the annotations of the leaf nodes of T with real values. We denote $I_A(x) = 1$ when $x \in A$, and $I_A(x) = 0$ otherwise.

3. *(Almost sure partitions.)* The partitions which we consider are collections of closed rectangles whose intersection may be nonempty. However, the intersection has zero Lebesgue measure. Density functions are defined only almost everywhere, and thus we may use such partitions to define piecewise constant density functions.

4. *(Evaluation trees in density estimation.)* In density estimation piecewise constant functions may appear at least in two ways: (a) we construct first a continuous density estimate, like a kernel density estimate, and then approximate this continuous function with a piecewise constant function, or (b) we construct directly a piecewise constant estimate, as in the case of CART-histograms and aggregated estimates.

   By the definition of the evaluation tree, the root node of this tree is annotated with a rectangle R_0 and the evaluation tree generates a partition of R_0. When a density estimate vanishes in some regions of R_0, we do not have to include rectangles corresponding to those regions in the evaluation tree. For example, the evaluation tree of Figure 13.1 does not have leaf nodes corresponding to all regions of $[0, 4] \times [0, 4]$.

5. *(Advantages of evaluation trees.)* There are two advantages we get from representing piecewise constant functions with binary trees.

   (a) We are able to find fast the value $\bar{f}(x)$ at any $x \in \mathbf{R}^d$, using the binary search algorithm. Thus we are also able to refine fast the values of the stored function. Refining may mean changing the value of the estimate at some region. In the case of the function approximation refining may mean splitting further some regions where the approximation was constant, thus giving a more accurate approximation of the function.

   (b) Siblings in the evaluation tree of the function correspond to spatially close sets. Thus we may apply a dynamic programming algorithm to find separated components of the level sets. We take a bottom-up trip on this tree, at which time we join the connected parts of level sets, thus solving the problem of finding separated components by solving a "dual" problem of joining the connected components. The algorithm DynaDecompose, described in page 357, is this kind of algorithm.

Example Figure 13.1 shows a partition-generating tree. Figure 13.2 shows the rectangles annotated with the nodes of this tree and the process of growing this tree. We start with the rectangle $[0, 4] \times [0, 4]$. We split first parallel to the y-axis making dyadic splits, and then parallel to the x-axis.

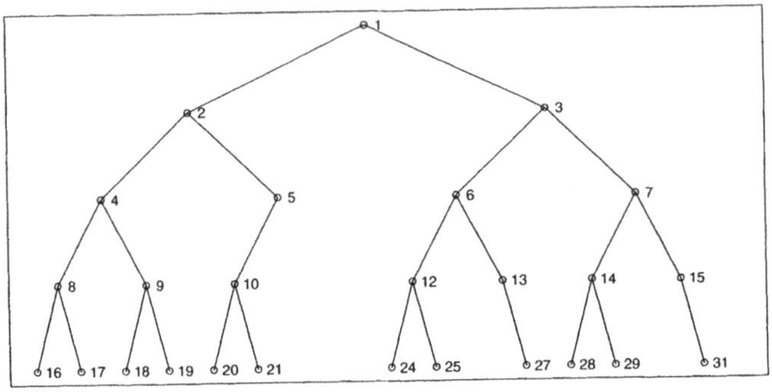

Figure 13.1 *(Partition-generating tree.)* Shown is a partition-generating tree. Figure 13.2 shows the annotations of the nodes with rectangles.

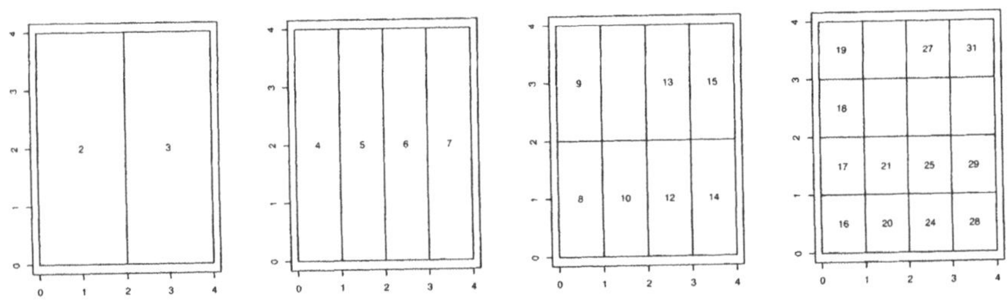

Figure 13.2 *(Partition-generating tree, continued.)* Shown are the rectangles annotated with the nodes of the tree of Figure 13.1. The process of growing this tree is illustrated simultaneously.

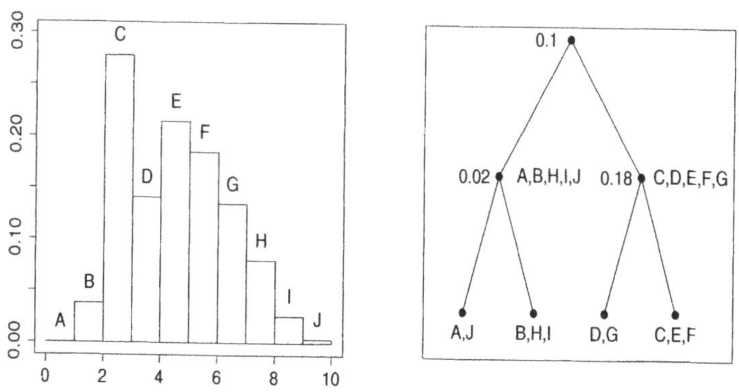

Figure 13.3 *(Range tree.)* Shown are a piecewise constant function and its range tree.

13.1.2 Range Trees

Evaluation trees are used to approximate functions by partitioning the domain to a union of rectangles. Range trees can be used when the function is approximated or represented by partitioning the domain to more general cells, for example to triangles or hexagons. Range trees organize the cells by the values of the function taken on the cells.

Figure 13.3 illustrates a range tree. The function is piecewise constant on 10 intervals and the intervals are arranged on the binary tree by splitting the range at every nonterminal node. The intervals where the function takes a smaller value than the split value are moved to the left child and the intervals where the function takes a larger value than the split value are moved to the right child. The split values are 0.1, 0.02, and 0.18.

In the general case we assume that a list of cells $c \subset \mathbf{R}^d$ is given, and each cell is associated with the interval

$$I_c = \left[\min_{x\in c} f(x), \max_{x\in c} f(x)\right] .$$

When the function is piecewise constant on the cells, the intervals I_c shrink to single values. The cells are arranged into the range tree by using the left or right boundary of the interval.

13.2 CONSTRUCTING VISUALIZATION TREES

We use the term *visualization tree* to mean a level set tree, a shape tree, or a tail tree. A single algorithm can be used for calculating all these trees. We call this algorithm the LeafsFirst algorithm. However, we discuss also an alternative to the LeafsFirst algorithm that is based on dynamic programming.

13.2.1 Leafs First

LeafsFirst algorithm calculates (upper) level set trees. The calculation of lower level set trees can be done analogously. The LeafsFirst algorithm was introduced in Klemelä (2006). We restrict ourselves to the case where the domain is a subset of $\mathbf{R}^d$. The input for the algorithm is a grid $x_1, \ldots, x_n \in \mathbf{R}^d$ and a set of scalar values $f(x_1), \ldots, f(x_n) \in \mathbf{R}$.

Interpolation The grid can be regular or scattered. Each algorithm assumes implicitly an interpolation scheme for the scalar values. When the grid is a regular rectilinear (rectangular) grid, we have available at least the following interpolation schemes:

1. The nearest neighbor interpolation takes every point x_i to be a center point of a rectangle, and the function constructed by the interpolation is rectangularwise constant, taking the value $f(x_i)$ on the rectangle whose center is x_i.

2. The cells can be split into a number of simplices (triangles in 2D, tetrahedra in 3D, ...) to fit a linear interpolant within each simplex.

3. Linear interpolation can be used to interpolate the values measured on the grid to all points between the grid points. For example, the linear interpolation in 2D means that (a) we construct 1D linear interpolants in the x-direction along the top and bottom edges of the cell, and (b) we calculate an 1D linear interpolant in the y-direction, passing through the (x, y) position of the point to be interpolated. In 3D a linear interpolation is also called a trilinear intrepolation. Note that this method does not lead to a globally piecewise linear function.

In the case of a scattered grid we can connect points in a triangulation and then fit a linear interpolant within each simplex. The Delaunay triangulation is a well-known example. (In 3D a triangulation is also called a tetrahedralisation.)

LeafsFirst Algorithm The LeafsFirst algorithm assumes the nearest neighbor interpolating scheme, and the rectangle whose center is x_i is denoted with R_i.

ALGORITHM: **LeafsFirst**

1. Order the values $f(x_1), \ldots, f(x_n)$, starting with the largest. From now on, assume that $f(x_1) > \cdots > f(x_n)$.

2. For $i = 1$ to n: consider grid point x_i. Create a new node to the level set tree whose level is $f(x_i)$. Find which sets, associated with the current root nodes (those nodes which do not yet have a parent) touch rectangle R_i. At this step we will apply the bounding box technique, which is explained below.

(a) If rectangle R_i touches sets $C_1, \ldots, C_M$ associated with the current root nodes, then create a parent to the nodes that are touched. The set of this parent node is $R_i \cup C_1 \cup \cdots \cup C_M$.

(b) If rectangle R_i does not touch any sets associated with the current root nodes, then create a new leaf node to the tree. The set associated with this node is R_i.

The sorting in line 1 of the algorithm takes $O(n \log n)$ steps. In the worst case the line 2 of the algorithm requires the pairwise comparison of all rectangles in $R_1, \ldots, R_n$ to find which rectangles touch, and this takes $O(dn^2)$ steps, since we need $O(d)$ steps to calculate whether two rectangles touch. Thus the worst case complexity of the algorithm is

$$O\left(n \log n + dn^2\right) = O\left(dn^2\right).$$

However, we have enhanced the algorithm with the *bounding box technique*. In the bounding box technique we associate the nodes with the bounding box of the set associated with a node. The bounding box of set $A \subset \mathbf{R}^d$ is the smallest rectangle containing A, such that the sides of the rectangle are parallel to the coordinate axes. In step 2 of the algorithm we find which rectangles, associated with the current root nodes, are touched by rectangle R_i. If rectangle R_i does not touch the bounding box of those rectangles, then it does not touch any rectangles inside the bounding box. Only if it does touch the bounding box, will we have to travel further toward the leaf nodes to find whether R_i touches any of the smaller bounding boxes. With the bounding box enhancement the worst case complexity of line 2 is still $O(dn^2)$, but with this technique we achieve considerable improvements in typical cases.

The Bounding Box Enhancement of Algorithm LeafsFirst In the bounding box technique we associate the nodes of a level set tree with the bounding box of the set associated with a node. The bounding box of set $A \subset \mathbf{R}^d$ is the smallest rectangle containing A such that the sides of the rectangle are parallel to the coordinate axes.

1. **Input** is rectangle R and the collection of the current root nodes of the level set tree (we are building the level set tree starting from the leaf nodes and thus the current root nodes of the unfinished level set tree are the nodes without parent).

2. **Output** is the list of those current root nodes that are touched by rectangle R (list of those current root nodes whose associated set is touched by rectangle R).

3. An additional internal data structure of the algorithm consists of the annotations of each node of the (unfinished) shape tree with the bounding box of the set associated with the node.

ALGORITHM: **Bounding Box Enhancement** (of algorithm LeafsFirst)

1. **answer=emptyList** (start by assuming that there are no touches);

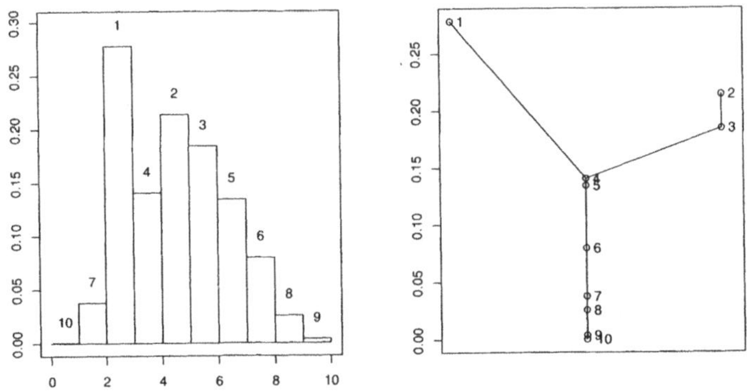

(a) (b)

Figure 13.4 *(Illustration of the* LeafsFirst *algorithm; level set tree.)* Shown are a 1D piecewise constant function and its level set tree.

2. **loop 1**: **go** through the current root nodes; assume that the current root node m_r was encountered;

3. **loop 2**: **go** through the nodes of the tree whose root is m_r; start with node m_r and go through the nodes in such a way that a parent is always encountered before children;

   assume that node m associated with set A_m and bounding box $\mathcal{B}_m$ was encountered;

   (a) **if** rectangle R does not touch bounding box $\mathcal{B}_m$, **then** (conclude that R does not touch the set associated with node m, and thus does not touch the set associated with m_r), **goto loop 1**;

   (b) **else** (R touches bounding box $\mathcal{B}_m$) **if** R touches the set A_m, **then** (conclude that R touches the set associated with node m, and thus touches the set associated with node m_r) **concatenate** m_r to the **answer**, **goto loop 1**;

   (c) **else** (R touches bounding box $\mathcal{B}_m$, but does not touch set A_m) **continue loop 2**;

4. **return answer**.

Illustration of the LeafsFirst Algorithm

1. *(1D case.)* Figure 13.4 illustrates LeafsFirst algorithm in the 1D case. The function is piecewise constant in 10 rectangles. Panel (a) shows the function and the ordering of the intervals according to the values of the function on the intervals. Panel (b) shows the level set tree of the function.

2. *(2D case.)* Figure 13.5 illustrates LeafsFirst algorithm in the 2D case. The function is a piecewise constant approximation on a grid of 16^2 points of the function shown in Figure 2.7. The algorithm proceeds by ordering the rectangles according to the value the function takes at the rectangles; the first rectangle is the one where the function takes the highest value. Panel (d) shows the 28 first rectangles; at this stage the level set consists of three separated components. Panel (e) shows the stage where two of the components are joined together, and panel (f) shows the stage where all three components are joined together.

3. *(Shape tree.)* Figure 13.6 illustrates the LeafsFirst algorithm used in calculating a shape tree for the set in Figure 5.3(a). The set is approximated on a grid of 16^2 rectangles and the reference point is the origin. The algorithm proceeds by ordering the rectangles according to the distance from a center point; the first rectangle is the one that is farthest away from the reference point. Panel (a) shows the 38 first rectangles; at this stage the tail region consists of three separated components. Panel (b) shows the stage where two of the components are joined together, and panel (c) shows the stage where all three components are joined together.

13.2.2 Roots First

We can build a level set tree by first decomposing the lowest level set of the function. The separated regions of the lowest level set correspond to the root nodes of the trees; then we travel toward upper level sets always decomposing regions of the level sets. The RootsFirst algorithm is based on the algorithm DynaDecompose, which was introduced in Klemelä (2005).

RootsFirst Algorithm RootsFirst is a typical tree-creating algorithm that starts by creating the roots of the tree, and then travels toward the leafs of the tree. When there is a branching of the tree, it travels to one of the branches and will stack those nodes that are the root nodes of the other branches. As a leaf node is reached,the algorithm takes a a node from the stack and starts again traveling toward the leafs. See Exercise 13.6.

Decomposing a Set We want to decompose a set $B \subset \mathbf{R}^d$ to connected components that are mutually separated. We assume that set B is a union of small sets (atoms),

$$B = \cup_{A \in \mathcal{A}} A \tag{13.4}$$

where $\mathcal{A}$ is a collection of connected sets. Thus sets in $\mathcal{A}$ are "atomic" in the sense that they cannot be decomposed to two separated components. For example, $\mathcal{A}$ could be a collection of small rectangles, or a collection of small balls. Since sets $A \in \mathcal{A}$ are connected, our task is to find the partition

$$\mathbb{P} = \{\mathcal{A}_1, \ldots, \mathcal{A}_M\} \tag{13.5}$$

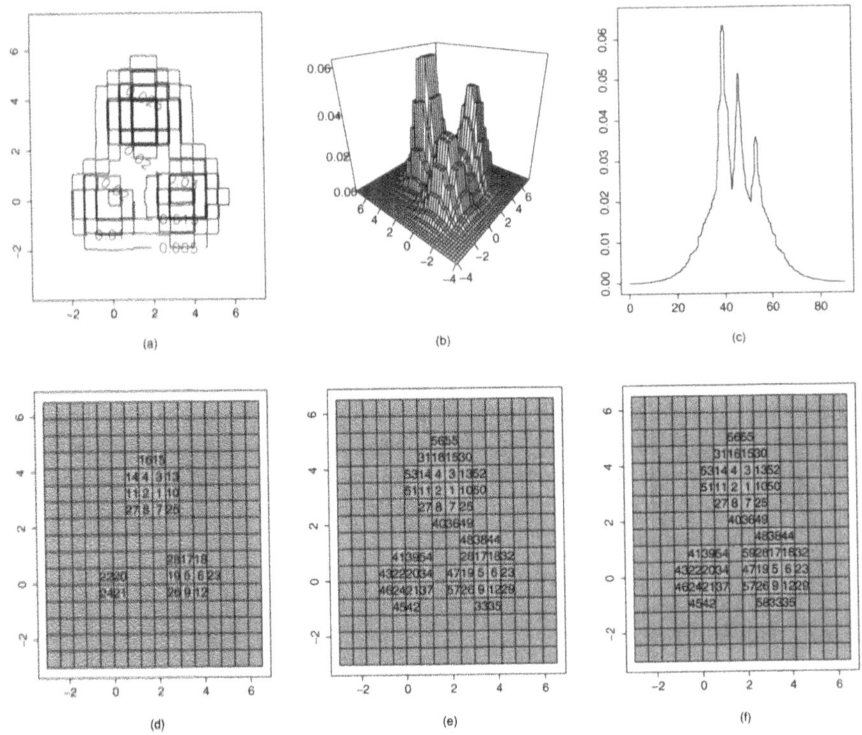

Figure 13.5 *(Illustration of the* LeafsFirst *algorithm; level set tree.)* The upper row shows a 2D piecewise constant function: (a) A contour plot, (b) perspective plot, and (c) volume plot. The lower row shows the order of going through the rectangles as the level set tree is calculated of the function shown in the upper row.

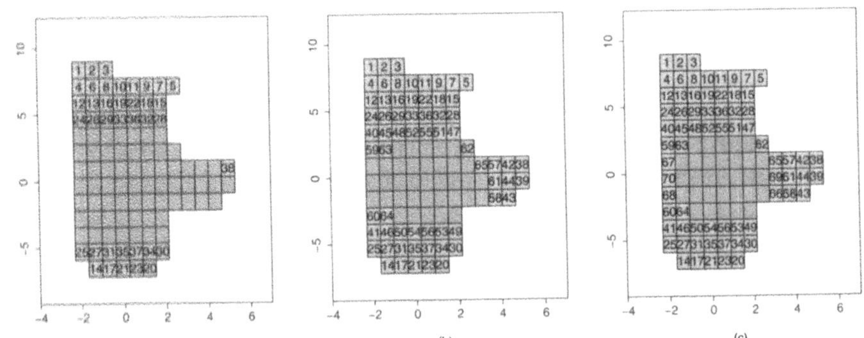

Figure 13.6 *(Illustration of the* LeafsFirst *algorithm; shape tree.)* Shown are three steps of the algorithm in the calculation of a shape tree.

of $\mathcal{A}$ so that sets

$$B_i = \cup_{A \in \mathcal{A}_i} A, \qquad i = 1, \ldots, M,$$

form the required decomposition of B, that is, the finite maximal decomposition to connected components, as in Definition 4.4. Thus the following conditions are met:

1. $B = \cup_{i=1}^{M} B_i$, that is, $\mathcal{A} = \cup_{i=1}^{M} \mathcal{A}_i$.

2. For $i, j = 1, \ldots, M$, $i \neq j$, B_i and B_j are separated.

3. For $i = 1, \ldots, M$, B_i is connected.

Condition 2 says that the members of the partition are pairwise separated. Condition 2 implies that for $i, j = 1, \ldots, M$, $i \neq j$, $B_i \cap B_j = \emptyset$, and $\mathcal{A}_i \cap \mathcal{A}_j = \emptyset$. Thus condition 1 and condition 2 together imply that a partition of set B is made to sets B_i, $i = 1, \ldots, M$. Condition 3 says that the members of the partition are connected. Condition 3 implies that the partition is maximal in the sense that trying to split some member of the partition will lead to a violation of condition 2. Since we assumed that collection $\mathcal{A}$ is finite and sets $A \in \mathcal{A}$ are connected, we can always find a decomposition satisfying conditions 1 through 3.

A straightforward algorithm compares all sets $A \in \mathcal{A}$ in the representation (13.4), with each other, to find which sets A touch each other. This algorithm needs $O((\#\mathcal{A})^2)$ steps, where $\#\mathcal{A}$ is the cardinality of set $\mathcal{A}$. When this cardinality is very large, this is not a feasible algorithm. See Exercise 13.4.

There exists a dynamic programming algorithm for the decomposition of a set, in the case where the set is a union of rectangles and the set is constructable by a recursive partition. That is, we assume that the set is associated with a partition generating tree in the sense of (13.2) when a partition-generating tree is defined as in Definition 13.1. The algorithm proceeds by traveling the partition-generating tree, starting from the leaf nodes of the tree. We utilize the fact that the sets that are spatially close are close in the tree structure of the partition tree.

Pseudo-code for the Algorithm DynaDecompose We give a pseudo-code for the algorithm DynaDecompose that decomposes a set to maximally separated components. This algorithm takes as an input the set associated with a partition-generating tree, as defined in (13.2).

1. **Input** of the algorithm are a partition-generating tree as defined in Definition 13.1, and a list $\mathcal{L}$ that labels the terminal nodes of the partition-generating tree. The terminal nodes are associated with sets in $\mathcal{A}$, and the set that we decompose is given by (13.2) or by (13.4).

2. **Output** of the algorithm is a partition $\mathbb{P}$ of $\mathcal{A}$ as in (13.5). This is equivalent to giving a partition of list $\mathcal{L}$.

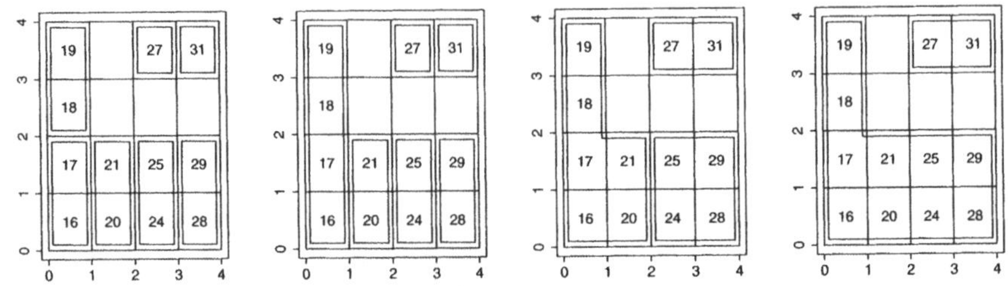

Figure 13.7 *(Illustration of the algorithm DynaDecompose.)* Illustrated is the process of finding the two separated regions of a set. The corresponding partition-generating tree is shown in Figure 13.1.

ALGORITHM: **DynaDecompose**

1. **travel** the evaluation tree, starting from the terminal nodes labeled by the list $\mathcal{L}$; traveling has to proceed so that the children are encountered before the parent;

   (a) assume that you have encountered node m;

   (b) **if** node m is a leaf node annotated with set A, **then** annotate node m with the partition $\{\{A\}\}$;

   (c) **else** (m is not a leaf) check whether there are connections with the sets in the partitions that are annotated with left and right child, and annotate node m with the partition joining the two partitions;

2. **end travel**

3. **return** the partition that is annotated with the root node

Especially step 1(c) is rather complex, requiring to make a pairwise comparison of all "current components". See Exercise 13.5.

Example Figure 13.7 illustrates the evolution of the algorithm. We assume that the partition-generating tree of the function is given in Figure 13.1. We want to decompose the union of the rectangles that are annotated with the leaf nodes. The first window shows how we have formed 1×2 rectangles. We join subsets inside 1×4 rectangles to get window 2. Next we join subsets inside 2×4 rectangles to get window 3. Then we join subsets inside the 4×4 rectangle to get the 4th window, which shows that the support of the function consists of two separated and connected components.

13.2.3 Bibliographic Notes

The first algorithm for constructing Reeb graphs is given in Shinagawa and Kunii (1991). The algorithm works for 2D functions and has the worst case complexity $O(n^2)$. In van Kreveld, van Oostrum, Bajaj, Pascucci and Schikore (1997) an algorithm is designed with the complexity $O(m \log m)$ for 2D data and $O(m^2)$ for general data where m is the number of simplices in a simplicial mesh. Tarasov and Vyalyi (1998) improved the time complexity to $O(m \log m)$ in the 3D case, using an algorithm with three passes of data: they calculated an upper level set tree, a lower level set tree, and then combined these to a contour tree by a merge procedure. Carr et al. (2003) simplified the algorithm of Tarasov and Vyalyi (1998) and designed it for any dimension. The complexity of the algorithm in Carr et al. (2003) is $O(m + n \log n)$. Improved algorithms include Carr, Snoeyink and van de Panne (2004) and Chiang, Lenz, Lu and Rote (2005).

EXERCISES

13.1 Sketch an algorithm for evaluating a multivariate function $f : \mathbf{R}^d \to \mathbf{R}$ on a regular grid, when we have an algorithm for evaluating the function at each point $x \in \mathbf{R}^d$. The main difficulty is to design an algorithm that takes as input functions with different dimensions d; the algorithm has to work for all values of d.

13.2 We have defined shape trees in Definition 5.3 in terms of level set trees: a shape tree is a level set tree of a distance function. However, we have defined shape trees in Definition 5.4 directly, without using the concept of a level set tree. Tail trees were defined in Definition 6.2 in terms of level set trees and directly in Definition 6.3. Generalize Definition 5.4 and define the concept of a *visualization tree*, that includes level set trees, shape trees, and tail trees as a special case.

13.3 Design an algorithm for calculating the distance of a point to the boundary of a rectangle: calculate $\inf_{y \in R} \|x - y\|$, where $R \subset \mathbf{R}^d$ is a rectangle whose sides are parallel to the coordinate axes and $x \in \mathbf{R}^d$. The algorithm should have $O(d)$ flops; calculating the distance of a point to the center of the rectangle is not faster, since it takes also $O(d)$ flops. The algorithm is needed when one uses the LeafsFirst algorithm to calculate shape trees.

13.4 Sketch an algorithm PairwiseComparison for decomposing a set $B \subset \mathbf{R}^d$ to mutually separated and connected components. Assume that (13.4) is satisfied. Use pairwise comparison to write an algorithm which takes $O((\#\mathcal{A})^2)$ steps. This algorithm could illustrate the use of stacks.

13.5 Give a more precise pseudo-code for the algorithm DynaDecompose than the pseudo-code given in Section 13.2.2, page 358.

13.6 Write a pseudo-code for the RootsFirst algorithm, The algorithm should call procedure "decompose" for partitioning a set of atoms. This procedure could be given by either DynaDecompose, or the PairwiseComparison of Exercise 13.4.

PART III

TOOLBOX OF DENSITY ESTIMATORS

CHAPTER 14

LOCAL AVERAGING

Local averaging is one of the basic methods of density estimation in the one-dimensional case. The generalization of one-dimensional local averaging to multivariate cases is straightforward. However, even in the two-dimensional case one has to take into account multivariate features. For example, multivariate function may have anisotropic smoothness (variability may be different in different directions).

A prototype of local averaging may be defined as a procedure where density f is estimated at point $x \in \mathbf{R}^d$ by choosing a local neighborhood $N(x) \subset \mathbf{R}^d$ of x and estimating

$$f(x) \approx \hat{f}(x) = \frac{n_{N(x)}/n}{\text{volume}(N(x))}, \tag{14.1}$$

where $n_{N(x)}$ is the number of observations in $N(x)$. This idea of estimating densities is related to the Lebesgue density theorem, which says that for almost all $x \in \mathbf{R}^d$ the density $f(x)$ at x is the limit

$$f(x) = \lim_{r \to 0} \frac{P_f(B_r(x))}{\text{volume}(B_r(x))},$$

where P_f is the probability distribution corresponding to f and $B_r(x) = \{y : \|y - x\| \le r\}$; see Wheeden and Zygmund (1977). We could apply some other decreasing sequences of sets than the sequence of balls with decreasing radius. The estimator in (14.1) is a kernel estimator with a uniform kernel. Histograms, kernel estimators with general kernels, and nearest neighborhood estimators, defined in Sections 14.2, 14.3, and 14.4, amount to various modifications of this idea of local averaging.

Local averaging estimators belong to the class of "nonparametric estimators". The qualification "nonparametric" refers to the fact that these estimators are meant to be applied in cases where we are not postulating an underlying probability model that is characterized by a finite dimensional parameter. An alternative term is "model-free estimator", although strictly speaking there is always a model behind it; at least the existence of the density function has to be assumed.

14.1 CURSE OF DIMENSIONALITY

Local averaging suffers from the curse of dimensionality. Bellman (1961) coined the phrase "curse of dimensionality" while discussing the computational complexity in optimization over many dimensions. A multivariate grid that has 10 gridpoints in one direction has 10^d total gridpoints. The exhaustive enumeration strategy in optimization cannot work over such large grids, and he suggested dynamic programming algorithms.

A particular manifestation of the curse of dimensionality is the fact that local averaging does not work for high-dimensional data. This is because local neighborhoods are empty of observations and neighborhoods that are not empty are not local. For example, consider the case where the density is the uniform distribution on the rectangle $[0, 1]^d$. Say the histogram estimator has 10 bins for every dimension. This histogram estimator has a total of 10^d bins, and thus only about $0.1^d \times 100\%$ of the observations is contained in each bin. The number of observations in each bin will diminish rapidly as the dimension of the sample space grows. Simonoff (1996) gives a detailed discussion of the concept of the curse of dimensionality.

We mentioned in page 315 two approaches to the theory statistical estimation. (1) Postulate a density class and ask what would be the best possible estimator when it is known that the true density belongs to this class. (2) Construct an estimator and ask what are the statistical properties of the estimator, when the true density belongs to a given test bed. The curse of dimensionality arises in the first approach because a seemingly restrictive class like a Sobolev class turns out to be so large in multivariate cases that even the best estimators have a very slow rate of convergence. The curse of dimensionality arises in the second approach because a seemingly reasonable estimator (kernel estimator) that works in low-dimensional cases turns out to perform badly in high-dimensional cases.

Asymptotic theory indicates that the kernel estimator suffers from the curse of dimensionality because the rate of convergence of the kernel estimator is $n^{s/(2s+d)}$ where s is the smoothness index of the density. This rate increasingly slows as d increases.

As a response to the curse of dimensionality, we will construct other estimators than estimators based on local averaging, and we will analyze estimators in other test beds than the Sobolev classes and Hölder classes.

14.2 HISTOGRAMS

14.2.1 Definition of Histogram

A histogram is an estimate of the form

$$\hat{f}(x, \mathcal{P}) = \sum_{R \in \mathcal{P}} \frac{n_R}{n\, \text{volume}(R)} I_R(x), \qquad x \in \mathbf{R}^d, \tag{14.2}$$

where $\mathcal{P}$ is a finite collection of almost everywhere disjoint rectangles, n_R is the number of observations $X_1, \ldots, X_n \in R$, $I_R(x) = 1$ when $x \in R$, and $I_R(x) = 0$ otherwise, and volume$(R) = \int_R dx$. Typically $\mathcal{P}$ is a partition to rectangles of a rectangle which is estimated to contain the support of the true density and whose sides are parallel to the coordinate axes. Other shapes than rectangles may also be considered.

14.2.2 Average Shifted Histogram

The average shifted histogram (ASH) is an average of histograms whose partitions are shifted versions of each other. The partitions have the same binwidths $h = (h_1, \ldots, h_d)$, and they are shifted by the amount h/m to get m shifted histograms. When the number m of histograms increases, the ASH approximates a kernel estimator with a triangular kernel. An ASH may be computationally more effective than a kernel estimator but statistically more accurate than a histogram. ASH is piecewise constant on $(2m-1)^d$ rectangles. The average shifted histogram was introduced by Scott (1985).

Let $\hat{f}_j$, $j = 1, \ldots, m$, be one-dimensional histograms with equal sized bins $A_{j,k}$, $k \in \mathbf{Z} = \{\ldots, -1, 0, 1, \ldots\}$, where the width of the bins is $h > 0$. Then the average shifted histogram is

$$\hat{f}(x) = \frac{1}{m} \sum_{j=1}^{m} \sum_{k \in \mathbf{Z}} \frac{n_{j,k}}{n \cdot \text{volume}(A_{j,k})} I_{A_{j,k}}(x), \qquad x \in \mathbf{R},$$

where $n_{j,k} = \#\{X_i : X_i \in A_{j,k}\}$. Define small bins B_l, $l \in \mathbf{Z}$, where the width of the bins is h/m. Each $A_{j,k}$ is divided to m small bins B_l. Let $k(x)$ be such that $x \in A_{j,k(x)}$. Let us denote by $B_j^{(x)}, \ldots, B_{j+2m-1}^{(x)}$ such intervals in $(B_l)_{l \in \mathbf{Z}}$ that

$$A_{j,k(x)} = \bigcup_{l=0}^{m-1} B_{j+l}^{(x)},$$

and let $\nu_{j+l}^{(x)} = \#\{X_i : X_i \in B_{j+l}^{(x)}\}$. Then

$$n_{j,k(x)} = \sum_{l=0}^{m-1} \nu_{j+l}^{(x)}.$$

Also, volume$(A_{j,k}) = h$, which gives

$$\begin{aligned}
\hat{f}(x) &= \frac{1}{m} \sum_{j=1}^{m} \frac{\sum_{l=0}^{m-1} \nu_{j+l}^{(x)}}{nh} \\
&= \frac{1}{mnh} \left(\nu_1^{(x)} + 2\nu_2^{(x)} + \cdots + m\nu_m^{(x)} + \cdots + 2\nu_{2m-2}^{(x)} + \nu_{2m-1}^{(x)} \right) \\
&= \frac{1}{nh} \sum_{j=1-m}^{m-1} \left(1 - \frac{|j|}{m}\right) \nu_{j+m}^{(x)}.
\end{aligned}$$

14.3 KERNEL ESTIMATORS

14.3.1 Definitions of Kernel Estimators

Basic Form of the Kernel Estimator The kernel estimator based on observations $X_1, \ldots, X_n \in \mathbf{R}^d$ is defined as

$$\hat{f}(x) = n^{-1} \sum_{i=1}^{n} K_h(x - X_i), \qquad x \in \mathbf{R}^d, \tag{14.3}$$

where

$$K_h(x) = h^{-d} K(x/h) \tag{14.4}$$

is the scaled kernel, $h > 0$ is the smoothing parameter, and $K : \mathbf{R}^d \to \mathbf{R}$ is the kernel function. When the kernel function K is a density, kernel estimates are densities.

Anisotropic Kernel Estimator We may add flexibility to the basic form of the kernel estimator by using a different smoothing parameter in each direction. We define the anisotropically scaled kernel as

$$K_h(x) = \left(\Pi_{i=1}^{d} h_i^{-1}\right) K(x_1/h_1, \ldots, x_d/h_d), \qquad x \in \mathbf{R}^d, \tag{14.5}$$

where $h = (h_1, \ldots, h_d)$ is a vector of the smoothing parameters. The kernel estimator with the scaled kernel as in (14.5) achieves the optimal rates of convergence in anisotropic Besov classes when $p_i \geq 2$, $i = 1, \ldots, d$, for the L_2 loss function. This is proved in Section 14.3.2.

Spatially Adaptive Kernel Estimator We may let the smoothing parameter depend on the point $x \in \mathbf{R}^d$ where the density is estimated. The kernel estimator with local smoothing parameters is defined as

$$\hat{f}(x) = n^{-1} \sum_{i=1}^{n} K_{h(x)}(x - X_i), \qquad x \in \mathbf{R}^d, \tag{14.6}$$

where $h : \mathbf{R}^d \to (0,\infty)$ is a smoothing parameter function and K_h is defined in (14.4). This estimator is a generalization of the nearest neighbor estimator to be defined in (14.54). Estimator (14.6) may achieve the optimal rates of convergence in an isotropic Besov space also when $p < 2$. Such a choice for $h : \mathbf{R}^d \to (0,\infty)$ was defined in Lepski, Mammen and Spokoiny (1997) for the one-dimensional case. However, when the Besov space is anisotropic and some $p_i < 2$, we would need to generalize estimator (14.6) to obtain different smoothing parameters in the different directions. Kerkyacharian, Lepski and Picard (2001) and Kerkyacharian, Lepski and Picard (2007) considered such estimator.

Time-Localized Kernel Estimator We define the time-localized kernel estimator by associating weights p_i, $i = 1, \ldots, n$, to the observations, replacing the usual weights n^{-1}. The *time-localized kernel estimator* is defined by

$$\hat{f}(x) = \sum_{i=1}^{n} p_i K_h(x - X_i), \qquad x \in \mathbf{R}^d, \tag{14.7}$$

where $p_i \geq 0$, $\sum_{i=1}^{n} p_i = 1$. The kernel estimator in (14.3) is a special case of estimator (14.7) when $p_i = n^{-1}$ for $i = 1, \ldots, n$. The time-localized kernel estimator is used when the observations are not identically distributed but the distribution is changing continuously as a function of time. We use the term "time", but the observations may also be spatially indexed and not be a time series. The weights may take the form

$$p_i = L_g(\delta(t_0, t(i))), \qquad i = 1, \ldots, n,$$

where $t : \{1, \ldots, n\} \to \mathbf{T}$ maps the indexes to a metric space $\mathbf{T}$ with distance δ, we want to estimate the density at point $t_0 \in \mathbf{T}$, $L : [0,\infty) \to [0,\infty)$, $L_g(t) = c(g)L(t/g)$, and $c(g)$ is the normalization constant. For example, let $X_1, \ldots, X_n$ be a time series so that X_i corresponds to the previous observation rather than X_j when $i < j$. When we want to estimate the density of the latest observation, we may take the Gaussian weights

$$p_i = \frac{q_{g,i}}{\sum_{i=1}^{n} q_{g,i}}, \qquad q_{g,i} = \exp\left\{-(n-i)^2/g^2\right\},$$

$i = 1, \ldots, n$. We give more weight to the recent observations and gradually decrease the weights for the more distant observations. Parameter $g > 0$ is the time localization parameter so that, when g is small, the estimate is effectively based only on recent observations. We choose g to reflect the rate of change of the density as the function of time. Time-localized kernel estimators are studied in Klemelä (2008*a*).

Kernels Often, in choosing the kernel one has to strike a balance between computational efficiency and statistical efficiency. The standard Gaussian density leads to smooth estimates. The Bartlett–Epanechnikov product kernel is defined by

$$K(x) = (3/4)^d \, \Pi_{i=1}^{d} (1 - x_i^2)_+, \qquad x \in \mathbf{R}^d,$$

where $(a)_+ = \max\{0, a\}$. The Bartlett–Epanechnikov kernel has often spurious local maxima at the tails of the estimate, but it is computationally effective when one applies algorithms of Section 14.3.4, since this kernel is a polynomial and has a compact support.

Examples Figure 14.1 illustrates smoothing parameter selection with the help of volume functions. Volume functions are defined in Section 4.2. A kernel estimator has two parameters: the kernel function and the smoothing parameter. We illustrate in the figure the influence of these parameters on the kernel estimator. The upper row shows kernel estimates with the Bartlett–Epanechnikov product kernel and the lower row shows kernel estimates with the standard Gaussian kernel. The underlying density is that shown in Figure 2.7 and the sample size is $n = 200$. In the upper row, the kernel estimates have the smoothing parameters $h = 0.3$, $h = 0.6$, and $h = 0.9$. In the lower row, the kernel estimates have the smoothing parameters $h = 0.2$, $h = 0.3$, and $h = 0.45$. Notice that the optimal smoothing parameters are different for the two kernels. Figure 14.1 shows that the estimates with the standard Gaussian kernel have heavier and smoother tails. When the smoothing parameter is small, the supports of the estimates with the Bartlett–Epanechnikov kernel tend to split to a larger number of separated tail regions than the estimates with the standard Gaussian kernel.

Bibliographic Notes Rosenblatt (1956*b*) defined the kernel estimator using a uniform kernel. General kernel estimators were introduced by Parzen (1962). Cacoullos (1966) considered multivariate kernel density estimation with a scalar smoothing parameter, and Epanechnikov (1969) considered a vector of smoothing parameters. Deheuvels (1977) contributed to the discussion on a matrix of smoothing parameters.

14.3.2 Rates of Convergence

We prove here that kernel estimators can achieve the optimal rates of convergence in anisotropic Besov classes $B_{spq}(L)$ for locally identically distributed dependent observations.

The Basic Form of the Kernel Estimator and i.i.d. Observations For the mean integrated squared error we have the bias-variance decomposition (11.9). The expectation of the kernel estimator (14.3) may be written under the assumption of identically distributed observations as

$$E\hat{f}(x) = \int_{\mathbf{R}^d} f(x) K_h(x - y)\, dy \stackrel{def}{=} f * K_h(x), \tag{14.8}$$

and the variance may be written under the assumption of i.i.d. observations as

$$\mathrm{Var}(\hat{f}(x)) = \frac{1}{nh^d}\left[f * (K^2)_h(x) - (f * K_h(x))^2 \right]. \tag{14.9}$$

Thus the consistency may be derived from the convergence of the convolutions as stated in Lemma B.1.

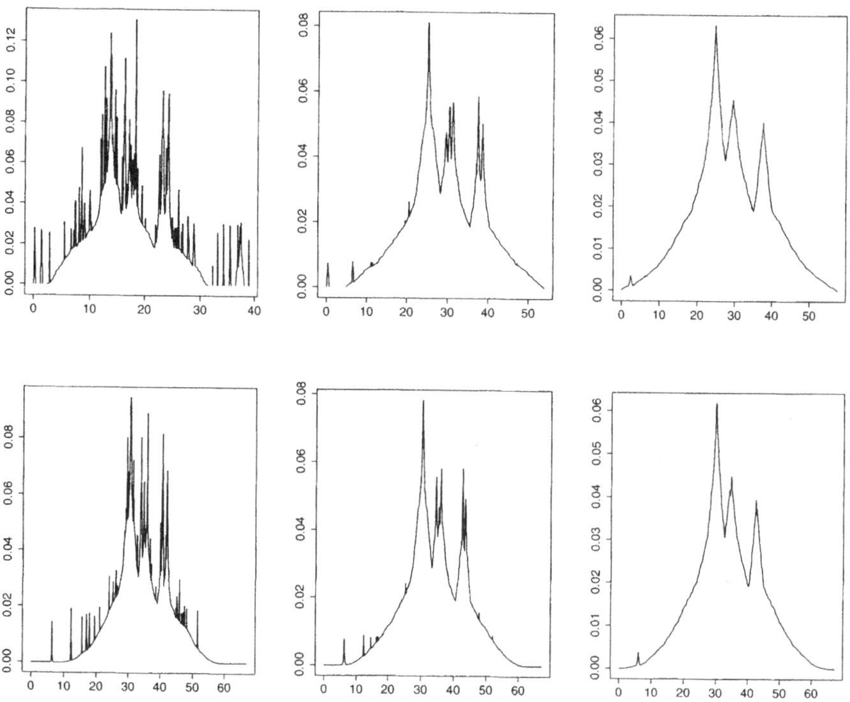

Figure 14.1 *(Volume functions of kernel estimates.)* Illustrated is smoothing parameter selection aided by volume functions. The kernel estimates in the upper row have the Bartlett–Epanechnikov product kernel and the smoothing parameters $h = 0.3$, $h = 0.6$, and $h = 0.9$. The kernel estimates in the lower row have the standard Gaussian kernel and the smoothing parameters $h = 0.2$, $h = 0.3$, and $h = 0.45$.

Lemma 14.1 *Let $X_1, \ldots, X_n$ be i.i.d. with density f, and assume that $\|f\|_2 < \infty$. Let $\hat{f}$ be the kernel estimator defined in (14.3), with the isotropic scaling as in (14.4). Assume that the kernel function K satisfies $\|K\|_1 < \infty$, $\|K\|_2 < \infty$, $\int_{\mathbf{R}^d} K = 1$, the support of K is compact, and smoothing parameter $h = h_n$ is such that $\lim_{n\to\infty} h_n = 0$ and $\lim_{n\to\infty} nh_n^d = \infty$. Then*

$$\lim_{n\to\infty} E\|f - \hat{f}\|_2 = 0.$$

Proof. The lemma follows from the bias-variance decomposition (11.9), from the bias formula (14.8), from the variance formula (14.9), and from an application of Lemma B.1. □

We may write the bias of the kernel estimator under the assumption of identically distributed observations as

$$\begin{aligned} E\hat{f}(x) - f(x) &= \int_{\mathbf{R}^d} [f(x+hz) - f(x)]K(z)\,dz \\ &= \sum_{k=1}^{s-1} h^k \sum_{|\alpha|=k} \frac{1}{\alpha!} \int_{\mathbf{R}^d} z^\alpha K(z)\,dz\, D^\alpha f(x) \\ &\quad + h^s \sum_{|\alpha|=s} \frac{1}{\alpha!} \int_{\mathbf{R}^d} z^\alpha K(z) D^\alpha f(x + \theta_\alpha hz)\,dz, \end{aligned}$$

where $0 < \theta_\alpha < 1$, $\alpha = (\alpha_1, \ldots, \alpha_d)$ is a multi-index; we have applied the Taylor expansion given in (B.1). This gives the rate of convergence $O(h^s)$ for the bias, for densities that are s times continuously differentiable, when the kernel is chosen so that the lower order moments vanish: $\int_{\mathbf{R}^d} z^\alpha K(z)\,dz = 0$, for $|\alpha| = k$, $k = 1, \ldots, s-1$.

We study the rates of convergence for the test bed of anisotropic Besov classes, for locally identically distributed observations, and for approximately independent observations. We need to replace the basic form of the kernel estimator with the anisotropic time-localized kernel estimator. The smoothness assumption and the assumption that the observations are locally identically distributed is needed in the bias calculation. The assumption that the observations are approximately independent is needed in the variance calculation.

Convolutions in Anisotropic Besov Classes We postpone the stochastic issues and study first the convergence of the convolutions. The expectation of the basic form of the kernel estimator under the assumption of identically distributed observations is equal to the convolution, as written in (14.8). Also in the case of approximately identically distributed observations the calculation of the convergence of the bias will be reduced to the calculation of the convergence of convolutions. Let $s = (s_1, \ldots, s_d)$, $p = (p_1, \ldots, p_d)$, $L = (L_1, \ldots, L_d)$, $s_i, L_i > 0$, $2 \le p_i < \infty$,

$$B = \left\{ f : [0,1]^d \to \mathbf{R} \,\middle|\, \rho^{(i)}_{s_i p_i \infty}(f) \le L_i,\ i = 1, \ldots, d \right\},$$

where $\rho^{(i)}_{s_i p_i \infty}$ is the Besov seminorm defined in (10.38). We construct a kernel estimator that converges with the optimal rates when the true density belongs to B.

Let

$$K(x) = \Pi_{i=1}^{d} g_i(x_i), \tag{14.10}$$

where

$$g_i(t) = \sum_{j=1}^{l_i} (-1)^{j+l_i} \binom{l_i}{j} \frac{1}{j} g\left(\frac{t}{j}\right), \qquad t \in \mathbf{R}.$$

Here l_i is the smallest integer greater than s_i, and $g : \mathbf{R} \to \mathbf{R}$ is such that $\int_{-\infty}^{\infty} g = 1$, $g(t) = g(-t)$, and $\int_{-\infty}^{\infty} g^2 < \infty$. Note that $\int_{\mathbf{R}^d} K = 1$, since $\int_{\mathbf{R}} g = \sum_{j=1}^{l} (-1)^{j+1} \binom{l}{j} = 1$ due to $\sum_{j=0}^{l} (-1)^{j+1} \binom{l}{j} = 0$.

Lemma 14.2 *Let K_h be defined in (14.5) with the kernel (14.10). We have that*

$$\sup_{f \in B} \|K_h * f - f\|_2 \le C \sum_{i=1}^{d} L_i h_i^{s_i},$$

for a positive constant C.

Proof. We can write the lth order difference of a function, using (10.35), as

$$\Delta_y^l f(x) = (-1)^{l+1} \sum_{j=1}^{l} (-1)^{j+l} \binom{l}{j} f(x + jy) - f(x).$$

We have for $x \in \mathbf{R}^d$, denoting $h^{-1} = (h_1^{-1}, \ldots, h_d^{-1})$, $x \cdot y = (x_1 y_1, \ldots, x_d y_d)$,

$$E\hat{f}(x) = \Pi_{i=1}^{d} h_i^{-1} \int_{\mathbf{R}^d} K((h^{-1}) \cdot (x - y)) f(y) dy = \int_{\mathbf{R}^d} K(y) f(x + h \cdot y) dy$$

and

$$E\hat{f}(x) - f(x) = \int_{\mathbf{R}^d} K(y) \left[f(x + h \cdot y) - f(x) \right] dy.$$

We write for shortness

$$z_i(y, h) = \sum_{j=1}^{i} y_j h_j e_j \in \mathbf{R}^d$$

where $y, h, e_j \in \mathbf{R}^d$, and e_j are defined in (10.39). We write the difference of f by way of telescoping

$$f(x + h \cdot y) - f(x) = \sum_{i=1}^{d} \left[f\left(x + z_i(y, h)\right) - f\left(x + z_{i-1}(y, h)\right) \right], \tag{14.11}$$

where we denote $z_0(y, h) = 0$. For simplicity we consider only the first term in the sum above. We have that

$$\int_{\mathbf{R}^d} K(y) [f(x + y_1 h_1 e_1) - f(x)] \, dy$$

$$
\begin{aligned}
&= \int_{\mathbf{R}} g_1(y_1)[f(x+y_1h_1e_1) - f(x)]\,dy_1 \\
&= \int_{\mathbf{R}} \sum_{j=1}^{l_1} (-1)^{j+l_1} \binom{l_1}{j} \frac{1}{j} g\left(\frac{y_1}{j}\right) f(x+y_1h_1e_1)\,dy_1 - f(x) \\
&= \int_{\mathbf{R}} g(t)\,\Delta_{th_1e_1}^{l_1} f(x)\,dt. \qquad (14.12)
\end{aligned}
$$

Exercise 14.1 calls for proving the analogous formula for the ith term. Then

$$
E\hat{f}(x) - f(x) = \sum_{i=1}^{d} \int_{\mathbf{R}^{i-1}} \Pi_{k=1}^{i-1} dy_k g_k(y_k) \int_{\mathbf{R}} g_i\,(t)\,\Delta_{th_ie_i}^{l_i} f(x + z_{i-1}(y,h))\,dt
$$

and

$$
\begin{aligned}
&\|E\hat{f} - f\|_2 \\
&\le \sum_{i=1}^{d} \int_{\mathbf{R}^{i-1}} \Pi_{k=1}^{i-1} dy_k |g_k(y_k)| \int_{\mathbf{R}} |g_i\,(t)\,|\, \|\Delta_{th_ie_i}^{l_i} f(x + z_{i-1}(y,h))\|_{L_2(dx)}\,dt \\
&\le \sum_{i=1}^{d} \int_{\mathbf{R}^{i-1}} \Pi_{k=1}^{i-1} dy_k |g_k(y_k)| \int_{\mathbf{R}} |g_i\,(t)\,|\, \|\Delta_{th_ie_i}^{l_i} f(x + z_{i-1}(y,h))\|_{L_{p_i}(dx)}\,dt \\
&\le \sum_{i=1}^{d} L_i h_i^{s_i} \int_{\mathbf{R}} |t|^{s_i} |g_i\,(t)\,|\,dt \int_{\mathbf{R}^{i-1}} \Pi_{k=1}^{i-1} |g_k(y_k)|\,dy_k,
\end{aligned}
$$

where we have used the notation $\|\cdot\|_{L_2(dx)}$ to indicate that the integration is with respect to variable x. We used also the fact that for $f : [0,1]^d \to \mathbf{R}$, $\|f\|_2 \le \|f\|_{p_i}$, when $2 \le p_i < \infty$. This follows from Jensen's inequality; see (B.21). □

Variance We prove the bound for the variance under weaker than i.i.d. assumptions, assuming only approximately independent observations, and under the setting of approximately identically distributed observations. We denote with f_i the density of X_i, $i = 1, \ldots, n$.

Lemma 14.3 *Let $X_1, \ldots, X_n \in \mathbf{R}^d$ be locally identically distributed with rate $o(1)$ in the sense of Definition 9.1, Eq. (9.1), with rate function $r : [0,\infty) \to [0,\infty)$ that satisfies Assumption 9.1. Let either Assumption 9.2 on the β-mixing coefficients or Assumption 9.3 with (14.23)–(14.25) on the α-mixing coefficients be satisfied. Let $\hat{f}$ be the anisotropic time-localized kernel estimator defined in (14.7) and (14.5). Assume that weights $p_i = p_{i,n}$ satisfy*

$$
\limsup_{n\to\infty} \max_{i=1,\ldots,n} \frac{p_i}{\pi_i} < \infty, \qquad (14.13)
$$

where

$$
\pi_i = \frac{r[\delta(t(i),t_0)]}{\sum_{j=1}^{n} r[\delta(t(j),t_0)]}\,. \qquad (14.14)
$$

Let $h = (h_1, \ldots, h_d)$ be such that

$$\lim_{n\to\infty} \max_{i=1,\ldots,d} h_i = 0, \qquad \lim_{n\to\infty} n\,\Pi_{i=1}^d h_i = \infty.$$

Then

$$\int_{\mathbf{R}^d} \mathrm{Var}\left(\hat{f}\right) = O\left(\left(n\,\Pi_{i=1}^d h_i\right)^{-1}\right),$$

as $n \to \infty$.

Proof. By Assumption 9.1,

$$\limsup_{n\to\infty} \max_{i=1,\ldots,n} \pi_i \cdot \frac{n}{r[\delta(t(i), t_0)]} < \infty. \tag{14.15}$$

Denote

$$\mathrm{Cov}_{ij} = \int_{\mathbf{R}^d} \mathrm{Cov}\left(K_h(x - X_i), K_h(x - X_j)\right) dx.$$

We have

$$\begin{aligned}
\int_{\mathbf{R}^d} \mathrm{Var}\left(\hat{f}\right) &= \sum_{i,j=1}^n p_i p_j \mathrm{Cov}_{ij} \\
&\preccurlyeq \sum_{i,j=1}^n \pi_i \pi_j \mathrm{Cov}_{ij} && (14.16) \\
&\preccurlyeq n^{-2} \sum_{i,j=1}^n r[\delta(t(i), t_0)] \cdot r[\delta(t(j), t_0)]\ \mathrm{Cov}_{ij}, && (14.17)
\end{aligned}$$

where in (14.16) we applied assumption (14.13) on the weights and in (14.17) we applied (14.15). Applying Lemma B.1 obtains

$$\begin{aligned}
\mathrm{Cov}_{ii} &= \int_{\mathbf{R}^d} \mathrm{Var}\left(K_h(x - X_i)\right) dx \\
&\leq \int_{\mathbf{R}^d} dx \int_{\mathbf{R}^d} K_h^2(x - y) f_i(y)\, dy && (14.18) \\
&\sim \left(\Pi_{i=1}^d h_i\right)^{-1} \int_{\mathbf{R}^d} K^2, && (14.19)
\end{aligned}$$

as $n \to \infty$. It is left to calculate Cov_{ij} when $i \neq j$. The calculation is different for the β-mixing coefficients and for the α-mixing coefficients.

Covariance with β-Mixing We assume that Assumption 9.2 holds. We generalize Theorem 2.1 of Viennet (1997) to the nonstationary case. We have for $x \in \mathbf{R}^d$, $i, j = 1, \ldots, n$, $i \neq j$, after applying the Cauchy–Schwartz inequality,

$$\mathrm{Cov}\left(K_h(x - X_i), K_h(x - X_j)\right)$$

$$
\begin{aligned}
&= \int_{\mathbf{R}^d \times \mathbf{R}^d} K_h(x-y) K_h(x-z)\, d\left(P_{i,j} - P_i \otimes P_j\right)(y,z) \qquad (14.20) \\
&\le \left(\int_{\mathbf{R}^d \times \mathbf{R}^d} K_h^2(x-y)\, d\left|P_{i,j} - P_i \otimes P_j\right|(y,z) \right)^{1/2} \\
&\quad \times \left(\int_{\mathbf{R}^d \times \mathbf{R}^d} K_h^2(x-z)\, d\left|P_{i,j} - P_i \otimes P_j\right|(y,z) \right)^{1/2},
\end{aligned}
$$

where P_i is the distribution of X_i, P_j is the distribution of X_j, and $P_{i,j}$ is the distribution of (X_i, X_j). Denote by $Q_{i,j}^{(i)}$ the first marginal distribution of $|P_{i,j} - P_i \otimes P_j|$: $Q_{i,j}^{(i)}(A) = \int_{A \times \mathbf{R}^d} d\left|P_{i,j} - P_i \otimes P_j\right|(y,z)$, and denote by $Q_{i,j}^{(j)}$ the second marginal distribution. Denote by $g_{i,j}^{(l)} : \mathbf{R}^d \to \mathbf{R}$ the Radon–Nikodym density of $Q_{i,j}^{(l)}$ with respect to P_l, $l = i, j$:

$$
g_{i,j}^{(l)} = \frac{dQ_{i,j}^{(l)}}{dP_l}, \qquad l = i, j.
$$

Then we have

$$
\int_{\mathbf{R}^d \times \mathbf{R}^d} K_h^2(x-y)\, d\left|P_{i,j} - P_i \otimes P_j\right|(y,z) = \int_{\mathbf{R}^d} K_h^2(x-y) g_{i,j}^{(i)}(y) f_i(y)\, dy,
$$

and

$$
\int_{\mathbf{R}^d \times \mathbf{R}^d} K_h^2(x-z)\, d\left|P_{i,j} - P_i \otimes P_j\right|(y,z) = \int_{\mathbf{R}^d} K_h^2(x-z) g_{i,j}^{(j)}(z) f_j(z)\, dz.
$$

Thus

$$
\begin{aligned}
&\mathrm{Cov}\left(K_h(x-X_i), K_h(x-X_j)\right) \\
&\quad \le \int_{\mathbf{R}^d} K_h^2(x-y) \left(g_{i,j}^{(i)}(y) f_i(y) + g_{i,j}^{(j)}(y) f_j(y) \right) dy.
\end{aligned}
$$

We next obtain

$$
\int_{\mathbf{R}^d} g_{i,j}^{(i)} f_i = \int_{\mathbf{R}^d \times \mathbf{R}^d} d\left|P_{i,j} - P_i \otimes P_j\right| = 2 \left\| P_{i,j} - P_i \otimes P_j \right\|_{tot} = 2\beta(X_i, X_j)
$$

and also $\int_{\mathbf{R}^d} g_{i,j}^{(j)} f_j = 2\beta(X_i, X_j)$, where $\beta(X_i, X_j)$ is defined in (9.8). Now applying Lemma B.1 yields

$$
\mathrm{Cov}_{ij} \lesssim (\Pi_{i=1}^d h_i)^{-1}\, 4\, \beta(X_i, X_j) \int_{\mathbf{R}^d} K^2, \qquad (14.21)
$$

where $a_n \lesssim b_n$ means that $\limsup_{n \to \infty} a_n / b_n \le 1$. Equation (14.21) holds also for $i = j$, as noted in (14.19). Applying (14.17) and (14.21), we get

$$
\int_{\mathbf{R}^d} \mathrm{Var}\left(\hat{f}\right)
$$

$$
\begin{aligned}
&\preccurlyeq (n\Pi_{i=1}^d h_i)^{-1} n^{-1} \sum_{i,j=1}^{n} r[\delta(t(i), t_0)] \cdot r[\delta(t(j), t_0)] \; \beta(X_i, X_j) \\
&\preccurlyeq O\left((n\Pi_{i=1}^d h_i)^{-1}\right) \\
&= o(1) \qquad (14.22)
\end{aligned}
$$

as $n \to \infty$. Here we applied also Assumption 9.2 on the β-mixing coefficients and the assumption $\lim_{n\to\infty} n\Pi_{i=1}^d h_i = \infty$.

Covariance with α-Mixing In the case of α-mixing we have to make the following technical assumptions, which are similar to the assumptions in Bosq (1998, thm. 2.1). Let the densities f_t have compact supports contained in a fixed compact set:

$$
\text{supp}(f_i) \subset A, \qquad i = 1, 2, \ldots, \qquad \text{for some compact set } A. \qquad (14.23)
$$

For $i \neq j$ and for some $p > 2$,

$$
\|d\,(P_{i,j} - P_i \otimes P_j)\|_p \leq C_p, \qquad (14.24)
$$

where P_i is the distribution of X_i, P_j is the distribution of X_j, $P_{i,j}$ is the distribution of (X_i, X_j), and C_p is a positive constant. Constant a in Assumption 9.3 satisfies

$$
a > 2\frac{p-1}{p-2}. \qquad (14.25)
$$

We modify the proof of Theorem 2.1 in Bosq (1998) for the nonstationary case. We apply Billingsley's inequality:

$$
\text{Cov}(X, Y) \leq 4\|X\|_\infty \|Y\|_\infty \alpha(X, Y), \qquad (14.26)
$$

which is proved in Doukhan (1994, lmm. 3, p. 10), and Bosq (1998, cor. 1.1), Corollary 1.1. We will use the shorthand notation

$$
H^d = \prod_{i=1}^{d} h_i.
$$

Equation (14.26) implies that

$$
\text{Cov}\left(K_h(x - X_i), K_h(x - X_i)\right) \leq 4H^{-2d}\|K\|_\infty^2 \alpha(X_i, X_j).
$$

On the other hand, from (14.20) and by Hölder's inequality for $1/p + 1/q = 1$, we get

$$
\text{Cov}\left(K_h(x - X_i), K_h(x - X_j)\right) \leq \|d\,(P_{i,j} - P_i \otimes P_j)\,\|_p \cdot H^{-2d/p}\|K\|_q^2,
$$

since $\|K_h(x - \cdot)\|_q = H^{-d}H^{d/q}\|K\|_q$. Since f_i have compact support (assumption (14.23)), the volume of the support of the kernel estimator can be bounded by a constant C_s (we will assume from now on that $h_i \leq 1$). Thus we have

$$
\text{Cov}_{ij} \leq C_s \cdot \max\left\{C_p H^{-2d/p}\|K\|_q^2, 4H^{-2d}\|K\|_\infty^2 \alpha(X_i, X_j)\right\},
$$

where C_s is a positive constant and we used also assumption (14.24). Let

$$m = \lceil H^{-2d/(qa)} \rceil.$$

The integer m solves asymptotically $mH^{-2d/p} \approx m^{1-a}H^{-2d}$. Now, continuing from (14.17), and applying Assumption 9.3, we have

$$\begin{aligned}
& n^{-2} \sum_{i,j=1,i\neq j}^{n} r[\delta(t(i),t_0)] \cdot r[\delta(t(j),t_0)] \,\mathrm{Cov}_{ij} \qquad (14.27) \\
&= \\
&\leq C_s n^{-2} \left[\sum_{i=1}^{n} \sum_{j=1,1\leq|j-i|\leq m}^{n} r[\delta(t(i),t_0)] \cdot r[\delta(t(i),t_0)] \, C_p H^{-2d/p} \|K\|_q^2 \right. \\
&\qquad \left. + \sum_{i=1}^{n} \sum_{j=1,|j-i|>m}^{n} r[\delta(t(i),t_0)] \cdot r[\delta(t(j),t_0)] \, 4H^{-2d} \|K\|_\infty^2 \alpha(X_i,X_j) \right] \\
&\leq CC_s n^{-1} \left[C_p H^{-2d/p} \|K\|_q^2 \cdot m + 4H^{-2d} \|K\|_\infty^2 \cdot m^{1-a} \right] \\
&= O\left(\left(nH^d\right)^{-1} \right), \qquad (14.28)
\end{aligned}$$

since

$$mH^{-2d/p} < H^{-2d[1/(qa)+1/p]} < H^{-d},$$

since $1/(qa) < (p-2)/(2p)$ (remember, $q = p/(p-1)$ and $a > 2(p-1)/(p-2)$ by assumption (14.25)), and

$$m^{1-a}H^{-2d} < H^{2d(a-1)/(qa)} < H^{-d},$$

since $(a-1)/a > q/2$ (note $1 - 1/a > p/[2(p-1)] = q/2$). Combining (14.17), (14.19), and (14.28) gives

$$\int_{\mathbf{R}^d} \mathrm{Var}\left(\hat{f}\right) = O\left((nH^d)^{-1}\right) = o(1), \qquad (14.29)$$

which proves the lemma. □

Rates of Convergence of the MISE We combine the variance and bias calculations to get the rates of convergence of the mean integrated squared error (MISE) for the anisotropic time-localized kernel estimator. The case $p_i < 2$ for some $i = 1, \ldots, d$ is considered in Section 16.4.2.

Let $s = (s_1, \ldots, s_d)$, $p = (p_1, \ldots, p_d)$, $L = (L_1, \ldots, L_d)$, $s_i, L_i > 0$, $2 \leq p_i < \infty$ and

$$\mathcal{F} = \left\{ f : [0,1]^d \to \mathbf{R} \;\middle|\; \rho^{(i)}_{s_i p_i \infty}(f) \leq L_i, \; i = 1, \ldots, d \right\} \cap \mathcal{F}_{den}, \qquad (14.30)$$

where $\rho^{(i)}_{s_i p_i \infty}$ is the Besov seminorm defined in (10.38) and $\mathcal{F}_{den}$ is the class of densities.

Theorem 14.4 *Let $X_1, \ldots, X_n \in \mathbf{R}^d$ be locally identically distributed with rate $O(n^{-1/2})$ in the sense of Definition 9.1, Eq. (9.2). Assume otherwise the conditions of Lemma 14.3. Let $\hat{f}$ be the time-localized kernel estimator defined in (14.7), with the anisotropic scaling as in (14.5), with the kernel (14.10). Let the smoothing parameters be*

$$h_i = n^{-r_i}, \qquad r_i = \frac{\sigma}{s_i(2\sigma+1)}, \tag{14.31}$$

for $i = 1, \ldots, d$, where

$$\sigma = \left(\sum_{i=1}^{d} s_i^{-1} \right)^{-1}.$$

Then for $t_0 \in \mathbf{T}$,

$$\limsup_{n\to\infty} \sup_{f_1,\ldots,f_n \in \mathcal{F}} n^{2\sigma/(2\sigma+1)} E\|f - \hat{f}\|_2^2 < \infty,$$

where f_i is the density of X_i and f is the density associated to time point t_0 in Definition 9.1.

Proof. We give first a bound for the bias of the estimator and then apply Lemma 14.3 to get a bound for the variance.

Bias We prove that

$$\sup_{f_1,\ldots,f_n \in \mathcal{F}} \left\| E\hat{f} - f \right\|_2^2 = O\left(\sum_{i=1}^{d} h_i^{2s_i} + n^{-1} \right), \tag{14.32}$$

as $n \to \infty$. We have for $x \in \mathbf{R}^d$ that

$$\begin{aligned} E\hat{f}(x) - f(x) &= \sum_{i=1}^{n} p_i \int_{\mathbf{R}^d} K_h(x-y) f_i(y)\, dy - f(x) \\ &= \sum_{i=1}^{n} p_i \left(K_h * f_i(x) - f_i(x) \right) \\ &\quad + \sum_{i=1}^{n} p_i (f_i(x) - f(x)) \\ &\stackrel{def}{=} A_1(x) + A_2(x). \end{aligned} \tag{14.33}$$

Term A_1 Lemma 14.2 gives the bound

$$\sup_{f_1,\ldots,f_n \in \mathcal{F}} \|A_1\|_2 = O\left(\sum_{i=1}^{d} L_i h_i^{2s_i} \right).$$

Term A_2 Applying assumption (14.13) on the weights and (14.15) obtains

$$\begin{aligned} \|A_2\|_2 &\preccurlyeq \sum_{i=1}^{n} \pi_i \, \|f_i - f\|_2 \\ &\preccurlyeq n^{-1} \sum_{i=1}^{n} r[\delta(t(i), t_0)] \, \|f_i - f\|_2 \, , \end{aligned} \tag{14.34}$$

where we denote $a_n \preccurlyeq b_n$ when $\limsup_{n\to\infty} a_n/b_n < \infty$. Thus, applying the assumption (9.2) of the locally identically distribution with rate $O(n^{-1/2})$, we have

$$\sup_{f_1,\ldots,f_n \in \mathcal{F}} \|A_2\|_2^2 = O(n^{-1}).$$

Collecting the Results The theorem follows from the bias-variance decomposition (11.9), from the bias bound in (14.32), and from the variance bound in Lemma 14.3, when we plug in the formula for h_i in (14.31). □

Consistency We give conditions for the consistency of the time-localized kernel estimator.

Theorem 14.5 *Assume that the conditions of Lemma 14.3 hold. (In particular, we need only the rate $o(1)$ in the definition of locally identically distributed observations.) Assume that for all $i = 1, \ldots, n$,*

$$\|f_i\|_2 \le C, \tag{14.35}$$

for a positive constant C. Let $\hat{f}$ be the time-localized kernel estimator defined in (14.7), with the isotropic scaling as in (14.4). Let $h = h_n > 0$ be such that $\lim_{n\to\infty} h = 0$, $\lim_{n\to\infty} nh^d = \infty$. Then for $t_0 \in \mathbf{T}$,

$$\lim_{n\to\infty} E\|\hat{f} - f\|_2 = 0,$$

where f is the density associated with time point t_0 in Definition 9.1.

Proof. We prove that

$$\left\| E\hat{f} - f \right\|_2^2 = o(1) \tag{14.36}$$

as $n \to \infty$. We apply the decomposition (14.33). The assumption in (14.35) and Lemma B.1 imply that

$$\max_{i=1,\ldots,n} \|K_h * f_i - f_i\|_2 \to 0$$

as $n \to \infty$. This and the assumption $\max_{i=1,\ldots,d} h_i \to 0$ imply that $\|A_1\|_2 = o(1)$. Equation (14.34) and the assumption (9.1) of the locally identically distribution with rate $o(1)$ imply that $\|A_2\|_2 = o(1)$. The theorem follows from the bias-variance decomposition (11.9), from the bias bound in (14.36), and from the variance bound in Lemma 14.3. □

14.3.3 Inverse Problems

We show how kernel estimation can be used to solve two statistical inverse problems: the problem of measurement errors in observations and the problem of emission tomography.

Measurement Errors We consider a model of measurement errors as in (9.19) and (9.20). We observe i.i.d. random vectors $Y_1, \ldots, Y_n \in \mathbf{R}^d$, where the observations Y_i are contaminated with measurement errors. We write

$$Y_i = X_i + \epsilon_i, \qquad i = 1, \ldots, n,$$

where X_i and ϵ_i are independent. The density of Y_i is $Af = a * f$, where $a : \mathbf{R}^d \to \mathbf{R}$ is a known density of the measurement errors ϵ_i, and f is the density of X_i. We want to estimate density f. Let kernel estimator $\hat{f}$ be defined in (14.3), where kernel function $K(\,\cdot\,; h) : \mathbf{R}^d \to \mathbf{R}$, defined in terms of its Fourier transform by

$$FK(\omega; h) = \frac{L(\|\omega\|)}{Fa(\omega/h)}, \qquad \omega \in \mathbf{R}^d, \tag{14.37}$$

where $h > 0$ is the smoothing parameter and L is chosen as

$$L(t) = L_r(t) = \begin{cases} (1 - t^r) I_{[0,1]}(t), & 0 < r < \infty, \\ I_{[0,1]}(t), & r = \infty, \end{cases} \qquad t \geq 0. \tag{14.38}$$

Regularized Inverse The kernel estimator is based on the fact that we have constructed a regularized inverse for the convolution operator A. Define operator $B_h : \{g : \mathbf{R}^d \to \mathbf{R}\} \to \{f : \mathbf{R}^d \to \mathbf{R}\}$ by

$$B_h g = g * K_h(\,\cdot\,; h),$$

where $g : \mathbf{R}^d \to \mathbf{R}$ and K_h is defined by (14.37). We call operator B_h a regularized inverse because for $f : \mathbf{R}^d \to \mathbf{R}$,

$$\lim_{h \to 0} \|B_h A f - f\|_2 = 0. \tag{14.39}$$

Equation (14.39) follows from (14.42) below. Note that Lemma 9.1 gives the inverse of the convolution operator.

Rates of Convergence The rate of convergence of the kernel estimator depends on the smoothness on the density a. The smoothness is imposed as an assumption on the tails of the Fourier transform Fa of density a:

$$Fa(\omega) \geq B\|\omega\|^{-b}, \tag{14.40}$$

where B and b are positive constants. Let us denote with $\mathcal{F} = \mathcal{F}_s(L, B_\infty)$ the following class of densities $f : \mathbf{R}^d \to \mathbf{R}$:

$$\mathcal{F} = \{f : \rho_s(f) \leq L\} \cap \{f : \|f\|_\infty \leq B_\infty\} \cap \mathcal{F}_{den},$$

where ρ_s is the Sobolev seminorm defined in (10.27), B_∞ is a positive constant, and $\mathcal{F}_{den}$ is the class of densities.

Theorem 14.6 *Let kernel K of $\hat{f}$ be defined in (14.37) and (14.38) with $r \geq s$. Let the smoothing parameter h satisfy*

$$h = h_n \asymp n^{-1/(2s+2b+d)}. \tag{14.41}$$

Then

$$\limsup_{n\to\infty} n^{2s/(2s+2b+d)} \sup_{f\in\mathcal{F}} E\left\|\hat{f} - f\right\|_2^2 < \infty.$$

Proof. We have that for $f \in \mathcal{F}$, for $h > 0$,

$$\begin{aligned}
\left\|E\hat{f} - f\right\|_2^2 &= \|(Af) * K_h(\cdot\,; h) - f\|_2^2 \\
&= (2\pi)^{-d}\|F(a * f * K_h(\cdot\,; h)) - Ff\|_2^2 \\
&= (2\pi)^{-d}\|Fa \cdot Ff \cdot FK_h(\cdot\,; h) - Ff\|_2^2 \\
&= (2\pi)^{-d}\int_{\mathbf{R}^d} \{(Ff)(\omega)\cdot[L(h\|\omega\|) - 1]\}^2\, d\omega \\
&\leq (2\pi)^{-d}\int_{\mathbf{R}^d} \|h\omega\|^{2s}|(Ff)(\omega)|^2\, d\omega \\
&\leq L^2h^{2s},
\end{aligned} \tag{14.42}$$

where we used (B.6) and (B.7), the fact $FK_h(\omega; h) = (FL)(h\omega)/(Fa)(\omega)$, and the fact,

$$\begin{aligned}
|L(\|h\omega\|) - 1| &\leq \begin{cases} \|h\omega\|^s, & \text{when } \|h\omega\| \leq 1 \\ 1, & \text{when } \|h\omega\| \geq 1 \end{cases} \\
&\leq \|h\omega\|^s,
\end{aligned} \tag{14.43}$$

because $r \geq s$. Using (14.9), we get

$$\begin{aligned}
\int_{\mathbf{R}^d} \mathrm{Var}\left(\hat{f}\right) &\leq n^{-1}\|f\|_\infty \int_{\mathbf{R}^d} K_h^2(\cdot\,; h) \\
&= n^{-1}\|f\|_\infty(2\pi)^{-d}\int_{\mathbf{R}^d} L^2(h\|\omega\|)/(Fa)^2(\omega)\, d\omega \\
&= n^{-1}h^{-d}\|f\|_\infty(2\pi)^{-d}\int_{\mathbf{R}^d} L^2(\|\omega\|)/(Fa)^2(\omega/h)\, d\omega \\
&\leq n^{-1}h^{-d-2r}\|f\|_\infty(2\pi)^{-d}B^{-2}\int_{\mathbf{R}^d} \|\omega\|^{2r}L^2(\|\omega\|)\, d\omega.
\end{aligned}$$

The theorem follows from the bias-variance decomposition (11.9), from the bias bound in (14.42), and from the variance bound (14.44) when we plug in the formula for h in (14.41). □

Tomography We consider a model of emission tomography as in (9.27) and (9.29). We observe realizations of $Y_1, \ldots, Y_n \in \mathbf{Y}$, that are i.i.d. random variables with density Af. Operator A is a constant times the Radon transform, as defined in (9.29). Sample space is $\mathbf{Y} = \mathbf{S}_{d-1} \times [0, \infty)$, and we write $Y_i = (S_i, U_i)$. We define the estimator of the density $f : \mathbf{R}^d \to \mathbf{R}$ by

$$\hat{f}(x) = \frac{1}{n} \sum_{i=1}^{n} K\left(S_i^T x - U_i; h\right), \qquad x \in \mathbf{R}^d, \tag{14.44}$$

where kernel $K(\,\cdot\,; h) : \mathbf{R} \to \mathbf{R}$ is defined by giving its Fourier transform

$$FK(t; h) = C_K |t|^{d-1} L(h|t|), \qquad t \in \mathbf{R}, \tag{14.45}$$

with

$$C_K = \frac{1}{2} (2\pi)^{-d+1} \mu(\mathbf{S}_{d-1}). \tag{14.46}$$

Here $h > 0$ is the smoothing parameter, and $L : [0, \infty) \to \mathbf{R}$ is a kernel function defined in (14.38). We have

$$K(u; h) = 2(2\pi)^{-1} C_K \int_0^\infty \cos(tu) t^{d-1} L(ht)\, dt, \qquad u \in \mathbf{R}.$$

In the case $d = 2$ and $L = I_{[0,1]}$ we have

$$K(u; h) = 2(2\pi)^{-1} C_K h^{-1} \cdot \left(\frac{\cos(u/h)}{(u/h)^2} + \frac{\sin(u/h)}{u/h} - \frac{1}{(u/h)^2} \right).$$

Figure 14.2 shows the kernel function $K(\,\cdot\,; h)$ for the smoothing parameter values $h = 0.3, 0.5, 0.6$. Panel (a) shows the case where $d = 2$ and $r = 2$, panel (b) shows the case where $d = 2$ and $r = \infty$, and panel (c) shows the case where $d = 4$ and $r = 2$. The kernel is decreasing with the oscillating decoration. The amplitude and frequency of the oscillation increases as the smoothing parameter decreases, but the kernel does not localize. When the parameter r is increased, the amplitude of the oscillation increases.

Regularized Inverse The kernel estimator is based on the fact that we can construct a regularized inverse B_h for the operator A. Define operator $B_h : \{g : \mathbf{S}_{d-1} \times \mathbf{R} \to \mathbf{R}\} \to \{f : \mathbf{R}^d \to \mathbf{R}\}$ by

$$(B_h g)(x) = \int_{\mathbf{S}_{d-1}} d\mu(s) \int_{-\infty}^{\infty} K(s^T x - u; h) g(s, u)\, du, \qquad x \in \mathbf{R}^d, \tag{14.47}$$

where $g : \mathbf{S}_{d-1} \times \mathbf{R} \to \mathbf{R}$ and $K(\,\cdot\,; h)$ is defined by (14.45). We may call operator B_h a regularized inverse because for $f : \mathbf{R}^d \to \mathbf{R}$,

$$\lim_{h \to 0} \|B_h A f - f\|_2 = 0. \tag{14.48}$$

We proof (14.48) in Lemma 14.8 below, where also the rate of convergence is given. Note that the inverse of the Radon transform is given in Lemma 9.3.

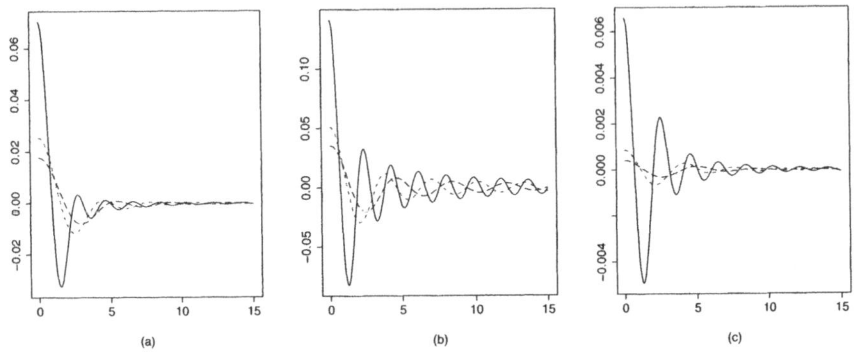

Figure 14.2 *(Kernels of tomography.)* Shown are the kernel function $K(\cdot\,; h)$ when $h = 0.3, 0.5, 0.6$: (a) $d = 2$ and $r = 2$; (b) $d = 2$ and $r = \infty$; (c) $d = 4$ and $r = 2$. The solid line corresponds to the case $h = 0.3$, the dashed line corresponds to the case $h = 0.5$, and the line with longer dashes corresponds to the case $h = 0.6$.

Rates of Convergence The following theorem gives the rate of convergence of the estimator $\hat{f}$, defined in (14.44). Let $\mathcal{F}$ be the collection of densities $f : \mathbf{R}^d \to \mathbf{R}$ defined by

$$\mathcal{F} = \{f : \rho_s(f) \le L\} \cap \{f : \|f\|_\infty \le B_\infty,\ \text{supp}(f) \subset A\} \cap \mathcal{F}_{den},$$

where ρ_s is the Sobolev semi-norm defined in (10.27), $A \subset \mathbf{R}^d$ with volume$(A) < \infty$, and $\mathcal{F}_{den}$ is the class of densities.

Theorem 14.7 *Let kernel L be defined in (14.38) with $r \ge s$. Let the smoothing parameter h of $\hat{f}$ satisfy*

$$h = h_n \asymp n^{-1/(2s+2d-1)}. \tag{14.49}$$

Then

$$\limsup_{n\to\infty} n^{2s/(2s+2d-1)} \sup_{f\in\mathcal{F}} E\left\|\hat{f} - f\right\|_2^2 < \infty.$$

Proof. The theorem follows from the bias-variance decomposition (11.9), from the bias bound in Lemma 14.8 below and from the variance bound in Lemma 14.9 below when we plug in the formula for h in (14.49). □

Lemma 14.8 *For $f \in \mathcal{F}$ and $h > 0$,*

$$\left\|E\hat{f} - f\right\|_2^2 \le L^2 h^{2s}.$$

Proof. We have for $x \in \mathbf{R}^d$,

$$E\hat{f}(x) \quad = \quad \frac{1}{n}\sum_{i=1}^{n} E\left(E\left[K\left(S_i^T x - U_i; h\right) \mid S_i\right]\right)$$

$$= \frac{1}{n}\sum_{i=1}^{n} E\left(\int_{-\infty}^{\infty} K(S_i^T x - u; h) Rf(S_i, u)\, du\right) \qquad (14.50)$$

$$= B_h Af(x), \qquad (14.51)$$

where we applied (9.28) in (14.50) and B_h is defined in (14.47). For fixed $s \in \mathbf{S}_{d-1}$, we denote with $F_1(Rf)(s,t)$ the Fourier transform of $u \mapsto Rf(s,u)$. We have

$$F_1(Rf)(s,t) = (Ff)(ts), \qquad t \in \mathbf{R}. \qquad (14.52)$$

For a proof of (14.52), see the projection theorem in Appendix B.11. We have that for fixed $s \in \mathbf{S}_{d-1}$ and fixed $x \in \mathbf{R}^d$, the Fourier transform of $u \mapsto K(s^T x - u; h)$ is

$$t \mapsto FK(t;h)\exp\{-its^T x\},$$

where $FK(\,\cdot\,; h)$ is the Fourier transform of $K(\,\cdot\,; h)$. Thus for fixed $s \in \mathbf{S}_{d-1}$ and fixed $x \in \mathbf{R}^d$,

$$\begin{aligned} &\int_{-\infty}^{\infty} K(s^T x - u; h) Rf(s,u)\, du \\ &= (2\pi)^{-1}\int_{-\infty}^{\infty} FK(t;h)\exp\{-its^T x\} F_1(Rf)(s,t)\, dt \\ &= (2\pi)^{-1} C_K \int_{-\infty}^{\infty} |t|^{d-1} L(h|t|)\exp\{-its^T x\}(Ff)(ts)\, dt. \end{aligned}$$

Thus

$$\begin{aligned} &B_h Af(x) \\ &= 2(2\pi)^{-1}\frac{C_K}{\mu(\mathbf{S}_{d-1})}\int_{\mathbf{S}_{d-1}} d\mu(s)\int_0^{\infty} t^{d-1} L(ht)\exp\{-its^T x\}(Ff)(ts)\, dt \\ &= 2(2\pi)^{-1}\frac{C_K}{\mu(\mathbf{S}_{d-1})}\int_{\mathbf{R}^d} L(h\|\omega\|)\exp\{-i\omega^T x\}(Ff)(\omega)\, d\omega, \qquad (14.53) \end{aligned}$$

where we used the integration rule in (B.2). Combining (14.51) and (14.53), and using the definition of C_K in (14.46), gives

$$\begin{aligned} &E\hat{f}(x) - f(x) \\ &= B_h Af(x) - f(x) \\ &= (2\pi)^{-d}\int_{\mathbf{R}^d} \left[L(h\|\omega\|) - 1\right](Ff)(\omega)\exp\{-ix^T\omega\}\, d\omega, \end{aligned}$$

where we used the formula for the Fourier inverse in (B.5). Thus using (14.43) gives

$$\begin{aligned} \|E\hat{f} - f\|_2^2 &= (2\pi)^{-d}\int_{\mathbf{R}^d} \left[L(h\|\omega\|) - 1\right]^2 |(Ff)(\omega)|^2\, d\omega \\ &\le (2\pi)^{-d}\int_{\mathbf{R}^d} \|h\omega\|^{2s} |(Ff)(\omega)|^2\, d\omega \\ &\le h^{2s} L^2. \end{aligned}$$

We have proved the lemma. □

Lemma 14.9 *Let the support of f be contained in set $A \subset \mathbf{R}^d$ that satisfies* volume$(A) < \infty$. *Let f be bounded: $\|f\|_\infty \le B_\infty$. Then, for $h > 0$,*

$$\int_{\mathbf{R}^d} \mathrm{Var}\left(\hat{f}\right) \le C n^{-1} h^{-2d+1},$$

for a positive constant C.

Proof. We have for $x \in \mathbf{R}^d$,

$$\begin{aligned}
\mathrm{Var}\left(\hat{f}(x)\right) &\le n^{-1} E\left(E\left[K^2\left(S_i^T x - U_i; h\right) \mid S_i\right]\right) \\
&= n^{-1} \int_{\mathbf{S}_{d-1}} d\mu(s) \int_{-\infty}^{\infty} K^2(s^T x - u; h) Rf(s,u)\, du \\
&\le n^{-1} \mathrm{volume}(A) B_\infty \mu(\mathbf{S}_{d-1}) \int_{-\infty}^{\infty} K^2(u; h)\, du \\
&= n^{-1} \mathrm{volume}(A) B_\infty \mu(\mathbf{S}_{d-1}) \\
&\quad \times (2\pi)^{-1} C_K^2 \int_{-\infty}^{\infty} |t|^{2d-2} L^2(h|t|)\, dt.
\end{aligned}$$

We have proved the lemma. □

14.3.4 Algorithms for Computing Kernel Estimates

Algorithm Using Evaluation Trees We describe an algorithm for creating an evaluation tree for a kernel estimate. The evaluation tree is defined in Definition 13.2. We evaluate the kernel estimate on the grid by going through the observations, finding which gridpoints belong to the support of the kernel centered at the current observation, and updating the value of the estimate at those grid points. Because we create a binary search tree from the grid points where the estimate is positive, we are able to quickly find whether a given grid point already exists in the tree, and if it exists, we are able to update fast the value of the estimate at this grid point.

1. Input of the algorithm are the observations $X_1, \ldots, X_n \in \mathbf{R}^d$, kernel function $K : \mathbf{R}^d \to \mathbf{R}$ that has a compact support, smoothing parameter $h > 0$, and a collection of gridpoints.

2. Output of the algorithm is an evaluation tree, as defined in Definition 13.2.

ALGORITHM: **EvaKer** (Evaluation tree of a kernel density estimate)

1. Find the smallest rectangle containing the support of the kernel estimate (14.3) whose sides are parallel to the coordinate axis. The root node of the evaluation

tree will be annotated with this rectangle. We will evaluate the kernel estimate at gridpoints that are inside the rectangle.

2. **for** i=1 to n (go through observations $X_1, \ldots, X_n$);
   (a) denote with $y_1, \ldots, y_m \in \mathbf{R}^d$ the gridpoints that lie inside the support of function $x \mapsto K((x - X_i)/h)$, $x \in \mathbf{R}^d$;
   (b) **for** j=1 to m (go through gridpoints $y_1, \ldots, y_m$);
      i. **if** the evaluation tree of the kernel estimate already contains grid point y_j, **then** add the value $(nh^d)^{-1}K((y_j - X_i)/h)$ to the current value at this grid point;
      ii. **else** create the node for grid point y_j and store to this node the value $(nh^d)^{-1}K((y_j - X_i)/h)$
   (c) **end for**
3. **end for**

Remarks We can make some observations about the algorithm.

1. In step 2(a) we have to find the gridpoints lying in the support of function $x \mapsto K((x - X_i)/h)$. This can be done fast when the grid is regular.
2. In step 2(b)i we may apply binary search algorithm to find out whether the evaluation tree already contains the grid point y_j.
3. Algorithm EvaKer evaluates the kernel estimate at a predetermined grid. We have two possibilities for evaluating the kernel estimate at arbitrary points: (a) we may apply binning, as explained below, or (b) we may evaluate kernel estimate at the knots of a grid and then use interpolation at the other points. Besides piecewise constant interpolation we can apply quadratic or cubic interpolation. Interpolation schemes are discussed in page 352.

Bibliographic Notes The following proposals have been made for the fast computation of kernel estimates:

1. *(Binning.)* We can make a regular binning of $\mathbf{R}^d$, consisting of bins

$$B_\mu = \prod_{i=1}^{d} [(\mu_i - 1/2)\delta, (\mu_i + 1/2)\delta),$$

where $\mu \in \mathbf{Z}^d$ and $\mathbf{Z} = \{\ldots, -1, 0, 1, \ldots\}$. We let $n_\mu = \#\{X_i \in B_\mu\}$ be the number of observations in B_μ. We can define the binned kernel density estimator as

$$\bar{f}(x) = \sum_{\mu \in \mathbf{Z}^d} \frac{n_\mu}{n} K_h(x - \delta\mu), \qquad x \in \mathbf{R}^d.$$

Binning in the one-dimensional case is discussed in Fan and Marron (1994). Calculation of multivariate kernel estimates with binning is considered in Wand (1994) and the accuracy of binned kernel estimators is studied in Hall and Wand (1996). Accuracy and complexity of binning is studied by Holmström (2000).

We do not have to consider binning as an alternative to the method of evaluation trees because the evaluation tree is also a useful data structure for storing the result of binning. In this case every leaf node is annotated with a bin and the weight of this bin.

2. *(An updating method.)* Seifert, Brockmann, Engel and Gasser (1994) consider an updating method for polynomial kernel functions. They expand a kernel estimate $\hat{f}(x)$ in sums of powers of X_i, and update these sums when moving to nearby values of x.

14.4 NEAREST NEIGHBOR ESTIMATOR

14.4.1 Definition of Nearest Neighbor Estimator

For $k = 1, 2, \ldots$ the kth nearest neighbor density estimator is defined as

$$\hat{f}_k(x) = \frac{k}{n \cdot \text{volume}(B_{r_{n,k}(x)}(x))}, \qquad x \in \mathbf{R}^d, \tag{14.54}$$

where $B_r(x) = \{y : \|x - y\| \leq r\}$ is the ball with radius r and center x, and $r_{n,k}(x)$ is the minimal radius such that $B_{r_{n,k}}(x)$ contains k observations:

$$r_{n,k}(x) = \min \{r > 0 : \#\{X_i \in B_r(x)\} = k\}.$$

We have

$$\text{volume}(B_{r_{n,k}(x)}(x)) = r_{n,k}(x)^d \cdot \text{volume}(B_1),$$

where the volume of the unit ball is given in (B.10).

Remarks We can make some observations about the nearest neighbor estimator.

1. The nearest neighbor estimator is a special case of the locally adaptive kernel estimator defined in (14.6) when we choose the kernel function to be the indicator of the unit ball: $K = I_{B_1}/\text{volume}(B_1)$ and $h(x) = r_{n,k}(x)$.

2. When $k = 1$, the nearest neighbor estimator is not bounded at the points of observations.

3. The level sets of the nearest neighbor estimator with $k = 1$ are unions of balls centered at the observations:

$$\left\{x \in \mathbf{R}^d : \hat{f}_1(x) \geq \lambda\right\} = \bigcup_{i=1}^{n} B_{r(\lambda)}(X_i), \qquad r(\lambda) = (n\lambda \cdot \text{volume}(B_1))^{-1/d}.$$

This follows from the fact that $\hat{f}_k(x) \ge \lambda$ is equivalent to

$$r_{n,k}(x) \le \left(\frac{k}{\lambda \, n \, \text{volume}(B_1)} \right)^{1/d},$$

which in the case $k = 1$ means that the distance of x to the nearest observation is less than $r(\lambda)$ so that $x \in B_{r(\lambda)}(X_i)$ for some $i = 1, \dots, n$.

4. The clusters of a single linkage clustering are the observations in disconnected components of a level set of the 1st nearest neighbor estimator $\hat{f}_1$.

14.4.2 Bibliographic Notes

Fix and Hodges (1951) defined multivariate nearest neighbor estimator in an unpublished technical report. Loftsgaarden and Quesenberry (1965) considered also multivariate nearest neighbor density estimation.

14.5 SERIES ESTIMATORS

14.5.1 Definition of Series Estimator

Density estimators, which can be written as

$$\hat{f}(x) = \sum_{g \in \mathcal{D}} w_g \hat{\theta}_g g(x), \qquad x \in \mathbf{R}^d, \tag{14.55}$$

are called series estimators. Here $\mathcal{D}$ is a collection of functions $g : \mathbf{R}^d \to \mathbf{R}$, $w_g \in [0, 1]$ are shrinking coefficients, and $\hat{\theta}_g$ is the arithmetical mean of g:

$$\hat{\theta}_g = \frac{1}{n} \sum_{i=1}^{n} g(X_i). \tag{14.56}$$

The class of series estimators is in fact a large class of estimators, and many of them cannot be classified as based on local averaging. Collection $\mathcal{D}$ can be chosen in many ways and the coefficients w_g can be also chosen in many ways. We will encounter series estimators again in Chapter 16, where wavelet estimators are considered, and in Chapter 18, where best basis selection is considered.

Remarks We can make some remarks about series estimators.

1. *(Kernel form.)* When we change the order of summation, the series estimator can be written in a kernel form as

$$\hat{f}(x) = \frac{1}{n} \sum_{i=1}^{n} K(x, X_i),$$

where

$$K(x,y) = \sum_{g\in\mathcal{B}} w_g g(x) g(X_i).$$

2. *(Coefficients w_g.)* The coefficients w_g can have two roles.

   a. If set $\mathcal{D}$ has infinite cardinality, then we use coefficients w_g to choose a finite subset of $\mathcal{D}$. When $w_g > 0$ only for a finitely many g, the summation in the definition of $\hat{f}$ is over a finite set.

   b. In addition, even when $\mathcal{D}$ has a finite cardinality, coefficients w_g can be used to further regularize the estimator.

3. *(Famous shrinking coefficients.)* If we can write $\mathcal{D} = \{g_1, g_2, \ldots\}$, then the series estimator can be written as

$$\hat{f}(x) = \sum_{j=1}^{\infty} w_j \hat{\theta}_j g_j(x). \tag{14.57}$$

   Famous shrinking coefficients in the enumerated case of (14.57) include the projection, Pinsker, and Tikhonov coefficients. The projection coefficients are equal to $w_j = I_{\{1,2,\ldots,N\}}(j)$, where N is an integer. Pinsker coefficients are defined as $w_j = \left(1 - \lambda j^{2s}\right)_+$, where $s > 0$ is a smoothness parameter and $\lambda > 0$ is the smoothing parameter. Tikhonov coefficients, or spline coefficients, are defined as $w_j = \left(1 + \lambda j^{2s}\right)^{-1}$.

4. *(Hard thresholding.)* We can choose a subset of $\mathcal{D}$ by using hard thresholding, in which case

$$w_g = \begin{cases} 1, & \text{when } |\hat{\theta}_g| > \lambda, \\ 0, & \text{when } |\hat{\theta}_g| \le \lambda, \end{cases}$$

   where $\lambda > 0$ is a smoothing parameter.

5. *(Soft thresholding.)* Series estimators can be written more generally as

$$\hat{f}(x) = \sum_{g\in\mathcal{D}} \delta(\hat{\theta}_g, \lambda)\, g(x), \qquad x \in \mathbf{R}^d,$$

   where $\delta(\hat{\theta}_g, \lambda)$ is the shrinking function and $\lambda > 0$ is a smoothing parameter. In soft thresholding we define

$$\delta(\hat{\theta}_g, \lambda) = \left(|\hat{\theta}_g| - \lambda\right)_+ \operatorname{sign}(\hat{\theta}_g),$$

   where $(x)_+ = \max\{x, 0\}$. Soft thresholding both chooses a subset of $\mathcal{D}$ and pulls each noisy coefficient toward zero by the amount λ.

6. *(Orthogonal series estimator.)* Suppose that $\mathcal{D} \subset \Phi \subset L_2(\mathbf{R}^d)$, where Φ is a basis of $L_2(\mathbf{R}^d)$. Then $f \in L_2(\mathbf{R}^d)$ can be written as a series expansion

$$f = \sum_{g\in\Phi} \theta_g g,$$

where

$$\theta_g \stackrel{def}{=} \int_{\mathbf{R}^d} fg.$$

Thus the estimator in (14.55) has been obtained by a twofold approximation.

(a) First we approximate the theoretical coefficients θ_g with the arithmetical means $\hat{\theta}_g$. This is natural, since θ_g is equal to the expectation of function g with respect to density f:

$$\theta_g = E_f g(X_1) \approx \frac{1}{n} \sum_{i=1}^{n} g(X_i) = \hat{\theta}_g,$$

where the approximation holds when $X_1, \ldots, X_n$ are identically distributed with density f.

(b) Second we approximate Φ by choosing a finite subset of Φ. Then we reach a series estimate where the summation is over a finite set. Note that there are two possibilities to choose a finite subset: (i) collection $\mathcal{D} \subset \Phi$ is chosen finite, and (ii) $\mathcal{D} = \Phi$ in which case the coefficients w_g choose a finite subset of $\mathcal{D}$.

The orthogonal series estimator appears already in Cencov (1963, 1972).

7. *(Reduction of density estimation to regression function estimation.)* Density estimation with a series estimator is related to the shrinkage methods of regression. We calculate the statistics $(\hat{\theta}_g)_{g \in D}$ from the original observation vector $(X_1, \ldots, X_n)$. When $X_1, \ldots, X_n$ are i.i.d., we have

$$E_f(\hat{\theta}_g) = \theta_g = \int_{\mathbf{R}^d} fg, \qquad \mathrm{Var}_f(\hat{\theta}_g) = \sigma_g^2 \stackrel{def}{=} \frac{1}{n} \left(\int_{\mathbf{R}^d} fg^2 - \theta_g^2 \right).$$

Thus we can write

$$\hat{\theta}_g = \theta_g + \xi_g,$$

where ξ_g is a random variable with $E_f(\xi_g) = 0$, $\mathrm{Var}(\xi_g) = \sigma_g^2$. This leads to the application of shrinkage methods of heteroscedastic regression function estimation in density estimation. We can approximate σ_g^2 by substituting the term $\int_{\mathbf{R}^d} fg^2$ with the estimate $n^{-1} \sum_{i=1}^{n} g^2(X_i)$ and θ_g with the estimate $\hat{\theta}_g$.

14.5.2 Singular Value Decomposition

We can design an orthogonal series estimator for inverse problems using the singular value decomposition. In (B.14) we write an expansion of $f : \mathbf{R}^d \to \mathbf{R}$ as

$$f = \sum_{j=1}^{\infty} b_j^{-1} \langle Af, \psi_j \rangle \phi_j,$$

where (ϕ_j, ψ_j, b_j) is the singular system of the operator A. Since we observe $Y_1, \ldots, Y_n$ whose density is Af, we may estimate

$$\langle Af, \psi_j \rangle \approx \frac{1}{n} \sum_{i=1}^{n} \psi_j(Y_i) \stackrel{def}{=} \hat{\theta}_j.$$

This leads to a series estimator that is similar to estimator in (14.57):

$$\hat{f}(x) = \sum_{j=1}^{\infty} w_j b_j^{-1} \hat{\theta}_j \phi_j(x), \qquad x \in \mathbf{R}^d.$$

EXERCISES

14.1 Prove an analogous formula to (14.12) for the ith term in (14.11), $i = 1, \ldots, d$.

CHAPTER 15

MINIMIZATION ESTIMATORS

Empirical risk minimization is one of the standard tools of statistical estimation. Least squares estimators of regression functions and maximum likelihood estimators are examples of empirical risk minimizers.

A recurrent theme in the analysis of estimators minimizing empirical risk is the reduction of a statistical error to an approximation error. The reduction appears in two ways. (1) Rate of convergence of minimization estimators is shown to depend on the entropy and capacity numbers of the underlying function class. (2) A bound for a complexity-penalized empirical risk is given in terms of the best complexity-penalized theoretical error.

Theory of minimization estimators gives a tool for deriving rates of convergence of estimators: if it can be shown that an estimator is a minimizer of a suitable empirical risk, then the rates of convergence follow from the general theory. On the other hand, general theorems on the rate of convergence of minimization estimators might have no application if there does not exist a reasonable algorithm for actually finding a minimizer of the empirical risk.

- In Section 15.1 two empirical risk functionals are defined: the L_2 empirical risk and the log-likelihood empirical risk. The theoretical results are given only for the L_2 empirical risk.

- In Section 15.2 a theory for δ-net estimators is developed. With the help of δ-net estimators we may complement the theory of lower bounds given in Chapter 11, and show that the optimal rate of convergence–a statistical concept–is determined by the entropy number–an approximation theoretic concept.

- In Section 15.3 minimization estimators are studied that are defined by "deep" minimization: the minimization is not restricted to a δ-net. The relaxation makes it possible to find better algorithms to calculate the minimizer, but the estimators behave well only for classes of smooth functions.

- In Section 15.4 series estimators are studied. The main application is the empirical choice of an orthonormal system. The concept of oracle inequality appears. An oracle inequality gives a bound for the estimator that is not essentially larger than a bound that is reachable by the complete knowledge of the underlying density. The results are applied to give a bound for the L_2 error of a dyadic histogram in Section 18.3.

- In Section 15.5 minimization over convex hulls is considered. The special structure of the problem leads to simplifications. The results are applied to give a bound for the L_2 error of a stagewise minimization estimator in Chapter 19.

In this chapter the setting of inverse problems is often considered, as in Section 9.3.4. We assume to have i.i.d. observations $Y_1, \ldots, Y_n \in \mathbf{Y}$ from the distribution of density $Af : \mathbf{Y} \to \mathbf{R}$, where $(\mathbf{Y}, \mathcal{Y}, \mu)$ is a Borel space. Operator A is assumed to be linear. When A is the identity operator, we use notation $X_1, \ldots, X_n \in \mathbf{R}^d$ for the observations.

We use extensively the pedagogical tool of the Gaussian white noise model. Many results are derived first for the Gaussian white noise and then extended to density estimation. The Gaussian white noise model is defined in Section 9.4.3, and the setting of inverse observations is specified in (9.47), where the signal function Af is observed.

15.1 EMPIRICAL RISK

15.1.1 Empirical Risk Functionals

We consider two different empirical risk functionals: L_2 *empirical risk* and *log-likelihood empirical risk*. These are defined in the case of density estimation by

$$\gamma_n\left(\hat{f}\right) = \begin{cases} -\frac{2}{n}\sum_{i=1}^n \hat{f}(X_i) + \|\hat{f}\|_2^2, & L_2 \text{ empirical risk,} \\ -\frac{1}{n}\sum_{i=1}^n \log \hat{f}(X_i), & \text{log-likelihood empirical risk,} \end{cases} \tag{15.1}$$

where $\hat{f} : \mathbf{R}^d \to \mathbf{R}$. We will define the empirical risk functionals both for the density estimation model and for the Gaussian white noise model, and in the case of the L_2 empirical risk we will consider also inverse problems. The empirical risk functionals will be defined using the notation of contrast function. We denote in the direct case the empirical risk of an estimator $\hat{f} : \mathbf{R}^d \to \mathbf{R}$ by

$$\gamma_n\left(\hat{f}\right) = \begin{cases} \frac{1}{n}\sum_{i=1}^n \gamma\left(\hat{f}, X_i\right), & \text{density estimation,} \\ \int_{\mathbf{R}^d} \gamma\left(\hat{f}, \cdot\right) dX_n, & \text{Gaussian white noise,} \end{cases} \tag{15.2}$$

where $\gamma(g, x)$ is a *contrast function*, for $g : \mathbf{R}^d \to \mathbf{R}$ and $x \in \mathbf{R}^d$. In the setting of inverse problems, introduced in Section 9.3.4, the empirical risk is written

$$\gamma_n\left(\hat{f}\right) = \begin{cases} \frac{1}{n}\sum_{i=1}^n \gamma\left(\hat{f}, Y_i\right), & \text{density estimation,} \\ \int_{\mathbf{Y}} \gamma\left(\hat{f}, \cdot\right) dY_n, & \text{Gaussian white noise.} \end{cases} \tag{15.3}$$

where $\gamma(g, y)$ is a contrast function, for $g : \mathbf{R}^d \to \mathbf{R}$ and $y \in \mathbf{Y}$.

***L_2* Empirical Risk** In the direct case the L_2 *contrast function* is defined by

$$\gamma(g, x) = -2g(x) + \|g\|_2^2, \qquad g \in L_2(\mathbf{R}^d), \ \ x \in \mathbf{R}^d. \tag{15.4}$$

In the inverse case, let $Q = (A^{-1})^*$ be the adjoint of the inverse of operator A:

$$\int_{\mathbf{R}^d} (A^{-1}h)g = \int_{\mathbf{Y}} h(Qg), \tag{15.5}$$

for $h \in L_2(\mathbf{Y})$, $g \in L_2(\mathbf{R}^d)$. The L_2 contrast function is now defined by

$$\gamma(g, y) = -2(Qg)(y) + \|g\|_2^2, \qquad g \in L_2(\mathbf{R}^d), \ \ y \in \mathbf{Y}. \tag{15.6}$$

Minimizing the empirical risk with this contrast function is approximately the same as minimizing the L_2 loss:

$$\begin{aligned} \left\|\hat{f} - f\right\|_2^2 - \|f\|_2^2 &= -2\int_{\mathbf{R}^d} \hat{f}f + \left\|\hat{f}\right\|_2^2 \\ &= -2\int_{\mathbf{Y}} (Q\hat{f})(Af) + \|\hat{f}\|_2^2 \\ &\approx -\frac{2}{n}\sum_{i=1}^n (Q\hat{f})(Y_i) + \left\|\hat{f}\right\|_2^2 \\ &= \gamma_n\left(\hat{f}\right). \end{aligned} \tag{15.7}$$

Term $\|f\|_2^2$ does not depend on $\hat{f}$ and thus may be subtracted from the objective function.

Log-likelihood Empirical Risk With the log-likelihood empirical risk we cannot utilize the L_2 structure, and we restrict ourselves to the direct case. The log-likelihood empirical risk is defined by using the *log-likelihood contrast function*

$$\gamma(g, x) = -\log g(x), \tag{15.8}$$

where $g : \mathbf{R}^d \to \mathbf{R}$. Minimizing the empirical risk with this contrast function is approximately the same as minimizing the Kullback–Leibler loss:

$$\begin{aligned}\int_{\mathbf{R}^d} f \log\left(\frac{f}{\hat{f}}\right) - \int_{\mathbf{R}^d} f \log f &= -\int_{\mathbf{R}^d} f \log \hat{f} \\ &\approx -\frac{1}{n}\sum_{i=1}^{n} \log \hat{f}(X_i) \\ &= \gamma_n\left(\hat{f}\right).\end{aligned}$$

Term $\int_{\mathbf{R}^d} f \log f$ does not depend on $\hat{f}$ and thus may be subtracted from the objective function.

Section 11.1.2 contains a discussion about the L_2 loss, Kullback–Leibler loss, and other loss functions.

15.1.2 Minimization Estimators

Empirical risk minimization requires regularization. A regularization may be achieved by choosing a small class $\mathcal{C}$ over which the minimizer is searched, or it may be achieved by minimizing a complexity-penalized empirical risk. Note that the empirical risk functionals contain already a certain regularization: taking a logarithm of the density penalizes from the high values of the density, and the L_2 empirical risk contains the square integral of the density, which penalizes also from the high values of the density.

Empirical Risk Minimizer Let $\mathcal{C} \subset \{g : \mathbf{R}^d \to \mathbf{R}\}$ be a class of functions. The minimization estimator $\hat{f}$ minimizes the empirical risk functional over class $\mathcal{C}$, up to $\epsilon > 0$: let $\hat{f} \in \mathcal{C}$ be such that

$$\gamma_n(\hat{f}) \le \inf_{g \in \mathcal{C}} \gamma_n(g) + \epsilon, \tag{15.9}$$

where $\gamma_n(\,\cdot\,)$ is defined in (15.3).

Examples We can choose $\mathcal{C}$ to be the class of functions inside a ball of a smoothness class:

$$\mathcal{C} = \{f : \rho_s(f) \le L\},$$

where $\rho_s(f)$ could the Sobolev seminorm as defined in (10.23). The radius L controls the smoothness of the estimate. When we let radius L increase as the sample size increases, we can call this method of estimation the method of sieves. In Section 15.2 we consider δ-net estimators. The δ-net estimators are empirical risk minimizers where the class $\mathcal{C}$ is a δ-net of a function class. For example, $\mathcal{C}$ could be a δ-net of a Sobolev ball.

Regularization When class $\mathcal{C}$ is too large, the empirical risk minimizer overfits the data. Figure 15.1 shows a density estimate that overfits the data.

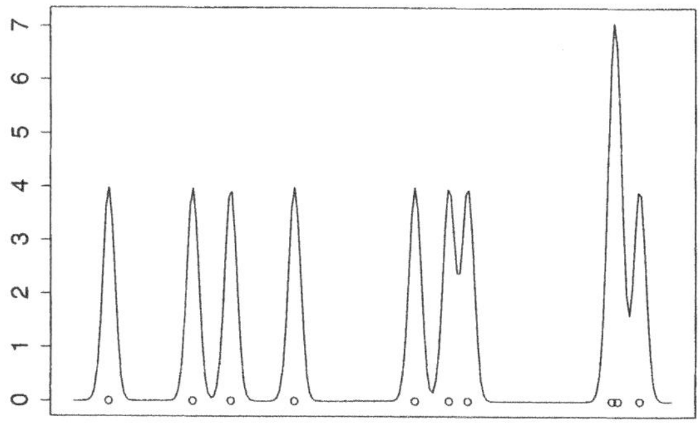

Figure 15.1 *(Overfitting.)* The circles correspond to the observations in the real line and the function shows a density estimate which overfits the data.

Complexity-Penalized Empirical Risk Minimizer We will consider the minimization of a complexity-penalized empirical risk, and control the complexity of the estimate by changing the amount of penalization. To define the complexity-penalized empirical risk, we write class $\mathcal{C}$ as a parametrized set:

$$\mathcal{C} = \{g_\kappa : \kappa \in \mathbb{K}\}, \tag{15.10}$$

where $g_\kappa : \mathbf{R}^d \to \mathbf{R}$ and $\mathbb{K}$ is a set of parameters. Let $D : \mathbb{K} \to [0, \infty)$ be a penalization function and define the complexity-penalized empirical risk as

$$\mathcal{E}_n(\kappa, \alpha) = \gamma_n(g_\kappa) + \alpha\, D(\kappa), \tag{15.11}$$

where $\gamma_n(\,\cdot\,)$ is as in (15.3) and $\alpha \geq 0$ is the smoothing parameter, controlling the amount of penalization. We assume that $D(\kappa)$ takes larger values for more complex g_κ. Let $\hat{\kappa}$ be such that

$$\mathcal{E}_n(\hat{\kappa}, \alpha) \leq \inf_{\kappa \in \mathbb{K}} \mathcal{E}_n(\kappa, \alpha) + \epsilon, \tag{15.12}$$

where $\epsilon > 0$, and define the minimization estimator by

$$\hat{f} = g_{\hat{\kappa}}. \tag{15.13}$$

Examples We can choose the class $\mathcal{C}$ to be the class of densities with a finite smoothness:

$$\mathcal{C} = \{f : \rho_s(f) < \infty\},$$

where $\rho_s(f)$ could the Sobolev seminorm as defined in (10.23). The complexity penalization can be defined using a seminorm of a smoothness class:

$$D(\kappa) = \rho_s^2(g_\kappa),$$

where ρ_s is defined in (10.23), for example.

Regularization The multiplier α is a regularization parameter. Too small values of α lead to estimates that overfit the data; see Figure 15.1. Parameter α has a similar kind of effect on the estimate as the smoothing parameter in kernel estimation: choosing α too small introduces much noise to the estimate, and choosing α too large introduces much bias to the estimate.

15.1.3 Bounds for the L_2 Error

The basic step in analyzing an empirical risk minimizer with the L_2 contrast function is to give a bound to the L_2 error of the estimate in terms of the L_2 error of an arbitrary member from the class. Similarly a complexity-penalized empirical risk minimizer is analyzed by giving a bound to the complexity-penalized L_2 error of the estimate in terms of the complexity-penalized L_2 error of an arbitrary member from the class.

A Bound for the Empirical Risk Minimizer We prove that the L_2 error of a minimization estimator may be bounded by the L_2 error of an arbitrary member of the class and by an additional stochastic term. This upper bound gives a certain kind of bias-variance decomposition.

Lemma 15.1 *Let $\mathcal{C} \subset L_2(\mathbf{R}^d)$. Let $\hat{f} \in \mathcal{C}$ be such that (15.9) holds with the L_2 contrast function defined in (15.6). Then for each $f^0 \in \mathcal{C}$,*

$$\left\|\hat{f} - f\right\|_2^2 \leq \|f^0 - f\|_2^2 + \epsilon + 2\nu_n\left[Q\left(\hat{f} - f^0\right)\right],$$

where f is the true density, and $\nu_n(g)$ is the centered empirical operator:

$$\nu_n(g) = \begin{cases} n^{-1}\sum_{i=1}^n g(Y_i) - \int_{\mathbf{Y}} g(Af), & \text{density estimation,} \\ \int g\, dY_n - \int_{\mathbf{Y}} g(Af), & \text{white noise model,} \end{cases} \tag{15.14}$$

where $g : \mathbf{R}^d \to \mathbf{R}$.

Proof. We have for $g = \hat{f}, g = f^0$,

$$\begin{aligned}&\|g - f\|_2^2 - \gamma_n(g) \\ &= \begin{cases} \|f\|_2^2 - 2\int_{\mathbf{R}^d} fg + 2n^{-1}\sum_{i=1}^n (Qg)(Y_i), & \text{density estimation,} \\ \|f\|_2^2 - 2\int_{\mathbf{R}^d} fg + 2\int (Qg)\, dY_n, & \text{white noise model.} \end{cases}\end{aligned}$$

We have $\int_{\mathbf{R}^d} fg = \int_{\mathbf{Y}} (Af)(Qg)$. Thus

$$\left\|\hat{f} - f\right\|_2^2 - \gamma_n\left(\hat{f}\right) + \gamma_n\left(f^0\right) - \|f^0 - f\|_2^2 = 2\nu_n\left[Q\left(\hat{f} - f^0\right)\right]. \tag{15.15}$$

Thus,

$$\left\|\hat{f} - f\right\|_2^2 - \|f^0 - f\|_2^2$$

$$
\begin{aligned}
&= \left\|\hat{f} - f\right\|_2^2 - \gamma_n(\hat{f}) + \gamma_n(\hat{f}) - \|f^0 - f\|_2^2 \\
&\le \left\|\hat{f} - f\right\|_2^2 - \gamma_n(\hat{f}) + \gamma_n(f^0) + \epsilon - \|f^0 - f\|_2^2 \qquad (15.16) \\
&= 2\nu_n\left[Q\left(\hat{f} - f^0\right)\right] + \epsilon. \qquad (15.17)
\end{aligned}
$$

In (15.16) we applied (15.9), and in (15.17) we applied (15.15). □

A Bound for the Complexity-Penalized Empirical Risk Minimizer Let class $\mathcal{C}$ be parametrized by (15.10) and define the complexity-penalized L_2 error by

$$K(f, \kappa, \alpha) = \|g_\kappa - f\|_2^2 + \alpha \cdot D(\kappa),$$

where f is the true density or the true signal function and $\alpha \ge 0$. We prove that the complexity-penalized L_2 error of a minimization estimator may be bounded by the complexity-penalized L_2 error of an arbitrary member of the class $\mathcal{C}$ and by an additional stochastic term. The term $K(f, \kappa, \alpha)$ consists itself from two terms: the L_2 error and the complexity penalization, and to find a minimum of $K(f, \kappa, \alpha)$ over κ one has to balance between these two terms.

Lemma 15.2 *Let $\mathcal{C} \subset L_2(\mathbf{R}^d)$ be parametrized by (15.10), and let $\hat{f} = g_{\hat{\kappa}} \in \mathcal{C}$ where $\hat{\kappa}$ satisfies (15.12). Then for each $f^0 = g_{\kappa^0} \in \mathcal{C}$,*

$$K(f, \hat{\kappa}, \alpha) \le K(f, \kappa^0, \alpha) + \epsilon + 2\nu_n\left[Q\left(\hat{f} - f^0\right)\right],$$

where f is the true density or the signal function, and $\nu_n(g)$ is the centered empirical operator defined in (15.14).

Proof. We have

$$
\begin{aligned}
&K(f, \hat{\kappa}, \alpha) - K(f, \kappa^0, \alpha) \\
&= K(f, \hat{\kappa}, \alpha) - \mathcal{E}_n(\hat{\kappa}, \alpha) + \mathcal{E}_n(\hat{\kappa}, \alpha) - K(f, \kappa^0, \alpha) \\
&\le K(f, \hat{\kappa}, \alpha) - \mathcal{E}_n(\hat{\kappa}, \alpha) + \mathcal{E}_n(\kappa^0, \alpha) + \epsilon - K(f, \kappa^0, \alpha) \qquad (15.18) \\
&= \left\|\hat{f} - f\right\|_2^2 - \gamma_n(\hat{f}) + \gamma_n(f^0) + \epsilon - \|f^0 - f\|_2^2 \\
&= 2\nu_n\left[Q\left(\hat{f} - f^0\right)\right] + \epsilon. \qquad (15.19)
\end{aligned}
$$

In (15.18) we applied (15.12), and in (15.19) we applied (15.15). □

15.1.4 Historical and Bibliographic Notes

Maximum Likelihood Maximum likelihood was invented by R. A. Fisher in 1930's. In maximum likelihood estimation we find the parameter $\hat{\theta}$ that maximizes the likelihood. That is, we suppose that the true density belongs to the collection

$$\{f(\cdot, \theta) : \theta \in \Theta\}, \qquad (15.20)$$

where Θ may be infinite dimensional, or $\Theta \subset \mathbf{R}^k$, and we define

$$\hat{\theta} = \text{argmax}_{\theta \in \Theta} \text{Lik}_n(\theta),$$

where

$$\text{Lik}_n(\theta) = \prod_{i=1}^{n} f(X_i, \theta) \tag{15.21}$$

is the likelihood function for i.i.d. observations. Maximizing $\text{Lik}_n(\theta)$ is the same as minimizing $-\log \text{Lik}_n(\theta)$, and we arrive at the contrast function (15.8). Exercise 15.1 shows that set Θ has to satisfy constraints (even when it may be infinite dimensional) in order the maximization problem to have a solution.

Bayesian Estimation In Bayesian estimation one postulates a prior density $p : \Theta \to \mathbf{R}$ on the parameter space Θ, with dominating measure μ, and constructs the posterior density

$$g(\theta; X_1, \ldots, X_n) = \frac{\text{Lik}_n(\theta)p(\theta)}{\int \text{Lik}_n(\theta)p(\theta)\, d\mu(\theta)}.$$

Then we choose a loss function $L : \Theta \times \Theta \to [0, \infty)$, and define the Bayesian risk at $\theta \in \Theta$:

$$R_n^B(\theta, \Theta) = \int_\Theta L(\theta, t)\, g(t; X_1, \ldots, X_n)\, d\mu(t).$$

We may define estimator $\hat{\theta}$ as the minimizer of the Bayesian risk:

$$\hat{\theta} = \text{argmin}_{\theta \in \Theta} R_n^B(\theta, \Theta).$$

For example, when $\Theta \subset \mathbf{R}^k$ and $L(\theta, t) = \|\theta - t\|^2$, the Bayesian estimator is the posterior mean $\hat{\theta} = \int_\Theta t\, g(t, X_1, \ldots, X_n)\, d\mu(t)$.

Ill-posed Problems Minimization of a complexity-penalized error has its origins in the theory of ill-posed problems. Assume that we have an operator equation

$$Af = g,$$

where $A : \mathcal{F} \to \mathcal{G}$ is an operator from a function space $\mathcal{F}$ to a function space $\mathcal{G}$. The inverse problem is the problem of finding f, given g. This problem is said to be ill-posed when it may happen that the distance $\|f_1 - f_2\|$, $f_1, f_2 \in \mathcal{F}$, is large, but the distance $\|Af_1 - Af_2\|$ is small. In practice, we observe disturbed g, say g_δ, where δ is the level of noise. Hadamard discovered in the early 1900's the existence of ill-posed inverse problems. Ivanov (1962), Philips (1962), and Tikhonov (1963) proposed to find f that minimizes

$$R(f) = \|Af - g_\delta\|^2 + \alpha(\delta)D(f), \tag{15.22}$$

where $D(f)$ is a penalization functional and $\alpha(\delta)$ is a constant depending on the noise level δ. The history of ill-posed problems was discussed in Vapnik (1995). The least squares estimator as defined in (15.22) would correspond to the choice of the contrast function

$$\tilde{\gamma}(g, y) = -2(Ag)(y) + \|Ag\|_{2,\mathbf{Y}}^2 \,. \tag{15.23}$$

Indeed, as in (15.7), $\|A\hat{f} - Af\|_{2,\mathbf{Y}}^2 - \|Af\|_{2,\mathbf{Y}}^2 \approx \tilde{\gamma}_n(\hat{f})$. This approach has been discussed for example in O'Sullivan (1986). Analogously to (15.23) one could consider the log-likelihood contrast function $\gamma(g, y) = -\log(Ag)(y)$. We have defined the contrast function for inverse problems in (15.6).

Penalization Terms Silverman (1982) studies the penalized log-likelihood empirical risk minimization in the 1D case with the penalization

$$\int_{-\infty}^{\infty} \left(\frac{\partial^3}{\partial t^3} \log f(t) \right)^2 \, dt.$$

This penalization term is zero for the Gaussian densities. More generally, in the multivariate case Silverman (1982) considers the penalization

$$\int_{\mathbf{R}^d} [D(\log f)]^2,$$

where

$$D(g) = \sum_{1 \le |\alpha| \le m} c_\alpha D^\alpha g,$$

with $c_\alpha \in \mathbf{R}$.

Minimum Distance Estimators In minimum distance estimation one tries to find a density $\hat{f} \in \mathcal{C}$ that minimizes some distance to the empirical measure $\mu_n(A) = \sum_{i=1}^n I_A(X_i)$, where $\mathcal{C}$ is some class of densities. The general principle was discussed by Wolfowitz (1957) and studied, for example, by Le Cam (1966), Beran (1977), and Yatracos (1985). Devroye and Lugosi (2001) proposed to use the distance

$$\Delta(g) = \sup_{A \in \mathcal{A}} \left| \int_A g - \mu_n(A) \right|,$$

where $\mathcal{A}$ is the Yatracos class of sets:

$$\mathcal{A} = \{\{x : f(x) \ge g(x)\} : f, g \in \mathcal{C}\}.$$

This choice leads to a method for universal smoothing parameter selection.

15.2 δ-NET ESTIMATOR

15.2.1 Definition of δ-Net Estimator

A δ-net estimator is a minimizer of an empirical risk over a δ-net of a function class. Let $\mathcal{F}$ be a class of densities or signal functions $\mathbf{R}^d \to \mathbf{R}$. Let $\mathcal{G}_\delta$ be a finite δ-net of

$\mathcal{F}$ in the L_2 metric, where $\delta > 0$. That is, for each $f \in \mathcal{F}$ there is $\phi \in \mathcal{G}_\delta$ such that $\|f - \phi\|_2 \le \delta$. Define estimator $\hat{f}$ by

$$\hat{f} = \text{argmin}_{\phi \in \mathcal{G}_\delta} \gamma_n(\phi),$$

where $\gamma_n(\phi)$ is defined in (15.3). We assume that $\mathcal{F}$ is bounded in the L_2 metric,

$$\sup_{g \in \mathcal{F}} \|g\|_2 \le B_2,$$

where $0 < B_2 < \infty$.

15.2.2 An Upper Bound to MISE

Theorem 15.3 below gives a bound for the mean integrated squared error of the estimate. We may identify the first term in the bound as a bias term and the second term as a variance term, which depends on the operator norm of Q over the δ-net $\mathcal{G}_\delta$, defined as

$$\varrho(Q, \mathcal{G}_\delta) = \max_{\phi, \phi' \in \mathcal{G}_\delta, \phi \ne \phi'} \frac{\|Q(\phi - \phi')\|_2}{\|\phi - \phi'\|_2}, \qquad \delta > 0, \tag{15.24}$$

where $Q = (A^{-1})^*$ is defined by (15.5). For the density estimation we need additional assumptions that $\varrho(Q, \mathcal{G}_\delta) \ge 1$ and that $A\mathcal{F}$ and $Q\mathcal{F}$ are bounded in the L_∞ metric:

$$\varrho(Q, \mathcal{G}_\delta) \ge 1, \qquad \sup_{f \in \mathcal{F}} \|Af\|_\infty \le B_\infty, \qquad \sup_{f \in \mathcal{F}} \|Qf\|_\infty \le B'_\infty, \tag{15.25}$$

where $0 < B_\infty, B'_\infty < \infty$.

Theorem 15.3 *We consider both density estimation with i.i.d. observations whose density is Af and the Gaussian white noise model (9.47). Operator A is linear. For the density estimation we assume that (15.25) is satisfied. We have for $f \in \mathcal{F}$,*

$$E_f \left\| \hat{f} - f \right\|_2^2 \le C_1 \delta^2 + C_2 \frac{\varrho^2(Q, \mathcal{G}_\delta) \cdot (\log_e(\#\mathcal{G}_\delta) + 1)}{n},$$

where

$$C_1 = (1 - 2\xi)^{-1}(1 + 2\xi), \tag{15.26}$$

$$C_2 = (1 - 2\xi)^{-1} \xi C_\tau, \tag{15.27}$$

$$C_\tau > 0, \tag{15.28}$$

and ξ is such that

$$\begin{cases} C_\tau^{-1}\left(4B'_\infty/3 + \sqrt{2\left[8(B'_\infty)^2/9 + C_\tau B_\infty\right]}\right) \le \xi < 1/2, & \textit{density estimation} \\ \sqrt{2/C_\tau} \le \xi < 1/2, & \textit{white noise.} \end{cases} \tag{15.29}$$

Proof. Let $f \in \mathcal{F}$ be the true density. Let $\phi^0 \in \mathcal{G}_\delta$. Denote

$$\zeta = C_1 \|\phi^0 - f\|_2^2 + C_2 n^{-1} \varrho^2(Q, \mathcal{G}_\delta) \log_e(\#\mathcal{G}_\delta),$$

where C_1 is defined in (15.26) and C_2 is defined in (15.27). We have that

$$\begin{aligned} E\|\hat{f} - f\|_2^2 &= \int_0^\infty P\left(\|\hat{f} - f\|_2^2 > t\right) dt \\ &\le \zeta + \int_\zeta^\infty P\left(\|\hat{f} - f\|_2^2 > t\right) dt \\ &= \zeta + C_2 n^{-1} \varrho^2(Q, \mathcal{G}_\delta) \int_0^\infty P(A_t)\, dt, \end{aligned} \tag{15.30}$$

where

$$A_t \stackrel{def}{=} \left(\|\hat{f} - f\|_2^2 > C_2 n^{-1} \varrho^2(Q, \mathcal{G}_\delta) t + \zeta\right).$$

Denote

$$\tau_n = C_\tau n^{-1} \varrho^2(Q, \mathcal{G}_\delta) \left(\log_e(\#\mathcal{G}_\delta) + t\right),$$

where C_τ is defined in (15.28). Then

$$\begin{aligned} A_t &= \left(\|\hat{f} - f\|_2^2 > C_1 \|\phi^0 - f\|_2^2 + C_2 C_\tau^{-1} \tau_n\right) \\ &= \Big((1 - 2\xi)^{-1} \|\hat{f} - f\|_2^2 \\ &\quad > 2\xi(1 - 2\xi)^{-1} \|\hat{f} - f\|_2^2 + C_1 \|\phi^0 - f\|_2^2 + C_2 C_\tau^{-1} \tau_n\Big) \\ &= \left(\|\hat{f} - f\|_2^2 > 2\xi \|\hat{f} - f\|_2^2 + (1 + 2\xi) \|\phi^0 - f\|_2^2 + \xi \tau_n\right). \end{aligned} \tag{15.31}$$

We have by Lemma 15.1,

$$\left\|\hat{f} - f\right\|_2^2 \le \|\phi^0 - f\|_2^2 + 2\nu_n \left[Q\left(\hat{f} - \phi^0\right)\right].$$

Denote

$$w(\phi) = \|\phi - f\|_2^2 + \|\phi^0 - f\|_2^2 + \tau_n/2.$$

We continue (15.31) with

$$\begin{aligned} A_t &\subset \left(\nu_n[Q(\hat{f} - \phi^0)] > \xi \|\hat{f} - f\|_2^2 + \xi \|\phi^0 - f\|_2^2 + \xi \tau_n/2\right) \\ &= \left(\nu_n[Q(\hat{f} - \phi^0)] > w(\hat{f}) \xi\right) \\ &\subset \left(\max_{\phi \in \mathcal{G}_\delta, \phi \ne \phi^0} \frac{\nu_n[Q(\phi - \phi^0)]}{w(\phi)} > \xi\right) \\ &\stackrel{def}{=} B_t, \end{aligned}$$

and thus

$$P(A_t) \le P(B_t). \tag{15.32}$$

We prove that

$$P(B_t) \le \exp(-t), \tag{15.33}$$

and this proves the theorem when we combine (15.30) and (15.32).

Proof of (15.33) Denote

$$\Phi = \left\{ \frac{Q(\phi - \phi^0)}{w(\phi)} : \phi \in \mathcal{G}_\delta, \ \phi \ne \phi^0 \right\}.$$

We have that

$$P(B_t) \le \sum_{g \in \Phi} P\left(\nu_n(g) > \xi\right). \tag{15.34}$$

Also

$$w(\phi) \ge \frac{1}{2}\left(\|\phi - \phi^0\|_2^2 + \tau_n \right) \ge \|\phi - \phi^0\|_2 \, \tau_n^{1/2},$$

and thus

$$v_0 \overset{def}{=} \max_{g \in \Phi} \|g\|_2^2 \le \frac{1}{\tau_n} \max_{\phi \in \mathcal{G}_\delta, \phi \ne \phi^0} \frac{\|Q(\phi - \phi_0)\|_2^2}{\|\phi - \phi_0\|_2^2} = \frac{\varrho^2(Q, \mathcal{G}_\delta)}{\tau_n}. \tag{15.35}$$

From now on, we consider the Gaussian white noise model and density estimation separately.

Gaussian White Noise When $W \sim N(0, \sigma^2)$, we have

$$P(W > \xi) \le 2^{-1} \exp\left\{ -\frac{\xi^2}{2\sigma^2} \right\}$$

for $\xi > 0$; see (12.4). We have that $\nu_n(g) \sim N(0, n^{-1}\|g\|_2^2)$. Thus

$$P\left(\nu_n(g) > \xi\right) \le 2^{-1} \exp\left\{ -\frac{n\xi^2}{2v_0} \right\} \le 2^{-1} \exp\left\{ -\frac{n\tau_n \xi^2}{2\varrho^2(Q, \mathcal{G}_\delta)} \right\}.$$

Thus, continuing from (15.34), denoting $C_\xi \overset{def}{=} \xi^2 C_\tau / 2$ obtains

$$\begin{aligned} P(B_t) &\le \#\mathcal{G}_\delta \cdot \exp\left\{ -\frac{n\tau_n \xi^2}{2\varrho^2(Q, \mathcal{G}_\delta)} \right\} \\ &= \#\mathcal{G}_\delta \cdot \exp\left\{ -C_\xi [\log_e(\#\mathcal{G}_\delta) + t] \right\} \\ &\le \exp(-t), \end{aligned}$$

since $C_\xi \ge 1$ by the choice of ξ.

Density Estimation Denote $v = \sup_{g \in \Phi} \mathrm{Var}_f(g(Y_1))$, and $b = \sup_{g \in \Phi} \|g\|_\infty$. We have

$$v \le \|Af\|_\infty v_0 \le B_\infty \frac{\varrho^2(Q, \mathcal{G}_\delta)}{\tau_n}, \tag{15.36}$$

by (15.35). Also

$$w(\phi) \geq \frac{\tau_n}{2}$$

and thus because $\varrho(Q, \mathcal{G}_\delta) \geq 1$,

$$b \leq 2B'_\infty \frac{2}{\tau_n} \leq 4B'_\infty \frac{\varrho^2(Q, \mathcal{G}_\delta)}{\tau_n}. \tag{15.37}$$

Applying Bernstein's inequality of Theorem 12.1, applying (15.36) and (15.37), leads to

$$\begin{aligned} P\left(\nu_n(g) > \xi\right) &\leq \exp\left\{\frac{-n\xi^2}{2(v + \xi b/3)}\right\} \\ &\leq \exp\left\{\frac{-n\xi^2\tau_n}{2\varrho^2(Q, \mathcal{G}_\delta)(B_\infty + 4B'_\infty\xi/3)}\right\}. \end{aligned}$$

Thus, continuing from (15.34), we write

$$\begin{aligned} P(B_t) &\leq \#\mathcal{G}_\delta \cdot \exp\left\{\frac{-n\xi^2\tau_n}{2\varrho^2(Q, \mathcal{G}_\delta)(B_\infty + 4B'_\infty\xi/3)}\right\} \\ &= \#\mathcal{G}_\delta \cdot \exp\left\{-C_\xi[\log_e(\#\mathcal{G}_\delta) + t]\right\} \\ &\leq \exp(-t), \end{aligned}$$

where

$$C_\xi \stackrel{def}{=} \frac{\xi^2 C_\tau}{2(B_\infty + 4B'_\infty\xi/3)},$$

and $C_\xi \geq 1$ by the choice of ξ. We have proved (15.33) and thus the theorem. □

15.2.3 Rates of Convergence of δ-Net Estimator

General Formula Theorem 15.3 shows that the δ-net estimator achieves the rate of convergence ψ_n when ψ_n is the solution of the equation

$$\psi_n^2 \asymp n^{-1}\varrho^2(Q, \mathcal{G}_{\psi_n}) \log(\#\mathcal{G}_{\psi_n}). \tag{15.38}$$

We calculate the rate under the assumptions that $\log(\#\mathcal{G}_\delta)$ and $\varrho(Q, \mathcal{G}_\delta)$ increase polynomially as δ decreases: we assume that one can find a δ-net whose cardinality satisfies

$$\log(\#\mathcal{G}_\delta) = C\delta^{-b} \tag{15.39}$$

for some constants $b, C > 0$, and we assume that

$$\varrho(Q, \mathcal{G}_\delta) = C'\delta^{-a} \tag{15.40}$$

for some $a, C' > 0$ (in the direct case $a = 0$ and $C' = 1$). Then (15.38) can be written as $\psi_n^2 \asymp n^{-1}\psi_n^{-2a-b}$, and the rate of the δ-net estimator is equal to

$$\psi_n \asymp n^{-1/[2(a+1)+b]}. \tag{15.41}$$

Smoothness Classes Let $\mathcal{F}$ be a set of s-smooth d-dimensional functions so that $b = d/s$; see (10.70). Then the rate is

$$\psi_n \asymp n^{-s/[2(a+1)s+d]}, \tag{15.42}$$

which gives in the direct case ($a = 0$) the classical rate $\psi_n \asymp n^{-s/(2s+d)}$.

Convex Hull Let $\mathcal{F}$ be the L_2 closure of the convex hull of a class $\mathbb{G}$, and assume that $N(\delta, \mathbb{G}, \|\cdot\|_2) \leq Cn^{-V}$ for $0 < V < \infty$ (there exists a δ-net for $\mathbb{G}$ whose cardinality is $\leq C\delta^{-V}$, for a positive constant C). Then (15.39) holds with $b = 2V/(2+V)$; see Theorem 10.13. Thus the rate is in the direct case

$$\psi_n \asymp n^{-(V+2)/[4(V+1)]}. \tag{15.43}$$

For $V = 0$ the rate would be $n^{-1/2}$, and for $V = \infty$ the rate would be $n^{-1/4}$.

Operator Norms We give examples where (15.40) holds. We calculate the operator norm $\varrho(Q, \mathcal{G}_\delta)$ when the function class $\mathcal{F}$ and its δ-net $\mathcal{G}_\delta$ are given by

$$\mathcal{F} = \left\{ \sum_{j=1}^{\infty} \theta_j \phi_j : \theta \in \Theta \right\}, \qquad \mathcal{G}_\delta = \left\{ \sum_{j=1}^{\infty} \theta_j \phi_j : \theta \in \Theta_\delta \right\}, \tag{15.44}$$

where Θ is the ellipsoid defined in (10.59) and Θ_δ is given in (10.63). We consider two cases.

1. Functions ϕ_j are the singular functions of the operator A, as given in (B.13). Let b_j be the singular values in (B.15). We assume that there exists positive constants q and C such that for all $j \in \{1, 2, \ldots\}$,

$$b_j \geq C \cdot j^{-q}. \tag{15.45}$$

The functions $Q\phi_j$ are orthogonal and $\|Q\phi_j\|_2 = b_j^{-1}$. Indeed $Q = (A^{-1})^*$, and thus

$$\langle Q\phi_j, Q\phi_l \rangle = \langle \phi_j, A^{-1}(A^{-1})^* \phi_l \rangle = b_l^{-2} \langle \phi_j, \phi_l \rangle,$$

where we used the fact

$$A^{-1}(A^{-1})^* \phi_l = A^{-1}(A^*)^{-1} \phi_l = (A^*A)^{-1} \phi_l = b_l^{-2} \phi_l.$$

Thus, when $g, g' \in \mathcal{G}_\delta$ and we write $g - g' = \sum_{j=1}^M (\theta_j - \theta'_j)\phi_j$,

$$\begin{aligned} \|Q(g-g')\|_2^2 &= \left\| \sum_{j=1}^{M} (\theta_j - \theta'_j)^2 Q\phi_j \right\|_2^2 \\ &= \sum_{j=1}^{M} (\theta_j - \theta'_j)^2 b_j^{-2} \\ &\leq CM^{2q} \sum_{j=1}^{M} (\theta_j - \theta'_j)^2. \end{aligned}$$

On the other hand, $\|g - g'\|_2 = \|\theta - \theta\|_{l_2}$. This gives the upper bound for the operator norm

$$\varrho(Q, \mathcal{G}_\delta) \le CM^q \le C'\delta^{-q/s}, \tag{15.46}$$

by the definition of M in (10.61). Thus (15.40) holds with $a = q/s$.

2. Functions ϕ_j are wavelet basis functions and we use the wavelet-vaguelette system as given in (B.17). The Besov body defined through the semi-norm in (10.37) is an ellipsoid when we take there $p = q = 2$. Let b_j be the quasi-singular values in (B.16). We assume that there exists positive constants q and C such that for all $j \in \{1, 2, \ldots\}$,

$$b_j \ge C \cdot 2^{-qj}. \tag{15.47}$$

Since $Q = (A^{-1})^*$, then $QA^* = (AA^{-1})^* = I$. From (B.16) we get $A^* u_{jk} = b_j \phi_{jk}$, where (u_{jk}) is a near-orthogonal system. Thus

$$\begin{aligned}\langle Q\phi_{jk}, Q\phi_{j'k'}\rangle &= b_j^{-1} b_{j'}^{-1} \langle QA^* u_{jk}, QA^* u_{j'k'}\rangle \\ &= b_j^{-1} b_{j'}^{-1} \langle u_{jk}, u_{j'k'}\rangle.\end{aligned}$$

Then, when $g, g' \in \mathcal{G}_\delta$ and we write $g - g' = \sum_{j=1}^{J} \sum_k (\theta_j - \theta_j')\phi_{jk}$, where $2^J \asymp \delta^{-1/s}$,

$$\begin{aligned}\|Q(g - g')\|_2^2 &= \left\| \sum_{j=1}^{J} \sum_k (\theta_{jk} - \theta_{jk}')\, Q\phi_{jk} \right\|_2^2 \\ &= \left\| \sum_{j=1}^{J} b_j^{-1} \sum_k (\theta_{jk} - \theta_{jk}')\, u_{jk} \right\|_2^2 \\ &\asymp \sum_{j=1}^{J} b_j^{-2} \sum_k (\theta_{jk} - \theta_{jk}')^2 \\ &\le C 2^{2Jq} \sum_{j=0}^{J} \sum_k (\theta_{jk} - \theta_{jk}')^2.\end{aligned}$$

On the other hand, $\|g - g'\|_2 = \|\theta - \theta\|_{l_2}$. This gives the upper bound for the operator norm

$$\varrho(Q, \mathcal{G}_\delta) \le C 2^{Jq} \le C'\delta^{-q/s}.$$

Thus (15.40) holds with $a = q/s$.

Sobolev Space and Convolution Let A be a convolution operator: $Af = a * f$ where $a : \mathbf{R}^d \to \mathbf{R}$ is a known function. Let the collection (ϕ_{jk}), $(j, k) \in \{0, 1, \ldots\}^d \times K_j$, of trigonometric polynomials be defined in (10.29). When the convolution kernel a is a 1-periodic function in $L_2([0, 1]^d)$, we can write the Fourier series of a with the Fourier coefficients b_{jk} as in (9.23), The functions ϕ_{jk} are the singular functions of the operator A, and the values b_{jk} are the corresponding singular values. We assume that the underlying function space is given in (11.50) and (11.51). We give the rate of convergence of the δ-net estimator. The corresponding result for a kernel estimator is given in Theorem 14.6.

Corollary 15.4 *Let $Y_1, \ldots, Y_n$ be i.i.d. with density $Af \in \mathcal{F}$, where $\mathcal{F}$ is defined in (11.50). We assume that the coefficients of the ellipsoid (11.51) satisfy*

$$a_{jk} \geq C|j|^s$$

for some $s > 0$ and $C > 0$. We assume that the convolution filter a is 1-periodic function in $L_2([0,1]^d)$ and that the Fourier coefficients of filter a satisfy

$$b_{jk} \geq C|j|^{-q}$$

for some $q \geq 0$, $C > 0$. Then

$$\limsup_{n \to \infty} n^{2s/(2s+2q+d)} \sup_{f \in \mathcal{F}} E \left\| \hat{f} - f \right\|_2^2 < \infty,$$

where $\hat{f}$ is the δ-net estimator.

Proof. We apply Theorem 15.3. Let $\mathcal{G}_\delta$ be the δ-net of $\mathcal{F}$ as constructed in (15.44). We showed in (15.46) that $\varrho(Q, \mathcal{F}_\delta) \leq C\delta^{-a}$, where $a = q/s$. We stated in (10.70) that the cardinality of the δ-net satisfies $\log(\#\mathcal{G}_\delta) \leq C\delta^{-b}$, where $b = d/s$. Thus we can apply (15.41) to get the rate

$$\psi_n = n^{-1/(2(a+1)+b)} = n^{-s/(2s+2q+d)}.$$

The corollary is proved. □

Sobolev Space and Radon Transform We consider the 2D Radon transform as defined in (9.42). The singular value decomposition is given in Section 9.3.4, page 248. We assume that the underlying function space is as given in (11.52) and (11.53). We give the rate of convergence of the δ-net estimator. The corresponding result for a kernel estimator is given in Theorem 14.7. Since the singular functions are given only for the two-dimensional case, the corollary holds only in the two-dimensional case.

Corollary 15.5 *Let $\mathcal{F}$ be the function class as defined in (11.52). We assume that the coefficients of the ellipsoid (11.53) satisfy*

$$a_{jk} \geq C|j|^s$$

for some $s > 0$ and $C > 0$. Then for $d = 2$,

$$\limsup_{n \to \infty} n^{2s/(2s+2d-1)} \sup_{f \in \mathcal{F}} E \left\| \hat{f} - f \right\|_2^2 < \infty,$$

where $\hat{f}$ is the δ-net estimator.

Proof. We apply Theorem 15.3. Let $\mathcal{G}_\delta$ be the δ-net of $\mathcal{F}$ as constructed in (15.44). We showed in (15.46) that $\varrho(Q, \mathcal{G}_\delta) \le C\delta^{-a}$, where $a = q/s$ and $q = 1/2$ (so that $a = (d-1)/(2s)$), since the singular values are given in (9.43). We have stated in (10.65) that the cardinality of the δ-net satisfies $\log(\#\mathcal{G}_\delta) \le C\delta^{-b}$, where $b = d/s$. Thus we may apply (15.41) to get the rate $\psi_n = n^{-s/(2s+2d-1)}$. □

15.3 DENSE MINIMIZER

A dense empirical risk minimizer minimizes the empirical risk over a function class. In contrast to the δ-net estimator the minimization is not restricted to a δ-net. We call this estimator a "dense minimizer" because it is defined as a minimizer over possibly an uncountable function class. The δ-net estimator is more universal: it may be applied also for unsmooth functions and for severely ill-posed operators. The dense empirical minimizer may be applied only for relatively smooth cases (the entropy integral has to converge), but without a restriction to a δ-net we have available a larger toolbox of algorithms for finding an approximate minimizer of the empirical risk.

We will consider the setting of inverse problems introduced in Section 9.3.4 for the density estimation and in (9.47) for the Gaussian white noise model. The definition of the estimator can be given by the same formula for the two models but the presentation of the MISE bounds shall be given separately.

15.3.1 Definition of Dense Minimizer

Let $\mathcal{F}$ be a collection of functions $\mathbf{R}^d \to \mathbf{R}$ and let the estimator $\hat{f}$ be a minimizer of the empirical risk over $\mathcal{F}$, up to $\epsilon > 0$:

$$\gamma_n(\hat{f}) \le \inf_{g\in\mathcal{F}} \gamma_n(g) + \epsilon,$$

where $\gamma_n(\cdot)$ is defined in (15.3). We assume that $\mathcal{F}$ is bounded in the L_2 metric:

$$\sup_{g\in\mathcal{F}} \|g\|_2 \le B_2,$$

where $0 < B_2 < \infty$.

15.3.2 Gaussian White Noise

We prove an upper bound to the mean integrated squared error. Let $\mathcal{G}_\delta$, $\delta > 0$, be a δ-net of $\mathcal{F}$, with respect to the L_2 norm. Let $\varrho(Q, \mathcal{G}_\delta)$ be the norm defined in (12.7), and let $G(\delta)$ be the entropy integral defined in (12.8).

Theorem 15.6 *We consider the Gaussian white noise model (9.47) with linear operator A. Assume that*

1. *the entropy integral* $G(\delta)$ *defined in (12.8) is finite,*
2. $G(\delta)/\delta^2$ *is decreasing on the interval* $(0, B_2]$,
3. $\varrho(Q, \mathcal{G}_\delta) = c\delta^{-a}$, *where* $0 \le a < 1$ *and* $c > 0$,
4. $\lim_{\delta \to 0} G(\delta)\delta^{a-1} = \infty$,
5. $\delta \mapsto \varrho(Q, \mathcal{G}_\delta)\sqrt{\log_e(\#\mathcal{G}_\delta)}$ *is decreasing on* $(0, B_2]$. *(This assumption is only needed to give an upper bound to a finite Riemann sum with the entropy integral.)*

Let ψ_n *be such that*

$$\psi_n^2 \ge C n^{-1/2} G(\psi_n) \tag{15.48}$$

for a positive constant C, *and assume that* $\lim_{n\to\infty} n\psi_n^{2(1+a)} = \infty$. *Then for* $f \in \mathcal{F}$,

$$E\left\|\hat{f} - f\right\|_2^2 \le C' \left(\psi_n^2 + \epsilon\right),$$

for a positive constant C', *for sufficiently large* n.

Proof. Let $f \in \mathcal{F}$ be the true density. Denote

$$\zeta = C_1\epsilon + C_2\psi_n^2,$$

where $C_1 = (1-2\xi)^{-1}$, $C_2 = 1 - 2\xi$, $0 < \xi \le (3-\sqrt{5})/4$. We have that

$$\begin{aligned} E\|\hat{f} - f\|_2^2 &= \int_0^\infty P\left(\|\hat{f} - f\|_2^2 > t\right) dt \\ &\le \zeta + \int_\zeta^\infty P\left(\|\hat{f} - f\|_2^2 > t\right) dt \\ &= \zeta + C_2\psi_n^2 \int_0^\infty P(A_t)\, dt, \end{aligned} \tag{15.49}$$

where

$$A_t \stackrel{def}{=} \left(\|\hat{f} - f\|_2^2 > C_2\psi_n^2 t + \zeta\right).$$

Denote

$$\tau_n = C_\tau \psi_n^2(1+t), \qquad C_\tau = \xi^{-1}(1-2\xi)^2.$$

Then

$$\begin{aligned} A_t &= \left(\|\hat{f} - f\|_2^2 > C_2 C_\tau^{-1}\tau_n + C_1\epsilon\right) \\ &= \Big((1-2\xi)^{-1}\|\hat{f} - f\|_2^2 \\ &\qquad > 2\xi(1-2\xi)^{-1}\|\hat{f} - f\|_2^2 + C_2 C_\tau^{-1}\tau_n + C_1\epsilon\Big) \\ &= \left(\|\hat{f} - f\|_2^2 > 2\xi\|\hat{f} - f\|_2^2 + \xi\tau_n + \epsilon\right). \end{aligned} \tag{15.50}$$

Choosing $f^0 = f$ in Lemma 15.1, we get

$$\left\|\hat{f} - f\right\|_2^2 \le 2\nu_n[Q(\hat{f} - f)] + \epsilon.$$

Denote

$$w(g) = \|g - f\|_2^2 + \tau_n/2.$$

Then we may continue (15.50) with

$$\begin{aligned} A_t &\subset \left(\nu_n[Q(\hat{f} - f)] > \xi\|\hat{f} - f\|_2^2 + \xi\tau_n/2\right) \\ &= \left(\nu_n[Q(\hat{f} - f)] > w(\hat{f})\xi\right) \\ &\subset P\left(\sup_{g\in\mathcal{F}} \frac{\nu_n[Q(g - f)]}{w(g)} > \xi\right) \\ &\stackrel{def}{=} B_t, \end{aligned}$$

and thus

$$P(A_t) \subset P(B_t). \tag{15.51}$$

We prove that

$$P(B_t) \le \exp(-t \cdot \log_e 2), \tag{15.52}$$

and this proves the theorem when we combine (15.49) and (15.51).

Proof of (15.52) We use the peeling device; see for example van de Geer (2000, p. 69). Denote

$$a_0 = \tau_n/2, \qquad a_j = 2^{2j} a_0, \qquad b_j = 2^2 a_j, \qquad j = 0, 1, \ldots$$

Let $\mathcal{H}_j$ be the set of functions

$$\mathcal{H}_j = \{g \in \mathcal{F} : a_j \le w(g) < b_j\}, \qquad j = 0, 1, \ldots,$$

and

$$\mathcal{F}_j = \left\{g \in \mathcal{F} : \|g - f\|_2^2 < b_j\right\}, \qquad j = 0, 1, \ldots.$$

We have

$$\mathcal{F} = \{g \in \mathcal{F} : w(g) \ge a_0\} = \bigcup_{j=0}^{\infty} \mathcal{H}_j.$$

Thus

$$\begin{aligned} P(B_t) &\le \sum_{j=0}^{\infty} P\left(\sup_{g\in\mathcal{H}_j} \frac{\nu_n[Q(g - f)]}{w(g)} > \xi\right) \\ &\le \sum_{j=0}^{\infty} P\left(\sup_{g\in\mathcal{F}, w(g)<b_j} \nu_n[Q(g - f)] > \xi a_j\right) \\ &\le \sum_{j=0}^{\infty} P\left(\sup_{g\in\mathcal{F}_j} \nu_n[Q(g - f)] > \xi a_j\right). \end{aligned} \tag{15.53}$$

By Assumption 4, $\tilde{G}(\psi_n) = 24\sqrt{2}G(\psi_n)$, where $\tilde{G}$ is defined in (12.9) for sufficiently large n. Thus by the choice of $C = \xi^{-1}4 \cdot 24\sqrt{2}$ in (15.48),

$$\psi_n^2 \geq n^{-1/2}\xi^{-1}4\tilde{G}(\psi_n).$$

By the choice of ξ we have that $C_\tau \geq 2$, and thus $a_0 = C_\tau \psi_n^2(1+t)/2 \geq \psi_n^2$. Since $G(\delta)/\delta^2$ is decreasing, by Assumption 2, then $\tilde{G}(\delta)/\delta^2$ is decreasing, and

$$\frac{\xi n^{1/2}}{4} \geq \frac{\tilde{G}(\psi_n)}{\psi_n^2} \geq \frac{\tilde{G}(a_0^{1/2})}{a_0} \geq \frac{\tilde{G}(b_j^{1/2})}{b_j},$$

that is,

$$\xi a_j = \xi b_j/4 \geq n^{-1/2}\tilde{G}(b_j^{1/2}). \tag{15.54}$$

We may apply Lemma 12.4 with (15.54) to get

$$\begin{aligned}
&P\left(\sup_{g\in\mathcal{F}_j} \nu_n[Q(g-f)] > \xi a_j\right) \\
&\leq \exp\left\{-\frac{n(\xi a_j)^2 C'}{c^2 b_j^{1-a}}\right\} \qquad (15.55)\\
&\leq \exp\left\{-C'' 2^{2j(a+1)} n\psi_n^{2(1+a)}(1+t)^{1+a}\right\} \\
&\leq \exp\left\{-C''(j+1) n\psi_n^{2(1+a)}(1+t)^{1+a}\right\}, \qquad (15.56)
\end{aligned}$$

where $C'' = C'c^{-2}\xi^2 2^{2(a-1)}(C_\tau/2)^{1+a}$, and we used the facts

$$a_j^2/b_j^{1-a} = 2^{2(a-1)}a_j^{1+a} = 2^{2(a-1)}(2^{2j}a_0)^{1+a} = 2^{2(a-1)}[2^{2j}C_\tau\psi_n^2(1+t)/2]^{1+a}$$

and $2^{2j(a+1)} \geq j+1$. When $0 \leq b \leq 1/2$, we have $\sum_{j=0}^{\infty} b^{j+1} = \sum_{j=1}^{\infty} b^j = b/(1-b) \leq 2b$; see B.20. When $n\psi_n^{2(1+a)} \geq (\log_e 2)/C''$, we have $\exp\{-C'' n\psi_n^{2(1+a)}(1+t)^{1+a}\} \leq 1/2$, and we combine (15.53) and (15.56) to get the upper bound

$$\begin{aligned}
2\exp\left\{-C'' n\psi_n^{2(1+a)}(1+t)^{1+a}\right\} &\leq 2\exp\left\{-C'' n\psi_n^{2(1+a)}(1+t)\right\} \\
&\leq \exp\left\{-t\log_e 2\right\}.
\end{aligned}$$

We have proved (15.52), and thus we have proved Theorem 15.6. □

15.3.3 Density Estimation

Let $\mathcal{G}_\delta$ be a δ-bracketing net of $\mathcal{F}$ with respect to the L_2 norm. A δ-bracketing net was defined in Definition 10.6. Let $\varrho(Q, \mathcal{G}_\delta)$ be the operator defined in (12.11), and let $G(\delta)$ be the entropy integral defined in (12.12).

Theorem 15.7 *Let $Y_1, \ldots, Y_n \in \mathbf{Y}$ be i.i.d. with density Af, where A is a linear operator. We make the Assumptions 1 through 5 of Theorem 15.6. In addition we assume that $\sup_{f \in \mathcal{F}} \|Af\|_\infty < \infty$ and $\sup_{g \in \mathcal{G}^L_{B_2} \cup \mathcal{G}^U_{B_2}} \|Qg\|_\infty < \infty$, where $\mathcal{G}^L_\delta = \{g^L : (g^L, g^U) \in \mathcal{G}_\delta\}$ and $\mathcal{G}^U_\delta = \{g^U : (g^L, g^U) \in \mathcal{G}_\delta\}$. We assume also that the operator Q preserves positivity ($g \geq 0$ implies that $Qg \geq 0$). Let ψ_n be such that*

$$\psi_n^2 \geq C n^{-1/2} G(\psi_n), \tag{15.57}$$

for a positive constant C, and assume that $\lim_{n\to\infty} n\psi_n^{2(1+a)} = \infty$. Then for $f \in \mathcal{F}$,

$$E\left\|\hat{f} - f\right\|_2^2 \leq C'\left(\psi_n^2 + \epsilon\right),$$

for a positive constant C', for sufficiently large n.

Proof. The proof goes similarly as the proof of Theorem 15.6 until step (15.55). At this step we apply Lemma 12.5 to get

$$\begin{aligned}
&P\left(\sup_{g \in \mathcal{F}_j} \nu_n[Q(g-f)] > \xi a_j\right) \\
&\leq \exp\left\{-\frac{n(\xi a_j)^2 C'}{c^2 b_j^{1-a}}\right\} + 2\#\mathcal{G}_{B_2} \exp\left\{-\frac{1}{12}\frac{n(\xi a_j)^2}{B_\infty c^2 b_j^{1-a} + 2\xi a_j B'_\infty/9}\right\}.
\end{aligned}$$

The first term in the right-hand side is handled similarly as in the Proof of Theorem 15.6. For the second term in the right-hand side we have, for sufficiently large n,

$$\begin{aligned}
\exp\left\{-\frac{1}{12}\frac{n(\xi a_j)^2}{B_\infty c^2 b_j^{1-a} + 2\xi a_j B'_\infty/9}\right\} &= \exp\left\{-\frac{1}{12}\frac{n\xi^2 a_j}{B_\infty c^2 a_0^{-a} + 2\xi B'_\infty/9}\right\} \\
&\leq \exp\left\{-\frac{1}{12}\frac{n\xi^2 a_j a_0^a}{B_\infty c^2 + 2\xi B'_\infty/9}\right\} \\
&= \exp\left\{-n\psi_n^{2(1+a)} 2^{2j}(1+t)^{1+a} C''\right\},
\end{aligned}$$

since $a_j^{-a} = (2^{2j}a_0)^{-a} \leq a_0^{-a}$ and $a_0^{-a} \geq 1$ for sufficiently large n, and we denote $C' = \xi^2 C_\tau^{1+a}/[2^{1+a}12(B_\infty c^2 + 2\xi B'_\infty/9)]$. The proof is finished the same as the proof of Theorem 15.6. □

15.3.4 Rates of Convergence of Dense Minimizer

Theorem 15.6 and Theorem 15.7 show that the rate of convergence of the dense minimizer is the solution of the equation

$$\psi_n^2 = n^{-1/2} G(\psi_n). \tag{15.58}$$

To get the optimal rate, the net $\mathcal{G}_\delta$ is chosen so that its cardinality is minimal. In the polynomial case one can find a δ-net whose cardinality satisfies

$$\log(\#\mathcal{G}_\delta) = C\delta^{-b}$$

for some constants $b, C > 0$ and the operator norm satisfies

$$\varrho(Q, \mathcal{G}_\delta) = C'\delta^{-a}$$

for some $a, C' > 0$. (In the direct case $a = 0$ and $C' = 1$.) Thus the entropy integral $G(\delta)$ is finite for $\int_0^\delta u^{-a-b/2}\, du < \infty$, which holds when

$$a + b/2 < 1. \tag{15.59}$$

Then (15.58) leads to $\psi_n^2 \asymp n^{-1/2}\psi_n^{-a-b/2+1}$, and the rate of the dense minimization estimator is

$$\psi_n \asymp n^{-1/[2(a+1)+b]}. \tag{15.60}$$

This is the same rate as the rate of the δ-net estimator given in (15.41).

1. *(Smoothness classes.)* Let $\mathcal{F}$ be a set of s-smooth d-dimensional functions, so that $b = d/s$; see (10.70). Then condition (15.59) may be written as a condition for the smoothness index s:

$$s > \frac{d}{2(1-a)}.$$

When the problem is direct, $a = 0$, and we have the classical condition $s > d/2$. The rate is given in (15.42).

2. *(Convex hull.)* Let $\mathcal{F}$ be the L_2 closure of the convex hull of class $\mathbb{G}$, and assume that the cardinality of a minimal δ-net is $N(\delta, \mathbb{G}, \|\cdot\|_2) \le C\delta^{-V}$. Then (15.39) holds with $b = 2V/(2+V)$; see Theorem 10.13. Condition (15.59) can be written as

$$V < \frac{2(1-a)}{a}. \tag{15.61}$$

The rate is given in (15.43).

15.4 SERIES ESTIMATORS

We have discussed series estimators in Section 14.5. We will give bounds to the mean integrated squared error of these estimates in this section.

15.4.1 An Orthogonal Series Estimator

We consider estimators of the form

$$\tilde{f}(x, W, \Theta) = \sum_{j=1}^{k} w_j \theta_j \phi_j(x), \qquad x \in \mathbf{R}^d,$$

where $W = (w_1, \ldots, w_k)$, $\Theta = (\theta_1, \ldots, \theta_k)$. We should choose at least Θ based on the data $X_1, \ldots, X_n$. As in (14.57) we can define a series estimator as

$$\tilde{f}(x, W, \hat{\Theta}) = \sum_{j=1}^{k} w_j \hat{\theta}_j \phi_j(x), \qquad x \in \mathbf{R}^d, \tag{15.62}$$

where

$$\hat{\Theta} = (\hat{\theta}_1, \ldots, \hat{\theta}_k), \qquad \hat{\theta}_j = \frac{1}{n} \sum_{i=1}^{n} \phi_j(X_i), \tag{15.63}$$

are the empirical coefficients. The estimator is called an orthogonal series estimator when $\{\phi_1, \ldots, \phi_k\}$ is an orthonormal system in $L_2(\mathbf{R}^d)$.

Orthogonal Series Estimator as a Minimization Estimator We can show that the estimator in (15.62) is a minimization estimator when we minimize over the coefficients θ_j. Namely the empirical coefficients minimize the L_2 empirical risk of an orthogonal series estimator when we minimize over the coefficients and fix the weights. Lemma 15.8 is analogous to the approximation theoretic result in Exercise 15.2.

Lemma 15.8 *Let*

$$\hat{f} = \operatorname{argmin}_{f \in \mathcal{C}} \gamma_n(f), \tag{15.64}$$

be a minimization estimator where $\gamma_n(\,\cdot\,)$ *is the empirical risk functional as defined in (15.1) for the* L_2 *error and*

$$\mathcal{C} = \{\tilde{f}(\cdot, W, \Theta) : \Theta \in \mathbf{R}^k, \|\Theta\| < \infty\}. \tag{15.65}$$

We have that

$$\hat{f} = \tilde{f}(\cdot, W, \hat{\Theta}),$$

when $\{\phi_1, \ldots, \phi_k\}$ *is an orthonormal system in* $L_2(\mathbf{R}^d)$.

Proof. Let $\gamma(\,\cdot\,)$ be the L_2 contrast function as defined in (15.4). By the orthonormality, we have that

$$\gamma\left(\tilde{f}(\cdot, W, \Theta), X_i\right) = \sum_{j=1}^{k} \left(w_j^2 \theta_j^2 - 2 w_j \theta_j \phi_j(X_i)\right).$$

Thus

$$\begin{aligned}\gamma_n\left(\tilde{f}(\,\cdot\,,W,\Theta)\right) &= \frac{1}{n}\sum_{i=1}^{n}\gamma(\tilde{f}(\,\cdot\,,W,\Theta),X_i)\\ &= \sum_{j=1}^{k}\left(w_j^2\theta_j^2 - 2w_j\theta_j\theta_{n,j}\right), \qquad (15.66)\end{aligned}$$

and it follows that

$$\mathrm{argmin}_{\Theta\in\mathbf{R}^k,\|\Theta\|<\infty}\gamma_n\left(\tilde{f}(\,\cdot\,,W,\Theta)\right) = \hat{\Theta}, \qquad (15.67)$$

as claimed. □

Hard Thresholding In estimator (15.62) the coefficient vector Θ is chosen based on the data $X_1,\ldots,X_n$. We want to choose also the vector W based on the data. Hard thresholding chooses the vector of weights by thresholding the empirical coefficients:

$$\hat{W}_{thre} = \left(I_{[\alpha,\infty)}(\hat{\theta}_j^2)\right)_{j=1,\ldots,k},$$

where $\alpha > 0$. The number k of the terms in the estimate may be chosen large and the weights are used to remove the less important terms. We prove that hard thresholding is equivalent to choosing the weights by minimizing a complexity-penalized empirical risk when the complexity is the number of terms in the estimate. Let

$$\hat{W}_{pena} = \mathrm{argmin}_{W\in\{0,1\}^k}\mathcal{E}_n(W,\alpha), \qquad (15.68)$$

where

$$\mathcal{E}_n(W,\alpha) = \gamma_n\left(\tilde{f}(\cdot,W,\Theta_n)\right) + \alpha\cdot\#\{w_j : j=1,\ldots,k,\ w_j>0\}, \qquad (15.69)$$

where $\gamma_n(\,\cdot\,)$ is the empirical risk functional as defined in (15.1) for the L_2 error.

Lemma 15.9 *We have that* $\hat{W}_{thre} = \hat{W}_{pena}$.

Proof. We may write, using (15.66),

$$\gamma_n\left(\tilde{f}(\,\cdot\,,W,\Theta_n)\right) = -\sum_{j=1}^{k} w_j\theta_{n,j}^2.$$

Thus for $W\in\{0,1\}^k$,

$$\mathcal{E}_n(W,\alpha) = \sum\left\{\alpha - w_j\theta_{n,j}^2 : j=1,\ldots,k,\ w_j=1\right\},$$

which is minimized over w_j by the hard thresholding weights. □

15.4.2 A General Series Estimator

We consider estimators that are linear combinations of functions in a finite dictionary of L_2 functions. Denote

$$\tilde{f}(x,\Lambda) = \sum_{\phi \in \mathcal{D}} \lambda_\phi \phi(x), \qquad x \in \mathbf{R}^d, \tag{15.70}$$

where $\mathcal{D} \subset L_2([0,1]^d)$ is a collection of functions, and we will assume that $\mathcal{D}$ has finite cardinality; $\Lambda = (\lambda_\phi)_{\phi \in \mathcal{D}} \in \mathbf{R}^{\mathcal{D}}$ gives the coefficients of the expansion.

Complexity-Penalized Empirical Risk We consider estimators that are minimizers of a complexity-penalized empirical risk. They are of the type (15.13). We apply definition (15.13) by taking the class of functions to be

$$\mathcal{C} = \{\tilde{f}(\cdot\,,\Lambda) : \Lambda \in \mathbb{K}\},$$

where $\mathbb{K} \subset \mathbf{R}^{\mathcal{D}}$. We consider subsets of $\mathbf{R}^{\mathcal{D}}$, since in later examples we impose several restrictions on the coefficients. The complexity penalization is taken to be the number of terms in the expansion:

$$D(\Lambda) = \#\{\lambda_\phi : \lambda_\phi \neq 0\}, \tag{15.71}$$

where $\Lambda = (\lambda_\phi)_{\phi \in \mathcal{D}} \in \mathbf{R}^{\mathcal{D}}$. As in (15.11) we define the complexity-penalized empirical risk as

$$\mathcal{E}_n(\Lambda,\alpha) = \gamma_n\left(\tilde{f}(\cdot,\Lambda)\right) + \alpha \cdot D(\Lambda), \tag{15.72}$$

where γ_n is defined in (15.2) with the L_2 contrast function defined in (15.4), and $\alpha \geq 0$.

Definition of the Estimator We define the empirical choice for vector Λ when we restrict the minimization to the set $\mathbb{K} \subset \mathbf{R}^{\mathcal{D}}$, as

$$\hat{\Lambda}_\alpha = \hat{\Lambda}_\alpha(\mathbb{K}) = \text{argmin}_{\Lambda \in \mathbb{K}} \mathcal{E}_n(\Lambda,\alpha). \tag{15.73}$$

We denote the minimization estimator with

$$\hat{f}_\alpha(x) = \tilde{f}\left(x, \hat{\Lambda}_\alpha\right), \qquad x \in \mathbf{R}^d. \tag{15.74}$$

Error Bound We give an upper bound for the mean integrated squared error of the minimization estimator in terms of the best approximation error. We define the best approximation error to be a minimum of the theoretical error-complexity

$$K(f,\Lambda,\alpha) = \left\|\tilde{f}(\cdot\,,\Lambda) - f\right\|_2^2 + \alpha \cdot D(\Lambda), \tag{15.75}$$

where f is the true density to be estimated, $\tilde{f}$ is defined in (15.70), $\alpha \geq 0$, and $D(\Lambda)$ is defined in (15.71). The minimization is over $\Lambda \in \mathbb{K}_0 \subset \mathbb{K}$, where $\mathbb{K}_0$ is defined in the case of density estimation as

$$\mathbb{K}_0 = \left\{\Lambda \in \mathbb{K} : \|\tilde{f}(\cdot,\Lambda)\|_\infty \leq 2B_\infty\right\}, \tag{15.76}$$

where $B_\infty > \|f\|_\infty$ is a positive constant. In the case of Gaussian white noise we take $\mathbb{K}_0 = \mathbb{K}$, where $\mathbb{K} \subset \mathbf{R}^{\mathcal{D}}$ is the same as in (15.73). That is, we have restricted ourselves to the bounded approximating functions $\tilde{f}(\cdot, \Lambda)$ in the case of density estimation.

Theorem 15.10 *We have for the estimator $\hat{f}_\alpha$ defined in (15.74), based on i.i.d. observations $X_1, \ldots, X_n$ from the distribution of a continuous density $f : [0,1]^d \to \mathbf{R}$, or based on an observation from a Gaussian white noise model with a continuous signal function, that*

$$E\left\|\hat{f}_{\alpha_n} - f\right\|_2^2 1_{\tilde{\Omega}} \leq C_1 \inf_{\Lambda \in \mathbb{K}_0} K(f, \Lambda, \alpha_n) + C_2 n^{-1}, \tag{15.77}$$

where

$$\alpha_n = C_L\, B_\infty \frac{\log_e(\#\mathcal{D})}{n}, \tag{15.78}$$

with B_∞ being the same as in (15.76) for density estimation and $B_\infty = 1$ for the Gaussian white noise, C_L, C_1, C_2 are positive constants, $1_{\tilde{\Omega}}$ is the indicator of the event

$$\tilde{\Omega} = \left(\|\hat{f}_{\alpha_n}\|_\infty \leq 2B_\infty\right) \tag{15.79}$$

in the case of density estimation and in the case of Gaussian white noise we take $\tilde{\Omega} = \Omega$ to be the whole probability space.

Proof of Theorem 15.10 Denote $\hat{f} = \hat{f}_{\alpha_n}$ and $\hat{\Lambda} = \hat{\Lambda}_{\alpha_n}$. We condition on the set $\tilde{\Omega}$ so that $\hat{\Lambda} \in \mathbb{K}_0$. Let f be the true density and let $\Lambda^0 \in \mathbb{K}_0$. Denote

$$\zeta = C_1 K(f, \Lambda^0, \alpha_n),$$

where C_1 is a positive constant to be chosen later. We have that

$$\begin{aligned} E\|\hat{f} - f\|_2^2 &= \int_0^\infty P\left(\|\hat{f} - f\|_2^2 > t\right) dt \\ &\leq \zeta + \int_\zeta^\infty P\left(\|\hat{f} - f\|_2^2 > t\right) dt \\ &= \zeta + C_2 n^{-1} \int_0^\infty P(A_t)\, dt, \end{aligned} \tag{15.80}$$

where

$$A_t \stackrel{def}{=} \left(\|\hat{f} - f\|_2^2 > C_2 n^{-1} t + \zeta\right)$$

and C_2 is a positive constant to be chosen later. Let $a > 0$. Now

$$A_t = \left((a+1)\|\hat{f} - f\|_2^2 > a\|\hat{f} - f\|_2^2 + C_1 K(f, \Lambda^0, \alpha_n) + C_2 n^{-1} t\right).$$

Lemma 15.2 implies that the theoretical error-complexity of the minimization estimator may be bounded by the theoretical error-complexity of $f^0 = \tilde{f}(\cdot, \Lambda^0)$, with the additional empirical term:

$$K(f, \hat{\Lambda}, \alpha_n) \le K(f, \Lambda^0, \alpha_n) + 2\nu_n\left(\hat{f} - f^0\right)$$
$$\Leftrightarrow \quad \|\hat{f} - f\|_2^2 \le K(f, \Lambda^0, \alpha_n) - \alpha_n D(\hat{\Lambda}) + 2\nu_n\left(\hat{f} - f^0\right).$$

Thus

$$\begin{aligned} A_t \quad &\subset \quad \left(2\nu_n\left(\hat{f} - f^0\right) > \frac{a}{a+1}\|\hat{f} - f\|_2^2 + \alpha_n D(\hat{\Lambda}) \right. \\ &\qquad \left. + \left(\frac{C_1}{a+1} - 1\right) K(f, \Lambda^0, \alpha_n) + \frac{C_2}{a+1} n^{-1} t\right) \\ &\subset \quad \left(\nu_n\left(\hat{f} - f^0\right) > w(\hat{\Lambda})\xi\right) \\ &\subset \quad \left(\sup_{\Lambda \in \mathbb{K}_0} \frac{\nu_n(\tilde{f}(\cdot, \Lambda)) - \nu_n(f^0)}{w(\Lambda)} > \xi\right) \\ &\stackrel{def}{=} \quad B_t, \end{aligned}$$

where we used the fact that on $\tilde{\Omega}$, $\hat{\Lambda} \in \mathbb{K}_0$, and we denote

$$\xi = \frac{1}{2} \min\left\{\frac{C_1}{a+1} - 1, \frac{a}{a+1}\right\},$$

$$w(\Lambda) = \left\|\tilde{f}(\cdot, \Lambda) - f\right\|_2^2 + \|f - f^0\|_2^2 + \frac{\tau(\Lambda)}{2n},$$

$$\tau(\Lambda) = C_\tau\left[n\alpha_n\left(C_{\tau,1} D(\Lambda^0) + C_{\tau,2} D(\Lambda)\right) + t\right], \qquad (15.81)$$

$$C_\tau = \frac{1}{\xi}\frac{C_2}{a+1}, \quad C_{\tau,1} = \frac{C_1 - a - 1}{C_2}, \quad C_{\tau,2} = \frac{a+1}{C_2}.$$

We need to choose C_1, C_2, and a so that $2C_\tau^{-1} \le \xi^2$. This inequality will be needed in (15.90). This choice is possible, since we can take $2(a+1)/C_2 \le \xi$. We need also $C_1/(a+1) - 1 > 0$ to guarantee $\xi > 0$. We have

$$P(A_t) \le P(B_t). \qquad (15.82)$$

We prove that

$$P(B_t) \le C \exp\{-t(C_L B_\infty)^{-1}\}, \qquad (15.83)$$

where C and C_L are positive constants. Eq. (15.83) proves the theorem, when we combine (15.80) and (15.82).

Proof of (15.83) For $\Phi \subset \mathcal{D}$, let $\mathbb{K}_{0,\Phi}$ be the set of coefficients in $\mathbb{K}_0$ that are nonzero exactly at the positions given by set Φ:

$$\mathbb{K}_{0,\Phi} = \{\Lambda \in \mathbb{K}_0 : \lambda_\phi \neq 0 \text{ if and only if } \phi \in \Phi\},$$

where we use again the notation $\Lambda = (\lambda_\phi)_{\phi \in \mathcal{D}}$. Let for $l = 1, 2, \ldots$,

$$\mathbb{D}_l = \{\Phi \subset \mathcal{D} : \#\Phi = l\}$$

be the set of subsets of $\mathcal{D}$ of cardinality l. We can write

$$\mathbb{K}_0 = \bigcup_{l=1}^{\infty} \bigcup_{\Phi \in \mathbb{D}_l} \mathbb{K}_{0,\Phi}.$$

That is, we can make a countable partition of $\mathbb{K}_0$ and each member of the partition is the set of vectors Λ with exactly l nonzero elements. We have

$$B_t \subset \bigcup_{l=1}^{\infty} \bigcup_{\Phi \in \mathbb{D}_l} B_\Phi,$$

where

$$B_\Phi = \left(\sup_{\Lambda \in \mathbb{K}_{0,\Phi}} \frac{\nu_n(\tilde{f}(\cdot, \Lambda)) - \nu_n(f^0)}{w(\Lambda)} > \xi \right).$$

For $\Phi \in \mathbb{D}_l$,

$$P(B_\Phi) \leq 2 \exp\{-(C_L B_\infty)^{-1}(t + lL)\}, \tag{15.84}$$

where C_L is a positive constant defined in (15.95) and

$$L = C_L B_\infty \log_e(\#\mathcal{D}).$$

We prove (15.84) below. We have

$$\#\mathbb{D}_l = \binom{\#\mathcal{D}}{l} \leq \left(\frac{e\#\mathcal{D}}{l}\right)^l. \tag{15.85}$$

The proof of (15.85) is left in Exercise 15.4. Thus

$$\begin{aligned} P(B_t) &\leq 2 \sum_{l=1}^{\infty} \sum_{\Phi \in \mathbb{D}_l} \exp\{-(C_L B_\infty)^{-1}(t + lL)\} \\ &\leq 2 \sum_{l=1}^{\infty} \left(\frac{e\#\mathcal{D}}{l}\right)^l \exp\{-(C_L B_\infty)^{-1}(t + lL)\} \\ &\leq C \exp\{-(C_L B_\infty)^{-1} t\}, \end{aligned} \tag{15.86}$$

by the choice of L. We have proved (15.83) up to proving (15.84).

Proof of (15.84) Denote

$$Z = \sup_{g \in \mathcal{G}} \nu_n(g),$$

where

$$\mathcal{G} = \mathcal{G}_\Phi = \left\{ \frac{\tilde{f}(\cdot, \Lambda) - f^0}{w(\Lambda)} : \Lambda \in \mathbb{K}_{0,\Phi} \right\}, \qquad \Phi \in \mathbb{D}_l,$$

and we denote $w = w_1$. Next denote

$$v_0 = \sup_{g \in \mathcal{G}} \|g\|_2^2 = \sup_{\Lambda \in \mathbb{K}_{0,\Phi}} \frac{\|\tilde{f}(\cdot, \Lambda) - f^0\|_2^2}{w^2(\Lambda)}$$

and

$$\tau = C_\tau \left[n\alpha_n \left(C_{\tau,1} D(\Lambda^0) + C_{\tau,2}\, l \right) + t \right].$$

We have for $\Phi \in \mathbb{D}_l$, $\Lambda \in \mathbb{K}_{0,\Phi}$, that $\tau(\Lambda) = \tau$, where $\tau(\Lambda)$ is defined in (15.81). Thus for $\Phi \in \mathbb{D}_l$ and $\Lambda \in \mathbb{K}_{0,\Phi}$,

$$w(\Lambda) \geq \frac{1}{2} \left(\left\| \tilde{f}(\cdot, \Lambda) - f^0 \right\|_2^2 + \frac{\tau}{n} \right) \geq \left\| \tilde{f}(\cdot, \Lambda) - f^0 \right\|_2 \left(\frac{\tau}{n} \right)^{1/2}. \tag{15.87}$$

Thus

$$v_0 \leq \frac{n}{\tau}. \tag{15.88}$$

We have

$$(EZ)^2 \leq \begin{cases} \left(l + D(\Lambda^0)\right) \tau^{-1}, & \text{Gaussian white noise,} \\ \|f\|_\infty \left(l + D(\Lambda^0)\right) \tau^{-1}, & \text{density estimation.} \end{cases} \tag{15.89}$$

We prove equation (15.89) below in page 421. Denote

$$\eta^2 = \tau^{-1}(t + C_{\tau,2} L l).$$

Then

$$\begin{aligned} (EZ + \eta)^2 &\leq 2\left[(EZ)^2 + \eta^2 \right] \\ &\leq 2\tau^{-1} \left[B_\infty \left(l + D(\Lambda^0) \right) + t + C_{\tau,2} L l \right] \\ &\leq 2C_\tau^{-1} \\ &\leq \xi^2 \end{aligned} \tag{15.90}$$

since

$$C_\tau^{-1} \tau \geq L \left(C_{\tau,1} D(\Lambda^0) + C_{\tau,2}\, l \right) + t \geq B_\infty D(\Lambda^0) + (B_\infty + C_{\tau,2} L)\, l + t,$$

where we used $n\alpha_n = L$ and $C_{\tau,1} C_L \log_e(\#\mathcal{D}) \geq 1$. Equation (15.90) implies that

$$P(B_\Phi) = P\left(Z \geq \xi\right) \leq P\left(Z \geq EZ + \eta\right). \tag{15.91}$$

Gaussian White Noise We have that $\nu_n(g) \sim N(0, n^{-1}\|g\|_2^2)$, and thus

$$\sup_{g\in\mathcal{G}} \mathrm{Var}(\nu_n(g)) = n^{-1} v_0.$$

Theorem 12.2 and (15.88) imply that

$$P(Z \geq EZ + \eta) \leq 2\exp\left\{-\frac{n\eta^2}{2v_0}\right\} \leq 2\exp\left\{-\frac{t+lL}{2}\right\}. \tag{15.92}$$

Thus (15.84) follows for the case of Gaussian white noise from (15.91) and (15.92).

Density Estimation Denote

$$b = \sup\{\|g\|_\infty : g \in \mathcal{G}\}$$

and

$$v = \sup\{\mathrm{Var}_f(g(X_1)) : g \in \mathcal{G}\}.$$

By Theorem 12.3,

$$P(Z \geq EZ + \eta) \leq \exp\left\{\frac{-n\eta^2}{2[v + 2bEZ + \eta b/3]}\right\}. \tag{15.93}$$

We have

$$v \leq \|f\|_\infty v_0 \leq \|f\|_\infty \frac{n}{\tau}.$$

Also for for $\Phi \in \mathbb{D}_l$ and $\Lambda \in \mathbb{K}_{0,\Phi}$, we have

$$w(\Lambda) \geq \frac{\tau}{2n}, \tag{15.94}$$

and thus

$$b \leq \frac{2n}{\tau} \sup_{\Lambda\in\mathbb{K}_{0,\Phi}} \|\tilde{f}(\cdot,\Lambda) - f^0\|_\infty \leq \frac{4n}{\tau} B, \qquad B = 2\|f\|_\infty,$$

by the definition of $\mathbb{K}_0$. Thus applying inequalities $EZ \leq \xi, \eta \leq \xi$, obtains

$$\begin{aligned} v + 2bEZ + \eta b/3 &\leq \frac{n}{\tau}\|f\|_\infty (1 + 8\cdot\xi(2+1/3)) \\ &\stackrel{def}{=} \frac{n}{\tau}\|f\|_\infty C_L/2. \end{aligned} \tag{15.95}$$

Thus

$$P(Z \geq EZ + \eta) \leq \exp\left\{-\frac{t+lL}{B_\infty C_L}\right\}. \tag{15.96}$$

Equation (15.84) follows for the case of density estimation from (15.91) and (15.96).

Proof of (15.89) Let $\Lambda \in \mathbb{K}_{0,\Phi}$ where $\Phi \in \mathbb{D}_l$. Denote $\mathcal{D}(\Lambda, \Lambda^0) = \{\phi \in \mathcal{D} : \lambda_\phi \neq 0 \text{ or } \lambda_\phi^0 \neq 0\}$. Let $\{\psi_1, \ldots, \psi_k\}$ be a basis of the span of $\mathcal{D}(\Lambda, \Lambda^0)$. We have $k \leq \#\mathcal{D}(\Lambda, \Lambda^0) \leq l + D(\Lambda^0)$. Applying (15.87) obtains

$$\sup_{\Lambda \in \mathbb{K}_{0,\Phi}} \frac{\|\tilde{f}(\cdot, \Lambda) - f^0\|_2^2}{w^2(\Lambda)} \leq \frac{n}{\tau}.$$

Thus we may apply Lemma 12.9 with $B_2^2 = n/\tau$ to get (15.89). □

15.4.3 Best Basis Estimator

The estimator in Section 15.4.2 is a quite general series estimator. We specialize to the practically interesting case where the estimator is a linear combination of functions in an orthonormal system. Denote

$$\tilde{f}(x, W, \Theta, \mathcal{B}) = I_{[0,1]^d}(x) + \sum_{\phi \in \mathcal{B}} w_\phi \theta_\phi \phi(x), \qquad x \in \mathbf{R}^d, \tag{15.97}$$

where $\mathcal{B}$ is an orthonormal system of functions in $L_2([0,1]^d)$, $W = (w_\phi)_{\phi \in \mathcal{B}} \in \{0,1\}^{\mathcal{B}}$, $\Theta = (\theta_\phi)_{\phi \in \mathcal{B}} \in \mathbf{R}^{\mathcal{B}}$. Vector W chooses a subset of $\mathcal{B}$ and vector Θ gives the coefficients of the expansion. We will assume that $\int_{[0,1]^d} \phi = 0$ for all $\phi \in \mathcal{B}$, and since we estimate densities we may include the indicator $I_{[0,1]^d}$ to all expansions. We assume to have a collection $\mathcal{L}$ of orthonormal systems.

Assumption 15.1 *Collection $\mathcal{L}$ of orthonormal systems is such that for each $\mathcal{B} \in \mathcal{L}$,*

1. *$\mathcal{B}$ is a finite orthonormal system of functions $\phi : [0,1]^d \to \mathbf{R}$,*
2. *for each $\phi \in \mathcal{B}$, $\int_{[0,1]^d} \phi = 0$.*

Complexity-Penalized Empirical Risk We consider estimators that minimize a complexity-penalized empirical risk. Namely, we apply definition (15.13) by taking the class of functions to be

$$\mathcal{C} = \{\tilde{f}(\cdot, W, \Theta, \mathcal{B}) : (W, \Theta, \mathcal{B}) \in \mathbb{K}\},$$

where

$$\mathbb{K} = \left\{(W, \Theta, \mathcal{B}) : \mathcal{B} \in \mathcal{L},\ \ W \in \mathcal{W}(\mathcal{B}),\ \ \Theta \in \mathbf{R}^{\mathcal{B}}\right\},$$

where $\mathcal{W}(\mathcal{B}) \subset \{0,1\}^{\mathcal{B}}$. The complexity penalization is taken to be the number of terms in the expansion:

$$D(W, \Theta, \mathcal{B}) = D(W) = \#\{w_\phi : w_\phi = 1\} + 1, \tag{15.98}$$

where $W = (w_\phi)_{\phi \in \mathcal{B}} \in \{0,1\}^{\mathcal{B}}$. We added 1 in the definition of $D(W)$ because the function $I_{[0,1]^d}$ is also in the expansion (15.97). As in (15.11) we define the complexity-penalized empirical risk as

$$\mathcal{E}_n(W, \Theta, \mathcal{B}, \alpha) = \gamma_n\left(\tilde{f}(\cdot, W, \Theta, \mathcal{B})\right) + \alpha \cdot D(W), \tag{15.99}$$

where γ_n is defined in (15.2) with the L_2 contrast function as in (15.4), and $\alpha \geq 0$.

Definition of the Estimator We define an empirical choice for basis $\mathcal{B}$, for coefficients Θ, and for weights $W \in \{0,1\}^{\mathcal{B}}$. We restrict the minimization to the set $\mathcal{W}(\mathcal{B}) \subset \{0,1\}^{\mathcal{B}}$. We show in (15.109) that the coefficients Θ can in fact taken to be the empirical coefficients. Let

$$\left(\hat{\mathcal{B}}_\alpha, \hat{\Theta}_\alpha, \hat{W}_\alpha\right) = \text{argmin}_{\mathcal{B}\in\mathcal{L},\Theta\in\mathbf{R}^{\mathcal{B}},W\in\mathcal{W}(\mathcal{B})}\mathcal{E}_n\left(W,\Theta,\mathcal{B},\alpha\right). \tag{15.100}$$

We denote the best-basis minimization estimator by

$$\hat{f}_\alpha(x) = \tilde{f}\left(x, \hat{W}_\alpha, \hat{\Theta}_\alpha, \hat{\mathcal{B}}_\alpha\right), \qquad x \in \mathbf{R}^d. \tag{15.101}$$

Error Bound We give an upper bound for the mean integrated squared error of the minimization estimator in terms of the best approximation error. We define the best approximation error to be a minimum of the theoretical error-complexity

$$K(f, W, \Theta, \mathcal{B}, \alpha) = \left\|\tilde{f}(\,\cdot\,, W, \Theta, \mathcal{B}) - f\right\|_2^2 + \alpha \cdot D(W), \tag{15.102}$$

where f is the true density to be estimated, $\tilde{f}$ is defined in (15.97), $\alpha \geq 0$, and $D(W)$ is defined in (15.98). The minimization is over $(W,\Theta,\mathcal{B}) \in \mathbb{K}_0$, where $\mathbb{K}_0$ is defined for the density estimation as

$$\mathbb{K}_0 = \left\{(W,\Theta,\mathcal{B}) \in \mathcal{W}(\mathcal{B}) \times \mathbf{R}^{\mathcal{B}} \times \mathcal{L} : \|\tilde{f}(\cdot, W, \Theta, \mathcal{B})\|_\infty \leq 2B_\infty\right\}, \tag{15.103}$$

with $B_\infty > \|f\|_\infty$ as a positive constant. For the Gaussian white noise we define $\mathbb{K}_0 = \mathcal{W}(\mathcal{B}) \times \mathbf{R}^{\mathcal{B}} \times \mathcal{L}$. That is, we restrict ourselves to the bounded approximating functions $\tilde{f}(\cdot, W, \Theta, \mathcal{B})$ in the case of density estimation and to the set $\mathcal{W}(\mathcal{B})$ of weights.

Theorem 15.11 *We have for the estimator $\hat{f}_\alpha$ defined in (15.74), based on i.i.d. observations $X_1, \ldots, X_n$ from the distribution of a continuous density $f : [0,1]^d \to \mathbf{R}$, or based on an observation from a Gaussian white noise model with a continuous signal function, that*

$$E\left\|\hat{f}_{\alpha_n} - f\right\|_2^2 1_{\tilde{\Omega}} \leq C_1 \inf_{(W,\Theta,\mathcal{B})\in\mathbb{K}_0} K\left(f, W, \Theta, \mathcal{B}, \alpha_n\right) + C_2 n^{-1}, \tag{15.104}$$

where

$$\alpha_n = C_L\, B_\infty \frac{\log_e(\#\mathcal{D})}{n}, \tag{15.105}$$

with

$$\mathcal{D} = \bigcup_{\mathcal{B}\in\mathcal{L}} \mathcal{B}, \tag{15.106}$$

and B_∞ is the same as in (15.103) for the density estimation and $B_\infty = 1$ for the Gaussian white noise, C_L, C_1, C_2 are positive constants, $1_{\tilde{\Omega}}$ is the indicator of the event

$$\tilde{\Omega} = \left(\|\hat{f}_{\alpha_n}\|_\infty \leq 2B_\infty\right) \tag{15.107}$$

in the case of density estimation and in the case of Gaussian white noise we take $\tilde{\Omega} = \Omega$ to be the whole probability space.

Proof. Theorem 15.11 follows from Theorem 15.10 when we choose dictionary $\mathcal{D}$ as in (15.106) and the coefficients λ_ϕ are obtained by merging w_ϕ and θ_ϕ: $\lambda_\phi = w_\phi \theta_\phi$. The vectors $\Lambda = (\lambda_\phi)_{\phi \in \mathcal{D}}$ have to be such that components are nonzero only for a single basis $\mathcal{B}$: $\{\phi : \lambda_\phi \neq 0\} \subset \mathcal{B}$ for some $\mathcal{B} \in \mathcal{L}$. That is, we define the set of coefficients over which one minimizes to be

$$\mathbb{K} = \left\{\Lambda \in \mathbf{R}^{\mathcal{D}} : \Lambda \in \mathcal{W}(\mathcal{B}) \text{ for some } \mathcal{B} \in \mathcal{L}\right\}.$$

□

Empirical Coefficients The minimization problem in (15.100) can be written in a more transparent way, since the coefficient vector Θ can be chosen to be the vector of empirical coefficients. Define the vector of empirical coefficients as

$$\hat{\Theta} = \hat{\Theta}(\mathcal{B}) = \left(\hat{\theta}_\phi\right)_{\phi \in \mathcal{B}}, \qquad \hat{\theta}_\phi = \frac{1}{n} \sum_{i=1}^{n} \phi(X_i). \tag{15.108}$$

We have that $\hat{\Theta}_\alpha = \hat{\Theta}(\hat{\mathcal{B}}_\alpha)$ and

$$\left(\hat{\mathcal{B}}_\alpha, \hat{W}_\alpha\right) = \text{argmin}_{\mathcal{B} \in \mathcal{L}, W \in \mathcal{W}(\mathcal{B})} \mathcal{E}_n\left(W, \hat{\Theta}(\mathcal{B}), \mathcal{B}, \alpha\right). \tag{15.109}$$

This follows from Lemma 15.8. Lemma 15.8 implies that for every $(W, \Theta) \in \{0,1\}^{\mathcal{B}} \times \mathbf{R}^{\mathcal{B}}$,

$$\mathcal{E}_n(W, \hat{\Theta}(\mathcal{B}), \mathcal{B}, \alpha) \leq \mathcal{E}_n(W, \Theta, \mathcal{B}, \alpha).$$

Theoretical Error Complexity The minimal complexity-penalized approximation error can be bound by finding a good orthonormal system $\mathcal{B}^*$ and choosing the coefficients to be the coefficients of f in this orthonormal system as follows:

$$\inf_{(W,\Theta,\mathcal{B}) \in \mathbb{K}_0} K\left(f, W, \Theta, \mathcal{B}, \alpha_n\right) \leq \inf_{W \in \{0,1\}^{\mathcal{B}^*}} K\left(f, W, \Theta_f(\mathcal{B}^*), \mathcal{B}^*, \alpha_n\right),$$

where $\Theta_f(\mathcal{B})$ is defined by

$$\Theta_f(\mathcal{B}) = \left(\int_{\mathbf{R}^d} f\phi\right)_{\phi \in \mathcal{B}}. \tag{15.110}$$

It is left to minimize the complexity-penalized approximation error with respect to weights. We may use the following lemma to find the optimal weights:

Lemma 15.12 *Let $\mathcal{B}$ be a basis of $L_2([0,1]^d)$, and let $\mathcal{B}_0$ be an orthonormal system, such that $\mathcal{B}_0 \subset \mathcal{B}$. Denote $\theta_{f,\phi} = \int_{\mathbf{R}^d} f\phi$. Then*

$$\begin{aligned}
&\min_{W \in \{0,1\}^{\mathcal{B}_0}} K\left(f, W, \Theta_f(\mathcal{B}_0), \mathcal{B}_0, \alpha\right) \\
&= \alpha + \sum_{\phi \in \mathcal{B}_0} \min\left\{\theta_{f,\phi}^2, \alpha\right\} + \sum_{\phi \in \mathcal{B} \setminus \mathcal{B}_0} \theta_{f,\phi}^2.
\end{aligned}$$

Proof. We have

$$
\begin{aligned}
& K\left(f, W, \Theta_f(\mathcal{B}_0), \mathcal{B}_0, \alpha\right) \qquad (15.111)\\
&= \sum_{\phi \in \mathcal{B}_0} \theta_{f,\phi}^2 (1 - w_\phi)^2 + \sum_{\phi \in \mathcal{B} \setminus \mathcal{B}_0} \theta_{f,\phi}^2 + \alpha D(W) \\
&= \alpha + \sum_{\phi \in \mathcal{B}_0} \left(\theta_{f,\phi}^2 I_{\{\phi : w_\phi = 0\}}(\phi) + \alpha I_{\{\phi : w_\phi = 1\}}(\phi) \right) + \sum_{\phi \in \mathcal{B} \setminus \mathcal{B}_0} \theta_{f,\phi}^2 .
\end{aligned}
$$

Thus the lemma follows by our choosing $w_\phi = 1$ when $\theta_{f,\phi}^2 > \alpha$ and choosing $w_\phi = 0$ otherwise. □

15.5 MINIMIZATION OVER CONVEX HULLS

The main application of the results in this section wiill be made in Chapter 19, where stagewise minimization is considered. We define here in Definition 15.1 an estimator as a minimizer of an empirical risk. Chapter 19 considers estimators that are defined less implicitly and can actually be calculated.

15.5.1 Definition of the Estimator

We define an estimator as a minimizer of the empirical risk over the convex hull of a dictionary $\mathcal{D}$. At this point we do not fix the dictionary $\mathcal{D}$. We will derive error bounds under the assumption that $\mathcal{D}$ is a δ-net of a base class $\mathbb{G}$ and the true density belongs to the closure of the convex hull of $\mathbb{G}$: $f \in \bar{\text{co}}(\mathbb{G})$.

Definition 15.1 Minimization estimator over mixtures $\hat{f} : \mathbf{R}^d \to \mathbf{R}$, *with dictionary* $\mathcal{D}$ *of functions* $\mathbf{R}^d \to \mathbf{R}$, *and with the approximation bound* $\varepsilon > 0$, *satisfies*

$$
\gamma_n(\hat{f}) \le \inf_{g \in \text{co}(\mathcal{D})} \gamma_n(g) + \varepsilon, \qquad (15.112)
$$

where $\text{co}(\mathcal{D})$ *is the convex hull of* $\mathcal{D}$, *as defined in (10.50).*

As in (15.70) we denote

$$
\tilde{f}(x, \Lambda) = \tilde{f}(x, \Lambda, \mathcal{D}) = \sum_{\phi \in \mathcal{D}} \lambda_\phi \phi(x), \qquad x \in \mathbf{R}^d, \qquad (15.113)
$$

where $\Lambda = (\lambda_\phi)_{\phi \in \mathcal{D}} \in \mathcal{W}(\mathcal{D})$, and $\mathcal{W}(\mathcal{D})$ is the set of vectors of coefficients of finite convex combinations, as defined in (10.51). Then

$$
\text{co}(\mathcal{D}) = \{\tilde{f}(\cdot, \Lambda) : \Lambda \in \mathcal{W}(\mathcal{D})\}
$$

and

$$
\gamma_n(\hat{f}) \le \inf_{\Lambda \in \mathcal{W}(\mathcal{D})} \gamma_n(\tilde{f}(\cdot, \Lambda)) + \varepsilon .
$$

15.5.2 An Error Bound

We prove that the L_2 error of the minimization estimator over mixtures may be bounded by the optimal theoretical error and an additional stochastic term. Lemma 15.13 is analogous to Lemma 15.1 and Lemma 15.2.

Lemma 15.13 *Let $\hat{f}$ be defined as in Definition 15.1. Then for density f,*

$$\left\|\hat{f} - f\right\|_2^2 \leq \inf_{g \in \mathrm{co}(\mathcal{D})} \|g - f\|_2^2 + \varepsilon + 4 \sup_{\phi \in \mathcal{D}} |\nu_n(\phi)|,$$

where $\nu_n(\phi)$ is the centered empirical operator, as defined in (12.1).

Proof. Let $f^0 \in \mathrm{co}(\mathcal{D}) = \{f(\cdot, \Lambda) : \Lambda \in \mathcal{W}\}$. Let $\hat{\Lambda} = (\hat{\lambda}_\phi)_{\phi \in \mathcal{D}} \in \mathcal{W}$, $\Lambda^0 = (\lambda_\phi^0)_{\phi \in \mathcal{D}} \in \mathcal{W}$ be such that

$$\hat{f} = \tilde{f}(\cdot, \hat{\Lambda}), \qquad f^0 = \tilde{f}(\cdot, \Lambda^0).$$

Denote the vectors of the empirical and theoretical coefficients as

$$\hat{\Theta} = \left(\frac{1}{n}\sum_{i=1}^{n} \phi(X_i)\right)_{\phi \in \mathcal{D}}, \qquad \Theta_f = \left(\int_{\mathbf{R}^d} f\phi\right)_{\phi \in \mathcal{D}}.$$

Then

$$\begin{aligned} \nu_n\left(\hat{f} - f^0\right) &= \left(\hat{\Theta} - \Theta_f\right)^T \left(\hat{\Lambda} - \Lambda^0\right) \\ &\leq \left[\|\hat{\Lambda}\|_{l_1} + \|\Lambda^0\|_{l_1}\right] \left\|\hat{\Theta} - \Theta_f\right\|_{l_\infty} \\ &\leq 2\left\|\hat{\Theta} - \Theta_f\right\|_{l_\infty} \\ &= 2 \sup_{\phi \in \mathcal{D}} |\nu_n(\phi)|. \end{aligned} \tag{15.114}$$

Here we applied the notation $\|\Lambda\|_{l_1} = \sum_{\phi \in \mathcal{D}} |\lambda_\phi|$ and $\|\Lambda\|_{l_\infty} = \sup_{\phi \in \mathcal{D}} |\lambda_\phi|$, and the fact that $\|\hat{\Lambda}\|_{l_1}, \|\Lambda^0\|_{l_1} \leq 1$. The lemma follows from Lemma 15.1 and (15.114). □

15.5.3 MISE Bounds

Lemma 15.13 implies that when we want to get a bound to the uniform expected squared L_2 error

$$\sup_{f \in \mathcal{F}} E \left\|\hat{f} - f\right\|_2^2,$$

we need to give a bound to the variance term

$$E \sup_{\phi \in \mathcal{D}} |\nu_n(\phi)|.$$

Bounds for this variance term were given in Section 12.2. We need also a bound to the bias term

$$\sup_{f \in \mathcal{F}} \inf_{g \in \mathrm{co}(\mathcal{D})} \|g - f\|_2^2 .$$

We get a bound for this approximation error when we assume that $\mathcal{F} \subset \bar{\mathrm{co}}(\mathbb{G})$, where $\mathbb{G}$ is some base class of functions, $\bar{\mathrm{co}}(\mathbb{G})$ is the L_2 closure of the convex hull of $\mathbb{G}$, and the dictionary $\mathcal{D}$ is a δ-net for $\mathbb{G}$. We prove that when $\mathcal{D}$ is a δ-net for $\mathbb{G}$, then $\mathrm{co}(\mathcal{D})$ is a δ-net for $\bar{\mathrm{co}}(\mathbb{G})$.

Lemma 15.14 *Let $\delta > 0$ and let $\mathcal{D}$ be a δ net of $\mathbb{G}$. Then*

$$\sup_{f \in \bar{\mathrm{co}}(\mathbb{G})} \inf_{g \in \mathrm{co}(\mathcal{D})} \|g - f\|_2^2 \le \delta.$$

Proof. Applying notation (15.113), we need to show that

$$\sup_{f \in \bar{\mathrm{co}}(\mathbb{G})} \inf_{\Lambda \in \mathcal{W}(\mathcal{D})} \left\| f - \tilde{f}(\cdot, \Lambda, \mathcal{D}) \right\|_2 \le \delta. \tag{15.115}$$

We have clearly

$$\sup_{f \in \bar{\mathrm{co}}(\mathbb{G})} \inf_{\Lambda \in \mathcal{W}(\mathbb{G})} \left\| f - \tilde{f}(\cdot, \Lambda, \mathbb{G}) \right\|_2 = 0. \tag{15.116}$$

Equation (15.115) follows from (15.116) and (15.117) below. Equation (15.117) states that every convex combination of the functions in $\mathbb{G}$ can be approximated with some convex combination of the functions in dictionary $\mathcal{D}$ with accuracy δ:

$$\sup_{\Lambda \in \mathcal{W}(\mathbb{G})} \inf_{\Lambda' \in \mathcal{W}(\mathcal{D})} \left\| \tilde{f}(\cdot, \Lambda, \mathbb{G}) - \tilde{f}(\cdot, \Lambda', \mathcal{D}) \right\|_2 \le \delta. \tag{15.117}$$

Let Φ_ψ, $\psi \in \mathcal{D}$, be the collection of those $\phi \in \mathbb{G}$ for which ψ is the closest member of $\mathcal{D}$:

$$\Phi_\psi = \left\{ \phi \in \mathbb{G} : \psi = \mathrm{argmin}_{\psi' \in \mathcal{D}} \|\psi' - \phi\|_2 \right\} .$$

We can solve the ties arbitrarily to make a partition of $\mathbb{G}$:

$$\mathbb{G} = \cup_{\psi \in \mathcal{D}} \Phi_\psi, \qquad \Phi_\psi \cap \Phi_{\psi'}, \quad \psi \ne \psi'.$$

Let $\Lambda \in \mathcal{W}(\mathbb{G})$, $\Lambda = (\lambda_\phi)_{\phi \in \mathbb{G}}$. Let $\Lambda' \in \mathcal{W}(\mathcal{D})$, $\Lambda' = (\lambda'_\psi)_{\psi \in \mathcal{D}}$ be such that

$$\lambda'_\psi = \sum_{\phi \in \Phi_\psi} \lambda_\phi .$$

Now

$$\tilde{f}(\cdot, \Lambda, \mathbb{G}) = \sum_{\phi \in \mathbb{G}} \lambda_\phi \phi = \sum_{\psi \in \mathcal{D}} \sum_{\phi \in \Phi_\psi} \lambda_\phi \phi$$

and

$$\tilde{f}(\cdot, \Lambda', \mathcal{D}) = \sum_{\psi \in \mathcal{D}} \lambda'_\psi \psi = \sum_{\psi \in \mathcal{D}} \left(\sum_{\phi \in \Phi_\psi} \lambda_\phi \right) \psi.$$

Thus

$$\begin{aligned}\left\|\tilde{f}(\cdot,\Lambda,\mathbb{G})-\tilde{f}(\cdot,\Lambda',\mathcal{D})\right\|_2 &\le \sum_{\psi\in\mathcal{D}}\sum_{\phi\in\Phi_\psi}\lambda_\phi\|\phi-\psi\|_2\\ &\le \max_{\psi\in\mathcal{D}}\sup_{\phi\in\Phi_\psi}\|\phi-\psi\|_2\le\delta.\end{aligned}$$

We have proved (15.117) and thus the lemma. □

15.6 BIBLIOGRAPHIC NOTES

Inequalities, similar to those of Section 15.1.3, were used in Birgé and Massart (1997, lmm. 1). The results for the δ-net estimator in Section 15.2 and the dense minimizer in Section 15.3 were given in Klemelä and Mammen (2008). The contrast function in (15.6) was in deconvolution for projection estimators by Comte, Taupin and Rozenholc (2006). δ-net estimators were considered as well by van der Laan, Dudoit and van der Vaart (2004). The results in Section 15.4.2 are based on Klemelä (2009).

Oracle Inequalities Oracle inequalities appeared implicitly in the proofs of optimality of Mallows C_p method, cross-validation, and related data-driven methods, given by Li (1986), Li (1987), Polyak and Tsybakov (1990), and Polyak and Tsybakov (1992). Oracle inequalities are also implicit in the minimax adaptive constructions of Golubev (1987), Golubev (1992), and Golubev and Nussbaum (1990). Apparently first explicit uses appear in Donoho and Johnstone (1994a) and Kneip (1994). A very general exposition of model selection is given by Barron et al. (1999).

Smoothing Parameter Selection We have not discussed empirical risk minimization in smoothing parameter selection. Let $\mathcal{C}=\{\hat{f}_h : h\in[h_0,h_1]\}$ be a class of kernel density estimators, where $0<h_0<h_1<\infty$ and $\hat{f}_h$ is defined by (14.3). The estimator $\hat{f}_{h_0}$ minimizes the empirical risk over class $\mathcal{C}$, and we have to propose a better empirical risk to be applied in smoothing parameter selection. The jackknife method minimizes

$$\gamma_n^{(J)}(h)=\frac{1}{n}\sum_{i=1}^{n}\gamma\left(\hat{f}_{n,h,-i},X_i\right)$$

over h, where

$$\hat{f}_{n,h,-i}(x)=\frac{1}{n}\sum_{j=1,\dots,n,j\neq i}K_h(x-X_j)$$

and γ is either the L_2 contrast function or the log-likelihood contrast function.

EXERCISES

15.1 Let the set (15.20) be defined by

$$\Theta=\{(\mu,\sigma):\mu\in\mathbf{R},\ \ \sigma\in(0,\infty)\}$$

and

$$f(x,(\mu,\sigma)) = \frac{1}{2\sqrt{2\pi}\sigma}\exp\left\{-\frac{(x-\mu)^2}{2\sigma^2}\right\} + \frac{1}{2\sqrt{2\pi}}\exp\left\{-\frac{x^2}{2}\right\},$$

where $x \in \mathbf{R}$. Construct sequence $\theta_n = (\mu_n, \sigma_n)$ so that $\mathrm{Lik}_n(\theta_n) \to \infty$ as $n \to \infty$, where Lik_n is defined in (15.21).

15.2 Let $\{\phi_1, \ldots, \phi_k\}$ be an orthonormal system in $L_2(\mathbf{R}^d)$. Let $f \in L_2(\mathbf{R}^d)$. Prove that

$$\mathrm{argmin}_{\Theta\in\mathbf{R}^k, \|\Theta\|<\infty} \left\| f - \sum_{j=1}^{k} \theta_j \phi_j \right\|_2 = \Theta_f,$$

where

$$\Theta_f = \left(\int_{\mathbf{R}^d} f\phi_1, \ldots, \int_{\mathbf{R}^d} f\phi_k \right).$$

15.3 Prove Lemma 15.8 and Lemma 15.9 for the Gaussian white noise model.

15.4 Prove inequality (15.85).

CHAPTER 16

WAVELET ESTIMATORS

Wavelet estimators are a special case of orthogonal series estimators discussed in Section 14.5. We proved in Section 14.3.2 that a kernel estimator can achieve optimal rates of convergence in anisotropic Besov classes. In this section we prove similar results for wavelet estimators, but we consider also Besov spaces containing functions with high spatial variability and the space of dominating mixed derivatives, which were not considered in the case of kernel estimators. Adaptive estimation is not considered in this chapter, but in Chapter 18 where we construct an adaptive estimator that achieves nearly the optimal rates in anisotropic Besov classes.

We start in Section 16.1 by presenting the concepts of linear algebra needed in this chapter. Section 16.2 considers univariate wavelet bases. We define the concept of multiresolution analysis and then the setting of multiresolution analysis is specialized to bases constructed from univariate Haar wavelets. Section 16.3 constructs two multivariate wavelet bases. The first basis is a multiresolution basis, which is, however, not suitable to use in constructing estimators for functions with anisotropic smoothness, or for functions lying in the space of dominating mixed derivatives. Thus we construct an anisotropic basis that is able to handle these cases.

In Section 16.4 three wavelet estimators are constructed. All three estimators use the anisotropic basis. The first estimator is a linear estimator that achieves optimal rates of convergence in anisotropic Besov classes $B_{spq}(L)$ when $p_i \geq 2$ for all $i = 1, \ldots, d$. In order to estimate functions that belong to Besov spaces $B_{spq}(L)$ when $p_i < 2$ for some $i = 1, \ldots, d$, we consider also a wavelet estimator where the terms are chosen by hard thresholding. When $p_i < 2$ for some $i = 1, \ldots, d$, the functions may have high spatial variability. The third estimator is a linear estimator, but the anisotropic basis is now truncated so that functions lying in the space of dominating mixed derivatives can be efficiently estimated.

16.1 LINEAR ALGEBRA

We define first the concepts of linear algebra that we will need in this chapter.

We call $V \subset L_2([0,1]^d)$ a (linear) *subspace* if every linear combination of elements of V belongs to V. A collection of functions $\mathcal{B}$ is called an *orthonormal system* if for every $\phi, \psi \in \mathcal{B}$, $\int_{[0,1]^d} \phi\psi = 1$ when $\phi = \psi$, and 0 otherwise. An orthonormal system $\mathcal{B}$ is called an *orthonormal basis* in a subspace V of $L_2([0,1]^d)$ if any function $f \in V$ has a representation $f = \sum_{\phi \in \mathcal{B}} \theta_\phi \phi$ where coefficients $\theta_\phi \in \mathbf{R}$ satisfy $\sum_{\phi \in \mathcal{B}} \theta_\phi^2 < \infty$.

When $V_0, V_1, \ldots$ are subspaces, we denote $V_0 + V_1 = \{f_0 + f_1 : f_0 \in V_0, f_1 \in V_1\}$ and $\sum_{i=0}^{\infty} V_i = \cup_{m=0}^{\infty} \sum_{i=0}^{m} V_i$. We write $V = V_0 \oplus V_1$ when $V = V_0 + V_1$ and $V_0 \perp V_1$, where $V_0 \perp V_1$ means that for all $f_0 \in V_0$, $f_1 \in V_1$, $\int_{[0,1]^d} f_0 f_1 = 0$. Then we write also $V_0 = V_2 \ominus V_1$. We write $V = \bigoplus_{i=0}^{\infty} V_i$ when $V = \sum_{i=0}^{\infty} V_i$ and $V_i \perp V_j$ for $i \neq j$.

When $A \subset L_2([0,1]^d)$, we define $\mathrm{span}(A)$ as the set of all linear combinations of A and $\mathrm{cl}(A)$ as the closure of A in the L_2 metric. We use the following lemma to construct bases of $L_2([0,1]^d)$:

Lemma 16.1 *Let $U_0, U_1, \ldots$ be linear subspaces of $L_2([0,1]^d)$ and let $\mathcal{B}_i$ be a basis of U_i, $i = 0, 1, \ldots$. If*

$$L_2([0,1]^d) = \mathrm{cl}\left(\bigoplus_{i=0}^{\infty} U_i \right), \tag{16.1}$$

then

$$\cup_{i=0}^{\infty} \mathcal{B}_i \text{ is a basis of } L_2([0,1]^d).$$

Note that (16.1) means that

1. $\sum_{i=0}^{\infty} U_i$ is dense in $L_2([0,1]^d)$, and

2. $U_i \perp U_j$ for $i \neq j$.

16.2 UNIVARIATE WAVELET BASES

We start by defining univariate wavelet systems. The Haar wavelet basis is the only example that we consider. Multivariate wavelet bases can be constructed with the help of a univariate basis.

16.2.1 Multiresolution Analysis

Let $\phi \in L_2([0,1])$, $\|\phi\|_2 = 1$. Denote

$$\phi_{jk}(t) = 2^{j/2}\phi(2^j t - k), \qquad j = 0, 1, \ldots, \qquad k = 0, \ldots, 2^j - 1, \tag{16.2}$$

and define the linear subspaces

$$V_j = \text{span}\left(\{\phi_{jk} : k = 0, \ldots, 2^j - 1\}\right), \qquad j = 0, 1, \ldots. \tag{16.3}$$

Definition 16.1 (Multiresolution analysis.) *A multiresolution analysis of $L_2([0,1])$ is a sequence of spaces $(V_j)_{j=0,1,\ldots}$, as defined in (16.3), when the spaces are nested:*

$$V_j \subset V_{j+1}, \qquad j = 0, 1, \ldots, \tag{16.4}$$

and

$$\bigcup_{j=0}^{\infty} V_j \text{ is dense in } L_2([0,1]). \tag{16.5}$$

The function ϕ generating a multiresolution analysis is called the scaling function, *or the* father wavelet.

Let $(V_j)_{j=0,1,\ldots}$ be a multiresolution analysis and define

$$W_j = V_{j+1} \ominus V_j, \qquad j = 0, 1, \ldots. \tag{16.6}$$

Then, by telescoping, we get

$$V_{j+1} = V_0 \oplus \bigoplus_{l=0}^{j} W_l. \tag{16.7}$$

Since $V_j \subset V_{j+1}$, we get

$$\bigcup_{j=0}^{\infty} V_j = V_0 \oplus \bigoplus_{j=0}^{\infty} W_j.$$

It follows, by (16.5), that

$$L_2([0,1]) = \text{cl}\left(V_0 \oplus \bigoplus_{j=0}^{\infty} W_j\right). \tag{16.8}$$

Lemma 16.1 and (16.8) imply that any $f \in L_2([0,1])$ can be represented as a series

$$f(t) = \alpha_0\phi(t) + \sum_{j=0}^{\infty}\sum_{k=0}^{2^j-1} \beta_{jk}\psi_{jk}(t), \qquad t \in [0,1], \tag{16.9}$$

that converges in $L_2([0,1])$, α_0 and β_{jk} are real coefficients, and $\{\psi_{jk} : k = 0, \ldots, 2^j - 1\}$ is a basis for W_j. We call (16.9) a *multiresolution expansion* of f. If there exists such a function $\psi : [0,1] \to \mathbf{R}$ that

$$\psi_{jk}(t) = 2^{j/2}\psi(2^j t - k), \qquad t \in [0,1], \tag{16.10}$$

$j = 0, 1, \ldots$, $k = 0, \ldots, 2^j - 1$, then ψ is called a *wavelet*, or a *mother wavelet*. The space W_j is called the *resolution level* of the multiresolution analysis.

Wavelet System Construction The steps for the wavelet system construction are the following:

1. Find a scaling function, that is, function ϕ, such that $(V_j)_{j=0,1,\ldots}$ is a multiresolution analysis, as in Definition 16.1.

2. Find the wavelet, that is, function $\psi \in W_0$ with $\|\psi\|_2 = 1$, such that $\{\psi_{jk} : k = 0, \ldots, 2^j - 1\}$ is a basis in W_j, where ψ_{jk} satisfies (16.10).

Then any $f \in L_2([0,1])$ has a unique representation (16.9) as an $L_2([0,1])$-convergent series, where

$$\alpha_0 = \int_0^1 f\phi, \qquad \beta_{jk} = \int_0^1 f\psi_{jk}.$$

Inhomogeneous Wavelet Expansion We get more generally for $j_0 = 0, 1, \ldots,$ that

$$L_2([0,1]) = \mathrm{cl}\left(V_{j_0} \oplus \bigoplus_{j=j_0}^{\infty} W_j \right).$$

Then we get an expansion

$$f(t) = \sum_{k=0}^{2^{j_0}-1} \alpha_k \phi_{j_0 k}(t) + \sum_{j=j_0}^{\infty} \sum_{k=0}^{2^j-1} \beta_{jk}\psi_{jk}(t), \qquad t \in [0,1], \tag{16.11}$$

where $\alpha_k = \int_0^1 f\phi_{j_0 k}$. This is called an *inhomogeneous wavelet expansion* of f. The coefficients α_k summarize the general form of the function and the β_{jk} coefficients represent the innovations to this general form. The coefficients β_{jk} are also called *detail coefficients*.

16.2.2 The Haar Basis

The Haar basis is known since 1910. We take the scaling function

$$\phi = \eta^{(0)} = I_{[0,1]} \tag{16.12}$$

and the wavelet

$$\psi = \eta^{(1)} = I_{[1/2,1)} - I_{[0,1/2)}. \tag{16.13}$$

Define ϕ_{jk} by (16.2) and ψ_{jk} similarly, by (16.10). Then

$$\psi_{jk} = 2^{j/2}\left(I_{[(2k+1)/2^{j+1},(2k+2)/2^{j+1})} - I_{[2k/2^{j+1},(2k+1)/2^{j+1})} \right). \tag{16.14}$$

For example, $\psi_{00} = \psi$ and

$$\psi_{10} = 2^{1/2}\left(I_{[1/4,1/2)} - I_{[0,1/4)} \right), \qquad \psi_{11} = 2^{1/2}\left(I_{[3/4,1)} - I_{[1/2,3/4)} \right).$$

To prove that we have succeeded in the wavelet system construction, we have to show that the scaling function ϕ generates a multiresolution analysis and that the wavelet ψ generates a basis of W_j.

Lemma 16.2 *Sequence of spaces $(V_j)_{j=0,1,\ldots}$, defined by (16.3) with scaling function ϕ as in (16.12), is a multiresolution analysis.*

Proof. Space V_j consists of functions that are constant on intervals $[k/2^j, (k+1)/2^j)$. Thus the nestedness property (16.4) is obvious. One may easily prove the denseness property (16.5), and this is the objective of Exercise 16.1. □

Lemma 16.3 *Collection $(\psi_{jk})_{k=0,\ldots,2^j-1}$ is an orthonormal basis in W_j, for $j = 0, 1, \ldots$*

Proof. The lemma follows from the facts that:

1. $(\psi_{jk})_{k=0,\ldots,2^j-1}$ is an orthonormal system,
2. $(\psi_{jk})_{k=0,\ldots,2^j-1}$ is orthogonal to V_j, and
3. every $f \in V_{j+1}$ has a unique representation in terms of the joint system $(\phi_{jk}, \psi_{jk})_{k=0,\ldots,2^j-1}$.

Items 1 and 2 are obvious. To prove item 3, it suffices to note that

$$\phi_{j+1,2k} = 2^{-1/2}(\phi_{jk} - \psi_{jk}), \qquad \phi_{j+1,2k+1} = 2^{-1/2}(\phi_{jk} + \psi_{jk}),$$

$k = 0, \ldots, 2^j - 1$. □

When f is a density and we use the Haar wavelets, we have $\alpha_0 = 1$ in expansion (16.9), since densities integrate to one.

16.3 MULTIVARIATE WAVELET BASES

We may approximate multivariate functions with expansions whose members are products of univariate functions. In general, a multivariate basis can be constructed with the help of tensor products of univariate bases. For $V, W \subset \{f : \mathbf{R} \to \mathbf{R}\}$, define the tensor product

$$V \otimes W = \{f : \mathbf{R}^2 \to \mathbf{R} \mid f(x_1, x_2) = g(x_1)h(x_2),\ g \in V,\ h \in W\}.$$

Note that $V \otimes W \neq W \otimes V$. We will denote also

$$V^d = V^{d-1} \otimes V,\ d \geq 2, \qquad V^1 = V.$$

When $(V_j)_{j=0,1,\ldots}$ is a multiresolution analysis of $L_2([0,1])$, as in Definition 16.1, we have

$$\bigcup_{j=0}^{\infty} V_j^d \text{ is dense in } L_2([0,1]^d). \tag{16.15}$$

The proof of (16.15) is left as Exercise 16.2. We will use this fact to construct multivariate wavelet bases.

Haar Wavelet Notation We will restrict ourselves to bases that are constructed from the univariate Haar scaling function and the Haar wavelet. We will denote the product of univariate Haar scaling functions and Haar wavelets by

$$h^i_{jk}(x) = \eta^{(i_1)}_{j_1k_1}(x_1)\cdots\eta^{(i_d)}_{j_dk_d}(x_d), \qquad x = (x_1,\ldots,x_d) \subset [0,1]^d, \tag{16.16}$$

where $i \in \{0,1\}^d$, $j \in \{0,1,\ldots\}^d$, $k \in K_j$, with K_j as the set of translation coefficients corresponding to resolution index $j = (j_1,\ldots,j_d)$,

$$K_j = \{k = (k_1,\ldots,k_d) : k_l = 0,\ldots,2^{j_l}-1, l = 1,\ldots,d\}, \tag{16.17}$$

and where

$$\eta^{(i_l)}_{j_lk_l}(t) = 2^{j_l/2}\eta^{(i_l)}(2^{j_l}t - k_l),$$

with $\eta^{(0)}$ defined as in (16.12) and $\eta^{(1)}$ defined as in (16.13). We will denote the wavelet coefficients by

$$\theta^i_{jk} = \int_{[0,1]^d} h^i_{jk} f. \tag{16.18}$$

16.3.1 Multiresolution Basis

We will construct a basis $\mathcal{B}^M$ consisting of functions h^i_{jk}, defined in (16.16), where vectors of resolution indexes j are such that $j_1 = \cdots = j_d$. Basis $\mathcal{B}^M$ contains the indicator $I_{[0,1]^d}(x) = \eta^{(0)}_{00}(x_1)\cdots\eta^{(0)}_{00}(x_d)$, but other functions in the basis are products that contain a wavelet at least for one coordinate.

Definition 16.2 (Multiresolution basis.) *The* multiresolution basis $\mathcal{B}^M$ *is defined by*

$$\mathcal{B}^M = \{I_{[0,1]^d}\} \cup \bigcup_{m=0}^{\infty} \left[\{h^i_{j_m,k} : i \in \{0,1\}^d, k \in K_j\} \setminus \left\{h^{(0,\ldots,0)}_{j_m,k} : k \in K_j\right\}\right]$$

where $j_m = (m,\ldots,m) \in \{0,1,\ldots\}^d$, h^i_{jk} *is as defined in (16.16), and* K_j *is as defined in (16.17).*

We may write the basis compactly with

$$\mathcal{B}^M = \left\{h^i_{jk} : i \in \{0,1\}^d, j \in \mathcal{J}_i, k \in K_j\right\},$$

where

$$\mathcal{J}_i = \begin{cases} \{0\}^d, & \text{when } i = \{0\}^d, \\ \{1\}^d \cup \{2\}^d \cup \cdots, & \text{otherwise}. \end{cases}$$

Thus $(i_1,\ldots,i_d) = (0,\ldots,0)$ only when $(j_1,\ldots,j_d) = (0,\ldots,0)$.

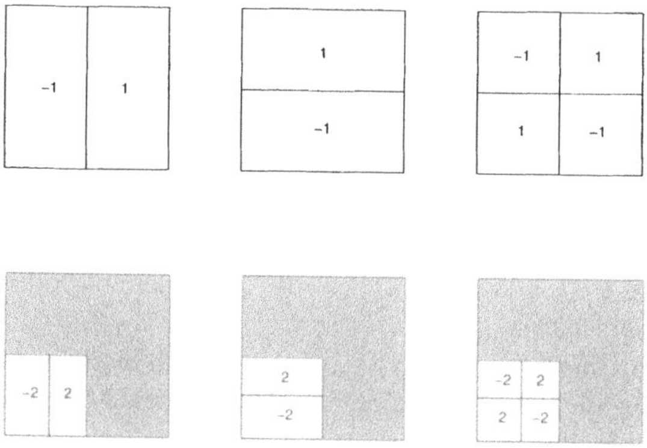

Figure 16.1 *(First basis functions of the 2D multiresolution basis.)* The upper row shows functions $h^{(1,0)}_{(0,0),(0,0)}$, $h^{(0,1)}_{(0,0),(0,0)}$, $h^{(1,1)}_{(0,0),(0,0)}$. The lower row shows functions $h^{(1,0)}_{(1,1),(0,0)}$, $h^{(0,1)}_{(1,1),(0,0)}$, $h^{(1,1)}_{(1,1),(0,0)}$. The values of the functions are also given in the figure.

Figure 16.1 illustrates the multiresolution basis in the two-dimensional case. The upper row shows the basis functions $h^{(1,0)}_{(0,0),(0,0)}$, $h^{(0,1)}_{(0,0),(0,0)}$, $h^{(1,1)}_{(0,0),(0,0)}$. These are the functions whose support is the whole rectangle $[0,1]^2$. For example, basis function $h^{(1,0)}_{(0,0),(0,0)}$ is a multiplication of the indicator in the y-direction and the step function in the x-direction. The lower row shows the basis functions $h^{(1,0)}_{(1,1),(0,0)}$, $h^{(0,1)}_{(1,1),(0,0)}$, $h^{(1,1)}_{(1,1),(0,0)}$. The support of these functions is the lower rectangle $[0,1/4] \times [0,1/4]$.

We prove that $\mathcal{B}^M$ is a basis.

Lemma 16.4 *Collection $\mathcal{B}^M$, defined in Definition 16.2, is a basis of $L_2([0,1]^d)$.*

Proof. Define

$$U_j = V^d_{j+1} \ominus V^d_j. \tag{16.19}$$

Then, by telescoping, we get $V^d_{j+1} = V^d_0 \oplus \bigoplus_{l=0}^j U_l$. We have that $V^d_j \subset V^d_{j+1}$ and

$$\bigcup_{j=0}^{\infty} V^d_j = V^d_0 \oplus \bigoplus_{j=0}^{\infty} U_j. \tag{16.20}$$

We get, by (16.15) and (16.20), that

$$L_2([0,1]^d) = \mathrm{cl}\left(V^d_0 \oplus \bigoplus_{j=0}^{\infty} U_j \right). \tag{16.21}$$

Decomposition (16.21) is analogous to the univariate decomposition in (16.8). Second, we have defined W_j in (16.6) so that $V_{j+1} = W_j \oplus V_j$, and thus by (16.19),

$$\begin{aligned} U_j &= (W_j \oplus V_j)^d \ominus V_j^d \\ &= \left(\bigoplus_{(i_1,\ldots,i_d)\in\{0,1\}^d} V_j^{(i_1)} \otimes \cdots \otimes V_j^{(i_d)} \right) \ominus V_j^d, \end{aligned} \tag{16.22}$$

where we denote $V_j^{(0)} = V_j$ and $V_j^{(1)} = W_j$. The lemma follows from (16.21), Lemma 16.1, and (16.22). □

We may approximate functions $f \in L_2([0,1]^d)$ with expansions whose elements are basis functions in $\mathcal{B}^M$. For example, a function could be approximated with a finite number of basis functions

$$\mathcal{B}_J^M = \{h_{jk}^i \in \mathcal{B}^M : j_l \le J,\ l = 1,\ldots,d\},$$

where $J \ge 1$. Note that

$$\#\mathcal{B}_J^M = 1 + \sum_{m=1}^{J} \left[(2\cdot 2^m)^d - 2^{md}\right] = O\left(2^{Jd}\right).$$

For resolutions $m = 1,\ldots,J$ we have to choose for each direction $l = 1,\ldots,d$ either $i_l = 0$ or $i_l = 1$, and we have to choose over locations $k_l = 0,\ldots,2^m - 1$. This gives $(2\cdot 2^m)^d$ choices for resolution m. However, we remove 2^{md} functions $h_{jk}^{(0,\ldots,0)}$ but add the function $I_{[0,1]^d}$.

The large number $O(2^{Jd})$ of basis functions is a manifestation of the curse of dimensionality. We would need a method for finding expansions with a small number of terms. Since the coefficient of each basis function has to be estimated with data, a large number of basis functions will increase the variance of the estimator.

When the function that we want to approximate has anisotropic smoothness, the collection $\mathcal{B}_J^M$ contains some basis functions that are not needed to get an accurate approximation. Indeed we need to choose the maximal resolution index J to cope with the direction where the the function has the least smoothness, and this introduces unnecessarily fine resolution in the other directions.

16.3.2 Anisotropic Basis

The anisotropic basis contains basis functions that have different resolution indexes for different coordinates.

Definition 16.3 (Anisotropic basis.) *The* anisotropic basis $\mathcal{B}^A$ *is defined by*

$$\mathcal{B}^A = \left\{h_{jk}^i : i \in \{0,1\}^d, j \in \mathcal{J}_i^A, k \in K_j\right\},$$

where h_{jk}^i *is as defined in (16.16),*

$$\mathcal{J}_i^A = \mathcal{J}_{i_1} \times \cdots \times \mathcal{J}_{i_d},$$

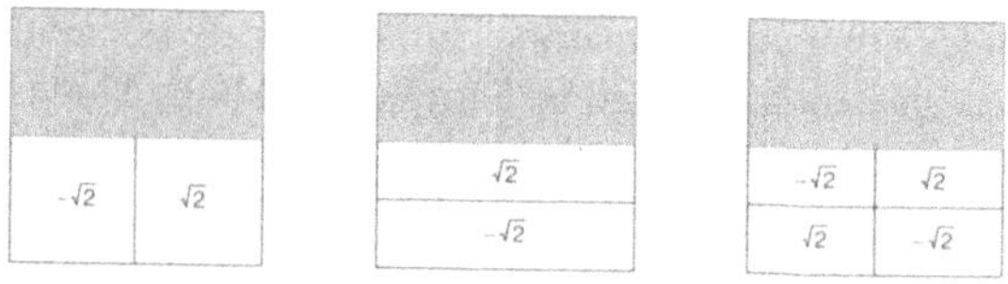

Figure 16.2 *(Some anisotropic basis functions of the 2D multiresolution basis.)* Shown are functions $h^{(1,0)}_{(0,1),(0,0)}$, $h^{(0,1)}_{(0,1),(0,0)}$, $h^{(1,1)}_{(0,1),(0,0)}$, with their values.

$$\mathcal{J}_{i_l} = \begin{cases} \{0\}, & \textit{when } i_l = 0, \\ \{0, 1, \ldots\}, & \textit{when } i_l = 1, \end{cases}$$

and K_j is as defined in (16.17).

The basis functions h^i_{jk} in $\mathcal{B}^A$ are such that if $i_l = 0$, then $j_l = 0$. That is, we take scaling functions only for the resolution index $j_l = 0$. Figure 16.2 illustrates the anisotropic basis in the two-dimensional case: we show the functions $h^{(1,0)}_{(0,1),(0,0)}$, $h^{(0,1)}_{(0,1),(0,0)}$, $h^{(1,1)}_{(0,1),(0,0)}$. The support of these basis functions is the rectangle $[0,1] \times [0,1/4]$. Unlike the multiresolution basis, the anisotropic basis has basis functions that contain indicator functions and step functions of different lengths.

We prove that $\mathcal{B}^A$ is a basis.

Lemma 16.5 *Collection $\mathcal{B}^A$ defined in Definition 16.3 is a basis of $L_2([0,1]^d)$.*

Proof. From (16.7) we get that

$$V^d_{j+1} = \bigoplus_{j_1,\ldots,j_d=-1}^{j} W_{j_1} \otimes \cdots \otimes W_{j_d},$$

where $W_{-1} = V_0$. By (16.15),

$$L_2([0,1]^d) = \mathrm{cl}\left(\bigoplus_{j_1,\ldots,j_d=-1}^{\infty} W_{j_1} \otimes \cdots \otimes W_{j_d} \right).$$

The lemma follows by an application of Lemma 16.1. □

16.4 WAVELET ESTIMATORS

We will consider three different estimators. The first estimator in Section 16.4.1 is a linear orthogonal series estimator that achieves the optimal rates of convergence for anisotropic Besov spaces $B_{spq}(L)$ when $p_l \geq 2$ for all $l = 1, \ldots, d$. In order to achieve optimal rates of convergence for the cases where $p_l < 2$ for some $l = 1, \ldots, d$, we need to consider nonlinear estimators. In Section 16.4.2 we define the second estimator, which is an estimator based on hard thresholding. This estimator achieves the optimal rates of convergence for anisotropic Besov spaces $B_{spq}(L)$ when possibly $p_l < 2$ for some $l = 1, \ldots, d$. In Section 16.4.3 we define the third estimator, which is again a linear orthogonal series estimator, but the truncation of the basis is done in such a way that the estimator achieves the optimal rates of convergence in the spaces $D_s(L)$ of dominating mixed derivatives. Adaptive estimation is not considered in this section. Adaptation will be considered in Chapter 18, where we construct an adaptive estimator which achieves nearly the optimal rates in anisotropic Besov classes.

16.4.1 Linear Estimator

We study the performance of the linear estimator in anisotropic Besov spaces $B_{spq}(L)$ defined in (10.41). We make the restriction $p_l \geq 2$ for all $l = 1, \ldots, d$. Let

$$\mathcal{F} = B_{sp\infty}(L) \cap \mathcal{F}_{den}(B_\infty),$$

where $s = (s_1, \ldots, s_d) \in (0, 1]^d$, $p = (p_1, \ldots, p_d) \in [2, \infty]^d$, $0 < L < \infty$, and $\mathcal{F}_{den}(B_\infty)$ is the collection of densities f with $\|f\|_\infty \leq B_\infty$, where B_∞ is a positive constant.

We construct an orthogonal series estimator with the anisotropic basis $\mathcal{B}^A$, defined in Definition 16.3. We choose a finite subcollection of $\mathcal{B}^A$ by limiting the maximal resolution of basis functions. Each direction will have its own resolution bound. We write the estimator with the index notation of Definition 16.3, where the basis functions $h^i_{jk} \in \mathcal{B}^A$ have the index set

$$\mathbb{J} = \left\{(ijk) : i \in \{0,1\}^d,\ j \in \mathcal{J}^A_i,\ k \in K_j\right\}. \tag{16.23}$$

The estimator is defined as

$$\hat{f}(x) = \sum_{(ijk) \in \mathbb{J}_{j^*}} \hat{\theta}^i_{jk} h^i_{jk}(x), \qquad x \in \mathbf{R}^d,$$

where

$$\mathbb{J}_{j^*} = \{(ijk) \in \mathbb{J} : j_l \leq j^*_l,\ l = 1, \ldots, d\}$$

and $j^* = (j^*_1, \ldots, j^*_d)$, where the maximal resolution indexes $j^*_l \geq 1$ are such integers that

$$j^*_l \asymp \frac{\sigma}{s_l} \frac{1}{2\sigma + 1} \log_2 n, \qquad l = 1, \ldots, d, \tag{16.24}$$

as $n \to \infty$, and the anisotropic smoothness index is defined by

$$\sigma = \left(\sum_{i=1}^d s_i^{-1} \right)^{-1}. \tag{16.25}$$

The empirical coefficients are defined for the observations $X_1, \ldots, X_n$ as

$$\hat{\theta}^i_{jk} = \sum_{i=1}^{n} h^i_{jk}(X_i). \tag{16.26}$$

Theorem 16.6 *Let $X_1, \ldots, X_n$ be i.i.d. observations from the distribution of the density $f \in \mathcal{F}$. Let estimator $\hat{f}$ be defined in (16.4.1) with the maximal resolution indexes as in (16.24). Then*

$$\limsup_{n\to\infty} n^{2\sigma/(2\sigma+1)} \sup_{f\in\mathcal{F}} E\int_{[0,1]^d} \left(f - \hat{f}\right)^2 < \infty.$$

Proof. We have that

$$E\int_{[0,1]^d} \left(f - \hat{f}\right)^2 = \sum_{(ijk)\in\mathbb{J}_{j^*}} E\left(\theta_{ijk} - \hat{\theta}_{ijk}\right)^2 + \sum_{(ijk)\in\mathbb{J}\setminus\mathbb{J}_{j^*}} (\theta^i_{jk})^2,$$

where we denote the wavelet coefficients of function $f \in L_2([0,1]^d)$ by

$$\theta^i_{jk} = \int_{[0,1]^d} h^i_{jk} f,$$

with h^i_{jk} a product of univariate Haar scaling functions or Haar wavelets as defined in (16.16). Now

$$E\left(\theta^i_{jk} - \hat{\theta}^i_{jk}\right)^2 = n^{-1}\mathrm{Var}\left(\hat{\theta}^i_{jk}\right) \le n^{-1}\|f\|_\infty.$$

Thus

$$\begin{aligned}
\sum_{(ijk)\in\mathbb{J}_{j^*}} E\left(\theta^i_{jk} - \hat{\theta}^i_{jk}\right)^2 &\le n^{-1}\|f\|_\infty \,\#\mathbb{J}_{j^*} \\
&= O\left(n^{-1}\sum_{j_1=0}^{j_1^*}\cdots\sum_{j_d=0}^{j_d^*} 2^{|j|}\right) \\
&= O\left(n^{-1}2^{|j^*|}\right) \\
&= O\left(n^{-2\sigma/(2\sigma+1)}\right).
\end{aligned}$$

Denote

$$\mathbb{J}^c_{j^*,l} = \{j \in \mathbb{J} : j_l > j_l^*\}.$$

Then applying (10.46) obtains

$$\sum_{(ijk)\in\mathbb{J}\setminus\mathbb{J}_{j^*}} (\theta^i_{jk})^2 = O\left(\sum_{l=1}^{d}\sum_{i\in\{0,1\}^d}\sum_{j\in\mathbb{J}^c_{j^*,l}}\sum_{k\in K_j}(\theta^i_{jk})^2\right)$$

$$
\begin{aligned}
&= O\left(\sum_{l=1}^{d} \sum_{i\in\{0,1\}^d} \sum_{j\in\mathbb{J}^c_{j^*,l}} 2^{-2j_l s_l}\right) \\
&= O\left(\sum_{l=1}^{d} 2^{-2j_l^* s_l}\right) \\
&= O\left(n^{-2\sigma/(2\sigma+1)}\right).
\end{aligned}
$$

We have proved the result. □

16.4.2 Nonlinear Estimator

We study the performance of the nonlinear estimator in anisotropic Besov spaces $B_{spq}(L)$, where possibly $p_l < 2$ for some $l = 1, \ldots d$. Let

$$\mathcal{F} = B_{sp\infty}(L)$$

where $B_{sp\infty}(L)$ is defined in (10.41), $s = (s_1, \ldots, s_d) \in (0,1]^d$, $p = (p_1, \ldots, p_d) \in [1,\infty]^d$, $0 < L < \infty$.

We apply hard thresholding of the empirical coefficients. The estimator is defined by

$$\hat{f}(x) = \sum_{(ijk)\in\mathbb{J}} w^i_{jk}\hat{\theta}^i_{jk}h^i_{jk}(x), \qquad x \in \mathbf{R}^d, \tag{16.27}$$

where

$$w^i_{jk} = \begin{cases} 1, & \text{when } |\hat{\theta}^i_{jk}| \geq \lambda_j, \\ 0, & \text{when } |\hat{\theta}^i_{jk}| < \lambda_j, \end{cases}$$

and

$$\lambda_j = n^{-1/2}\kappa\sqrt{\max_{1\leq l\leq d}\{s_l(j_l - j_l^*)_+\}\sigma^{-1}}, \tag{16.28}$$

with $\kappa > \sqrt{2\log_e 2}$ and $(x)_+ = \max\{0, x\}$. The indexes j_l^* are defined in (16.24), and the smoothness index σ is defined in (16.25). We will prove the result for the Gaussian white noise model, so we define the empirical coefficients as

$$\hat{\theta}^i_{jk} = \int h^i_{jk}\, dX_n,$$

where X_n is the Gaussian white noise model in (9.45).

Theorem 16.7 *Let X_n be a realization of the Gaussian white noise process with the signal function $f \in \mathcal{F}$. Let estimator $\hat{f}$ be as defined in (16.27) with the smoothing parameter (16.28). Assume that*

$$\sigma > \frac{1}{p_l} - \frac{1}{2}, \qquad l = 1, \ldots, d. \tag{16.29}$$

Then,

$$\limsup_{n\to\infty} n^{2\sigma/(2\sigma+1)} \sup_{f\in\mathcal{F}} E\int_{[0,1]^d}\left(f-\hat{f}\right)^2 < \infty.$$

Proof. We have

$$E\int_{[0,1]^d}\left(f-\hat{f}\right)^2 = \sum_{(ijk)\in\mathbb{J}} E\left(\theta^i_{jk} - w^i_{jk}\hat{\theta}^i_{jk}\right)^2.$$

We can in the following ignore the summation over i With an abuse of the notation we shall consider the sum

$$\sum_{j\in\{0,1,\ldots\}^d}\sum_{k\in K_j} E\left(\theta_{jk} - w_{jk}\hat{\theta}_{jk}\right)^2.$$

Denote

$$\mathcal{J} = \left\{j\in\{0,1,\ldots\}^d : j_l \le j_l^*,\ l=1,\ldots,d\right\}$$

and

$$\mathcal{J}^c = \left\{j\in\{0,1,\ldots\}^d : j_l > j_l^* \text{ for some } l=1,\ldots,d\right\}.$$

When $j\in\mathcal{J}$, we have $\lambda_j = 0$, $w_{jk}=1$ and

$$E\left(\theta_{jk} - w_{jk}\hat{\theta}_{jk}\right)^2 = n^{-1}\mathrm{Var}\left(\hat{\theta}_{jk}\right) = n^{-1}.$$

Thus

$$\begin{aligned}
\sum_{j\in\mathcal{J}}\sum_{k\in K_j} E\left(\theta_{jk} - w_{jk}\hat{\theta}_{jk}\right)^2 &\le \sum_{j\in\mathcal{J}}\sum_{k\in K_j} n^{-1} \\
&= O\left(n^{-1}\sum_{j_1=0}^{j_1^*}\cdots\sum_{j_d=0}^{j_d^*} 2^{|j|}\right) \\
&= O\left(n^{-1}2^{|j^*|}\right) \\
&= O\left(n^{-2\sigma/(2\sigma+1)}\right).
\end{aligned}$$

We have

$$E\left(\theta_{jk} - w_{jk}\hat{\theta}_{jk}\right)^2 \le C\left(n^{-1}\left(\frac{\lambda_j}{n^{1/2}}+1\right)\phi\left(\frac{\lambda_j}{n^{1/2}}\right) + \min\left\{\lambda_j^2, \theta_{jk}^2\right\}\right),$$

where C is a positive constant and ϕ is the density of the standard Gaussian distribution. This is proved in Appendix E.3. We can write

$$\sum_{j\in\mathcal{J}^c}\sum_{k\in K_j} E\left(\theta_{jk} - w_{jk}\hat{\theta}_{jk}\right)^2 \le C(A+B),$$

where

$$A = \sum_{j \in \mathcal{J}^c} \sum_{k \in K_j} n^{-1} \left(\frac{\lambda_j}{n^{1/2}} + 1 \right) \phi \left(\frac{\lambda_j}{n^{1/2}} \right)$$

and

$$B = \sum_{j \in \mathcal{J}^c} \sum_{k \in K_j} \min \left\{ \lambda_j^2, \theta_{jk}^2 \right\}.$$

Term A Denote

$$\mathcal{J}_{i,\iota} = \{(j_1, \ldots, j_{i-1}, \iota, j_{i+1}, \ldots, j_d) : 0 \le j_l \le \iota s_i / s_l, \ \ l \ne i \ \ l = 1, \ldots, d\}.$$

We have

$$\begin{aligned}
\#\{(j,k) : j \in \mathcal{J}_{i,\iota}, \ \ k \in K_j\} &= \sum_{l=1, l \ne i}^{d} \sum_{j_l=0}^{\iota s_i / s_l} \sum_{k \in K_j} \\
&= \sum_{l=1, l \ne i}^{d} \sum_{j_l=0}^{\iota s_i / s_l} 2^{|j| - \iota} \\
&= O\left(2^{\iota s_i \sigma^{-1}} \right) \\
&= O\left(2^{|j^*|} 2^{(\iota - j_i^*) s_i \sigma^{-1}} \right),
\end{aligned}$$

where we used the fact $|j^*| = j_i^* s_i \sigma^{-1}$. Now

$$\begin{aligned}
A &= \sum_{i=1}^{d} \sum_{j_i = j_i^*}^{\infty} \sum_{j \in \mathcal{J}_{i,j_i}} \sum_{k \in K_j} n^{-1} \left(\frac{\lambda_j}{n^{1/2}} + 1 \right) \phi \left(\frac{\lambda_j}{n^{1/2}} \right) \\
&= O\left(n^{-2\sigma/(2\sigma+1)} \right)
\end{aligned}$$

because for fixed $i \in \{1, \ldots, d\}$ and $j_i \ge j_i^*$,

$$\begin{aligned}
&\sum_{j \in \mathcal{J}_{i,j_i}} \sum_{k \in K_j} n^{-1} \left(\frac{\lambda_j}{n^{1/2}} + 1 \right) \phi \left(\frac{\lambda_j}{n^{1/2}} \right) \\
&= O\left(n^{-1} 2^{|j^*|} \right) O\left(2^{(j_i - j_i^*) s_i \sigma^{-1}} \sqrt{j_i - j_i^*} \exp\left\{ -\frac{1}{2} \kappa^2 (j_i - j_i^*) s_i \sigma^{-1} \right\} \right) \\
&= O\left(n^{-2\sigma/(2\sigma+1)} \right) O\left(\sqrt{j_i - j_i^*} \exp\left\{ (\log_e 2 - \kappa^2/2)(j_i - j_i^*) s_i \sigma^{-1} \right\} \right),
\end{aligned}$$

where $\log_e 2 - \kappa^2/2 < 0$.

Term B It holds that $\min \left\{ \theta_{jk}^2, \lambda_j^2 \right\} \le \lambda_j^{2 - \tilde{p}_l} |\theta_{jk}|^{\tilde{p}_l}$. Indeed, when $\theta_{jk}^2 \le \lambda_j^2$, we have $\lambda_j^{2-\tilde{p}_l} |\theta_{jk}|^{\tilde{p}_l} \ge \theta_{jk}^2$ and when $\theta_{jk}^2 > \lambda_j^2$, we have $\lambda_j^{2-\tilde{p}_l} |\theta_{jk}|^{\tilde{p}_l} \ge \lambda_j^2$. Thus for

$l \in \{1, \ldots, d\}$ and $j_l \geq j_l^*$,

$$\begin{aligned}
&\sum_{j \in \mathcal{J}_{l,j_l}} \sum_{k \in K_j} \min\left\{\theta_{jk}^2, \lambda_j^2\right\} \\
&\quad \leq \lambda_j^{2-\tilde{p}_l} \sum_{j \in \mathcal{J}_{l,j_l}} \sum_{k \in K_j} |\theta_{jk}|^{\tilde{p}_l} \\
&\quad \leq \lambda_j^{2-\tilde{p}_l} \sum_{j \in \mathcal{J}_{l,j_l}} 2^{-\tilde{p}_l j_l s_l} 2^{-\tilde{p}_l |j|(1/2 - 1/\tilde{p}_l)} && (16.30) \\
&\quad \leq \lambda_j^{2-\tilde{p}_l} \sum_{j \in \mathcal{J}_{l,j_l}} 2^{-\tilde{p}_l j_l s_l} 2^{\tilde{p}_l j_l s_l (1/\tilde{p}_l - 1/2)\sigma^{-1}} && (16.31) \\
&\quad = O\left(n^{\tilde{p}_l/2-1} (j_l - j_l^*)^{1-\tilde{p}_l/2} 2^{-\tilde{p}_l j_l s_l} 2^{\tilde{p}_l j_l s_l (1/\tilde{p}_l - 1/2)\sigma^{-1}}\right) && (16.32) \\
&\quad = O\left(n^{-1} 2^{|j^*|}\right) O\left(n^{\tilde{p}_l/2} 2^{-\tilde{p}_l j_l^* s_l} 2^{-\tilde{p}_l |j^*|/2}\right) \\
&\qquad \times O\left((j_l - j_l^*)^{1-\tilde{p}_l/2} 2^{-\tilde{p}_l (j_l - j_l^*) s_l} 2^{\tilde{p}_l j_l s_l (1/\tilde{p}_l - 1/2)\sigma^{-1}} 2^{\tilde{p}_l |j^*|(1/2 - 1/\tilde{p}_l)}\right),
\end{aligned}$$

where in (16.30) we applied Lemma 10.3, in (16.31) we applied that for $j \in \mathcal{J}_{l,j_l}$, $|j| \leq j_l + \sum_{i=1, i \neq l}^{d} j_l s_l / s_i = j_l s_l \sigma^{-1}$, in (16.32) we applied that for $j \in \mathcal{J}_{l,j_l}$, $(j_k - j_k^*) s_k \leq (j_l - j_l^*) s_l$ (because $j_k^* s_k$ does not depend on k). We have

$$O\left(n^{\tilde{p}_l/2} 2^{-\tilde{p}_l j_l^* s_l} 2^{-\tilde{p}_l |j^*|/2}\right) = O(1)$$

because $2^{2 j_l^* s_l} = n^{2\sigma/(2\sigma+1)}$ and $2^{|j^*|} = n^{1/(2\sigma+1)}$. Also

$$\begin{aligned}
&\sum_{l=1}^{d} \sum_{j_l = j_l^*}^{\infty} (j_l - j_l^*)^{1-\tilde{p}_l/2} 2^{-\tilde{p}_l (j_l - j_l^*) s_l} 2^{\tilde{p}_l j_l s_l (1/\tilde{p}_l - 1/2)\sigma^{-1}} 2^{\tilde{p}_l |j^*|(1/2 - 1/\tilde{p}_l)} \\
&\qquad = \sum_{l=1}^{d} \sum_{j_l = j_l^*}^{\infty} (j_l - j_l^*)^{1-\tilde{p}_l/2} 2^{-\tilde{p}_l (j_l - j_l^*) s_l [1 - (1/\tilde{p}_l - 1/2)\sigma^{-1}]} = O(1)
\end{aligned}$$

because $|j^*| = j_l^* s_l \sigma^{-1}$ and because (16.29) implies

$$1 - (1/\tilde{p}_l - 1/2)\sigma^{-1} > 0.$$

Thus

$$B = O\left(n^{-1} 2^{|j^*|}\right) = O\left(n^{-2\sigma/(2\sigma+1)}\right).$$

We have proved the theorem. □

16.4.3 Dominating Mixed Derivatives

We study the performance of the estimator defined in this section in the space $D_s(L)$ of dominating mixed derivatives. Let

$$\mathcal{F} = D_s(L) \cap \mathcal{F}_{den}(B_\infty),$$

where $D_s(L)$ is defined in (10.48), $s \geq 1$ integer, $L > 0$, and $\mathcal{F}_{den}(B_\infty)$ is the collection of densities f with $\|f\|_\infty \leq B_\infty$, where B_∞ is a positive constant.

We construct an orthogonal series estimator with the anisotropic basis $\mathcal{B}^A$, defined in (16.3). The estimator is defined as

$$\hat{f}(x) = \sum_{(ijk)\in\mathbb{J}_J} \hat{\theta}^i_{jk} h^i_{jk}(x), \qquad x \in \mathbf{R}^d, \tag{16.33}$$

where h^i_{jk} are the basis functions in $\mathcal{B}^A$,

$$\mathbb{J}_J = \{(ijk) \in \mathbb{J} : |j| \leq J\},$$

$\mathbb{J}$ is defined in (16.23), and we choose the maximal resolution index J by

$$2^J \asymp \left(\frac{n}{(\log n)^{d-1}}\right)^{1/(2s+1)}, \tag{16.34}$$

as $n \to \infty$. The empirical coefficients $\hat{\theta}^i_{jk}$ are defined for the observations $X_1, \ldots, X_n$ as in (16.26).

Theorem 16.8 *Let $X_1, \ldots, X_n$ be i.i.d. observations from the distribution of the density $f \in \mathcal{F}$. Let estimator $\hat{f}$ be defined in (16.33) with the maximal resolution index as in (16.34). Then*

$$\limsup_{n\to\infty} \left(\frac{n}{(\log n)^{d-1}}\right)^{2s/(2s+1)} \sup_{f\in\mathcal{F}} E \int_{\mathbf{R}^d} \left(f - \hat{f}\right)^2 < \infty.$$

Proof. We have that

$$E \int_{[0,1]^d} \left(f - \hat{f}\right)^2 = \sum_{(ijk)\in\mathbb{J}_J} E\left(\theta_{ijk} - \hat{\theta}_{ijk}\right)^2 + \sum_{(ijk)\in\mathbb{J}\setminus\mathbb{J}_J} (\theta^i_{jk})^2,$$

where we denote the wavelet coefficients of function $f \in L_2([0,1]^d)$ with $\theta^i_{jk} = \int_{[0,1]^d} h^i_{jk} f$, where h^i_{jk} is a product of univariate Haar scaling functions or Haar wavelets defined in (16.16). Now

$$E\left(\theta^i_{jk} - \hat{\theta}^i_{jk}\right)^2 = n^{-1}\mathrm{Var}\left(\hat{\theta}^i_{jk}\right) \leq n^{-1}\|f\|_\infty.$$

Thus

$$\begin{aligned}
\sum_{(ijk)\in\mathbb{J}_J} E\left(\theta^i_{jk} - \hat{\theta}^i_{jk}\right)^2 &\leq n^{-1}\|f\|_\infty \,\#\mathbb{J}_J \\
&= O\left(n^{-1}2^J J^{d-1}\right) \\
&= O\left(\left(n(\log n)^{1-d}\right)^{2s/(2s+1)}\right).
\end{aligned}$$

Applying the equivalent semi-norm in (10.49) obtains

$$
\begin{aligned}
\sum_{(ijk)\in\mathbb{J}\setminus\mathbb{J}_J} (\theta_{jk}^i)^2 &= O\left(\sum_{i\in\{0,1\}^d}\sum_{j\in\mathbb{J}\setminus\mathbb{J}_J}\sum_{k\in K_j}(\theta_{jk}^i)^2\right)\\
&= O\left(\sum_{i\in\{0,1\}^d}\sum_{j\in\mathbb{J}\setminus\mathbb{J}_J} 2^{-2|j|s}\right)\\
&= O\left(2^{-2Js}\right)\\
&= O\left(\left(n\left(\log n\right)^{1-d}\right)^{2s/(2s+1)}\right).
\end{aligned}
$$

We have proved the result. □

16.5 BIBLIOGRAPHIC NOTES

Multiresolution analysis was introduced by Mallat (1989), Meyer (1990), and Meyer (1991). We have followed in Section 16.2.1 the exposition of Härdle et al. (1998).

Univariate density estimation is considered in Donoho, Johnstone, Kerkyacharian and Picard (1996), who proved adaptive minimax results in Besov spaces for various loss functions. A multivariate extension of multiresolution analysis was introduced by Mallat (1989). Computation of two-dimensional wavelet estimators was considered by Nason and Silverman (1994) and Ogden (1997). Tribouley (1995) considers cross-validation in multivariate isotropic Besov spaces. Delyon and Juditsky (1996) generalized this result. Neumann and von Sachs (1997) and Neumann (2000) cover the anisotropic Besov classes. Section 16.4.2 and Section 16.4.3 are based on Neumann (2000).

EXERCISES

16.1 Prove the denseness property (16.5) for the case of the Haar scaling function.

16.2 Prove the denseness property (16.15): when $(V_j)_{j=0,1,\ldots}$ is a multiresolution analysis of $L_2([0,1])$, then $\bigcup_{j=0}^{\infty} V_j^d$ is dense in $L_2([0,1]^d)$.

16.3 Show that a wavelet estimator with the multiresolution Haar basis is a histogram with a regular dyadic partition.

CHAPTER 17

MULTIVARIATE ADAPTIVE HISTOGRAMS

We have defined histograms in (14.2) as setwise constant functions whose value in each set is the empirical probability divided by the volume of the set. Histograms are uniquely determined by a partition on a set containing the observations. The main problem is to choose the partition defining the histogram in an optimal way. We call an adaptive histogram such a histogram whose partition is chosen in a data-dependent way. The partition is usually not regular: typically bins in the tail regions should be chosen larger and bins in the high density areas should be chosen smaller.

Histograms with equispaced bins are not able to adapt to spatially varying smoothness. This problem appears already in the one-dimensional case. In the multivariate case the density to be estimated may further have anisotropic smoothness: the density function may vary more in one direction than in the other directions. We should choose bins to be thinner in the direction where the density varies more.

By partitioning in a flexible way, we are not so vulnerable to the curse of dimensionality. Indeed in high-dimensional cases accurate estimation may be possible if the "effective dimension" of the density is small. Effective dimension could mean, for example, the number of coordinate directions with respect which the density has variability. Indeed an extreme case of anisotropic smoothness occurs when the density is almost constant on its support with respect to some variables. These types of densities could be estimated well if we had a method of choosing the partition of the histogram economically: we should choose a partition that only delineates the support with respect to directions that have no variation.

We will consider three types of adaptive histograms.

1. *Greedy histograms* are histograms whose partition is grown in a greedy way. Greedy histograms are studied in Section 17.1.

   We start with the trivial partition consisting from a single rectangle containing the observations, and then continue growing the partition by splitting partition members one at a time so that the empirical risk is minimized. A stopping rule completes the growth at some point.

2. *CART histograms* are based on greedy growth followed by a complexity-penalized empirical risk minimization. CART histograms are studied in Section 17.2.

   CART histograms implement a familiar strategy familiar from CART regressograms (classification and regression trees): construct first a fine partition with myopic splitting and then choose the final partition by minimizing a complexity-penalized empirical risk.

3. *Dyadic histograms* are based on a complexity-penalized empirical risk minimization over a set of recursive dyadic partitions. Dyadic histograms are studied in Chapter 18.

   Dyadic histograms are based on finding an exact solution to the problem of minimizing the complexity-penalized empirical risk. We consider, however, a very restrictive class of partitions over which we search for the minimizer. By restricting the class of partitions, we make the problem of finding the minimizing partition feasible.

It is possible to enhance adaptive histograms with bootstrap aggregation. In bootstrap aggregation one generates several bootstrap samples from the original sample, constructs an adaptive histogram based on each bootstrap sample, and defines the estimate as the pointwise arithmetic mean over those estimates. Bootstrapping increases the granularity of estimates. We study bootstrap aggregation of greedy histograms and CART histograms in Section 17.3.

Another use of adaptive histograms comes in the connection of stagewise minimization estimators. A stagewise minimization estimator is an estimator that minimizes empirical risk in stages as it tries to find an optimal convex combination of functions in a dictionary. We may choose the dictionary to be the class of adaptive histograms. Stagewise minimization estimators are discussed in Chapter 19, and adaptive histograms in the stagewise minimization are discussed in Section 19.3.2.

17.1 GREEDY HISTOGRAMS

Greedy (myopic) optimization refers to an optimization strategy that does not guarantee that an optimal solution would be found (which would take too much time) but instead apply stepwise optimization by looking only one step ahead.

17.1.1 Definition

A greedy histogram is constructed by the following rules. We find a partition that consist of M rectangles and this partition defines the greedy histogram.

1. Find the smallest rectangle containing the observations, whose sides are parallel to the coordinate axes.

2. Repeat the next step until the partition has M rectangles: split a rectangle in the current partition so that the empirical risk of the corresponding histogram is minimized.

Pool of Split Points We define a pool of split points over which we search for the best splits. One may either (1) construct a regular equispaced grid for each coordinate or (2) one may construct an empirical grid for each coordinate from the midpoints of the coordinates of the observations. In both cases we start with the smallest rectangle containing the observations, whose sides are parallel to the coordinate axis, and construct a pool of split points inside the boundary of this rectangle. Let us denote the pool of split points by

$$\mathcal{G} = \mathcal{G}_1 \times \cdots \times \mathcal{G}_d, \tag{17.1}$$

where $\mathcal{G}_k$ is the grid of split points in direction k. In the case of the empirical grid we have

$$\mathcal{G}_k = \{Z_1^k, \ldots, Z_{n-1}^k\}, \qquad k = 1, \ldots, d,$$

where Z_i^k is the midpoint of $X_{(i)}^k$ and $X_{(i+1)}^k$: $Z_i^k = X_{(i)}^k + (X_{(i+1)}^k - X_{(i)}^k)/2$, where $X_{(1)}^k, \ldots, X_{(n)}^k$ is the order statistic of the kth coordinate of the observations $X_1, \ldots, X_n$.

Partition Growth We say that partition $\mathcal{P}$ is grown if it is replaced by partition

$$\mathcal{P}_{R,k,s} = \mathcal{P} \setminus \{R\} \cup \left\{R_{k,s}^{(0)}, R_{k,s}^{(1)}\right\}, \tag{17.2}$$

where rectangle $R \in \mathcal{P}$ is splitted in direction $k = 1, \ldots, d$ through the point

$$s \in S_{R,k} \stackrel{def}{=} \mathcal{G}_k \cap \mathrm{int}(\mathrm{proj}_k(R)), \tag{17.3}$$

where $\mathcal{G}_k$ are the grid points in the kth direction, as defined by (17.1), $\mathrm{proj}_k(R) = R_k$, when $R = R_1 \times \cdots \times R_d$, $\mathrm{int}(R_k)$ is the interior of set R_k,

$$R_{k,s}^{(0)} = \{(x_1, \ldots, x_d) \in R : x_k \le s\} \tag{17.4}$$

and

$$R_{k,s}^{(1)} = \{(x_1, \ldots, x_d) \in R : x_k > s\}. \tag{17.5}$$

Above we take $\mathrm{int}(R_k)$ instead of R_k to exclude the case where a split is made at the boundary of R_k.

Greedy Partition We define a greedy partition for a given cardinality bound $M \geq 1$. It may happen that M is so large that we have to stop the partition's growth before reaching the cardinality M. Once we have reached the finest resolution level, defined by the pool of split points in (17.1), we cannot grow the partition any further. Also, growth of the partition can be reasonably restricted not to split rectangles that contain fewer observations than a given threshold. This does not exclude the case where some rectangles of the partition contain no observations.

The partition is grown by minimizing an empirical risk, which is typically that defined by (15.2), with the contrast function γ as in (15.4) or (15.8). However, we apply also a third kind of empirical risk in Section 19.3.2, where we use greedy histograms in stagewise minimization.

Definition 17.1 (Greedy partition.) *The* greedy partition, *with empirical risk $\tilde{\gamma}_n$, with cardinality bound $M \geq 1$, with minimal observation number $m \geq 1$, is defined recursively by the following rules:*

1. *Start with the partition $\mathcal{P}_1 = \{R_0\}$, where R_0 is the smallest rectangle that contains the observations and whose sides are parallel to the coordinate axes.*

2. *For $L = 1, \ldots, M-1$, assume that we have constructed partition $\mathcal{P}_L$ of cardinality L.*

   (a) *Partition $\mathcal{P}_L$ is the final partition when $I = \emptyset$, where*

   $$I = \{(R, k, s) \;:\; R \in \mathcal{P}_L,\ \#\{X_i \in R\} \geq m, \\ k = 1, \ldots, d,\ s \in S_{R,k}\},$$

   and where $S_{R,k}$ is the set of split points defined in (17.3). That is, we stop the growth when no rectangles R exist that contain at least m observations and for which the finest resolution level is not reached in each direction.

   (b) *Otherwise, if partition $\mathcal{P}_L$ is not the final partition, we construct new partition $\mathcal{P}_{\hat{R},\hat{k},\hat{s}}$, where*

   $$\left(\hat{R}, \hat{k}, \hat{s}\right) = \mathrm{argmin}_{(R,k,s) \in I} ERR\left(\mathcal{P}_{R,k,s}\right), \tag{17.6}$$

   where $\mathcal{P}_{R,k,s}$ is the partition defined in (17.2),

   $$ERR(\mathcal{P}) = \tilde{\gamma}_n\left(\hat{f}(\,\cdot\,, \mathcal{P})\right), \tag{17.7}$$

   and $\hat{f}$ is histogram defined in (14.2).

Definition 17.2 (Greedy histogram.) *Let $\hat{\mathcal{P}}_M$ be the greedy partition defined in Definition 17.1. The greedy histogram corresponding to $\hat{\mathcal{P}}_M$ is defined by*

$$\hat{f}_M = \hat{f}(\cdot, \hat{\mathcal{P}}_M)$$

where $\hat{f}$ is defined in (14.2).

Illustrations We illustrate greedy partitions and greedy histograms.

1. Figure 17.1 illustrates the growth of a greedy partition. We have generated 600 observations from the density shown in Figure 2.7 and the log-likelihood contrast function is applied. The partitions of cardinality 2, 3, 4, 7, 15, and 20 are shown. The first partitions delineate the support, and the three modes are detected by the partition of size 15.

2. Figure 17.2 shows volume functions of the level set trees of the histograms in the lower row of Figure 17.1 (7, 15, and 20 rectangles in the partition).

3. Figure 17.3 shows a perspective plot of the last histogram of Figure 17.1 (the partition has cardinality 20).

17.1.2 Contrast Functions

We study the minimization problem in (17.6) for the two contrast functions in (15.4) and (15.8). In (17.6) we minimize the quantity

$$\begin{aligned}
&\frac{1}{n}\sum_{i=1}^{n}\gamma\left(\hat{f}(\cdot, \mathcal{P}_{R,k,s}), X_i\right) \\
&= \frac{1}{n}\sum_{X_i \in R^c}\gamma\left(\hat{f}(\cdot, \mathcal{P}_{R,k,s}), X_i\right) + \frac{1}{n}\sum_{X_i \in R}\gamma\left(\hat{g}_{R,k,s}, X_i\right), \qquad (17.8)
\end{aligned}$$

over sets $R \in \mathcal{P}_L$, over directions $k = 1, \ldots, d$, and over split points $s \in S_{R,k}$, where

$$\begin{aligned}
\hat{g}_{R,k,s}(x) &= \hat{f}(x, \mathcal{P}_{R,k,s}) I_R(x) \\
&= \frac{n_1/n}{\text{volume}(R_{k,s}^{(0)})} I_{R_{k,s}^{(0)}}(x) + \frac{n_2/n}{\text{volume}(R_{k,s}^{(1)})} I_{R_{k,s}^{(1)}}(x),
\end{aligned}$$

where n_j is the number of observations in sets $R_{k,s}^{(j)}$, $j = 0, 1$, defined in (17.4) and (17.5). The first summand in the right-hand side of (17.8) does not depend on k or s, since the summation is over observations not in R. In the second summand on

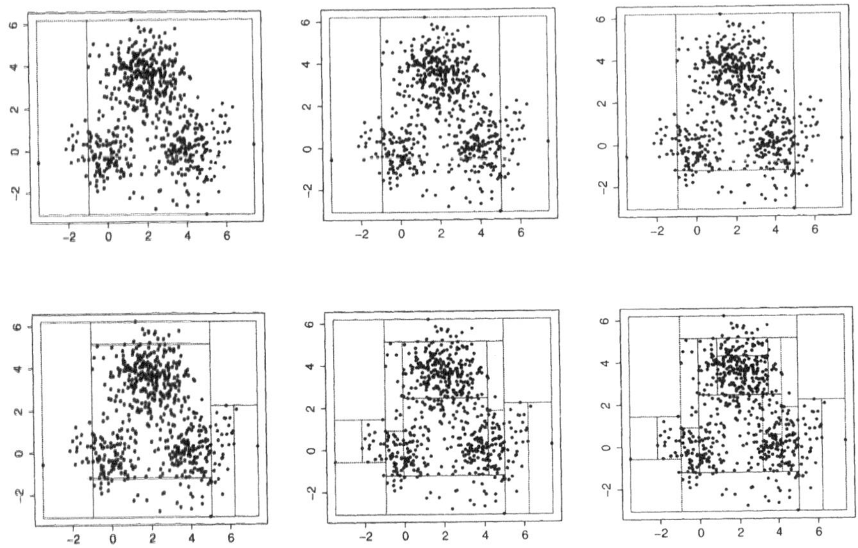

Figure 17.1 *(Growing of a greedy partition.)* The upper row shows partitions with 2 to 4 rectangles and the lower row shows partitions with 7, 15, and 20 rectangles. Observations ($n = 600$) are generated from the density shown in Figure 2.7. The log-likelihood contrast function is applied.

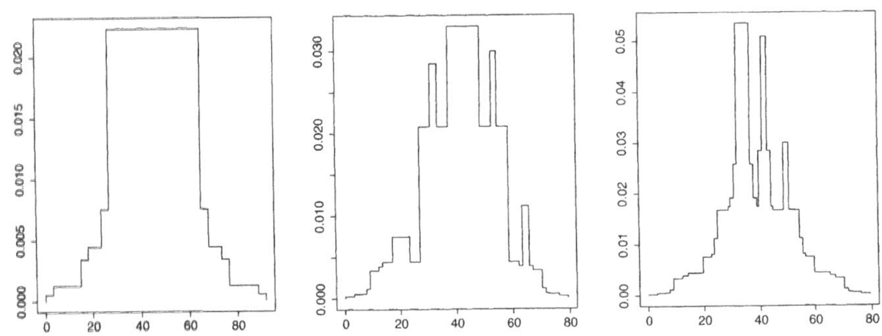

Figure 17.2 *(Volume functions of greedy histograms.)* Shown are volume functions of level set trees of the histograms corresponding to the partitions in the lower row of Figure 17.1 (partitions have cardinalities 7, 15, 20).

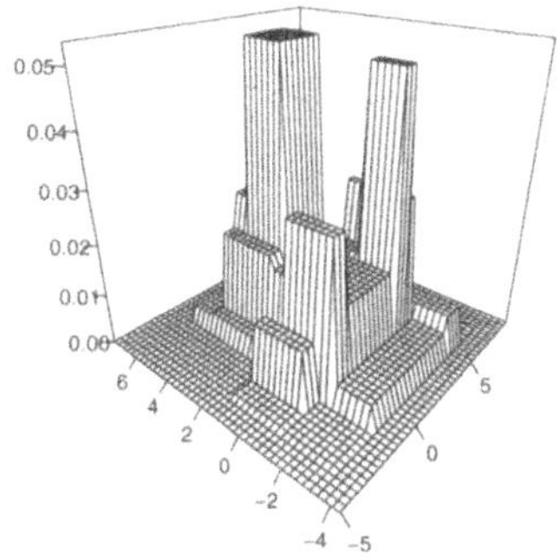

Figure 17.3 *(Perspective plot of a greedy histogram.)* Shown is a perspective plot of the histogram corresponding to the last partition in Figure 17.1. The partition has cardinality 20.

the right-hand side of (17.8) we have localized the summation to rectangle R. The second summand may be written as

$$\frac{1}{n}\sum_{X_i\in R}\gamma(\hat{g}_{R,k,s},X_i)$$
$$= \begin{cases} -\frac{n_1}{n}\log\left(\frac{n_1/n}{\text{volume}(R^{(0)}_{k,s})}\right) - \frac{n_2}{n}\log\left(\frac{n_2/n}{\text{volume}(R^{(1)}_{k,s})}\right) \\ \left(\frac{n_1^2/n^2}{\text{volume}(R^{(0)}_{k,s})} + \frac{n_2^2/n^2}{\text{volume}(R^{(1)}_{k,s})}\right)\left(\frac{n_1+n_2}{n}-2\right) \end{cases} \quad (17.9)$$

depending on the choice of γ either as log-likelihood contrast in (15.8) or as L_2-error contrast in (15.4). Above we use the convention $0\cdot(-\infty)=0$. Note that term $(n_1+n_2)/n-2$ is always negative, since $n_1+n_2\le n$. Term $(n_1+n_2)/n-2$ does not affect the choice of split point s and direction k in a given rectangle, but it may have an effect on the choice for the rectangle to be splitted.

We study how the split point will be chosen when the rectangle and the direction is fixed. Let

$$G(v,p) = \begin{cases} -p\log\left(\frac{p}{v}\right) - (p_{max}-p)\log\left(\frac{p_{max}-p}{v_{max}-v}\right), \\ -\frac{p^2}{v} - \frac{(p_{max}-p)^2}{v_{max}-v}, \end{cases} \quad (17.10)$$

where $0\le v\le v_{max}<\infty, 0\le p\le p_{max}\le 1$. We have defined G by modifying the formulas in (17.9) by taking $p=n_1/n$, $p_{max}=(n_1+n_2)/n$, $v=\text{volume}(R^{(0)}_{l,k,s})$, and $v_{max}=\text{volume}(R^{(0)}_{l,k,s})+\text{volume}(R^{(1)}_{l,k,s})$. When the split point s is varied over its range from left to right, the volume v_s varies from 0 to v_{max} and the (empirical)

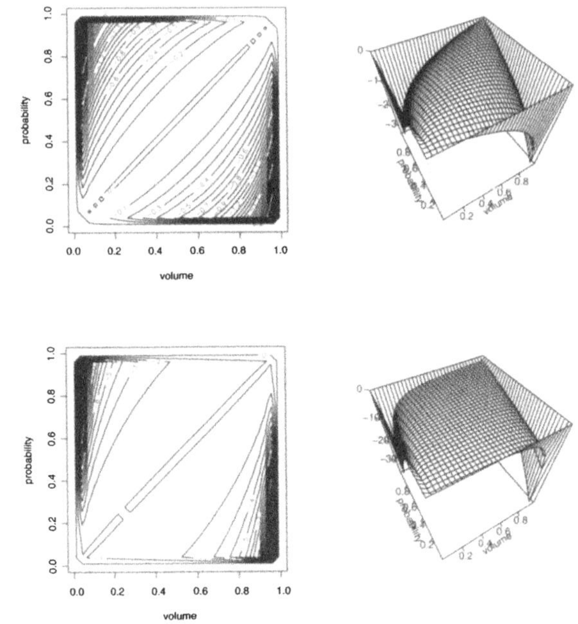

Figure 17.4 *(Comparison of the contrast functions.)* The upper row shows the function G in (17.10) for the log-likelihood contrast. We show both the contour plot and the perspective plot. The lower row shows the function G for the L_2 contrast function. We have $v_{max} = p_{max} = 1$.

probability p_s varies from 0 to p_{max}. We have to find the split point minimizing

$$g(s) = G(v_s, p_s). \tag{17.11}$$

This function is different for different (empirical) marginal distributions.

1. Figure 17.4 draws function G in (17.10) both for the log-likelihood contrast and for the L_2 contrast. We chose $v_{max} = p_{max} = 1$. For the L_2 contrast the function G is flatter.

2. Figure 17.5 shows the shape of function g in (17.11) for a collection of marginal distributions.

   Panel (a) shows three marginal densities (solid line, dashed line, line with long dashes): a Gaussian density and two equal mixtures of Gaussians, with $\sigma = 0.07$.

   Panel (b) shows the distribution functions together with a contour plot of function G, with the log-likelihood contrast (the upper row in Figure 17.4).

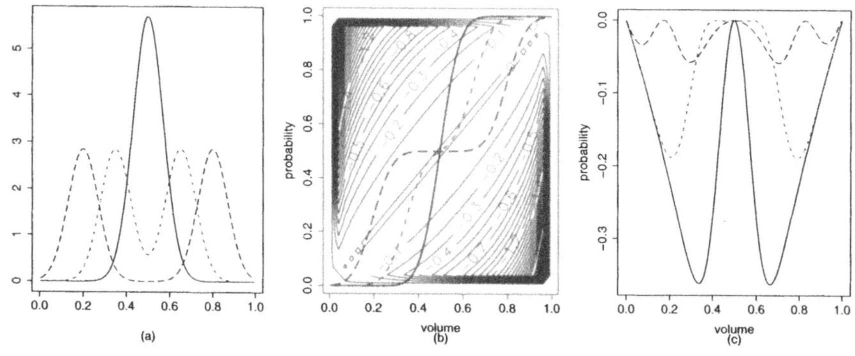

Figure 17.5 *(Log-likelihood contrast, search of splitting points.)* (a) Three density functions, (b) the corresponding distribution functions and the contour plot of function G for the log-likelihood contrast, and (c) function g for the log-likelihood criterion.

Panel (c) shows functions g for the three marginal densities, for the log-likelihood contrast. We have chosen $v_s = s$ and $s \in [0, 1]$ for simplicity. The best split point minimizes function g.

For the solid and dashed densities the best split points (the split points minimizing g) delineate the support from the left and right. For the density with long dashes (the density where the mixture centers are most well separated) the best split points separate the modes. It is not optimal to separate the modes at the middle but rather either at the right boundary of the left mixture component or at the left boundary of the right mixture component.

17.2 CART HISTOGRAMS

17.2.1 Definition

A CART histogram is constructed by making a greedy histogram whose partition has large cardinality (the greedy partition is grown "to the end"), and then choosing the final partition by minimizing a complexity-penalized error criterion. We define an overfitting partition as a partition that is grown in a greedy way but differently than the greedy partition defined in Definition 17.1. In Definition 17.1 the splits were chosen by choosing first the best rectangle and then choosing the best split inside the best rectangle. When we grow the overfitting partition, we do not have to choose the best rectangle, since all rectangles will be split as long as it is possible.

Definition 17.3 (Overfitting partition and the associated binary tree.) *An overfitting partition, with minimal observation number* $m \geq 1$, *and the associated binary tree are defined recursively by the following rules:*

1. *Start with the partition* $\mathcal{P}_1 = \{R_0\}$, *where* R_0 *is the smallest rectangle which contains the observations and whose sides are parallel to the coordinate axes. Rectangle* R_0 *corresponds to the root of the binary tree.*

2. *Let* $\mathcal{P}$ *be the current partition and let* $R \in \mathcal{P}$. *Let*

$$J = \{(k, s) : k = 1, \ldots, d, \ s \in S_{R,k}\} .$$

where $S_{R,k}$ *is the set of split points defined in (17.3).*

   (a) *If* $\#\{X_i : i = 1, \ldots, n, \ X_i \in R\} < m$ *or* $J = \emptyset$, *then rectangle* R *will not be split and* R *corresponds to a leaf node of the binary tree. That is, we do not split* R *if it does not contain at least* m *observations and if the finest resolution level is reached in this rectangle for each direction.*

   (b) *Otherwise, we find*

$$\left(\hat{k}, \hat{s}\right) = \text{argmin}_{(k,s)\in J} ERR\left(\mathcal{P}_{R,k,s}\right),$$

where $\mathcal{P}_{R,k,s}$ *is partition defined in (17.2) and* $ERR(\mathcal{P})$ *is defined in (17.7). The rectangle* $R^{(0)}_{\hat{k},\hat{s}}$ *is the left child node of the rectangle* R *and* $R^{(1)}_{\hat{k},\hat{s}}$ *is the right child node, where* $R^{(0)}_{k,s}$ *and* $R^{(1)}_{k,s}$ *are defined in (17.4) and (17.5).*

Remarks

1. The binary tree associated with an overfitting partition is a partition generating tree in the sense of Definition 13.1. We may make an evaluation tree in the sense of Definition 13.2 from this tree by the histogram rule: every leaf node is associated with the value that is equal to the empirical probability of the rectangle divided be the volume of the rectangle.

2. If the pool of split points is a regular grid on R_0, and has $n - 1$ split points for each direction, then we cannot split a rectangle R whose volume is equal to volume$(R_0)/n^d$, since then $S_{R,k}$ is empty for all $k = 1, \ldots, d$.

Complexity-Penalized Empirical Risk Let $0 \leq \alpha < \infty$, and define the complexity-penalized error criterion as

$$COPERR(\mathcal{P}, \alpha) = ERR(\mathcal{P}) + \alpha \cdot \#(\mathcal{P}), \tag{17.12}$$

where $ERR(\mathcal{P})$ is defined in (17.7). Note that

$$ERR(\mathcal{P}) = \begin{cases} -\sum_{R\in\mathcal{P}} \frac{n_R}{n} \log\left(\frac{n_R/n}{\text{volume}(R)}\right), \\ -\sum_{R\in\mathcal{P}} \frac{n_R^2/n^2}{\text{volume}(R)}, \end{cases} \tag{17.13}$$

corresponding respectively to the choices of contrast function as in (15.4) or as in (15.8), where n_R is the number of observations in R. Above we use the convention $0 \cdot (-\infty) = 0$.

Collection of Partitions We write $T' \preceq T$ to mean that T' is a subtree of T obtained by pruning away some number of branches of T; see (D.1). Let T_0 be the binary tree associated to the overfitting partition, as constructed in Definition 17.3. To any $T \preceq T_0$ corresponds a partition $\mathcal{P}(T)$ of R_0 that is the collection of rectangles associated with the leaf nodes of T. The available collection of partitions from which we search the best partition is defined by

$$\mathcal{L}(T_0) = \{\mathcal{P}(T) : T \preceq T_0\} . \tag{17.14}$$

Definition 17.4 (CART histogram.) *Define the partition corresponding to parameter* $\alpha \geq 0$ *as*

$$\hat{\mathcal{P}}_\alpha = \text{argmin}_{\mathcal{P}\in\mathcal{L}(T_0)} COPERR(\mathcal{P}, \alpha), \tag{17.15}$$

where $\mathcal{L}(T_0)$ *is defined in (17.14), and define the CART histogram by*

$$\hat{f}_\alpha = \hat{f}(\,\cdot\,, \hat{\mathcal{P}}_\alpha), \tag{17.16}$$

where $\hat{f}$ *is defined in (14.2).*

Illustrations We illustrate CART partitions and CART histograms.

1. Figure 17.6 shows CART partitions and CART histograms for the same data as in Figure 17.1. The upper row shows CART partitions and the lower row shows volume plots of the corresponding CART histograms.

   First, we show the overfitting partition with the minimal observation number 5. Second, we show the CART partition that is pruned from the overfitting partition to have cardinality 16. Third, we show the CART partition, which has 10 rectangles.

2. Figure 17.7 shows a perspective plot of the CART histogram corresponding to the partition in Figure 17.6. This partition has cardinality 16.

17.2.2 Pruning Algorithms

Single α We need algorithms for solving the minimization problem (17.15). The problem may be solved by the dynamic programming algorithm "prune" given in Appendix D.4. Indeed the complexity-penalized error defined in (17.12) and in (17.13) is a sum of a function f over the leafs of a subtree of the overfitting tree, where f is defined by

$$f(R) = -\frac{n_R}{n}\log\left(\frac{n_R/n}{\text{volume}(R)}\right) + \alpha$$

for the log-likelihood criterion, where we identified the nodes with rectangles R associated with the nodes. We search for a pruned tree which minimizes the sum of f over the leafs.

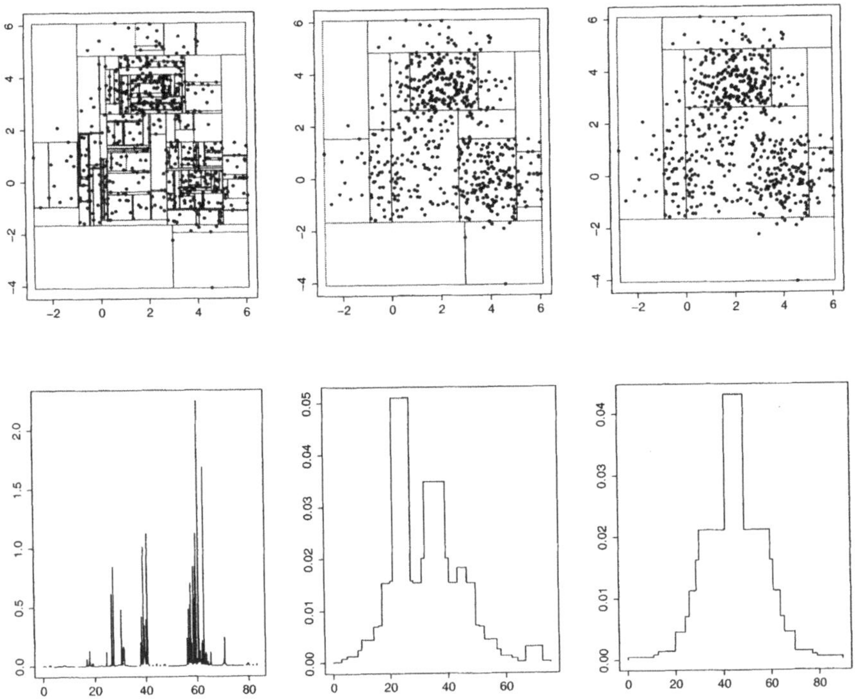

Figure 17.6 *(CART partitions.)* The upper row shows CART partitions, and the lower row shows volume functions of the histograms corresponding to the partitions. The first panel shows the overfitting partition for the same data as in Figure 17.1. The second panel shows the CART partition, which is pruned from the overfitting partition to have 16 rectangles. The third panel shows the CART partition, which has 10 rectangles.

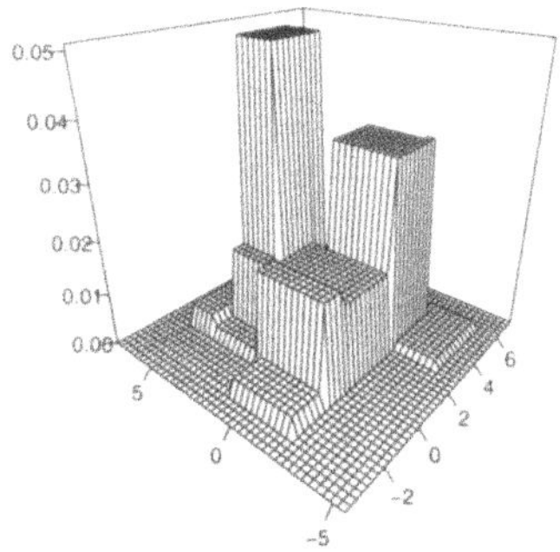

Figure 17.7 *(CART histogram.)* Shown is a perspective plot of the histogram corresponding to the partition in Figure 17.6 that has 16 rectangles (the partition in the middle).

The dynamic programming algorithm utilizes a bottom-up travel of the overfitting tree. We formulate a lemma that formalizes the idea that we may solve the global minimization problem (17.15) by first solving localized subproblems, and building the global solution from the previously solved local problems.

Lemma 17.1 *Let T_0 be the overfitting tree defined in Definition 17.3. Let t_0 be some nonterminal node of T_0, t_1 left child of t_0, and t_2 right child of t_0. Let T_{0,t_i}, $i = 0, 1, 2$, be the subtrees of T_0 where the root of tree T_{0,t_i} is t_i. Let the collection of partitions defined by the terminal nodes of subtree T_{0,t_i} be denoted by*

$$\mathcal{L}^{(i)}(T_0) = \{\mathcal{P}(T) : T \preceq T_{0,t_i}\}, \qquad i = 1, 2, 3,$$

where relation $\preceq$ is defined in (D.1). Denote the partition minimizing the complexity-penalized error criterion by

$$\widehat{\mathcal{P}}_\alpha^{(i)} = \text{argmin}_{\mathcal{P} \in \mathcal{L}^{(i)}(T_0)} COPERR(\mathcal{P}, \alpha), \qquad i = 1, 2, 3.$$

Then

$$\widehat{\mathcal{P}}_\alpha^{(0)} = \begin{cases} \{R_{t_0}\}, \text{ when } COPERR(\{R_0\}, \alpha) \leq \\ \qquad\qquad COPERR\left(\widehat{\mathcal{P}}_\alpha^{(1)}, \alpha\right) + COPERR\left(\widehat{\mathcal{P}}_\alpha^{(2)}, \alpha\right) \\ \widehat{\mathcal{P}}_\alpha^{(1)} \cup \widehat{\mathcal{P}}_\alpha^{(2)}, \text{ otherwise,} \end{cases}$$

where R_{t_0} is the rectangle associated to node t_0.

Proof. When $\mathcal{P}_i \in \mathcal{L}^{(i)}(T_0)$, $i = 1, 2$, we have

$$COPERR(\mathcal{P}_1 \cup \mathcal{P}_2, \alpha) = COPERR(\mathcal{P}_1, \alpha) + COPERR(\mathcal{P}_2, \alpha).$$

This follows directly from (17.13) and from the fact that $\#(\mathcal{P}_1 \cup \mathcal{P}_2) = \#\mathcal{P}_1 + \#\mathcal{P}_2$, where we used the fact that $\mathcal{P}_1$ and $\mathcal{P}_2$ are partitions of disjoint rectangles. On the other hand,

$$\mathcal{L}^{(0)}(T_0) = \{\{R_{t_0}\}\} \cup \left\{\mathcal{P}_1 \cup \mathcal{P}_2 : \mathcal{P}_i \in \mathcal{L}(T_0)^{(i)},\ i = 1, 2\right\}.$$

We have proved the lemma. □

In particular, $\widehat{\mathcal{P}}_\alpha^{(0)} = \widehat{\mathcal{P}}_\alpha$ is the global solution defined in (17.15) when we choose t_0 in Lemma 17.1 to be the root of tree T_0.

The Sequence of α When $\alpha = 0$, then $\hat{\mathcal{P}}_\alpha$ is equal to $\mathcal{P}(T_0)$, where T_0 is the binary tree associated to the overfitting partition constructed in Definition 17.2. For some large value of α the complexity dominates the complexity-penalized error criterion and the partition $\hat{\mathcal{P}}_\alpha$ is the trivial partition consisting of the root node R_0 of tree T_0: $\hat{\mathcal{P}}_\alpha = \{R_0\}$. Thus, when α increases from 0 to infinity, there exists only a finite number of values for α, say $\{0 = \alpha_0 < \cdots < \alpha_k\}$, for which the corresponding partitions are different. When $0 \le \alpha < \alpha_1$, the partition is $\mathcal{P}(T_0)$; when $\alpha_1 \le \alpha_2$, the partition is smaller, and so on. There exists an algorithm for finding this α-sequence and the corresponding sequence of partitions. We present an algorithm in Appendix D.5 that is a modification of the algorithm of Breiman, Friedman, Olshen and Stone (1984, sec. 10.3).

17.3 BOOTSTRAP AGGREGATION

Adaptive histograms (greedy and CART histograms) give piecewise constant estimates of low granularity. It is possible to increase the granularity of these estimates by bootstrap aggregation. Average shifted histograms discussed in Section 14.2.2 are a related method to increase the granularity of regular non-adaptive histograms. The following procedure will be called a bootstrap aggregation:

1. Generate B bootstrap samples from the original sample.

2. Calculate adaptive histogram $\hat{f}_j$, $j = 1, \ldots, B$, based on each bootstrap sample. We may choose the adaptive histograms to be greedy histograms defined in Section 17.1 or CART histograms defined in Section 17.2.

3. Define estimator $\hat{f}$ as the arithmetic mean of estimators $\hat{f}_j$,

$$\hat{f}(x) = \frac{1}{B} \sum_{j=1}^{B} \hat{f}_j(x), \qquad x \in \mathbf{R}^d.$$

Breiman (2001) considers a class of procedures that includes random split selection from the set of best splits, and a random perturbation of observations. Bootstrap aggregation has been regarded as a method of decreasing variance of an unstable estimator, like an adaptive histogram.

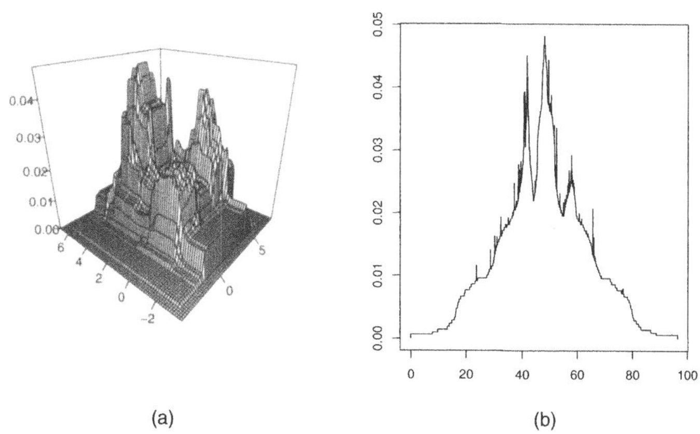

Figure 17.8 *(Bootstrap aggregation with CART histograms.)* (a) A perspective plot and (b) a volume function of a bootstrap aggregated estimate. The estimate is an average of $B = 50$ CART histograms. The cardinality of the partitions of the CART histograms is 15. We applied $n/2$-out-of-n without replacement bootstrap. The data of size 600 is the same as in Figure 17.1.

There exist several bootstrapping schemes and we mention two of them:

1. We may apply n-out-of-n with replacement bootstrap. This means that we take B bootstrap samples of size n with replacement from the empirical distribution $P_n(A) = \frac{1}{n}\sum_{i=1}^{n} I_A(X_i)$. Some observations may appear more than once in the bootstrap sample and some observations may be missing.

2. We may apply $n/2$-out-of-n without replacement bootstrap. This means that we take B bootstrap samples of size $[n/2]$ without replacement. Every observation appears at most once in the bootstrap sample.

The original proposal in Breiman (1996*a*) suggested n-out-of-n bootstrap with replacement.

Illustrations We illustrate bootstrap aggregation.

1. Figure 17.8 shows an estimate that is obtained by bootstrap aggregation for the same data as in Figure 17.1. The estimate is an average of 50 CART histograms that have a partition of cardinality 15. The overfitting partition is grown with the minimal observation number 5. The $n/2$-out-of-n without replacement bootstrap is applied. Panel (a) shows a perspective plot and panel (b) shows a volume function of the estimate. The volume function shows that the estimate has a large number of local maxima, but the estimate has three main modes.

2. *(Level sets.)* CART histograms can find qualitative features like the number of modes of the density or the number of modes of a level set. Bootstrap

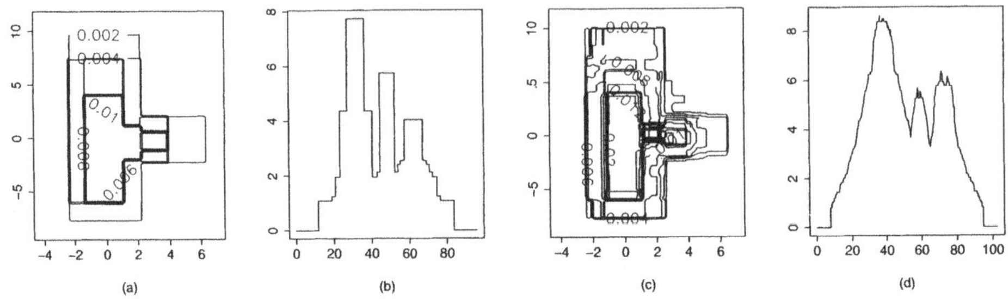

Figure 17.9 *(Radius functions.)* (a) A contour plot of a CART histogram, (b) a radius function of the 0.002 level set of the CART histogram, (c) a contour plot of a bootstrap aggregated estimate, and (d) a radius function of the 0.002 level set of the bootstrap aggregated estimate. The underlying density is shown in Figure 5.5(a).

aggregation can give a more accurate estimate of the shape of the function. We illustrate the estimation of the shape of a level set with an CART histogram and an bootstrap aggregated estimate. We generated a sample of size 1000 from the density of Figure 5.6(a).

Figure 17.9(a) shows a contour plot of a CART histogram and panel (b) shows a radius function of the 0.002 level set of the histogram. Radius functions are defined in Section 5.3.1. The partition of the CART histogram has 15 rectangles.

Figure 17.9(c) shows a contour plot of a bootstrap aggregated estimate, and panel (b) shows a radius function of the 0.002 level set of the estimate. The bootstrap aggregated estimate is an average of 20 CART histograms whose partition has cardinality 15. The CART histograms are constructed using $n/2$-out-of-n without replacement bootstrap.

17.4 BIBLIOGRAPHIC NOTES

Univariate Histograms Scott (1979), Freedman and Diaconis (1981), and Rudemo (1982) derived expressions for optimal interval widths in fixed-interval univariate histograms. In the one-dimensional case histogram estimation based on irregular partitions was considered by Kogure (1987), who extends the rule of Rudemo (1982) to a histogram with locally equispaced cells. He considers the L_2 loss. Taylor (1987) studied histograms with a regular partition and considered Kullback–Leibler loss, applying a general penalized maximum likelihood model selection procedure proposed by Akaike (1973). Kanazawa (1992) defines an irregular partition with knots among the order statistics. He considers Hellinger loss. A model selection procedure for one-dimensional histograms was given in Catoni (2002).

Multivariate Histograms Hüsemann and Terrell (1991) consider the problem of optimal fixed and variable cell dimensions in bivariate histograms. Scott (1992, p. 82), suggested a normal reference rule to choose the bin widths differently in different directions. Lugosi and Nobel (1996) present L_1 consistency results on density estimators based on data-dependent partitions. Heikkinen and Arjas (1998) present a histogram type of Bayesian estimator for a spatial Poisson intensity. Barron et al. (1999) constructed a multivariate histogram that achieves asymptotic minimax rates over anisotropic Hölder classes for the L_2 loss. Their histograms had different number of bins in different directions, but in the single direction bins were equispaced. A modified Akaike criterion for histogram estimation with irregular splits was studied by Castellan (1999), who gives oracle inequalities for the Kullback–Leibler and Hellinger losses.

CART Histograms Breiman et al. (1984) introduced CART (classification and regression trees) as a method for estimating classification and regression functions with piecewise constant estimates. Density estimation with CART-type methods was considered by Shang (1994), Sutton (1994), and Ooi (2002). Quinlan (1993) considers C4.5 algorithm that is related to CART. An overview of tree structured classification can be found from Ripley (1996). Yu and Lambert (2000) fit trees to functional data.

Pruning Algorithms Chou, Lookabaugh and Gray (1989) present a fast pruning algorithm for finding the sequence of subtrees. The algorithm needs $O(\#T \log(\#T))$ flops, where $\#T$ is the number of leafs of the tree. Besides CART pruning, some related instances of dynamic programming algorithms are the "Best-Ortho-Basis" algorithm of Coifman and Wickerhauser (1992), Fast Algorithm for Dyadic CART of Donoho (1997), and Optimal Decorated Quadtree algorithm of Donoho (1999).

Bootstrap Aggregation Bagging was introduced by Breiman (1996*a*) and Breiman (1996*b*). Boosting and bagging in density estimation was considered by Ridgeway (2002). Bühlmann and Yu (2002) define the unstability of the base learners as the lack of convergence when the sample size increases.

EXERCISES

17.1 Write a pseudo-code for an algorithm that calculates a weighted average of two functions when the functions are represented with evaluation trees.

CHAPTER 18

BEST BASIS SELECTION

Best basis selection shall serve as an example of the theory of adaptive estimation. We consider best basis selection in the collection of dyadic histograms. The partition of the histograms is chosen data dependently, and this is equivalent to choosing empirically a basis of multivariate Haar basis functions.

Dyadic histograms are histograms whose partition is chosen as a minimizer of a complexity-penalized empirical risk over a set of partitions defined by dyadic recursive splitting sequences. As the error criterion we take the empirical risk with the L_2 contrast function, and the complexity of the histogram is defined to be the number of sets in the partition. The partition may have a different bin width in each direction, and also in a single direction the bin widths may have variable lengths.

Dyadic histograms may be written as series estimators with a basis of multivariate Haar functions. Instead of thresholding the empirical coefficients in a fixed basis, the method chooses empirically a basis where the thresholding is performed. The thresholds are not simple hard thresholds but satisfy certain constraints.

There is a similarity between CART histograms and dyadic histograms: both histograms are found as minimizers of a complexity-penalized empirical risk over a set of partitions. In the case of CART histograms, the set of partitions is constructed empirically and in the case of dyadic histograms, the set of partitions is a fixed set of recursive dyadic partitions.

The estimator achieves the minimax rates up to a logarithmic factor over a scale of anisotropic smoothness classes, for the L_2 loss. We consider histograms with unequal bin widths in every direction, and thus we nearly achieve the minimax rates over smoothness classes containing functions with considerable spatially varying smoothness. To apply the estimator we have to choose a bound for the maximal fineness of the partitions being considered. We may increase the flexibility of the estimator by choosing the maximal allowed resolution to be fine. On the other hand, this will increase the computational complexity of the estimator. The bound for the maximal fineness depends on the spatial inhomogeneity of the density, and also the computational complexity depends on this bound for the maximal fineness. The method may be seen as based on presmoothing the data, since the estimator uses only the frequencies on the partition defined by the finest resolution level.

18.1 ESTIMATORS

18.1.1 Dyadic Histogram

We defined in (14.2) a histogram

$$\hat{f}(x, \mathcal{P}) = \sum_{R \in \mathcal{P}} \frac{n_R}{n \, \mathrm{volume}(R)} I_R(x), \qquad x \in \mathbf{R}^d, \tag{18.1}$$

where $\mathcal{P}$ is a partition and $n_R = \#\{X_i \in R : i = 1, \ldots, n\}$ are the frequencies for the sets of the partition. We will define a set of partitions from among which the optimal partition will be searched.

Collection of Partitions We need to define a collection of trees that generate dyadic partitions. Dyadic partition-generating trees are a special case of the partition-generating trees defined in Definition 13.1. Dyadic partition-generating trees have three constraints, as compared to general partition-generating trees: (1) splits are made only at the midpoints of the sides of the rectangles, (2) a dyadic partition-generating tree is a binary tree where each node has either 0 or 2 children, (3) a bound is given for the directionwise split number.

Remember, in Definition 17.1 we defined a single partition and in Definition 17.3 we defined a single partition and a partition-generating tree. Now we define a collection of partition-generating trees. The optimal partition will be searched over the collection of partitions generated by these partition-generating trees.

Definition 18.1 (Dyadic partition-generating tree.) *A collection of dyadic partition-generating trees $\mathbb{T}(R_0, J)$, associated with a rectangle $R_0 \subset \mathbf{R}^d$, whose sides are parallel to the coordinate axes, and with a bound for split numbers $J = (J_1, \ldots, J_d) \in \{0, 1, \ldots\}^d$, consists of binary trees whose each node is annotated with a rectangle, and each non-leaf node is annotated with a splitting direction in $\{1, \ldots, d\}$.*

1. *The root node is annotated with* R_0.

2. *Let a non-leaf node be annotated with rectangle* $R = \prod_{m=1}^{d}[c_m, d_m]$ *and direction* $l \in \{1, \ldots, d\}$. *Let the split point be*

$$s = \frac{d_l - c_l}{2}.$$

Denote

$$R_{l,s}^{(0)}(R) = \{x \in R : x_l < s\}, \qquad R_{l,s}^{(1)}(R) = \{x \in R : x_l \geq s\}.$$

The left child of the node is annotated with $R_{l,s}^{(0)}(R)$, *and the right child is annotated with* $R_{l,s}^{(1)}(R)$.

3. *At most* J_l *splits will be made in direction* l, $l = 1, \ldots, d$.

We make some observations concerning the definition.

- A set of dyadic partition-generating trees is completely determined by the initial rectangle R_0 and by the splitting directions; since the splits are always made at the midpoints of the sides of the rectangles, the annotation of the nodes with rectangles is redundant.

- The simplest dyadic partition-generating tree is the tree that consists only of the root node, and this tree is the single member of $\mathbb{T}(R_0, 0)$.

- The bound J for the number of splits implies a bound for the depth of the tree: the depth is at most $|J| = \sum_{i=1}^{d} J_i$. (We define the depth of a tree to be equal to the largest depth among the depths of its nodes and we stipulate that the depth of the root node is 0 and the depth of the children of the root is 1, etc.).

- A tree that generates a dyadic partition may be an unbalanced tree; some terminal nodes may have their depths equal to $|J|$, but the depths of some other terminal nodes may be less than $|J|$.

Each tree in the set $\mathbb{T}(R_0, J)$ generates a partition: the partition is the collection of the rectangles associated with the leaf nodes of the tree. This is the content of the net definition.

Definition 18.2 (Collection of dyadic partitions.) *The dyadic partition associated to tree* $\mathcal{T} \in \mathbb{T}(R_0, J)$, *where* $\mathbb{T}(R_0, J)$ *is defined in Definition 18.1, is*

$$\mathcal{P}(\mathcal{T}) = \{R(t) : t \in \text{Ter}(\mathcal{T})\}, \tag{18.2}$$

where $\text{Ter}(\mathcal{T})$ *is the set of terminal nodes of* $\mathcal{T}$, *and* $R(t)$ *is the rectangle annotated to node* t. The collection of dyadic partitions $\mathbb{P} = \mathbb{P}(R_0, J)$, *with base rectangle* R_0 *and with depth bound* J, *is defined by*

$$\mathbb{P}(R_0, J) = \{\mathcal{P}(\mathcal{T}) : \mathcal{T} \in \mathbb{T}(R_0, J)\}. \tag{18.3}$$

Complexity-Penalized Error Criterion We apply the empirical risk $\gamma_n(\hat{f}(\cdot, \mathcal{P}))$, defined in (15.2). We take the contrast function to be the L_2 contrast function defined in (15.4). The estimator is uniquely defined through its partition, and we use the notation

$$ERR_n(\mathcal{P}) = \gamma_n\left(\hat{f}(\cdot, \mathcal{P})\right). \tag{18.4}$$

We have that

$$ERR_n(\mathcal{P}) = -\sum_{R \in \mathcal{P}} \frac{n_R^2}{n^2 \text{volume}(R)} = -\left\|\hat{f}^2(\cdot, \mathcal{P})\right\|_2^2. \tag{18.5}$$

The complexity of a histogram is taken to be the number of sets in the partition of the histogram. Let $0 \le \alpha < \infty$, and define the complexity-penalized error criterion as

$$COPERR_n(\mathcal{P}, \alpha) = ERR_n(\mathcal{P}) + \alpha \cdot \#\mathcal{P}. \tag{18.6}$$

Definition of a Dyadic Histogram A dyadic histogram is defined as a minimizer of the complexity-penalized empirical risk, as we minimize the complexity-penalized empirical risk over the set of dyadic partitions.

Definition 18.3 (Dyadic histogram.) *Define the partition corresponding to parameter α as*

$$\hat{\mathcal{P}}_\alpha = \text{argmin}_{\mathcal{P} \in \mathbb{P}(R_0, J)} COPERR_n(\mathcal{P}, \alpha), \tag{18.7}$$

where $\mathbb{P}(R_0, J)$ is defined in (18.3). A dyadic histogram *is defined as*

$$\hat{f} = \hat{f}(\,\cdot\,, \hat{\mathcal{P}}_\alpha) \tag{18.8}$$

where $\hat{f}(\,\cdot\,, \mathcal{P})$ is defined in (18.1).

We make some observations concerning the definition.

- A dyadic histogram is a minimization estimator, as defined in (15.13), when we choose the parameter of the estimate as $\kappa = \mathcal{P}$, which lies in the set $\mathbb{K} = \mathbb{P}(R_0, J)$, and define the complexity penalization as $D(\kappa) = D(\mathcal{P}) = \#\mathcal{P}$. Furthermore $\epsilon = 0$: we find the exact minimizer.

- Besides depending on the smoothing parameter α, dyadic histogram estimator depends also on the maximal directionwise split numbers J and on the initial rectangle R_0. Theorem 18.2 gives conditions for the choice of these parameters. In particular, α and J will depend on the sample size n. In Theorem 18.2 we take $R_0 = [0, 1]^d$ but in practice we would estimate R_0. A reasonable choice is to define R_0 as the smallest rectangle containing the observations whose sides are parallel to the coordinate axes.

18.1.2 Series Estimator

We define a series estimator using a basis of Haar wavelets. The estimator is identical with a dyadic histogram. The representation of the estimator as a series estimator is useful when we want to analyze the asymptotic properties of the estimator, but the representation of the estimator as a dyadic histogram is useful when we want to find algorithms for the calculation of the estimates.

Multivariate Haar Wavelets Denote

$$\phi_{j,k}^{(l)}(x) = \eta_{j_l,k_l}^{(1)}(x_l) \prod_{m=1,m\neq l}^{d} \eta_{j_m,k_m}^{(0)}(x_m), \qquad x = (x_1,\ldots,x_d) \in \mathbf{R}^d, \quad (18.9)$$

where $l \in \{1,\ldots,d\}$, $j = (j_1,\ldots,j_d) \in \{0,1,\ldots\}^d$, and $k = (k_1,\ldots,k_d) \in K_j$, where K_j is defined in (16.17), as the set of translation coefficients corresponding to resolution index $j = (j_1,\ldots,j_d)$: $K_j = \{k = (k_1,\ldots,k_d) : k_l = 0,\ldots,2^{j_l}-1, l = 1,\ldots,d\}$. Here we define the univariate Haar scaling function $\eta^{(0)} = I_{[0,1]}$ as in (16.12) and the univariate Haar wavelet $\eta^{(1)} = I_{[1/2,1)} - I_{[0,1/2)}$ as in (16.13). We denote

$$\eta_{j_m,k_m}^{(\iota)}(t) = 2^{j_m/2}\eta^{(\iota)}(2^{j_m}t - k_m), \qquad t \in [0,1],$$

where $\iota \in \{0,1\}$, $j_m \in \{0,1,\ldots\}$, and $k_m \in \{0,\ldots,2^{j_m}-1\}$. Compare notation $\phi_{j,k}^{(l)}(x)$ with the notation h_{jk}^{i} in (16.16). Function $\prod_{m=1}^{d} \eta_{j_m,k_m}^{(0)}(x_m)$ is (a constant times) the indicator of the rectangle R_{jk} (defined below) but in (18.9) we have multiplied it with Haar wavelet $\eta_{j_l,k_l}^{(1)}(x_l)$.

Dyadic Rectangles Define the rectangle corresponding to the pair of multi-indexes $(j,k) \in \{0,1,\ldots\}^d \times K_j$ as

$$R_{jk} = \prod_{l=1}^{d} \left[\frac{k_l}{2^{j_l}}, \frac{k_l+1}{2^{j_l}}\right). \quad (18.10)$$

We have defined in Definition 18.1 a collection of dyadic partition-generating trees. When the root node is annotated with rectangle $[0,1)^d$, every node of the tree is annotated with a dyadic rectangle. We have a bijective correspondence between dyadic rectangles and pairs of multi-indexes, as defined by (18.10). We denote with $\mathcal{I}(t)$ the pair of multi-indexes associated with node t; that is, when a node is annotated with rectangle R_{jk}, we have $\mathcal{I}(t) = (j,k)$. Exercise 18.1 requires the reader to give a direct definition of the annotation with multi-indexes.

Collection of Pre-Bases Definition 18.4 of a collection of pre-bases is the counterpart of Definition 18.2 of a collection of partitions. A difference is that now we take the initial rectangle $R_0 = [0,1)^d$. We defined in Definition 18.2 a collection of partitions generated by a collection of partition generating trees. We define analogously a collection of pre-bases generated by a collection of partition-generating trees.

In (18.2) we defined the partition associated with a partition-generating tree. We define analogously a pre-basis $\mathcal{B}(\mathcal{T})$ annotated with a partition-generating tree. Collection $\mathcal{B}(\mathcal{T})$ is a finite orthonormal system and $\int_{[0,1]^d} \phi = 0$ for each $\phi \in \mathcal{B}(\mathcal{T})$. We call these collections "pre-bases" because it is possible to extend these to be bases of $L_2([0,1]^d)$.

Definition 18.4 (Collection of pre-bases.) *When $\mathcal{T} \in \mathbb{T}([0,1]^d, J)$ is a dyadic partition-generating tree, where $\mathbb{T}([0,1]^d, J)$ is defined in Definition 18.1, and t is a node of $\mathcal{T}$, let $s(t) \in \{1, \ldots, d\}$ be the direction annotated with t and let $\mathcal{I}(t)$ be the pair of multi-indexes annotated with t. Denote with $NT(\mathcal{T})$ the set of nonterminal nodes of $\mathcal{T}$. The* pre-basis associated to tree $\mathcal{T}$ *is*

$$\mathcal{B}(\mathcal{T}) = \left\{ \phi_{\mathcal{I}(t)}^{(s(t))} : t \in NT(\mathcal{T}) \right\}, \tag{18.11}$$

where $\phi_{j,k}^{(l)}$ is defined in (18.9). The collection of pre-bases $\mathcal{L}(\mathcal{J})$, *with depth bound $J = (J_1, \ldots, J_d)$, is*

$$\mathcal{L}(J) = \left\{ \mathcal{B}(\mathcal{T}) : \mathcal{T} \in \mathbb{T}([0,1]^d, J) \right\}. \tag{18.12}$$

Collection of Tree Weights We define a series estimator whose terms are a subset of a pre-basis $\mathcal{B}(\mathcal{T})$. The series estimator is defined with the help of 0–1-weights by which a subset of the pre-basis is chosen. In order the series estimator to be equivalent with a dyadic histogram, we need to make a restriction on the weights of the series estimator. The pre-basis $\mathcal{B}(\mathcal{T})$ is associated with tree $\mathcal{T}$, and we require the weights to be such that they correspond to a pruning of the associated tree. The *collection of treeweights* $\mathcal{W}_{tree,J} = \mathcal{W}_{tree,J}(\mathcal{B})$, associated with $\mathcal{B} \in \mathcal{L}(J)$, is the set of vectors $W = (w_\phi)_{\phi \in \mathcal{B}} \in \{0,1\}^{\mathcal{B}}$, that satisfy the condition that a weight can be zero only when all the "ancestor" weights are zero at the coarser resolution levels. Define

$$\mathcal{W}_{tree,J} = \left\{ (w_\phi)_{\phi \in \mathcal{B}} \in \{0,1\}^{\mathcal{B}} : \text{ if } w_\phi = 0 \text{ then } w_{\phi'} = 0 \text{ for all } \phi' \subset \phi \right\}. \tag{18.13}$$

Here $\phi' \subset \phi$ means for $\phi = \phi_{\mathcal{I}(t)}^{(s(t))}, \phi' = \phi_{\mathcal{I}(t')}^{(s(t'))} \in \mathcal{B}$, that $R_{\mathcal{I}(t')} \subset R_{\mathcal{I}(t)}$, where R_{jk} is defined in (18.10).

When $\phi' \subset \phi$, we say that ϕ' is a child of ϕ. The tree condition says that if $w_\phi = 0$, then $w_{\phi'} = 0$ for all children ϕ' of ϕ. Choosing a subset of $\mathcal{B}(\mathcal{T})$ with the help of weights $W \in \mathcal{W}_{tree,J}(\mathcal{B}(\mathcal{T}))$ is equivalent to the pruning of tree $\mathcal{T} \in \mathbb{T}([0,1]^d, J)$.

Definition of the Series Estimator The series estimator f^* is a minimization estimator where the search is for a best pre-basis $\mathcal{B}_\alpha^*$ and a best subset of $\mathcal{B}_\alpha^*$ under the tree condition.

Definition 18.5 (Dyadic series estimator.) *The empirical choice for basis $\mathcal{B}$ and for vector W is defined as*

$$(\mathcal{B}_\alpha^*, W_\alpha^*) = \text{argmin}_{\mathcal{B} \in \mathcal{L}(J), W \in \mathcal{W}_{tree,J}(\mathcal{B})} \mathcal{E}_n(W, \Theta_n(\mathcal{B}), \mathcal{B}, \alpha), \tag{18.14}$$

where $\mathcal{E}_n(W, \Theta, \mathcal{B}, \alpha)$ is as defined in (15.99) and $\Theta_n(\mathcal{B}) = (\hat{\theta}_\phi)_{\phi \in \mathcal{B}}$ is the vector of empirical coefficients as defined in (15.108). The dyadic series estimator *is defined by*

$$f^*(x) = \tilde{f}\left(x, W^*_\alpha, \Theta_n(\mathcal{B}^*_\alpha), \mathcal{B}^*_\alpha\right), \qquad x \in \mathbf{R}^d, \tag{18.15}$$

where $\tilde{f}(\cdot, W, \Theta, \mathcal{B})$ is as defined in (15.97).

The series estimator is a minimization estimator, as defined in (15.13), when we choose the parameter of the estimate as $\kappa = (\mathcal{B}, W)$, which lies in the set

$$\mathbb{K} = \{(\mathcal{B}, W) : \mathcal{B} \in \mathcal{L}(J), W \in \mathcal{W}_{tree,J}(\mathcal{B})\},$$

and the complexity penalization is defined by $D(\kappa) = D(W) = \#\{w_\phi \neq 0\} + 1$. In addition $\epsilon = 0$, so we find the exact minimizer.

18.1.3 Equivalence Between the Estimators

We prove that dyadic histograms and series estimators are equal.

Lemma 18.1 *We have that $\hat{f} = f^*$ when $\hat{f}$ is as defined in (18.8) with the initial rectangle $R_0 = [0,1)^d$ and f^* is as defined in (18.15).*

Proof. Estimator $\hat{f}$ is defined as a minimizer of the $COPERR_n$ criterion among the partitions in $\mathbb{P}(R_0, J)$. Estimator f^* is defined as a minimizer of $\mathcal{E}_n$ criterion among the bases in $\mathcal{L}(J)$ and the weights in $\mathcal{W}_{tree,J}$. There exists a natural bijection $G : \mathcal{L}(J) \times \mathcal{W}_{tree,J} \to \mathbb{P}(J)$. Indeed a pre-basis $\mathcal{B} \in \mathcal{L}(J)$ is defined with the help of a tree $\mathcal{T} \in \mathbb{T}(J)$, where $\mathbb{T}(J)$ is the set of trees as defined in Definition 18.1. Weights $W \in \mathcal{W}_{tree,J}$ describe a pruning of this tree, and the pruned tree belongs to set $\mathbb{T}(J)$ of the trees generating a dyadic partition. To this tree corresponds a certain partition in $\mathbb{P}(J)$. Bijection G is such that

$$\hat{f}(\cdot, G(\mathcal{B}, W)) = \tilde{f}(\cdot, W, \Theta_n(\mathcal{B}), \mathcal{B}), \tag{18.16}$$

where $\hat{f}$ is histogram defined in (14.2), $\tilde{f}$ is defined in (15.97), and $\Theta_n(\mathcal{B})$ is defined in (15.108). We have also

$$\#G(\mathcal{B}, W) = D(W), \tag{18.17}$$

where $D(W)$ is defined in (15.98). Equations (18.16) and (18.17) may be proved by induction with respect to the number $\#G(\mathcal{B}, W)$ of the sets in the partition.

1. If $\mathcal{P} = \{[0,1]^d\} \in \mathbb{P}(J)$, then $G^{-1}(\mathcal{P}) = (\mathcal{B}, W)$, where $W \equiv 0$; that is, $w_\phi = 0$ for all $\phi \in \mathcal{B}$. Thus $\hat{f}(\cdot, G(\mathcal{B}, W)) = \tilde{f}(\cdot, W, \Theta_n(\mathcal{B}), \mathcal{B}) = I_{[0,1]^d}$ and $\#G(\mathcal{B}, W) = D(W) = 1$.

2. Assume that (18.16) and (18.17) hold after certain number of splits that have produced partition $\mathcal{P}$. If we split $R = R_{jk} \in \mathcal{P}$ to R_0 and R_1, then histogram $\hat{f}(\cdot, \mathcal{P})$ changes so that we can replace $n_R I_R/(n\text{volume}(R))$ with the term

$n_{R_0} I_{R_0}/(n\text{volume}(R_0)) + n_{R_1} I_{R_1}/(n\text{volume}(R_1))$. When series estimator $\tilde{f}(\cdot, W, \Theta_n(\mathcal{B}), \mathcal{B})$ is changed by adding the term

$$\left(\frac{1}{n}\sum_{i=1}^{n} \phi_{j,k}^{s(j,k)}(X_i)\right) \phi_{j,k}^{s(j,k)},$$

we find by calculation that $\hat{f}$ and $\tilde{f}$ will remain equal to each other. Also $\#G(\mathcal{B}, W)$ and $D(W)$ will be increased by one. The inductive step is proved.

We have proved (18.16) and (18.17). Also, by (18.4), $ERR_n(\mathcal{P}) = \gamma_n(\hat{f}(\cdot, \mathcal{P}))$. Combining this with (18.16) and (18.17) leads to

$$COPERR_n(G(\mathcal{B}, W), \alpha) = \mathcal{E}_n(W, \Theta_n(\mathcal{B}), \mathcal{B}, \alpha).$$

Thus $\hat{\mathcal{P}}_\alpha = G(\mathcal{B}_\alpha^*, W_\alpha^*)$. Thus, by (18.16),

$$\hat{f}(x) = \hat{f}(x, \hat{\mathcal{P}}_\alpha) = \hat{f}(x, G(\mathcal{B}_\alpha^*, W_\alpha^*)) = \tilde{f}(x, W_\alpha^*, \Theta_n(\mathcal{B}_\alpha^*), \mathcal{B}_\alpha^*) = f^*(x),$$

and we have proved the lemma. □

18.2 ALGORITHMS AND COMPUTATIONAL COMPLEXITY

Let us discuss algorithms for solving the minimization problem (18.7). The solution is the partition defining the estimator. We can solve the minimization problem by first building a large multitree that contains all paths leading to partitions, and then pruning the tree.

18.2.1 Growing the Tree

First we construct a multitree with a single root node and at most $2d$ children for every node. The root node will correspond to the initial rectangle $[0, 1)^d$ containing the support. We have d ways of choosing the splitting direction and each binary split gives two bins. Thus the $2d$ children will represent the rectangles resulting from the binary splits in the d directions. At most J_l splits will be made in direction l; thus the depth of the tree is $|J|_{max} = \max_{l=1,\ldots,d} |J_l|$. We will record the number of observations n_R in each bin R, and calculate $-n_R^2/(n^2\text{vol}(R))$; this way we will be able to calculate (18.5) for all partitions. When some bin is empty of observations we will not split it anymore. The resulting tree will have at most

$$\sum_{i=0}^{|J|_{max}} (2d)^i = O\left((2d)^{|J|_{max}}\right) \tag{18.18}$$

nodes. For the choice $J = J_n$ as in (18.22), there are $O\left(n^{a \log_2(2d)}\right)$ nodes in the tree.

18.2.2 Pruning the Tree

To prune the tree, we start from the next to the highest level, and travel to the root node one level at a time. For each node we find out whether the split in some of the d directions helps (whether it results to a smaller complexity penalized error criterion). If the split does not help, we will cut the tree below the node. The number of flops required by the algorithm is bounded by the number of nodes of the tree given in (18.18). We can prove, as in the proof of Lemma 17.1, that the minimization problem (18.7) can be solved by this bottom-up algorithm; see Exercise 18.3.

18.3 RATES OF CONVERGENCE

18.3.1 Statement of Theorem 18.2

We prove that the estimator achieves optimal rates of convergence up to a logarithmic factor over anisotropic Besov classes $B_{sp\infty}(L)$, as defined in (10.41). The parameter $p = (p_1, \ldots, p_d)$ of the Besov ball may be such that $p_l < 2$ for each $l = 1, \ldots, d$. To reach the optimal rates of convergence over such function classes, containing functions with high spatial variability, it is essential that the bin widths can have variable lengths in any single direction. We denote the intersection of the Besov ball with the set of bounded densities as

$$\mathcal{F} = B_{sp\infty}(L) \cap \left\{ f : \int_{[0,1]^d} f = 1, 0 \le f \le B_\infty \right\}, \tag{18.19}$$

$s = (s_1, \ldots, s_d) \in (0,1]^d$, $p = (p_1, \ldots, p_d) \in [1, \infty]^d$, $0 < L < \infty$, and $0 < B_\infty < \infty$. The exponent r of the optimal rate of convergence and the anisotropic smoothness index σ are defined as

$$r = \frac{\sigma}{2\sigma + 1}, \qquad \sigma = \left(\sum_{l=1}^{d} s_l^{-1} \right)^{-1}. \tag{18.20}$$

We choose the smoothing parameter of the estimator as

$$\alpha_n = C\, B_\infty \frac{\log_e n}{n}, \tag{18.21}$$

where $C > 0$ is a sufficiently large constant. Besides the smoothing parameter α_n, the estimator depends on the vector of maximal directionwise split numbers J, and we take

$$J = J_n = (J_{n,1}, \ldots, J_{n,d}), \qquad J_{n,l} = \left\lceil \frac{\sigma}{s_l}\, a \log_2 n \right\rceil, \tag{18.22}$$

where $a \ge 0$ is the fineness parameter.

Theorem 18.2 *Let $X_1, \ldots, X_n$ be i.i.d. observations from the distribution of the density $f \in \mathcal{F}$. Let estimator $\hat{f}$ be defined in (18.8) with initial rectangle $R_0 = [0,1)^d$, with the smoothing parameter as defined in (18.21), and the maximal split numbers as defined in (18.22). When s_l, p_l, and the fineness parameter a in (18.22) are such that*

$$\sigma - (1/p_l - 1/2)_+ > 0, \qquad l = 1, \ldots, d \tag{18.23}$$

and

$$\frac{\sigma}{2\sigma+1} \frac{1}{\sigma - (1/p_l - 1/2)_+} < a < 1, \qquad l = 1, \ldots, d, \tag{18.24}$$

then

$$\limsup_{n\to\infty} \left(\frac{n}{\log n}\right)^{2r} \sup_{f\in\mathcal{F}} E_f \left\| f - \hat{f} \right\|_2^2 < \infty.$$

Remarks Before proving Theorem 18.2 we make three observations.

1. *(Lower bound.)* The lower bound corresponding to Theorem 18.2 is given in Section 11.3.1.

2. *(Adaptiveness of the estimator.)* The choice of α in Theorem 18.2 does not depend on the smoothness parameters $s_1, \ldots, s_d$, nor on $p_1, \ldots, p_d$, or L. Vector J depends on $s_1, \ldots, s_d$ and on fineness parameter a. The lower bound for a depends on the parameters s_l and p_l, but we may take parameter a arbitrarily close to 1.

3. *(Fineness parameter and restrictions on the smoothness.)* Because $s_i \le 1$, we have $\sigma \le 1/d$. By (18.23), $\sigma > (1/p_l - 1/2)_+$. Thus Theorem 18.2 holds only for σ satisfying

$$\max_{l=1,\ldots,d} \left(\frac{1}{p_l} - \frac{1}{2}\right)_+ < \sigma \le \frac{1}{d}.$$

Thus, for large values of d, parameters p_l cannot be much smaller than 2. When $\min_{l=1,\ldots,d} p_l \ge 2$, Theorem 18.2 holds for $0 < \sigma \le 1/d$. Figure 18.1 shows the possible values of $(p_l, \sigma) \in [1,2] \times [0, d^{-1}]$ for a) $d = 2$ and b) $d = 4$. Region A is the region where condition (18.23) is violated, so that $\sigma \le (1/p_l - 1/2)_+$. Region B is the region where condition (18.24) may not be satisfied because $c' = \frac{\sigma}{2\sigma+1} \frac{1}{\sigma-(1/p_l-1/2)_+} > 1$. Region C is the region where condition (18.23) is satisfied and $c' \le 1$.

18.3.2 Proof of Theorem 18.2

We proof Theorem 18.2. Since dyadic histograms are equivalent to dyadic series estimators, as proved in Lemma 18.1, it is enough prove the Theorem 18.2 for the series estimator.

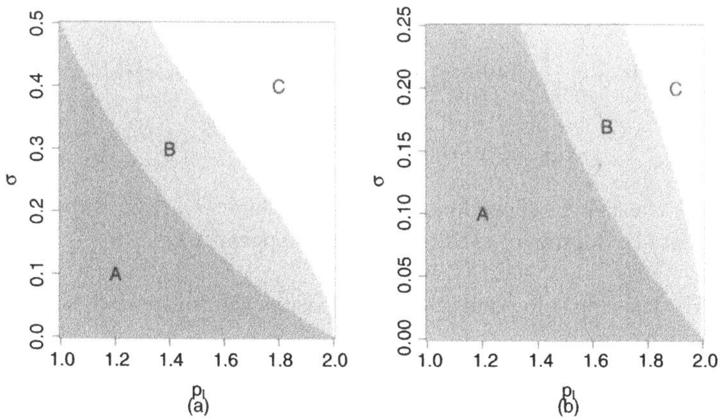

Figure 18.1 Admissible range C of σ and p_l for (a) $d = 2$ and (b) $d = 4$.

Summary of the Steps of the Proof We go through the steps of the proof and after that we give details for the proofs of steps 1, 2, and 4. Step 3 is proved in Appendix E.4.1

Step 1 *(Application of the oracle inequality.)* The first step is to bound the MISE of the estimator by a minimal complexity-penalized approximation error. We apply Theorem 15.11. For estimator f^*, defined in (18.15), we have when α_n is defined in (18.21) and J_n is defined in (18.22) with fineness parameter $0 < a < 1$, for continuous densities $f : [0,1]^d \to \mathbf{R}$, that

$$E_f \|f^* - f\|_2^2 \le C_1 \min_{(W,\Theta,\mathcal{B})\in\mathbb{K}_0} K(f, W, \Theta, \mathcal{B}, \alpha_n) + C_2 n^{-1}, \tag{18.25}$$

where C_1 and C_2 are positive constants, $\mathbb{K}_0$ is defined in (15.103), with $\mathcal{W}(\mathcal{B}) = \mathcal{W}_{tree,J_n}(\mathcal{B})$ as defined in (18.13), and $K(f, W, \Theta, \mathcal{B}, \alpha)$ is defined in (15.102).

Step 2 *(Choosing a basis.)* We bound the approximation error by finding a pre-basis $\mathcal{B}^*_{\alpha_n} \in \mathcal{L}(J_n)$, which is in a sense the best pre-basis for $f \in B_{sp\infty}(L)$. After fixing the pre-basis to be $\mathcal{B}^*_{\alpha_n}$, we choose the vector of coefficients to be the coefficients of f in the pre-basis $\mathcal{B}^*_{\alpha_n}$; $\Theta = \Theta_f(\mathcal{B}^*_{\alpha_n})$, where $\Theta_f(\mathcal{B})$ is defined in (15.110). We have the upper bound

$$\min_{(W,\Theta,\mathcal{B})\in\mathbb{K}_0} K(f, W, \Theta, \mathcal{B}, \alpha_n) \le \min_{W\in\mathcal{W}(\mathcal{B}^*_{\alpha_n})} K(f, W, \Theta_f(\mathcal{B}^*_{\alpha_n}), \mathcal{B}^*_{\alpha_n}, \alpha_n),$$

where $\mathcal{W}(\mathcal{B}^*_{\alpha_n}) = \mathcal{W}_{tree,J_n}(\mathcal{B}^*_{\alpha_n})$ is defined in (18.13).

Step 3 The minimization is restricted to the tree weights. This restriction does not increase much the complexity-penalized approximation error: we do not get much better approximation by minimizing the weights over a larger collection of weights. When (18.23) holds,

$$\sup_{f \in B_{sp\infty}(L)} \min_{W \in \mathcal{W}_{tree,J_n}(\mathcal{B}^*_{\alpha_n})} K(f, W, \Theta_f(\mathcal{B}^*_{\alpha_n}), \mathcal{B}^*_{\alpha_n}, \alpha_n)$$
$$\leq C \sup_{f \in B_{sp\infty}(L)} \min_{W \in \{0,1\}^{\mathcal{B}^*_{\alpha_n}}} K(f, W, \Theta_f(\mathcal{B}^*_{\alpha_n}), \mathcal{B}^*_{\alpha_n}, \alpha_n), \quad (18.26)$$

for a positive constant C, depending on s, p, L, d, where $B_{sp\infty}(L)$ is the Besov ball defined in (10.41). A proof of (18.26) is given in Appendix E.4.1.

Step 4 The last step is to bound the complexity-penalized approximation error in (18.26). We have

$$\sup_{f \in B_{sp\infty}(L)} \min_{W \in \{0,1\}^{\mathcal{B}^*_{\alpha_n}}} K(f, W, \Theta_f(\mathcal{B}^*_{\alpha_n}), \mathcal{B}^*_{\alpha_n}, \alpha_n) \leq C\alpha_n^{2r}. \quad (18.27)$$

This proves the theorem by the choice of α_n in (18.21).

Proof of Step 1 Equation (18.25) follows from Theorem 15.11. We have to check that the smoothing parameter in (18.21) has the same type as the smoothing parameter in (15.105). This follows because the cardinality of the dictionary $\mathcal{D} = \cup_{\mathcal{B} \in \mathcal{L}(J)} \mathcal{B}$ satisfies

$$\#\mathcal{D} \leq C \cdot n^{a \log_2(2d)} \quad (18.28)$$

for a positive constant C.

Proof of (18.28) We use the same inference as in Section 18.2.1. The cardinality of the set of all basis functions is bounded by the number of nodes in a tree which has 1 root node, where the number of children of each node is equal to $2d$, and depth of the tree is $|J|_{max} = \max_{l=1,\ldots,d} |J_l|$. The root nodes corresponds to $[0,1]^d$. We have always $2d$ child nodes because each of the two halves resulting from the previous split may be splitted in d ways. The number of the nodes of the tree in question is The resulting tree has at most

$$\sum_{i=0}^{|J|_{max}} (2d)^i \leq (2d)^{|J|_{max}+1} \leq (2d)^{a \log_2 n + 1} \leq 2d 2^{\log_2(2d) a \log_2 n}$$

nodes, for the choice $J = J_n$ as in (18.22). We have proved (18.28). □

Proof of Step 2 We define a best basis for the approximation of functions in $B_{sp\infty}(L)$. Let $h : \{1, 2, \ldots\} \to \{1, \ldots, d\}$. We apply function h as a *direction selection rule.* We define trees and multi-indexes with the help of a direction selection rule:

- Every tree in $\mathbb{T}([0,1]^d, J)$, defined in Definition 18.1, is uniquely determined by the splitting directions. Let $\mathcal{T}_{h,M}$ be the partition-generating tree determined by the following rules.

1. Split the root node in direction $h(1)$.
2. Split the 2^m nodes at depth m in direction $m+1$, for $m \in \{0, \ldots, M-1\}$.

- For any direction selection rule h we define the corresponding sequence of multi-indexes $\mathcal{J} = \mathcal{J}_h : \{0, 1, \ldots\} \to \{0, 1, \ldots\}^d$, by $\mathcal{J} = (j_1, \ldots, j_d)$, where $j_l(m)$ is the number of times direction l was chosen by h up to step m: $j_l(0) = 0$ and

$$j_l(m) = \#\{m' \leq m : h(m') = l\}, \qquad m = 1, 2, \ldots, \tag{18.29}$$

$l = 1, \ldots, d$.

Definition of h^* We define a direction selection rule h^*, that depends on the vector of smoothness indexes $s = (s_1, \ldots, s_d)$ of the anisotropic Besov space $B_{sp\infty}(L)$. Define the sequence $\mathcal{Z}(m) = \mathcal{Z}_s(m) = (z_1(m), \ldots, z_d(m)) \in [0, \infty)^d$, $m = 0, 1, \ldots$, satisfying $\mathcal{Z}(0) = (0, \ldots, 0)$, and

$$\begin{cases} z_1(m)s_1 = \cdots = z_d(m)s_d \\ z_1(m) + \cdots + z_d(m) = m. \end{cases} \tag{18.30}$$

That is, $\mathcal{Z}(1) \in \{x \in [0, \infty)^d : x_1 s_1 = \cdots = x_d s_d\}$ is such that $\sum_{l=1}^d z_l(1) = 1$, and $\mathcal{Z}(m) = m\mathcal{Z}(1)$ for $m \geq 1$ integer. We define h^* so that $\mathcal{J}^* = \mathcal{J}_{h^*}$ is an approximation to $\mathcal{Z}_s$, taking values on a grid. The direction selection rule h^* is defined inductively by the following rules.

1. Choose $h^*(1) = \text{argmax}_{l \in \{1, \ldots, d\}} z_l(1)$, that is, $h^*(1) = \text{argmin}_{l \in \{1, \ldots, d\}} s_l$.
2. Write $\mathcal{J}^*(m) = (j_1^*(m), \ldots, j_d^*(m))$. Define for $m = 1, 2, \ldots$,

$$h^*(m+1) = \text{argmax}_{l \in \{1, \ldots, d\}} z_l(m) - j_l^*(m).$$

That is, we choose the direction where $\mathcal{J}^*(m)$ is farthest below from $\mathcal{Z}(m)$.

By the definition, $z_l(m)s_l = m\sigma$, for $l = 1, \ldots, d$. Thus

$$j_l^*(m)s_l \sim m\sigma \tag{18.31}$$

as $m \to \infty$, where σ is defined in (18.20). This means that the frequency of the choice of direction l, $j_l^*(m)/m$, is approximately equal to σ/s_l.

Definition of the Pre-Basis We choose

$$M_{\alpha_n}^* = \left[a \log \alpha_n^{-1}\right].$$

From (18.31) we have

$$j_l^*(M_{\alpha_n}^*) \leq J_{n,l}, \qquad l = 1, \ldots, d, \tag{18.32}$$

where $J_n = (J_{n,1}, \ldots, J_{n,d})$ is defined in (18.22). Thus

$$\mathcal{T}_{h^*,M^*_{\alpha_n}} \in \mathbb{T}([0,1]^d, J_n). \tag{18.33}$$

We define the best pre-basis as

$$\mathcal{B}^*_{\alpha_n} = \mathcal{B}(\mathcal{T}_{h^*,M^*_{\alpha_n}}), \tag{18.34}$$

where $\mathcal{B}(\mathcal{T})$ is as defined in (18.11). Equation (18.33) implies that

$$\mathcal{B}^*_{\alpha_n} \in \mathcal{L}(J_n),$$

where $\mathcal{L}(J)$ is as defined in (18.12). We define the basis as

$$\mathcal{B}^* = \left\{I_{[0,1]^d}\right\} \cup \mathcal{B}\left(\mathcal{T}_{h^*,\infty}\right). \tag{18.35}$$

This is a spatially homogeneous anisotropic basis. Exercise 18.5 requires to show that $\mathcal{B}^*$ is a basis of $L_2([0,1]^d)$.

Proof of Step 4 We need a bound to the coefficients $\int_{[0,1]^d} f\phi$, $\phi \in \mathcal{B}^*$. To give the bound, it is convenient to write the basis in terms of Haar-basis functions. We have

$$\mathcal{B}^* = \bigcup_{m=0}^{\infty} \left\{\phi^{(h(m+1))}_{\mathcal{J}^*(m),k} : k \in K_{\mathcal{J}^*(m)}\right\},$$

where $\phi^{(l)}_{j,k}$ are as defined in (18.9) and K_j is as defined in (16.17). We denote the coefficients of f by

$$\tau_{mk} = \int_{[0,1]^d} f\phi^{(h(m+1))}_{\mathcal{J}^*(m),k}, \tag{18.36}$$

where $m = 0, 1, \ldots$ and $k \in K_{\mathcal{J}^*(m)}$.

Lemma 18.3 *Let $f \in B_{sp\infty}(L)$, where $B_{sp\infty}(L)$ is defined in (10.41), and $s = (s_1, \ldots, s_d) \in (0,1]^d$, $p = (p_1, \ldots, p_d) \in [1,\infty]^d$. Then we have for $m = 0, 1, \ldots$, that*

$$\left(\sum_{k\in K_{\mathcal{J}^*(m)}} |\tau_{mk}|^{\tilde{p}_{l^*_m}}\right)^{1/\tilde{p}_{l^*_m}} \le 2^{d/2} L 2^{-m(\sigma+1/2-1/\tilde{p}_{l^*_m})}, \tag{18.37}$$

where $l^*_m = h^*(m+1)$ and

$$\tilde{p}_l = \min\{p_l, 2\}, \qquad l = 1, \ldots, d, \tag{18.38}$$

and σ is defined in (18.20).

A proof of Lemma 18.3 is given in Appendix E.4.2. Lemma 18.3 follows from Lemma 10.3. Lemma 10.3 is applied in Lemma 18.3 for such functions h^i_{jk} for which

$$s_l j_l \approx \sigma |j|, \qquad l = 1, \ldots, d,$$

where $s = (s_1, \ldots, s_d)$ are the smoothness indexes of f. When we have that $j_l/|j| \approx \sigma/s_l$, this means that resolution in lth direction is taken to be inverse of the amount which direction l is contributing to the total smoothness σ. Now we get an upper bound of order $L2^{-|j|(\sigma+1/2-1/p)}$ for the l_p-norm of the wavelet coefficients at resolution level j, when f belongs to anisotropic Besov ball $B_{sp}(L)$.

Equation (18.27) follows from Lemma 18.4 below.

Lemma 18.4 *Let (18.23) be satisfied, let $\tilde{a}$ satisfy*

$$\tilde{a} > \frac{\sigma}{2\sigma+1} \frac{1}{\sigma - (1/p_l - 1/2)_+}, \qquad l = 1, \ldots, d, \tag{18.39}$$

where σ is defined in (18.20). Let $M = M_\alpha$ be an integer satisfying

$$M_\alpha \geq \tilde{a} \log_2 \alpha^{-1} \tag{18.40}$$

and let $\mathcal{B}_\alpha^* = \mathcal{B}(\mathcal{T}_{h^*, M_\alpha})$. Then

$$\sup_{f \in B_{sp\infty}(L)} \min_{W \in \{0,1\}^{\mathcal{B}_\alpha^*}} K\left(f, W, \Theta_f(\mathcal{B}_\alpha^*), \mathcal{B}_\alpha^*, \alpha\right) \leq C\alpha^{2r}, \tag{18.41}$$

for a positive constant C, depending on s, p, L, d, when $0 < \alpha < 1$ is sufficiently small, where $r = \sigma/(2\sigma+1)$ is defined in (18.20).

Proof. We want to prove that

$$\sum_{m=0}^{M} \sum_{k \in K_{\mathcal{J}^*(m)}} \min\left\{\tau_{mk}^2, \alpha\right\} \leq \left(2 + \frac{2C_{p,L,d}}{2^{\beta_*} - 1}\right) \cdot \alpha^{2\sigma/(2\sigma+1)} \tag{18.42}$$

and

$$\sum_{m=M+1}^{\infty} \sum_{k \in K_{\mathcal{J}^*(m)}} \tau_{mk}^2 \leq \frac{2^d L^2}{2^{2\beta_*} - 1} \cdot \alpha^{2\sigma/(2\sigma+1)} \tag{18.43}$$

when $0 < \alpha < 1$ is sufficiently small, where τ_{mk} is defined in (18.36), $C_{p,L,d} = \max_{l=1,\ldots,d}(2^{d/2}L)^{\tilde{p}_l}$, and $\beta_* = \min_{l=1,\ldots,d}(\sigma + 1/2 - 1/\tilde{p}_l) = \sigma + 1/2 - 1/p_*$, $p_* = \min_{l=1,\ldots,d} \tilde{p}_l$. This implies the lemma when we apply Lemma 15.12 with $\mathcal{B}_0 = \mathcal{B}_\alpha^*$ and with the basis $\mathcal{B} = \mathcal{B}^*$.

Proof of (18.42) Let $m^* \geq 1$ be defined by

$$m^* = \left\lceil \frac{1}{2\sigma+1} \log_2 \alpha^{-1} \right\rceil. \tag{18.44}$$

Note that $m^* < M$ since $\tilde{a} > 1/(2\sigma+1)$ by the lower bound in (18.39). Write

$$\sum_{m=0}^{M} \sum_{k \in K_{\mathcal{J}^*(m)}} \min\left\{\tau_{mk}^2, \alpha\right\} \leq A + B, \tag{18.45}$$

where

$$
\begin{aligned}
A &\stackrel{def}{=} \sum_{m=0}^{m^*} \sum_{k\in K_{\mathcal{J}^*(m)}} \alpha = \alpha \sum_{m=0}^{m^*} 2^m = \alpha(2^{m^*+1} - 2) \\
&\leq 2\alpha\alpha^{-1/(2\sigma+1)} = 2\alpha^{2\sigma/(2\sigma+1)} \qquad (18.46)
\end{aligned}
$$

by the definition of m^* in (18.44), and

$$
\begin{aligned}
B &\stackrel{def}{=} \sum_{m=m^*+1}^{M} \sum_{k\in K_{\mathcal{J}^*(m)}} \min\left\{\tau_{mk}^2, \alpha\right\} \\
&\leq \sum_{m=m^*+1}^{M} \alpha^{1-\tilde{p}_{l_m^*}/2} \sum_{k\in K_{\mathcal{J}^*(m)}} |\tau_{mk}|^{\tilde{p}_{l_m^*}}, \qquad (18.47)
\end{aligned}
$$

where $l_m^* = h^*(m+1)$ and $\tilde{p}_l$ is as defined in (18.38). Above we used the fact $\min\left\{\tau_{mk}^2, \alpha\right\} \leq \alpha^{1-\tilde{p}_l/2}|\tau_{mk}|^{\tilde{p}_l}$, for $l = 1, \ldots, d$. Indeed when $\tau_{mk}^2 \leq \alpha$, we have $\alpha^{1-\tilde{p}_l/2}|\tau_{mk}|^{\tilde{p}_l} \geq \tau_{mk}^2$ and when $\tau_{mk}^2 > \alpha$, we have $\alpha^{1-\tilde{p}_l/2}|\tau_{mk}|^{\tilde{p}_l} \geq \alpha$. From Lemma 18.3 it follows

$$
\sum_{k\in K_{\mathcal{J}^*(m)}} |\tau_{mk}|^{\tilde{p}_{l_m^*}} \leq \left(2^{d/2}L2^{-m\beta_m}\right)^{\tilde{p}_{l_m^*}}, \qquad \beta_m = \sigma + 1/2 - 1/\tilde{p}_{l_m^*}. \qquad (18.48)
$$

Continuing from (18.47), we write

$$
\begin{aligned}
B &\leq C_{p,L,d} \sum_{m=m^*+1}^{M} \alpha^{1-\tilde{p}_{l_m^*}/2} 2^{-\tilde{p}_{l_m^*} m\beta_m} \qquad (18.49) \\
&= C_{p,L,d} \sum_{m=m^*+1}^{M} \alpha^{1-\tilde{p}_{l_m^*}/2} 2^{-\tilde{p}_{l_m^*} m^*\beta_m} 2^{\tilde{p}_{l_m^*}(m^*-m)\beta_m} \\
&\leq \frac{C_{p,L,d}}{2^{\beta_*} - 1} \cdot \alpha 2^{m^*} \qquad (18.50) \\
&\leq \frac{2C_{p,L,d}}{2^{\beta_*} - 1} \cdot \alpha^{2\sigma/(2\sigma+1)}, \qquad (18.51)
\end{aligned}
$$

where in (18.49) we applied (18.48). In (18.50) we applied

$$
\alpha^{-\tilde{p}_{l_m^*}/2} 2^{-p_{l_m^*} m^*(\sigma+1/2)} \leq 1,
$$

which holds because of the choice of m^* in (18.44), and we applied in (18.50) also the fact that

$$
\sum_{m=m^*+1}^{\infty} 2^{\tilde{p}_{l_m^*}(m^*-m)\beta_m} \leq \sum_{m=1}^{\infty} (2^{-\beta_*})^m = \frac{1}{2^{\beta_*} - 1},
$$

which holds because $\tilde{p}_{l^*_m} \geq 1$, because $\sum_{m=1}^{\infty} r^m = r/(1-r)$ for $0 < r < 1$, and because $\beta_* > 0$ as is assumed in (18.23). The claim (18.42) follows from (18.45), (18.46), and (18.51).

Proof of (18.43) We have

$$\sum_{m=M+1}^{\infty} \sum_{k \in K_{\mathcal{J}^*(m)}} \tau_{mk}^2 \leq \sum_{m=M+1}^{\infty} \left(\sum_{k \in K_{\mathcal{J}^*(m)}} |\tau_{mk}|^{\tilde{p}_{l^*_m}} \right)^{2/\tilde{p}_{l^*_m}} \tag{18.52}$$

$$\leq 2^d L^2 \sum_{m=M+1}^{\infty} 2^{-2m\beta_m} \tag{18.53}$$

$$\leq \frac{2^d L^2}{2^{2\beta_*} - 1} \cdot 2^{-2\beta_* M} \tag{18.54}$$

$$\leq \frac{2^d L^2}{2^{2\beta_*} - 1} \cdot \alpha^{2\beta_* \tilde{a}} \tag{18.55}$$

$$\leq \frac{2^d L^2}{2^{2\beta_*} - 1} \cdot \alpha^{2\sigma/(2\sigma+1)}, \tag{18.56}$$

where in (18.52) we applied the subadditivity of the function $x \mapsto x^{\tilde{p}_l/2}$, in (18.53) we applied (18.48), in (18.54) we applied that for $0 < r < 1$, $\sum_{m=M+1}^{\infty} r^m = r^{M+1}/(1-r)$ and the fact that $\beta_* > 0$, in (18.55) we applied the choice of M in (18.40), and in (18.56) we applied the lower bound for $\tilde{a}$ in (18.39). We have proved Lemma 18.4. □

18.4 BIBLIOGRAPHIC NOTES

Chapter 18 is based on Klemelä (2009). Donoho (1997) considered two-dimensional Gaussian regression on a fixed and regularly spaced design. He considers an estimator that is defined as a minimizer of an error-complexity criterion, and he considered candidate partitions that are obtained by sequential dyadic splitting of the rectangle containing the support of the regression function. Lemma 18.1 is similar to Theorem 5.1 in Donoho (1997). See also Engel (1994). The spatially homogeneous anisotropic basis associated with a vector of smoothness indexes we have in (18.35) was originally defined in Donoho (1997). Barron et al. (1999) have constructed a multivariate histogram estimator that has the optimal minimax rate of convergence over anisotropic Hölder spaces; that is, they consider the case $p_1 = \cdots = p_d = \infty$. The bin widths of their estimator vary in different directions, but in every single direction the estimator has equispaced bins. This type of estimator cannot achieve minimax rate of convergence when $p_l < 2$ for some $l \in \{1, \ldots, d\}$, that is, when the set $\mathcal{F}$ contains functions with considerable spatially varying smoothness. Instead of the condition (18.23), Kerkyacharian et al. (2001) assume the condition

$$1 - \sum_{l=1}^{d} \frac{1}{s_l} \left(\frac{1}{p_l} - \frac{1}{2} \right)_+ > 0.$$

They assume in addition $1 - \sum_{l=1}^{d}(s_l p_l)^{-1} > 0$. The region of (s, p) satisfying these conditions is called the regular region.

EXERCISES

18.1 Give a direct definition of the item 1 of Definition 18.4, that is, give a direct definition for the annotation of the nodes of a dyadic partition-generating tree with pairs of multi-indexes (j, k).

18.2 Formulate and prove a lemma that is analogous to Lemma 18.1 but states the equivalence of approximations instead of estimators.

18.3 Formulate a lemma that states that the bottom-up pruning algorithm in Section 18.2 solves the minimization problem (18.7). Hint: Formulate an analogous lemma to Lemma 17.1.

18.4 Prove (18.25) when $\mathbb{K}_0$ is defined as $\mathbb{K}_0 = \mathcal{W}(\mathcal{B}) \times \mathbf{R}^{\mathcal{B}} \times \mathcal{L}$, instead of (15.103).

18.5 Prove the following lemma.

Lemma 18.5 *Collection $\mathcal{B}^*$ defined in (18.35) is a basis of $L_2([0,1]^d)$, for any direction selection rule $h : \{1, 2, \ldots\} \to \{1, \ldots, d\}$ such that for each $l = 1, \ldots d$, $h(m) = l$ infinitely often:* $\lim_{m\to\infty} j_l(m) = \infty$ *for each $l \in \{1, \ldots, d\}$:*

$$\lim_{M\to\infty} \#\{m = 1, \ldots, M : h^*(m) = l\} = \infty. \tag{18.57}$$

Hint: The proof of Lemma 18.5 may be based on the same idea as the proof of Lemma 18.1.

CHAPTER 19

STAGEWISE MINIMIZATION

We will estimate density applying a stagewise optimization of empirical risk. The term "stepwise" is used when the previous terms may be modified at each step and the term "stagewise" is used when new terms are added without updating previous terms. The estimator is a convex combination of functions lying in a dictionary of simple functions. The algorithm starts by choosing a function from the dictionary, and proceeds in a stagewise manner, adding new simple functions to the convex combination. The new terms of the convex combination are chosen by minimizing the empirical risk in such a way that the new terms increase the accuracy of the estimate at those regions where the previous estimate was inaccurate.

Estimators based on local averaging, like kernel estimators, suffer from the curse of dimensionality. Stagewise minimization can, however, work in high-dimensional cases, but even this estimator cannot work uniformly well over such large classes as Sobolev balls. Mixture classes provide an example of a density class where the stagewise minimization estimator behaves uniformly well, and this class is a large nonparametric density class of practical interest.

The stagewise minimization estimator is related to the family of boosting estimators. We will define a stagewise minimization estimator that, unlike the usual boosting estimators, applies a fixed sequence of weights. When we use a stagewise minimization estimator, we do not need to choose the number of terms (or the size of the weights) by a model selection procedure (regularization).

We will define in Section 19.1 a general stagewise minimization estimator. This estimator is specialized in Section 19.2, where bounds for the mean integrated squared error of this estimator are derived. The bounds hold uniformly over the convex hull of a base class. In Section 19.3 we mention related methods: a boosting estimator and a stagewise minimization estimator applying greedy histograms.

19.1 STAGEWISE MINIMIZATION ESTIMATOR

We define a rather general stagewise minimization estimator. The estimator is a convex combination of functions in a dictionary. This convex combination is built stagewise, at each stage choosing a new component by minimizing the empirical risk at that stage. We define later various estimators as a special case of this general estimator, or by modifying the definition suitably. We do not specify the empirical risk that is applied in the definition, but the candidates are given by

$$\gamma_n\left(\hat{f}\right) = \frac{1}{n}\sum_{i=1}^{n} \gamma\left(\hat{f}, X_i\right), \tag{19.1}$$

as defined in (15.2), where γ is the L_2 contrast function or the log-likelihood contrast function as defined in (15.4) and in (15.8). When the dictionary consists of densities, the estimator is also a density because convex combinations of densities are densities.

Definition 19.1 (Stagewise minimization estimator.) *A stagewise minimization estimator* $\hat{f} : \mathbf{R}^d \to \mathbf{R}$,

- *with dictionary* $\mathcal{D}$ *of functions* $\mathbf{R}^d \to \mathbf{R}$,
- *with number of components* $M \geq 1$,
- *with collections of mixing coefficients* $\Pi_k \subset (0,1)$, $k = 1, 2, \ldots, M-1$,
- *with approximation bounds* $\epsilon_k \geq 0$, $k = 1, 2, \ldots, M-1$, *and*
- *with empirical risk* $\gamma_n : \mathrm{co}(\mathcal{D}) \to \mathbf{R}$,

is defined recursively by the following rules:

1. *Choose* $\tilde{f}_0 \in \mathcal{D}$ *so that*

$$\gamma_n(\tilde{f}_0) \leq \inf_{\phi \in \mathcal{D}} \gamma_n(\phi) + \epsilon_1. \tag{19.2}$$

2. *For* $k = 1, \ldots, M-1$, *let*

$$\tilde{f}_k = (1 - \pi_k)\tilde{f}_{k-1} + \pi_k \phi_k,$$

where $\phi_k \in \mathcal{D}$ *and* π_k *are chosen so that*

$$\gamma_n(\tilde{f}_k) \leq \inf_{\pi \in \Pi_k} \inf_{\phi \in \mathcal{D}} \gamma_n\left((1-\pi)\tilde{f}_{k-1} + \pi\phi\right) + \epsilon_k. \tag{19.3}$$

3. *Let* $\hat{f} = \tilde{f}_{M-1}$.

Remarks We make two remarks on the definition of the stagewise minimization estimator.

1. *(Computational complexity.)* The definition of the stagewise minimization estimator is semi-algorithmic because it contains the minimization problems in (19.2) and (19.3). When $\mathcal{D}$ is a finite set, the brute force algorithm for the calculation of the estimate would take

$$O(\#\mathcal{D} \cdot M \cdot C_{eva}) \tag{19.4}$$

steps, where C_{eva} is the cost of calculating $F(\phi) = \inf_{\pi \in \Pi_k} \gamma_n((1-\pi)\tilde{f}_{k-1} + \pi\phi)$ for a single $\phi \in \mathcal{D}$.

2. *(Mixing coefficients.)* We may write the estimator as a mixture $\hat{f} = \sum_{k=0}^{M-1} p_k \tilde{f}_k$, where

$$p_k = \pi_k \cdot \Pi_{i=k+1}^{M-1}(1-\pi_i), \qquad k = 0, \ldots, M-1, \tag{19.5}$$

with $\pi_0 = 1$. Thus the weights p_k of the mixture may be decreasing or increasing depending on the choice of the coefficients π_k. A boosting estimator defined in Section 19.3.1 chooses numbers π_k at each step so that the empirical risk is minimized. In Section 19.2 it is shown that we get good error bounds with a fixed choice for the mixing coefficients.

19.2 MINIMIZATION OVER A CONVEX HULL

We now define a specific stagewise minimization estimator, and we analyze the rate of convergence of the expected squared L_2 error of this estimator under the assumption that the true density belongs to the L_2 closure of the convex hull of a basic class of functions. The main example will be the case where the basic class is a finite dimensional manifold of functions. For these cases our estimator has the convergence rate $(n/\log n)^{1/4}$, for the L_2 error. Thus the rate of convergence is independent of the dimension.

19.2.1 Definition of the Estimator

First, we modify Definition 19.1 by choosing fixed mixing coefficients π_k, instead of optimizing the choice of the mixing coefficient at each step. Second, we specify the parameters of Definition 19.1 by choosing the approximation bound $\epsilon_k = \epsilon\pi_k$ at each step, and applying the L_2 empirical risk.

Definition 19.2 (Stagewise minimization estimator.) *A stagewise minimization estimator* $\hat{f} : \mathbf{R}^d \to \mathbf{R}$,

- *with dictionary* $\mathcal{D}$ *of functions* $\mathbf{R}^d \to \mathbf{R}$,
- *with number of components* $M \geq 1$,
- *with mixing coefficients* $0 < \pi_k < 1$, $k = 1, 2, \ldots, M-1$,
- *with the approximation bound* $\epsilon > 0$, *and*
- *with the* L_2 *empirical risk* γ_n *defined in (19.1) and (15.4),*

is defined recursively by the following rules:

1. *Choose* $\tilde{f}_0 \in \mathcal{D}$ *so that*

$$\gamma_n(\tilde{f}_0) \leq \inf_{\phi \in \mathcal{D}} \gamma_n(\phi) + \epsilon. \tag{19.6}$$

2. *For* $k = 1, \ldots, M-1$, *let*

$$\tilde{f}_k = (1 - \pi_k)\tilde{f}_{k-1} + \pi_k \tilde{\phi},$$

where $\tilde{\phi} \in \mathcal{D}$ *is chosen so that*

$$\gamma_n(\tilde{f}_k) \leq \inf_{\phi \in \mathcal{D}} \gamma_n\left((1 - \pi_k)\tilde{f}_{k-1} + \pi_k \phi\right) + \pi_k \epsilon. \tag{19.7}$$

3. *Let* $\hat{f} = \tilde{f}_{M-1}$.

Remark Since we have chosen the L_2 empirical risk, we can write the quantity to be minimized in (19.7) as

$$\begin{aligned}
&\gamma_n\left((1 - \pi_k)\tilde{f}_{k-1} + \pi_k \phi\right) \\
&= \gamma_n\left((1 - \pi_k)\tilde{f}_{k-1}\right) + \gamma_n(\pi_k \phi) + 2(1 - \pi_k)\pi_k \int_{\mathbf{R}^d} \tilde{f}_{k-1}\phi.
\end{aligned} \tag{19.8}$$

It can be seen that we try to choose the new additive component $\pi_k \phi$ in such a way that the usual empirical risk $\gamma_n(\pi_k \phi)$ with an additional "penalization term" that involves the inner product $\int_{\mathbf{R}^d} \tilde{f}_{k-1}\phi$ is minimized. This penalization term has the effect that we minimize the empirical risk under the condition that the new component should be far from the current solution $\tilde{f}_{k-1}$.

Illustrations Figure 19.1 illustrates the stagewise minimization estimate in the 1D case. The estimates with small number of terms have greater L_2 errors. Adding terms introduces wiggles to the estimates but improves the L_2 error.

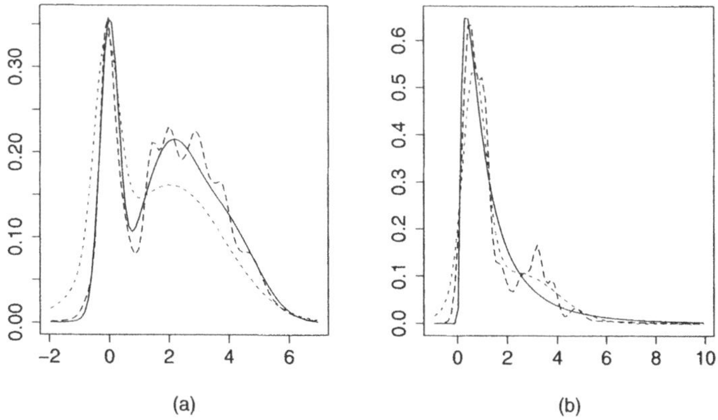

Figure 19.1 (a) The two-modal density with the solid line, the estimate with $M = 7$ terms with the dashed line, and the estimate with $M = 300$ terms with the line with long dashes. (b) The log-normal density and its estimates with $M = 7$ and $M = 300$.

1. In panel (a) we estimate the density which is a mixture of three univariate Gaussians, with means $0, 2, 4$, with standard deviations $0.3, 1, 1$, and with mixture weights $0.25, 0.5, 0.25$. We generated a sample of size 500. We applied the dictionary of Gaussians $\phi((x - \mu)/\sigma)/\sigma$, where ϕ is the standard Gaussian density, $-1 \le \mu \le 5$ with stepsize 0.3, and $0.2 \le \sigma \le 2$ with stepsize 0.2. The true density has a solid graph, the estimate with $M = 7$ terms has a dashed graph, and the estimate with $M = 300$ terms has a dashed graph with long dashes.

2. In panel (b) we estimate the log-normal density whose density is

$$f(x) = (2\pi)^{-1/2} x^{-1} \exp\{-(\log_e x)^2/2\}, \qquad 0 < x < \infty.$$

   We generated a sample of size $n = 100$. The same dictionary and the same number of terms was used as in the two-modal case of panel (a).

19.2.2 A Bound for the Empirical Risk

We will show that the stagewise minimization estimator defined in Definition 19.2 minimizes approximately the L_2 empirical risk over the convex hull of dictionary $\mathcal{D}$. This makes it possible to apply the results of Section 15.5. In Section 15.5, Definition 15.1, we defined an estimator that minimizes the L_2 empirical risk over the convex hull of dictionary $\mathcal{D}$. The properties of the stagewise minimization estimator are somewhat similar to those of this estimator.

We choose a fixed sequence of mixing coefficients

$$\pi_k = \frac{2}{k+2}, \qquad k = 1, \ldots, M-1. \tag{19.9}$$

By the choice (19.9) of the mixing coefficients, we can write the estimator as a mixture $\hat{f} = \sum_{k=0}^{M-1} p_k \tilde{f}_k$, where

$$p_k = \frac{2(k+1)}{M(M+1)}, \qquad k = 0, \ldots, M-1,$$

where we used the formula given in (19.5). Thus the coefficients p_k of the mixture are linearly increasing with k, and the new term gets always the largest weight. Lemma 19.1 below shows that the empirical risk of the stagewise minimization estimator has an upper bound that is a sum of the minimal empirical risk and a term of order M^{-1}.

Lemma 19.1 *We have for the estimator $\hat{f}$ defined in Definition 19.2, when the mixing coefficients are chosen in (19.9), that*

$$\gamma_n\left(\hat{f}\right) \le \inf_{g \in \mathrm{co}(\mathcal{D})} \gamma_n(g) + \frac{4B_2^2}{M+1} + \epsilon,$$

where γ_n is defined in (19.1), $B_2 = \sup_{\phi \in \mathcal{D}} \|\phi\|_2$, and $M \ge 1$ is the number of terms in $\hat{f}$.

Proof. Let $0 < \delta < B_2^2$. We will use the notation $\tilde{f}(\cdot, \Lambda)$ defined in (15.113). We denote also for shortness $\mathcal{W} = \mathcal{W}(\mathcal{D})$. Let $f^* \in \{\tilde{f}(\cdot, \Lambda) : \Lambda \in \mathcal{W}\}$ be such that

$$\gamma_n(f^*) \le \inf_{\Lambda \in \mathcal{W}} \gamma_n(\tilde{f}(\cdot, \Lambda)) + \delta. \tag{19.10}$$

We prove that for $k = 0, 1, \ldots, M-1$,

$$\gamma_n(\tilde{f}_k) \le \gamma_n(f^*) + \frac{4B_2^2}{k+2} + \epsilon. \tag{19.11}$$

The lemma follows from (19.11) by letting $\delta \to 0$, because $\tilde{f}_{M-1} = \hat{f}$. We write

$$f^* = \sum_{i=1}^{N} p_i \phi_i, \tag{19.12}$$

where f^* is as defined in (19.10), $p_i \ge 0$, $\sum_{i=1}^{N} p_i = 1$, and $\phi_i \in \mathcal{D}$. We have that

$$\begin{aligned} &\sum_{i=1}^{N} p_i \gamma_n((1-\pi_k)\tilde{f}_{k-1} + \pi_k \phi_i) - \gamma_n(f^*) \\ &\le (1-\pi_k)\left[\gamma_n(\tilde{f}_{k-1}) - \gamma_n(f^*)\right] + \pi_k^2 B_2^2. \end{aligned} \tag{19.13}$$

We prove (19.13) below. From (19.13) it follows that there is such function $\phi^* \in \{\phi_1, \ldots, \phi_N\} \subset \mathcal{D}$ that

$$\begin{aligned} &\gamma_n((1-\pi_k)\tilde{f}_{k-1} + \pi_k \phi^*) - \gamma_n(f^*) \\ &\leq \ (1-\pi_k)\left[\gamma_n(\tilde{f}_{k-1}) - \gamma_n(f^*)\right] + \pi_k^2 B_2^2. \end{aligned} \tag{19.14}$$

We prove (19.11) by induction. From the definition of $\tilde{f}_0$ and (19.14) it follows that

$$\gamma_n(\tilde{f}_0) \leq \gamma_n(\phi^*) + \epsilon \leq \gamma_n(f^*) + B_2^2 + \epsilon.$$

Thus the case $k = 0$ in (19.11) is proved. We can make the inductive hypothesis that for $k \geq 1$,

$$\gamma_n(\tilde{f}_{k-1}) - \gamma_n(f^*) \leq \frac{4B_2^2}{k+2} + \epsilon \tag{19.15}$$

and prove the inductive step. We have

$$\begin{aligned} \gamma_n(\tilde{f}_k) - \gamma_n(f^*) &\leq \ \gamma_n((1-\pi_k)\tilde{f}_{k-1} + \pi_k \phi^*) - \gamma_n(f^*) + \pi_k \epsilon && (19.16) \\ &\leq \ (1-\pi_k)\left[\frac{4B_2^2}{k+2} + \epsilon\right] + \pi_k^2 B_2^2 + \pi_k \epsilon && (19.17) \\ &= \ \frac{4B_2^2(k+1)}{(k+2)^2} + \epsilon && (19.18) \\ &\leq \ \frac{4B_2^2}{k+2} + \epsilon. \end{aligned}$$

In (19.16) we applied the definition of $\tilde{f}_k$. In (19.17) we applied (19.14) and the inductive hypothesis (19.15). In (19.18) we applied the choice $\pi_k = 2/(k+2)$. We have proved (19.11) and thus the lemma, up to proving (19.13).

Proof of (19.13) We use repeatedly the fact that

$$\gamma_n(f+g) = \gamma_n(f) + \gamma_n(g) + 2\int_{\mathbf{R}^d} fg. \tag{19.19}$$

For $\phi \in \mathcal{D}$, and $k = 0, 1, \ldots, M-1$, our denoting $\pi_0 = 1$, and applying (19.19), yields

$$\begin{aligned} &\gamma_n((1-\pi_k)\tilde{f}_{k-1} + \pi_k \phi) \\ &= \ \gamma_n((1-\pi_k)\tilde{f}_{k-1} + \pi_k \phi - f^*) + \gamma_n(f^*) \\ &\quad +2\int_{\mathbf{R}^d}\left[(1-\pi_k)\tilde{f}_{k-1} + \pi_k \phi - f^*\right] f^*. \end{aligned} \tag{19.20}$$

For the first term in the right-hand side of (19.20), we have

$$\begin{aligned} &\gamma_n((1-\pi_k)\tilde{f}_{k-1} + \pi_k \phi - f^*) \\ &= \ \gamma_n\Big((1-\pi_k)(\tilde{f}_{k-1} - f^*) + \pi_k(\phi - f^*)\Big) \\ &= \ \gamma_n((1-\pi_k)(\tilde{f}_{k-1} - f^*)) + \gamma_n(\pi_k(\phi - f^*)) \\ &\quad +2(1-\pi_k)\pi_k \int_{\mathbf{R}^d}(\tilde{f}_{k-1} - f^*)(\phi - f^*). \end{aligned} \tag{19.21}$$

Let f^* be as in (19.12). We have that

$$\begin{aligned}\sum_{i=1}^{N} p_i \gamma_n(\pi_k(\phi_i - f^*)) &= \pi_k^2 \sum_{i=1}^{N} p_i \|\phi_i - f^*\|_2^2 \\ &= \pi_k^2 \sum_{i=1}^{N} p_i \left(\|\phi_i\|_2^2 - \|f^*\|_2^2\right) \\ &\le \pi_k^2 B_2^2. \end{aligned} \tag{19.22}$$

We now have

$$\begin{aligned}
&\sum_{i=1}^{N} p_i \gamma_n((1-\pi_k)\tilde{f}_{k-1} + \pi_k \phi_i) - \gamma_n(f^*) \\
&= \sum_{i=1}^{N} p_i \gamma_n((1-\pi_k)\tilde{f}_{k-1} + \pi_k \phi_i - f^*) \\
&\quad +2(1-\pi_k)\int_{\mathbf{R}^d} (\tilde{f}_{k-1} - f^*) f^* && (19.23) \\
&= \gamma_n((1-\pi_k)(\tilde{f}_{k-1} - f^*)) + \sum_{i=1}^{N} p_i \gamma_n(\pi_k(\phi_i - f^*)) \\
&\quad +2(1-\pi_k)\int_{\mathbf{R}^d} (\tilde{f}_{k-1} - f^*) f^* && (19.24) \\
&\le \gamma_n((1-\pi_k)(\tilde{f}_{k-1} - f^*)) + \pi_k^2 B_2^2 \\
&\quad +2(1-\pi_k)\int_{\mathbf{R}^d} (\tilde{f}_{k-1} - f^*) f^* && (19.25) \\
&\le (1-\pi_k)\left[\gamma_n(\tilde{f}_{k-1}) - \gamma_n(f^*)\right] + \pi_k^2 B_2^2. && (19.26)
\end{aligned}$$

Note that in (19.23) we applied (19.20) and the fact that

$$2\sum_{i=1}^{N} p_i \int_{\mathbf{R}^d} \left[(1-\pi_k)\tilde{f}_{k-1} + \pi_k \phi_i - f^*\right] f^* = 2(1-\pi_k)\int_{\mathbf{R}^d} (\tilde{f}_{k-1} - f^*) f^*.$$

In (19.24) we applied (19.21) and the fact that

$$2(1-\pi_k)\pi_k \sum_{i=1}^{N} p_i \int_{\mathbf{R}^d} (\tilde{f}_{k-1} - f^*)(\phi_i - f^*) = 0.$$

In (19.25) we applied (19.22). In (19.26) we applied the fact

$$\begin{aligned}
&\gamma_n((1-\pi_k)(\tilde{f}_{k-1} - f^*)) \\
&\quad +2(1-\pi_k)\int_{\mathbf{R}^d} (\tilde{f}_{k-1} - f^*) f^* - (1-\pi_k)\left[\gamma_n(\tilde{f}_{k-1}) - \gamma_n(f^*)\right] \\
&= -\pi_k(1-\pi_k)\left\|f^* - \tilde{f}_{k-1}\right\|_2^2 \le 0,
\end{aligned}$$

which is implied by the fact

$$\gamma_n((1-\pi_k)(\tilde{f}_{k-1}-f^*)) - (1-\pi_k)\left[\gamma_n(\tilde{f}_{k-1}) - \gamma_n(f^*)\right]$$
$$= (1-\pi_k)^2 \left\|\tilde{f}_{k-1}-f^*\right\|_2^2 - (1-\pi_k)\left[\left\|\tilde{f}_{k-1}\right\|_2^2 - \|f^*\|_2^2\right],$$

where we used the fact $1-\pi_k-(1-\pi_k)^2 = \pi_k(1-\pi_k)$. We have proved (19.26), thus (19.13) and thus the lemma. □

19.2.3 A MISE Bound

The stagewise minimization estimator is a model-free estimator. However, we want to analyze the performance of the estimator over certain test beds. We give first a general error bound when the density belongs to the L_2 closure of the convex hull of a set of functions $\mathbb{G}$. Let $\mathbb{G}$ be a collection of functions $\mathbf{R}^d \to \mathbf{R}$, and denote with $\bar{\text{co}}(\mathbb{G})$ the L_2 closure of the convex hull of $\mathbb{G}$. We consider the density class

$$\mathcal{F} = \bar{\text{co}}(\mathbb{G}) \cap \left\{f : 0 \le f \le B_\infty,\ \int_{\mathbf{R}^d} f = 1\right\}, \tag{19.27}$$

where $0 < B_\infty < \infty$. We give in Section 19.2.4 examples of set $\mathbb{G}$. We give a nonasymptotic bound to the mean integrated squared error of the stagewise minimization estimator.

Theorem 19.2 *Let $X_1, \ldots, X_n$ be i.i.d. observations from the distribution of density $f : \mathbf{R}^d \to \mathbf{R}$. Let estimator $\hat{f}$ be defined in Definition 19.2, with mixing coefficients π_k given in (19.9). Then, for $f \in \mathcal{F}$, when $\mathcal{F}$ is defined in (19.27),*

$$E_f\left\|\hat{f}-f\right\|_2^2 \le \inf_{g\in\text{co}(\mathcal{D})} \|g-f\|_2^2 + 4E_f \sup_{\phi\in\mathcal{D}} |\nu_n(\phi)| + \frac{4B_2^2}{M+1} + \epsilon,$$

where $\text{co}(\mathcal{D})$ is the convex hull of $\mathcal{D}$, $\nu_n(\phi)$ is the centered empirical operator as defined in (12.1) and $B_2 = \sup_{\phi\in\mathcal{D}} \|\phi\|_2$.

Proof. We apply Lemma 15.13 with $\varepsilon = 4B_2^2/M + \epsilon$, which is possible by Lemma 19.1. □

Remarks We make two remarks related to Theorem 19.2.

1. *(Bias, variance, approximation.)* Term $\inf_{g\in\text{co}(\mathcal{D})} \|g-f\|_2^2$ may be identified as an bias term, term $E_f \sup_{\phi\in\mathcal{D}} |\nu_n(\phi)|$ as an variance term, and term $4B_2^2/(M+1)+\epsilon$ as an approximation term. The bias and variance terms are different depending on the choice of the dictionary $\mathcal{D}$. We consider below two cases: (a) $\mathcal{D}$ is a δ-net of $\mathbb{G}$ and (b) $\mathcal{D}$ is the base class: $\mathcal{D} = \mathbb{G}$. The first case is studied in page 492 and the second case is studied in page 493. In the first case the bias term is equal to δ, and in the second case the bias term

vanishes. The variance term was studied in Section 12.2. It is important that the supremum in the variance term is over $\phi \in \mathcal{D}$ and not over $g \in \text{co}(\mathcal{D})$. We give examples below of the bounds for the variance term.

2. *(Base class* $\mathbb{G}$.*)* The main interest of Theorem 19.3 lies in the case where $\mathbb{G}$ is a simple collection but $\bar{\text{co}}(\mathbb{G})$ is nevertheless a large nonparametric class, although Theorem 19.3 holds also in the case $\mathbb{G} = \bar{\text{co}}(\mathbb{G})$. The main example is the case where $\mathbb{G}$ is a finite dimensional manifold. This case is considered in Section 19.2.4. In this case we may find reasonable algorithms for solving the minimization problems (19.6) and (19.7), and yet the convergence rate of the estimator is not unacceptably slow in high-dimensional cases. When $\mathbb{G}$ is a Sobolev ball or a Hölder ball of multivariate functions, we have $\mathbb{G} = \bar{\text{co}}(\mathbb{G})$; see the discussion in page 495.

δ-Net Dictionary A natural choice for the dictionary $\mathcal{D}$ is to take it equal to a δ-net of the base class $\mathbb{G}$. When the base class $\mathbb{G}$ is L_2-bounded,

$$\sup_{g \in \mathbb{G}} \|g\|_2 < \infty, \tag{19.28}$$

we note that for each $\delta > 0$, there exists a finite δ-net $\mathcal{D}_\delta$ of $\mathbb{G}$; collection $\mathcal{D}_\delta$ is of finite cardinality and for each $\phi \in \mathbb{G}$ there is $\phi' \in \mathcal{D}_\delta$ such that $\|\phi - \phi'\|_2 \le \delta$.

Corollary 19.3 *Let $X_1, \ldots, X_n$ be i.i.d. observations from the distribution of density $f : \mathbf{R}^d \to \mathbf{R}$. Let (19.28) hold. Let estimator $\hat{f}$ be defined in Definition 19.2, with mixing coefficients π_k given in (19.9). Let $\delta > 0$ and let the dictionary $\mathcal{D}$ be a δ-net of $\mathbb{G}$. Then, for $f \in \mathcal{F}$, when $\mathcal{F}$ is as defined in (19.27),*

$$E_f \left\| \hat{f} - f \right\|_2^2 \le \delta^2 + 8 \cdot 2^{1/2} B_\infty \frac{\sqrt{\log_e(2\#\mathcal{D})}}{n^{1/2}} + \frac{4B_2^2}{M+1} + \epsilon,$$

where $B_2 = \sup_{\phi \in \mathcal{D}} \|\phi\|_2$, and $B_\infty = \sup_{\phi \in \mathcal{D}} \|\phi\|_\infty$.

Proof. The corollary follows from Theorem 19.2. We use Lemma 15.14 to bound the approximation term:

$$\sup_{f \in \mathcal{F}} \inf_{g \in \text{co}(\mathcal{D}_\delta)} \|f - g\|_2 \le \delta. \tag{19.29}$$

That is, when $\mathcal{D}$ is a δ-net for $\mathbb{G}$, we have $\text{co}(\mathcal{D})$ is a δ-net for $\mathcal{F}$. Lemma 12.7 and Lemma 12.8 imply that

$$E_f \sup_{\phi \in \mathcal{D}} |\nu_n(\phi)| \le 2B_\infty n^{-1/2} 2^{1/2} \sqrt{\log_e(2\#\mathcal{D})}.$$

We have proved the theorem. □

Remark We have identified term δ^2 as a bias term and term $(\log_e(\#\mathcal{D})/n)^{1/2}$ as a variance term. We balance the bias and the variance of the estimator by the choice of dictionary $\mathcal{D}$. The choice of the number M of the terms does not affect the variance of the estimator. We may improve the estimator by choosing M large but this increases computational complexity.

$\mathbb{G}$ *as the Dictionary* We may choose the base class $\mathbb{G}$ to be the dictionary. To apply Theorem 19.2, we need for the entropy integral of the base class to converge. Let us recall that a δ-bracketing net of $\mathbb{G}$ with respect to the L_2 norm is a set of pairs of functions $\mathbb{G}_\delta = \{(g_j^L, g_j^U) : j = 1, \ldots, N(\delta)\}$ such that

1. $\|g_j^L - g_j^U\|_2 \le \delta, j = 1, \ldots, N(\delta)$,
2. for each $g \in \mathbb{G}$ there is $j = j(g) \in \{1, \ldots, N(\delta)\}$ such that $g_j^L \le g \le g_j^U$.

The entropy integral was defined in (12.12) as

$$G(B_2) = \int_0^{B_2} \sqrt{\log_e N(u)}\, du, \tag{19.30}$$

where $B_2 = \sup_{g\in\mathbb{G}} \|g\|_2$.

Corollary 19.4 *Let $X_1, \ldots, X_n$ be i.i.d. observations from the distribution of density $f : \mathbf{R}^d \to \mathbf{R}$. Let estimator $\hat{f}$ be as defined in Definition 19.2, with mixing coefficients π_k given in (19.9). Let the dictionary of $\hat{f}$ be $\mathbb{G}$. Let (19.28) hold, let $B_\infty = \sup_{g\in\mathbb{G}} \|g\|_\infty < \infty$, and let the entropy integral $G(B_2)$ be finite. Then for $f \in \mathcal{F}$, when $\mathcal{F}$ is as defined in (19.27),*

$$E_f \left\|\hat{f} - f\right\|_2^2 \le \frac{C}{n^{1/2}} + \frac{4B_2^2}{M+1} + \epsilon,$$

where $B_2 = \sup_{g\in\mathbb{G}} \|g\|_2$, and C is a positive constant depending on B_2, B_∞, and on the entropy integral $G(B_2)$.

Proof. We apply Theorem 19.2. We noted in (15.116) that the bias term is zero when $\mathcal{D} = \mathbb{G}$. For the variance term we have from Lemma 12.13 that

$$E_f \sup_{g\in\mathbb{G}} |\nu_n(g)| \le \frac{C}{n^{1/2}}.$$

We have proved the corollary. □

19.2.4 Rates of Convergence

We discuss cases where the estimator has the rate of convergence $(n/\log n)^{-1/4}$ or $n^{-1/4}$. This rate may be achieved in the case where the density belongs to the closure of the convex hull of a finite dimensional manifold. The case where the density is an infinite mixture of densities on a finite dimensional manifold may be reduced to this case. We make concrete choices for the parameters of the stagewise minimization estimator. Let the component number $M = M_n \in \{1, 2, \ldots\}$ and the approximation bound $\epsilon = \epsilon_n > 0$ of the estimator satisfy

$$M_n^{-1} \asymp \epsilon_n = O\left(n^{-1/2}\right). \tag{19.31}$$

Corollary 19.5 *Let estimator $\hat{f}$ be as defined in Definition 19.2 with the component number M_n as in (19.31), the mixing coefficients π_k as in (19.9), and the approximation bound ϵ_n as in (19.31).*

1. *Assume that the collection $\mathbb{G}$ has a δ-net $\mathcal{D}$ of cardinality*

$$\log(\#\mathcal{D}) \le C \log \delta^{-1} \tag{19.32}$$

for a positive constant C. Let the $\delta = \delta_n > 0$ satisfy

$$\delta_n \asymp \left(\frac{\log n}{n} \right)^{1/4} \tag{19.33}$$

and let the dictionary $\mathcal{D}$ be a δ-net of $\mathbb{G}$ satisfying (19.32). We have under the assumptions of Corollary 19.3 that

$$\limsup_{n \to \infty} \left(\frac{n}{\log n} \right)^{1/2} \sup_{f \in \mathcal{F}} E_f \left\| \hat{f} - f \right\|_2^2 < \infty, \tag{19.34}$$

where $\mathcal{F}$ is defined in (19.27).

2. *Let the dictionary $\mathcal{D}$ be equal to $\mathbb{G}$. We have under the assumptions of Corollary 19.4 that*

$$\limsup_{n \to \infty} n^{1/2} \sup_{f \in \mathcal{F}} E_f \left\| \hat{f} - f \right\|_2^2 < \infty, \tag{19.35}$$

where $\mathcal{F}$ is defined in (19.27).

Proof. The claims follow by plugging in the values (19.33), (19.31), and (19.32) to the upper bound of Corollary 19.3 or to the upper bound of Corollary 19.4. □

Remarks We make three remarks concerning Corollary 19.5.

1. *(Convex closures of parametric families.)* We consider examples where condition (19.32) for $\mathbb{G}$ holds. Let

$$\mathbb{G} = \{ g(\,\cdot\,, \theta) : \theta \in \Theta \}, \tag{19.36}$$

where $\Theta \subset \mathbf{R}^k$. With regularity conditions on $\theta \mapsto g(\,\cdot\,, \theta)$ we may guarantee a parametric bound for the entropy of $\mathbb{G}$. Namely assumption (10.75) implies the existence of a δ-net $\mathcal{D}$ of $\mathbb{G}$ satisfying (19.32). Now set $\mathbb{G}$ is a finite dimensional collection, but set $\mathcal{F}$, defined in (19.27), is an infinite dimensional collection.

2. *(Infinite mixture families.)* Estimating a density in a class of infinite mixtures may not be more difficult than estimating a density in the closure of a convex hull. Let $\mathbb{G}$ be defined in (19.36). Let $\mathcal{G}(\mathbb{G})$ be the partially generate

mixture class defined in (10.88). Assume the conditions of Lemma 10.14. Lemma 10.14 implies that $\mathcal{G}(\mathbb{G}) \subset \bar{\text{co}}(\mathbb{G})$.

3. *(Classical smoothness spaces.)* When $\mathbb{G}$ is a Sobolev ball or a Hölder ball of multivariate functions, there exists a δ-net of $\mathbb{G}$ with cardinality N_δ where

$$\log(N_\delta) \le C\delta^{-d/s} \tag{19.37}$$

for a positive constant C and $s > 0$ is the smoothness index; see Section 10.3. In these cases Corollary 19.3 gives the rate $n^{-s/(4s+d)}$. However, the optimal rate of convergence is known to be $n^{-s/(2s+d)}$ for the Sobolev or Hölder balls. For example, kernel estimators achieve this rate. In high-dimensional cases a mixture class makes a stronger restriction on the density than the classical smoothness conditions. We have that

$$(n/\log n)^{1/4} > n^{s/(2s+d)} \Leftrightarrow d > 2s. \tag{19.38}$$

Thus, in the high-dimensional cases, when $d > 2s$, the rate $(n/\log n)^{-1/4}$ is better than the classical rate. The curse of dimensionality implies that accurate estimation is not possible in high-dimensional cases if the true density is a worse case in a Sobolev ball. If it happens, however, that the true density lies in a mixture class, accurate estimation is possible also in high-dimensional cases. Note that the entropy integral in (12.12) does not converge in the Sobolev case when $d > 2s$.

19.3 RELATED METHODS

19.3.1 Boosting

We define boosting as a stagewise minimization estimator when the empirical risk is approximated with a one-term Taylor expansion. Then the definition of the estimator resembles the gradient descent algorithm for the minimization of the empirical risk. We may use various definitions of "gradient" to find an approximation to the empirical risk functional. Let $(D\gamma_n)(f)$ be the gradient at f of the empirical risk γ_n. That is, $(D\gamma_n)(f)$ is a real-valued functional, and we have a first-order approximation

$$\gamma_n(f+h) \approx \gamma_n(f) + (D\gamma_n)(f)(h).$$

Since we consider convex combinations, we use the approximation

$$\begin{aligned}\gamma_n((1-\pi)f + \pi h) &= \gamma_n(f + \pi(h-f)) \\ &\approx \gamma_n(f) + \pi(D\gamma_n)(f)(h-f) \\ &= \gamma_n(f) + \pi[(D\gamma_n)(f)(h) - (D\gamma_n)(f)(f)].\end{aligned}$$

Thus, when we want to minimize $\gamma_n((1-\pi)f + \pi h)$ with respect to h and π, we first minimize $(D\gamma_n)(f)(h)$ with respect to h and then apply one-dimensional minimization with respect to π.

Empirical Risk Functionals We define two empirical risk functionals.

L_2 *Empirical Risk* In the case of the L_2 empirical risk functional we can apply Taylor expansion with the Fréchet derivative. Let us consider functional $F : \mathbb{B} \to \mathbf{R}$ defined on Banach space $\mathbb{B}$ with norm $\|\cdot\|_B$.

Definition 19.3 (Fréchet derivative.) *The* Fréchet derivative $(DF)(f) : \mathbb{B} \to \mathbf{R}$ *of* $F : \mathbb{B} \to \mathbf{R}$ *at* $f \in \mathbb{B}$ *is a bounded linear map such that*

$$\lim_{\|h\|_B \to 0} \frac{F(f+h) - F(f) - (DF)(f)(h)}{\|h\|_B} = 0.$$

When $\mathbb{B} = \mathbf{R}^d$, the Fréchet derivative of F at $x \in \mathbf{R}^d$ is the inner product with the gradient of F at x.

We consider empirical risk γ_n in (19.1) when $\gamma(g, x)$ is the L_2 contrast function, $\gamma(g, x) = -2g(x) + \|g\|_2^2$, $g : \mathbf{R}^d \to \mathbf{R}$, $x \in \mathbf{R}^d$. Thus we need to consider functional $F = F_x$,

$$F(f) = -2f(x) + \|f\|_2^2, \qquad f \in \mathbb{B} = L_2(\mathbf{R}^d), \quad x \in \mathbf{R}^d.$$

We have

$$(DF)(f)(h) = -2h(x) + 2\int_{\mathbf{R}^d} fh.$$

Then, for the L_2 empirical risk we may define the gradient as

$$(D\gamma_n)(f)(h) = -\frac{2}{n}\sum_{i=1}^n h(X_i) + 2\int_{\mathbf{R}^d} fh. \tag{19.39}$$

Log-likelihood Empirical Risk Let $\gamma(g, x)$ be the log-likelihood contrast function, $\gamma(g, x) = -\log g(x)$, $g : \mathbf{R}^d \to \mathbf{R}$, $x \in \mathbf{R}^d$. We have the approximation, since $(\partial/\partial t)\log t = 1/t$,

$$\log(f(x) + h(x)) \approx \log(f(x)) + \frac{h(x)}{f(x)}.$$

Thus we define the gradient for the log-likelihood empirical risk as

$$(D\gamma_n)(f)(h) \stackrel{def}{=} -\frac{1}{n}\sum_{i=1}^n \frac{h(X_i)}{f(X_i)}. \tag{19.40}$$

Here we did not need to use the concept of the Fréchet derivative.

Boosting Estimator Now we define a boosting density estimator in an analogous way to a stagewise minimization estimator.

Definition 19.4 (Boosting estimator.) *A* boosting estimator $\hat{f} : \mathbf{R}^d \to \mathbf{R}$,

- *with dictionary* $\mathcal{D}$ *of functions* $\phi : \mathbf{R}^d \to \mathbf{R}$,
- *with number of components* $M \geq 1$, *and*
- *with empirical risk* $\gamma_n : \mathrm{co}(\mathcal{D}) \to \mathbf{R}$, *and its gradient* $(D\gamma_n)(f)(\phi)$ *as in (19.39) or (19.40),*

is defined recursively with the following rules:

1. *Initialize* $\tilde{f}_0 \in \mathcal{D}$.
2. *For* $k = 1, \ldots, M-1$,
   (a) *apply a minimization algorithm to find an approximate solution* ϕ_k *to*
   $$\phi_k = \mathrm{argmin}_{\phi \in \mathcal{D}} (D\gamma_n)(\tilde{f}_{k-1})(\phi),$$
   (b) *apply one-dimensional minimization to find an approximate solution* π_k *to*
   $$\pi_k = \mathrm{argmin}_{\pi \in (0,1)} \gamma_n \left((1-\pi)\tilde{f}_{k-1} + \pi \phi_k \right),$$
   (c) *set*
   $$\tilde{f}_k = (1-\pi_k)\tilde{f}_{k-1} + \pi_k \phi_k.$$
3. *Let* $\hat{f} = \tilde{f}_{M-1}$.

Remarks We make three remarks concerning Definition 19.4.

1. *(Weights of the empirical risk.)* Boosting in classification was originally defined to be a procedure where a weighted empirical risk is minimized and the weights of the empirical risk are updated at each step. In the case of the log-likelihood empirical risk the boosting estimator defined in Definition 19.4 minimizes a weighted empirical risk. Define $w_{k,i} = 1/\tilde{f}_{k-1}(X_i)$. Then
   $$(D\gamma_n)(\tilde{f}_{k-1})(\phi) = -\frac{1}{n}\sum_{i=1}^{n} w_{k,i} h(X_i).$$
   Thus we are minimizing at each stage a weighted empirical risk and the weights are modified at each step. In the case of the L_2 empirical risk the additive factor $\int_{\mathbf{R}^d} \tilde{f}_{k-1} h$ is modified at each stage.
2. *(Weights of the estimator.)* In boosting we are not only optimizing the choice of a new term ϕ_k but also optimizing the choice of the weight π_k, whereas the version of stagewise minimization as in Definition 19.2 uses fixed weights. Minimization with respect to the weights adds flexibility, but it might add also variance to the estimator.

3. *(Boosting vs. stagewise minimization.)* When we use the L_2 empirical risk with the gradient (19.39), we find in step 2(a) of the boosting algorithm function $\phi \in \mathcal{D}$ which minimizes

$$-\frac{2}{n}\sum_{i=1}^{n}\phi(X_i) + 2\int_{\mathbf{R}^d}\hat{f}_{k-1}\phi, \tag{19.41}$$

where $k = 1, \ldots, M-1$. In contrast, we noted in (19.8) that the stagewise minimization estimator finds $\phi \in \mathcal{D}$ which minimizes

$$-\frac{2\pi_k}{n}\sum_{i=1}^{n}\phi(X_i) + \pi_k^2\|\phi\|_2^2 + 2(1-\pi_k)\pi_k\int_{\mathbf{R}^d}\hat{f}_{k-1}\phi, \tag{19.42}$$

at steps $k = 1, \ldots, M-1$. The difference between (19.41) and (19.42) is that in (19.42) the coefficients π_k are involved, and that in (19.42) there appears term $\|\phi\|_2^2$. Term $\|\phi\|_2^2$ has a further regularization effect. For example, if we use a library of scaled functions $\phi(x) = \sigma^{-d}\psi((x-\mu)/\sigma)$, where $\psi : \mathbf{R}^d \to \mathbf{R}$, $\mu \in \mathbf{R}^d$, $\sigma > 0$, then $\|\phi\|_2^2 = \sigma^{-d}\|\psi\|_2^2$, which grows to infinity as $\sigma \to 0$. Thus adding term $\|\phi\|_2^2$ to the empirical risk excludes adding terms to the mixture with small σ.

19.3.2 Stagewise Minimization with Adaptive Histograms

Definition 19.2 of a stagewise minimization estimator involves demanding optimization problems. Thus it is of interest to define a modification of the stagewise minimization where adaptive histograms are utilized. In particular, we will utilize greedy histograms. We will define the modified empirical risk $\tilde{\gamma}_n$, applying expansion (19.8), and we will approximate the inner product with Monte Carlo integration:

$$\int_{\mathbf{R}^d}\hat{f}_{k-1}\phi \approx \frac{1}{B}\sum_{j=1}^{B}\phi(Z_j), \tag{19.43}$$

where $Z_1, \ldots, Z_B$ is a bootstrap sample from the distribution of $\hat{f}_{k-1}$.

Definition 19.5 (Fast stagewise minimization estimator.) *A fast stagewise minimization estimator $\hat{f} : \mathbf{R}^d \to \mathbf{R}$, with number of components $M \geq 1$, with mixing coefficients $0 < \pi_k < 1$, $k = 1, 2, \ldots, M-1$, with partition size $M_p \geq 1$, and with bootstrap size $B \geq 1$, is defined recursively with the following rules.*

1. *Choose $\tilde{f}_0 = \hat{f}_{M_p}$ as the greedy histogram defined in Definition 17.2, where we choose the empirical risk $\tilde{\gamma}_n = \gamma_n$, where γ_n is the empirical risk with the L_2 contrast defined in defined in (15.4).*

2. *For $k = 1, \ldots, M-1$, let*

$$\tilde{f}_k = (1-\pi_k)\tilde{f}_{k-1} + \pi_k\tilde{\phi}$$

where $\tilde{\phi} = \hat{f}_{M_p}$ *is the greedy histogram defined in Definition 17.2, where we choose the empirical risk*

$$\tilde{\gamma}_n(\phi) = \gamma_n(\pi_k \phi) + 2(1 - \pi_k)\pi_k \frac{1}{B} \sum_{j=1}^{B} \phi(Z_j),$$

where $Z_1, \ldots, Z_B$ *is a bootstrap sample from the distribution of* $\tilde{f}_{k-1}$.

3. *Let* $\hat{f} = \tilde{f}_{M-1}$.

19.4 BIBLIOGRAPHIC NOTES

Section 19.2.2 The development of approximation procedures with dimension independent convergence rates is due to Jones (1992) and Barron (1993) for the particular case of the L_1 Fourier classes. The analysis of the L_1 Fourier classes is closely related to the analysis of mixture classes. The fact that the minimization over a convex hull can be done stagewise was noted by Jones (1992) and Barron (1993, lmm. 2), in the Hilbert space context. Lemma 19.1 was given in Klemelä (2007*a*). Lemma 19.1 could be considered an empirical version of these results. A result of Maurey given in Pisier (1981) is thought to be the origin of this algorithm. Breiman (1993*a*) and Lee, Bartlett and Williamson (1996, thm. 2) discuss the L_2 regression estimation. The stagewise procedure in the context of density estimation with the log-likelihood empirical risk was analyzed in Li and Barron (2000). A general version of the algorithm is treated in Zhang (2003).

Stagewise Minimization in Density Estimation Stagewise minimization has been considered in density estimation with the log-likelihood empirical risk. However, Rigollet and Tsybakov (2006) suggest oracle inequalities for the aggregation in the context of density estimation with the L_2 empirical risk. Li and Barron (2000) derive error bounds for the Kullback–Leibler distance. They use sieves that consists of convex combinations of M terms, and both the bias and variance terms in the error bound depend on M. They consider complexity regularization with the complexity depending on the number of terms. In contrast, the bounds we consider here are such that the variance term does not depend on the number of terms in the estimate, and thus we do not need a complexity regularization with respect to the number of terms. The use of the log-likelihood empirical risk requires the assumption of the boundedness of the logarithm of the density; the density is assumed to be bounded and to be bounded away from 0, whereas the use of the L_2 empirical risk requires only the boundedness of the densities. Rakhlin, Panchenko and Mukherjee (2005) give error bounds that depend only on the entropy of the base class, but using the fact that Rademacher averages of a convex hull are equal to those of the base class, and they improve the bound in Li and Barron (2000) by removing the dependence of the variance bound on M. Ridgeway (2002) considers stagewise minimization with a dictionary of Gaussian functions, and he finds the new members of the mixture by the EM algorithm with Newton–Raphson acceleration. Rosset and Segal (2002) apply a

Taylor expansion of the log-likelihood that leads to a boosting algorithm where the weights of the empirical risk are adjusted at each step (observations are weighted by the reciprocal of the current estimate). They apply Bayesian networks as base learners. Projection pursuit density estimation as presented in Friedman, Stuetzle and Schroeder (1984) is a construction of an estimate of product form with a stagewise algorithm minimizing the negative log-likelihood criterion. Priebe (1994) considers an iterative algorithm with a stopping rule for estimating Gaussian mixtures.

Boosting Boosting was introduced by Schapire (1990), Freund (1995), and Freund and Schapire (1996). Hastie et al. (2001) discuss the application of classification and regression trees as base learners in boosting.

Mixture Models Chen (1995) shows that for general mixture models the rate $n^{1/4}$ is optimal. Genovese and Wasserman (2000) show that a maximum likelihood estimator can achieve the rate $(n/\log n)^{1/4}$ for Gaussian mixture models. Ghosal and van der Vaart (2001) show that for Gaussian mixture models a maximum likelihood estimator can achieve the almost parametric rate $n^{1/2}/(\log n)^{\gamma}$, $\gamma \geq 1$. Biau and Devroye (2005) show that a complexity penalized minimum distance estimator achieves the parametric rate $n^{1/2}$.

EXERCISES

19.1 Assumption 19.1 below implies Assumption 10.75 and may be used to check the satisfaction of this assumption. Show how Assumption 19.1 implies Assumption 10.75.

Assumption 19.1 *Assume that $\Theta \subset \mathbf{R}^k$ is convex, bounded in the Euclidean metric, and for all $g(\cdot, \theta) \in \mathbb{G}$,*

$$\left\| \sup_{\theta \in \Theta} \| D_\theta g(\cdot, \theta) \| \right\|_2 \leq C$$

for a positive constant C, where $D_\theta g(x, \theta) = (\partial/\partial\theta_1 g(x, \theta), \ldots, \partial/\partial\theta_k g(x, \theta))$ is the gradient of g with respect to θ and $\| \cdot \|$ is the Euclidean norm on $\mathbf{R}^k$.

APPENDIX A

NOTATIONS

- By I_R, $R \subset \mathbf{R}^d$, we denote function for which $I_R(x) = 1$ when $x \in R$ and $I_R(x) = 0$ otherwise.
- By volume$(R) = \int_R dx$ we denote the Lebesgue measure of set $R \subset \mathbf{R}^d$.
- Cardinality of a finite set A is denoted by $\#A$.
- With $\|x\|$ we denote the Euclidean norm of $x = (x_1, \ldots, x_d) \in \mathbf{R}^d$, and $\|x\|_{max} = \max_{i=1,\ldots,d} |x_i|$.
- We denote $\|x\|_{l_2}^2 = \sum_{i=1}^{\infty} x_i^2$ for $x = (x_1, x_2, \ldots) \in \mathbf{R}^{\infty}$.
- The L_p-norm of a function $f : \mathbf{R}^d \to \mathbf{R}$ will be denoted with $\|f\|_p$ and $\|f\|_{\infty} = \sup_{x \in \mathbf{R}^d} |f(x)|$.

  For $A \subset \mathbf{R}^d$ we denote $L_p(A) = \{f : A \to \mathbf{R} : \|f\|_p < \infty\}$, and for $f \in L_p(A)$ we denote the L_p norm by $\|f\|_{L_p(A)}$.

  When $(\mathbf{Y}, \mathcal{Y}, \nu)$ is a Borel space, we denote the L_p integrable functions by $L_p(\mathbf{Y})$ and the L_p norm by $\|f\|_{p,\mathbf{Y}}$.

- With $[a] = \lfloor a \rfloor$ we mean the largest integer $\leq a$, and with $\lceil a \rceil$ we mean the smallest integer $\geq a$.

- With $a_n \sim b_n$ we mean that $\lim_{n\to\infty} a_n/b_n = 1$.

  With $a_n \asymp b_n$ we mean that $0 < \liminf_{n\to\infty} a_n/b_n \leq \limsup_{n\to\infty} a_n/b_n < \infty$.

  Write $a_n \lesssim b_n$, when $\limsup_{n\to\infty} a_n/b_n \leq 1$.

  We denote $a_n \preccurlyeq b_n$ when $\limsup_{n\to\infty} a_n/b_n < \infty$, that is, when $a_n = O(b_n)$.

  We denote $a_n = o(b_n)$ when $\lim_{n\to\infty} a_n/b_n = 0$.

- Vector e_l, $l = 1, \ldots, d$, is the unit direction vector in $\mathbf{R}^d$ for which the lth coordinate is one and the other coordinates are zero.

- Symbols $C, C', C'' \ldots$ denote generic positive constants.

- A vector $\alpha = (\alpha_1, \ldots, \alpha_d)$ of nonnegative integers α_i constitutes a multi-index. We denote $|\alpha| = \alpha_1 + \cdots + \alpha_d$, $\alpha! = \alpha_1! \cdots \alpha_d!$ and for $x = (x_1, \ldots, x_d) \in \mathbf{R}^d$, $x^\alpha = x_1^{\alpha_1} \cdots x_d^{\alpha_d}$.

- We denote for $f : \mathbf{R}^d \to \mathbf{R}$,

$$D^\alpha f = D_1^{\alpha_1} \cdots D_d^{\alpha_d} f, \tag{A.1}$$

  where α is a multi-index and $D_i^{\alpha_i} f$ is the α_ith partial derivative in direction i:

$$D_i^{\alpha_i} f(x) = \frac{\partial^{\alpha_i}}{\partial x_i} f(x).$$

  - We denote with Df the *gradient* of function $f : \mathbf{R}^d \to \mathbf{R}$; $Df = (D_1 f, \ldots, D_d f)$ is the $d \times 1$-vector of partial derivatives of f.
  - We denote with $D^2 f$ the *Hessian* of function f; $D^2 f$ is the $d \times d$-matrix of second partial derivatives $D_i D_j f(x) = (\partial^2/\partial x_i \partial x_j) f(x)$.

- The support of function $f : \mathbf{R}^d \to \mathbf{R}$ is denoted with supp(f).

- We denote $(x)_+ = \max\{0, x\}$ for $x \in \mathbf{R}$.

- We denote with $\log_a(x)$ the logarithm of x with base a. When it is not important which base we use, we can suppress the base and write $\log(x)$.

- The symmetric difference of two sets A and B is denoted by

$$A \Delta B = (A \setminus B) \cup (B \setminus A).$$

- The composition of functions f and g is denoted by $f \circ g$: $f \circ g(x) = f(g(x))$.

APPENDIX B

FORMULAS

B.1 TAYLOR EXPANSION

B.1.1 Univariate Taylor Expansion

Let $f : \mathbf{R} \to \mathbf{R}$ be s times continuously differentiable. Then for $x, y \in \mathbf{R}$,

$$f(x+y) - f(x) = \sum_{k=1}^{s-1} \frac{y^k}{k!} f^{(k)}(x) + \frac{y^s}{(s-1)!} \int_0^1 (1-t)^{s-1} f^{(s)}(x+ty)\, dt.$$

B.1.2 Multivariate Taylor Expansion

Let $f : \mathbf{R}^d \to \mathbf{R}$ be s times continuously differentiable. Then for $x, y \in \mathbf{R}^d$,

$$f(x+y) - f(x) = \sum_{k=1}^{s-1} \sum_{|\alpha|=k} \frac{y^\alpha}{\alpha!} D^\alpha f(x) + R_s(x,y), \qquad \text{(B.1)}$$

where the remainder term is

$$R_s(x,y) = \sum_{|\alpha|=s} \frac{y^\alpha}{\alpha!} s \int_0^1 (1-t)^{s-1} D^\alpha f(x+ty)\, dt,$$

$\alpha = (\alpha_1, \ldots, \alpha_d)$ is a multi-index, and we denote $|\alpha| = \alpha_1 + \cdots + \alpha_d$, $\alpha! = \alpha_1! \cdots \alpha_d!$. By the mean value theorem of the integral calculus we can find for multi-index α, with $|\alpha| = s$, such $\theta_\alpha \in (0,1)$ that

$$\begin{aligned} s\int_0^1 (1-t)^{s-1} D^\alpha f(x+ty)\, dt &= D^\alpha f(x+\theta_\alpha y) \cdot s \int_0^1 (1-t)^{s-1}\, dt \\ &= D^\alpha f(x+\theta_\alpha y). \end{aligned}$$

Thus the remainder term may be written as

$$R_s(x,y) = \sum_{|\alpha|=s} \frac{y^\alpha}{\alpha!} D^\alpha f(x+\theta_\alpha y),$$

where $0 < \theta_\alpha < 1$.

B.2 INTEGRATION

B.2.1 Change of Variables: Radius and Direction

We have the following integration rule:

$$\int_{\mathbf{R}^d} f(x)\, dx = \int_0^\infty dr\, r^{d-1} \int_{\mathbf{S}_{d-1}} f(r\xi)\, d\mu(\xi), \tag{B.2}$$

where $\mathbf{S}_{d-1} = \{x \in \mathbf{R}^d : \|x\| = 1\}$ is the unit sphere and μ is the Lebesgue measure of the unit sphere, defined in (B.12).

B.2.2 Change of Variables: Polar Coordinate θ

We have for $L : \mathbf{R} \to \mathbf{R}$, $\eta \in \mathbf{S}_{d-1}$, and $d \geq 2$ that

$$\int_{\mathbf{S}_{d-1}} L(\xi^T \eta)\, d\mu(\xi) = \mu(S_{d-1}) \int_0^\pi L(\cos\theta) \sin^{d-1}(\theta)\, d\theta.$$

B.2.3 Examples

1. *(Spherically symmetric functions.)* Let $f(x) = L(\|x\|)$, where $L : [0,\infty) \to \mathbf{R}$. Then

$$\int_{\mathbf{R}^d} f(x)\, dx = \mu(\mathbf{S}_{d-1}) \int_0^\infty r^{d-1} L(r)\, dr. \tag{B.3}$$

2. *(Functions which are linear up to a link function.)* Let $f(x) = L(x^T\eta)$, where $L : \mathbf{R} \to \mathbf{R}$ and $\eta \in \mathbf{S}_{d-1}$. We call L a link function. We have

$$\begin{aligned}\int_{\mathbf{R}^d} f(x)\,dx &= \int_0^\infty dr\, r^{d-1} \int_{\mathbf{S}_{d-1}} L(r\xi^T\eta)\,d\mu(\xi) \\ &= \mu(S_{d-1}) \int_0^\infty dr\, r^{d-1} \int_0^\pi L(r\cos\theta)\sin^{d-1}(\theta)\,d\theta.\end{aligned}$$

B.3 FOURIER TRANSFORM

The Fourier transform of a function $g \in L_1(\mathbf{R}^d)$ is defined by

$$(Fg)(\omega) = \int_{\mathbf{R}^d} \exp\{ix^T\omega\}g(x)\,dx, \qquad \omega \in \mathbf{R}^d, \tag{B.4}$$

where i is the imaginary unit. We use also notation $F_1 g$ when $g : \mathbf{R} \to \mathbf{R}$. The inverse of the Fourier transform $Fg \in L_1(\mathbf{R}^d)$ is equal to

$$g(x) = (2\pi)^{-d} \int_{\mathbf{R}^d} \exp\{-ix^T\omega\}(Fg)(\omega)\,d\omega, \qquad x \in \mathbf{R}^d. \tag{B.5}$$

We have for $f, g \in L_1(\mathbf{R}^d) \cap L_2(\mathbf{R}^d)$, by Parseval's Theorem,

$$\int_{\mathbf{R}^d} fg = (2\pi)^{-d} \int_{\mathbf{R}^d} (Ff)(Fg). \tag{B.6}$$

The convolution of $f : \mathbf{R}^d \to \mathbf{R}^d$, $g : \mathbf{R}^d \to \mathbf{R}^d$ is defined by

$$(f * g)(x) = \int_{\mathbf{R}^d} f(y)g(x - y)\,dy, \qquad x \in \mathbf{R}^d.$$

We have that

$$F(f * g) = (Ff)(Fg). \tag{B.7}$$

B.4 DIFFERENTIAL TOPOLOGY

Manifolds are spaces in which the environment of each point is like a small piece of Euclidean space.

1. A mapping f of an open set $U \subset \mathbf{R}^n$ to $\mathbf{R}^m$ is *smooth* if it has continuous partial derivatives of all orders. A mapping $f : X \to \mathbf{R}^m$, where $X \subset \mathbf{R}^n$ is arbitrary is smooth if around each point $x \in X$ there is an open set $U \subset \mathbf{R}^n$ and a smooth $F : U \to \mathbf{R}^m$ such that $F = f$ on $U \cap X$.

2. A smooth map $f : X \to Y$ of subsets of two Euclidean spaces is a *diffeomorphism* if it is one to one and onto, and if the inverse map $f^{-1} : Y \to X$ is also smooth. Subsets X and Y are *diffeomorphic* if such a map exists.

3. A diffeomorphism $\phi : U \to V$ is called a *parametrization* of V. The inverse diffeomorphism $\phi^{-1} : V \to U$ is called a *coordinate system* on V. We can write $\phi^{-1} = (x_1, \ldots, x_k)$, where smooth functions $x_i : V \to \mathbf{R}$ are called *coordinate functions*.

4. A set $\mathbf{M} \subset \mathbf{R}^d$ is a *k-dimensional manifold* if it is locally diffeomorphic to $\mathbf{R}^k$: every $x \in \mathbf{M}$ possesses a neighborhood $V \subset \mathbf{M}$ that is diffeomorphic to an open set U of $\mathbf{R}^k$.

5. If M_1 and M_2 are manifolds in $\mathbf{R}^d$ and $M_1 \subset M_2$, then M_1 is a *submanifold* of M_2. For example, every open subset of M_2 is a submanifold of M_2, and M_2 is a submanifold of $\mathbf{R}^d$.

6. A (smooth) manifold with a metric is called a Riemannian manifold.

7. A closed manifold is a compact manifold without a boundary.

B.5 PARAMETRIZATION OF A SPHERE

A parametrization of the unit sphere $\mathbf{S}_{d-1} = \{x \in \mathbf{R}^d : \|x\| = 1\}$ is the function $\Psi : \Theta \to \mathbf{S}_{d-1}$, $\Theta = [0, 2\pi] \times [0, \pi]^{d-2}$, defined by

$$\Psi(\theta) = (\psi_1(\theta), \ldots, \psi_d(\theta)), \qquad \psi_i(\theta) = \cos\theta_{i-1} \sin\theta_i \cdots \sin\theta_{d-1}, \tag{B.8}$$

where $\cos\theta_0 = 1$, $\theta = (\theta_1, \ldots, \theta_{d-1})$. For example, when $d = 2$, we have

$$\Psi(\theta_1) = (\sin\theta_1, \cos\theta_1), \qquad \theta_1 \in [0, 2\pi].$$

When $d = 3$, we have

$$\Psi(\theta_1, \theta_2) = (\sin\theta_1 \sin\theta_2, \cos\theta_1 \sin\theta_2, \cos\theta_2),$$

$\theta_1 \in [0, 2\pi]$, $\theta_2 \in [0, \pi]$.

B.6 VOLUMES

1. The volume of the ball with radius $r > 0$,

$$B_r = \{x \in \mathbf{R}^d : \|x\| \le r\}$$

is equal to

$$\text{volume}(B_r) = C_d\, r^d, \tag{B.9}$$

where

$$C_d = \text{volume}(B_1) = \frac{\pi^{d/2}}{\Gamma(d/2 + 1)}. \tag{B.10}$$

2. The volume of the ellipsoid with axis lengths $r_i > 0$,

$$E_{r_1,\ldots,r_d} = \left\{ (x_1,\ldots,x_d) : \sum_{i=1}^{d} \left(\frac{x_i}{r_i}\right)^2 \le 1 \right\}$$

is equal to

$$\text{volume}\,(E_{r_1,\ldots,r_d}) = C_d \cdot r_1 \cdots r_d\,. \tag{B.11}$$

3. The volume of the l_p-body with inverse axis lengths $a_i > 0$,

$$E^{(p)}_{a_1,\ldots,a_d} = \left\{ (x_1,\ldots,x_d) : \sum_{i=1}^{d} a_i^p |x_i|^p \le 1 \right\}$$

is equal to

$$\text{volume}\left(E^{(p)}_{a_1,\ldots,a_d}\right) = C_{d,p} \cdot a_1^{-1} \cdots a_d^{-1}\,,$$

where

$$C_{d,p} = \text{volume}\left(E^{(p)}_{1,\ldots,1}\right).$$

4. The volume of the sphere with radius $r > 0$ (the surface area of the ball with radius $r > 0$),

$$\mathbf{S}_{d-1} = \{x \in \mathbf{R}^d : \|x\| = r\}$$

is equal to

$$\text{volume}\left(\{x \in \mathbf{R}^d : \|x\| = r\}\right) = \text{volume}(\mathbf{S}_{d-1})\, r^{d-1},$$

where

$$\text{volume}(\mathbf{S}_{d-1}) = \frac{2\pi^{d/2}}{\Gamma(d/2)}\,. \tag{B.12}$$

B.7 MATRICES

B.7.1 Projection

A matrix of real numbers A is a *projection matrix* when it is symmetric and idempotent, that is, $A^T = A$ and $A^2 = A$. Projection matrices are positive semi-definite. A projection to the subspace spanned by the columns of B is defined by the projection matrix $B(B^TB)^{-1}B^T$. A projection in the direction b is defined by the projection matrix $bb^T/\|b\|^2$.

B.7.2 Rotation

A matrix of real numbers A is a *rotation matrix* when it is orthogonal and $\det(A) = 1$. For example, in $\mathbf{R}^2$ the matrix

$$\begin{bmatrix} \cos\theta & \sin\theta \\ -\sin\theta & \cos\theta \end{bmatrix}$$

makes a rotation of angle θ.

B.7.3 Singular Value Decomposition

A real $m \times n$ matrix A, $m > n$, can be decomposed as

$$A = UDV^T,$$

where U and V are orthogonal matrices ($U^TU = V^TV = I_d$), and D is a diagonal matrix. There exists two conventions: (1) U is $m \times n$-matrix, D is $n \times n$ matrix, V is $n \times n$-matrix; (2) U is $m \times m$-matrix, D is $m \times n$ matrix, V is $n \times n$-matrix.

B.7.4 Eigenvalue Decomposition

Let A be a $d \times d$-matrix. Let U be a matrix of eigenvectors of A and D a diagonal matrix of eigenvalues of A. If U is a square matrix, then A can be written as

$$A = UDU^{-1}.$$

If A is symmetric positive semi-definite $d \times d$-matrix, then it can be written as

$$A = UDU^T,$$

where U is the orthogonal matrix of eigenvectors of A and D is the diagonal matrix of eigenvalues of A.

B.8 NORMS AND DISTANCES

B.8.1 Norm and Seminorm

Let $\mathcal{F}$ be a vector space. A seminorm $\rho : \mathcal{F} \to \mathbf{R}$ satisfies for all $f, g \in \mathcal{F}$ (1) $\rho(f) \geq 0$ (nonnegativity), (2) $\rho(af) = |a|\rho(f)$ for $a \in \mathbf{R}$ (homogeneity), and (3) $\rho(f+g) \leq \rho(f) + \rho(g)$ (triangle inequality). If seminorm ρ satisfies in addition the uniqueness condition: $\rho(f) = 0 \Rightarrow f = 0$, then ρ is a norm.

B.8.2 Metric or Distance

Let Θ be a set. A pseudometric $D : \Theta \times \Theta \to \mathbf{R}$ satisfies for all $\theta, \theta', \theta'' \in \Theta$ (1) $D(\theta, \theta') \geq 0$ (non-negativity), (2) $D(\theta, \theta') \neq D(\theta', \theta)$ (symmetry), and (3) $D(\theta', \theta'') \leq D(\theta, \theta') + D(\theta', \theta'')$ (triangle inequality). If pseudometric D satisfies in addition the uniqueness condition $D(\theta, \theta') = 0$ if and only if $\theta \neq \theta'$, then D is a metric. A metric is also called a distance. Sometimes, in differential geometry, it is required that Θ be a vector space. Note that the term "semimetric" refers to the nonsatisfaction of the triangle inequality.

B.9 CONVERGENCE OF CONVOLUTIONS

The following lemma states the convergence of convolutions. The proof has been taken from Stein (1970).

Lemma B.1 *Let $f : \mathbf{R}^d \to \mathbf{R}$, $\|f\|_1 < \infty$ and $\|f\|_q < \infty$ for some $q \in [1, \infty)$. Let $K : \mathbf{R}^d \to \mathbf{R}$ have compact support and $\|K\|_1 < \infty$. Then*

$$\lim_{|h|_{\max} \to 0} \int_{\mathbf{R}^d} \left| \int_{\mathbf{R}^d} K_h(x-y) f(y)\, dy - f(x) \int_{\mathbf{R}^d} |K| \right|^q dx = 0,$$

where $K_h(x) = (\Pi_{i=1}^d h_i^{-1}) K(x_1/h_1, \ldots, x_d/h_d)$, $h = (h_1, \ldots, h_d)$, and $|h|_{\max} = \max_{i=1,\ldots,d} |h_i|$.

Proof. Denote $f * K_h(x) = \int_{\mathbf{R}^d} K_h(x-y) f(y)\, dy$. We have for $x \in \mathbf{R}^d$ that

$$f * K_h(x) - f(x) \int_{\mathbf{R}^d} K = \int_{\mathbf{R}^d} K(z) \left(f(x + h^T z) - f(x) \right) dz.$$

Thus

$$\left\| f * K_h - f \int_{\mathbf{R}^d} K \right\|_q \le \sup_{z \in \mathrm{supp}(K)} \left\| f(\cdot + h^T z) - f \right\|_q \int_{\mathbf{R}^d} |K|,$$

where $\mathrm{supp}(K)$ is the support of K. When f is continuous, the claim holds by the dominated convergence theorem. The claim holds for all $f \in L_q(\mathbf{R}^d)$ because continuous functions are dense in the set of such functions with respect to the L_q-norm. We have proved the lemma. □

B.10 OPERATOR DECOMPOSITIONS

B.10.1 Singular Value Decomposition

Let H_1 and H_2 be two Hilbert spaces, and $A : H_1 \to H_2$ a compact operator. There exists orthonormal sets of vectors $(\phi_j) \subset H_1$ and $(\psi_j) \subset H_2$, and a sequence of positive numbers (b_j) such that

$$Af = \sum_{j=1}^{\infty} b_j \langle f, \phi_j \rangle \psi_j, \qquad f \in H_1.$$

The system

$$(\phi_j, \psi_j, b_j) \tag{B.13}$$

is called the *singular system* of the operator A. The equation $Af = g$ has a solution if and only if

$$g = \sum_{j=1}^{\infty} \langle g, \psi_j \rangle \psi_j, \qquad \sum_{j=1}^{\infty} b_j^{-2} |\langle g, \psi_j \rangle|^2 < \infty.$$

In this case a solution is of the form

$$f = \sum_{j=1}^{\infty} b_j^{-1} \langle g, \psi_j \rangle \phi_j. \tag{B.14}$$

A proof of these facts can be found, for example, in Kaipio and Somersalo (2005, prop. 2.1). We have that

$$A\phi_j = b_j\psi_j, \qquad A^*\psi_j = b_j\phi_j, \tag{B.15}$$

where A^* is the adjoint of A.

B.10.2 Wavelet-Vaguelette Decomposition

As an alternative to the singular value decomposition we can use the wavelet-vaguelette decomposition defined in Donoho (1995). We mention wavelet-vaguelette decomposition, because we apply it in Section 15.2. We assume that the underlying function space $\mathcal{F}$ consists of d-variate functions that are linear combinations of orthonormal wavelet functions (ϕ_{jk}), where j and k are multi-indexes: $j = (j_1, \ldots, j_d) \in \{0, 1, \ldots\}^d$, and $k \in K_j$, where

$$K_j = \{k = (k_1, \ldots, k_d) : k_l = 0, \ldots, 2^{j_l} - 1, l = 1, \ldots, d\}.$$

We have available the following three sets of functions: $(\phi_{jk})_{jk}$ is an orthogonal wavelet basis and $(u_{jk})_{jk}$ and $(v_{jk})_{jk}$ are near-orthogonal sets:

$$\left\| \sum_{jk} a_{jk} u_{jk} \right\|_2 \asymp \|(a_{jk})\|_{l_2}, \qquad \left\| \sum_{jk} a_{jk} v_{jk} \right\|_2 \asymp \|(a_{jk})\|_{l_2},$$

where $a \asymp b$ means that there exists positive constants C, C' such that $Cb \le a \le C'b$. The following quasi-singular relations hold:

$$A\phi_{jk} = b_j v_{jk}, \qquad A^* u_{jk} = b_j \phi_{jk}, \tag{B.16}$$

where b_j are quasi-singular values. The system

$$(\phi_{jk}, u_{jk}, v_{jk}, b_j) \tag{B.17}$$

may be called the wavelet-vaguelette system of the linear operator A.

B.11 PROJECTION THEOREM

The projection theorem is also called the "central slice theorem" or "projection-slice theorem". The Fourier transform $F(g)$ of a function $g : \mathbf{R}^d \to \mathbf{R}$ is defined in (B.4).

Lemma B.2 *Denote $g(u) = g_{f,\xi}(u) = (Rf)(\xi, u)$, $u \in \mathbf{R}$, where Rf is the Radon transform of f, defined in (9.26). Then*

$$(Fg)(t) = (Ff)(t\xi), \qquad t \in \mathbf{R}.$$

Proof. By a change of the coordinate system, we can write, for $\xi \in \mathbf{S}_{d-1}$, $h : \mathbf{R}^d \to \mathbf{R}$, $\int_{\mathbf{R}^d} h(x)\,dx = \int_{-\infty}^{\infty} \int_{\xi^\perp} h(u\xi + z)\,dz\,du$. Thus

$$\begin{aligned}(Fg)(t) &= \int_{-\infty}^{\infty} e^{iut}(Rf)(\xi,u)\,du \\ &= \int_{-\infty}^{\infty} du\, e^{iut} \int_{\xi^\perp} f(u\xi + z)\,dz \\ &= \int_{-\infty}^{\infty} du \int_{\xi^\perp} e^{it\xi^T(u\xi+z)} f(u\xi + z)\,dz \\ &= \int_{\mathbf{R}^d} e^{t\xi^T x} f(x)\,dx \\ &= (Ff)(t\xi).\end{aligned}$$

We have proved the lemma. □

B.12 MISCELLANEOUS

1. *(Stirling's formula.)* Stirling's formula states that

$$n! \sim n^{n+1/2} e^{-n} \sqrt{2\pi}, \tag{B.18}$$

as $n \to \infty$. We have also

$$n^{n+1/2} e^{-n+1/(12n+1)} \sqrt{2\pi} < n! < n^{n+1/2} e^{-n+1/(12n)} \sqrt{2\pi}. \tag{B.19}$$

These facts are proved in Feller (1968, pp. 50–53).

2. *(Geometric series.)* For $q \neq 1$ we have $\sum_{i=0}^{K} q^i = (q^{K+1} - 1)/(q - 1)$ and for $|q| < 1$ we have

$$\sum_{i=0}^{\infty} q^i = \frac{1}{1-q}. \tag{B.20}$$

3. *(Jensen's inequality.)* Jensen's inequality states that for convex functions $\phi : \mathbf{R} \to \mathbf{R}$,

$$\phi\left(\int_{\mathbf{R}^d} f\,dP\right) \le \int_{\mathbf{R}^d} \phi(f)\,dP, \tag{B.21}$$

where P is a probability measure on $\mathbf{R}^d$ and $f \in L_1(P)$.

4. Let $X \in \mathbf{R}$ be a random variable and $A : \mathbf{R} \to \mathbf{R}$ be a monotonic function. Denote by f_X the density of X and by $f_{A(X)}$ the density of $A(X)$. We have

$$\begin{aligned}f_{A(X)}(t) &= \frac{\partial}{\partial t} P(A(X) \le t) \\ &= \frac{\partial}{\partial t} P\left(X \le A^{-1}(t)\right)\end{aligned}$$

$$
\begin{aligned}
&= f_X\left(A^{-1}(t)\right) \cdot \frac{\partial}{\partial t} A^{-1}(t) \\
&= f_X\left(A^{-1}(t)\right) \cdot \frac{1}{A'\left(A^{-1}(t)\right)}, \qquad \text{(B.22)}
\end{aligned}
$$

where $A'(t) = \partial A(t)/\partial t$.

APPENDIX C

THE PARENT–CHILD RELATIONS IN A MODE GRAPH

We need to define the parent–child relations for a mode graph, as presented in Section 7.1.

Let $H = \{h_1, \ldots, h_L\}$, where $h_1 < \cdots < h_L$, be the set of smoothing parameters. Denote with $\mathcal{M}_h$ the set of mode locations corresponding to smoothing parameter $h \in H$:

$$\mathcal{M}_h = \{M_1^{(h)}, \ldots, M_{m_h}^{(h)}\}.$$

We give the steps for defining the parent–child relations for a mode graph. In those steps we apply a procedure $vectormatch$ to find for two finite sets of vectors $\mathbb{X}, \mathbb{Y} \subset \mathbf{R}^d$, $\#\mathbb{X} \leq \#\mathbb{Y}$, a mapping $vm : \mathbb{X} \to \mathbb{Y}$. Mapping vm is defined in such a way that x and $vm(x)$ are as close to each other as possible. Procedure $vectormatch$ is described in page 515.

1. The modes in $\mathcal{M}_{h_L}$, corresponding to the largest smoothing parameter h_L, are the root nodes. The modes in $\mathcal{M}_{h_1}$ are the leaf nodes.

2. We define child nodes for the modes in $\mathcal{M}_{h_i}$, $i = L, \ldots, 2$.

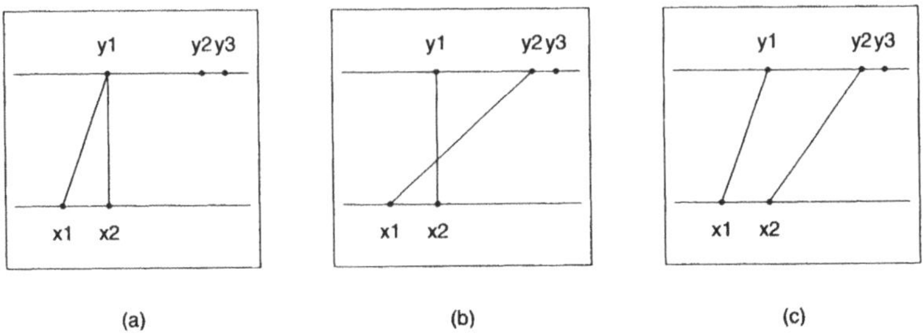

Figure C.1 *(Finding an injection $vm : \mathbb{X} \to \mathbb{Y}$, where $\mathbb{X} = \{x_1, x_2\}$ and $\mathbb{Y} = \{y_1, y_2, y_3\}$.)* (a) A conflict where x_1 has y_1 as its closest and x_2 has y_1 as its closest. (b) A suboptimal resolution of the conflict. (c) A better resolution of the conflict.

(a) Assume that $\#\mathcal{M}_{h_i} \leq \#\mathcal{M}_{h_{i-1}}$: the number of modes is at this step increasing as the smoothing parameter is decreasing. (This is the usual case.)
Let $\mathbb{X} = \mathcal{M}_{h_i}$ and $\mathbb{Y} = \mathcal{M}_{h_{i-1}}$.
If $vm(x) = y$, then y is a child of x. The color of y is the same as that of x.
Let M be a mode that had not a parent assigned to it: $M \in \mathbb{Y} \setminus \{vm(x) : x \in \mathbb{X}\}$. We choose the closest member in $\mathbb{X}$ to be the parent of M. We choose a new color for M.

(b) Assume that $\#\mathcal{M}_{h_i} > \#\mathcal{M}_{h_{i-1}}$: the number of modes is at this step decreasing as the smoothing parameter is decreasing. (This is the unusual case.)
Let $\mathbb{X} = \mathcal{M}_{h_{i-1}}$ and $\mathbb{Y} = \mathcal{M}_{h_i}$.
If $vm(x) = y$, then y is the parent of x. The color of x is the same as that of y.
Let M be a mode that had not a child assigned to it: $M \in \mathbb{Y} \setminus \{vm(x) : x \in \mathbb{X}\}$. Mode M is a leaf node of the mode tree.

It is left for us to describe the procedure *vectormatch* for finding mapping $vm : \mathbb{X} \to \mathbb{Y}$, $\#\mathbb{X} \leq \#\mathbb{Y}$. Mapping vm should be such that x and $vm(x)$ are close. In addition vm should be injective. The injectivity means that $vm(x_1) \neq vm(x_2)$ when $x_1 \neq x_2$. The number of all injections is

$$\#\mathbb{Y} \cdot (\#\mathbb{Y} - 1) \cdots (\#\mathbb{Y} - \#\mathbb{X} + 1),$$

which is so large number that one has to find a suboptimal solution. When we find for each $x \in \mathbb{X}$ the closest $y \in \mathbb{Y}$, we do not get an injective mapping vm in all cases. So we have to find a way to resolve the conflicts. If x_1 and x_2 are competing over y, that is, x_1 has y as its closest in $\mathbb{Y}$ and x_2 has y as its closest in $\mathbb{Y}$, then a simple way to resolve the conflict would be to take $vm(x_1) = y$ if $\|x_1 - y\| \leq \|x_2 - y\|$, and $vm(x_2) = y$ otherwise. This might lead to bad matching overall, see Figure C.1. We define the procedure *vectormatch* so that a conflict is resolved by looking such matching pairs that both x_1 and x_2 find a relatively good match. One fixes the better of these matches and continues by finding for each remaining $x \in \mathbb{X}$ the closest remaining $y \in \mathbb{Y}$. If there is a conflict, one resolves it as before; otherwise the procedure is done. The precise algorithm for *vectormatch* is given below.

Procedure *vectormatch* We give a pseudocode that describes the procedure *vectormatch* for finding vm.

1. For each $x \in \mathbb{X}$ find the closest $y \in \mathbb{Y}$ in the Euclidean metric, denote $vm_1(x) = y$.

2. If $\#\{x \in \mathbb{X} : vm_1(x) = y\} = 1$ for each $y \in \mathbb{Y}$, then return $vm = vm_1$.

3. Else

   (a) Set $\mathbb{X}_0 = \mathbb{X}$ and $\mathbb{Y}_0 = \mathbb{Y}$ to be the sets of available vectors.

   (b) Repeat until $\#A(y) = 1$ for each $y \in \mathbb{Y}_0$, where $A(y) = \{x \in \mathbb{X}_0 : vm_1(x) = y\}$.

      i. Set $B = \cup\{A(y) : y \in \mathbb{Y}_0,\ \#A(y) > 1\}$ to be the set of vectors $x \in \mathbb{X}_0$ that have competitors.
      For each $x \in B$, let $vm_2(x)$ be the second closest to x in $\mathbb{Y}_0$, after $vm_1(x)$.

      ii. We go through all ordered subsets (x, z) of size 2 from B, and calculate

      $$crit(x, z) = \|vm_1(x) - x\|^2 + \|vm_2(z) - z\|^2.$$

      That is, when $x, z \in B$, $x \neq z$, we calculate $crit(x, z)$ and $crit(z, x)$.

      iii. We find the minimal value of $crit(x, z)$ over all ordered subsets of B of size 2.
      When (x_0, z_0) achieves the minimum, set

      $$vm(x_0) = vm_1(x_0).$$

      iv. Set

      $$\mathbb{X}_0 = \mathbb{X}_0 \setminus \{x_0\}, \qquad \mathbb{Y}_0 = \mathbb{Y}_0 \setminus \{vm_1(x_0)\}.$$

      and for each $x \in \mathbb{X}_0$ we find the closest $y \in \mathbb{Y}_0$, and set $vm_1(x) = y$. The idea is that we do not simply choose from B the vector that is closest to y but take into account whether the choice allows further good choices: we take into account the distance to the second-best choice to guarantee that there remains a potential for further good choices.

4. Set $vm(x) = vm_1(x)$ for those x for which $vm(x)$ was not yet determined and return vm.

APPENDIX D

TREES

We start by defining trees as a subclass of graphs in Section D.1. In Section D.2 we discuss how trees can be implemented in computer programs. In Section D.3 we discuss segmentation and ordering of trees. In Section D.4 we discuss dynamic programming in the context of choosing an optimal subtree. In Section D.5 we describe an algorithm for finding a sequence of optimal CART partitions.

D.1 GRAPHS AND TREES

We may define trees as a certain subclass of graphs. Graph G is a pair (V, E), where V is a set and E is a set of pairs (v_0, v_1) where $v_0, v_1 \in V$. Set V is called the set of vertices (or nodes) and E is called the set of edges (or arcs). Thus graph is a relation in set V. Note that sometimes graphs are defined as relations between two different sets. For example, the graph of a function f is the set of all pairs $(x, f(x))$.

Tree is a connected and acyclic graph. We define the concepts of a connected and acyclic graph.

1. If it is possible to establish a path connecting any two vertices of the graph, the graph is said to be *connected*. That is, we say that graph is connected if for

each $v, v' \in V$, there is $u_1, \ldots, u_k \in V$ so that

$$(v, u_1), (u_1, u_2), \ldots, (u_{k-1}, u_k), (u_k, v') \in E$$

or

$$(v', u_1), (u_1, u_2), \ldots, (u_{k-1}, u_k), (u_k, v) \in E.$$

2. A cycle (or circuit) is a path that begins and ends with the same vertex. *Acyclic graph* is a graph with no cycles. That is, we say that graph is acyclic when for each $v_0 \in V$, there does not exist $u_1, \ldots, u_k \in V$ so that

$$(v_0, u_1), (u_1, u_2), \ldots, (u_{k-1}, u_k), (u_k, v_0) \in E,$$

where $v_0, u_1, \ldots, u_k$ are distinct.

We have defined a *free tree*, a tree whose nodes are treated equally and there is no root. A *rooted tree* is a tree with a distinguished vertex, which is called the root. If a tree is rooted, then it is common to regard the tree as a directed graph, with the edges oriented away from the root. (We say that a graph is *undirected* if the relation is symmetric; that is, if $(v_0, v_1) \in E$, then $(v_1, v_0) \in E$.) A *binary tree* is a rooted tree where each vertex has at most two children. An *ordered tree* is a rooted tree where the children of each node have been ordered.

D.2 IMPLEMENTATIONS

For the purpose of programming, a tree is a finite set of linked storage cells (nodes). There are many ways to define the abstract data type of a tree corresponding to various uses of a tree. For example, a tree may be specialized to the abstract data type of a binary search tree. Common operations on trees are enumerating all the items, searching for an item, adding a new item at a certain position on the tree, deleting an item, removing a whole section of a tree (pruning), adding a whole section to a tree (grafting), and finding the root for any node.

We implement trees with arrays. The implementation is similar to the implementation of linked lists. We define trees with the help of two types of arrays: arrays that store the data and arrays that contain the links between nodes. We discuss two basic implementations: (1) implementation with a pointer to the parent, (2) implementation with a pointer to a child and to a sibling. In the special case of binary trees the implementation with pointers to the children is a third possible implementation. The implementation with pointers to the parent may be enough if we want only to plot the tree. If we want to search the tree, then we need to implement the tree with links to children and to siblings.

D.2.1 Pointer to the Parent

A directed tree is uniquely defined by giving the parent node for each node. We need an array for the parent links and the arrays for the data. However, for many operations on the tree it is useful to introduce links to the children and siblings.

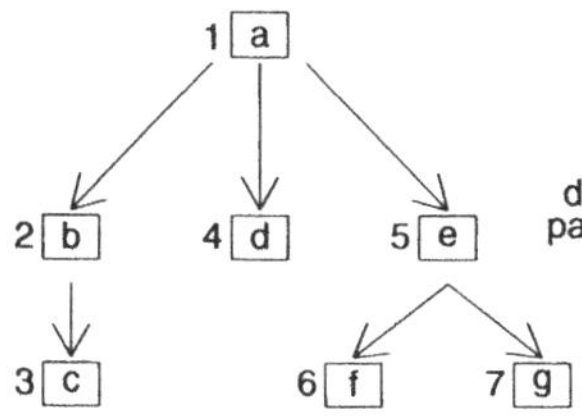

(a)

data	a	b	c	d	e	f	g	NA	NA	NA
parent	0	1	2	1	1	5	5	NA	NA	NA

(b)

Figure D.1 *(Parent links.)* (a) A tree and (b) an implementation of this tree with arrays and parent links. Each node of the tree is associated with data.

```
# create tree
maxsize<-10
data<-matrix(NA,maxsize,1)
parent<-matrix(NA,maxsize,1)

# links to the parents
root<-1             #we create the link to the root
parent[1]<-0        #we use 0 when there is no parent
parent[2]<-1
parent[3]<-2
parent[4]<-1
...

# store the items
data[1]<-''a''
data[2]<-''b''
data[3]<-''c''
...
```

Figure D.1(a) shows a tree. For each node we have associated data a–g and we have numerated the nodes with numbers 1–7. Panel (b) shows an implementation of the tree with parent links.

D.2.2 Pointer to a Child and to a Sibling

When we implement a directed tree using links to a child and to a sibling, we can conveniently implement several operations on the tree. Figure D.2(a) shows the tree of Figure D.1 with links to a child and to a sibling. Panel (b) shows an implementation of the tree using arrays "child" and "sibling".

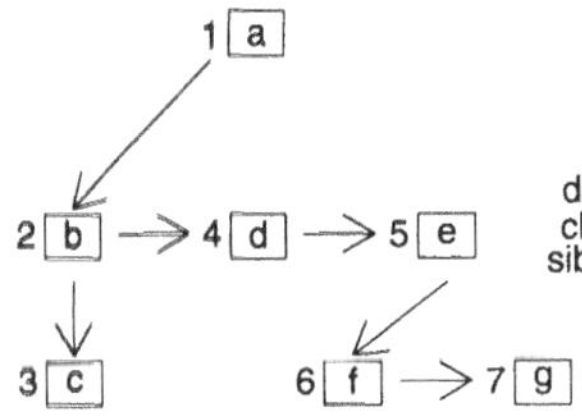

(a)

data	a	b	c	d	e	f	g	NA	NA	NA
child	2	3	0	0	6	0	0	NA	NA	NA
sibling	0	4	0	5	0	7	0	NA	NA	NA

(b)

Figure D.2 *(Child- and sibling-links.)* (a) The same tree as in Figure D.1 and (b) an implementation of this tree with arrays and child and sibling links.

We may additionally introduce the array of parent links to facilitate some operations. It may be possible to cope without parent links when we use the following implementation:

1. The root node is stored at the first position of the array.

2. The children are stored at later positions than their parent.

Then, as we travel the array starting from the last element, we will always meet children before the parent. This implementation allows the tree to grow, but pruning the tree becomes more difficult.

D.2.3 Binary Tree

A binary tree is a tree where each tree has either 0 or 2 children. A binary tree may be implemented with two arrays of links: we give a pointer to both the left child and the right child.

D.3 SEGMENTATION AND ORDERING

D.3.1 Segmentation

The segmentation of a tree to branches is a useful tool that has many applications: to define a coloring of a tree, to find unimodal components of a function (Section 4.4), to segment tail regions of a level set, and to segment tail regions of data (Section 6.4).

The segmentation of a tree to branches may be defined recursively: start at the root node by taking the root node to be the first node of the first segment (first branch). Add descendants of the root to the first segment until the tree is branching (there is a node which has several children). The node that has several children is the last node of the first segment. The children of this node are the root nodes of the new branches. Continue in this way, creating new segments when a node has several children.

1. The root node is the first node of the first branch.
2. Let a node of a branch be given.
   (a) If this node is a leaf node, then this node is the last node of its branch.
   (b) If this node has exactly 1 child node, then the child node belongs to the same branch.
   (c) If this node has more than 1 child, then this node is the last node of its branch, and the children of the node are the first nodes of new branches.

We define the segmentation of the nodes in a short way in the following definition.

Definition D.1 (Segmentation of a tree to branches.) *We define a partition of the nodes of a tree. The members of the partition are called branches. If node m belongs to branch B, then a child of node m belongs to branch B if and only if it has no siblings.*

D.3.2 Ordered Trees

We defined a function tree in Definition 3.3 as an ordered tree: the children of a given parent have to be ordered. Often there is no natural ordering for the nodes of a function tree, but if the tree is a spatial tree in the sense of Definition 3.1, then we may use the locations of the nodes to define an ordering that reflects the spatial location of the nodes.

We suggest two orderings for spatial trees. The first rule orders the siblings according to the closeness to a given reference point, say the origin. The second rule tries to reach an optimal ordering for the leaf nodes. Thus at each level the ordering is based on the locations of all the descendants of the nodes. We call this rule *descendant-based ordering*, We have used the descendant-based ordering rule of Definition D.3 in this book.

Reference Point-Based Ordering We fix the reference point to be the origin.

Definition D.2 (Origin-based ordering of a tree.) *The* origin-based ordering *of siblings in a tree, whose nodes are associated with vectors in the d-dimensional Euclidean space, is given by the following rule:*

1. *The first node among the siblings is the one with the largest Euclidean distance from the origin.*
2. *After that, the sibling nodes are ordered according to the distance to the first node; the second node is the one with the closest Euclidean distance from the first node. We continue the ordering by finding the distance to the first node.*

For example, level set trees are such that typically each node has 0, 1, or 2 nodes. Only when we use few levels to build a level set tree, it may happen that some node has 3 or more children. Thus, to define an ordering, we need to concentrate on finding the first sibling: the rule for ordering the remaining siblings is not important. In this case item 2 of Definition D.2 is less important than item 1.

Descendant-Based Ordering We may express useful information with the ordering if the ordering of the leaf nodes reflects somehow the spatial location of the leaf nodes. However, a tree may only be ordered by starting from the root nodes. To partially solve this problem we associate each node m with a vector $\mathrm{da}(m)$, that depends in a certain way on all the descendants of the node. When we order the siblings with the help of vectors $\mathrm{da}(m)$, the ordering takes into better account the location of the leaf nodes.

Definition D.3 (Descendant-based ordering of a tree.) Descendant-based ordering *of siblings in a tree whose nodes are associated with vectors in the d-dimensional Euclidean space is given by the following rule: annotate node m with that vector $da(m) \in \mathbf{R}^d$ of its descendants that has the largest Euclidean distance from the origin:*

$$\mathrm{da}(m) = \max\left\{\|\mathrm{vector}(m')\| : m' \textit{ is a descendant of } m\right\},$$

where $\mathrm{vector}(m')$ *is the vector associated with node m' and we always include the node itself to the set of its descendants.*

1. *Node m is the first node among the siblings if $\|\mathrm{da}(m)\|$ is larger than $\|\mathrm{da}(m')\|$ for other siblings m'.*
2. *Node m' is the second node if $\|\mathrm{da}(m') - \mathrm{da}(m)\|$ is smaller than $\|\mathrm{da}(m'') - \mathrm{da}(m)\|$ for other siblings m'', where m is the first node. The ordering is continued by finding the distance to the first node.*

Again, in many cases item 1 of Definition D.3 is the important part. Note that Definition D.3 is equivalent to Definition D.2 if we define $\mathrm{da}(m) = \mathrm{vector}(m)$. The descendant-based ordering is not essentially more complex computationally than the origin-based ordering, when one calculates the vectors $\mathrm{da}(m)$ for each node m as the tree is grown.

D.4 MINIMIZATION OVER SUBTREES

D.4.1 Dynamic Programming

Dynamic programming is called also backward induction. For example, Sedgewick (1988, ch. 42) gives an introduction to dynamic programming. The term "programming" refers to the process of formulating the constraints of a problem so that the method is applicable. In dynamic programming we solve smaller problems and store the answers to be used later in solving larger problems. We may apply dynamic programming when any solution to a small subproblem remains a good one when that subproblem becomes a piece of some larger problem.

We do not have a general characterization of the applicability of the technique. There may appear two problems: (1) it is not possible to combine the solutions of smaller problems to form the solution of a larger one and (2) the number of small problems to solve may be unacceptably large. Applications of dynamic programming can have quite different time and memory requirements depending on the amount of information about the small subproblems.

D.4.2 Minimization over Subtrees

As an example of a dynamic programming algorithm we present a solution to the minimization problem of finding a subtree of a binary tree so that the subtree minimizes the sum of a cost function over its leafs. This minimization problem finds its application in Section 17.2.2 where we search for an adaptive histogram minimizing a complexity-penalized error criterion.

For trees T and T', we write

$$T' \preceq T \tag{D.1}$$

if T' has been constructed from T by pruning away some number of branches; that is, for some collection of nodes $\{t_1, \ldots, t_l\}$ of T, we have made these nodes leaf nodes by removing the branches whose roots are t_i, $i = 1, \ldots, l$. Let T be a binary tree where each node t is associated with a real value $f(t)$. We want to solve the minimization problem

$$\min \left\{ \sum_{t \in \text{leafs}(T')} f(t) : T' \preceq T \right\},$$

where leafs(T') is the collection of leafs of T'.

We use the following data structures.

- Input: Binary tree T is represented through vectors $left$, and $right$, which are pointers to the right and left child of a node. These are arrays of size $maxnode$. Function f is an array of size $maxnode$. We implement the binary tree so that the index of a child is larger than the index of the parent. (The root has thus the smallest index, for example the root could have index 1.)

- Output: Modified vectors $right$ and $left$ that define binary tree $T' \preceq T$.

- Internal data structure: Array S that stores the costs of subtrees. Every node t is associated with value $S[t]$, which is the sum of the values $f[t']$ over the leafs t' of the subtree whose root is t.

ALGORITHM **prune** (left, right, f, maxnode)

1. **for** t=maxnode to 1

(a) **if** t is leaf **then** S[t]=f[t]

(b) **else** (t is not leaf)

i. **if** (f[t] $\le$ S[left[t]]+S[right[t]]) **then**

A. S[t]=f[t]

B. left[t]=0, right[t]=0

ii. **else**

A. S[t]=S[left[t]]+S[right[t]]

2. **return**(left, right)

The pruning algorithm is fast because it allows us to find a solution by traveling through the nodes of tree only once.

D.5 PRUNING ALGORITHM

We give an algorithm for finding the α-sequence and the corresponding sequence of partitions, as discussed in Section 17.2.2, page 460.

Let T_0 be the binary tree associated to the overfitting partition constructed in Definition 17.2. Let us denote by $\{0 = \alpha_0 < \cdots < \alpha_K\}$ the α-values for which the corresponding partitions are different. Denote the corresponding trees by $T_0, \ldots, T_K$. Let us denote for nodes t of T_0,

$$\mathrm{err}(t) = \begin{cases} -\frac{n_{R_t}}{n} \log\left(\frac{n_{R_t}}{n\,\mathrm{volume}(R_t)}\right), \\ -\frac{n_{R_t}^2}{n^2\,\mathrm{volume}(R_t)}, \end{cases} \tag{D.2}$$

where R_t is the rectangle associated with node t. Denote by $\mathrm{ERR}(t)$ the sum of $\mathrm{err}(t')$ over the leafs t' of the subtree T_t whose root is t. We have for every nonterminal node t of T_0 that

$$\mathrm{ERR}(t) < \mathrm{err}(t).$$

Denote by $\#T_t$ the number of leaves in the subtree T_t. As long as

$$\mathrm{ERR}(t) + \alpha \cdot \#T_t < \mathrm{err}(t) + \alpha \tag{D.3}$$

the branch T_t has a smaller error-complexity than the single node $\{t\}$, but at some critical value of α, the two error-complexities become equal. At this point the subbranch $\{t\}$ is smaller than T_t, has the same error-complexity, and is therefore preferable. To find this α, solve (D.3) to get

$$\alpha < \frac{\mathrm{err}(t) - \mathrm{ERR}(t)}{\#T_t - 1}.$$

The algorithm is based on finding the "weakest links", which are the nodes minimizing

$$g_k(t) = \begin{cases} \frac{\mathrm{err}(t) - \mathrm{ERR}(t)}{\#T_t - 1}, & t \text{ is not a leaf in } T_k \\ \infty, & t \text{ is a leaf in } T_k, \end{cases} \tag{D.4}$$

$k = 1, \ldots, K$. Let $t_1 = \mathrm{argmin}_{t \in T_0} g_0(t)$. Then we have $\alpha_1 = g_0(t_1)$ and T_1 is the subtree of T_0 obtained by making t_1 a leaf node. We continue in this way: $\alpha_k = g_{k-1}(t_k)$ for $k = 1, \ldots, K$. If at any stage there is a multiplicity of weakest links, for example, $g_{k-1}(t_k) = g_{k-1}(t'_k)$, then define $T_k = T_{k-1} - T_{t_k} - T_{t'_k}$. We get a sequence where $\alpha_1 = 0$, and for $\alpha_k \le \alpha < \alpha_{k+1}$, the corresponding partition is $\mathcal{P}(T_k)$.

Pseudocode Algorithm PruneSeq is described below. We represent the overfitting tree T_0, defined in Definition 17.2, as a collection of arrays, as explained in Section D.2. We apply the binary tree representation so that the tree is defined by arrays "left" and "right".

- Input:
  - Overfitting tree T_0. We denote with right$[t]$ and left$[t]$ the pointers to the left and right child of node t. We denote with parent$[t]$ the parent of t.
  - Vector ERR$[t]$ whose values are equal to the sum of err(t') over the leafs t' of the subtree T_t whose root is t, where err is defined in (D.2).
  - Vector leafs$[t]$ is the number of leaves in the tree T_t whose root is t.
  - Vector $g[t] = (\mathrm{err}[t] - \mathrm{ERR}[t])/(\#T_t - 1)$, defined in (D.4).
  - Vector

$$G[t] = \begin{cases} \min\{g[t], G[\mathrm{left}[t]], G[\mathrm{right}[t]]\}, & t \text{ is not a leaf} \\ \infty, & t \text{ is a leaf}. \end{cases}$$

    Note that always $G[t] > 0$.

- Output:
  - Vector alpha$[k]$ that gives the sequence of distinct smoothing parameter values. We always have alpha$[1] = 0$.
  - Matrix delnodes$[k]$[jnode] that gives pointers to the nodes to be pruned. That, is rows $1 \le i \le k$ of delnodes give pointers to the nodes that are to be pruned when $\alpha = \mathrm{alpha}[k]$.

ALGORITHM **PruneSeq**(left, right, parent, ERR, leafs, g, G)

1. **if** leafs$[1] = 1$ **then** the tree is trivial **else**
2. k=1, α=0
3. **while** leafs$[1] > 1$
   (a) **if** $G[1] > \alpha$ **then**
      i. alpha$[k] = \alpha$,
      ii. $\alpha = G[1]$, $k = k + 1$, jnode $= 0$

(b) **end if**

(c) $t = 1$ (Now we start from the root and find a node t which is such that we may cut below this node.)

(d) **while** $G[t] < g[t]$ (We continue until g achieves its minimum.)

  i. **if** $G[t] = G[\text{left}[t]]$ **then** $t = \text{left}[t]$ **else** $t = \text{right}[t]$

(e) **end while**

(f) $\text{leafs}[t] = 1, \text{ERR}[t] = \text{err}[t], G[t] = \infty$ (We make t a leaf.)

(g) $\text{jnode} = \text{jnode} + 1$, $\text{delnodes}[k][\text{jnode}] = t$ (When $\alpha = \text{alpha}[k]$, then prune below t.)

(h) **while** $t > 1$ (We return to the root updating ERR, leafs, g, and G.)

  0. $t = \text{parent}[t], m = \min\{\text{err}[t], \text{ERR}[\text{left}[t]] + \text{ERR}[\text{right}[t]]\}$

  i. **if case** $\text{err}[t] = m$ **then**

    A. $\text{left}[t] = 0, \text{right}[t] = 0,$

    B. $\text{ERR}[t] = \text{err}[t], \text{leafs}[t] = 1, g[t] = \infty, G[t] = \infty$

  ii. **if case** $\text{ERR}[\text{left}[t]] + \text{ERR}[\text{right}[t]] = m$ **then**

    A. $\text{ERR}[t] = \text{ERR}[\text{left}[t]] + \text{ERR}[\text{right}[t]]$

    B. $\text{leafs}[t] = \text{leafs}[\text{left}[t]] + \text{leafs}(\text{right}[t]]$

    C. $g[t] = (\text{err}[t] - \text{ERR}[t])/(\text{leafs}[t] - 1)$

    D. $G[t] = \min\{g[t], G[\text{left}[t]], G[\text{right}[t]]\}$

(i) **end while**

4. **end while** (in 3)

5. **end if** (in 1)

6. **return**(alpha, delnodes)

APPENDIX E

PROOFS

E.1 PROOFS FOR CHAPTER 10

We prove Lemma 10.3. Thus we have to prove (10.43), (10.44), and (10.46).

E.1.1 Proofs of (10.43) and (10.44)

Denote by $R_{j_l k_l} = [k_l/2^{j_l}, (k_l + 1)/2^{j_l})$ a dyadic interval, and with $R_{j_l k_l 0} = [2k_l/2^{j_l+1}, (2k_l + 1)/2^{j_l+1})$, $R_{j_l k_l 1} = [(2k_l + 1)/2^{j_l+1}, (2k_l + 2)/2^{j_l+1})$ the left and the right half of $R_{j_l k_l}$. Then

$$\eta^{(1)}_{j_l k_l} = 2^{j_l/2}(I_{R_{j_l k_l 1}} - I_{R_{j_l k_l 0}})$$

by (16.14). We have for $x = (x_1, \ldots, x_d) \in [0, 1]^d$,

$$\begin{aligned} \int_0^1 f(x)\eta^{(1)}_{j_l k_l}(x_l)\, dx_l &= 2^{j_l/2}\left(\int_{R_{j_l k_l 1}} f(x)dx_l - \int_{R_{j_l k_l 0}} f(x)dx_l\right) \\ &= 2^{j_l/2}\int_{R_{j_l k_l 0}} (f(x + \mu e_l) - f(x))\, dx_l \end{aligned}$$

$$= \quad 2^{j_l/2} \int_{R_{j_l k_l 0}} \Delta_{\mu e_l} f(x)\, dx_l, \tag{E.1}$$

where $\mu = 2^{-j_l-1}$, $\Delta_y f(x) = f(x+y) - f(x)$, and $e_l \in \mathbf{R}^d$ is the unit direction vector the unit direction vector with $e_{li} = 1$ when $i = l$ and 0 otherwise. Denote $Q_{jkl} = R_{j_l k_l 0} \times \Pi_{m=1, m\neq l}^d R_{j_m k_m}$, so that

$$\text{volume}(Q_{jkl}) = 2^{-j_l-1} \Pi_{m=1,m\neq l}^d 2^{-j_m} = 2^{-(|j|+1)}. \tag{E.2}$$

We have from (E.1),

$$\begin{aligned} \theta_{jk}^i &= \int_{[0,1]^d} f(x)\, \eta_{j_1 k_1}^{(i_1)}(x_1) \cdots \eta_{j_d k_d}^{(i_d)}(x_d)\, dx \\ &= \left(2^{j_1} \cdots 2^{j_d}\right)^{1/2} \int_{[0,1]^d} \Delta_{\mu e_l} f(x) I_{Q_{jkl}}(x)\, dx. \end{aligned}$$

We consider the cases $1 < p_l < \infty$, $p_l = \infty$, and $p_l = 1$ separately.

1. When $1 < p_l < \infty$ by Hölder's inequality, by (E.2),

$$\begin{aligned} \left|\theta_{jk}^i\right|^{p_l} &\leq \left(2^{j_1} \cdots 2^{j_d}\right)^{p_l/2} \|\Delta_{\mu e_l} f\|_{L_{p_l}(Q_{jkl})}^{p_l} \cdot \left\|I_{Q_{jkl}}\right\|_{L_{p_l'}}^{p_l} \\ &= \left(2^{j_1} \cdots 2^{j_d}\right)^{p_l/2} \|\Delta_{\mu e_l} f\|_{L_{p_l}(Q_{jkl})}^{p_l} (\text{volume}(Q_{jkl}))^{p_l-1} \\ &= 2^{1-p_l} 2^{(1-p_l/2)|j|} \|\Delta_{\mu e_l} f\|_{L_{p_l}(Q_{jkl})}^{p_l}, \end{aligned} \tag{E.3}$$

where $p_l' = p_l/(p_l-1)$.

2. When $p_l = \infty$,

$$\begin{aligned} \left|\theta_{jk}^i\right| &\leq 2^{|j|/2} \|\Delta_{\mu e_l} f\|_{L_\infty(Q_{jkl})} \cdot \text{volume}(Q_{jkl}) \\ &= 2^{-1} 2^{-|j|/2} \|\Delta_{\mu e_l} f\|_{L_\infty(Q_{jkl})}. \end{aligned}$$

3. When $p_l = 1$, then (E.3) holds also, by Jensen's inequality.

Then for $1 \leq p_l < \infty$,

$$\begin{aligned} \sum_{k \in K_j} \left|\theta_{jk}^i\right|^{p_l} &\leq 2^{1-p_l} 2^{(1-p_l/2)|j|} \sum_{k \in K_j} \|\Delta_{\mu e_l} f\|_{L_{p_l}(Q_{jkl})}^{p_l} \\ &= 2^{1-p_l} 2^{(1-p_l/2)|j|} \|\Delta_{\mu e_l} f\|_{L_{p_l}(A_{l,\mu})}^{p_l} \\ &\leq 2^{1-p_l} 2^{(1-p_l/2)|j|} L^{p_l} \mu^{s_l p_l}, \end{aligned}$$

where $A_{l,\mu}$ is defined in (10.40), and for $p_l = \infty$,

$$\begin{aligned} \max_{k \in K_j} \left|\theta_{jk}^i\right| &\leq 2^{-1} 2^{-|j|/2} \max_{k \in K_j} \|\Delta_{\mu e_l} f\|_{L_\infty(Q_{jkl})} \\ &= 2^{-1} 2^{-|j|/2} \|\Delta_{\mu e_l} f\|_{L_\infty(A_{l,\mu})} \\ &\leq 2^{-1} 2^{-|j|/2} L \mu^{s_l}. \end{aligned}$$

The claims follow, since $\mu^{s_l} = 2^{-(j_l+1)s_l} \leq 2^{-j_l s_l}$. □

E.1.2 Proof of (10.46)

We consider separately the cases $1 \le p_l \le 2$, $2 < p_l < \infty$, and $p_l = \infty$.

1. When $1 \le p_l \le 2$, the lemma follows from (10.43).

2. When $2 < p_l < \infty$, we have by Hölder's inequality, by applying (10.43), because $\#K_j = 2^{|j|}$,

$$\begin{aligned}
\sum_{k\in K_j} \left|\theta^i_{jk}\right|^2 &\le \left(\sum_{k\in K_j} \left|\theta^i_{jk}\right|^{p_l}\right)^{2/p_l} \cdot (\#K_j)^{(p_l-2)/p_l} \\
&\le L^2 2^{-2j_l s_l} 2^{-2|j|(1/2-1/p_l)} \cdot 2^{2|j|(1/2-1/p_l)} \\
&= L^2 2^{-2j_l s_l}.
\end{aligned}$$

3. When $p_l = \infty$, we have by (10.44),

$$\begin{aligned}
\sum_{k\in K_j} \left|\theta^i_{jk}\right|^2 &\le \#K_j \cdot \max_{k\in K_j} \left|\theta^i_{jk}\right|^2 \\
&\le 2^{|j|} \cdot L^2 2^{-2j_l s_l} 2^{-|j|} \\
&= L^2 2^{-2j_l s_l}.
\end{aligned}$$

We have proved Lemma 10.3. □

E.2 PROOFS FOR CHAPTER 12

E.2.1 Proof of Theorem 12.1

We prove Theorem 12.1 (Bernstein's inequality). The proof of Bernstein's inequality is based on Chernoff's method. We follow Lugosi (2002). See also Theorem 2.8 in Petrov (1995), for example, for a proof of Bernstein's inequality.

Preliminary Lemma We have noted in (12.18) that in the Gaussian case, when $W \sim N(0, \sigma^2)$, then for $s > 0$,

$$Ee^{sW} \le e^{s^2\sigma^2/2}.$$

We use the following bound on the moment generating function in the possibly non-Gaussian cases.

Lemma E.1 *Let $EY = 0$ and $|Y| \le b$. Then for $s > 0$,*

$$Ee^{sY} \le \exp\left\{s^2 \mathrm{Var}(Y) C_s\right\},$$

where $C_s = (e^{sb} - 1 - sb)/(sb)^2$.

Proof. The exponential function may be written $e^s = \sum_{k=0}^{\infty} \frac{s^k}{k!}$. Thus we have that

$$Ee^{sY} = \sum_{k=0}^{\infty} \frac{s^k EY^k}{k!} = 1 + s^2 \mathrm{Var}(Y) \sum_{k=2}^{\infty} \frac{s^{k-2} EY^k}{k! \mathrm{Var}(Y)}.$$

We have that $EY^k \le b^{k-2}\mathrm{Var}(Y)$, which gives

$$\sum_{k=2}^{\infty} \frac{s^{k-2} EY^k}{k! \mathrm{Var}(Y)} \le \sum_{k=2}^{\infty} \frac{(sb)^{k-2}}{k!} = (sb)^{-2}(e^{sb} - 1 - sb) = C_s.$$

Thus

$$Ee^{sY} \le 1 + s^2 \mathrm{Var}(Y) C_s \le \exp\left\{ s^2 \mathrm{Var}(Y) C_s \right\},$$

since $1 + t \le e^t$. We have proved the lemma. □

Proof of Bernstein's Inequality By Markov's inequality, we have for a random variable $Y \in \mathbf{R}$,

$$P(Y \ge \zeta) = P(e^{sY} \ge e^{s\zeta}) \le e^{-s\zeta} Ee^{sY},$$

where $s, \zeta > 0$. Thus

$$\begin{aligned} P\left(\sum_{i=1}^{n} (X_i - EX_1) \ge \zeta \right) &\le e^{-s\zeta} E\left(\exp\left\{ s \cdot \sum_{i=1}^{n} (X_i - EX_1) \right\} \right) \\ &= e^{-s\zeta} \prod_{i=1}^{n} E\left(\exp\left\{ s(X_i - EX_1) \right\} \right), \end{aligned} \tag{E.4}$$

where we used independence. Using (E.4) and Lemma E.1, we get for $s, \zeta > 0$,

$$P\left(\sum_{i=1}^{n} (X_i - EX_1) \ge \zeta \right) \le e^{-s\zeta} \exp\left\{ s^2 n \sigma^2 C_s \right\}.$$

By the optimal choice of s we get the upper bound

$$\exp\left\{ -\frac{n\sigma^2}{b^2} h\left(\frac{b\zeta}{n\sigma^2} \right) \right\}, \tag{E.5}$$

where $h : [0, \infty) \to \mathbf{R}$ is defined by $h(u) = (1 + u)\log_e(1 + u) - u$. The inequality with the upper bound (E.5) is called Bennett's inequality. The Bernstein's inequality follows from the inequality $h(u) \ge u^2/(2 + 2u/3)$ and the substitution $\zeta = n\xi$. □

E.2.2 Proof of Theorem 12.4

We prove Theorem 12.4. The basic difference to the direct case is visible in Eq. (E.7). Let us denote $R_k = 2^{-k}R$, $N_k = \#\mathcal{G}_{R_k}$, $H_k = \log_e N_k$, and $T_k = \varrho(Q, \mathcal{G}_{R_k})$, where $k = 0, 1, \ldots$. For each $g \in \mathcal{G}$, let h_g^k be a member of R_k covering set of $\mathcal{G}$ such that $\|g - h_g^k\|_2 \le R_k$. We may write every $g \in \mathcal{G}$ with telescoping as

$$g = \sum_{k=1}^{\infty} \left(h_g^k - h_g^{k-1}\right),$$

where $h_g^0 \equiv 0$ and the convergence is in L_2. Let $\eta_k > 0$ be such that $\sum_{k=1}^{\infty} \eta_k \le 1$. We will define η_k in (E.9). Then

$$P\left(\sup_{g\in\mathcal{G}} \nu_n(Qg) \ge \xi\right) \le \sum_{k=1}^{\infty} P\left(\sup_{g\in\mathcal{G}} \nu_n\left(Q(h_g^k - h_g^{k-1})\right) \ge \xi\eta_k\right). \tag{E.6}$$

We have

$$\#\left\{h_g^k - h_g^{k-1} : g \in \mathcal{G}\right\} \le N_k N_{k-1} \le N_k^2.$$

We have

$$\begin{aligned} \max\left\{\left\|Q(h_g^k - h_g^{k-1})\right\|_2 : g \in \mathcal{G}\right\} &\le T_k \max\left\{\left\|h_g^k - h_g^{k-1}\right\|_2 : g \in \mathcal{G}\right\} \\ &\le 3T_k R_k, \end{aligned} \tag{E.7}$$

where we used the fact

$$\|h_g^k - h_g^{k-1}\|_2 \le \|h_g^k - g\|_2 + \|h_g^{k-1} - g\|_2 \le 2^{-k}R + 2^{-k+1}R = 3R_k.$$

When $W \sim N(0, \sigma^2)$, $\xi > 0$, we have $P(W > \xi) \le 2^{-1}\exp\{-\xi^2/(2\sigma^2)\}$; see (12.4). We have next

$$\nu_n(Q(h_g^k - h_g^{k-1})) \sim N\left(0, n^{-1}\|Q(h_g^k - h_g^{k-1})\|_2^2\right)$$

and thus

$$P\left(\sup_{g\in\mathcal{G}} \nu_n\left(Q(h_g^k - h_g^{k-1})\right) \ge \xi\eta_k\right) \le N_k^2 2^{-1} \exp\left\{-\frac{1}{2}\frac{n\xi^2\eta_k^2}{3^2T_k^2R_k^2}\right\}. \tag{E.8}$$

Now we choose

$$\eta_k = 3T_k R_k \max\left\{\frac{8^{1/2}H_k^{1/2}}{n^{1/2}\xi}, c^{-1}R^{a-1}(C'k)^{1/2}2\right\}, \tag{E.9}$$

where C' is defined in (12.10). Then we may apply (E.8) to continue (E.6) with an upper bound

$$2^{-1}\sum_{k=1}^{\infty} \exp\left\{2H_k - \frac{1}{2}\frac{n\xi^2\eta_k^2}{3^2T_k^2R_k^2}\right\}$$

$$\le 2^{-1}\sum_{k=1}^{\infty} \exp\left\{-\frac{1}{4}\frac{n\xi^2\eta_k^2}{3^2T_k^2R_k^2}\right\} \tag{E.10}$$

$$\le 2^{-1}\sum_{k=1}^{\infty} \exp\left\{-\frac{n\xi^2C'k}{c^2R^{2-2a}}\right\} \tag{E.11}$$

$$\le \exp\left\{-\frac{n\xi^2C'}{c^2R^{2-2a}}\right\}. \tag{E.12}$$

In (E.10) we applied (E.9), which implies $2H_k \le n\xi^2\eta_k^2/(4\cdot 3^2 T_k^2 R_k^2)$, when we apply the first term in the maximum. In (E.11) we applied also (E.9), which implies $\eta_k^2/(4\cdot 3^2 T_k^2 R_k^2) \ge C'k/[c^2 R^{2-2a}]$ where we applied the second term in the maximum. In (E.12) we applied that for $0 \le b \le 1/2$; $\sum_{k=1}^{\infty} b^k = b/(1-b) \le 2b$, which follows from (B.20). Here we need that $\exp\left\{-n\xi^2 C'/[c^2 R^{2-2a}]\right\} \le 1/2$, that is, $\xi \ge cR^{1-a}\left[\log_e 2/(nC')\right]^{1/2}$, which is implied by (12.9). We need to check that $\sum_{k=1}^{\infty} \eta_k \le 1$. Since $\delta \mapsto \varrho(Q,\mathcal{G}_\delta)\sqrt{\log_e(\#\mathcal{G}_\delta)}$ is decreasing,

$$\sum_{k=1}^{\infty} T_k R_k H_k^{1/2} = 2\sum_{k=1}^{\infty} 2^{-k-1} R T_{2^{-k}R}\sqrt{\log_e(\#\mathcal{G}_{2^{-k}R})} \le 2G(R). \tag{E.13}$$

We apply the assumption that $T_k = \varrho(Q,\mathcal{G}_{2^{-k}R}) = cR^{-a}2^{ak}$ to get

$$\begin{aligned}\sum_{k=1}^{\infty} k^{1/2} T_k R_k &= cR^{1-a}\sum_{k=1}^{\infty} k^{1/2} 2^{-(1-a)k} \\ &= cR^{1-a}\lim_{K\to\infty} K^{3/2}\int_0^1 t^{1/2} 2^{-(1-a)Kt}\,dt \\ &= cR^{1-a}(1-a)^{-3/2}\int_0^{\infty} u^{1/2} 2^{-u}\,du \\ &= cR^{1-a}C'',\end{aligned} \tag{E.14}$$

where C'' is defined in (12.10). We have from (E.13) and (E.14) that

$$\sum_{k=1}^{\infty} \eta_k \le \frac{8^{1/2} 6 G(R)}{n^{1/2}\xi} + 6\sqrt{C'}C'' \le \frac{1}{2} + \frac{1}{2} = 1,$$

when $\xi \ge 2 8^{1/2} 6 G(R) n^{-1/2}$, which is guaranteed by (12.9), and C' is chosen as in (12.10). The lemma follows from (E.6), (E.8), and (E.12). □

E.2.3 Proof of Theorem 12.5

We prove Theorem 12.5. The basic difference from the direct case is visible in (E.21) and (E.28). Let us denote $R_k = 2^{-k}R$, $N_k = \#\mathcal{G}_{R_k}$, $H_k = \log_e N_k$, $T_k = \varrho(Q,\mathcal{G}_{R_k})$, where $\varrho(Q,\mathcal{G}_\delta)$ is defined in (12.11) and $k = 0, 1, \ldots$. For each $g \in \mathcal{G}$, let $(h_g^{k,L}, h_g^{k,U})$ be the member of the bracketing net $\mathcal{G}_{R_k}$ such that $h_g^{k,L} \le g \le h_g^{k,U}$. Then write every $g \in \mathcal{G}$ with telescoping as

$$g = g - h_g^{\kappa_g,L} + \sum_{k=1}^{\kappa_g}\left(h_g^{k,L} - h_g^{k-1,L}\right) + h_g^{0,L},$$

where

$$\kappa_g = \begin{cases} \min\left\{0 \le k \le K-1 : Q\Delta_g^k \ge \beta_k\right\}, & \\ \qquad \text{if } Q\Delta_g^k \ge \beta_k \text{ for some } 0 \le k \le K-1, & \\ K, \quad \text{otherwise,} & \end{cases}$$

where $K \geq 1$ is defined in (E.29),

$$\Delta_g^k = h_g^{k,U} - h_g^{k,L},$$

and

$$\beta_k = \frac{12B_\infty T_k^2 R_k^2}{\xi}. \tag{E.15}$$

Then

$$\begin{aligned}
& P\left(\sup_{g\in\mathcal{G}} \nu_n(Qg) \geq \xi\right) \\
\leq \quad & P\left(\sup_{g\in\mathcal{G}} \sum_{k=1}^{\kappa_g} \nu_n\left(Q(h_g^{k,L} - h_g^{k-1,L})\right) \geq \frac{\xi}{3}\right) \\
& + P\left(\sup_{g\in\mathcal{G}} \nu_n\left(Q(g - h_g^{\kappa_g,L})\right) \geq \frac{\xi}{3}\right) \\
& + P\left(\sup_{g\in\mathcal{G}} \nu_n\left(Qh_g^{0,L}\right) \geq \frac{\xi}{3}\right) \\
\stackrel{def}{=} \quad & P_I + P_{II} + P_{III}.
\end{aligned} \tag{E.16}$$

Term P_I We have

$$\begin{aligned}
& \sup_{g\in\mathcal{G}} \sum_{k=1}^{\kappa_g} \nu_n\left(Q(h_g^{k,L} - h_g^{k-1,L})\right) \\
= \quad & \sup_{g\in\mathcal{G}} \sum_{k=1}^{K} I_{\{1,\ldots,\kappa_g\}}(k)\nu_n\left(Q(h_g^{k,L} - h_g^{k-1,L})\right) \\
\leq \quad & \sum_{k=1}^{K} \sup_{g\in\mathcal{G}} I_{\{1,\ldots,\kappa_g\}}(k)\nu_n\left(Q(h_g^{k,L} - h_g^{k-1,L})\right).
\end{aligned}$$

Let us denote

$$\begin{aligned}
\eta_k \quad = \quad & (9^2 + 96 \cdot 2^{-2a})^{1/2} T_k R_k \\
& \times \max\left\{\frac{8^{1/2} B_\infty^{1/2} H_k^{1/2}}{n^{1/2}\xi}, c^{-1} R^{a-1} (C'k)^{1/2} 2\right\},
\end{aligned} \tag{E.17}$$

where C' is defined by

$$C' = 4^{-2}(C'')^{-2}(9^2+96\cdot 2^{-2a})^{-1},\ C'' = (1-a)^{-3/2}\Gamma(3/2)(\log_e 2)^{-3/2}. \tag{E.18}$$

We have defined η_k in (E.17) so that $\eta_k > 0$ and $\sum_{k=1}^{\infty} \eta_k \leq 1$, which is proved in (E.40). Then

$$P_I \leq \sum_{k=1}^{K} P\left(\sup_{g\in\mathcal{G}} I_{\{1,\ldots,\kappa_g\}}(k)\, \nu_n\left(Q(h_g^{k,L} - h_g^{k-1,L})\right) \geq \frac{\eta_k \xi}{3}\right). \tag{E.19}$$

We have

$$\#\left\{h_g^{k,L} - h_g^{k-1,L} : g \in \mathcal{G}\right\} \le N_k N_{k-1} \le N_k^2. \tag{E.20}$$

Also

$$\begin{aligned}
&\max\left\{E\left|Q(h_g^{k,L} - h_g^{k-1,L})\right|^2 : g \in \mathcal{G}\right\} \\
\le\ & B_\infty \max\left\{\left\|Q(h_g^{k,L} - h_g^{k-1,L})\right\|_2^2 : g \in \mathcal{G}\right\} \\
\le\ & B_\infty T_k^2 \max\left\{\left\|h_g^{k,L} - h_g^{k-1,L}\right\|_2^2 : g \in \mathcal{G}\right\} \\
\le\ & B_\infty 3^2 T_k^2 R_k^2
\end{aligned} \tag{E.21}$$

because

$$\begin{aligned}
\left\|h_g^{k,L} - h_g^{k-1,L}\right\|_2 &\le \left\|h_g^{k,L} - g\right\|_2 + \left\|h_g^{k-1,L} - g\right\|_2 \\
&\le 2^{-k}R + 2^{-k+1}R = 3R_k.
\end{aligned}$$

When $k \le \kappa_g$, we have

$$Q(h_g^{k,L} - h_g^{k-1,L}) \le Q\Delta_g^{k-1} \le \beta_{k-1},$$

which implies

$$\left|Q(h_g^{k,L} - h_g^{k-1,L}) - EQ(h_g^{k,L} - h_g^{k-1,L})\right| \le 2\beta_{k-1}. \tag{E.22}$$

Thus, applying (E.20), (E.21), (E.22), by Bernstein's inequality, obtains

$$\begin{aligned}
&P\left(\sup_{g\in\mathcal{G}} I_{\{1,\ldots,\kappa_g\}}(k)\,\nu_n\left(Q(h_g^{k,L} - h_g^{k-1,L})\right) \ge \xi\eta_k/3\right) \\
\le\ & N_k^2 \exp\left\{-\frac{1}{2}\frac{n(\xi\eta_k/3)^2}{3^2 B_\infty T_k^2 R_k^2 + 2\beta_{k-1}\xi\eta_k/9}\right\} \\
\le\ & \exp\left\{2H_k - \frac{1}{2}\frac{n(\xi\eta_k)^2}{3^2(3^2 + 24\cdot 2^{2(1-a)}/9)B_\infty T_k^2 R_k^2}\right\} &\text{(E.23)} \\
\le\ & \exp\left\{-\frac{1}{4}\frac{n(\xi\eta_k)^2}{(9^2 + 96\cdot 2^{-2a})B_\infty T_k^2 R_k^2}\right\} &\text{(E.24)} \\
\le\ & \exp\left\{-\frac{n\xi^2 C' k}{c^2 B_\infty R^{2-2a}}\right\}. &\text{(E.25)}
\end{aligned}$$

In (E.23) we applied the fact that $\beta_{k-1}\xi\eta_k \le 12B_\infty 2^{2(1-a)}T_k^2 R_k^2$. This fact follows because $T_k R_k = cR_k^{1-a} = 2^{1-a}T_{k+1}R_{k+1}$, which implies that

$$\beta_k \le \frac{12B_\infty 2^{2(1-a)}T_{k+1}^2 R_{k+1}^2}{\eta_{k+1}\xi}, \tag{E.26}$$

since $0 < \eta_k \le 1$, where η_k is defined in (E.17). In (E.24) we applied the first term in the maximum in (E.17), which implies that $2H_k \le 4^{-1}n(\xi\eta_k)^2/[(9^2 + 96\cdot$

$2^{-2a})B_\infty T_k^2 R_k^2]$. In (E.25) we applied the second term in the maximum in (E.17), which implies that $\eta_k^2/[4\cdot(9^2+96\cdot 2^{-2a})T_k^2 R_k^2] \geq C'k/[c^2 R^{2-2a}]$. We may continue (E.19) with an upper bound

$$\sum_{k=1}^{\infty} \exp\left\{-\frac{n\xi^2 C' k}{c^2 B_\infty R^{2-2a}}\right\} \leq 2\exp\left\{-\frac{n\xi^2 C'}{c^2 B_\infty R^{2-2a}}\right\}. \tag{E.27}$$

We applied the fact that for $0 \leq a \leq 1/2$, $\sum_{k=1}^{\infty} a^k = a/(1-a) \leq 2a$; see (B.20). Here we need that $\exp\left\{-n\xi^2 C'/[c^2 B_\infty R^{2-2a}]\right\} \leq 1/2$, that is, $\xi \geq [\log_e 2/(nC')]^{1/2}\, cB_\infty^{1/2} R^{1-a}$ which is implied by (12.13).

Term P_{II} We have

$$g - h_g^{k,L} \leq \Delta_g^k,$$

and thus

$$\nu_n\left(Q(g - h_g^{\kappa_g,L})\right) \leq \nu_n\left(Q\Delta_g^{\kappa_g}\right) + 2E\left|Q\Delta_g^{\kappa_g}\right|.$$

Here we used the assumption that operator Q preserves positivity ($g \geq 0$ implies $Qg \geq 0$). We have for $k = 0, \ldots, K$,

$$\begin{aligned}
\max\left\{E\left|Q\Delta_g^k\right|^2 : g \in \mathcal{G}\right\} &\leq B_\infty \max\left\{\left\|Q\Delta_g^k\right\|_2^2 : g \in \mathcal{G}\right\} \\
&\leq B_\infty T_k^2 \max\left\{\left\|\Delta_g^k\right\|_2^2 : g \in \mathcal{G}\right\} \\
&\leq B_\infty T_k^2 R_k^2.
\end{aligned} \tag{E.28}$$

When $\kappa_g = k$, then $Q\Delta_g^{\kappa_g} \geq \beta_k$, for $k = 0, \ldots, K-1$. Thus, for $\kappa_g = k$, $k = 0, \ldots, K-1$, using (E.15), we have

$$E|Q\Delta_g^{\kappa_g}| \leq \beta_k^{-1} E|Q\Delta_g^k|^2 \leq \beta_k^{-1} B_\infty T_k^2 R_k^2 \leq \frac{\xi}{12},$$

and for $\kappa_g = K$,

$$E|Q\Delta_g^{\kappa_g}| \leq \left(E|Q\Delta_g^K|^2\right)^{1/2} \leq B_\infty^{1/2} T_K R_K \leq \frac{\xi}{12},$$

when we choose

$$K = \min\left\{k \geq 1 : 12 B_\infty^{1/2} T_k R_k < \xi\right\}. \tag{E.29}$$

Thus

$$P\left(\sup_{g\in\mathcal{G}} 2E|Q\Delta_g^{\kappa_g}| > \xi/6\right) = 0.$$

Define

$$\mathcal{G}^{(k)} = \{g \in \mathcal{G} : \kappa_g = k\}, \qquad k = 0, \ldots, K,$$

so that $\mathcal{G} = \bigcup_{k=0}^{K} \mathcal{G}^{(k)}$. Then

$$\begin{aligned} P_{II} &\le P\left(\sup_{g\in\mathcal{G}} \nu_n\left(Q\Delta_g^{\kappa_g}\right) \ge \xi/6\right) \\ &\le \sum_{k=0}^{K} P\left(\sup_{g\in\mathcal{G}^{(k)}} \nu_n\left(Q\Delta_g^{k}\right) \ge \xi/6\right) \\ &= P_{II}^{(0)} + P_{II}^{(1)}, \end{aligned} \tag{E.30}$$

where

$$P_{II}^{(0)} = P\left(\sup_{g\in\mathcal{G}^{(0)}} \nu_n\left(Q\Delta_g^{0}\right) \ge \xi/6\right),$$

$$P_{II}^{(1)} = \sum_{k=1}^{K} P\left(\sup_{g\in\mathcal{G}^{(k)}} \nu_n\left(Q\Delta_g^{k}\right) \ge \xi/6\right).$$

We now have

$$\#\left\{\Delta_g^k : g \in \mathcal{G}^{(k)}\right\} \le \#\left\{\Delta_g^k : g \in \mathcal{G}\right\} \le N_k. \tag{E.31}$$

It holds that

$$\left|Q\Delta_g^0 - EQ\Delta_g^0\right| \le 4B'_\infty. \tag{E.32}$$

Using (E.28), (E.31), (E.32), by Bernstein's inequality, we have

$$P_{II}^{(0)} \le N_0 \exp\left\{-\frac{1}{2}\frac{n(\xi/6)^2}{B_\infty T_0^2 R_0^2 + 2B'_\infty \xi/9}\right\}. \tag{E.33}$$

Let us turn to $P_{II}^{(1)}$. For $\kappa_g = k$ (i.e., when $g \in \mathcal{G}^{(k)}$), for $k = 1, \ldots, K$,

$$Q\Delta_g^k \le Q\Delta_g^{k-1} \le \beta_{k-1},$$

which implies that

$$\left|Q\Delta_g^k - EQ\Delta_g^k\right| \le 2\beta_{k-1}. \tag{E.34}$$

Thus, using (E.28), (E.31), (E.34), the fact that $0 < \eta_k \le 1$, where η_k is defined in (E.17), by Bernstein's inequality, for $k = 1, \ldots, K$, we have

$$\begin{aligned} &P\left(\sup_{g\in\mathcal{G}^{(k)}} \nu_n(Q\Delta_g^k) \ge \xi/6\right) \\ &\le P\left(\sup_{g\in\mathcal{G}^{(k)}} \nu_n(Q\Delta_g^k) \ge \xi\eta_k/6\right) \\ &\le N_k \exp\left\{-\frac{1}{2}\frac{n(\xi\eta_k/6)^2}{B_\infty T_k^2 R_k^2 + \beta_{k-1}\xi\eta_k/9}\right\} \\ &\le \exp\left\{H_k - \frac{1}{2}\frac{n(\xi\eta_k)^2}{6^2(1+2^{2(2-a)}/3)B_\infty T_k^2 R_k^2}\right\} \end{aligned} \tag{E.35}$$

$$
\begin{aligned}
&\leq \exp\left\{-\frac{1}{4}\frac{n(\xi\eta_k)^2}{(6^2+48\cdot 2^{-2a})B_\infty T_k^2 R_k^2}\right\} && \text{(E.36)}\\
&\leq \sum_{k=1}^{\infty}\exp\left\{-\frac{n\xi^2 C' k}{B_\infty c^2 R^{2-2a}}\right\}. && \text{(E.37)}
\end{aligned}
$$

In (E.35) we applied the fact that $\beta_{k-1}\xi\eta_k \leq 12B_\infty 2^{2(1-a)}T_k^2R_k^2$, which follows by using (E.26). In (E.36) we applied the first term in the maximum in (E.17), which implies that $H_k \leq 4^{-1}n(\xi\eta_k)^2/[(6^2+48\cdot 2^{-2a})B_\infty T_k^2 R_k^2]$, since $2^{-1}(6^2+48\cdot 2^{-2a}) \leq 9^2+96\cdot 2^{-2a}$. In (E.37) we applied the second term in the maximum in (E.17), which implies that $\eta_k^2/[4\cdot(6^2+48\cdot 2^{-2a})T_k^2R_k^2] \geq C'k/[c^2R^{2-2a}]$. We get

$$
P_{II}^{(1)} \leq \sum_{k=1}^{\infty}\exp\left\{-\frac{n\xi^2C'k}{B_\infty c^2R^{2-2a}}\right\} \leq 2\exp\left\{-\frac{n\xi^2C'}{B_\infty c^2R^{2-2a}}\right\}. \tag{E.38}
$$

In (E.38) we applied that for $0\leq a\leq 1/2$, $\sum_{k=1}^{\infty}a^k = a/(1-a)\leq 2a$. Here we need that $\exp\left\{-n\xi^2C'/[B_\infty c^2R^{2-2a}]\right\}\leq 1/2$, that is, $\xi \geq \left[\log_e 2/(nC')\right]^{1/2}B_\infty cR^{1-a}$ which is implied by (12.13).

Term P_{III} We have first,

$$
\#\left\{h_g^{0,L} : g\in\mathcal{G}\right\} \leq N_0,
$$

second,

$$
\sup_{g\in\mathcal{G}} E\left|Qh_g^{0,L}\right|^2 \leq B_\infty \sup_{g\in\mathcal{G}}\left\|Qh_g^{0,L}\right\|_2^2 \leq B_\infty T_0^2R_0^2 = B_\infty c^2R^{2-2a},
$$

and third,

$$
\sup_{g\in\mathcal{G}}\left\|Qh_g^{0,L}\right\|_\infty \leq B'_\infty.
$$

Thus, by Bernstein's inequality

$$
P_{III} \leq N_0\exp\left\{-\frac{1}{2}\frac{n(\xi/3)^2}{B_\infty T_0^2R_0^2+\xi 2B'_\infty/9}\right\}. \tag{E.39}
$$

Finishing the Proof The lemma follows from (E.16), (E.27), (E.33), (E.38), and (E.39), after we check some final facts. We need to check that $\sum_{k=1}^{\infty}\eta_k\leq 1$. Applying the calculations in (E.13) and (E.14), we get

$$
\begin{aligned}
\sum_{k=1}^{\infty}\eta_k &\leq (9^2+96\cdot 2^{-2a})^{1/2}\left(\frac{8^{1/2}B_\infty^{1/2}2G(R)}{n^{1/2}\xi}+2\sqrt{C'}C''\right)\\
&\leq \frac{1}{2}+\frac{1}{2}=1 && \text{(E.40)}
\end{aligned}
$$

when $\xi \geq 2\cdot 8^{1/2}6G(R)n^{-1/2}(9^2+96\cdot 2^{-2a})^{1/2}B_\infty^{1/2}$, which is guaranteed by (12.13), and C' is chosen as in (E.18). □

E.2.4 Proof of Lemma 12.6

We prove Lemma 12.6. We may continue (E.33) by

$$
\begin{aligned}
P_{II}^{(1)} &\le N_0 \exp\left\{-\frac{1}{2}\frac{n(\xi/6)^2}{B_\infty T_0^2 R_0^2 + 2B'_\infty \xi/9}\right\} \\
&\le \exp\left\{H_0 - \frac{1}{2}\frac{n\xi^2}{6^2(1+2/9)B_\infty c^2 R^{2(1-a)}}\right\} \qquad \text{(E.41)} \\
&\le \exp\left\{-\frac{1}{4}\frac{n\xi^2}{44 B_\infty c^2 R^{2(1-a)}}\right\}. \qquad \text{(E.42)}
\end{aligned}
$$

In (E.41) we applied the upper bound in (12.15) and the fact $T_0 R_0 = cR^{1-a}$. In (E.42) we applied the lower bound in (12.15), which implies that

$$H_0 \le 4^{-1} n\xi^2 / [44 B_\infty c^2 R^{2(1-a)}].$$

Also we may continue (E.39) by

$$
\begin{aligned}
P_{III} &\le N_0 \exp\left\{-\frac{1}{2}\frac{n(\xi/3)^2}{B_\infty T_0^2 R_0^2 + \xi 2B'_\infty/9}\right\} \\
&\le \exp\left\{H_0 - \frac{1}{2}\frac{n\xi^2}{3^2(1+2/9)B_\infty c^2 R^{2-2a}}\right\} \qquad \text{(E.43)} \\
&\le \exp\left\{-\frac{1}{4}\frac{n\xi^2}{11 B_\infty c^2 R^{2-2a}}\right\}. \qquad \text{(E.44)}
\end{aligned}
$$

In (E.43) we applied the upper bound in (12.15). In (E.44) we applied the lower bound in (12.15), which implies that $H_0 \le 4^{-1} n\xi^2/(11 B_\infty c^2 R^{2-2a})$.

E.2.5 Proof of Lemma 12.7

We prove Lemma 12.7. We follow Lugosi (2002). By Jensen's inequality, for all $s > 0$,

$$
\begin{aligned}
\exp\left\{sE \max_{1\le i\le k} Y_i\right\} &\le E\exp\left\{s \max_{1\le i\le k} Y_i\right\} = E \max_{1\le i\le k} \exp\{sY_i\} \\
&\le \sum_{i=1}^{k} E\exp\{sY_i\} \le k\exp\left\{\frac{s^2\sigma^2}{2}\right\}.
\end{aligned}
$$

Thus

$$E \max_{1\le i\le k} Y_i \le \frac{\log_e k}{s} + \frac{s\sigma^2}{2}.$$

Taking $s = \sqrt{2\log_e(k)/\sigma^2}$ yields

$$E \max_{1\le i\le k} Y_i \le \sigma\sqrt{2\log_e k}. \qquad \text{(E.45)}$$

We have that $\max_{1\le i\le k}|Y_i| = \max\{Y_1, -Y_1, \ldots, Y_k, -Y_k\}$, and thus the lemma follows from (E.45). □

E.2.6 Proof of Lemma 12.10

We prove Lemma 12.10. We follow Pollard (1989). Let us denote (with an abuse of notation) δ-nets by $\mathcal{G}_k = \mathcal{G}_{\delta_k}$, $k = 0, 1, \ldots$, where $\delta_k = \delta/2^k$, $\delta = R$. Denote also $T_k = \varrho(Q, \mathcal{G}_k)$. Now we have $\#\mathcal{G}_0 = 1$. (The optimal choice is to choose $\mathcal{G}_0 = \{g_0\}$ where $g_0 \in \mathcal{G}$ minimizes $\sup_{g\in\mathcal{G}} \|g - g_0\|_2$, and choose $\delta = \sup_{g\in\mathcal{G}} \|g - g_0\|_2$. We may assume that $0 \in \mathcal{G}$ and take $g_0 = 0$.) For $f \in \mathcal{G}_k$, let $g_f \in \mathcal{G}_{k-1}$ be such that $\|f - g_f\|_2 \le \delta_{k-1}$. Then, by the triangle inequality, we have

$$E \max_{f\in\mathcal{G}_k} |\nu_n(Qf)| \le E \max_{f\in\mathcal{G}_k} |\nu_n(Qg_f) - \nu_n(Qf)| + E \max_{g\in\mathcal{G}_{k-1}} |\nu_n(Qg)|. \tag{E.46}$$

We have

$$\nu_n(Qg_f) - \nu_n(Qf) \sim N\left(0, n^{-1}\|Q(g_f - f)\|_2^2\right),$$

and thus, by Lemma 12.7, we can write

$$E \max_{f\in\mathcal{G}_k} |\nu_n(Qg_f) - \nu_n(Qf)| \le n^{-1/2}\sqrt{2\log_e(2\#\mathcal{G}_k)} \max_{f\in\mathcal{G}_k} \|Q(g_f - f)\|_2. \tag{E.47}$$

Next we have

$$\max_{f\in\mathcal{G}_k} \|Q(g_f - f)\|_2 \le T_{k-1} \cdot \delta_{k-1}. \tag{E.48}$$

Thus

$$\begin{aligned}
& n^{1/2} E\left(\max_{f\in\mathcal{G}_k} |\nu_n(Qf)|\right) \\
\le\ & \sqrt{2\log_e(2\#\mathcal{G}_k)} \max_{f\in\mathcal{G}_k} \|Q(g_f - f)\|_2 \\
& + E\left(\max_{f\in\mathcal{G}_{k-1}} |\nu_n(Qf)|\right) && \text{(E.49)} \\
\le\ & \sqrt{2} T_{k-1}\delta_{k-1}\sqrt{\log_e(2\#\mathcal{G}_k)} + E\left(\max_{f\in\mathcal{G}_{k-1}} |\nu_n(Qf)|\right) && \text{(E.50)} \\
\le\ & 4\sqrt{2}\delta \sum_{i=0}^{k-1} 2^{-i-1} T_i \sqrt{\log_e(2\#\mathcal{G}_i)} && \text{(E.51)} \\
\le\ & 4\sqrt{2}\int_0^\delta \varrho(Q, \mathcal{G}_\epsilon)\sqrt{\log_e(2\#\mathcal{G}_\epsilon)}\, d\epsilon, && \text{(E.52)}
\end{aligned}$$

where in (E.49) we applied (E.46) and (E.47), in (E.50) we applied (E.48), in (E.51) we applied iteration, and in (E.52) we considered the sum as a lower bound to the Riemann integral, which is possible since the function $\epsilon \mapsto \varrho(Q, \mathcal{G}_\epsilon)\sqrt{\log \#\mathcal{G}_\epsilon}$ is decreasing on the interval $(0, \delta]$. □

E.2.7 Proof of Lemma 12.11

We prove Lemma 12.11. We follow Pollard (1989). We have by a Gaussian symmetrization that

$$E \sup_{g\in\mathcal{G}} |\nu_n(Qg)| \le 2\gamma^{-1} n^{-1/2} E \sup_{g\in\mathcal{G}} |Z_n(Qg, \mathbb{Y}_n)|, \tag{E.53}$$

where

$$Z_n(h, \mathbb{Y}_n) = n^{-1/2} \sum_{i=1}^{n} z_i \, h(Y_i), \qquad h : \mathcal{Y} \to \mathbf{R},$$

$z_1, \ldots, z_n$ are i.i.d. $N(0,1)$ random variables, independent from Y_i, $\gamma = E|z_i| = (2/\pi)^{1/2}$, and we denote $\mathbb{Y}_n = (Y_1, \ldots, Y_n)$.

Let $\mathcal{G}_{\delta_k}$, $k = 0, 1, \ldots$, be $\delta_k = \delta/2^k$ nets of $\mathcal{G}$ with respect to the metric $\|h\|_{2,n}^2 = n^{-1} \sum_{i=1}^n h^2(Y_i)$, where $\mathcal{G}_\delta = \{g_0\}$ for some $g_0 \in \mathcal{G}$, and $\delta = 2B_\infty$. Note that $\sup_{g \in \mathcal{G}} \|g - g_0\|_{2,n} \le 2B_\infty$. For $f \in \mathcal{G}_{\delta_k}$, let $g_f \in \mathcal{G}_{\delta_{k-1}}$ be such that $\|f - g_f\|_2 \le \delta_{k-1}$. Then, by the triangle inequality, we have

$$\begin{aligned} E \max_{f \in \mathcal{G}_{\delta_k}} |Z_n(Qf, \mathbb{Y}_n)| &= E \max_{f \in \mathcal{G}_{\delta_k}} |Z_n(Qg_f, \mathbb{Y}_n) - Z_n(Qf, \mathbb{Y}_n)| \\ &\quad + E \max_{g \in \mathcal{G}_{\delta_{k-1}}} |Z_n(Qg, \mathbb{Y}_n)|. \end{aligned} \tag{E.54}$$

Random variable $Z_n(Qg_f, \mathbb{Y}_n) - Z_n(Qf, \mathbb{Y}_n)$ is Gaussian conditionally on $\mathbb{Y}_n$:

$$Z_n(Qg_f, \mathbb{Y}_n) - Z_n(Qf, \mathbb{Y}_n) \mid \mathbb{Y}_n \sim N(0, \sigma_n^2),$$

where

$$\sigma_n^2 = \|Q(g_f - f)\|_{2,n}^2.$$

We have by Lemma 12.7 and by (12.18) (using Gaussianity) that

$$E\left[\max_{f \in \mathcal{G}_{\delta_k}} |Z_n(Qg_f, \mathbb{Y}_n) - Z_n(Qf, \mathbb{Y}_n)| \,\middle|\, \mathbb{Y}_n \right] \le \sigma_n \sqrt{2 \log_e(2\#\mathcal{G}_{\delta_k})}. \tag{E.55}$$

Denote $T_\delta = \varrho(Q, \mathcal{G}_\delta)$. Now

$$\sigma_n^2 \le T_{\delta_{k-1}}^2 \cdot \delta_{k-1}^2. \tag{E.56}$$

Thus

$$\begin{aligned} & E\left[\max_{f \in \mathcal{G}_{\delta_k}} |Z_n(Qf, \mathbb{Y}_n)| \,\middle|\, \mathbb{Y}_n \right] \\ &\le \sigma_n \sqrt{2 \log_e(2\#\mathcal{G}_{\delta_k})} + E\left[\max_{f \in \mathcal{G}_{\delta_{k-1}}} |Z_n(Qf, \mathbb{Y}_n)| \,\middle|\, \mathbb{Y}_n \right] && \text{(E.57)} \\ &\le T_{\delta_{k-1}} \delta_{k-1} \sqrt{2 \log_e \#(2\mathcal{G}_{\delta_k})} + E\left[\max_{f \in \mathcal{G}_{\delta_{k-1}}} |Z_n(Qf, \mathbb{Y}_n)| \,\middle|\, \mathbb{Y}_n \right] && \text{(E.58)} \\ &\le 2^{1/2} \cdot 4\delta \sum_{i=0}^{k-1} 2^{-i-1} T_{\delta_i} \sqrt{\log_e(2\#\mathcal{G}_{\delta_i})} && \text{(E.59)} \\ &\le 2^{1/2} \cdot 4 \int_0^\delta T_\epsilon \sqrt{\log_e(2\#\mathcal{G}_\epsilon)} \, d\epsilon, && \text{(E.60)} \end{aligned}$$

where in (E.57) we applied (E.54) and (E.55), in (E.58) we applied (E.56), in (E.59) we applied iteration, and in (E.60) we considered the sum as a lower bound to the Riemann integral, which is possible since the function $\epsilon \mapsto T_\epsilon \sqrt{\log_e \#\mathcal{G}_\epsilon}$ is decreasing on the interval $(0, \delta]$. □

E.2.8 Proof of Lemma 12.12

We prove Lemma 12.12. Let us denote $Z = \sup_{g \in \mathcal{G}} \nu_n(Qg)$,

$$\zeta = n^{-1/2}\tilde{G}(R), \qquad \alpha = \frac{C'}{c^2 R^{2-2a}} .$$

Now applying Lemma 12.4 obtains

$$\begin{aligned} EZ &= \int_0^\infty P(Z > u)\, du \\ &\le \zeta + \int_\zeta^\infty P(Z > u)\, du \\ &\le \zeta + \int_\zeta^\infty \exp\left\{-nu^2 \cdot \alpha\right\}\, du \\ &= \zeta + (2\alpha)^{-1/2}(\pi/2)^{1/2} n^{-1/2}, \end{aligned}$$

since $\int_0^\infty \exp\left\{-u^2/2\right\}\, du = (\pi/2)^{1/2}$. □

E.2.9 Proof of Lemma 12.13

We prove Lemma 12.13. Let us denote $Z = \sup_{g \in \mathcal{G}} \nu_n(g)$,

$$\zeta = n^{-1/2}\tilde{G}(R),$$

$$\alpha = \frac{C'}{B_\infty c^2 R^{2-2a}}, \qquad \beta_1 = 12 B_\infty c^2 R^{2-2a}, \qquad \beta_2 = \frac{24 B'_\infty}{9} .$$

Applying Lemma 12.5 obtains

$$\begin{aligned} &EZ \\ &= \int_0^\infty P(Z > u)\, du \\ &\le \zeta + \int_\zeta^\infty P(Z > u)\, du \\ &= \zeta + 4\int_\zeta^\infty \exp\left\{-nu^2 \cdot \alpha\right\}\, du + 2\#\mathcal{G}_R \int_\zeta^\infty \exp\left\{-\frac{nu^2}{\beta_1 + \beta_2 u}\right\}\, du \\ &\le \zeta + 4\int_\zeta^\infty \exp\left\{-nu^2 \cdot \alpha\right\}\, du \\ &\quad + 2\#\mathcal{G}_R \int_\zeta^{\max\{\zeta, \beta_1/\beta_2\}} \exp\left\{-\frac{nu^2}{2\beta_1}\right\}\, du \\ &\quad + 2\#\mathcal{G}_R \int_{\max\{\zeta, \beta_1/\beta_2\}}^\infty \exp\left\{-\frac{nu}{2\beta_2}\right\}\, du \\ &= \zeta + 4(2\alpha)^{-1/2}(\pi/2)^{1/2} n^{-1/2} + 2\#\mathcal{G}_R \beta_1^{1/2}(\pi/2)^{1/2} n^{-1/2} + 4\#\mathcal{G}_R \beta_2 n^{-1} \\ &= n^{-1/2}(2\pi)^{1/2}\left(\tilde{G}(R) + 2^{1/2}\alpha^{-1/2} + \#\mathcal{G}_R \beta_1^{1/2}\right) + 4\#\mathcal{G}_R \beta_2 n^{-1}, \end{aligned}$$

since $\int_0^\infty \exp\left\{-u^2/2\right\}\, du = (\pi/2)^{1/2}$, $\int_0^\infty \exp\left\{-u\right\}\, du = 1$. □

E.3 PROOFS FOR CHAPTER 16

Let

$$\hat{\theta} = \theta + n^{-1/2}\xi,$$

where $\theta \in \mathbf{R}$ and $\xi \sim N(0,1)$. Let $\lambda \geq 0$ and

$$w = \begin{cases} 1, & \text{when } |\hat{\theta}| \geq \lambda, \\ 0, & \text{when } |\hat{\theta}| < \lambda. \end{cases}$$

The following lemma was proved in Donoho and Johnstone (1994*b*, lmm. 1).

Lemma E.2 *When $\lambda > n^{-1/2}$, we have*

$$E\left(\theta - w\hat{\theta}\right)^2 \leq C\left(n^{-1}\left(\frac{\lambda}{n^{1/2}} + 1\right)\phi\left(\frac{\lambda}{n^{1/2}}\right) + \min\left\{\lambda^2, \theta^2\right\}\right),$$

where C is a positive constant and ϕ is the density of the standard univariate Gaussian distribution.

Proof. Below we denote $\epsilon = n^{-1/2}$.

1. When $|\theta| > \lambda$, then

$$\theta - w\hat{\theta} = \begin{cases} \theta = \hat{\theta} - \epsilon\xi, & |\hat{\theta}| \leq \lambda, \\ \epsilon\xi, & |\hat{\theta}| > \lambda. \end{cases}$$

   Thus $(\theta - w\hat{\theta})^2 \leq 2(\lambda^2 + \epsilon^2\xi^2)$ and

$$E(\theta - w\hat{\theta})^2 \leq 2(\lambda^2 + \epsilon^2) \leq 4\lambda^2$$

   when $|\theta| > \lambda \geq \epsilon$.

2. Denote $\rho_\epsilon(\lambda,\theta) = E(\theta - w\hat{\theta})^2$. We have that

$$\rho_\epsilon(\lambda,\theta) = \theta^2 P_\phi\left(\left[\frac{-\lambda-\theta}{\epsilon}, \frac{\lambda-\theta}{\epsilon}\right]\right) + \epsilon^2 \int_{\{y:|\epsilon y+\theta|>\lambda\}} y^2\phi(y)\, dy,$$

   where ϕ is the standard Gaussian density and P_ϕ is the corresponding probability measure. Let $|\theta| \leq \lambda$. Then we have for $\theta \geq 0$,

$$\begin{aligned} \rho_\epsilon(\lambda,\theta) &\leq \rho_\epsilon(\lambda,0) + \frac{\theta^2}{2}\sup_{t>0}\frac{\partial^2}{\partial t^2}\rho_\epsilon(\lambda,t) \\ &\leq \rho_\epsilon(\lambda,0) + 2\theta^2. \end{aligned}$$

When $\lambda > \epsilon$, we have

$$\rho_\epsilon(\lambda, 0) = \epsilon^2 \int_{\{y:|\epsilon y|>\lambda\}} y^2\phi(y)\,dy \le 2\epsilon^2\phi\left(\frac{\lambda}{\epsilon}\right)\left(\frac{\lambda}{\epsilon}+1\right).$$

By symmetry, we get the bounds also for $\theta > 0$. The inequalities can be checked numerically, or by using the calculations in Donoho and Johnstone (1994*b*, lmm. 1).

We get the claim by combining the items 1 and 2. □

E.4 PROOFS FOR CHAPTER 18

E.4.1 Proof of (18.26)

We prove (18.26). We prove that the best value for the error-complexity criterion, with the tree constraints $\mathcal{W}_{tree,J_n}(\mathcal{B}^*_{\alpha_n})$ for the weights, can be bounded by the best value for the error-complexity criterion without the tree constraints. We restate (18.26) as Lemma E.3. This lemma is a modification of Lemma 10.1 in Donoho (1997).

Lemma E.3 *If (18.23) holds, then*

$$\sup_{f\in B_{sp\infty}(L)} \min_{W\in\mathcal{W}_{tree,J_n}(\mathcal{B}^*_{\alpha_n})} K(f, W, \Theta_f(\mathcal{B}^*_{\alpha_n}), \mathcal{B}^*_{\alpha_n}, \alpha_n)$$
$$\le C \sup_{f\in B_{sp\infty}(L)} \min_{W\in\{0,1\}^{\mathcal{B}^*_{\alpha_n}}} K(f, W, \Theta_f(\mathcal{B}^*_{\alpha_n}), \mathcal{B}^*_{\alpha_n}, \alpha_n),$$

for a positive constant C, depending on s, p, L, d, where $B_{sp\infty}(L)$ is the Besov ball defined in (10.41), $\mathcal{W}_{tree,J}(\mathcal{B})$ is defined in (18.13), pre-basis $\mathcal{B}^*_{\alpha_n}$ is defined in (18.34), $K(f, W, \Theta, \mathcal{B}, \alpha)$ is defined in (15.102), and Θ_f is defined in (15.110).

Preliminary Lemma Before the proof we formulate a lemma that states that the equations in (15.111) imply an alternative expression for $K(f, W, \Theta, \mathcal{B}, \alpha)$ and they give the minimizer of $K(f, W, \Theta, \mathcal{B}, \alpha)$ over W.

Lemma E.4 *Let $\mathcal{B}$ be a basis of $L_2([0,1]^d)$, and let $\mathcal{B}_0 \subset \mathcal{B}$ be an orthonormal system. For each $W = (w_\phi)_{\phi\in\mathcal{B}_0} \in \{0,1\}^{\mathcal{B}_0}$,*

$$K(f, W, \Theta_f(\mathcal{B}_0), \mathcal{B}_0, \alpha) = \sum_{\phi\in\mathcal{B}_0\setminus\mathcal{S}(W)} \theta^2_{f,\phi} + \alpha\cdot(\#\mathcal{S}(W)+1) + \sum_{\phi\in\mathcal{B}\setminus\mathcal{B}_0} \theta^2_{f,\phi}, \tag{E.61}$$

where

$$\mathcal{S}(W) = \{\phi\in\mathcal{B}_0 : w_\phi = 1\}. \tag{E.62}$$

The minimum of $K(f, W, \Theta_f(\mathcal{B}_0), \mathcal{B}_0, \alpha)$ over $W \in \{0,1\}^{\mathcal{B}_0}$ is achieved by

$$W_{f,\alpha} = (w_{f,\phi})_{\phi\in\mathcal{B}_0}, \qquad w_{f,\phi} = w_{f,\phi,\alpha} = \begin{cases} 1, & \text{when } \theta^2_{f,\phi} > \alpha,\ \phi\in\mathcal{B}_0, \\ 0, & \text{otherwise,} \end{cases} \tag{E.63}$$

where $\theta_{f,\phi} = \int_{\mathbf{R}^d} f\phi$.

Proof of Lemma E.3 We prove Lemma E.3. Let W_{f,α_n} be defined in (E.63) and $\mathcal{S}(W_{f,\alpha_n})$ be defined in (E.62), for $\mathcal{B} = \mathcal{B}^*_{\alpha_n}$. Denote by $\mathcal{S}^*(W_{f,\alpha_n})$ the hereditary cover of $\mathcal{S}(W_{f,\alpha_n})$:

$$\mathcal{S}^*(W_{f,\alpha_n}) = \bigcup_{\phi \in \mathcal{S}(W_{f,\alpha_n})} \{\phi' : \phi \subset \phi'\}.$$

That is, for each $\phi \in \mathcal{S}(W_{f,\alpha_n})$ we form $\mathcal{S}^*(W_{f,\alpha_n})$ by adding all ancestors ϕ' of ϕ. We recall from Section 18.1.2, page 470 that ϕ' is an ancestor of ϕ if $\phi \subset \phi'$, and this means for $\phi = \phi^{(s(t))}_{\mathcal{I}(t)}, \phi' = \phi^{(s(t'))}_{\mathcal{I}(t')} \in \mathcal{B}^*_{\alpha_n}$, that $R_{\mathcal{I}(t')} \subset R_{\mathcal{I}(t)}$, where R_{jk} is defined in (18.10). Define the weights corresponding to locations $\mathcal{S}^*(W_{f,\alpha_n})$ by

$$W^*_{f,\alpha_n} = \left(I_{\mathcal{S}^*(W_{f,\alpha_n})}(\phi)\right)_{\phi \in \mathcal{B}^*_{\alpha_n}}.$$

We have that

$$W^*_{f,\alpha_n} \in \mathcal{W}_{tree,J_n}(\mathcal{B}^*_{\alpha_n}). \tag{E.64}$$

Thus, with basis $\mathcal{B}^*$ as defined in (18.35),

$$\begin{aligned}
&\min_{W \in \mathcal{W}_{tree,J_n}(\mathcal{B}^*_{\alpha_n})} K(f, W, \Theta_f(\mathcal{B}^*_{\alpha_n}), \mathcal{B}^*_{\alpha_n}, \alpha_n) \\
&\le K(f, W^*_{f,\alpha_n}, \Theta_f(\mathcal{B}^*_{\alpha_n}), \mathcal{B}^*_{\alpha_n}, \alpha_n) \\
&= \sum_{\phi \in \mathcal{B}^*_{\alpha_n} \setminus \mathcal{S}^*(W_{f,\alpha_n})} \theta^2_{f,\phi} + \alpha_n \cdot (\#\mathcal{S}^*(W_{f,\alpha_n}) + 1) + \sum_{\phi \in \mathcal{B}^* \setminus \mathcal{B}^*_{\alpha_n}} \theta^2_{f,\phi},
\end{aligned}$$

where we applied (E.61) and the fact that $\mathcal{S}(W^*_{f,\alpha_n}) = \mathcal{S}^*(W_{f,\alpha_n})$. When we prove

$$\sup_{f \in B_{sp\infty}(L)} \#\mathcal{S}^*(W_{f,\alpha_n}) \le C \sup_{f \in B_{sp\infty}(L)} \#\mathcal{S}(W_{f,\alpha_n}) \tag{E.65}$$

for a positive constant C, then the lemma follows by one more application of (E.61). We prove, first, that

$$\sup_{f \in B_{sp\infty}(L)} \#\mathcal{S}^*(W_{f,\alpha_n}) \le C' \alpha_n^{-1/(2\sigma+1)} \tag{E.66}$$

and second, that

$$\sup_{f \in B_{sp\infty}(L)} \#\mathcal{S}(W_{f,\alpha_n}) \ge C'' \alpha_n^{-1/(2\sigma+1)}. \tag{E.67}$$

The inequality (E.65) follows from (E.66) and (E.67).

Proof of (E.66) Denote

$$\begin{aligned}
\mathcal{S}(W_{f,\alpha_n})_m &= \left\{\phi^{(h(m+1))}_{\mathcal{J}_h(m),k} \in \mathcal{S}(W_{f,\alpha_n}) : k \in K_{\mathcal{J}_h(m)}\right\} \\
&= \left\{\phi^{(h(m+1))}_{\mathcal{J}_h(m),k} : k \in K_{\mathcal{J}_h(m)}, |\tau_{mk}| > \alpha_n^{1/2}\right\},
\end{aligned}$$

where τ_{mk} is defined in (18.36). Define also

$$\mathcal{S}^*(W_{f,\alpha_n})_m = \left\{\phi^{(h(m+1))}_{\mathcal{J}_h(m),k} \in \mathcal{S}^*(W_{f,\alpha_n}) : k \in K_{\mathcal{J}_h(m)}\right\}.$$

From Lemma 18.3 we get for $f \in B_{sp\infty}(L)$,

$$\#\mathcal{S}(W_{f,\alpha_n})_m \cdot \alpha_n^{\tilde{p}_{h(m)}/2} \le \sum_{k \in K_{\mathcal{J}_h(m)}} |\tau_{mk}|^{\tilde{p}_{h(m)}} \le \Lambda_m^{\tilde{p}_{h(m)}},$$

where

$$\Lambda_m = 2^{d/2} L 2^{-m(\sigma+1/2-1/\tilde{p}_{h(m)})}.$$

Thus

$$\#\mathcal{S}(W_{f,\alpha_n})_m \le \left(\frac{\Lambda_m}{\alpha_n^{1/2}}\right)^{\tilde{p}_{h(m)}}. \tag{E.68}$$

On the other hand, for $\mathcal{J}_h(m) = (j_1(m), \ldots, j_d(m))$,

$$\#\mathcal{S}^*(W_{f,\alpha_n})_m \le \#K_{\mathcal{J}_h(m)} = 2^{j_1(m)} \cdots 2^{j_d(m)} = 2^m, \tag{E.69}$$

since by the definition of $\mathcal{J}_h(m)$, $j_1(m) + \cdots + j_d(m) = m$. Let m^* be an integer such that

$$m^* \sim \frac{1}{2\sigma+1} \log_2 \alpha_n^{-1} \tag{E.70}$$

as $\alpha_n \to 0$. Note that $\mathcal{S}^*(W_{f,\alpha_n})_m$ consists of $\mathcal{S}(W_{f,\alpha_n})_m$ and of all ancestors of all $\phi \in \mathcal{S}(W_{f,\alpha_n})_{m'}$, $m' = m+1, \ldots$, that is,

$$\#\mathcal{S}^*(W_{f,\alpha_n})_m \le \sum_{m'=m}^{\infty} \#\mathcal{S}(W_{f,\alpha_n})_{m'}.$$

Thus

$$\sum_{m=m^*+1}^{\infty} \#\mathcal{S}^*(W_{f,\alpha_n})_m \le \sum_{m'=m^*+1}^{\infty} (m'-m^*)\#\mathcal{S}(W_{f,\alpha_n})_{m'}. \tag{E.71}$$

Denoting $\beta_m = \sigma + 1/2 - 1/\tilde{p}_{h(m)}$, we have,

$$\begin{aligned}
&\#\mathcal{S}^*(W_{f,\alpha_n}) \\
&\le \sum_{m=0}^{\infty} \#\mathcal{S}^*(W_{f,\alpha_n})_m && \text{(E.72)} \\
&\le \sum_{m=0}^{m^*} 2^m + \sum_{m=m^*+1}^{\infty} (m-m^*)\#\mathcal{S}(W_{f,\alpha_n})_m && \text{(E.73)} \\
&\le 2^{m^*+1} + \sum_{m=m^*+1}^{\infty} (m-m^*)\left(2^{d/2}L2^{-m\beta_m}\alpha_n^{-1/2}\right)^{\tilde{p}_{h(m)}} && \text{(E.74)}
\end{aligned}$$

$$\leq \ 2^{m^*+1} + C_{p,L,d}$$
$$\times \sum_{m=m^*+1}^{\infty} \left(\alpha_n^{-1/2} 2^{-m^*\beta_m}\right)^{\tilde{p}_{h(m)}} (m - m^*) 2^{\tilde{p}_{h(m)}(m^*-m)\beta_m} \quad \text{(E.75)}$$
$$\leq \ C2^{m^*}. \quad \text{(E.76)}$$

In (E.72) we applied the fact that

$$\mathcal{S}^*(W_{f,\alpha_n}) \subset \bigcup_{m=0}^{\infty} \mathcal{S}^*(W_{f,\alpha_n})_m.$$

In (E.73) we applied (E.69) and (E.71). In (E.74) we applied (E.68). In (E.75) we used the constant $C_{p,L,d}$ defined after display (18.43). In (E.76) we applied the fact that for $m \geq 1$,

$$\left(\alpha_n^{-1/2} 2^{-m^*(\sigma+1/2)}\right)^{\tilde{p}_{h(m)}} \sim 1$$

due to the choice of m^* in (E.70), and in (E.76) we applied also the fact that, denoting $\beta_* = \min_{l=1,\ldots,d}(\sigma + 1/2 - 1/\tilde{p}_l)$,

$$\sum_{m=m^*+1}^{\infty} (m - m^*) 2^{\tilde{p}_{h(m)}(m^*-m)\beta_m} \leq \sum_{m=1}^{\infty} m(2^{-\beta_*})^m < \infty,$$

which holds because $\tilde{p}_{h(m)} \geq 1$, and because $\beta_* > 0$ as is assumed in (18.23). We have proved (E.66).

Proof of (E.67) Let us remind that by the definitions

$$\mathcal{S}(W_{f,\alpha_n}) = \left\{\phi \in \mathcal{B}^*_{\alpha_n} : \theta^2_{f,\phi} > \alpha_n\right\}, \quad \text{(E.77)}$$

where $\theta_{f,\phi} = \int_{[0,1]^d} f\phi$. Let $m^* = m^*_{\alpha_n}$ be defined in (E.70); that is, m^* is an integer such that

$$2^{m^*} \sim \alpha_n^{-1/(2\sigma+1)}$$

when $\alpha_n \to 0$. Let $\psi \in B_{sp\infty}(L')$ where

$$L' = L/(2C_0), \qquad C_0 > \left(\min_{l=1,\ldots,d} \int_{\mathbf{R}^d} \psi\phi^{(l)}_{00}\right)^{-1},$$

$\phi^{(l)}_{00}$ is defined in (18.9), and ψ is such that $\int_{[0,1]^d} \psi\phi^{(l)}_{00} > 0$ for $l = 1, \ldots, d$. Define the periodization of ψ as

$$\psi_{per}(x) = \sum_{k \in K_{\mathcal{J}(m^*)}} \psi(x-k) I_{[0,1]^d}(x-k), \qquad x \in [0, 2^{j_1(m^*)}) \times \cdots \times [0, 2^{j_d(m^*)}).$$

Define

$$f_0(x) = \alpha_n^{1/2} C_0 2^{m/2} \psi_{per}(2^{\mathcal{J}(m)} x), \qquad x \in [0,1]^d,$$

where we denote $2^j x = (2^{j_1}x_1, \ldots, 2^{j_d}x_d)$ for vectors $j = (j_1, \ldots, j_d)$ and $x = (x_1, \ldots, x_d)$, and we denote $\mathcal{J} = (j_1, \ldots j_d)$. Let $\mathcal{A}_{\alpha_n} \subset \mathcal{B}^*_{\alpha_n}$ be defined by

$$\mathcal{A}_{\alpha_n} = \left\{ \phi^{(h(m^*+1))}_{\mathcal{J}(m^*),k} : k \in K_{\mathcal{J}(m^*)} \right\}.$$

For all $\phi = \phi^{(h(m^*+1))}_{\mathcal{J}(m^*),k} \in \mathcal{A}_{\alpha_n}$, by a change of integration,

$$\theta_{f_0,\phi} = \alpha_n^{1/2} C_0 \int_{\mathbf{R}^d} \psi \phi^{(h(m^*+1))}_{00} > \alpha_n^{1/2}. \tag{E.78}$$

Thus, applying (E.77) and (E.78), we have

$$\mathcal{A}_{\alpha_n} \subset \mathcal{S}(W_{f_0,\alpha_n}).$$

Also

$$\#\mathcal{A}_{\alpha_n} = \#\mathcal{K}_{\mathcal{J}(m^*)} = 2^{j_1(m^*)} \cdots 2^{j_d(m^*)} = 2^{m^*}.$$

We need to proof that

$$f_0 \in B_{sp\infty}(L). \tag{E.79}$$

Then, collecting the results, we get

$$\sup_{f \in B_{sp\infty}(L)} \#\mathcal{S}(W_{f,\alpha_n}) \geq \#\mathcal{S}(W_{f_0,\alpha_n}) \geq \#\mathcal{A}_{\alpha_n} \sim \alpha_n^{-1/(2\sigma+1)}$$

for sufficiently large n, and we have proved (E.67).

Proof of (E.79) Let $h \in (0,1)$. We have for $l = 1, \ldots, d$, when $1 \leq p_l < \infty$,

$$\begin{aligned}
&\|D^l_h f_0\|^{p_l}_{L_p(A^l_h)} \\
&= \int_{A^l_h} |f_0(x + he_l) - f_0(x)|^{p_l}\, dx \\
&= \sum_{k \in K_{\mathcal{J}(m^*)}} \int_{R_{\mathcal{J}(m^*),k} \cap A^l_h} |f_0(x + he_l) - f_0(x)|^{p_l}\, dx \\
&= \sum_{k \in K_{\mathcal{J}(m^*)}} \int_{A^l_h} (\alpha_n^{1/2} C_0 2^{m^*/2})^{p_l} \left| \psi_{per}(y + 2^{j_l(m^*)} he_l) - \psi_{per}(y) \right|^{p_l} 2^{-m^*}\, dy \\
&= \int_{A^l_h} (\alpha_n^{1/2} C_0 2^{m^*/2})^{p_l} \left| \psi_{per}\left(y + 2^{j_l(m^*)} he_l\right) - \psi_{per}(y) \right|^{p_l}\, dy \\
&\leq (\alpha_n^{1/2} C_0 2^{m^*/2})^{p_l} \cdot [L'(2^{j_l(m^*)} h)^{s_l}]^{p_l} \\
&\leq (Lh^{s_l})^{p_l}
\end{aligned}$$

because $\psi_{per} \in B_{sp\infty}(L')$, $\alpha_n^{1/2} 2^{m^*/2} \sim \alpha_n^{\sigma/(2\sigma+1)}$ and $2^{j_l(m^*)s_l} \leq 2^{\sigma m^* + 1} \sim 2\alpha_n^{-\sigma/(2\sigma+1)}$, where we used the fact that $j_l(m)s_l \leq m\sigma + 1$ following from (E.81) and (E.83). The proof of the case $p_l = \infty$ is even easier. We have proved (E.79) and thus (E.67). □

E.4.2 Proof of Lemma 18.3

We prove Lemma 18.3 Let $\mathcal{Z} = (z_1, \ldots, z_d)$ let defined in (18.30). Write $\mathcal{J} = \mathcal{J}^*$, $\mathcal{J} = (j_1, \ldots, j_d)$, where $\mathcal{J}^*$ is defined by (18.29) with $h = h^*$. Assume that we have

$$j_l(m) \ge z_l(m) - \frac{d}{2}, \qquad l \in \{1, \ldots, d\}, m \in \{0, 1, \ldots\}. \tag{E.80}$$

By the definition of $\mathcal{Z}$ we have

$$z_l(m) s_l = m\sigma. \tag{E.81}$$

Thus

$$j_l(m)\, s_l \ge \left(z_l(m) - \frac{d}{2} \right) s_l = m\sigma - \frac{ds_l}{2}$$

and also

$$2^{-j_l(m)s_l} \le 2^{-m\sigma + ds_l/2} \le 2^{-m\sigma + d/2}. \tag{E.82}$$

Lemma 18.3 follows from Lemma 10.3 and (E.82). It is left to prove (E.80).

Proof of (E.80) We prove in fact a slightly stronger property

$$z_l(m) - \frac{d}{2} \le j_l(m) \le z_l(m) + 1, \qquad l \in \{1, \ldots, d\},\ m \in \{0, 1, \ldots\}. \tag{E.83}$$

The case $m = 0$ holds because $j_l(0) = z_l(0) = 0$. Assume that (E.83) holds for m, $m \ge 0$. We prove (E.83) for $m + 1$. Now, at least for one $l_0 \in \{1, \ldots, d\}$,

$$j_{l_0}(m) \le z_{l_0}(m), \tag{E.84}$$

because $\sum_{l=1}^d j_l(m) = \sum_{l=1}^d z_l(m) = m$. On the other hand, only for one $l_0 \in \{1, \ldots, d\}$ we may have that

$$j_{l_0}(m) < z_{l_0}(m) - \frac{d}{2} + 1. \tag{E.85}$$

Indeed, if (E.85) would hold for $l_0, l_1, l_0 \ne l_1$, then applying the induction hypothesis, we get

$$\begin{aligned} m = \sum_{l=1}^d j_l(m) &< z_{l_0}(m) + z_{l_1}(m) - d + 2 + \sum_{l \in \{1,\ldots,d\} \setminus \{l_0, l_1\}} (z_l(m) + 1) \\ &= \sum_{l=1}^d z_l(m) = m. \end{aligned}$$

Assume that $h(m+1) = l_0$, that is, $j_{l_0}(m+1) = j_{l_0}(m) + 1$ and $j_l(m+1) = j_l(m)$ for $l \ne l_0$. First, l_0 has to satisfy (E.84) because l_0 must be a direction such that $j_{l_0}(m)$

is farthest below from $z_{l_0}(m)$. Thus, applying also the property $z_l(m) = mz_l(1)$ so that $z_l(m) = z_l(m+1) - z_l(1)$, we get

$$j_{l_0}(m+1) = j_{l_0}(m)+1 \le z_{l_0}(m)+1 = z_{l_0}(m+1) - z_l(1)+1 \le z_{l_0}(m+1)+1.$$

On the other hand, applying induction hypothesis, we have

$$\begin{aligned} j_{l_0}(m+1) &= j_{l_0}(m)+1 \ge z_{l_0}(m) - \frac{d}{2} + 1 = z_{l_0}(m+1) - z_l(1) - \frac{d}{2} + 1 \\ &\ge z_{l_0}(m+1) - \frac{d}{2}. \end{aligned}$$

For other indexes $l \in \{1, \ldots, d\} \setminus \{l_0\}$, which are not chosen at step $m+1$, we know that (E.85) does not hold; that is, $j_l(m) \ge z_l(m) - d/2 + 1$ because, if (E.85) would hold for l, then it would have to hold for l_0 (because l_0 was chosen and thus it is a direction for which $j_l(m)$ is farthest below from $z_l(m)$). But we know that (E.85) holds at most for one direction. Thus

$$j_l(m+1) = j_l(m) \ge z_l(m) - \frac{d}{2} + 1 = z_l(m+1) - z_l(1) - \frac{d}{2} + 1 \ge z_l(m+1) - \frac{d}{2}$$

and, applying the induction hypothesis, we get

$$j_l(m+1) = j_l(m) \le z_l(m) + 1 = z_l(m+1) - z_l(1) + 1 \le z_l(m+1) + 1.$$

We have proved (E.83), thus we have proved (E.80), and thus we have proved the lemma. □

PROBLEM SOLUTIONS

SOLUTIONS FOR CHAPTER 10

10.1 For notational convenience we take $d = 2$. Let F be the distribution function of (X_1, X_2), G be the distribution function of $(h_1(X_1), h_2(X_2))$, and C_G be the copula. Then

$$\begin{aligned} G(x_1, x_2) &= P(h_1(X_1) \leq x_1, h_2(X_2) \leq x_2) \\ &= P\left(X_1 \leq h_1^{-1}(x_1), X_2 \leq h_2^{-1}(x_2)\right) \\ &= F\left(h_1^{-1}(x_1), h_2^{-1}(x_2)\right), \end{aligned}$$

$$G_1(x_1) = F_1\left(h_1^{-1}(x_1)\right), \qquad G_2(x_2) = F_2\left(h_2^{-1}(x_2)\right),$$

and

$$G_1^{-1}(u_1) = h_1\left(F_1^{-1}(u_1)\right), \qquad G_2^{-1}(u_2) = h_2\left(F_2^{-1}(u_2)\right).$$

Thus

$$\begin{aligned} C_G(u_1, u_2) &= G\left(G_1^{-1}(u_1), G_2^{-1}(u_2)\right) \\ &= F\left(F_1^{-1}(u_1), F_2^{-1}(u_2)\right) \\ &= C(u_1, u_2). \end{aligned}$$

10.2 For notational convenience we take $d = 2$. By the triangle inequality, we get

$$|C(u) - C(v)| \leq |C(u) - C(v_1, u_2)| + |C(v_1, u_2) - C(v)|.$$

Taking $u_2 = 1$ obtains

$$|C(u) - C(v_1, u_2)| \leq |u_1 - v_1|,$$

and taking $v_1 = 1$ obtains

$$|C(v_1, u_2) - C(v)| \leq |u_2 - v_2|.$$

10.3 We prove the claim for $d = 2$. Choose $v_1 = v_2 = 1$ in (10.10), which gives

$$C(u_1, u_2) \geq u_1 + u_2 - 1.$$

In addition, any copula is nonnegative, which proves the lower bound. Choose $u_1 = 0$ and $v_2 = 1$ in (10.10), which gives

$$C(v_1, u_2) \leq v_1.$$

Choose $u_2 = 0$ and $v_1 = 1$ in (10.10), which gives

$$C(u_1, v_2) \leq v_2.$$

This proves the upper bound.

10.5 The proof uses just the fact that the integral in the definition of $g \in \mathcal{G}(\mathbb{G})$ may be approximated with a Riemann sum. Let $f \in \mathcal{G}(\mathbb{G})$,

$$f = \int_{\Theta^{(2)}} g\left(\cdot, \theta^{(1)}, \theta^{(2)}\right) dQ\left(\theta^{(2)}\right).$$

Let $\mathcal{P}_m$, $m = 1, 2, \ldots$, be a regular collection of rectangles covering $\Theta^{(2)}$ eventually and with vanishing volumes:

$$\Theta^{(2)} \subset \bigcup_{m=1}^{\infty} \bigcup_{R \in \mathcal{P}_m} R, \qquad \lim_{m \to \infty} \max_{R \in \mathcal{P}_m} \mathrm{vol}(R) = 0.$$

Let $\theta_R \in R$ be the center of $R \in \mathcal{P}_m$. For each $x \in \mathbf{R}^d$, $\theta^{(1)} \in \Theta^{(1)}$, $g(x, \theta^{(1)}, \cdot)$ is Riemann integrable. Thus, by the definition of the Riemann integral, we write

$$\begin{aligned} \lim_{m \to \infty} \sum_{R \in \mathcal{P}_m} g\left(x, \theta^{(1)}, \theta_R\right) Q(R) &= \int_{\Theta^{(2)}} g\left(x, \theta^{(1)}, \theta^{(2)}\right) dQ\left(\theta^{(2)}\right) \\ &= f(x). \end{aligned}$$

Jensen's inequality, Fubini's lemma, and assumption (10.89) imply that f is square integrable. Thus, by the dominated convergence theorem, we have

$$\lim_{m \to \infty} \int_{\mathbf{R}^d} \left(f(x) - \sum_{R \in \mathcal{P}_m} g\left(x, \theta^{(1)}, \theta_R\right) Q(R) \right)^2 dx = 0.$$

We have proved the lemma, since $\sum_{R\in\mathcal{P}_m} g\left(x,\theta^{(1)},\theta_R\right) Q(R) \in \mathrm{co}(\mathbb{G})$, for all $m = 1, 2, \ldots$

10.6 We can write the kernel of the Bessel potential as

$$G_s(x) = C_{s,d} \int_0^\infty e^{-\sigma^2} \sigma^{s-d-1} e^{-\|x\|^2/\sigma^2} \, d\sigma, \qquad x \in \mathbf{R}^d.$$

When $f \in L_{s,p}$, we have

$$f(x) = \int_0^\infty \int_{\mathbf{R}^d} \phi(\mu) \psi_{s,d}(\sigma) e^{-\|x-\mu\|^2/\sigma^2} \, d\mu \, d\sigma,$$

where $\psi_{s,d}(\sigma) = e^{-\sigma^2} \sigma^{s-d-1}$ and $\|\phi\|_p < \infty$. When $s > d$, then $\psi_{s,d}$ is integrable.

10.7 Let $f \in \mathcal{F}$. We have for $x \in \mathbf{R}^d$,

$$f(x) = (2\pi)^{-d} \int_{\mathbf{R}^d} e^{-i\omega^T x} F(f)(\omega) d\omega = \int_{\mathbf{R}^d} g_f(x,\omega) Q_f(d\omega),$$

where

$$g_f(x,\omega) = \frac{C_f}{\|\omega\|^s} e^{-i\omega^T x - i\theta_f(\omega)} = \frac{C_f}{\|\omega\|^s} \cos(\omega^T x + \theta_f(\omega)),$$

where $\theta_h(\omega)$ is the phase of $F(h)(\omega)$: $F(h)(\omega) = e^{i\theta_h(\omega)} |F(h)(\omega)|$,

$$C_h = (2\pi)^{-d} \int_{\mathbf{R}^d} \|\omega\|^s |F(h)(\omega)| d\omega,$$

and Q_h is the probability measure with $Q_h(A) = C_h^{-1} (2\pi)^{-d} \int_A \|\omega\|^s |F(h)(\omega)| d\omega$ for $A \subset \mathbf{R}^d$. We have $0 \le C_h \le L$. Thus there exists $\theta^{(1)} \in \Theta^{(1)}$ and a probability measure Q on $\Theta^{(2)}$ so that

$$f(x) = \int_{\Theta^{(2)}} g(x, \theta^{(1)}, \theta^{(2)}) \, dQ(\theta^{(2)}), \qquad x \in \mathbf{R}^d,$$

where $g \in \mathbb{G}$ and $\mathbb{G}$ is defined in the statement of the lemma.

10.8 Denote $\mathcal{G} = \{f : \|Ff\|_\infty \le B_\infty\}$. Denote $q_1 = \beta_1 - d/2$ and $q_2 = 2\beta_2$. Inclusion (10.91) follows from the facts that for $f \in \mathcal{F}_{\beta_1,1}(L') \cap \mathcal{G}$,

$$\int_{\mathbf{R}^d} \|\omega\|^{q_2} |(Ff)(\omega)|^2 \, d\omega \le B_\infty \int_{\mathbf{R}^d} \|\omega\|^{q_2} |(Ff)(\omega)| \, d\omega$$

and $q_1 \ge q_2 \Leftrightarrow \beta_1 \ge 2\beta_2 + d/2$. To prove (10.92), we apply the Cauchy inequality for $f \in \mathcal{F}_{\beta_2,2}(L'') \cap \mathcal{F}_{\beta_1 - d/2,2}(L''')$:

$$\begin{aligned}\int_{\mathbf{R}^d} \|\omega\|^{q_1} |(Ff)(\omega)| \, d\omega \quad \le \quad & \left[\int_{\mathbf{R}^d} (1 + \|\omega\|^{2\beta_2'})^{-1} d\omega \right]^{1/2} \\ & \times \left[\int_{\mathbf{R}^d} (1 + \|\omega\|^{2\beta_2'}) \|\omega\|^{2q_1} |(Ff)(\omega)|^2 \, d\omega \right]^{1/2},\end{aligned}$$

where $\beta_2' = \beta_2 - \beta_1 + d/2$. The first integral on the right-hand side is finite, since $\beta_2' > d/2$. The second integral on the right-hand side is finite, since $\rho_{\beta_1 - d/2,2}(f) < \infty$, $\rho_{\beta_2,2}(f) < \infty$, and $2(\beta_2' + q_1) = q_2$.

10.9 We prove that $\mathcal{H}_\delta(f_0) \subset S_s(L)$, where isotropic Sobolev ball $\mathcal{S}_s(L)$ is defined in (10.23) and (10.24). Let $f_a \in \mathcal{H}_\delta(f_0)$. We have

$$\rho_s(f_a) \le \rho_s(f_0) + \delta^{(2s+d)/(2s)} \rho_s\left(\sum_{k\in I} a_k G_k\right) \tag{5.86}$$

and

$$\rho_s^2\left(\sum_{k\in I} a_k G_k\right) = \sum_{|\alpha|=s} \left\| \sum_{k\in I} a_k D^\alpha G_k \right\|_2^2 .$$

For $|\alpha| = s$, $k \in I$,

$$D^\alpha G_k(x) = m^{s+d/2}(D^\alpha G)(mx - k).$$

Function G_k has the support on the cell S_k where cells S_k constitute an almost sure partition of $[0,1]^d$: $[0,1]^d = \cup_{k\in I} S_k$, where S_k is defined in (10.82). Thus, for $|\alpha| = s$,

$$\begin{aligned}
\left\| \sum_{k\in I} a_k D^\alpha G_k \right\|_2^2 &= \sum_{k\in I} \int_{S_k} (D^\alpha G_k)^2 \\
&= m^{2s+d} \sum_{k\in I} \int_{S_k} ((D^\alpha G)(mx - k))^2 \\
&= m^{2s}(\#I) \int_{[0,1]^d} (D^\alpha G)^2 \\
&\le \delta^{-(2s+d)/s} \int_{[0,1]^d} (D^\alpha G)^2 ,
\end{aligned}$$

since $\#I = m^d$ and $m \le \delta^{-1/s}$. Thus

$$\rho_s^2\left(\sum_{k\in I} a_k G_k\right) \le \delta^{-(2s+d)/s} \rho_s^2(G)$$

and

$$\rho_s(f_a) \le \rho_s(f_0) + \rho_s(G).$$

To obtain that $f_a \in \mathcal{S}_s(L)$, we need that $\rho_s(f_0) + \rho_s(G) \le L$. For example, we can choose f_0 so that $\rho_s(f_0) = 0$ ($f_0 = I_{[0,1]^d}$) and then let $g_i : [-\infty, \infty] \to \mathbf{R}$ belong to the univariate Sobolev balls, $g_i \in \mathcal{S}_s(L_i)$, where $\prod_{i=1}^d L_i = L$, $L_i > 0$.

10.10 Let us consider the case where $\mathcal{F}_s$ is an anisotropic Besov ball $\mathcal{B}_{s\infty\infty}(L)$ defined in (10.41). Let $f_a \in \mathcal{H}_\delta(f_0)$. Let l_i be the smallest integer $\geq s_i$. We have

$$\rho_s\left(\sum_{k\in I} a_k G_k\right) = \sum_{i=1}^{d} \sup_{|h|\leq 1} |h|^{-s_i} \left\|\sum_{k\in I} a_k \Delta_{he_i}^{l_i} G_k\right\|_{L_\infty(A_{i,hl_i})} .$$

Now, for $i \in \{1,\ldots,d\}$, $|h| < 1$, $k \in I$,

$$\Delta_{he_i}^{l_i} G_k(x) = m_1^{1/2}\cdots m_d^{1/2}(\Delta_{hm_ie_i}^{l_i} G)(m^T x - k)$$

as can be seen from (10.35). Then, for $i \in \{1,\ldots,d\}$, $|h| < 1$,

$$\begin{aligned}
\left\|\sum_{k\in I} a_k \Delta_{he_i}^{l_i} G_k\right\|_{L_\infty(A_{i,hl_i})} &= \max_{k\in I} \sup_{x\in A_{i,hl_i}\cap S_k} \left|\Delta_{he_i}^{l_i} G_k(x)\right| \\
&= m_1^{1/2}\cdots m_d^{1/2} \sup_{x\in A_{i,hl_i}} \left|\Delta_{hm_ie_i}^{l_i} G(x)\right| \\
&= m_1^{1/2}\cdots m_d^{1/2}\cdot L_i|hm_i|^{s_i} \\
&\leq \delta^{-(2\sigma+1)/(2\sigma)}\cdot L_i|h|^{s_i},
\end{aligned}$$

since $m_1^{1/2}\cdots m_d^{1/2} \leq \delta^{-1/(2\sigma)}$ and $m_i^{s_i} \leq \delta^{-1}$, and we have assumed that g_i belongs to the univariate Besov class $B_{s_i\infty\infty}(L_i)$. Thus

$$\rho_s\left(\sum_{k\in I} a_k G_k\right) \leq \delta^{-(2\sigma+1)/(2\sigma)} \sum_{i=1}^{d} L_i,$$

and using (5.86), we write

$$\rho_s(f_a) \leq \rho_s(f_0) + \sum_{i=1}^{d} L_i.$$

To obtain that $f_a \in \mathcal{B}_{s\infty\infty}(L)$, we need that $\rho_s(f_0) + \rho_s(G) \leq L$. For example, we can choose f_0 so that $\rho_s(f_0) = 0$ ($f_0 = I_{[0,1]^d}$) and then let $\sum_{i=1}^d L_i = L$.

SOLUTIONS FOR CHAPTER 11

11.1 Let $B = (f > g)$. The upper bound holds because

$$\int_{\mathbf{R}^d} |f-g| = 2\int_B (f-g) \leq 2\sup_A \left|\int_A f - \int_A g\right|.$$

The lower bound holds because for $A \subset \mathbf{R}^d$,

$$\left|\int_A f - \int_A g\right| = \left|\int_{A\cap B}(f-g) + \int_{A\cap B^c}(f-g)\right|$$

$$
\begin{aligned}
&\leq \max\left\{\int_{A\cap B}(f-g), \int_{A\cap B^c}(g-f)\right\} \\
&\leq \max\left\{\int_{B}(f-g), \int_{B^c}(g-f)\right\} \\
&= \frac{1}{2}\int_{\mathbf{R}^d}|f-g|.
\end{aligned}
$$

11.2 For $a, b \geq 0$, $(a-b)^2 \leq |a-b|\cdot|a+b| = |a^2-b^2|$. Thus

$$
\int_{\mathbf{R}^d}\left(\sqrt{f}-\sqrt{g}\right)^2 \leq \int_{\mathbf{R}^d}\left|\sqrt{f}-\sqrt{g}\right|\cdot\left|\sqrt{f}+\sqrt{g}\right| = \int_{\mathbf{R}^d}|f-g|.
$$

Second, by the Cauchy-Schwarz inequality,

$$
\begin{aligned}
\left(\int_{\mathbf{R}^d}|f-g|\right)^2 &\leq \int_{\mathbf{R}^d}\left(\sqrt{f}-\sqrt{g}\right)^2\int_{\mathbf{R}^d}\left(\sqrt{f}+\sqrt{g}\right)^2 \\
&= D_H^2(f,g)\left(2+2\int_{\mathbf{R}^d}\sqrt{fg}\right) \\
&= D_H^2(f,g)\left(4-D_H^2(f,g)\right) \\
&\leq 4D_H^2(f,g).
\end{aligned}
$$

11.3 We may write

$$
\int_{\mathbf{R}^d}(f-g)^2 = \int_{\mathbf{R}^d}\left(\sqrt{f}-\sqrt{g}\right)^2\left(\sqrt{f}+\sqrt{g}\right)^2.
$$

SOLUTIONS FOR CHAPTER 12

12.1 We have that

$$
\begin{aligned}
\int_t^\infty \exp\{-\psi(u)\}\,du &\leq \int_t^\infty \frac{\psi'(u)}{\psi'(t)}\exp\{-\psi(u)\}\,du \\
&= \frac{-1}{\psi'(t)}\int_t^\infty d(\exp\{-\psi(u)\})\,du \\
&= \frac{\exp\{-\psi(t)\}}{\psi'(t)}.
\end{aligned}
$$

This solution was given in Devroye (1987, p. 92).

12.2 Denote $\hat{f} = \hat{f}(\cdot,\mathcal{P})$, $\bar{f} = \bar{f}(\cdot,\mathcal{P})$ and

$$
\mathcal{G} = \left\{\frac{1}{\mathrm{vol}(R)}I_R : R \in \mathcal{P}\right\}.
$$

By Theorem 12.1 (Bernstein's inequality),

$$\begin{aligned} P\left(\|\hat{f}-\bar{f}\|_\infty > \xi\right) &= P\left(\max_{g\in\mathcal{G}} |\nu_n(g)| > \xi\right) \\ &\le \sum_{g\in\mathcal{G}} P\left(|\nu_n(g)| > \xi\right) \\ &\le \#\mathcal{G} \max_{g\in\mathcal{G}} P\left(|\nu_n(g)| > \xi\right) \\ &\le \#\mathcal{P}\cdot \exp\left\{-\frac{n\xi^2}{2}\cdot T\right\}, \end{aligned}$$

where

$$T = \frac{1}{A_2 + \xi A_\infty/3},$$

$$A_2 \stackrel{def}{=} \max_{g\in\mathcal{G}} \mathrm{Var}(g(X^1)) \le \max_{R\in\mathcal{P}} \frac{\int_R f}{\mathrm{vol}(R)^2} \le \|f\|_\infty \max_{R\in\mathcal{P}} \mathrm{vol}(R)^{-1}$$

and

$$A_\infty \stackrel{def}{=} \max_{g\in\mathcal{G}} \|g\|_\infty = \max_{R\in\mathcal{P}} \mathrm{vol}(R)^{-1}.$$

Thus

$$T \ge \left[\min_{R\in\mathcal{P}} \mathrm{vol}(R)\cdot(\|f\|_\infty + \xi/3)\right]^{-1}.$$

We have proved the lemma.

SOLUTIONS FOR CHAPTER 13

13.1 An evaluation tree may be an evaluation tree of a piecewise constant function whose partition is regular (equispaced) or irregular. In both cases we can construct an underlying (fine) equispaced partition: choose the support

$$[t,u] = [t_1,u_1]\times\cdots\times[t_d,u_d]$$

and the number of intervals for each direction, given by the vector $N = (N_1,\ldots,N_d)$, where $N_i \ge 1$ are integers. Then the underlying finest partition consists of the sets indexed with multi-indexes $\alpha = (\alpha_1,\ldots,\alpha_d)$,

$$A_\alpha = [a_\alpha, b_\alpha], \qquad \alpha \in \{0,\ldots,N_1-1\}\times\cdots\times\{0,\ldots,N_d-1\}, \tag{5.87}$$

where

$$[a_\alpha, b_\alpha] = [a_{\alpha_1}, b_{\alpha_1}]\times\cdots\times[a_{\alpha_d}, b_{\alpha_d}]$$

and

$$a_{\alpha_k} = t_k + \alpha_k\cdot\delta_k, \qquad b_{\alpha_k} = a_{\alpha_k} + \delta_k,$$

where the stepsize is $\delta_k = (u_k - t_k)/N_k, = 1, \ldots, d$. When we build an evaluation tree, it is enough to identify split points as integers $s \in \{1, \ldots, N_k - 1\}$, where the direction is $k \in \{1, \ldots, d\}$. The true split point is found by

$$s \leftrightarrow t_k + s \cdot \delta_k.$$

A multivariate rectangle is specified by giving a rectangle whose vertices are integers:

$$[\alpha, \beta] = [\alpha_1, \beta_1] \times \cdots \times [\alpha_d, \beta_d].$$

The true rectangle $[a, b]$ is found by

$$a_k = t_k + \alpha_k \cdot \delta_k, \qquad b_k = a_k + \delta_k.$$

Going through a Set Indexed with a Multi-Index To make a piecewise constant approximation of a function $f : \mathbf{R}^d \to \mathbf{R}$, we choose a grid point $x_\alpha \in A_\alpha$ in each rectangle A_α (the midpoint of A_α), where A_α is defined in (5.87), and evaluate $f(x_\alpha)$. Thus we evaluate f on the set

$$G = \{x_\alpha : \alpha = (\alpha_1, \ldots, \alpha_d), \ \alpha_i = 0, \ldots, N_i - 1\}.$$

The cardinality of the set G is $\#G = \Pi_{i=1}^d N_i$. We use the following procedure to calculate values $f(x)$ for $x \in G$:

- for $i = 0$ to $\#G - 1$,
  - find multi-index $\alpha = \text{multi}(i)$ corresponding to index i;
  - find $x \in G$ corresponding to multi-index α; if these are the mid-points of the rectangles then $x_k = t_k + \alpha_k \cdot \delta_k + \delta_k/2$; and
  - calculate f(x).

We need to define the procedure $\text{multi}(i)$. Consider the mapping $\alpha \mapsto i$ defined by

$$i = \alpha_1 + \alpha_2 N_1 + \alpha_3 N_1 N_2 + \cdots + \alpha_{d-1} N_1 \cdots N_{d-2} + \alpha_d N_1 \cdots N_{d-1}.$$

The inverse of this mapping is $i \mapsto \text{multi}(i) = \alpha$ defined by

$$\begin{aligned}
\alpha_d &= \left\lfloor i / \Pi_{j=1}^{d-1} N_j \right\rfloor \\
\alpha_{d-1} &= \left\lfloor \left(i - \alpha_d \Pi_{j=1}^{d-1} N_j\right) / \Pi_{j=1}^{d-2} N_j \right\rfloor \\
&\vdots \\
\alpha_2 &= \left\lfloor (i - v_2) / N_1 \right\rfloor, \quad v_2 = i - \alpha_d \Pi_{j=1}^{d-1} N_j - \cdots - \alpha_3 N_1 N_2 \\
\alpha_1 &= i - v_1, \quad v_1 = i - \alpha_d \Pi_{j=1}^{d-1} N_j - \cdots - \alpha_2 N_1.
\end{aligned}$$

13.2 Level set trees, shape trees, and tail trees are all based on a same type of idea where we grow a tree by finding separated components of each set in a decreasing

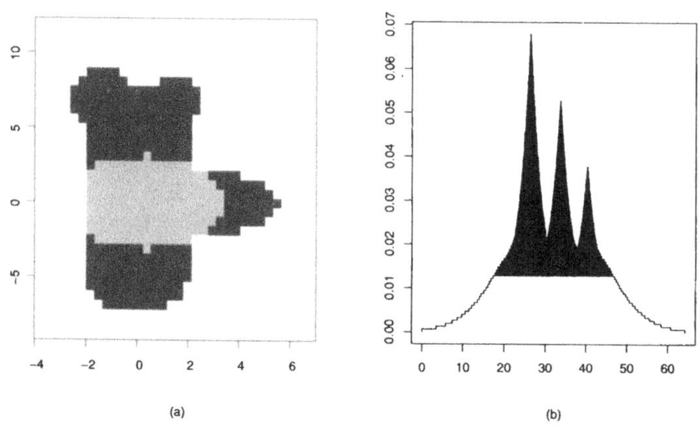

Figure 5.1 *(Illustration of the growing of a visualization tree.)* (a) In the case of a shape tree we look at the complements of balls intersected with the set; (b) in the case of a level set tree, we look at the half spaces intersected with the graph of the function.

sequence of sets. This idea of the definition of a visualization tree is illustrated in Figure 5.1. We take a shape tree as a basic concept, and we get the definition of a level set tree by making a shape tree of the graph of the function. The graph of the function is a connected set in the $d + 1$ dimensional Euclidean space. The decreasing sequence of sets consists of the half spaces intersected with the graph of the function.

The definition of a visualization tree uses the concepts of a ρ-connected set and ρ-separated sets, defined in Definition 6.1.

Definition 5.1 (Visualization tree.) A visualization tree *of* $A \subset \mathbf{M}$ *is associated with a resolution threshold* $\tau \geq 0$, *and with an increasing sequence* $\mathcal{R} = \{R_0 \subset \cdots \subset R_L\}$ *of radius sets, where* $R_l \subset \mathbf{M}$, $l = 0, \ldots, L$. *The nodes of the visualization tree are associated with subsets of* A *and sets in* $\mathcal{R}$ *in the following way:*

1. *Write*

$$A \setminus R_0 = A_1 \cup \cdots \cup A_K,$$

where sets A_i *are pairwise* τ*-separated, and each is* τ*-connected. The visualization tree has* K *root nodes which are associated with sets* A_i, $i = 1, \ldots, K$, *and each root node is associated with the same set* R_0.

2. *Let node* m *be associated with set* $B \subset A$ *and with set* $R_l \in \mathcal{R}$, $0 \leq l < L$.

   (a) *If* $B \setminus R_{l+1} = \emptyset$, *then node* m *is a leaf node.*

   (b) *Otherwise, write*

$$B \setminus R_{l+1} = C_1 \cup \cdots \cup C_M, \tag{5.88}$$

where sets C_i are pairwise separated, and each is connected, with the resolution threshold τ. Then node m has M children, which are associated with sets C_i, $i = 1, \ldots, M$, and each child is associated with the same set R_{l+1}.

We show how level set trees, shape trees, and tail trees are special cases of a visualization tree.

1. A level set tree of a function $f : \mathbf{R}^d \to \mathbf{R}$ is defined in Definition 4.2.

   0. We choose set A to be the graph of $f : \mathbf{R}^d \to \mathbf{R}$:

   $$A = \left\{ z = (x, y) \in \mathbf{R}^{d+1} : x \in \mathbf{R}^d, y \le f(x) \right\}.$$

   1. The level set tree is associated with the resolution threshold $\tau = 0$ (graphs are connected for $\rho = 0$ if the support of the function is connected). In Definition 4.2 the level set tree was associated with the sequence $\lambda_0, \ldots, \lambda_L$ of levels, and this defines a sequence of half spaces

   $$R_l = \{ z = (x, y) \in \mathbf{R}^{d+1} : x \in \mathbf{R}^d, y < \lambda_l \},$$

   $l = 0, \ldots, L$, which gives the increasing sequence of sets used in the definition of a visualization tree.

   2. In Definition 4.2 the nodes of a level set trees were annotated with levels and separated subsets of level sets. Associating the nodes with sets R_l is equivalent to associating the nodes with the levels λ_l. Associating the nodes with subsets of A leads to the annotation of the nodes with the separated subsets of level sets if we define for each $B \subset A$, $\Lambda(B) = \{ x \in \mathbf{R}^d : (x, y) \in B \}$. Set $\Lambda(B)$ is a separated subset of a level set corresponding to the separated subset B of the graph A.

2. A shape tree of a connected set $A \subset \mathbf{R}^d$ is defined in Definition 5.3.

   1. The shape tree is associated with the resolution threshold $\tau = 0$. In Definition 5.3 the shape tree was associated with the sequence $r_0, \ldots, r_L$ of radii, and this defines an increasing sequence of balls $R_l = B_{r_l}(\mu)$, where μ is a center point, which gives the increasing sequence of sets used in the definition of a visualization tree. We choose $r_0 = 0$ so that $R_0 = \emptyset$.

   2. Associating the nodes with sets R_l is equivalent to associating the nodes with the radii r_l.

3. A tail tree of a ρ-connected set $A = \{x_1, \ldots, x_n\} \subset \mathbf{R}^d$ is defined in Definition 6.3.

   1. The increasing sequence of sets used in the definition of a visualization tree is chosen to be the increasing sequence of balls $R_l = B_{r_l}(\mu)$, where

μ is a center point, and the grid of radii $0 = r_0 < r_1 < \ldots < r_{n-1}$ is chosen by $r_i = \|x_i - \mu\|$, $i = 1, \ldots, n-1$, where we assume that the points x_i are labeled so that $r_1 < \cdots < r_{n-1}$.

2. In Definition 6.3 the nodes of a tail trees were annotated with single observations. The observation annotated with a node is found by taking the intersection with the set $B \subset \{x_1, \ldots, x_n\}$ and set R_l associated with the node.

13.3 We may calculate the distance to the boundary of the rectangle with the following algorithm, which calculates the distance of vector **point** to the boundary of rectangle **rec**, where rectangle $[a_1, b_1] \times \cdots \times [a_d, b_d]$ is represented as vector $(a_1, b_1, \ldots, a_d, b_d)$.

```
dist<-0
for (i in 1:d){
   if (point[i]>rec[2*i])
        dist<-dist+(point[i]-rec[2*i])^2
   else if (point[i]<rec[2*i-1])
        dist<-dist+(point[i]-rec[2*i-1])^2
}
return(sqrt(dist))
```

13.4 We give the algorithm PairwiseComparison for the decomposition of a set.

1. **Input** of the algorithm is a list $\mathcal{A}$ of connected sets.

2. **Output** of the algorithm is a partition $\mathbb{P}$ of $\mathcal{A}$ as in (13.5) ($\mathbb{P}$ is a list of lists of sets in $\mathcal{A}$). This partition is initialized to be empty.

3. **Internal data structures** are stack S of sets, which is initialized to be empty, and a 0/1-labeling of each set $A \in \mathcal{A}$. We initialize label$(A) = 0$ for all $A \in \mathcal{A}$

ALGORITHM **PairwiseComparison**

1. **first loop**: go through sets in $\mathcal{A}$, consider set $A \in \mathcal{A}$;

2. **if** label$(A) = 0$ (set A is not already a member of some set in $\mathbb{P}$);

   (a) create empty list $\mathcal{C}$;

   (b) **push** A to the stack S, **set** label$(A) = 1$ (stack S will contain sets whose union is the connected component whose one member is A);

   (c) **while** stack S is not empty;

      i. **pop** set B from stack S;
      ii. **insert** and **store** B to the current component $\mathcal{C}$;
      iii. **second loop**: go through sets in $\mathcal{A}$, consider set $C \in \mathcal{A}$;

A. **if** C touches set B and label$(C) = 0$ (C is not already in set $\mathcal{C}$), **then push** C to the stack S, **set** label$(C) = 1$;

iv. **goto second loop**

(d) **goto while**

(e) **insert** and **store** the current component $\mathcal{C}$ to $\mathbb{P}$

3. **end if**

4. **goto first loop**

5. **return** $\mathbb{P}$

13.5 We give a more precise pseudo-code for the algorithm **DynaDecompose** than the pseudo-code given in Section 13.2.2, page 358.

The algorithm proceeds by joining those sets in the partitions annotated with the two children that are touching each other. (Partitions are represented as lists of lists of terminal nodes of the evaluation tree.) To find which sets are touching each other we have to travel through the boundaries of the sets. We have to consider four cases corresponding to whether left and right children exist, and whether the boundaries of the left and right children are empty or not.

1. **Input** and **Output** of the algorithm are given in Section 13.2.2, page 358.

2. **Internal data structures** of the algorithm are the following. We associate to every node m of the evaluation tree three lists of lists of terminal nodes of the evaluation tree: a list of lists $DC(m)$ of nodes in the separated components, a list of lists $LB(m)$ of nodes in the left boundaries of the separated components, and a list of lists $RB(m)$ of nodes in the right boundaries of the separated components.

ALGORITHM **DynaDecompose**

1. **travel** the evaluation tree, starting from the terminal nodes labeled by the list $\mathcal{L}$; traveling has to proceed so that the children are encountered before the parent;

2. assume we have encountered node m;

3. **if** node m is a leaf node, **then** we annotate node m with the partition consisting of the single rectangle that is annotated with this leaf node;

4. **else** (m is not a leaf node)

   (a) **if** left child of m does not exist, **then**

      i. the separated parts of m are inherited from the right child rm (which exists): $DC(m) = DC(rm)$;

      ii. the left boundary of m is empty: $LB(m)$ is empty list;

   iii. the right boundary of m is the right boundary of the right child: $RB(m) = RB(rm)$;

(b) **else, if** the right child does not exist, **then**

   i. the separated parts of m are inherited from the left child lm (which exists): $DC(m) = DC(lm)$;

   ii. the left boundary of m is the left boundary of the left child: $LB(m) = LB(lm)$;

   iii. the right boundary of m is is empty: $RB(m)$ is empty list;

(c) **else, if** the left boundary of the right child is empty **or** the right boundary of the left child is empty (so that there exists no connections between the sets of the left child and the sets of the right child), **then**

   i. the separated parts of m are inherited from the children, no parts are joined: $DC(m)$ is the concatenation of $DC(lm)$ and $DC(rm)$;

   ii. the left boundary of m is the left boundary of the left child (possibly empty): $LB(m) = LB(lm)$;

   iii. the right boundary of m is the right boundary of the right child (possibly empty): $RB(m) = RB(rm)$;

(d) **else**, (both children exist and both boundaries are nonempty);

   i. the separated parts are the result of joining the separated components of the left child and the right child: $DC(m)$ is formed by pairwise comparison of boundaries of components in $DC(lm)$ and $DC(rm)$ to find which components touch each other, and then concatenating the components which touch each other;

   ii. the left boundaries of m are the left boundaries of the left child (possibly some are joined): take $LB(m) = LB(lm)$ and then concatenate those lists for which the corresponding lists in $DC(lm)$ were concatenated in step 4(d)i to get $DC(m)$;

   iii. the right boundaries of m are the right boundaries of the right child (possibly some are joined): take $RB(m) = RB(rm)$ and then concatenate those lists for which the corresponding lists in $DC(rm)$ were concatenated in step 4(d)i to get $DC(m)$;

(e) **end if**

5. **end if**

6. **end travel**

7. **return** $DC(m)$ where m is the root node of the evaluation tree

The step 4(d)i is the most complex part of the algorithm. At this step we make pairwise comparisons of sets to find which sets touch each other. Note that in step 4(d)i we do not need always run through the complete boundaries, since we may stop immediately if we find a connection. Steps 4(a) through 4(c) of the algorithm are only bookkeeping.

We apply algorithm **PairwiseComparison**, designed in Exercise 13.4, in step 4(d)i to find which components of the left and the right child touch each other. The input is collection of sets $\mathcal{A}$, and in our application this is the collection of the intersections of the disconnected components, of the left and the right child, with respective boundaries. Furthermore these intersections are unions of rectangles.

13.6 We give a pseudo-code for algorithm RootsFirst. The algorithm calls a procedure "decompose" for partitioning a set of atoms. This procedure could be either given by DynaDecompose or by PairwiseComparison of Exercise 13.4.

1. **Input** for the algorithm is a function. Also an input is a grid of levels of the level set tree to be calculated: $0 < \lambda_1 < \cdots < \lambda_L$.

2. **Output** of the algorithm is a level set tree. With every node of the level set tree there is associated a set of grid points (separated components of level sets) and a level. The tree structure of the tree is uniquely determined by giving the parent for every node. See Appendix D.2.

3. An internal data structure of the algorithm is a stack S of nodes of the level set tree.

ALGORITHM **RootsFirst** for creating a level set tree

1. **decompose** the level set with level λ_1 to minimally separated components, these separated components are the root nodes of level set tree LST

2. put these components of level λ_1 to stack S

3. **if** we have more than one level ($L > 1$)

   (a) **while** stack S is not empty

      i. take a component C from stack S, assume that the level of this component is λ_l, $l \in \{1, \ldots, L\}$

      ii. make intersection with component C and the level set of level λ_{l+1}

      iii. **if** component C is the same as this intersection, **then**

         A. change the level of component C to be λ_{l+1}

         B. **if** $\lambda_{l+1} < \lambda_L$, **then** put component C back to stack S

      iv. **else** (intersection is a genuine subset of component C)

         A. **decompose** the intersection to separated components, assign the parent of these new components to be component C, add these new components to level set tree LST

         B. **if** $\lambda_{l+1} < \lambda_L$, **then** put new components to stack S

      v. **end if**

(b) **end while**

4. **end if**

5. **return** LST

Remarks We make two remarks concerning the algorithm.

1. In step 3(a)iii.A we have economized the level set tree by not taking such nodes to the tree whose associated set is the same but only the level is changing. That is, we create a level set tree without pillars, as defined in Definition 4.14.

2. In step 1 and step 3(a)iv.A we have called a procedure **decompose** that is the heart of the algorithm, and it is given either by **PairwiseComparison** or by **DynaDecompose**.

SOLUTIONS FOR CHAPTER 14

14.1 We have that

$$
\begin{aligned}
&\int_{\mathbf{R}^d} K(y)\,[f(x+z_i(y,h)) - f(x+z_{i-1}(y,h))]\,dy \\
&= \int_{\mathbf{R}^i} \Pi_{k=1}^{i} dy_k g_k(y_k)\,[f(x+z_i(y,h)) - f(x+z_{i-1}(y,h)] \\
&= \int_{\mathbf{R}^{i-1}} \Pi_{k=1}^{i-1} dy_k g_k(y_k) \int_{\mathbf{R}} g_i(y_i)\,[f(x+z_i(y,h)) - f(x+z_{i-1}(y,h))]\,dy_i \\
&= \int_{\mathbf{R}^{i-1}} \Pi_{k=1}^{i-1} dy_k g_k(y_k) \left\{ \int_{\mathbf{R}} g_i(y_i) f(x+z_i(y,h))\,dy_i - f(x+z_{i-1}(y,h)) \right\} \\
&= \int_{\mathbf{R}^{i-1}} \Pi_{k=1}^{i-1} dy_k g_k(y_k) \left\{ \sum_{j=1}^{l_i} (-1)^{j+l_i} \binom{l_i}{j} \frac{1}{j} \int_{\mathbf{R}} g\left(\frac{y_i}{j}\right) \right. \\
&\qquad \left. f(x+z_i(y,h))\,dy_i - f(x+z_{i-1}(y,h)) \right\} \\
&= \int_{\mathbf{R}^{i-1}} \Pi_{k=1}^{i-1} dy_k g_k(y_k) \\
&\qquad \left\{ \sum_{j=1}^{l} (-1)^{j+l_i} \binom{l_i}{j} \int_{\mathbf{R}} g(t)\, f(x+z_{i-1}(y,h)+jth_ie_i)\,dt \right. \\
&\qquad \left. - f(x+z_{i-1}(y,h)) \right\} \\
&= \int_{\mathbf{R}^{i-1}} \Pi_{k=1}^{i-1} dy_k g_k(y_k) \int_{\mathbf{R}} g(t)\,\Delta_{th_ie_i}^{l_i} f(x+z_{i-1}(y,h))\,dt.
\end{aligned}
$$

SOLUTIONS FOR CHAPTER 15

15.2 Let us denote with $\mathcal{B}$ a basis of $L_2(\mathbf{R}^d)$ such that $\mathcal{D} = \{\phi_1, \ldots, \phi_k\} \subset \mathcal{B}$. We have that

$$\int_{\mathbf{R}^d} \left(f - \sum_{j=1}^{k} \theta_j \phi_j \right)^2 = \sum_{j=1}^{k} \left(\theta_{f,\phi_j} - \theta_j \right)^2 + \sum_{g \in \mathcal{B} \setminus \mathcal{D}} \theta_{f,g}^2,$$

where $\theta_{f,g} = \int_{\mathbf{R}^d} fg$. The minimum is achieved by choosing $\theta_j = \theta_{f,\phi_j}$.

15.4 The inequality holds when $l = 1$. We assume that the inequality holds for l and prove that it holds for $l+1$:

$$\begin{aligned} \binom{k}{l+1} &= \binom{k}{l} \frac{k-l}{l+1} \\ &\le \left(\frac{ek}{l}\right)^l \frac{k-l}{l+1} \qquad (5.89) \\ &= \left(\frac{ek}{l+1}\right)^{l+1} \frac{1}{e} \left(\frac{l+1}{l}\right)^l \frac{k-l}{k} \\ &\le \left(\frac{ek}{l+1}\right)^{l+1}, \qquad (5.90) \end{aligned}$$

where in (5.89) we used the induction hypothesis and in (5.90) we used the fact that $((l+1)/l)^l = (1+1/l)^l \le e$.

SOLUTIONS FOR CHAPTER 16

16.1 We prove the fact for $d \ge 1$. Let $\bar{f}(\cdot, \mathcal{P})$ be the piecewise constant approximation of f defined in (12.21), where $\mathcal{P}$ is a partition of $[0,1]^d$. We may assume that f is continuous, since continuous functions are dense in $L_2([0,1]^d)$. Let $\mathcal{P}_M$ be a regular dyadic partition of $[0,1]^d$ of size M, and let $2^{-j(M)}$ be the side length of the rectangles. By the dominated convergence theorem it is enough to prove that for all $x \in [0,1]^d$,

$$\lim_{M \to \infty} \bar{f}(x, \mathcal{P}_M) = f(x).$$

For $x \in [0,1]^d$, let $R_M(x) \in \mathcal{P}_M$ be such that $x \in R_M(x)$. Let $\mu_M(x)$ be the center of $R_M(x)$. We have by the continuity of f that

$$\lim_{M \to \infty} f(\mu_M(x)) = f(x).$$

It is left to prove that

$$\lim_{M \to \infty} \left| \bar{f}(x, \mathcal{P}_M) - f(\mu_M(x)) \right| = 0.$$

We have that

$$\bar{f}(x, \mathcal{P}_M) = \frac{\int_{R_M(x)} f}{\text{vol}(R_M(x))}.$$

Then, denoting $\kappa = (2^{-j(M)}, \ldots, 2^{-j(M)}) \in \mathbf{R}^d$,

$$\begin{aligned}
&\bar{f}(x, \mathcal{P}_M) - f(\mu_M(x)) \\
&= \frac{1}{\text{vol}(R_M(x))} \int_{R_M(x)} [f - f(\mu_M(x))] \\
&= \int_{[0,1]^d} [f(\mu_R(x) + \kappa^T z) - f(\mu_R(x))] \, dz \longrightarrow 0,
\end{aligned}$$

as $M \to \infty$, where we applied the continuity of f and the dominated convergence theorem.

16.3 We may write

$$\begin{aligned}
&1 + \left(\frac{1}{n} \sum_{i=1}^{n} \psi_{0,0}(X_i) \right) \psi_{0,0} \\
&= 1 + \frac{1}{n} (n_{R_1} - n_{R_2})(I_{R_1} - I_{R_2}) \\
&= \frac{1}{n} (n + n_{R_1} - n_{R_2}) I_{R_1} + \frac{1}{n} (n - n_{R_1} + n_{R_2}) I_{R_2} \\
&= \frac{2n_{R_1}}{n} I_{R_1} + \frac{2n_{R_2}}{n} I_{R_2},
\end{aligned}$$

where $R_1 = [0, 1/2)$ and $R_2 = [1/2, 1)$.

SOLUTIONS FOR CHAPTER 17

17.1 We give a pseudo-code of algorithm **treeadd** that calculates a weighted average of two evaluation trees. We travel through the leaf nodes of $tr1$, and partition the sets associated with these leaf nodes by inducing the partition generated by $tr2$ to these sets.

Note that in the definition of an evaluation tree only leaf nodes were annotated with values of the function. In the following algorithm we have annotated also other nodes, see step 3.(b).ii.

1. **Input** of the algorithm are evaluation trees $tr1$ and $tr2$, and $p \in (0, 1)$.

2. **Output** of the algorithm is an evaluation tree that represents function $(1 - p)f_1 + pf_2$, where f_1 and f_2 are the functions represented by trees $tr1$ and $tr2$. We construct the output by further growing tree $tr1$.

3. **Internal data structures** of the algorithm are stacks $Sadd$ and $Str2$. Stack $Sadd$ will contain nodes that will be added to $tr1$ in order to get the final tree, and stack $Str2$ will contain nodes of $tr2$.

ALGORITHM **treeadd**($tr1, tr2, p$)

1. **go** through the leafs of tree $tr1$.
2. consider leaf $curleaf$ of tree $tr1$. We grow $tr1$ by making $curleaf$ the root node of a subtree of $tr1$. Initialize stacks: $Sadd[1] = curleaf$ and $Str2[1]$=root of $tr2$.
3. **while** stacks $Sadd$ and $Str2$ are not empty (the stacks will become empty simultaneously)
   (a) take from stack $Sadd$ node $curleaf$ of new tree and from stack $Str2$ node $nodeoftr2$ of tr2
   (b) **while** $nodeoftr2$ is not a leaf
      i. let $s \in \mathbf{R}$ be the split point and $k \in \{1, \ldots, d\}$ the direction annotated with node $nodeoftr2$
      ii. **if** s splits the set of the current node $curleaf$, **then**
         A. create left child $newleft$ and right child $newright$ for $curleaf$
         B. denote val = value of $curleaf$,
         $val.left$ = value of left child of $nodeoftr2$,
         $val.right$ = value of right child of $nodeoftr2$
         C. calculate mean: annotate left child $newleft$ with $(1-p) \cdot val + p \cdot val.left$ and annotate right child $newright$ with $(1-p) \cdot val + p \cdot val.right$
         D. put right child $newright$ to the stack $Sadd$ and the right child of node $nodeoftr2$ to the stack $Str2$
         E. set $nodeoftr2$=left child of node $nodeoftr2$
      iii. **else if** s is smaller than the left endpoint of the rectangle of the current node $curleaf$, in direction k, **then** set $nodeoftr2$=left child of node $nodeoftr2$
      iv. **else** (if s is larger than the right endpoint of the rectangle of the current node $curleaf$, in direction k, then) set $nodeoftr2$=right child of node $nodeoftr2$
   (c) **end while**
4. **end while**
5. **end go**
6. **return** $tr1$

SOLUTIONS FOR CHAPTER 18

18.1 1. Annotate the root with $(0, 0)$, where $0 \in \mathbf{R}^d$.

2. Assume a nonterminal node is annotated with direction $l \in \{1, \ldots, d\}$ and with pair of multi-indexes (j, k). Then the left child of the node is annotated with (j^0, k^0), where

$$\begin{cases} (j_l^0, k_l^0) = (j_l + 1, 2k_l), \\ (j_{l'}^0, k_{l'}^0) = (j_{l'}, k_{l'}), \quad l' \in \{1, \ldots, d\} \setminus \{l\} \end{cases}$$

The right child of the node is annotated with (j^1, k^1), where

$$\begin{cases} (j_l^1, k_l^1) = (j_l + 1, 2k_l + 1), \\ (j_{l'}^1, k_{l'}^1) = (j_{l'}, k_{l'}), \quad l' \in \{1, \ldots, d\} \setminus \{l\}. \end{cases}$$

18.2 *A series approximation* Define

$$\left(\mathcal{B}_{f,\alpha}^{00}, W_{f,\alpha}^{00}\right) = \text{argmin}_{\mathcal{B} \in \mathcal{L}(J), W \in \mathcal{W}_{tree,J}} K(f, W, \Theta_f(\mathcal{B}), \mathcal{B}, \alpha), \tag{5.91}$$

where the complexity-penalized theoretical error $K(f, W, \Theta, \mathcal{B}, \alpha)$ is defined in (15.102), $\mathcal{W}_{tree,J}$ is defined in (18.13), $\Theta_f(\mathcal{B})$ is the vector of coefficients of f defined in (15.110). Define

$$f_{f,\alpha}^{00}(x) = \tilde{f}\left(x, W_{f,\alpha}^{00}, \Theta_f(\mathcal{B}_{f,\alpha}^{00}), , \mathcal{B}_{f,\alpha}^{00}, \right), \qquad x \in \mathbf{R}^d, \tag{5.92}$$

where $\tilde{f}(\cdot, W, \Theta, \mathcal{B})$ is defined in (15.97).

A Theoretical Histogram Let $\bar{f}(\cdot, \mathcal{P})$ be the piecewise constant approximation to f defined in (12.21). Let $0 \leq \alpha < \infty$, and define the complexity-penalized theoretical error criterion

$$THEOCOPERR(f, \mathcal{P}, \alpha) = \left\|f - \bar{f}(\cdot, \mathcal{P})\right\|_2^2 + \alpha \cdot \#\mathcal{P}.$$

Define the partition corresponding to α as

$$\bar{\mathcal{P}}_{f,\alpha} = \text{argmin}_{\mathcal{P} \in \mathbb{P}(J)} THEOCOPERR(f, \mathcal{P}, \alpha), \tag{5.93}$$

where $\mathbb{P}(J)$ is defined in Definition 18.2. Define the approximation corresponding to α as

$$\bar{f}_{n,\alpha} = \bar{f}(\cdot, \bar{\mathcal{P}}_{f,\alpha}). \tag{5.94}$$

Equivalence The equivalence between optimal approximations is analogous to Lemma 18.1 stating the equivalence between estimators.

Lemma 5.5 *We have that $\bar{f}_{n,\alpha} = f_{f,\alpha}^{00}$ where $\bar{f}_{n,\alpha}$ is defined in (5.94) and $f_{f,\alpha}^{00}$ is defined in (5.92).*

Proof. To prove Lemma 5.5 we have to modify the proof of Lemma 18.1 only at the inductive step. Indeed we have only to note that if we split $R = R_{jk} \in \mathcal{P}$ to R_0 and R_1, then $\bar{f}(\cdot, f, \mathcal{P})$ changes so that we replace $I_R \int_R f/\text{vol}(R)$ with the term

$I_{R_0} \int_{R_0} f/\mathrm{vol}(R_0) + I_{R_1} \int_{R_1} f/\mathrm{vol}(R_1)$. When function $\tilde{f}(\cdot, W, \Theta_f)$ is changed by adding the term $\left(\int \phi_{j,k}^{s(j,k)} f\right) \phi_{j,k}^{s(j,k)}$, we see, by calculation, that the functions remain the same. We have proved the lemma. □

18.3 Lemma 5.6 *Let T be the tree grown in Section 18.2.1. Let t be some nonterminal node of T and t_{il}, $i = 1, 2$, $l = 1, \ldots, d$, be the children of t. Denote by R_t and R_{il} the rectangles annotated with these nodes. Denote the partition minimizing the complexity-penalized error criterion when we localize to rectangle R, annotated to a node of T, by*

$$\widehat{\mathcal{P}}_{n,\alpha}(R) = \mathrm{argmin}_{\mathcal{P} \in \tilde{\mathbb{P}}(R)} COPERR_n(\mathcal{P}, \alpha),$$

where $\tilde{\mathbb{P}}(R)$ is the set of partitions $\mathbb{P}(R, J')$, defined in Definition 18.2 and $J' = J - \mathrm{depth}(R)$, where $\mathrm{depth}(R)$ is the depth of the node with which R is annotated. Let

$$\begin{aligned} \mathcal{M} &= \min\left(\{COPERR_n(\{R_t\}, \alpha)\} \cup \right. \\ &\quad \left\{COPERR_n\left(\widehat{\mathcal{P}}_{n,\alpha}(R_{1l}), \alpha\right) + COPERR_n\left(\widehat{\mathcal{P}}_{n,\alpha}(R_{2l}), \alpha\right) : \right. \\ &\quad \left. l = 1, \ldots, d\}\right). \end{aligned}$$

Then

$$\widehat{\mathcal{P}}_{n,\alpha}(R_t) = \begin{cases} \{R_t\}, \text{ when } \mathcal{M} = COPERR_n(\{R_t\}, \alpha), \\ \widehat{\mathcal{P}}_\alpha(R_{1l}) \cup \widehat{\mathcal{P}}_\alpha(R_{2l}), \text{ when} \\ \mathcal{M} = COPERR_n\left(\widehat{\mathcal{P}}_{n,\alpha}(R_{1l}), \alpha\right) + COPERR_n\left(\widehat{\mathcal{P}}_{n,\alpha}(R_{2l}), \alpha\right). \end{cases}$$

Proof. When $\mathcal{P}_{il} \in \tilde{\mathbb{P}}(R_{il})$, $i = 1, 2$, $l = 1, \ldots, d$, we have

$$\begin{aligned} &COPERR_n(\mathcal{P}_{1l} \cup \mathcal{P}_{2l}, \alpha) \\ &= COPERR_n(\mathcal{P}_{1l}, \alpha) + COPERR_n(\mathcal{P}_{2l}, \alpha). \end{aligned} \tag{5.95}$$

Indeed $\mathcal{P}_{1l}$ and $\mathcal{P}_{2l}$ are partitions of disjoint rectangles, and thus (5.95) follows from (18.5) and from the fact that $\#(\mathcal{P}_{1l} \cup \mathcal{P}_{2l}) = \#\mathcal{P}_{1l} + \#\mathcal{P}_{2l}$. On the other hand,

$$\tilde{\mathbb{P}}(R_t) = \{\{R_t\}\} \cup \left\{\mathcal{P}_{1l} \cup \mathcal{P}_{2l} : \mathcal{P}_{il} \in \tilde{\mathbb{P}}(R_{il}), i = 1, 2, l = 1, \ldots, d\right\}.$$

We have proved the lemma. □

In particular, when we choose t in Lemma 17.1 to be the root of tree T, then $R_t = R_0$ and $\widehat{\mathcal{P}}_{n,\alpha}(R_t) = \widehat{\mathcal{P}}_{n,\alpha}$ is the global solution defined in (18.7).

18.4 We need two facts. First, for all samples,

$$\|f^*\|_\infty \leq n^\kappa \tag{5.96}$$

for some $\kappa > 0$. We may prove (5.96) by noting that by Lemma 18.1, $\|f^*\|_\infty = \|\hat{f}\|_\infty$ and $\|\hat{f}\|_\infty \le 2^{dJ}$, since 2^{-dJ} is the minimal volume of the rectangles in the partition of histogram $\hat{f}$. Second, we need that for $\tilde{\Omega}$ defined in (15.79), for sufficiently large n,

$$P\left(\tilde{\Omega}^c\right) \le \delta_n^* \stackrel{def}{=} n^{\kappa'} \exp\left\{-n^{1-a}\frac{3\|f\|_\infty}{8}\right\} \tag{5.97}$$

for some $\kappa' > 0$, where $0 < a < 1$ is the fineness parameter in (18.22). Equation (5.97) is proved below. (Note that also in the proof of (5.97) we apply Lemma 18.1.) Now the lemma follows from Theorem 15.10 and the bound

$$\begin{aligned} E_f \|f^* - f\|_2^2 1_{\tilde{\Omega}^c} &\le (\|f^*\|_\infty + \|f\|_\infty)^2 P\left(\tilde{\Omega}^c\right) \\ &\le \delta_n \stackrel{def}{=} (n^\kappa + \|f\|_\infty)^2 \delta_n^*, \end{aligned} \tag{5.98}$$

where we applied the fact that $\|g\|_2^2 \le \|g\|_\infty^2$ when the support of g is contained in $[0,1]^d$, κ is as in (5.96), and δ_n^* is defined in (5.97). We have proved the claim up to the proving (5.97). □

Proof of (5.97) Denote $B = 2\|f\|_\infty$. We have that

$$P\left(\|\hat{f}\|_\infty > B\right) \le P\left(\|\hat{f} - \bar{f}\|_\infty > \frac{B}{2}\right) + P\left(\|\bar{f}\|_\infty > \frac{B}{2}\right),$$

where $\hat{f} = \hat{f}(\cdot, \hat{\mathcal{P}})$ is the histogram with partition $\hat{\mathcal{P}}$, and $\bar{f} = \bar{f}(\cdot, \hat{\mathcal{P}})$, where $\bar{f}(\cdot, \mathcal{P})$ is as defined in (12.21). For each partition $\mathcal{P}$, $\|\bar{f}(\cdot, \mathcal{P})\|_\infty \le \|f\|_\infty$ and thus $P\left(\|\bar{f}\|_\infty > B/2\right) = 0$ because $B = 2\|f\|_\infty$. Finally, we apply Lemma 12.14. We choose $\xi = B/2 = \|f\|_\infty$. We have that $\min_{R \in \cup_{\mathcal{P}\in\mathbb{P}} \mathcal{P}} \mathrm{vol}(R) \ge n^{-a}$. We have that $\max_{\mathcal{P}\in\mathbb{P}} \#\mathcal{P} \le n^\kappa$ for some $\kappa > 0$; see Lemma 18.28. We have proved (5.97). □

18.5 Define the linear subspaces

$$V_{j_l} = \mathrm{span}\left(\{\eta_{j_l k_l}^{(0)} : k_l = 0, \ldots, 2^{j_l} - 1\}\right), \qquad j_l = 0, 1, \ldots, \tag{5.99}$$

where $\eta_{j_l k_l}^{(0)}(t) = \eta^{(0)}(2^{j_l} t - k_l)$, $\eta^{(0)} = I_{[0,1]}$. Define

$$W_{j_l} = \mathrm{span}\left(\left\{\eta_{j_l k_l}^{(1)} : k_l = 0, \ldots, 2^{j_l} - 1\right\}\right), \qquad j_l = 0, 1, \ldots, \tag{5.100}$$

where $\eta_{j_l k_l}^{(1)}(t) = \eta^{(1)}(2^{j_l} t - k_l)$, $\eta^{(1)} = I_{[1/2,1)} - I_{[0,1/2)}$. Note that $W_{j_l} = V_{j_l+1} \ominus V_{j_l}$. It holds that

$$\bigcup_{j_l=0}^{\infty} V_{j_l}^d \text{ is dense in } L_2([0,1]^d). \tag{5.101}$$

Indeed (5.101) holds because continuous functions are dense in $L_2([0,1]^d)$ and piecewise constant functions are dense among the set of continuous functions $[0,1]^d \to \mathbf{R}$.

We denote $V_{j_l}^{(0)} = V_{j_l}$ and $V_{j_l}^{(1)} = W_{j_l}$. Define the linear subspaces $U_0 = V_0^d$ and

$$U_{m+1} = \bigotimes_{l=1}^{d} V_{j_l(m)}^{(e_{h(m+1),l})}, \qquad m = 0, 1, \ldots,$$

where $e_l = (e_{l,1}, \ldots, e_{l,d})$, $l = 1, \ldots, d$, is the unit direction vector. For example, let $d = 2$ and $h(1) = 1, h(2) = 1, h(3) = 2, h(4) = 1, h(5) = 1, h(6) = 2$, then $U_0 = V_0^2$, $U_1 = W_0 \otimes V_0$, $U_2 = W_1 \otimes V_0$, $U_3 = V_2 \otimes W_0$, $U_4 = W_2 \otimes V_1$, $U_5 = W_3 \otimes V_1, U_6 = V_4 \otimes W_1, U_7 = V_4 \otimes W_2$. It holds that for $M \geq 0$,

$$\bigotimes_{l=1}^{d} V_{j_l(M)} = \sum_{m=0}^{M} U_m. \tag{5.102}$$

We prove (5.102) below. Note that by (18.57) and (5.102), and by the fact that $V_{j_1} \otimes \cdots \otimes V_{j_d} \subset V_{j_1'} \otimes \cdots \otimes V_{j_d'}$ when $j_l' \geq j_l, l = 1, \ldots, d$,

$$\bigcup_{j_l=0}^{\infty} V_{j_l}^d \subset \sum_{m=0}^{\infty} U_m.$$

Thus, applying (5.101), we find that $\sum_{m=0}^{\infty} U_m$ is dense in $L_2([0,1]^d)$. Also $U_m \perp U_{m'}$ for $m \neq m'$. Thus Lemma 18.5 follows from Lemma 16.1 because a basis of U_0 is $\{I_{[0,1]^d}\}$ and a basis of U_m, $m = 1, 2, \ldots$, is

$$\left\{\phi_{\mathcal{J}(m),k}^{(h(m+1))} : k \in K_{\mathcal{J}(m)}\right\}.$$

It is left to prove (5.102).

Proof of (5.102) We prove (5.102) by induction. We have that $V_0^d = U_0$ by the definition. We assume that (5.102) holds for M and aim to prove that it holds for $M + 1$. By the induction hypothesis we need to prove

$$\bigotimes_{l=1}^{d} V_{j_l(M+1)} = \bigotimes_{l=1}^{d} V_{j_l(M)} + U_{M+1}. \tag{5.103}$$

Let $l_0 \in \{1, \ldots, d\}$ be such that $h(M+1) = l_0$. Then $j_{l_0}(M+1) = j_{l_0}(M) + 1$ and $j_l(M+1) = j_l(M)$, for $l \neq l_0$. We have the following three facts.

1. A basis of $\bigotimes_{l=1}^{d} V_{j_l(M+1)}$ is

$$\left\{h_{\mathcal{J}(M+1),k}^{(0)} : k \in K_{\mathcal{J}(M+1)}\right\},$$

where

$$\begin{aligned} h_{\mathcal{J}(M+1),k}^{(0)}(x) &= \prod_{l=1}^{d} \eta_{j_l(M+1),k_l}^{(0)}(x_l), \\ &= \eta_{j_{l_0}(M)+1,k_{l_0}}^{(0)}(x_{l_0}) \prod_{l=1,l\neq l_0}^{d} \eta_{j_l(M),k_l}^{(0)}(x_l), \end{aligned}$$

$x = (x_1, \ldots, x_d)$.

2. A basis of $\bigotimes_{l=1}^{d} V_{j_l(M)}$ is

$$\left\{ h^{(0)}_{\mathcal{J}(M),k} : k \in K_{\mathcal{J}(M)} \right\},$$

where

$$h^{(0)}_{\mathcal{J}(M),k}(x) = \prod_{l=1}^{d} \eta^{(0)}_{j_l(M),k_l}(x_l), \qquad x = (x_1, \ldots, x_d).$$

3. A basis of U_{M+1} is

$$\left\{ \phi^{(l_0)}_{\mathcal{J}(M),k} : k \in K_{\mathcal{J}(M)} \right\}, \tag{5.104}$$

where

$$\phi^{(l_0)}_{\mathcal{J}(M),k}(x) = \eta^{(1)}_{j_{l_0}(M),k_{l_0}}(x_{l_0}) \prod_{l=1, l \neq l_0}^{d} \eta^{(0)}_{j_l(M),k_l}(x_l), \qquad x = (x_1, \ldots, x_d).$$

Thus functions $h^{(0)}_{\mathcal{J}(M+1),k}$, $h^{(0)}_{\mathcal{J}(M),k}$, and $\phi^{(l_0)}_{\mathcal{J}(M),k}$ differ only in the l_0th coordinate. We have

$$\eta^{(0)}_{j_l+1,k_l}(x_l) = 2^{-1/2} \left(\eta^{(0)}_{j_l,k_l/2}(x_l) - \eta^{(1)}_{j_l,k_l/2}(x_l) \right)$$

and

$$\eta^{(0)}_{j_l+1,k_l+1}(x_l) = 2^{-1/2} \left(\eta^{(0)}_{j_l,k_l/2}(x_l) + \eta^{(1)}_{j_l,k_l/2}(x_l) \right).$$

Thus

$$\begin{aligned} &\left\{ h^{(0)}_{\mathcal{J}(M+1),k} : k \in K_{\mathcal{J}(M+1)} \right\} \\ &= \left\{ 2^{-1/2} \left(h^{(0)}_{\mathcal{J}(M),k} - \phi^{(l_0)}_{\mathcal{J}(M),k} \right), 2^{-1/2} \left(h^{(0)}_{\mathcal{J}(M),k} + \phi^{(l_0)}_{\mathcal{J}(M),k} \right) : k \in K_{\mathcal{J}(M)} \right\}. \end{aligned}$$

We have proved (5.103) and thus (5.102). □

SOLUTIONS FOR CHAPTER 19

19.1 We have by the mean value theorem, Cauchy–Schwartz inequality, and convexity of Θ that for $\theta, \theta' \in \Theta$,

$$|g(x,\theta) - g(x,\theta')| = \left| (\theta - \theta')^T D_\theta\, g(x,\theta)|_{\theta=\tau} \right| \leq \|\theta - \theta'\| \cdot \sup_{\theta \in \Theta} \|D_\theta g(x,\theta)\|,$$

where $\tau \in \{\theta + t(\theta' - \theta) : 0 \leq t \leq 1\}$. Thus

$$\|g(\cdot,\theta) - g(\cdot,\theta')\|_2 \leq \|\theta - \theta'\| \cdot \left\| \sup_{\theta \in \Theta} \|Dg(\cdot,\theta)\| \right\|_2.$$

We have proved the claim.

REFERENCES

Akaike, H. (1973), Information theory and an extension of the maximum likelihood principle, *in* P. Petrov and F. Csaki, eds, 'Proc. 2nd Int. Symp. Information Theory', Akademia Kiado, Budapest, pp. 267–281.

Anderson, E. (1960), 'A semi-graphical method for the analysis of complex problems', *Technometrics* **2**, 287–292.

Andrews, D. (1972), 'Plots of high-dimensional data', *Biometrika* **28**, 125–136.

Ankerst, M., Breunig, M. M., Kriegel, H.-P. and Sander, J. (1999), OPTICS: ordering points to identify the clustering structure, *in* 'Proc. ACM SIGMOID'99 Int. Conf. on Management of Data', Philadelphia, pp. 49–60.

Asimov, D. (1985), 'The grand tour: a tool for viewing multidimensional data', *SIAM J. of Scient. and Statist. Comp.* **6**, 128–143.

Assouad, P. (1983), 'Deux remarques sur l'estimation', *Comptes Rendus Acad. Sci. Paris (A)* **296**, 1021–1024.

Azzalini, A. (2005), 'The skew-normal distribution and related multivariate families', *Scand. J. Statist.* **32**(2), 159–188.

Barnett, V. (1976), 'The ordering of multivariate data', *J. Roy. Statist. Soc. Ser. A* **139**, 319–354.

Barron, A. (1993), 'Universal approximation bounds for superpositions of a sigmoidal function', *IEEE Trans. Inform. Theory* **39**, 930–945.

Barron, A., Birgé, L. and Massart, P. (1999), 'Risk bounds for model selection via penalization', *Probab. Theory Relat. Fields* **113**, 301–413.

Bass, R. F. (1985), 'Law of the iterated logarithm for set-indexed partial sum processes with finite variance', *Z. Wahrsch. Verw. Gebiete* **65**, 181–237.

Becker, R., Clark, L. A. and Lambert, D. (1994), 'Cave plots: A graphical technique for comparing time series', *J. Comp. Graph. Stat.* **3**(3), 277–283.

Bellman, R. E. (1961), *Adaptive Control Processes*, Princeton University Press, Princeton, NJ.

Beran, R. (1977), 'Minimum Hellinger distance estimates for parametric models', *Ann. Statist.* **5**, 445–463.

Bergh, J. and Löfström, J. (1976), *Interpolation Spaces: An Introduction*, Springer, Berlin.

Bertin, J. (1967), *Semiologie Graphique*, Gauthier Villars, Paris.

Bertin, J. (1981), *Graphics and Graphic Information-Processing*, de Gruyter, Berlin.

Biau, G. and Devroye, L. (2005), 'Density estimation by the penalized combinatorial method', *J. Multivariate Anal.* **94**, 196–208.

Bickel, P. J., Klaassen, C. A. J., Ritov, Y. and Wellner, J. A. (1993), *Efficient and Adaptive Estimation for Semiparametric Models*, Johns Hopkins University Press, Baltimore.

Birgé, L. (1983), 'Approximation dans les espaces métriques et théorie de l'estimation', *Z. Wahrsch. Verw. Gebiete* **65**, 181–237.

Birgé, L. (2001), A new look at an old result: Fano's lemma, Prépublication 632, Universités Paris 6 et 7.

Birgé, L. and Massart, P. (1993), 'Rates of convergence for minimum contrast estimators', *Probab. Theory Relat. Fields* **97**, 113–150.

Birgé, L. and Massart, P. (1997), From model selection to adaptive estimation, *in* D. Pollard, E. Torgerson and G. Yang, eds, 'Festschrift for Lucien Le Cam: Research Papers in Probability and Statistics', Springer, Berlin, pp. 55–87.

Bock, H. H. (1974), *Automatische Klassifikation*, Vandenhoeck and Ruprecht.

Bosq, D. (1998), *Nonparametric Statistics for Stochastic Processes*, Vol. 110 of *Lecture Notes in Statistics*, Springer, Berlin.

Bouchaud, J.-P. and Potters, M. (2003), *Theory of Financial Risks*, Cambridge University Press, Cambridge.

Bousquet, O. (2002), 'A Bennett concentration inequality and its application to suprema of empirical processes', *C. R. Acad. Sci. Paris I* **334**, 495–500.

Breiman, L. (1993*a*), 'Hinging hyperplanes for regression, classification, and function approximation', *IEEE Trans. Inform. Theory* **39**(3), 999–1013.

Breiman, L. (1993*b*), *Probability*, 2nd edn, SIAM, Philadelphia, PA.

Breiman, L. (1996*a*), 'Bagging predictors', *Machine Learning* **24**, 123–140.

Breiman, L. (1996*b*), 'Heuristics of instability and stabilization in model selection', *Ann. Statist.* **24**, 2350–2383.

Breiman, L. (2001), 'Random forests', *Machine Learning* **45**, 5–32.

Breiman, L., Friedman, J., Olshen, R. and Stone, C. J. (1984), *Classification and Regression Trees*, Chapman and Hall, New York.

Bretagnolle, J. and Huber, C. (1979), 'Estimation des densités: Risque minimax', *Z. Wahrsch. Verw. Gebiete* **47**, 119–137.

Brown, L. and Low, M. (1996), 'A constrained risk inequality with applications to nonparametric estimation', *Ann. Statist.* **24**, 2524–2535.

Bühlmann, P. and Yu, B. (2002), 'Analyzing bagging', *Ann. Statist.* **30**, 927–961.

Cacoullos, T. (1966), 'Estimation of a multivariate density', *Ann. Inst. Statist. Math.* **18**, 179–189.

Card, S. K., Mackinlay, J. D. and Shneiderman, B. (1999), *Readings in information visualization: using vision to think*, Morgan Kaufmann, San Francisco.

Carl, B. (1997), 'Metric entropy of convex hulls in Hilbert spaces', *Bull. London Math. Soc.* **29**, 452–458.

Carl, B., Kyrezi, I. and Pajor, A. (1999), 'Metric entropy of convex hulls in Banach spaces', *J. London Math. Soc.* **60**(2), 871–896.

Carlsson, G. (2009), 'Topology and data', *Bulletin Am. Math. Soc.* **46**(2), 255–308.

Carr, H., Snoeyink, J. and Axen, U. (2003), 'Computing contour trees in any dimension', *Comput. Geometry: Theory Appl.* **24**(2), 75–94.

Carr, H., Snoeyink, J. and van de Panne, M. (2004), Simplifying flexible isosurfaces with local geometric measures, *in* 'VIS '04: Proc. Visualization '04', IEEE Computer Society, Washington, DC, pp. 497–504.

Carroll, R. J. and Hall, P. (1988), 'Optimal rates of convergence for deconvolving a density', *J. Amer. Statist. Assoc.* **83**, 1184–1186.

Castellan, G. (1999), Modified Akaike's criterion for histogram density estimation, Prépublication d'Orsay 61, Université Paris-Sud.

Catoni, O. (2002), Data compression and adaptive histograms, *in* F. Cucker and J. M. Rojas, eds, 'Foundations of Computational Mathematics, Proc. Smalfest 2000', World Scientific, Singapore, pp. 35–60.

Cencov, N. N. (1963), 'Evaluation of an unknown distribution density from observations', *Soviet Math.* **4**, 1559–1562.

Cencov, N. N. (1972), *Statistical Decision Rules and Optimal Inference*, Nauka, Moscow.

Chaudhuri, P. (1996), 'On a geometric notion of quantiles for multivariate data', *J. Amer. Statist. Assoc.* **91**, 862–872.

Chaudhuri, P. and Marron, J. S. (1999), 'Sizer for exploration of structures in curves', *J. Amer. Statist. Assoc.* **94**, 807–823.

Chaudhuri, P. and Marron, J. S. (2000), 'Scale space view of curve estimation', *Ann. Statist.* **28**, 408–428.

Chen, C. (2004), *Information Visualization: Beyond the Horizon*, Springer, Berlin. (1st ed. 1999).

Chen, J. (1995), 'Optimal rate of convergence for finite mixture models', *Ann. Statist.* **23**(1), 221–233.

Chernoff, H. (1973), 'Using faces to represent points in k-dimensional space graphically', *J. Amer. Statist. Assoc.* **68**, 361–368.

Chiang, Y.-J., Lenz, T., Lu, X. and Rote, G. (2005), 'Simple and optimal output-sensitive construction of contour trees using monotone paths', *Comput. Geometry: Theory Appl.* **30**, 165–195.

Chou, P., Lookabaugh, T. and Gray, R. M. (1989), 'Optimal pruning with applications to tree-structured source coding and modeling', *IEEE Trans. Inform. Theory* **37**, 31–42.

Clayton, D. G. (1978), 'A model for association in bivariate life tables and its application in epidemiological studies of familial tendency in chronic disease incidence', *Biometrika* **65**, 141–151.

Cleveland, W. S. (1993*a*), 'A model for studying display methods of statistical graphics', *J. Comput. Graph. Statist.* **2**(4), 323–343.

Cleveland, W. S. (1993*b*), *Visualizing Data*, Hobart Press, NJ.

Cleveland, W. S. (1994), *The Elements of Graphing Data*, Hobart Press, NJ.

Cleveland, W. S., McGill, M. E. and McGill, R. (1988), 'The shape parameter of a two-variable graph', *J. Amer. Statist. Assoc.* **33**, 289–300.

Coifman, R. R. and Wickerhauser, M. V. (1992), 'Entropy-based algorithms for best-basis selection', *IEEE Trans. Inform. Theory* **38**, 713–718.

Comte, F., Taupin, M.-L. and Rozenholc, Y. (2006), 'Penalized contrast estimator for density deconvolution', *Canadian J. Statist.* **34**, 431–452.

Cook, D., Buja, A. and Cabrera, J. (1993), 'Projection pursuit indexes based on orthonormal function expansions', *J. Comput. Graph. Statist.* **2**(3), 225–250.

Cook, D., Buja, A., Cabrera, J. and Hurley, C. (1995), 'Grand tour and projection pursuit', *J. Comput. Graph. Statist.* **4**(3), 155–172.

Dahlhaus, R. (1997), 'Fitting time series models to nonstationary processes', *Ann. Statist.* **25**, 1–37.

Davies, P. L. and Kovac, A. (2001), 'Local extremes, runs, strings and multiresolution', *Ann. Statist.* **29**, 1–65.

Deans, S. R. (1983), *The Radon Transform and some of its Applications*, Wiley, New York.

Deheuvels, P. (1977), 'Estimation non parametrique de la densité par histogrammes generalisés (II)', *Publications de l'Institut Statistique de l'Université de Paris* **22**, 1–23.

Delyon, B. and Juditsky, A. (1996), 'On minimax wavelet estimators', *J. Appl. Comput. Harmonic Anal.* **3**, 215–228.

Devroye, L. (1987), *A Course in Density Estimation*, Birkhäuser, Boston.

Devroye, L. (1989), 'Consistent deconvolution in density estimation', *Canad. J. Statist.* **7**, 235–239.

Devroye, L. and Györfi, L. (1985), *Density Estimation: The L^1 View*, Wiley, New York.

Devroye, L. and Lugosi, G. (2001), *Combinatiorial Methods in Density Estimation*, Springer, Berlin.

Dharmadhikari, S. and Joag-dev, K. (1988), *Unimodality, Convexity, and Applications*, Academic Press, San Diego.

Donoho, D. L. (1988), 'One-sided inference about functionals of a density', *Ann. Statist.* **16**, 1390–1420.

Donoho, D. L. (1995), 'Nonlinear solutions of linear inverse problems by wavelet-vaguelette decomposition', *J. Appl. Comput. Harmonic Anal.* **2**, 101–126.

Donoho, D. L. (1997), 'Cart and best-ortho-basis: A connection', *Ann. Statist.* **25**, 1870–1911.

Donoho, D. L. (1999), 'Wedgelets: Nearly minimax estimation of edges', *Ann. Statist.* **27**, 859–897.

Donoho, D. L. and Johnstone, I. M. (1994*a*), 'Ideal denoising in an orthonormal basis chosen from a library of bases', *C. R. Acad. Sci. Paris Sér. I Math.* **319**, 1317–1322.

Donoho, D. L. and Johnstone, I. M. (1994*b*), 'Ideal spatial adaptation by wavelet shrinkage', *Biometrika* **81**, 425–455.

Donoho, D. L., Johnstone, I. M., Kerkyacharian, G. and Picard, D. (1996), 'Density estimation by wavelet thresholding', *Ann. Statist.* **24**(2), 508–539.

Donoho, D. L. and Liu, R. C. (1991), 'Geometrizing rates of convergence III', *Ann. Statist.* **19**, 668–701.

Donoho, D. L. and Low, M. (1992), 'Renormalization exponents and optimal pointwise rates of convergence', *Ann. Statist.* **20**, 944–970.

Doukhan, P. (1994), *Mixing: Properties and Examples*, Vol. 85 of *Lecture Notes in Statistics*, Springer, Berlin.

Dudley, R. M. (1999), *Uniform Central Limit Theorems*, Cambridge University Press, Cambridge.

Edelsbrunner, H., Letscher, D. and Zomorodian, A. (2000), Topological persistence and simplification, *in* 'Proc. 41st Ann. IEEE Sympos. Found Comput. Sci.', IEEE Computer Society, Washington, DC, pp. 454–463.

Edwards, A. W. F. and Cavalli-Sforza, L. L. (1964), Reconstruction of evolutionary trees, *in* V. H. Heywood and J. McNeill, eds, 'Phenetic and Phylogenetic Classification', Systematics Assoc., London, pp. 67–76.

Efromovich, S. Y. (1999), *Nonparametric Curve Estimation: Methods, Theory, and Applications*, Springer.

Einmahl, J. H. and Mason, D. M. (1992), 'Generalized quantile processes', *Ann. Statist.* **20**, 1062–1078.

Embrechts, P., McNeil, A. J. and Straumann, D. (2002), Correlation and dependence in risk management: Properties and pitfalls, *in* M. A. H. Dempster, ed., 'Risk Management: Value at Risk and Beyond', Cambridge University Press, Cambridge, pp. 176–223.

Engel, J. (1994), 'A simple wavelet approach to nonparametric regression from recursive partitioning schemes', *J. Multivariate Anal.* **49**, 242–254.

Epanechnikov, V. A. (1969), 'Non-parametric estimation of a multivariate probability density', *Theory Probab. Appl.* **14**, 153–158.

Ermakov, M. S. (1989), 'Minimax estimation of the solution of an ill-posed convolution type problem', *Probl. Inform. Transmission* **25**, 191–200.

Everitt, B. (1974), *Cluster Analysis*, Heinemann, London.

Fan, J. (1991), 'On the optimal rates of convergence for nonparametric deconvolution problems', *Ann. Statist.* **10**, 1257–1272.

Fan, J. and Marron, J. S. (1994), 'Fast implementations of nonparametric curve estimators', *J. Comput. Graph. Statist.* **3**, 35–56.

Fano, R. M. (1952), Transmission of information, Class notes Course 6.574, MIT Cambridge.

Farrell, R. H. (1972), 'On the best obtainable asymptotic rates of convergence in estimation of a density function at a point', *Ann. Statist.* **43**, 170–180.

Feiner, S. and Beshers, C. (1990), Worlds within worlds: metaphors for exploring n-dimensional virtual worlds, *in* 'UIST '90: Proceedings 3rd annual ACM SIGGRAPH symposium on User interface software and technology', ACM, New York, pp. 76–83.

Feller, W. (1968), *An Introduction to Probability Theory and its Applications*, Vol. 1, 3rd edn, Wiley.

Ferguson, T. S. (1967), *Mathematical Statistics: A Decision Theoretic Approach*, Academic Press, New York.

Fisher, N. I., Mammen, E. and Marron, J. S. (1994), 'Testing for multimodality', *Comput. Statist. Data Anal.* **18**, 499–512.

Fix, E. and Hodges, J. L. (1951), Discriminatory analysis—nonparametric discrimination: Consistency properties, Report 4, USAF School of Aviation Medicine, Randolph Field, TX. Published in Internat. Statist. Rev., 1989, **57**, 238–247.

Flury, B. and Riedwyl, H. (1981), 'Graphical representation of multivariate data by means of asymmetrical faces', *Amer. Statist. Assoc.* **76**, 757–765.

Flury, B. and Riedwyl, H. (1988), *Multivariate Statistics: A Practical Approach*, Cambridge University Press, Cambridge.

Foley, J. D., van Dam, A., Feiner, S. K. and Hughes, J. F. (1997), *Computer Graphics: Principles and Practice*, Addison-Wesley, Reading, MA.

Fomenko, A. T. and Kunii, T. L., eds (1997), *Topological Modeling for Visualization*, Springer, Berlin.

Franke, J., Härdle, W. and Hafner, C. M. (2004), *Statistics of Financial Markets*, Springer, Berlin.

Freedman, D. and Diaconis, P. (1981), 'On the histogram as a density estimator: L^2 theory', *Z. Wahrsch. Verw. Gebiete* **57**, 453–476.

Freund, Y. (1995), 'Boosting a weak learning algorithm by majority', *Inform. Comput.* **121**, 256–285.

Freund, Y. and Schapire, R. (1996), Experiments with a new boosting algorithm, *in* 'Machine Learning: Proceedings of the Thirteenth International Conference', Morgan Kaufman, San Fransisco, pp. 148–156.

Friedman, H. P. and Rubin, J. (1967), 'On some invariant criterion for grouping data', *J. Amer. Statist. Assoc.* **62**, 1159–1178.

Friedman, J. H. and Rafsky, L. C. (1981), 'Graphics for the multivariate two-sample problem', *J. Amer. Statist. Assoc.* **7**(374), 277–287.

Friedman, J. H., Stuetzle, W. and Schroeder, A. (1984), 'Projection pursuit density estimation', *J. Amer. Statist. Assoc.* **79**, 599–608.

Friedman, J. H. and Tukey, J. (1974), 'A projection pursuit algorithm for exploratory data analysis', *IEEE Trans. Comput.* **C-23**, 881–889.

Furnas, G. W. and Buja, A. (1994), 'Prosection views: Dimensional inference through sections and projections (with discussion)', *J. Comput. Graph. Statist.* **3**(4), 323–385.

Genovese, C. R. and Wasserman, L. (2000), 'Rates of convergence for the Gaussian mixture sieve', *Ann. Statist.* **28**(4), 1105–1127.

Ghosal, S. and van der Vaart, A. W. (2001), 'Entropies and rates of convergence for maximum likelihood and Bayes estimation for mixtures of normal densities', *Ann. Statist.* **29**(5), 1233–1263.

Gnedenko, B. V. and Kolmogorov, A. N. (1954), *Limit distributions for sums of independent random variables*, Addison Wesley, Reading, MA. English transl. K. L. Chung.

Godtliebsen, F., Marron, J. S. and Chaudhuri, P. (2002), 'Significance in scale space for bivariate density estimation', *J. Comput. Graph. Statist.* **11**, 1–22.

Golubev, G. K. (1987), 'Adaptive asymptotically minimax estimation estimates of smooth signals', *Probl. Inform. Transmiss.* **23**, 57–67.

Golubev, G. K. (1991), 'LAN in problems of nonparametric estimation of functions and lower bounds for quadratic risks', *Theory Probab. Appl.* **63**, 152–157.

Golubev, G. K. (1992), 'Nonparametric estimation of smooth probability densities in L_2', *Probl. Inform. Transmiss.* **28**, 44–54.

Golubev, G. K. and Nussbaum, M. (1990), 'Adaptive spline estimates for nonparametric regression models', *Theory Probab. Appl.* **37**, 521–529.

Good, I. J. and Gaskins, R. A. (1980), 'Density estimation and bumb-hunting by the penalized likelihood method exemplified by the scattering and meteorite data (with discussion)', *J. Amer. Statist. Assoc.* **75**, 42–73.

Gordon, A. D. (1981), *Classification*, Chapman and Hall, London.

Granger, C. W., Maasoumi, E. and Racine, J. (2004), 'A dependence metric for possibly nonlinear processes', *J. Time Series Anal.* **25**(5), 649–669.

Guillemin, V. and Pollack, A. (1974), *Differential Topology*, Prentice-Hall, Englewood Cliffs, NJ.

Hall, P. and Wand, M. P. (1996), 'On the accuracy of binned kernel density estimators', *J. Multivariate Anal.* **56**, 165–184.

Han, T. and Verdú, S. (1994), 'Generalizing the Fano inequality', *IEEE Trans. Information Theory* **40**, 1247–1251.

Härdle, W., Kerkyacharian, G., Picard, D. and Tsybakov, A. B. (1998), *Wavelets, Approximation, and Statistical Applications*, Vol. 129 of *Lecture Notes in Statistics*, Springer, Berlin.

Härdle, W. and Scott, D. (1992), 'Smoothing by weighted averaging of rounded points', *Comput. Statist.* **7**, 97–128.

Härdle, W. and Simar, L. (2003), *Applied Multivariate Statistical Analysis*, Cambridge University Press, Cambridge.

Hartigan, J. A. (1975*a*), *Clustering Algorithms*, Wiley, New York.

Hartigan, J. A. (1975*b*), 'Printer graphics for clustering', *J. Statist. Comput. Simulation* **4**, 187–213.

Hartigan, J. A. (1987), 'Estimation of a convex density cluster in two dimensions', *J. Amer. Statist. Assoc.* **82**, 267–270.

Hartigan, J. A. and Hartigan, P. (1985), 'The DIP test of unimodality', *Ann. Statist.* **13**, 70–84.

Hartigan, J. A. and Mohanty, S. (1992), 'The RUNT test for multimodality', *J. Classification* **9**, 63–70.

Hasminskii, R. Z. (1978), 'A lower bound on the risk on non-parametric estimates of densities in the uniform metric', *Theory Probab. Appl.* **23**, 1053–1055.

Hasminskii, R. Z. and Ibragimov, I. A. (1990), 'On density estimation in the view of Kolmogorov's ideas in approximation theory', *Ann. Statist.* **18**, 999–1010.

Hastie, T., Tibshirani, R. and Friedman, J. (2001), *The Elements of Statistical Learning: Data Mining, Inference, and Prediction*, Springer, Berlin.

Haussler, D. (1991), 'Sphere packing numbers for subsets of the Boolean n-cube with bounded Vapnik-Chervonenkis dimension', *J. Combin. Theory* **A 69**, 217–232.

Heikkinen, J. and Arjas, E. (1998), 'Non-parametric Bayesian estimation of a spatial Poisson intensity', *Scand. J. Stat.* **25**, 435–450.

Hodges, J. (1955), 'A bivariate sign test', *Ann. Math. Statist.* **26**, 523–527.

Hoeffding, W. (1963), 'Probability inequalities for sums of bounded random variables', *J. Amer. Statist. Assoc.* **58**, 13–30.

Holmström, L. (2000), 'The error and the computational complexity of a multivariate binned kernel density estimator', *J. Multivariate Anal.* **72**, 264–309.

Huber, P. J. (1985), 'Projection pursuit', *Ann. Statist.* **13**(2), 435–475.

Hüsemann, J. A. and Terrell, G. R. (1991), 'Optimal parameter choice for error minimization in bivariate histograms', *J. Multivariate Anal.* **37**, 85–103.

Ibragimov, I. A. (2004), 'Estimation of multivariate regression', *Theory Probab. Appl.* **48**(2), 256–272.

Ibragimov, I. A. and Hasminskii, R. Z. (1977), 'A problem of statistical estimation in Gaussian white noise', *Dokl. Akad. Nauk SSSR* **236**, 1053–1055.

Ibragimov, I. A. and Hasminskii, R. Z. (1978), 'On the capacity of transmission by means of smooth signals', *Dokl. Akad. Nauk SSSR* **242**, 32–35.

Ibragimov, I. A. and Hasminskii, R. Z. (1980), 'On estimate of the density function', *Zap. Nauchn. Semin. LOMI* **98**, 61–85.

Ibragimov, I. A. and Hasminskii, R. Z. (1981*a*), 'On the non-parametric density estimates', *Zap. Nauchn. Semin. LOMI* **108**, 73–89.

Ibragimov, I. A. and Hasminskii, R. Z. (1981*b*), *Statistical Estimation: Asymptotic Theory*, Springer, Berlin. Originally published in Russian in 1979.

Inselberg, A. (1985), 'The plane with parallel coordinates', *Visual Computer* **1**, 69–91.

Inselberg, A. (1997), Multidimensional detective, *in* 'Proc. IEEE Information Visualization'97', IEEE, Washington, DC, pp. 100–107.

Inselberg, A. and Dimsdale, B. (1990), Parallel coordinates: A tool for visualizing multidimensional geometry, *in* 'Proc. IEEE Information Visualization'90', IEEE, Washington, DC, pp. 361–378.

Ivanov, V. V. (1962), 'On linear problems which are not well posed', *Soviet Math. Dokl.* **3**(4), 981–983.

Johnson, B. and Shneiderman, B. (1991), Tree-maps: A space-filling approach to the visualization of hierarchical information structures, *in* 'Proc. Information Visualization'91', IEEE, Washington, DC, pp. 284–291.

Johnstone, I. M. and Silverman, B. W. (1990), 'Speed of estimation in positron emission tomography and related inverse problems', *Ann. Statist.* **18**, 251–280.

Jones, L. K. (1992), 'A simple lemma on greedy approximation in Hilbert space and convergence rate for projection pursuit regression', *Ann. Statist.* **20**(1), 608–613.

Juditsky, A. and Nemirovski, A. (2000), 'Functional aggregation for nonparametric regression', *Ann. Statist.* **28**(3), 681–712.

Kaipio, J. and Somersalo, E. (2005), *Statistical and Computational Inverse Problems*, Springer, Berlin.

Kanazawa, Y. (1992), 'An optimal variable cell histogram based on the sample spacings', *Ann. Statist.* **20**, 291–304.

Keim, D. A. and Kriegel, H.-P. (1994), 'VisDB: Database exploration using multidimensional visualization', *IEEE Computer Graphics Appl.* **14**(5), 40–49.

Kerkyacharian, G., Lepski, O. and Picard, D. (2001), 'Nonlinear estimation in anisotropic multi-index denoising', *Probab. Theory Relat. Fields* **121**, 137–170.

Kerkyacharian, G., Lepski, O. and Picard, D. (2007), 'Nonlinear estimation in anisotropic multi-index denoising. sparse case.', *Teor. Veroyatn. Primen.* **52**(1), 150–171. Transl. in Theory Probab. Appl, (2008), **52**(1), 58–77.

Kleiner, B. and Hartigan, J. A. (1981), 'Representing points in many dimensions by trees and castles', *J. Amer. Statist. Assoc.* **76**(374), 260–269.

Klemelä, J. (2003), 'Lower bounds for the asymptotic minimax risk with spherical data', *J. Statist. Plann. Inference* **113**, 113–136.

Klemelä, J. (2004), 'Visualization of multivariate density estimates with level set trees', *J. Comput. Graph. Statist.* **13**(3), 599–620.

Klemelä, J. (2005), 'Algorithms for the manipulation of level sets of nonparametric density estimates', *Comput. Statist.* **20**, 349–368.

Klemelä, J. (2006), 'Visualization of multivariate density estimates with shape trees', *J. Comput. Graph. Statist.* **15**(2), 372–397.

Klemelä, J. (2007*a*), 'Density estimation with stagewise optimization of the empirical risk', *Machine Learning* **67**, 169–195.

Klemelä, J. (2007*b*), 'Visualization of multivariate data with tail trees', *Inform. Visualization* **6**, 109–122.

Klemelä, J. (2008*a*), 'Density estimation with locally identically distributed data and with locally stationary data', *J. Time Series Anal.* **29**(1), 125–141.

Klemelä, J. (2008*b*), 'Mode trees for multivariate data', *J. Comput. Graph. Statist.*. To appear.

Klemelä, J. (2009), 'Multivariate histograms with data-dependent partitions', *Statistica Sinica* **19**(1), 159–176.

Klemelä, J. and Mammen, E. (2008), 'Empirical risk minimization in inverse problems'. Submitted.

Kneip, A. (1994), 'Ordered linear smoothers', *Ann. Statist.* **22**, 835–866.

Kogure, A. (1987), 'Asymptotically optimal cells for a histogram', *Ann. Statist.* **15**, 1023–1030.

Kohonen, T. (2001), *Self-organizing Maps*, Springer, Berlin.

Kolmogorov, A. N. and Rozanov, Y. A. (1961), 'On the strong mixing conditions for stationary gaussian sequencies', *Theory Probab. Appl.* **5**, 204–207.

Kolmogorov, A. N. and Tikhomirov, V. M. (1961), 'ϵ-entropy and ϵ-capacity of sets in function spaces', *Transl. Am. Math. Soc.* **17**, 277–364.

Koltchinskii, V. I. (1997), 'M-Estimation, convexity and quantiles', *Ann. Stat.* **25**, 435–477.

Koo, J. Y. (1993), 'Optimal rates of convergence for nonparametric statistical inverse problems', *Ann. Statist.* **21**, 590–599.

Korobov, N. M. (1963), *Number Theoretic Methods of Approximate Analysis*, Fiznatgiz, Moscow.

Korostelev, A. P. and Tsybakov, A. B. (1991), 'Optimal rates of convergence of estimators in a probabilistic setup of tomography problem', *Probl. Inform. Transmiss.* **27**, 73–81.

Korostelev, A. P. and Tsybakov, A. B. (1993), *Minimax Theory of Image Reconstruction*, Vol. 82 of *Lecture Notes in Statistics*, Springer, Berlin.

Kunii, T. L. and Shinagawa, Y., eds (1992), *Modern Geometric Computing for Visualization*, Springer, Berlin.

Lambert, J. H. (1779), *Pyrometrie*, Babey Houde und Spener, Berlin.

Lamping, J. and Rao, R. (1994), Laying out and visualising large trees using a hyperbolic space, *in* 'Proc. UIST '94', ACM, New York, pp. 13–14.

Lamping, J. and Rao, R. (1996), 'The hyperbolic browser: A focus-context technique for visualizing large hierarchies', *J. Visual Languages and Computing* **7**(1), 33–55.

Lamping, J., Rao, R. and Pirolli, P. (1995), A focus-context technique based on hyperbolic geometry for visualizing large hierarchies, *in* 'Proc. CHI '95', ACM, pp. 401–408.

Le Cam, L. (1966), Likelihood functions for large number of independent observations, *in* F. N. David, ed., 'Festschrift for J. Neyman', Wiley, New York, pp. 167–187.

Le Cam, L. (1973), 'Convergence of estimates under dimensionality restrictions', *Ann. Statist.* **1**, 38–53.

Lee, W. S., Bartlett, R. C. and Williamson, R. C. (1996), 'Efficient agnostic learning of neural networks with bounded fan-in', *IEEE Trans. Inform. Theory* **42**(6), 2118–2132.

Lepski, O., Mammen, E. and Spokoiny, V. (1997), 'Optimal spatial adaptation to inhomogeneous smoothness: An approach based on kernel estimates with variable bandwidth selectors', *Ann. Statist.* **25**(3), 929–947.

Li, J. and Barron, A. (2000), Mixture density estimation, *in* S. Solla, T. Leen and K. Müller, eds, 'Advances in Neural Information Processing systems', Vol. 12, MIT Press, Cambridge.

Li, K.-C. (1986), 'Asymptotic optimality of C_L and generalized cross-validation in ridge regression with applications to spline smoothing', *Ann. Statist.* **14**, 1101–1112.

Li, K.-C. (1987), 'Asymptotic optimality of C_P, C_L, cross-validation and generalized cross-validation: Discrete index set', *Ann. Statist.* **15**, 958–976.

Liu, G. L. (1968), *Introduction to Combinatorial Mathematics*, McGraw-Hill, New York.

Liu, R. Y. (1990), 'On a notion of data depth based on random simplices', *Ann. Statist.* **18**, 405–414.

Liu, R. Y., Parelius, J. M. and Singh, K. (1999), 'Multivariate analysis by data depth: descriptive statistics, graphics and inference', *Ann. Statist.* **27**(3), 783–858.

Loftsgaarden, D. O. and Quesenberry, C. P. (1965), 'A nonparametric estimate of a multivariate density function', *Ann. Math. Statist.* **36**, 1049–1051.

Low, M. (1993), 'Renormalization upper and lower bounds for integrated risk in the white noise model', *Ann. Statist.* **21**, 577–589.

Lugosi, G. (2002), Pattern classification and learning theory, *in* L. Györfi, ed., 'Principles of Nonparametric Learning', Springer, Berlin, pp. 1–56.

Lugosi, G. and Nobel, A. (1996), 'Consistency of data-driven histogram methods for density estimation and classification', *Ann. Statist.* **24**, 687–706.

Macnaughton Smith, P., Williams, W., Dale, M. and Mockett, L. (1965), 'Dissimilarity analysis: a new technique of hierarchical subdivision', *Nature* **202**, 1034–1035.

Malevergne, Y. and Sornette, D. (2005), *Extreme Financial Risks: From Dependence to Risk Management*, Springer, Berlin.

Mallat, S. G. (1989), 'A theory for multiresolution signal decomposition: the wavelet representation', *IEEE Trans. Pattern Anal. Mach. Intell.* **11**, 674–693.

Mammen, E., Marron, J. S. and Fisher, N. I. (1992), 'Some asymptotics for multimodality tests based on kernel density estimates', *Probab. Theory Relat. Fields* **91**, 115–132.

Mandelbrot, B. (1963), 'The variation of certain speculative prices', *J. Business* **36**(4), 394–419.

Mandelbrot, B. and Hudson, R. L. (2004), *The (Mis)behavior of Markets : A Fractal View of Risk, Ruin, and Reward*, Basic Books, New York.

Marchette, D. J. and Wegman, E. J. (1997), 'The filtered mode tree', *J. Comput. Graph. Statist.* **6**(2), 143–159.

Marr, D. (1992), *Vision*, Freeman, San Francisco.

Marriot, F. H. C. (1971), 'Practical problems in a method of cluster analysis', *Biometrics* **27**, 501–514.

Matsumoto, Y. (2000), *An Introduction to Morse Theory*, Vol. 208 of *Translations of Mathematical Monographs*, AMS, Providence, RI. Originally published 1997 in Japanese.

Mendelson, S. (2002), 'On the size of convex hulls of small sets', *J. Mach. Learn. Res.* **2**, 1–18.

Meucci, A. (2005), *Risk and Asset Allocation*, Springer, Berlin.

Meyer, Y. (1990), *Ondelettes et opérateurs*, Hermann, Paris.

Meyer, Y. (1991), 'Ondelettes sur l'intervalle', *Rev. Mat. Iberoamericana* **7**, 115–133.

Michailidis, G. and de Leeuw, J. (2001), 'Data visualization through graph drawing', *Computational Statist.* **16**, 435–450.

Mihalisin, T., Timlin, J. and Schwegler, J. (1991), 'Visualizing multivariate functions, data, and distributions', *IEEE Comput. Graphics Appl.* **11**(13), 28–35.

Miller, J. J. and Wegman, E. J. (1991), Construction of line densities for parallel coordinate plots, *in* A. Buja and P. Tukey, eds, 'Computing and Graphics in Statistics', Springer, New York, pp. 107–123.

Milnor, J. (1963), *Morse Theory*, Princeton University Press, Princeton, NJ.

Minnotte, M. C. (1997), 'Nonparametric testing of the existence of modes', *Ann. Statist.* **25**, 1646–1660.

Minnotte, M. C., Marchette, D. J. and Wegman, E. J. (1998), 'The bumpy road to the mode forest', *J. Comput. Graph. Statist.* **7**, 239–251.

Minnotte, M. C. and Scott, D. W. (1993), 'The mode tree: A tool for visualization of nonparametric density features', *J. Comput. Graph. Statist.* **2**, 51–68.

Minnotte, M. C. and West, B. W. (1999), The data image: A tool for exploring high dimensional data sets, *in* '1998 Proc. ASA Section on Statistical Graphics', pp. 25–33.

Müller, D. W. and Sawitzki, G. (1991), 'Excess mass estimates and tests of multimodality', *J. Amer. Statist. Assoc.* **86**, 738–746.

Nason, G. and Silverman, B. W. (1994), 'The discrete wavelet transform in S', *J. Comput. Graph. Statist.* **3**, 163–191.

Natterer, F. (2001), *The Mathematics of Computerized Tomography*, Vol. 32 of *Classics in Applied Mathematics*, SIAM, Philadelphia, PA.

Nelsen, R. B. (1999), *An Introduction to Copulas*, Springer, Berlin.

Neumann, M. H. (2000), 'Multivariate wavelet thresholding in anisotropic function spaces', *Stat. Sin.* **10**, 399–431.

Neumann, M. H. and von Sachs, R. (1997), 'Wavelet thresholding in anisotropic function classes and applications to adaptive estimation of evolutionary spectra', *Ann. Stat.* **25**, 38–76.

Nikol'skii, S. M. (1975), *Approximation of Functions of Several Variables and Imbedding Theorems.*, Springer, Berlin.

Nolan, D. (1992), 'Asymptotics for multivariate trimming', *Stochast. Process. Appl.* **42**(1), 157–169.

Nussbaum, M. (1986), 'On nonparametric estimation of a regression function being smooth on a domain in R^k', *Theory Probab. Appl.* **31**, 108–115.

Nussbaum, M. (1996), 'Asymptotic equivalence of density estimation and Gaussian white noise', *Ann. Statist.* **24**, 2399–2430.

Ogden, T. (1997), *Essential Wavelets for Statistical Applications and Data Analysis*, Birkhäuser, Basel.

Oja, H. (1983), 'Descripitive statistics for multivariate distributions', *Statist. Probab. Lett.* **1**, 327–332.

Ooi, H. (2002), 'Density visualization and mode hunting using trees', *J. Comput. Graph. Statist.* **11**, 328–347.

Ossiander, M. (1987), 'A central limit theorem under metric entropy with L_2 bracketing', *Ann. Probab.* **15**, 897–919.

O'Sullivan, F. (1986), 'A statistical perspective on ill-posed inverse problems', *Statist. Science* **1**(4), 502–527.

Parzen, E. (1962), 'On estimation of a probability density function and mode', *Ann. Math. Statist.* **33**, 1065–1076.

Peetre, J. (1976), *New Thoughts on Besov Spaces*, Duke University Press, Durham, NC.

Petrov, V. V. (1995), *Limit Theorems of Probability Theory*, Clarendon Press, Oxford.

Philips, D. Z. (1962), 'A technique for numerical solution of certain integral equation of the first kind', *J. Assoc. Comput. Mach.* **9**, 84–96.

Pisier, G. (1981), Remarque sur un resultat non publie de B. Maurey, *in* 'Seminaire d'analyse fonctionelle 1980–1981', Ecole Polytechnique, Palaiseau, pp. 1–12.

Playfair, W. (1786), *The Commercial and Political Atlas*, Corry, London.

Playfair, W. (1801), *Statistical Breviary*, Wallis, London.

Pollard, D. (1989), 'Asymptotics via empirical processes', *Statist. Science* **4**(4), 341–366.

Polonik, W. (1999), 'Concentration and goodness-of-fit in higher dimensions: (asymptotically) distribution-free methods', *Ann. Statist.* **27**(4), 1210–1229.

Polyak, B. T. and Tsybakov, A. B. (1990), 'Asymptotic optimality of the C_p-test for the orthogonal series estimation of regression', *Theory Probab. Appl.* **35**, 293–306.

Polyak, B. T. and Tsybakov, A. B. (1992), 'A family of asymptotic optimal methods for choosing the estimate order in orthogonal series regression', *Theory Probab. Appl.* **37**, 471–481.

Pötzelberger, K. and Strasser, H. (2001), 'Clustering and quantization by MSP-partitions', *Stat. Decis.* **19**(4), 331–371.

Priebe, C. E. (1994), 'Adaptive mixtures', *J. Amer. Statist. Assoc.* **89**, 796–806.

Priestley, M. B. (1965), 'Evolutionary spectra and non-stationary processes', *J. Roy. Statist. Soc. Ser. B* **27**, 204–237.

Quinlan, J. R. (1993), *C4.5: Programs for Machine Learning*, Morgan Kaufmann, San Mateo.

Rakhlin, A., Panchenko, D. and Mukherjee, S. (2005), 'Risk bounds for mixture density estimation', *ESAIM: Probab. and Statist.* **9**, 220–229.

Reeb, G. (1946), 'Sur les points singuliers d'une forme de pfaff completement integrable ou d'une fonction numerique', *Comptes Rendus Acad. Sci. Paris* **222**, 847–849.

Ridgeway, G. (2002), 'Looking for lumps: Boosting and bagging for density estimation', *Comput. Statist. Data Anal.* **38**, 379–392.

Rigollet, P. and Tsybakov, A. B. (2006), Linear and convex aggregation of density estimators, Technical report, Université Paris 6.

Ripley, B. D. (1996), *Pattern Recognition and Neural Networks*, Cambridge University Press, Cambridge.

Robertson, G. G., Mackinlay, J. D. and Card, S. K. (1991), Cone trees: Animated 3D visualisations of hierarchical information, *in* 'Proc. CHI'95', ACM, New, York, pp. 189–194.

Robins, V., Abernethy, J., Rooney, N. and Bradley, E. (2004), 'Topology and intelligent data analysis', *Intell. Data Anal.* **8**(5), 505–515.

Rock, I. (1984), *Perception*, Scientific American Books, New York.

Rosenblatt, M. (1956*a*), 'A central limit theorem and a strong mixing condition', *Proc. Natl. Acad. Sci. USA* **42**, 43–47.

Rosenblatt, M. (1956*b*), 'Remarks on some nonparametric estimates of a density function', *Ann. Math. Statist.* **27**, 832–837.

Rosset, S. and Segal, E. (2002), Boosting density estimation, *in* 'Proc. 16th Intl. Conf. Neural Information Processing Systems (NIPS)'.

Rudemo, M. (1982), 'Empirical choice of histograms and kernel density estimators', *Scand. J. Statist.* **9**, 65–78.

Samarov, A. M. (1976), 'Minimax bound on the risk of nonparametric density estimates', *Probl. Inform. Transmiss.* **12**, 242–244.

Schapire, R. (1990), 'The strength of weak learnability', *Machine Learning* **5**, 197–227.

Schmeisser, H.-J. and Triebel, H. (1987), *Topics in Fourier Analysis and Function Spaces*, Geest and Portig, Leipzig.

Schweizer, B. and Wolff, E. F. (1981), 'On nonparametric measures of dependence for random variables', *Ann. Statist.* **9**(4), 879–885.

Scott, D. and Szewczyk, W. F. (2000), 'The stochastic mode tree and clustering'. Manuscript.

Scott, D. W. (1979), 'On optimal and data-based histograms', *Biometrika* **66**, 605–610.

Scott, D. W. (1985), 'Average shifted histograms: effective nonparametric density estimators in several dimensions', *Ann. Statist.* **13**, 1024–1040.

Scott, D. W. (1992), *Multivariate Density Estimation: Theory, Practice, and Visualization*, Wiley, New York.

Sedgewick, R. (1988), *Algorithms*, 2nd edn, Addison-Wesley, Reading, MA.

Seifert, B., Brockmann, M., Engel, J. and Gasser, T. (1994), 'Fast algorithms for nonparametric curve estimation', *J. Comput. Graph. Statist.* **3**, 192–213.

Serfling, R. (2002*a*), A depth function and a scale curve based on spatial quantiles, *in* Y. Dodge, ed., 'Statistical Data Analysis Based on the L1-Norm and Related Methods', Birkhäuser, Boston, pp. 25–38.

Serfling, R. (2002*b*), 'Quantile functions for multivariate analysis: Approaches and applications', *Stat. Neerl.* **56**(2), 214–232.

Shang, N. (1994), Tree-structured density estimation and dimensionality reduction, *in* 'Proc. 26rd Symp. on the Interface', pp. 172–176.

Shinagawa, Y. and Kunii, T. L. (1991), 'Constructing a Reeb graph automatically from cross sections', *IEEE Computer Graphics Appl.* **11**, 44–51.

Shneiderman, B. (1992), 'Tree visualization with treemaps: A 2-dimensional space filling approach', *ACM Trans. Graphics* **11**(1), 92–99.

Silverman, B. W. (1981), 'Using kernel density estimates to investigate multimodality', *J. Roy. Statist. Soc. Ser. B* **43**, 97–99.

Silverman, B. W. (1982), 'On the estimation of a probability density function by the maximum penalized likelihood method', *Ann. Statist.* **10**(3), 795–810.

Silverman, B. W. (1986), *Density Estimation for Statistics and Data Analysis*, Chapman and Hall, London.

Simonoff, J. S. (1996), *Smoothing Methods in Statistics*, Springer, Berlin.

Singh, G., Memoli, F. and Carlsson, G. (2007), Topological methods for the analysis of high dimensional data sets and 3D object recognition, *in* M. Botsch, R. Pajarola, B. Chen and M. Zwicker, eds, 'Symp. Point Based Graphics', Eurographics Association, Prague, pp. 91–100.

Sklar, A. (1959), 'Fonctions de répartition à n dimensions et leurs marges', *Publ. Inst. Statist. Univ. Paris* **8**, 229–231.

Sornette, D. (2003), *Why Stock Markets Crash*, Princeton University Press, Princeton.

Spence, I. and Lewandowsky, S. (1990), Graphical perception, *in* J. Fox and J. Scott Long, eds, 'Modern methods of data analysis', Sage Publications, Thousand Oaks, CA, pp. 13–57.

Spence, R. (2001), *Information Visualization*, Addison-Wesley, Reading, MA.

Spence, R. and Tweedie, L. (1998), 'The attribute explorer: Information synthesis via exploration', *Interact. Computers* **11**, 137–146.

Stefanski, L. A. and Carroll, R. J. (1990), 'Deconvoluting kernel density estimators', *Statistics* **2**, 169–184.

Stein, E. M. (1970), *Singular Integrals and Differentiability Properties of Functions*, Princeton University Press, Princeton.

Stărică, C. and Granger, C. (2005), 'Nonstationarities in stock returns', *Rev. Economics Statist.* **87**(3), 503–522.

Stuetzle, W. (2003), 'Estimating the cluster tree of a density by analyzing the minimal spanning tree of a sample', *J. Classification* **20**(5), 25–47.

Sutton, C. D. (1994), 'Tree structured density estimation', *Comput. Sci. Statist.* **26**, 167–171.

Talagrand, M. (2005), *The Generic Chaining: Upper and Lower Bounds of Stochastic Processes*, Springer, Berlin.

Tapia, R. A. and Thompson, J. R. (1978), *Nonparametric Probability Density Estimation*, Johns Hopkins University Press, Baltimore.

Tarasov, S. P. and Vyalyi, M. N. (1998), 'Construction of contour trees in 3D in $O(n \log n)$ steps', *ACM Symp. Comp. Geometry* pp. 68–75.

Taylor, C. (1987), 'Akaike's information criterion and the histogram', *Biometrika* **74**, 636–639.

Tikhonov, A. N. (1963), 'On solving ill-posed problem and method of regularization', *Doklady Akademii Nauk USSR* **153**, 501–504.

Tollis, I. G., Battista, G., Eades, P. and Tamassia, R. (1998), *Graph Drawing: Algorithms for the Visualization of Graphs*, Prentice Hall, Englewood Cliffs, NI.

Tribouley, K. (1995), 'Practical estimation of multivariate densities using wavelet methods', *Statist. Neerlandica* **49**, 41–62.

Triebel, H. (1992), *Theory of Function Spaces II*, Birkhäuser, Basel.

Troutt, M. D. (1991), 'A theorem on the density of the density ordinate and an alternative interpretation of the Box-Muller method', *Statistics* **3**, 463–466.

Tsybakov, A. B. (1998), 'Pointwise and sup-norm sharp adaptive estimation of functions on the Sobolev classes', *Ann. Statist.* **26**, 2420–2469.

Tsybakov, A. B. (2004), *Introduction à l'Estimation Non-paramétrique*, Vol. 41 of *Mathématiques et Applications*, Springer, Berlin.

Tufte, E. R. (1983), *The Visual Display of Quantitative Information*, Graphics Press, San Francisco.

Tufte, E. R. (1990), *Envisioning Information*, Graphics Press, San Francisco.

Tufte, E. R. (1997), *Visual Explanations*, Graphics Press, San Francisco.

Tukey, J. (1975), Mathematics and picturing data, *in* 'Proc. 1975 Intl. Congr. Mathematics', Vol. 2, pp. 523–531.

Tukey, J. (1977), *Exploratory Data Analysis*, Addison-Wesley, Reading, MA.

Tweedie, L., Spence, R., Williams, D. and Bhogal, R. (1994), The attribute explorer, *in* 'Video Proc. CHI '94', ACM, New York.

van de Geer, S. A. (2000), *Empirical Processes in M-Estimation*, Cambridge University Press, Cambridge.

van der Laan, M. J., Dudoit, S. and van der Vaart, A. W. (2004), The cross-validated adaptive epsilon-net estimator, Working paper series, U.C. Berkeley Division of Biostatistics.

van der Vaart, A. D. and Wellner, J. A. (1996), *Weak Convergence and Empirical Processes*, Springer, Berlin.

van Kreveld, M. J., van Oostrum, R., Bajaj, C. L., Pascucci, V. and Schikore, D. (1997), 'Contour trees and small seed sets for isosurface traversal', *ACM Symp. Comp. Geometry* pp. 212–220.

Vapnik, V. V. (1995), *The Nature of Statistical Learning Theory*, Springer, Berlin.

Vapnik, V. V. (1998), *Statistical Learning Theory*, Wiley, New York.

Viennet, G. (1997), 'Inequalities for absolutely regular sequences: Application to density estimation', *Probab. Theory Relat. Fields* **107**, 467–492.

Volkonskii, V. A. and Rozanov, Y. A. (1959), 'Some limit theorems for random functions, Part I', *Theory Probab. Appl.* **4**, 178–197.

Wand, M. P. (1994), 'Fast computation of multivariate kernel estimators', *J. Comput. Graph. Statist.* **3**, 433–445.

Wand, M. P. and Jones, M. C. (1995), *Kernel Smoothing*, Chapman and Hall, London.

Watt, A. (2000), *3D Computer Graphics*, Pearson, Addison-Wesley, Reading, MA.

Weber, G., Bremer, P.-T. and Pascucci, V. (2007), 'Topological landscapes: A terrain metaphor for scientific data', *IEEE Trans. Visualization Computer Graphics* **13**(6), 1416–1423.

Wegman, E. J. (1990), 'Hyperdimensional data analysis using parallel coordinates', *J. Amer. Statist. Assoc.* **85**(411), 664–675.

Wegman, E. J. and Luo, Q. (2002), 'On methods of computer graphics for visualizing densities', *J. Comput. Graph. Statist.* **11**(1), 137–162.

Wegman, E. J. and Solka, J. L. (2002), 'On some mathematics for visualizing high dimensional data', *Sankhyā, Ser. A* **64**, 429–452.

Weston, J., Gammerman, A., Stitson, M. O., Vapnik, V., Vovk, V. and Watkins, C. (1998), Support vector density estimation, *in* B. Schölkopf, B. Burges and A. Smola, eds, 'Advances in Kernel Methods–Support Vector Learning', MIT Press, Cambridge.

Wheeden, R. L. and Zygmund, A. (1977), *Measure and Integral*, Marcel Dekker, New York.

Wolfowitz, J. (1957), 'The minimum distance method', *Ann. Math. Statist.* **28**, 75–88.

Yan, J. (2007), 'Enjoy the joy of copulas: with a package copula', *Journal of Statistical Software* **21**(4), 1–21.

Yang, Y. and Barron, A. (1999), 'Information-theoretic determination of minimax rates of convergence', *Ann. Statist.* **27**, 1564–1599.

Yatracos, Y. G. (1985), 'Rates of convergence of minimum distance estimators and Kolmogorov's entropy', *Ann. Statist.* **13**(2), 768–774.

Yu, Y. and Lambert, D. (2000), 'Fitting trees to functional data, with an application to time of day patterns', *J. Comp. Graph. Stat.* **8**, 749–762.

Zhang, T. (2003), 'Sequential greedy approximation for certain convex optimization problems', *IEEE Trans. Information Theory* **49**, 682–691.

Zuo, Y. and Serfling, R. (2000), 'General notions of statistical depth function', *Ann. Statist.* **28**(2), 461–482.

AUTHOR INDEX

Smoothing of Multivariate Data: Density Estimation and Visualization. By Jussi Klemelä

TOPIC INDEX[1]

[1]The bold page numbers refer to the pages where a definition of a concept occurs and the underlined page numbers refer to the pages where a figure related to the concept occur.

WILEY SERIES IN PROBABILITY AND STATISTICS

ESTABLISHED BY WALTER A. SHEWHART AND SAMUEL S. WILKS

The ***Wiley Series in Probability and Statistics*** is well established and authoritative. It covers many topics of current research interest in both pure and applied statistics and probability theory. Written by leading statisticians and institutions, the titles span both state-of-the-art developments in the field and classical methods.

Reflecting the wide range of current research in statistics, the series encompasses applied, methodological and theoretical statistics, ranging from applications and new techniques made possible by advances in computerized practice to rigorous treatment of theoretical approaches.

This series provides essential and invaluable reading for all statisticians, whether in academia, industry, government, or research.

† ABRAHAM and LEDOLTER · Statistical Methods for Forecasting
AGRESTI · Analysis of Ordinal Categorical Data
AGRESTI · An Introduction to Categorical Data Analysis, *Second Edition*
AGRESTI · Categorical Data Analysis, *Second Edition*
ALTMAN, GILL, and McDONALD · Numerical Issues in Statistical Computing for the Social Scientist
AMARATUNGA and CABRERA · Exploration and Analysis of DNA Microarray and Protein Array Data
ANDĚL · Mathematics of Chance
ANDERSON · An Introduction to Multivariate Statistical Analysis, *Third Edition*
* ANDERSON · The Statistical Analysis of Time Series
ANDERSON, AUQUIER, HAUCK, OAKES, VANDAELE, and WEISBERG · Statistical Methods for Comparative Studies
ANDERSON and LOYNES · The Teaching of Practical Statistics
ARMITAGE and DAVID (editors) · Advances in Biometry
ARNOLD, BALAKRISHNAN, and NAGARAJA · Records
* ARTHANARI and DODGE · Mathematical Programming in Statistics
* BAILEY · The Elements of Stochastic Processes with Applications to the Natural Sciences
BALAKRISHNAN and KOUTRAS · Runs and Scans with Applications
BALAKRISHNAN and NG · Precedence-Type Tests and Applications
BARNETT · Comparative Statistical Inference, *Third Edition*
BARNETT · Environmental Statistics
BARNETT and LEWIS · Outliers in Statistical Data, *Third Edition*
BARTOSZYNSKI and NIEWIADOMSKA-BUGAJ · Probability and Statistical Inference
BASILEVSKY · Statistical Factor Analysis and Related Methods: Theory and Applications
BASU and RIGDON · Statistical Methods for the Reliability of Repairable Systems
BATES and WATTS · Nonlinear Regression Analysis and Its Applications
BECHHOFER, SANTNER, and GOLDSMAN · Design and Analysis of Experiments for Statistical Selection, Screening, and Multiple Comparisons

*Now available in a lower priced paperback edition in the Wiley Classics Library.
†Now available in a lower priced paperback edition in the Wiley–Interscience Paperback Series.

BELSLEY · Conditioning Diagnostics: Collinearity and Weak Data in Regression
† BELSLEY, KUH, and WELSCH · Regression Diagnostics: Identifying Influential Data and Sources of Collinearity
BENDAT and PIERSOL · Random Data: Analysis and Measurement Procedures, *Third Edition*
BERRY, CHALONER, and GEWEKE · Bayesian Analysis in Statistics and Econometrics: Essays in Honor of Arnold Zellner
BERNARDO and SMITH · Bayesian Theory
BHAT and MILLER · Elements of Applied Stochastic Processes, *Third Edition*
BHATTACHARYA and WAYMIRE · Stochastic Processes with Applications
BILLINGSLEY · Convergence of Probability Measures, *Second Edition*
BILLINGSLEY · Probability and Measure, *Third Edition*
BIRKES and DODGE · Alternative Methods of Regression
BISWAS, DATTA, FINE, and SEGAL · Statistical Advances in the Biomedical Sciences: Clinical Trials, Epidemiology, Survival Analysis, and Bioinformatics
BLISCHKE AND MURTHY (editors) · Case Studies in Reliability and Maintenance
BLISCHKE AND MURTHY · Reliability: Modeling, Prediction, and Optimization
BLOOMFIELD · Fourier Analysis of Time Series: An Introduction, *Second Edition*
BOLLEN · Structural Equations with Latent Variables
BOLLEN and CURRAN · Latent Curve Models: A Structural Equation Perspective
BOROVKOV · Ergodicity and Stability of Stochastic Processes
BOULEAU · Numerical Methods for Stochastic Processes
BOX · Bayesian Inference in Statistical Analysis
BOX · R. A. Fisher, the Life of a Scientist
BOX and DRAPER · Response Surfaces, Mixtures, and Ridge Analyses, *Second Edition*
* BOX and DRAPER · Evolutionary Operation: A Statistical Method for Process Improvement
BOX and FRIENDS · Improving Almost Anything, *Revised Edition*
BOX, HUNTER, and HUNTER · Statistics for Experimenters: Design, Innovation, and Discovery, *Second Editon*
BOX, JENKINS, and REINSEL · Time Series Analysis: Forcasting and Control, *Fourth Edition*
BOX, LUCEÑO, and PANIAGUA-QUIÑONES · Statistical Control by Monitoring and Adjustment, *Second Edition*
BRANDIMARTE · Numerical Methods in Finance: A MATLAB-Based Introduction
† BROWN and HOLLANDER · Statistics: A Biomedical Introduction
BRUNNER, DOMHOF, and LANGER · Nonparametric Analysis of Longitudinal Data in Factorial Experiments
BUCKLEW · Large Deviation Techniques in Decision, Simulation, and Estimation
CAIROLI and DALANG · Sequential Stochastic Optimization
CASTILLO, HADI, BALAKRISHNAN, and SARABIA · Extreme Value and Related Models with Applications in Engineering and Science
CHAN · Time Series: Applications to Finance
CHARALAMBIDES · Combinatorial Methods in Discrete Distributions
CHATTERJEE and HADI · Regression Analysis by Example, *Fourth Edition*
CHATTERJEE and HADI · Sensitivity Analysis in Linear Regression
CHERNICK · Bootstrap Methods: A Guide for Practitioners and Researchers, *Second Edition*
CHERNICK and FRIIS · Introductory Biostatistics for the Health Sciences
CHILÈS and DELFINER · Geostatistics: Modeling Spatial Uncertainty
CHOW and LIU · Design and Analysis of Clinical Trials: Concepts and Methodologies, *Second Edition*
CLARKE · Linear Models: The Theory and Application of Analysis of Variance

*Now available in a lower priced paperback edition in the Wiley Classics Library.
†Now available in a lower priced paperback edition in the Wiley–Interscience Paperback Series.

CLARKE and DISNEY · Probability and Random Processes: A First Course with Applications, *Second Edition*
* COCHRAN and COX · Experimental Designs, *Second Edition*
CONGDON · Applied Bayesian Modelling
CONGDON · Bayesian Models for Categorical Data
CONGDON · Bayesian Statistical Modelling
CONOVER · Practical Nonparametric Statistics, *Third Edition*
COOK · Regression Graphics
COOK and WEISBERG · Applied Regression Including Computing and Graphics
COOK and WEISBERG · An Introduction to Regression Graphics
CORNELL · Experiments with Mixtures, Designs, Models, and the Analysis of Mixture Data, *Third Edition*
COVER and THOMAS · Elements of Information Theory
COX · A Handbook of Introductory Statistical Methods
* COX · Planning of Experiments
CRESSIE · Statistics for Spatial Data, *Revised Edition*
CSÖRGŐ and HORVÁTH · Limit Theorems in Change Point Analysis
DANIEL · Applications of Statistics to Industrial Experimentation
DANIEL · Biostatistics: A Foundation for Analysis in the Health Sciences, *Eighth Edition*
* DANIEL · Fitting Equations to Data: Computer Analysis of Multifactor Data, *Second Edition*
DASU and JOHNSON · Exploratory Data Mining and Data Cleaning
DAVID and NAGARAJA · Order Statistics, *Third Edition*
* DEGROOT, FIENBERG, and KADANE · Statistics and the Law
DEL CASTILLO · Statistical Process Adjustment for Quality Control
DeMARIS · Regression with Social Data: Modeling Continuous and Limited Response Variables
DEMIDENKO · Mixed Models: Theory and Applications
DENISON, HOLMES, MALLICK and SMITH · Bayesian Methods for Nonlinear Classification and Regression
DETTE and STUDDEN · The Theory of Canonical Moments with Applications in Statistics, Probability, and Analysis
DEY and MUKERJEE · Fractional Factorial Plans
DILLON and GOLDSTEIN · Multivariate Analysis: Methods and Applications
DODGE · Alternative Methods of Regression
* DODGE and ROMIG · Sampling Inspection Tables, *Second Edition*
* DOOB · Stochastic Processes
DOWDY, WEARDEN, and CHILKO · Statistics for Research, *Third Edition*
DRAPER and SMITH · Applied Regression Analysis, *Third Edition*
DRYDEN and MARDIA · Statistical Shape Analysis
DUDEWICZ and MISHRA · Modern Mathematical Statistics
DUNN and CLARK · Basic Statistics: A Primer for the Biomedical Sciences, *Third Edition*
DUPUIS and ELLIS · A Weak Convergence Approach to the Theory of Large Deviations
EDLER and KITSOS · Recent Advances in Quantitative Methods in Cancer and Human Health Risk Assessment
* ELANDT-JOHNSON and JOHNSON · Survival Models and Data Analysis
ENDERS · Applied Econometric Time Series
† ETHIER and KURTZ · Markov Processes: Characterization and Convergence
EVANS, HASTINGS, and PEACOCK · Statistical Distributions, *Third Edition*
FELLER · An Introduction to Probability Theory and Its Applications, Volume I, *Third Edition,* Revised; Volume II, *Second Edition*
FISHER and VAN BELLE · Biostatistics: A Methodology for the Health Sciences

*Now available in a lower priced paperback edition in the Wiley Classics Library.
†Now available in a lower priced paperback edition in the Wiley–Interscience Paperback Series.

FITZMAURICE, LAIRD, and WARE · Applied Longitudinal Analysis
* FLEISS · The Design and Analysis of Clinical Experiments
FLEISS · Statistical Methods for Rates and Proportions, *Third Edition*
† FLEMING and HARRINGTON · Counting Processes and Survival Analysis
FULLER · Introduction to Statistical Time Series, *Second Edition*
† FULLER · Measurement Error Models
GALLANT · Nonlinear Statistical Models
GEISSER · Modes of Parametric Statistical Inference
GELMAN and MENG · Applied Bayesian Modeling and Causal Inference from Incomplete-Data Perspectives
GEWEKE · Contemporary Bayesian Econometrics and Statistics
GHOSH, MUKHOPADHYAY, and SEN · Sequential Estimation
GIESBRECHT and GUMPERTZ · Planning, Construction, and Statistical Analysis of Comparative Experiments
GIFI · Nonlinear Multivariate Analysis
GIVENS and HOETING · Computational Statistics
GLASSERMAN and YAO · Monotone Structure in Discrete-Event Systems
GNANADESIKAN · Methods for Statistical Data Analysis of Multivariate Observations, *Second Edition*
GOLDSTEIN and LEWIS · Assessment: Problems, Development, and Statistical Issues
GREENWOOD and NIKULIN · A Guide to Chi-Squared Testing
GROSS, SHORTLE, THOMPSON, and HARRIS · Fundamentals of Queueing Theory, *Fourth Edition*
GROSS, SHORTLE, THOMPSON, and HARRIS · Solutions Manual to Accompany Fundamentals of Queueing Theory, *Fourth Edition*
* HAHN and SHAPIRO · Statistical Models in Engineering
HAHN and MEEKER · Statistical Intervals: A Guide for Practitioners
HALD · A History of Probability and Statistics and their Applications Before 1750
HALD · A History of Mathematical Statistics from 1750 to 1930
† HAMPEL · Robust Statistics: The Approach Based on Influence Functions
HANNAN and DEISTLER · The Statistical Theory of Linear Systems
HARTUNG, KNAPP, and SINHA · Statistical Meta-Analysis with Applications
HEIBERGER · Computation for the Analysis of Designed Experiments
HEDAYAT and SINHA · Design and Inference in Finite Population Sampling
HEDEKER and GIBBONS · Longitudinal Data Analysis
HELLER · MACSYMA for Statisticians
HINKELMANN and KEMPTHORNE · Design and Analysis of Experiments, Volume 1: Introduction to Experimental Design, *Second Edition*
HINKELMANN and KEMPTHORNE · Design and Analysis of Experiments, Volume 2: Advanced Experimental Design
HOAGLIN, MOSTELLER, and TUKEY · Fundamentals of Exploratory Analysis of Variance
* HOAGLIN, MOSTELLER, and TUKEY · Exploring Data Tables, Trends and Shapes
* HOAGLIN, MOSTELLER, and TUKEY · Understanding Robust and Exploratory Data Analysis
HOCHBERG and TAMHANE · Multiple Comparison Procedures
HOCKING · Methods and Applications of Linear Models: Regression and the Analysis of Variance, *Second Edition*
HOEL · Introduction to Mathematical Statistics, *Fifth Edition*
HOGG and KLUGMAN · Loss Distributions
HOLLANDER and WOLFE · Nonparametric Statistical Methods, *Second Edition*
HOSMER and LEMESHOW · Applied Logistic Regression, *Second Edition*

*Now available in a lower priced paperback edition in the Wiley Classics Library.
†Now available in a lower priced paperback edition in the Wiley–Interscience Paperback Series.

HOSMER, LEMESHOW, and MAY · Applied Survival Analysis: Regression Modeling of Time-to-Event Data, *Second Edition*

† HUBER and RONCHETTI · Robust Statistics, *Second Edition*

HUBERTY · Applied Discriminant Analysis

HUBERTY and OLEJNIK · Applied MANOVA and Discriminant Analysis, *Second Edition*

HUNT and KENNEDY · Financial Derivatives in Theory and Practice, *Revised Edition*

HURD and MIAMEE · Periodically Correlated Random Sequences: Spectral Theory and Practice

HUSKOVA, BERAN, and DUPAC · Collected Works of Jaroslav Hajek—with Commentary

HUZURBAZAR · Flowgraph Models for Multistate Time-to-Event Data

IMAN and CONOVER · A Modern Approach to Statistics

† JACKSON · A User's Guide to Principle Components

JOHN · Statistical Methods in Engineering and Quality Assurance

JOHNSON · Multivariate Statistical Simulation

JOHNSON and BALAKRISHNAN · Advances in the Theory and Practice of Statistics: A Volume in Honor of Samuel Kotz

JOHNSON and BHATTACHARYYA · Statistics: Principles and Methods, *Fifth Edition*

JOHNSON and KOTZ · Distributions in Statistics

JOHNSON and KOTZ (editors) · Leading Personalities in Statistical Sciences: From the Seventeenth Century to the Present

JOHNSON, KOTZ, and BALAKRISHNAN · Continuous Univariate Distributions, Volume 1, *Second Edition*

JOHNSON, KOTZ, and BALAKRISHNAN · Continuous Univariate Distributions, Volume 2, *Second Edition*

JOHNSON, KOTZ, and BALAKRISHNAN · Discrete Multivariate Distributions

JOHNSON, KEMP, and KOTZ · Univariate Discrete Distributions, *Third Edition*

JUDGE, GRIFFITHS, HILL, LÜTKEPOHL, and LEE · The Theory and Practice of Econometrics, *Second Edition*

JUREČKOVÁ and SEN · Robust Statistical Procedures: Aymptotics and Interrelations

JUREK and MASON · Operator-Limit Distributions in Probability Theory

KADANE · Bayesian Methods and Ethics in a Clinical Trial Design

KADANE AND SCHUM · A Probabilistic Analysis of the Sacco and Vanzetti Evidence

KALBFLEISCH and PRENTICE · The Statistical Analysis of Failure Time Data, *Second Edition*

KARIYA and KURATA · Generalized Least Squares

KASS and VOS · Geometrical Foundations of Asymptotic Inference

† KAUFMAN and ROUSSEEUW · Finding Groups in Data: An Introduction to Cluster Analysis

KEDEM and FOKIANOS · Regression Models for Time Series Analysis

KENDALL, BARDEN, CARNE, and LE · Shape and Shape Theory

KHURI · Advanced Calculus with Applications in Statistics, *Second Edition*

KHURI, MATHEW, and SINHA · Statistical Tests for Mixed Linear Models

KLEIBER and KOTZ · Statistical Size Distributions in Economics and Actuarial Sciences

KLEMELÄ · Smoothing of Multivariate Data: Density Estimation and Visualization

KLUGMAN, PANJER, and WILLMOT · Loss Models: From Data to Decisions, *Third Edition*

KLUGMAN, PANJER, and WILLMOT · Solutions Manual to Accompany Loss Models: From Data to Decisions, *Third Edition*

KOTZ, BALAKRISHNAN, and JOHNSON · Continuous Multivariate Distributions, Volume 1, *Second Edition*

KOVALENKO, KUZNETZOV, and PEGG · Mathematical Theory of Reliability of Time-Dependent Systems with Practical Applications

*Now available in a lower priced paperback edition in the Wiley Classics Library.
†Now available in a lower priced paperback edition in the Wiley–Interscience Paperback Series.

KOWALSKI and TU · Modern Applied U-Statistics
KRISHNAMOORTHY and MATHEW · Statistical Tolerance Regions: Theory, Applications, and Computation
KROONENBERG · Applied Multiway Data Analysis
KVAM and VIDAKOVIC · Nonparametric Statistics with Applications to Science and Engineering
LACHIN · Biostatistical Methods: The Assessment of Relative Risks
LAD · Operational Subjective Statistical Methods: A Mathematical, Philosophical, and Historical Introduction
LAMPERTI · Probability: A Survey of the Mathematical Theory, *Second Edition*
LANGE, RYAN, BILLARD, BRILLINGER, CONQUEST, and GREENHOUSE · Case Studies in Biometry
LARSON · Introduction to Probability Theory and Statistical Inference, *Third Edition*
LAWLESS · Statistical Models and Methods for Lifetime Data, *Second Edition*
LAWSON · Statistical Methods in Spatial Epidemiology
LE · Applied Categorical Data Analysis
LE · Applied Survival Analysis
LEE and WANG · Statistical Methods for Survival Data Analysis, *Third Edition*
LePAGE and BILLARD · Exploring the Limits of Bootstrap
LEYLAND and GOLDSTEIN (editors) · Multilevel Modelling of Health Statistics
LIAO · Statistical Group Comparison
LINDVALL · Lectures on the Coupling Method
LIN · Introductory Stochastic Analysis for Finance and Insurance
LINHART and ZUCCHINI · Model Selection
LITTLE and RUBIN · Statistical Analysis with Missing Data, *Second Edition*
LLOYD · The Statistical Analysis of Categorical Data
LOWEN and TEICH · Fractal-Based Point Processes
MAGNUS and NEUDECKER · Matrix Differential Calculus with Applications in Statistics and Econometrics, *Revised Edition*
MALLER and ZHOU · Survival Analysis with Long Term Survivors
MALLOWS · Design, Data, and Analysis by Some Friends of Cuthbert Daniel
MANN, SCHAFER, and SINGPURWALLA · Methods for Statistical Analysis of Reliability and Life Data
MANTON, WOODBURY, and TOLLEY · Statistical Applications Using Fuzzy Sets
MARCHETTE · Random Graphs for Statistical Pattern Recognition
MARDIA and JUPP · Directional Statistics
MASON, GUNST, and HESS · Statistical Design and Analysis of Experiments with Applications to Engineering and Science, *Second Edition*
McCULLOCH, SEARLE, and NEUHAUS · Generalized, Linear, and Mixed Models, *Second Edition*
McFADDEN · Management of Data in Clinical Trials, *Second Edition*
* McLACHLAN · Discriminant Analysis and Statistical Pattern Recognition
McLACHLAN, DO, and AMBROISE · Analyzing Microarray Gene Expression Data
McLACHLAN and KRISHNAN · The EM Algorithm and Extensions, *Second Edition*
McLACHLAN and PEEL · Finite Mixture Models
McNEIL · Epidemiological Research Methods
MEEKER and ESCOBAR · Statistical Methods for Reliability Data
MEERSCHAERT and SCHEFFLER · Limit Distributions for Sums of Independent Random Vectors: Heavy Tails in Theory and Practice
MICKEY, DUNN, and CLARK · Applied Statistics: Analysis of Variance and Regression, *Third Edition*
* MILLER · Survival Analysis, *Second Edition*
MONTGOMERY, JENNINGS, and KULAHCI · Introduction to Time Series Analysis and Forecasting

*Now available in a lower priced paperback edition in the Wiley Classics Library.
†Now available in a lower priced paperback edition in the Wiley–Interscience Paperback Series.

MONTGOMERY, PECK, and VINING · Introduction to Linear Regression Analysis, *Fourth Edition*
MORGENTHALER and TUKEY · Configural Polysampling: A Route to Practical Robustness
MUIRHEAD · Aspects of Multivariate Statistical Theory
MULLER and STOYAN · Comparison Methods for Stochastic Models and Risks
MURRAY · X-STAT 2.0 Statistical Experimentation, Design Data Analysis, and Nonlinear Optimization
MURTHY, XIE, and JIANG · Weibull Models
MYERS, MONTGOMERY, and ANDERSON-COOK · Response Surface Methodology: Process and Product Optimization Using Designed Experiments, *Third Edition*
MYERS, MONTGOMERY, and VINING · Generalized Linear Models. With Applications in Engineering and the Sciences
† NELSON · Accelerated Testing, Statistical Models, Test Plans, and Data Analyses
† NELSON · Applied Life Data Analysis
NEWMAN · Biostatistical Methods in Epidemiology
OCHI · Applied Probability and Stochastic Processes in Engineering and Physical Sciences
OKABE, BOOTS, SUGIHARA, and CHIU · Spatial Tesselations: Concepts and Applications of Voronoi Diagrams, *Second Edition*
OLIVER and SMITH · Influence Diagrams, Belief Nets and Decision Analysis
PALTA · Quantitative Methods in Population Health: Extensions of Ordinary Regressions
PANJER · Operational Risk: Modeling and Analytics
PANKRATZ · Forecasting with Dynamic Regression Models
PANKRATZ · Forecasting with Univariate Box-Jenkins Models: Concepts and Cases
* PARZEN · Modern Probability Theory and Its Applications
PEÑA, TIAO, and TSAY · A Course in Time Series Analysis
PIANTADOSI · Clinical Trials: A Methodologic Perspective
PORT · Theoretical Probability for Applications
POURAHMADI · Foundations of Time Series Analysis and Prediction Theory
POWELL · Approximate Dynamic Programming: Solving the Curses of Dimensionality
PRESS · Bayesian Statistics: Principles, Models, and Applications
PRESS · Subjective and Objective Bayesian Statistics, *Second Edition*
PRESS and TANUR · The Subjectivity of Scientists and the Bayesian Approach
PUKELSHEIM · Optimal Experimental Design
PURI, VILAPLANA, and WERTZ · New Perspectives in Theoretical and Applied Statistics
† PUTERMAN · Markov Decision Processes: Discrete Stochastic Dynamic Programming
QIU · Image Processing and Jump Regression Analysis
* RAO · Linear Statistical Inference and Its Applications, *Second Edition*
RAUSAND and HØYLAND · System Reliability Theory: Models, Statistical Methods, and Applications, *Second Edition*
RENCHER · Linear Models in Statistics
RENCHER · Methods of Multivariate Analysis, *Second Edition*
RENCHER · Multivariate Statistical Inference with Applications
* RIPLEY · Spatial Statistics
* RIPLEY · Stochastic Simulation
ROBINSON · Practical Strategies for Experimenting
ROHATGI and SALEH · An Introduction to Probability and Statistics, *Second Edition*
ROLSKI, SCHMIDLI, SCHMIDT, and TEUGELS · Stochastic Processes for Insurance and Finance
ROSENBERGER and LACHIN · Randomization in Clinical Trials: Theory and Practice
ROSS · Introduction to Probability and Statistics for Engineers and Scientists
ROSSI, ALLENBY, and McCULLOCH · Bayesian Statistics and Marketing

*Now available in a lower priced paperback edition in the Wiley Classics Library.
†Now available in a lower priced paperback edition in the Wiley–Interscience Paperback Series.

† ROUSSEEUW and LEROY · Robust Regression and Outlier Detection
* RUBIN · Multiple Imputation for Nonresponse in Surveys
RUBINSTEIN and KROESE · Simulation and the Monte Carlo Method, *Second Edition*
RUBINSTEIN and MELAMED · Modern Simulation and Modeling
RYAN · Modern Engineering Statistics
RYAN · Modern Experimental Design
RYAN · Modern Regression Methods, *Second Edition*
RYAN · Statistical Methods for Quality Improvement, *Second Edition*
SALEH · Theory of Preliminary Test and Stein-Type Estimation with Applications
* SCHEFFE · The Analysis of Variance
SCHIMEK · Smoothing and Regression: Approaches, Computation, and Application
SCHOTT · Matrix Analysis for Statistics, *Second Edition*
SCHOUTENS · Levy Processes in Finance: Pricing Financial Derivatives
SCHUSS · Theory and Applications of Stochastic Differential Equations
SCOTT · Multivariate Density Estimation: Theory, Practice, and Visualization
† SEARLE · Linear Models for Unbalanced Data
† SEARLE · Matrix Algebra Useful for Statistics
† SEARLE, CASELLA, and McCULLOCH · Variance Components
SEARLE and WILLETT · Matrix Algebra for Applied Economics
SEBER · A Matrix Handbook For Statisticians
† SEBER · Multivariate Observations
SEBER and LEE · Linear Regression Analysis, *Second Edition*
† SEBER and WILD · Nonlinear Regression
SENNOTT · Stochastic Dynamic Programming and the Control of Queueing Systems
* SERFLING · Approximation Theorems of Mathematical Statistics
SHAFER and VOVK · Probability and Finance: It's Only a Game!
SILVAPULLE and SEN · Constrained Statistical Inference: Inequality, Order, and Shape Restrictions
SMALL and McLEISH · Hilbert Space Methods in Probability and Statistical Inference
SRIVASTAVA · Methods of Multivariate Statistics
STAPLETON · Linear Statistical Models
STAPLETON · Models for Probability and Statistical Inference: Theory and Applications
STAUDTE and SHEATHER · Robust Estimation and Testing
STOYAN, KENDALL, and MECKE · Stochastic Geometry and Its Applications, *Second Edition*
STOYAN and STOYAN · Fractals, Random Shapes and Point Fields: Methods of Geometrical Statistics
STREET and BURGESS · The Construction of Optimal Stated Choice Experiments: Theory and Methods
STYAN · The Collected Papers of T. W. Anderson: 1943–1985
SUTTON, ABRAMS, JONES, SHELDON, and SONG · Methods for Meta-Analysis in Medical Research
TAKEZAWA · Introduction to Nonparametric Regression
TAMHANE · Statistical Analysis of Designed Experiments: Theory and Applications
TANAKA · Time Series Analysis: Nonstationary and Noninvertible Distribution Theory
THOMPSON · Empirical Model Building
THOMPSON · Sampling, *Second Edition*
THOMPSON · Simulation: A Modeler's Approach
THOMPSON and SEBER · Adaptive Sampling
THOMPSON, WILLIAMS, and FINDLAY · Models for Investors in Real World Markets
TIAO, BISGAARD, HILL, PEÑA, and STIGLER (editors) · Box on Quality and Discovery: with Design, Control, and Robustness
TIERNEY · LISP-STAT: An Object-Oriented Environment for Statistical Computing and Dynamic Graphics

*Now available in a lower priced paperback edition in the Wiley Classics Library.
†Now available in a lower priced paperback edition in the Wiley–Interscience Paperback Series.

TSAY · Analysis of Financial Time Series, *Second Edition*
UPTON and FINGLETON · Spatial Data Analysis by Example, Volume II: Categorical and Directional Data
† VAN BELLE · Statistical Rules of Thumb, *Second Edition*
VAN BELLE, FISHER, HEAGERTY, and LUMLEY · Biostatistics: A Methodology for the Health Sciences, *Second Edition*
VESTRUP · The Theory of Measures and Integration
VIDAKOVIC · Statistical Modeling by Wavelets
VINOD and REAGLE · Preparing for the Worst: Incorporating Downside Risk in Stock Market Investments
WALLER and GOTWAY · Applied Spatial Statistics for Public Health Data
WEERAHANDI · Generalized Inference in Repeated Measures: Exact Methods in MANOVA and Mixed Models
WEISBERG · Applied Linear Regression, *Third Edition*
WELSH · Aspects of Statistical Inference
WESTFALL and YOUNG · Resampling-Based Multiple Testing: Examples and Methods for *p*-Value Adjustment
WHITTAKER · Graphical Models in Applied Multivariate Statistics
WINKER · Optimization Heuristics in Economics: Applications of Threshold Accepting
WONNACOTT and WONNACOTT · Econometrics, *Second Edition*
WOODING · Planning Pharmaceutical Clinical Trials: Basic Statistical Principles
WOODWORTH · Biostatistics: A Bayesian Introduction
WOOLSON and CLARKE · Statistical Methods for the Analysis of Biomedical Data, *Second Edition*
WU and HAMADA · Experiments: Planning, Analysis, and Parameter Design Optimization
WU and ZHANG · Nonparametric Regression Methods for Longitudinal Data Analysis
YANG · The Construction Theory of Denumerable Markov Processes
YOUNG, VALERO-MORA, and FRIENDLY · Visual Statistics: Seeing Data with Dynamic Interactive Graphics
ZACKS · Stage-Wise Adaptive Designs
ZELTERMAN · Discrete Distributions—Applications in the Health Sciences
* ZELLNER · An Introduction to Bayesian Inference in Econometrics
ZHOU, OBUCHOWSKI, and McCLISH · Statistical Methods in Diagnostic Medicine

*Now available in a lower priced paperback edition in the Wiley Classics Library.
†Now available in a lower priced paperback edition in the Wiley–Interscience Paperback Series.

Printed and bound by CPI Group (UK) Ltd, Croydon, CR0 4YY

16/09/2026

14963415-0004